SIXTH EDITION

FUNDAMENTALS OF DIMENSIONAL METROLOGY

My gratitude to

Charley Dotson, and Robert and Mary Bills—

You encourage and inspire me.

And in memory of

Earl Bills, for my first set of micrometers.

SIXTH EDITION

FUNDAMENTALS OF DIMENSIONAL METROLOGY

CONNIE L. DOTSON

Australia • Brazil • Japan • Korea • Mexico • Singapore • Spain • United Kingdom • United States

***Fundamentals of Dimensional Metrology*,**
Sixth Edition
Connie L. Dotson

SVP, GM Skills & Global Product Management: **Dawn Gerrain**

Product Team Manager: **Erin Brennan**

Associate Product Manager: **Nicole Sgueglia**

Senior Director, Development: **Marah Bellegarde**

Senior Product Development Manager: **Larry Main**

Senior Content Developer: **Sharon Chambliss**

Product Assistant: **Jason Koumourdas**

Vice President, Marketing Services: **Jennifer Ann Baker**

Marketing Manager: **Jennifer Barbic**

Senior Production Director: **Wendy Troeger**

Production Director: **Andrew Crouth**

Content Project Management: **S4 Carlisle**

Cover image(s): **Steven Peters: Getty Images**

Library of Congress Control Number: 2014947679

Book Only ISBN: 978-1-133-60089-3

Cengage Learning
20 Channel Center Street
Boston, MA 02210
USA

Cengage Learning is a leading provider of customized learning solutions with office locations around the globe, including Singapore, the United Kingdom, Australia, Mexico, Brazil, and Japan. Locate your local office at: **www.cengage.com/global.**

Cengage Learning products are represented in Canada by Nelson Education, Ltd.

To learn more about Cengage Learning, visit **www.cengage.com.**

Purchase any of our products at your local college store or at our preferred online store **www.cengagebrain.com.**

Printed at CLDPC, USA, 11-21

CONTENTS

PREFACE

The sixth edition of *Fundamentals of Dimensional Metrology* has been reorganized for better content flow with added clarification for calibration and traceability. Drawing from feedback from metrology course instructors, the content moves from an introduction to the language of metrology, to the use of simple tools for direct measurement, then to the importance of calibration and traceability. The content is expanded to the more complex measurement by comparison

<table>
<tr><td>1</td><td>Measurement and Metrology</td><td rowspan="4">Introduction to the Language of Metrology</td></tr>
<tr><td>2</td><td>Language and Systems of Measurement</td></tr>
<tr><td>3</td><td>Measurement and Tolerances</td></tr>
<tr><td>4</td><td>Statistics and Metrology</td></tr>
<tr><td>5</td><td>Measurement with Graduated Scales and Scaled Instruments</td><td rowspan="4">Basic Metrology Instruments</td></tr>
<tr><td>6</td><td>Vernier Instruments</td></tr>
<tr><td>7</td><td>Micrometer Instruments</td></tr>
<tr><td>8</td><td>Development and Use of Gage Blocks</td></tr>
<tr><td>9</td><td>Calibration</td><td>Traceability</td></tr>
<tr><td>10</td><td>Measurement by Comparison</td><td rowspan="4">Measurement by Comparison</td></tr>
<tr><td>11</td><td>Reference Planes</td></tr>
<tr><td>12</td><td>Angle Measurement</td></tr>
<tr><td>13</td><td>Surface Measurement</td></tr>
<tr><td>14</td><td>High-Amplification Comparators</td><td rowspan="5">Multiple-Scale Instruments</td></tr>
<tr><td>15</td><td>Pneumatic Measurement</td></tr>
<tr><td>16</td><td>Optical Flats and Optical Alignment</td></tr>
<tr><td>17</td><td>Optical Metrology</td></tr>
<tr><td>18</td><td>Coordinate Measuring Machines</td></tr>
</table>

studies, and combines the earlier concepts with a study of multiple scale instruments. The chapters for each topic are shown in the table on previous page.

While many tools are available with digital readout, and the ability to keep, record, and transfer data, there is a continued need for students to understand and be able to use basic vernier instruments, as shown in the earlier chapters. Batteries die, tools get dropped, and digital instruments are more expensive instruments to purchase. Most labs have a supply of basic vernier tools available.

The earlier content can then be applied to more complex measurement concepts and tools, where measurement by comparison and multiple scale instruments are introduced. The text continues to use both metric and English measurements throughout, as students still need to be proficient in both systems.

A new lab manual with expanded practical exercises is available to assist teachers and course instructors. The lab manual contains applications to reinforce the learning objectives, practice good metrology procedures, and promote writing skills that enhance a student's marketability.

This text can be used as a metrology course textbook or as a reference in the field of dimensional metrology. It is also intended that, as an instructor, you can apply the procedures and concepts to any educational level, and across disciplines. Machine shop, tool and die, and quality control courses and programs should find equal application of this text.

Although the principles of metrology have not changed, the impact of applying these tools correctly is progressively more important. Increasingly, employers are looking for quality control personnel, inspectors, engineers, and designers with a thorough understanding of metrology practices.

FEATURES OF THE SIXTH EDITION

- Better organizational content
- Increased traceability and calibration knowledge
- Updated drawings and illustrations
- Reduced theoretical material, replaced by more practical applications
- A new lab manual to accompany this text

INSTRUCTOR RESOURCES

The Instructor Resources section, found on www.cengagebrain.com, was developed to assist educators in planning and implementing their instructional programs. It includes an image gallery of the images from the book, an instructor's guide, and a student lab manual.

Also available:

Cengage Learning Testing Powered by Cognero a flexible, online system that allows you to:

- author, edit, and manage test bank content from multiple Cengage Learning solutions
- create multiple test versions in an instant
- deliver tests from your LMS, your classroom or wherever you want

ACKNOWLEDGMENTS

The revision of this text is not the author's alone. Many individuals in the metrology field were generous with their time and information. I would particularly like to thank Steven M. Hastings, Sr. Lab Coordinator at Arizona State University Polytechnic, and Sharon Chambliss at Cengage Learning. Their suggestions and review contributed extensively to the improvement of this edition.

I would also like to thank the following companies for their time, effort, and product information: AMETEK Taylor Hobson, Apex Tool Group, Apollo Research, ASME, BC Ames, Chicago Dial Indicator, Hexagon Metrology Inc., the International Bureau of Weights and Measures (BIPM), Mahr Federal, Micro-Radian Instruments, Mitutoyo America Corporation, Optical Gaging Products, Quality Vision International, Renishaw Plc., L. S. Starrett, Universal Gage, and Carl Zeiss, Inc.

ABOUT THE AUTHOR

Connie Dotson is a professional trainer and a journeyman tool-and-die maker, with strong technical and metrology experience.

She holds a Master of Business Administration degree, with emphasis on Adult Learning. Over the last 20 years, Connie has taught technical, apprentice, software, and metrology courses in colleges and private industry. The courses include Blueprint Reading, Geometric Dimensioning and Tolerancing, Drafting, Tool and Die Design, Fixturing, Machinery's Handbook, and Coordinate Measuring Machine Operation. She has also been a speaker at national conferences, including the American Society for Training and Development, and has taught Train-the-Trainer concepts and led interactive teaching skills workshops. Connie has also volunteered with the junior achievement program, as well as for high school and trade school career workshops.

CHAPTER ONE

MEASUREMENT AND METROLOGY

All things manufactured need to be measured. With the exception of one-of-a-kind artisan pieces, products need to conform to standards, federal or state regulations, and engineering drawings. The bumper on a car needs to fit and meet automotive regulations. A shirt collar needs to be measured for the proper amount of material. The wing of an airplane needs the correct shape for airlift. Metrology helps us determine whether a part meets its required form, fit, or function.

LEARNING OBJECTIVES

- Define metrology.
- Explain why measurement is relative.
- Define traceability.
- Explain the importance of traceability.
- Explain why measurement is similar to a language and essential for communication in industry.
- Describe how measurement is essential at many levels, such as skilled craftsmanship, production manufacturing, and scientific research.
- Explain why the principles of measurement are stressed in this text rather than the care of instruments.
- Explain the role of metrology in national and international trade.

OVERVIEW

We can trace measurement to the early Phoenicians when they were trading around the Mediterranean Sea. They had to develop methods of equating a quantity of a product to an amount of currency. All measurements are relative in that they are comparisons of some standard to the item being measured. Traceability is the comparison of any measuring tool or system to a standard of greater accuracy (see Figure 1–1). Without traceability, measurements are meaningless at best, and could be misleading.

We have all heard the old saying "Measure it twice and cut it once." From an economic standpoint, this comment makes a great deal of sense. The measurements we make with accuracy and precision will reduce the waste of materials, and will further contribute to the production of high-quality items. Costs affect the way companies compete for sales and customers. The competition is both global and keen. Thus, a simple error in measurement when setting up a machining center could cost a company a contract, loss of work, and, if sufficient, the loss of jobs and closure of the company.

Background

In our modern industrial society, we need to be able to produce manufactured goods made to exacting standards, repair them using interchangeable parts, and be consistent in their production so costs can be reduced.

FIGURE 1–1 Whenever you make a measurement, its accuracy, or lack of it, may be traced to the International Bureau of Weights and Measures in Sevres, France. *Reproduced with permission of the BIPM, which retains full internationally protected copyright (Photograph courtesy of the BIPM)*

1-1 MEASUREMENT AS THE LANGUAGE OF SCIENCE

Metrology is the science of measurement, and measurement is the language of science. It is the language we use to communicate size, quantity, position, condition, and time (see Figure 1–2).

A language consists of grammar and composition. Grammar is a science; composition, an art. This book unfolds the grammar of measurement, but only experience will develop the art. The language of measurement is much easier than French or Russian, because we already have considerable skill in it. Properly applied, this skill can decrease the effort given to, and improve the results of, work, hobbies, and continued studies. The accompanying lab manual exercises for this chapter provide you with some practice in technical writing and the language of measurement.

There are three reasons we all need measurements. First, we need measurements to make things, whether the things we make are of our own designs or somebody else's. This applies to all skilled workers and artisans. Second, we need measurements to control the way other people make things. This applies to ordering an engagement ring, fencing a yard, or producing a million spark plugs. Third, we need measurements for scientific description. It would be impossible to give definite information to someone else about aircraft design, electron mobility, or the plans for a birthday party without measurements.

A course in strength of materials will not make a person a bridge designer, but a full grasp of the course could make a bridge designer exceptional. So it is in measurement. The principles in this text are useful only when related to specific measurement situations. These are as diversified as the needs of mankind.

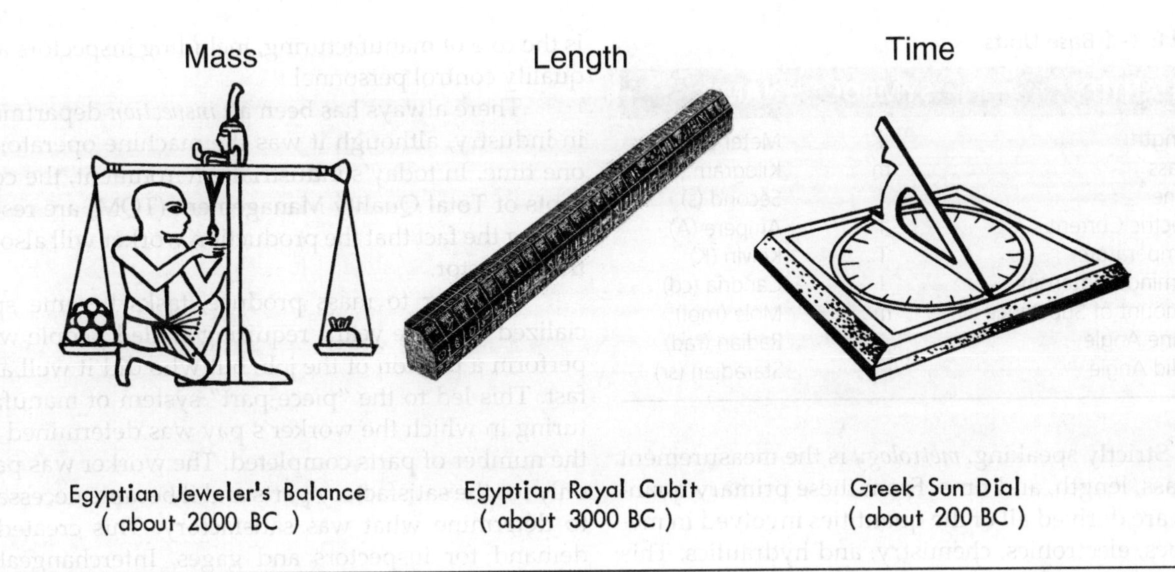

FIGURE 1–2 Early examples of metrology.

1–2 THE USES OF MEASUREMENT

Measurement to Make Things

In the late 1700s, Scotsman James Watt (1736–1819) was jubilant that Englishman John Wilkinson (1728–1808) had perfected the horizontal boring machine with such great precision that it could bore the cylinders for Watt's steam engine to a tolerance of "one thin shilling." Wilkinson's machine could bore a 144.78 cm diameter (57 in.) cylinder to an accuracy of about 1.59 mm (1/16 in.). This unheard-of accuracy made Watt's dream of the steam engine a reality. This same accuracy in today's automobile cylinders would make it an intolerable oil burner, if it would run at all.

Apart from the need for measurement in mass production, measurement is necessary whenever we make anything. In making a flint arrow, a Native American had an approximate idea of the size needed dependent on his target. This is true in all the crafts and skilled trades.

Without numerical values, things that must fit together can be made only by trial and error. Even today, some fitting is expected on one-of-a-kind jobs such as making a complex die, outfitting an ocean liner, or rebuilding an automobile engine. Measurement skill reduces hand fitting. The better the ability to measure, the faster skilled jobs are completed. Of course, the measurements required in one field may be very different from those required in another. However, for craftspeople there is no greater skill than the ability to measure.

The International System of Units has seven base units and two supplementary units. From these base units, several derived units with special names are extracted. The base units are listed in Table 1–1.

Plane angle and solid angle are additional units. The meter was defined in 1983 by the International Committee on Weights and Measures as the "distance light travels in a vacuum during a period of 1/299,792,458 m/s." This is the same as saying light travels at 299,792,458 meters per second.

TABLE 1–1 Base Units

QUANTITY	SYMBOL	NAME
Length	l	Meter (m)
Mass	m	Kilogram (kg)
Time	t	Second (s)
Electric Current	I	Ampere (A)
Temperature	T	Kelvin (K)
Luminous Intensity	I	Candria (cd)
Amount of Substance	m	Mole (mol)
Plane Angle		Radian (rad)
Solid Angle		Staradian (sr)

Strictly speaking, *metrology* is the measurement of mass, length, and time. From these primary quantities are derived all of the quantities involved in mechanics, electronics, chemistry, and hydraulics. This text is restricted to *mensuration.* Mensuration is the branch of applied geometry that is concerned with finding the length of lines, areas of surfaces, and volumes of solids from certain simple data pertaining to lines and angles. This includes those measurements required to use tools and instruments for designing, building, operating, and maintaining material objects, whether they be refrigerators or cyclotrons. This is the reason for the term *dimensional metrology*.

It is difficult for anyone to achieve experience in all of the special roles of measurement. Measurement practices across industries and countries varied greatly in the past. Standards in metrology have come a long way, but each industry must comply with its own regulations. As you might imagine, environment and safety regulations for medical manufacturers are different than those of aircraft manufacturers. There are three general areas to which the basic principles may be applied: communications about measurement, acts and applications of measurement, and codification of measurement.

Measurement to Control Manufacture

An extension of an individual making things is the control of the manufacturing process by others. This is the role of manufacturing, including inspectors and quality control personnel.

There always has been an *inspection* department in industry, although it was the machine operator at one time. In today's industrial environment, the concepts of Total Quality Management (TQM) are resurrecting the fact that the production worker will also be the inspector.

In order to mass produce, tasks became specialized over the years, requiring skilled people who perform a portion of the job, but who did it well and fast. This led to the "piece-part" system of manufacturing in which the worker's pay was determined by the number of parts completed. The worker was paid only for the satisfactory parts and it became necessary to determine what was satisfactory. This created a demand for inspectors and gages. Interchangeable manufacture does not stop at the factory walls. Parts are made in widely scattered plants and even in different countries. The result is a refinement of the language of measurement. Not only is the ability to measure to very small dimensions required, but the measurements must be based on an accepted standard or the parts will not interchange. Standard of length is as important as the means of measuring.

Measurement for Progress

Measurement is truly a universal language. Just as every other communication must be translated, industries today recognize the same standards of length, and convert in and out of each other's systems of measurement.

This has been due largely to industrial progress, but it is needed as much in pure science as it is in applied science. There is no way one research worker can repeat the work of another without specific measurements. This is quite true throughout all branches of science, from astronomy to biology. (In fact, the micrometer was invented for use in astronomy, not shop work.) For many years, progress was slowed and goods were made more expensive by the *"zero of ignorance."* This refers to the

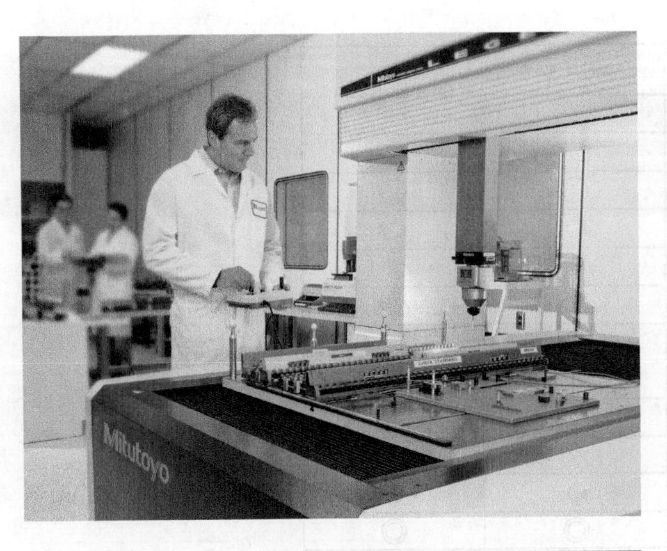

FIGURE 1–3 The computer has vastly expanded the potential for metrology, but the results are no better than the data fed into it. Multiple axis machines, such as the one shown here, benefits in particular using the computer's ability to manipulate metrological data. (*Courtesy of Mitutoyo American Corporation*)

extra decimal place added to the tolerance of a part because the designer was not sure how accurate the dimension really had to be. While computer-operated machines can provide faster, more complex data, they are only as reliable as the data that is input (see Figure 1-3).

As never before, the ability of an engineer, chemist, biologist, or physicist (and the technicians who support them) to test ideas, hinges on an understanding of measurement.

1–3 COMMUNICATIONS ABOUT MEASUREMENT

Measurement depends on communication. Furthermore, it is a social activity, because unless there is a need to communicate, there is no need for measurement. The only legitimacy that measurement units have, whether inches, meters, or archines (a Russian standard length equal to 28 in. or 71 cm), lies in their acceptance and usefulness. This acceptance requires the cooperation of other people; hence, the social aspect.

Communication requires language. Thus, we must first determine that all parties understand the meanings of certain terms. Two terms are central to understanding metrology: *precision* and *accuracy*. Roughly, precision pertains to the degree of fineness, whereas accuracy pertains to conformity with an accepted standard.

It is generally agreed that people are more efficient when they understand what they are doing. Therefore, the basic language of measurement is always expanding to meet new requirements. In the twentieth century, an entire vocabulary for expressing *dimensions* and *tolerances* has come into use. Along with it have come the symbols that represent the terms (see Figure 1–4).

Communications are subject to many serious distortions. Personal as well as cultural biases add to the propaganda and what is called "disinformation." In measurement, we know these as errors. For data to be useful, we must recognize the errors and quantify them.

A study of such errors is beneficial beyond the gathering of data for production or research, because we are bombarded with similar errors in all areas of our daily lives. As in measurement, some are easily recognized, but others are more obscure. We know that visible and invisible dirt or loose clamps will impair measurements, but it requires conscious effort to keep in mind the equally damaging effect of parallax, temperature fluctuations, cosine error, and hysteresis to name a few. Error always exists. The only questions are: How much? Where? To what effect? and, What can be done about it?

The precision of measurements often may be checked by *repeatability*. The accuracy can only be checked by comparison with a higher standard.

SYMBOL FOR:	ASME Y14.5M	ISO
STRAIGHTNESS	—	—
FLATNESS	▱	▱
CIRCULARITY	○	○
CYLINDRICITY	⌭	⌭
PROFILE OF A LINE	⌒	⌒
PROFILE OF A SURFACE	⌓	⌓
ALL AROUND	↙○	↙○ (proposed)
ANGULARITY	∠	∠
PERPENDICULARITY	⊥	⊥
PARALLELISM	//	//
POSITION	⌖	⌖
CONCENTRICITY (concentricity and coaxiality in ISO)	◎	◎
SYMMETRY	⌯	⌯
CIRCULAR RUNOUT	•↗	↗
TOTAL RUNOUT	•⌰	⌰
AT MAXIMUM MATERIAL CONDITION	Ⓜ	Ⓜ
AT LEAST MATERIAL CONDITION	Ⓛ	Ⓛ
REGARDLESS OF FEATURE SIZE	NONE	NONE
PROJECTED TOLERANCE ZONE	Ⓟ	Ⓟ
TANGENT PLANE	Ⓣ	Ⓣ (proposed)
FREE STATE	Ⓕ	Ⓕ
DIAMETER	⌀	⌀
BASIC DIMENSION (theoretically exact dimension in ISO)	[50]	[50]
REFERENCE DIMENSION (auxiliary dimension in ISO)	(50)	(50)
DATUM FEATURE	•[A]	• or •[A]

• MAY BE FILLED OR NOT FILLED

FIGURE 1–4 Revisions to the American Society of Mechanical Engineers Dimensioning and Tolerancing Standard (ASME Y14.5M-2009) are intended to improve national and international standardization. *Reprinted from ASME Y14.5-2009, by permission of The American Society of Mechanical Engineers.*

Because of the importance of the convertibility and interchangeability of measured data, the methods of traceability to the international standard have been formalized by governments and by trade groups. The major considerations are summarized in Figure 1–5.

1–4 ACTS AND APPLICATIONS OF MEASUREMENT

The applications of linear measurement are so diverse that it is surprising that they utilize so few basic acts. Those acts, of course, are the basis for this text.

Discipline of Measurement

1. The accuracy of a measurement can never be as great as that of the standard.
2. The accuracy is diminished by errors.
3. Calibration, comparison to a higher standard, is the test for accuracy.
4. Repeatability is the usual test for precision.
5. Every measurement alters the object being measured and the measurement system.
6. The potential errors are reduced by:

   Elimination of separate measurement acts.

   Elimination of separate parts of the measurement system.

   Elimination of separate fluctuating conditions.

   Elimination of separate positional variables.

7. The least positional error results when the line of measurement, standard, and axis of the comparison instrument are all in line.

FIGURE 1–5 These considerations are of extreme importance and must be understood and practiced.

An overview of the applications begins with *plate work*. Although often considered old fashioned, it offers the most immediate application of the basic principles and the greatest temptations to compromise.

Coordinate measurement is a technique for speeding up layout, machining, and inspection by having all part features dimensioned from the three rectangular coordinates: x, y, and z. A bolt circle, for example, would have each hole located by two dimensions, one from the vertical axis and the other from the horizontal axis. No angles and radii would be shown. It is important to know both when and how to use this technique.

When many parts must be made to close tolerances and be inspected by semiskilled persons, *gages* are substituted for direct measurement. Gages interface readily with computers. This facilitates the wonders of *statistical quality control*. It also clouds the line between the act of measurement and the effective use of measurement. Thus, such aids as the microprocessor and computer with their popular electronic digital displays call for greater caution, not less. See Figure 1–6.

A concern arising in metrology is which measurement system to apply when making a measurement—the SI (International System of Units [metric]) or the English system. Conversion tables are readily available and agreements have been reached on what the international conversion standards will be (25.4 mm equals 1 in.).

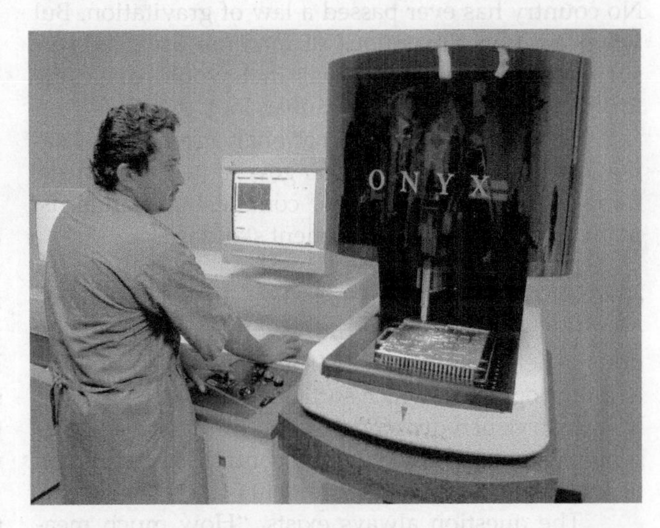

FIGURE 1–6 There is little, if anything, in today's high-tech society that has not benefited from the progress in metrology. Some fields such as solid-state electronics would have been impossible without it. (*Courtesy of CE Johansson*)

The only aspects of measurement that come immediately to mind in which the inch system cannot be converted readily with the SI (metric system) are screw threads and gears. These involve some of the most sophisticated problems in measurement, which are usually solved "by rule and by rote." This is fine until an error is made. Unlike solving problems by deductive reasoning, the error does not call attention to itself, and escapes unnoticed. In industry, at least, an understanding of the fundamentals of thread-and-gear measurement is invaluable.

The chances are that a technician in any of the branches of mechanical engineering will encounter cam measurement. The constantly changing curves relate to calculus. In actual practice, measurement shortcuts reliably reduce a cam to a series of measurable points.

1-5 CODIFICATION OF MEASUREMENT

When things derive from natural phenomena, we need little in the way of laws, statutes, and agreements. No country has ever passed a law of gravitation. But when arbitrary inventions of man are involved, we surround them with bastions that would make the fortifications of Monsieur Vauban look puny.

Thus, there are international agreements, national standards, various other government standards, trade association standards, commercial standards, plant standards, and department standards.

All too often, statisticians have extracted conclusions to the tenth decimal place from data reliable only to the third. The result can obscure the tremendous benefits that the statistical method can provide to any program in which sufficient data are available. In industry such decisions are the result of *quality assurance*, formerly called quality control. Its results can be no better than the measurements.

The question always exists, "How much measurement is required?" This cannot be answered without considering costs, production methods, and results. When the number of parts, the speed of operation, and the value of the results increase, the importance of inspection as well as its cost increases logarithmically. It is the function of the reliability people to balance results against the other factors. The answer to "How much measurement is necessary?" becomes a team decision and what is required by the product to meet the requirements of the customer. We must understand customer compliance.

1-6 A LOOK AHEAD

As shown, there are three reasons for measurement. Although each is different, they all use the same system of measurement and the same instruments. Moreover, they overlap.

1. The skilled diemaker might use a micrometer to measure a given part to one-half thousandth of an inch (0.0005 in. or 0.0127 mm). His experience and skill may enable him to do so with a high degree of *reliability*.
2. If the production of that same part is to be controlled to the same reliability, a much more *sensitive* instrument must be chosen to replace the skill of the diemaker.
3. The physicist might be able to use the same micrometer, but may also have to determine the atmospheric pressure, temperature, or some other variable before the measurement will be suitable for his purpose.

The following chapters will ensure that each of the preceding three groups means the same thing when they refer to their measurements. The standard measuring instruments will be discussed in the order in which they probably will be encountered, from the least precise to the most precise. Far more important, some rules will be developed as a guide for the selection of the best instrument for any given dimensional measurement. See Figure 1-7.

Principles are emphasized in this text, rather than products, for several reasons. Except for the simplest instruments, all differ somewhat. The separate manufacturers' models number in the thousands. It would be impossible to describe them all in detail. Fortunately, this is not necessary because manufacturers

FIGURE 1–7 Optical metrology is a good choice where non-contact inspection is desired. (*Photo courtesy of Optical Gaging Products*)

supply operating manuals for all except the simplest instruments. In all cases, these manuals should be used for specific operating and maintenance instructions.

Principles, rather than products, are emphasized for still another reason. Industry and science finally are aware of their dependency on precision measurement. Improvements in measuring instruments are rapidly being made. Tomorrow's instruments will not look much like today's. However, the principles will remain as true tomorrow as when they evolved long ago.

SUMMARY

- Measurements support a wide range of products and processes across science, medicine, manufacturing, health and safety, and government.
- Communication of measurements within and across industries is essential to both manufacturing and service companies.
- The quality of goods and services offered is highly dependent on reliable measurement.
- Traceability is the cornerstone of quality control programs. Without traceability, measurement is meaningless.
- Skills in linear measurement are readily marketable. They are far more widely needed as an aid for performing other work, because almost every technical pursuit involves measurement.
- Metrology requires communications for its use. It has its own terms and symbolism. These are analogous to all attempts to communicate without error and distortion. Part of the communication is codification. Codification ensures an orderly process that is understood by those working in a given area of metrology.
- Familiarity with the basic metrology principles provides a discipline for logical thought, a benefit that applies to all activities.

END-OF-CHAPTER QUESTIONS

1. Accuracy can only be checked by comparison with:
   a. another technician or metrologist
   b. a trade group
   c. a higher standard
   d. your boss
2. Measurement is most important to which one of the following applications?
   a. Mass production of automotive parts
   b. Mass production of aircraft parts
   c. One-of-a-kind handicrafts
   d. All manufacturing
   e. Manufacturing of parts from prepared plans
3. Measurement involves fundamental qualities. Select three qualities from the following list.
   a. Roundness
   b. Mass
   c. Flatness
   d. Time
   e. Weight
   f. Modules of elasticity
   g. Length
   h. Width
   i. Hardness

4. Which of the following best defines "dimensional metrology?"
   a. The measurements required to manufacture products
   b. Measurements found by the use of scales and other measuring instruments
   c. The measurement of lines, areas, volumes, and angles
   d. The measurement of real things, whether they be steel, wood, plastic, or any other material
   e. The measurement of lines, circles, and angles
5. Select one or more of the following that are *not* considered to be dimensional metrology.
   a. Bolt circle spacing
   b. The selection of lubricants for a given bearing allowance
   c. The torque requirement for a bolted assembly
   d. The tolerance required for a shaft in a bearing
   e. The size limits of a mass-produced replacement part
6. The role of the inspector in industry emerged as the result of which one of the following circumstances?
   a. Weapons were first mass produced.
   b. Gunpowder changed the threat from weapons.
   c. Foremen could not keep up with mass production rates.
   d. The piece-part system was introduced.
   e. Precision measurement tools such as the micrometer were introduced to industry.
7. Of the following characteristics, select the ones that affect quality control.
   a. Employee morale
   b. Pride of product
   c. Dimensional measurements
   d. Environmental conditions
   e. Product specifications
8. For every act of dimensional metrology, one or more of the following variables may apply. Identify the single most important variable.
   a. Number of decimal places
   b. Metric or customary system
   c. Reliability
   d. Repeatability
   e. Sensitivity
9. In this text, metrological methods or principles are emphasized rather than the measuring instruments themselves. There are several reasons for this. Which of the following is the most important reason?
   a. There is no complete agreement about the instruments.
   b. Instrument manufacturers provide detailed information.
   c. Principles are more interesting than methods of use.
   d. Principles and methods are easily adaptable to all instruments.

Discussion Question

10. Small errors can cost a company billions of dollars or have dire consequences. Discuss the potential impact that small errors can have in the following industries:
    - Automotive
    - Aerospace
    - Biotech/Medical

Additional Research

National Institute of Standards and Technology (http://museum.nist.gov/exhibits/ex1/index.html):

America Before Standard Weights and Measures

The Founding of the National Bureau of Standards in 1901

A Social History of Weights and Measures

CHAPTER TWO

LANGUAGE AND SYSTEMS OF MEASUREMENT

The Congress shall have Power . . . To regulate Commerce . . . among the several states . . . to fix the Standard of Weights and Measures . . .

Constitution of the United States, Section 8, Article 1

LEARNING OBJECTIVES

- Define a line of measurement.
- Define the terms *precision, accuracy,* and *reliability.*
- Describe the two systems in general use—English and metric.
- List the strengths and disadvantages of each measurement system.
- Explain how the choice of system to be utilized is made.
- Identify a standards body.

OVERVIEW

A child reportedly asked President Lincoln how long his legs were. The president answered, "Long enough to reach from my body to the ground." A humorous response, but the clever use of language does not provide an accurate response to the child's question.

As often happens today, language obscures the meaning of the questions we ask and the responses we receive. For example, if an automobile engine is bored for larger pistons and tested for acceleration time, the 0 to 96 kph (60 mph) time might be cut by 2 seconds. Then, in production, when someone asks how "accurately" the new pistons must fit for the same level

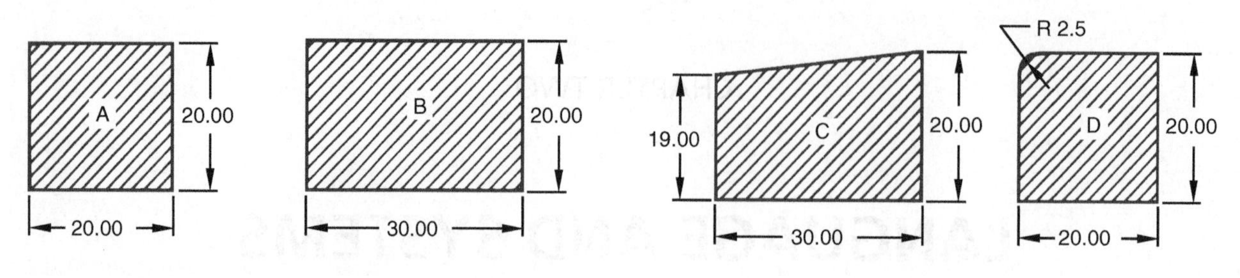

FIGURE 2–1 "A" is a 20 cm square and "B" is a 20 mm × 30 mm rectangle. What are the sizes of "C" and "D"? Only by dimensions can size be accurately defined.

of performance, how should we answer? One person says, "For a good fit." Another says, "The maximum clearance cannot exceed 0.050 mm (0.002 in.) or be less than 0.025 mm (0.001 in.)." One of these answers probably contains the information we need, but the *right* answer depends on the meaning of the word *accuracy* in the question.

Consider another everyday word: *size*. We all know what "400 mm^2 (0.62 in. sq.)" means (see Figure 2–1). We also understand a rectangle specified at 20 × 30 mm (0.75 × 1.2 in.). But what is the "size" of C or D in Figure 2–1? Furthermore, in A and B, are the lengths of the sides and the general shape enough to specify size, or do we also need to know how square or how rectangular the shape is?

Different interpretations of common terms like *accuracy* and *size* can cost manufacturers millions of dollars of waste each year. To reduce waste and improve measurement communication, manufacturers adhere to quality standards, like those developed by the International Organization for Standardization (ISO). The ISO is a worldwide federation of national standards bodies, with representatives from over 100 countries. The American National Standards Institute (ANSI), for example, is a member of ISO.

Founded in 1947, they have published more than 19,000 International Standards. According to the ISO, the term "ISO" is not an acronym but is derived from the Greek *isos* meaning "equal." "iso" is the prefix in a number of terms meaning equal or alike, such as "isometric" (equal measure or dimension), "isodiametric" (having equal diameters or axes), and "isonomy" (the equality of laws, or of people before the law). Outside of the metrology laboratory, manufacturers have only needed a few, all-inclusive terms. *Accurate* and *precise* were used interchangeably and were often synonymous with *reliable* and *repeatable. Measure, inspect,* and *gage* meant the same thing. *Standardize* meant a private agreement was reached with the boss that certain consistent levels would be reached in production from that moment forward.

In the complicated global economy of today, dimensional measurement requires the use of electronics, optical instruments, and interferometry, as well as time-honored mechanical methods of measurement. Adherence to measurement standards like ANSI's Y14.5 Geometric Dimensioning and Tolerancing Standard (GDT) help reduce waste and improve global communication.

When this text's first edition was published in 1964, the metric versus English measurements debate was at its height. Although English proponents have never attempted to outlaw metric use, a few well-intentioned but overly zealous metric proponents would not accommodate both measurement systems. Their "voluntary" compliance really meant "compulsory." This unfortunate disagreement took attention away from the fundamental principles of metrology and the need for standardized meanings that encourage international trade.

Anyone engaged in metrology—in fact, any member of an industrial society—must be familiar with the current accepted measurement system, because the ramifications of a standardized system of measurement extend into every aspect of our lives. This chapter and the accompanying lab manual exercises are designed to familiarize you with these important measurement systems.

2–1 COMMUNICATIONS CONSIDERATIONS

The usefulness of any measurement depends in part on your ability to communicate your results to other people. You must be familiar with the terminology of all the measurement systems commonly used today so you can understand what others are telling you and so others can understand what you are telling them.

Like all language, some of the terms used in measurement make sense, and others have no logical explanation. *Foot* is derived from the body part used originally to measure it, but *inch* does not have a similar correlation.

The prefixes used with the term *meter* are derived from the languages of ancient Greece and Rome, as are the terms used in the decimal-inch system (see Figure 2–2). For example, one-tenth of one thousandth of an inch is *point one mil* or *one hundred microinches* from the Greek and symbolized by µ (mu). The terminology of the decimal-inch system is shown in Figure 2–3.

2–2 HOW BIG?

If you can visualize the sizes involved, as in the comparisons in Figure 2–4, it is easier to understand the

Linear Measurement Units in the Inch-Pound and Metric Systems Compared

Linear Measure	Microinch (.000001y) µin.	Ten Thousandths (.0001y)	Thousandths (.001y)	Inch (y) or (z/12) in.	Foot (12y) or (z) ft.	Yard (3z) yd.	Mile (5280z) mi.
Nanometer nm (.000000001x)	0.0394 **25.4**						
Micrometer* µ (.000001x)	39.37 **0.0254**	0.3937 **2.54**	**25.4**				
Millimeter mm (.001x)		393.7	39.37 **0.0254**	0.0394 **25.4**	**304.8**		
Centimeter cm (.01x)		3937.	393.7	0.3937 **2.54**	0.0328 **30.48**	**91.44**	
Meter m (x)				39.37 **0.0254**	3.2808 **0.3048**	1.0936 **0.9144**	**1609.3**
Kilometer km (1000x)					3280.8	1093.6	0.6214 **1.6093**

*Micrometer is the official name but micron is widely used.

FIGURE 2–2 To use the English/metric conversions tables: enter a horizontal or vertical column; combine either the lightface or boldface units or numbers; and read its corresponding value in terms of one unit of the opposing scale. For example: 1.0936 yards = 1 meter.

Examples of Decimal-Inch Terminology		
Write		Say
For type-set material	On drawings	(Preferred form is first)
0.002″ or 2 mil	.002	two mil
0.012″ or 12 mil	.012	twelve mil
0.02″ or 20 mil	.02	twenty mil point zero two inch
0.2″	.20	point two inch
2.005″	2.005	two inch five mil two point zero zero five inch
2.00″	2.00	two inch
0.000005″ or 5 × 10–6″ or 5	.000005 or 5	five micro inch
0.00002″ or 20	.00002 or 20	twenty micro inch
0.0002	.0002	point two mil point zero zero zero two inch two tenths*
0.0025″	.0025	two point five mil point zero zero two five inch
2.000005″	2.000005	two inches and five micro inch
2.0005″	2.005	two inch point five mil two point zero zero zero five inch

*Colloquialism — not recommended

FIGURE 2–3 This is terminology from the American National Standards Institute (ANSI) and is used wherever appropriate in this text. *(From ANSI B87, American Standards Institute, Table 1)*

terminology. A graduation on the average machinist's rule is approximately *76* μm (0.003 in.) wide—*76 micrometers,* or *76 microns.* Your eye can see light through a crack as small as 2.5 μm (0.0001 in.)—*one tenth* of one thousandth of an inch, commonly called "one tenth." By comparison, 2.5 mm (0.10 in.) would be enormous.

We can eliminate any confusion by using decimal-inch terminology, which calls 0.0001 inch *point one mil.* In decimal terminology, your eye can see through an opening 2.5 μm (0.0001 in.) wide, or 2.5 millionths of a meter, usually read as 2.5 micrometers. Thus, by consistently using decimal terms, you can eliminate much of the confusion in both the English and SI systems of measurement.

2–3 HOW FAR APART?

Dimensional measurements are used daily in designing, building, and operating the objects that surround us and for communicating about past, present, and future objects.

The principal dimensional measurement is *length;* secondary measurements are *angle* and *curvature.* An expression of each value is a *dimension.* Using these measurements, we can describe surface, finish, flatness, and angular relationships among features—or *size.* Unless a product is sized by volume or weight, you can describe its shape without describing its size, but you cannot describe its size without knowing its shape.

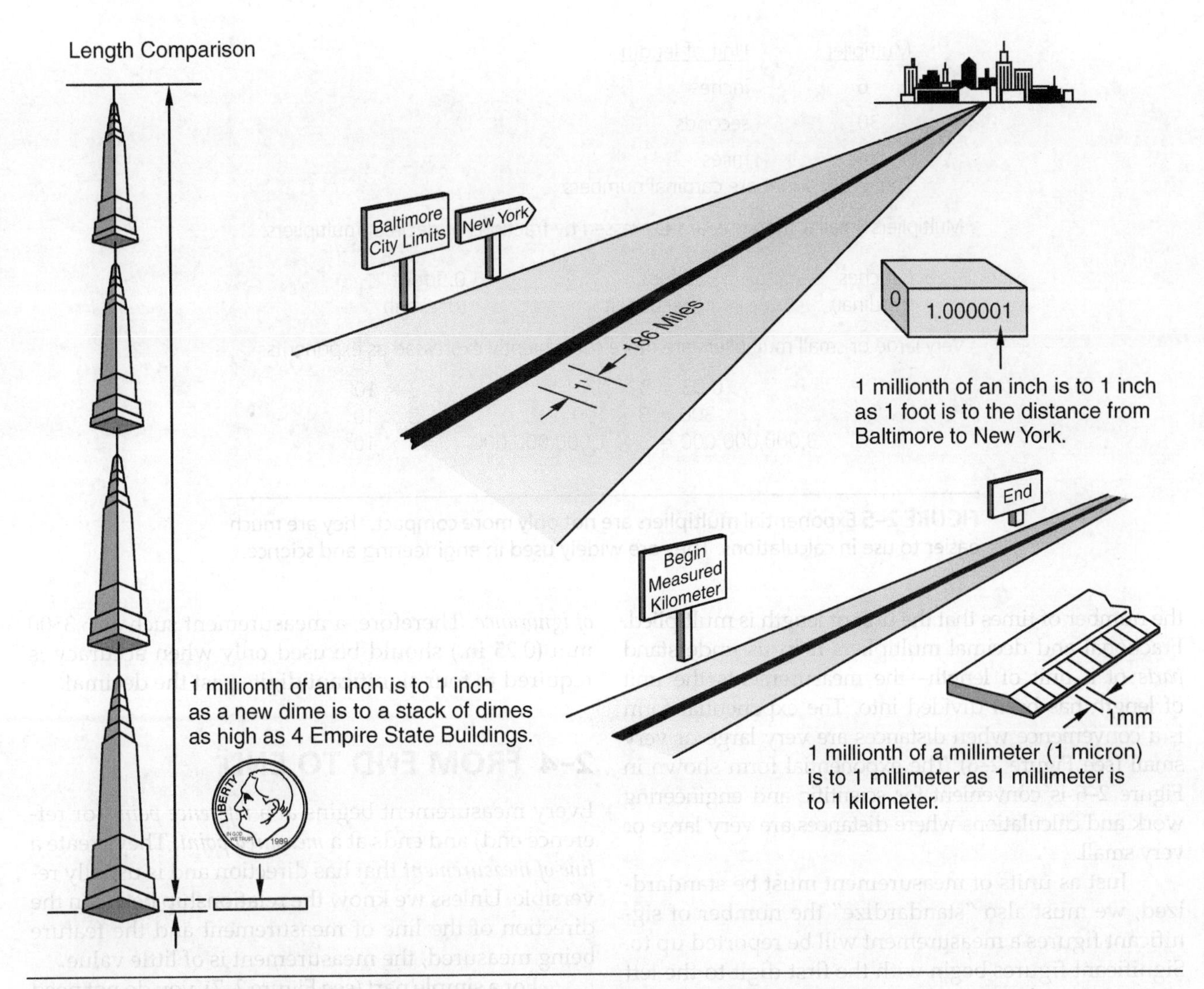

FIGURE 2–4 One millionth is to 1 inch as 1 inch is to 16 miles. Or in the metric system 1 micrometer is to 1 meter as 1 meter is to 1,000 kilometers. These examples show what that means in actual practice.

A linear measurement expresses the distance separating two points. Remember, it is not the distance; it is just a way to talk or write about the distance, and this expression of distance allows us to reconstruct it. *A measurement enables a distance to be reproduced.* A distance measured by "the left feet of the first sixteen men to leave church on a certain Sunday" sounds absurd, but it is a standard of measurement that can be reproduced—and was actually used in the sixteenth century. The micrometer, one thousandth of a millimeter, may be beyond imagination, but it represents distance and can be reproduced.

These measurements have two things in common: a *unit of length* and a *multiplier*. A measurement using a specific unit of length is easy to understand as long as everyone agrees on a *standard* for the unit of length. Standards preserve accepted units of length.

Multipliers can be cardinal, fractional, decimal, or exponential numbers (see Figure 2–5). *Cardinal* numbers are whole numbers (1, 2, 3, and so forth) and simply state

Multiplier	Unit of length
6	inches
30	seconds
4	miles

The 6, 30, and 4 are cardinal numbers.

Multipliers smaller than one are expressed by fractions or decimal multipliers:

6 inches (cardinal)	½ foot (fractional)	0.5 foot (decimal)

Very large or small multipliers are more conveniently expressed as exponents:

$$0.03 = 3 \times 0.1 = 3 \times 10^{-1}$$
$$300 = 3 \times 100 = 3 \times 10^{2}$$
$$3{,}000{,}000{,}000 = 3 \times 1{,}000{,}000{,}000 = 3 \times 10^{9}$$

FIGURE 2–5 Exponential multipliers are not only more compact, they are much easier to use in calculations. They are widely used in engineering and science.

the number of times that the unit of length is multiplied. Fractional and decimal multipliers help us understand *parts* of a unit of length—the measurements the unit of length has been divided into. The exponential form is a convenience when distances are very large or very small (see Figure 2–6). The exponential form shown in Figure 2–6 is convenient for scientific and engineering work and calculations where distances are very large or very small.

Just as units of measurement must be standardized, we must also "standardize" the number of significant figures a measurement will be reported up to. Significant figures begin with the first digit to the left of the decimal point that is not zero and end with the last digit to the right that is *correct*—*not* an approximation. That digit may be zero if it is the correct value.

Therefore, the precision demanded of a measurement increases as zeros are added to the right of the decimal: 0.25 in. is less precise than 0.250 in. or 0.2500 in. If a measurement of 0.250 in. is reported, it must be correct. The third number to the right is significant, and if it had been 0.251 in., that value would have been reported. Conversely, a 0.001 in. difference in a measurement requiring only two numbers past the decimal, such as 0.25 in., would not be significant. When in doubt, we often add an extra zero to the right of the requested significant digit called the *zero of ignorance*. Therefore, a measurement such as 6.3500 mm (0.25 in.) should be used only when accuracy is required to four significant digits past the decimal.

2-4 FROM END TO END

Every measurement begins at a *reference point* (or reference end) and ends at a *measured point*. They create a *line of measurement* that has direction and is usually reversible. Unless we know the relationship between the direction of the line of measurement and the feature being measured, the measurement is of little value.

For a simple part (see Figure 2–7), you do not need to differentiate between the reference point and the measured point. They are both *references*. At the design stage, these points create a dimension, a *feature*, that is the perfect separation between items either on one part or between parts. Manufacturing a perfect part is not possible, so the designer includes a range for these measurements to fall within. This range is the allowed "tolerance" or "limits." These limits represent acceptable deviation from the ideal dimensions (see Chapter 3).

The Act of Measurement

Generally, the act of measurement is a comparison of the standard of length or the distance to be

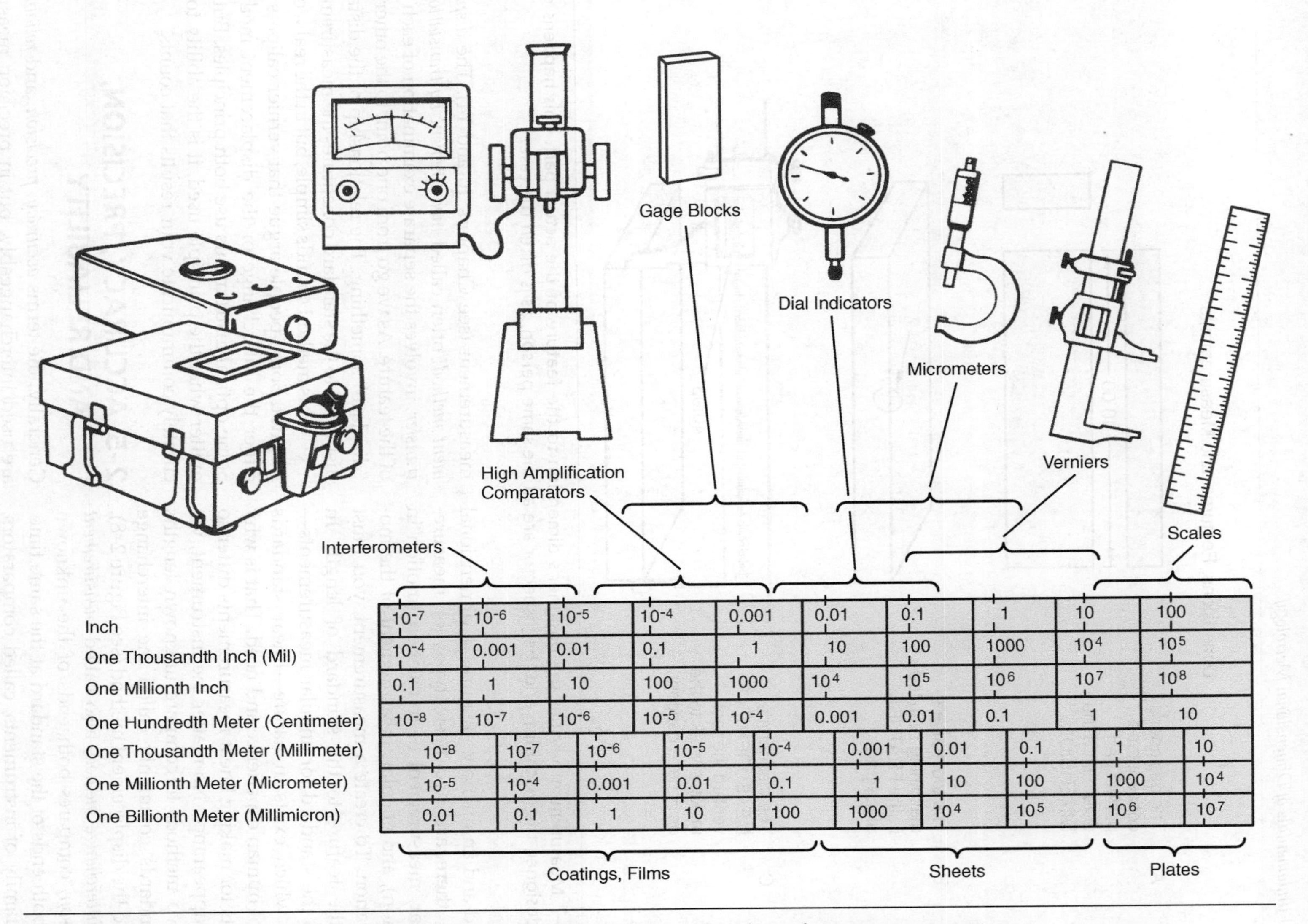

FIGURE 2–6 Exponential multipliers simplify the expression of very small and very large measurements.

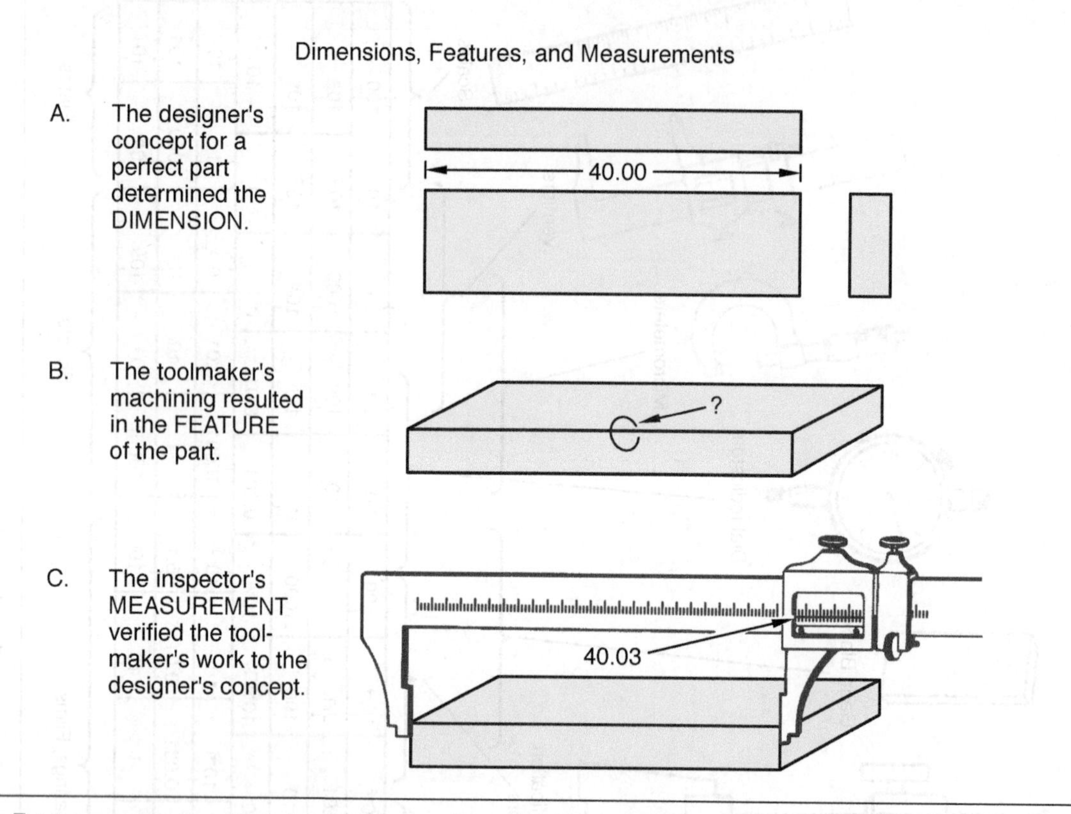

FIGURE 2–7 Measurement verifies the designer's dimension to the feature of the actual part. This happens even when the designer, the machinist, and the inspector are all the same person, as is often the case.

reproduced and an unknown feature. In dimensional metrology there are two basic types of measurement: linear, measurement of translation (motion in one direction), and angular, measurement of the motion of rotation. To create a measurement, you must compare the feature to the standard of length. In contrast to the standard for angular measurements—the circle, which exists in nature—linear standards grow out of human experience and need. That is why instruments for making linear measurements differ so greatly in appearance. No matter the instrument, we still use two methods to compare unknown lengths to the standards commonly called the interchange method and the displacement method (see Figure 2–8).

The *interchange method,* also called *measurement by comparison,* compares both ends of the unknown feature to both ends of the standard at the same time. An entire family of instruments, called comparators, has been created based on the interchange method of measurement (see Chapters 10 and 14). The *displacement method,* often called *measurement by translation or transfer,* involves the separate examination of each end of the feature. As we go from one point to the other, we "displace" something. The relationship of the distance displaced to the standard constitutes the measurement.

This method sounds simple, but in the real world, it is not. Some people argue that vernier calipers use either the interchange or the displacement method. Some optical instruments use both principles. But, no matter what the principle used, it is the ability to accurately communicate your results that counts.

2-5 ACCURACY, PRECISION, AND RELIABILITY

Generally, the terms *accuracy, precision,* and *reliability* are used interchangeably, but in precision measurement for manufacturing, the use of the wrong term

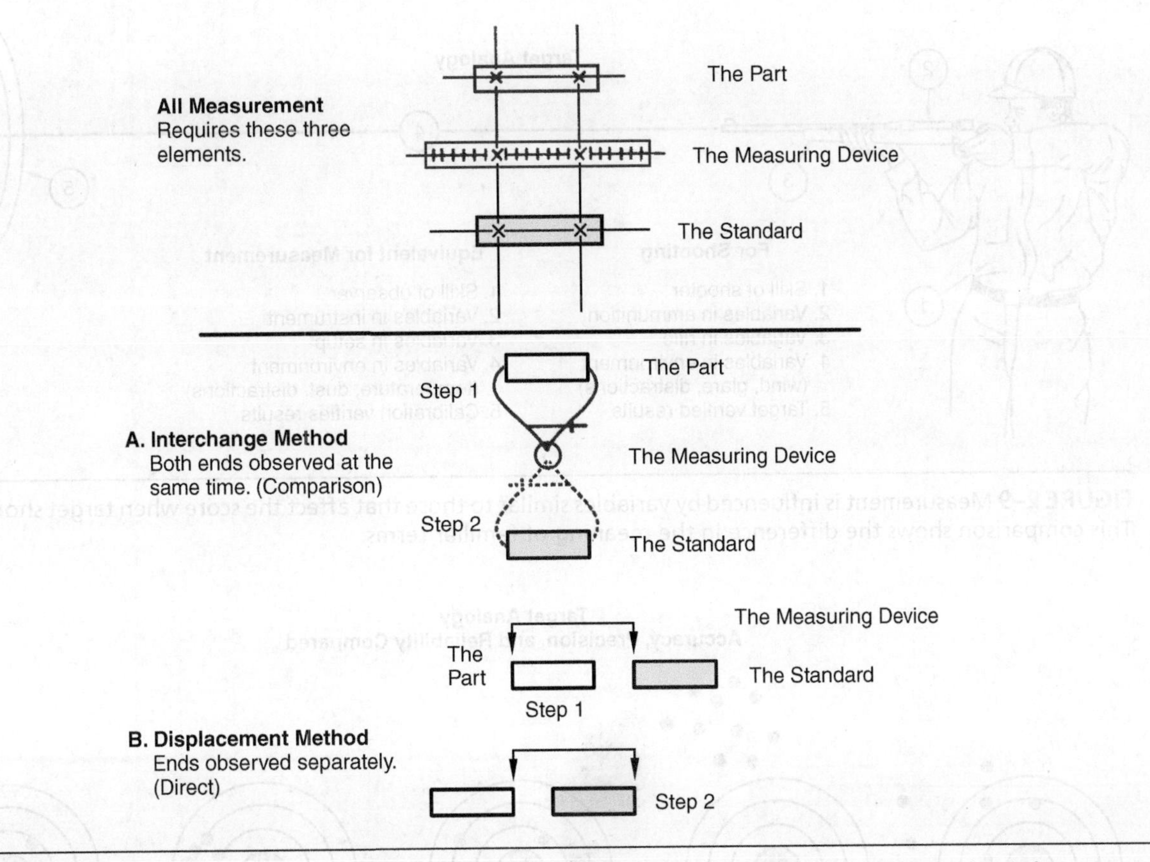

FIGURE 2–8 All measurements consist of the comparison of the unknown with a known. The methods for comparison vary but fall into one of two groups: interchange or displacement.

affects the choice of measuring instrument and the efforts of the quality control team. The incorrect use of one of these terms could mean failure to meet quality control standards, excess production, excess materials use, and more.

To help clarify the meaning of accuracy, precision, and reliability, we will compare the scores of five people in a shooting match. In Figure 2–9, all the components necessary to place a shot on the target and measure the shooter's score are listed, including subdivisions of these variables that affect the shooting and scoring process. The score of one shot can most easily be explained by luck, so we will record ten scores for each shooter. Our target is a simple circle: shots within the circle count; those outside the circle do not count. The results are shown in Figure 2–10.

Shooter A has five good and five bad shots. In comparison, Shooter B's grouping is close together. B shoots more precisely (more shots in the same area) but less accurately (fewer "good" shots) than A. Shooter C's shots are all good and are in a similar pattern to A's. Although the shooting is not as precise as B's, C's shots are more accurate than A and B.

In terms of their scores, Shooters C, D, and E all shoot as well as each other; however, there are significant differences among their patterns. C's group is not as precise as D's tighter grouping, and E's dead-center shots are the most precise and the most accurate, even though they score the same as the others.

Obviously, we need another term to describe E's superior shooting. That term is *reliability*. E's tight, precise pattern and dead-center, accurate shooting

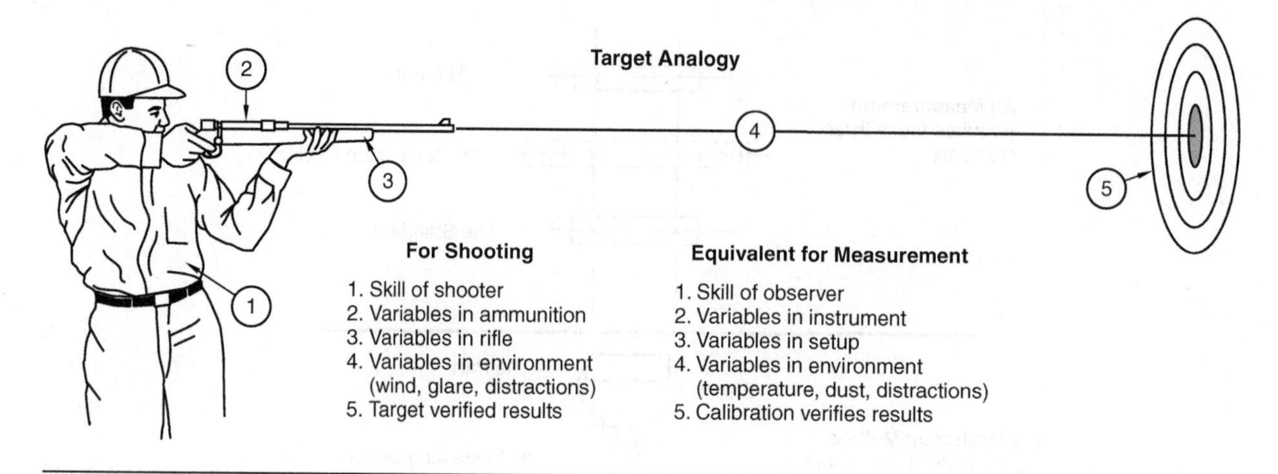

FIGURE 2–9 Measurement is influenced by variables similar to those that affect the score when target shooting. This comparison shows the difference in the meaning of familiar terms.

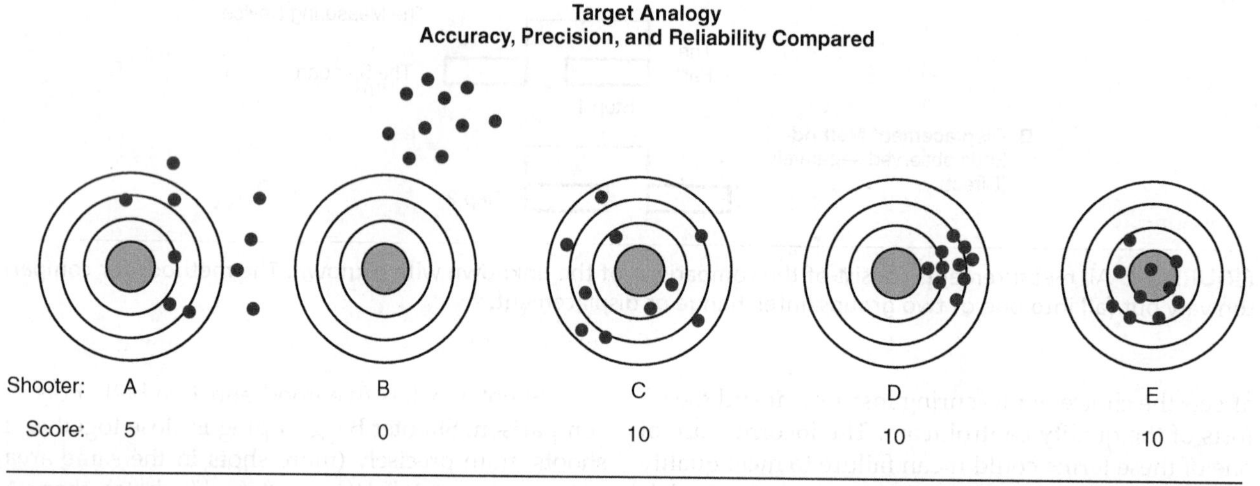

FIGURE 2–10 Which of these targets represents accurate shooting? Precise shooting? Reliable shooting?

have been consistently demonstrated in the ten shots. We can rely on Shooter E's performance.

Even as conditions change, we can count on Shooter E's reliability (see Figure 2–11). The addition of a crosswind causes scores to decrease in all cases except E's, because the "reliability" of E's shooting allows a comfortable margin around the grouping. There is a greater probability that Shooter E will perform as planned under less than perfect conditions.

Based on this shooting example, we can see that accuracy is a comparison: the desired result is compared with the actual measurement. Thus, accuracy is also frequently called the "quality of conformity." Precision reports the dispersement of results or the degree of repeatability within the manufacturing and measurement systems. Therefore, it is called "the quality of refinement." Reliability shows the relationship between the predicted results and the actual results

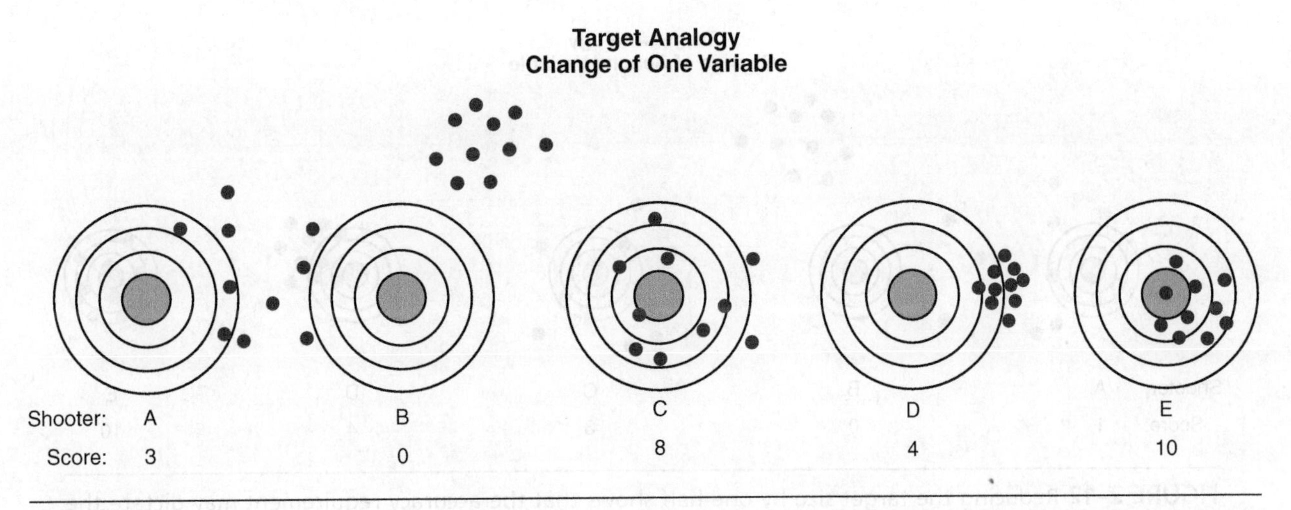

FIGURE 2–11 A change in one variable, such as wind, alters the results as shown. Does this show which shooting was most reliable?

and whether or not we actually *can* predict what will happen.

These definitions, when applied to a specific measurement such as our shooters' relative skill, provide a basis for "educated guesses" about future performances. These guesses are called "probabilities" before a shooting competition and "facts" afterward. Probabilities must have a range; facts are specific and can alter the total picture, either by causing us to upgrade or downgrade our expectation of future performance.

Accuracy, precision, and reliability exist in different time frames. If a steel cylinder is a precise part or an accurate part, we know that historically it measures within a range of permissible sizes. If it is reliable, we expect it to perform a certain way in the future.

Reliability can refer to the manufacturing or inspection process or to the part's expected future performance. As we have seen in other areas, it is vital that all parties involved understand exactly what was measured when reporting on reliability. For example, we might speak of an accurately inspected cylinder, a reliably designed cylinder, a cylinder with a precision tolerance, and so forth.

Precision answers the question *how much?* It provides fineness for the range of sizes allowable for one part or many parts, or it defines the fineness to which an instrument can read. Precision may be vague—*high precision*, or specific—*within 0.025 μm* (1.0 μin.) Accuracy answers the question *good enough?* It indicates whether the object complies with a standard, is too big or small, or is in or out of tolerance. Accuracy may be used in place of precision, but precision never designates accuracy. For example, in Figure 2–12 the target's size has been cut in half; therefore, you have a smaller area for "good" shots—the range of accuracy has been narrowed. You can shoot precisely without being accurate like Shooter B, whose shots all fall outside the acceptable area. Or you can shoot like E, whose shots are both accurate and precise. Notice, too, that there is no safety margin left. Outside variables could even mar our sharpshooter E's performance. Whenever requirements are made stricter, problems with reliability increase.

Without precision, we cannot have accuracy or reliability. Without skilled measurement specialists, we cannot meet the increasingly higher demand for accuracy, precision, and reliability in manufacturing.

As the National Institute of Standards and Technology (NIST) continues to divide our reference standard into smaller parts, we rise to higher levels of precision. But these are just abstractions until the users of measurement in science and industry make practical applications. These applications lead to improved accuracy, higher reliability, and new products. They may even change our knowledge of the universe itself.

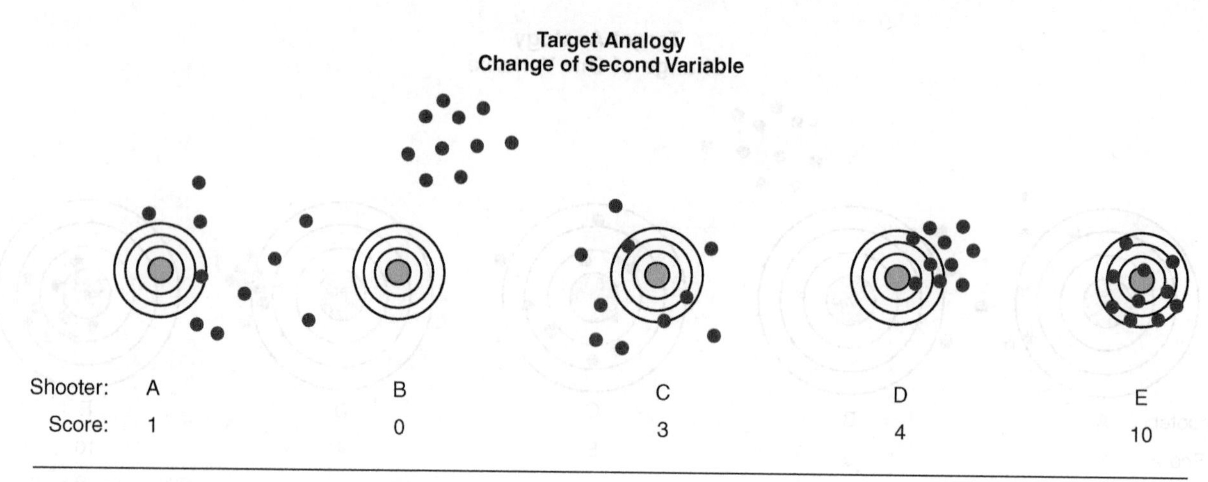

FIGURE 2–12 Reducing the target size by one-half shows that the accuracy requirement may dictate the precision requirement.

2-6 ACCURACY VERSUS PRECISION

Accuracy may mean many different things, but all of its meanings are related (see Figure 2–13). In industry, *accurate* is used commonly to mean precise; if you are not sure which word to use, we recommend you use *precise*, because it means you are looking for the higher standard. Precision can be used as an adjective—"a precision instrument"—or a noun—"the precision of your work."

These terms are your tools, just like your set of gage blocks or a microscope. We have started with basic definitions here, and we will refine and add to the meanings of these words as we work toward greater precision in our communication about measurement.

2-7 THE EVOLUTION OF STANDARDS

From our beginnings as creatures, we have struggled to harness our natural resources and know how much we had. We almost instinctively created systems of measurement, using body parts for length standards, for example, to answer this vital question. As we created cities and governments, we needed better refined standards for manufacturing and bartering with other cultures. After all, not everyone's forearm, for example, is the same length. The ruler of a certain city or state would usually eliminate conflict by choosing his own body parts as the standards. In the process, he or she vested the responsibility for determining measurement standards in the state. Even today, differences in measurement standards can create conflict among nations: for example, during the World Wars, some parts manufactured for military systems designed jointly by the United States and England would not fit because of the different measurement standards used.

The earliest recorded standard is the Egyptian cubit (see Figure 2–14). The Egyptians were very serious about ensuring that everyone was using the same measurement standard. Failure to calibrate the working cubit to the royal cubit at each full moon was punishable by death. But wonders of Egyptian architecture like the pyramids could not have been built without their strictly enforced system of measurement.

	Precision	Accuracy	Reliability
General Meaning	Mechanical or scientific exactness	Correctness	Probability of achieving accuracy
Measures	Fineness of readings	Ratio of correct/incorrect readings	Reliability of correct readings
Method of Stating	Within a 3-in circle Plus or minus one thousandth inch	5 out of 10 50% of full scale	90% reliable
Specific Meaning	The lower the standard deviation of measurement, the higher the precision	The number of measurements within a specified standard as compared with those outside	The probability of performing without failure—a specific function (measurement) under given conditions, for a specified period of time

FIGURE 2–13 These definitions fit most measurements, but many exceptions can be found. A good rule is to use the most precise term that the listener can understand easily.

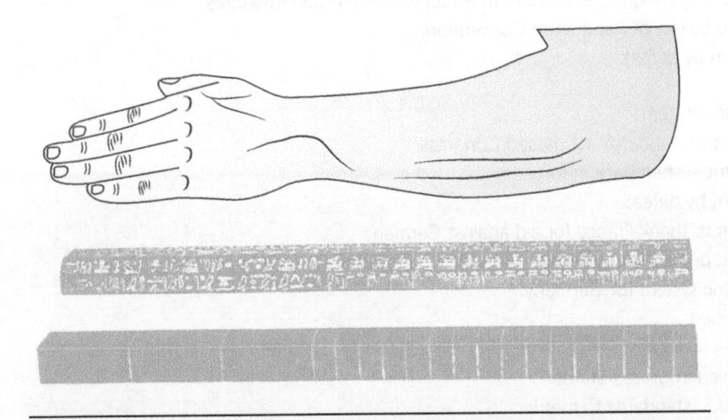

FIGURE 2–14 The Egyptian cubit shows recognition of metrology fundamentals that still apply today. The unit of length was the Pharaoh's forearm. The standard was the royal cubit, and to this standard the working cubits were added.

As use of the cubit spread throughout the ancient world, each culture modified the standard. The historical use of the cubit and its parallel measurements are not directly related, however. *Whenever humanity has developed civilization, we have created a measuring system, and these systems have always been very similar.* For example, the ancient Roman ounce and modern English ounce are nearly identical.

So we can see the similarity between the cubit of Rameses II (1324–1258 B.C.) and the English yard of King Henry I (1068–1135). In about 1130 A.D., Henry established the distance from his nose to the tip of his thumb when his arm was extended as a standard unit of length, *the iron ulna.* Of course, other cultures had used this body length as a measurement, but making the English yard a standard was a major advance in the Dark Ages. Around the same time, the inch, defined as one thumb-breadth, and the foot, were commonly in use (see Figure 2–15). It sounds funny today, but in the sixteenth century, the English rod was defined as the combined lengths of the left feet of the first 16 men to leave church on a particular Sunday.

Each culture creates its own standards for measurement, but these measurements do not necessarily convert easily—even among measurement systems of the same culture. These discrepancies allowed "sharp operators" to take unfair advantage of customers and allowed rulers to use overgenerous measures when collecting taxes.

Our inch-pound system is based on traditional English standards, but this system has had its problems, too. During the late eighteenth century, the English commonly used at least three different "miles"

Milestones in Measurement

Year	Place	Unit	Contribution
before 4000 B.C.E.	Chaldea	circle	First recorded standard of measurement.
about 4000 B.C.E.	Egypt	cubit	Length of Pharaoh's forearm.
4000 to 2000 B.C.E.	Egypt	span, palm	Outstretched hand.
		digit,	Middle of middle finger.
		meridian, mile	Equal to 400 cubits or 1000 fathoms.
		fathom	Length of outstretched arms, about 6 feet.
500 B.C.E.	Greece	stadia	1/10 of meridian mile, borrowed from Egypt.
		mile	1000 paces, similar to today's mile.
	Rome	thumb-breadth	Divided foot into 12 parts. First inch.
Ca. 180–Ca. 125 B.C.E.	Hipparchus of Nicaea		"Father of Trigonometry."
849–901	England	foot	King Alfred established foot as the measure of a cubical vessel containing 1000 Roman ounces of water.
1068–1135	England	yard	King Henry I established yard and made an iron standard.
1084–1153	Scotland	inch	King David made inch average measure of three men's thumbs.
1284–1327	England	inch	King Edward II made inch length of three barley-corns.
1776	U.S.		Articles of Confederation gave Congress power to establish weights and measures.
1790	France	meter	Metric system imposed by law of Republican Convention.
1795	Holland and Belgium		Forced to metric system by defeat.
1769–1821	France		Napoleon relaxed metric system.
1828	U.S.		First effective weights and standards act passed Congress.
1773–1850	France		Louis Philippe put metric system back in force.
1859	Italy		Forced to metric system by defeat.
1859	Austria		Adopted metric system to thank France for aid against Germany.
1866	U.S.		Metric system legalized but not compulsory.
1815–1898	Germany		Bismarck adopted metric system for Germany.
1878	U.S.		Metric Convention ratified.
	England		Yard defined.
1890	U.S.		Metric standards received in United States
1893	U.S.		Mendenhall Act links U.S. standards to metric.
1927	France	light wave-lengths	7th International Conference stated the meter was to be equal to 1,553,164.13 wavelengths of cadmium-red light.
1959			English-speaking countries agree on an International inch.
1960	France	meter	Extensive revision of system. Adoption of name Le Système International d'Unités and abbreviation SI.
1960	France		Length of meter expressed in reference to wavelength of krypton 86 radiation.
1964	France		Conferences for further improvements and clarification of SI.
1965	England		Adopted metric system.
1968	U.S.		Congress passed the Metric Study Act.
1975	U.S.		President Ford signed Metric Conversion Act.
1978	England		The compulsory metrication law was appealed.
1983		meter	1/299,792,458 second
1994	U.S.	Dimensioning Standard	ASME Y14.5M Dimensioning and Tolerancing Standard revised.

FIGURE 2–15 As this table shows, developments continue. Since 1927, the improvements in interferometry have allowed the meter to be defined with even greater accuracy. In 1960, it was defined in terms of an isotope of krypton gas. Then, in 1983, what may be the ultimate breakthrough was made. The meter was defined in terms of time. It is the length of time that light requires to travel one meter in a vacuum: 1/299,792,458 of a second.

and two units for subdivisions. Eventually, the English adopted the Imperial Standard Yard as the basis for all linear measurement. Unfortunately, the original standard was destroyed by the House of Parliament fire in 1834. When the new one was constructed, it was slightly different from the original.

The example of the Imperial Standard Yard demonstrates the two axioms of metrology. First, in order to measure, there must be a standard; second, *that standard must be reproducible.*

2–8 THE ORIGIN OF THE METRIC SYSTEM

Abuses of measurement were among the causes of the French Revolution, so the French Republicans addressed the problem of standardizing measurement early in setting up the new government. In 1790, they established the meter: the distance between the North Pole and the equator passing through Paris, divided into ten million parts. But the earth is difficult to use as a standard of length, so a metal standard, the Meter of the Archives, was made and adopted as the official standard in 1799. The Republicans also set up new units for mass and time.

It was hard for the people to get used to 110-day weeks and days divided into 10 hours of 100 minutes each. They were even more reluctant to buy produce in 10s and 20s instead of dozens. Despite the beheading of Antoine L. Lavoisier (1743–1794), the principal member of the Metric Committee, the new metric system was enforced. By 1799, the metric system was so unpopular that Napoleon Bonaparte I (1769–1821) won popular support by relaxing metric regulations and finally permitting old standards to be used again in 1812. The metric system, however, remained the *legal standard* of France.

In 1870, the first of a series of international conferences to establish the metric system worldwide was held, with 48 delegates representing 25 countries, including France.

In 1889, a general conference in Paris approved the work of the committee. Thirty prototype meters and 40 prototype kilograms were constructed of a platinum-iridium alloy and calibrated with each other. The one new standard most nearly equal to the Meter of the Archives was selected as the International Prototype Meter, and it is now located at the International Bureau of Weights and Measures near Paris.

The remaining 29 prototypes were distributed and became the national standards of the participating countries. Periodically, they are returned to the International Bureau for checking. This plan might have created an accepted international standard, but each country had its own interpretation of the metric system. In 1960, the metric system was revised worldwide, and the system was renamed to distinguish it from the other metric systems. Its new name: *Le Système International d'Unités*, abbreviated SI. As with all measurement systems, the people using it are still adapting it, identifying problem areas, and revising the standards. The terms, symbols, and abbreviations used in SI, summarized in Figure 2–16 and Appendix D, were established by an international committee and adopted by the National Institute of Standards and Technology (NIST).

As with all forms of language, the terminology of SI has its own quirks. For example, many of the units are derived from the names of famous scientists, but these units are not capitalized. As you become familiar with and use these terms, their conventions will come easily to you.

SI Prefixes

Factor	Prefix	Symbol	Factor	Prefix	Symbol
10^{24}	yotta	Y	10^{-1}	deci	d
10^{21}	zetta	Z	10^{-2}	centi	c
10^{18}	exa	E	10^{-3}	milli	m
10^{15}	peta	P	10^{-6}	micro	μ
10^{12}	tera	T	10^{-9}	nano	n
10^{9}	giga	G	10^{-12}	pico	p
10^{6}	mega	M	10^{-15}	femto	f
10^{3}	kilo	k	10^{-18}	atto	a
10^{2}	hecto	h	10^{-21}	zepto	z
10^{1}	deka	da	10^{-24}	yocto	y

FIGURE 2–16 These are the prefixes used in SI to show magnitude. A centimeter, for example, is one hundredth of a meter.

2-9 THE LEGALITY OF THE METRIC SYSTEM IN THE UNITED STATES

The early measurement standard established for the United States by Congress was an 82-inch brass bar that was prepared in London. This standard was brought to the United States in 1813, and it defined the "yard" as the distance between its 28th- and 64th-inch graduations.

In 1866, an act of Congress legalized the use of the metric system but did not make its use mandatory. This act also established the ratios between corresponding units of the measurement systems in the United States and the metric system; for example, the yard was defined as 3600/3937 meter. When the metric standards were created, the United States received two, Numbers 27 and 21, which were received in 1890. Three years later, Congress passed the Mendenhall Act, which established the metric International Prototype Meter as the legal standard and standardized its relationship to the yard. However, the Mendenhall Act did not specifically legalize the inch-pound system that was popularly in use. So the United States has a *legal* system of measurement not widely used and a *popular* system that never has been legalized.

2-10 THE INTERNATIONAL INCH

Popular measurement systems are also revised as governments change. The Mendenhall Act defined the U.S. inch as 25.4000508 mm, but the British Imperial Standard Inch was 25.399978 mm. In 1922, the British revised their standard to 29.399956 mm, increasing the discrepancy between measurements made in the United States and Great Britain. In 1951, Canada also revised their inch to exactly 25.4 mm. So, at one time, there were three "inches" commonly in use.

The three "inches" were reconciled to each other in 1959, when all three governments agreed that one "international" inch would equal 25.4 mm. It was accepted by general agreement but without specific congressional legislation.

Soon afterwards, light waves, a consistent, natural phenomenon, replaced metal bars, which were man-made, in the measurement of international standards. Scientists had started working on using light for standards back in 1892, but light waves were not accepted as the basis for standards until 1960. Light waves allow greater precision in measurement and work well for any measurement system.

2-11 FUNDAMENTAL CRITERIA

In order to evaluate possible measurement systems, we must understand as much as possible about each system and be as unbiased as possible.

Naturalness of the Systems

By now, you should be able to recognize that any measurement system is an arbitrary human invention: nature has no need for quantitative measurement. Even if a system is based on a natural phenomenon, such as the length of a forearm or the distance from pole to equator, practical application will compromise the pure, theoretical accuracy of the measurement system.

In addition, the subdivision of measurement standards is arbitrary. We may have 10 digits on both hands, but dividing a measurement into 10 sections is not necessarily a "natural" thing. By now, most of us think that it is natural to have a day of 24 hours, an hour of 60 minutes, a minute of 60 seconds, and so forth. But remember the French Republicans and their 10-hour day. The dozen and gross are used extensively in trade, but they are measurement standards that can be subdivided easily into a variety of smaller packages to meet the needs of a world of customers.

Economic Considerations

In the past, manufacturers have argued that it was too expensive to completely overhaul their facilities for a new measurement standard. Machines, tools, and manuals were replaced only when the old equipment was obsolete.

Today, however, manufacturers are competing in a global economy. For example, parts manufactured in the Far East must be able to fit precisely with parts made in Europe, and these parts must be able to be replaced by parts made in the United States. Manufacturers must recognize international demands and adapt manufacturing processes to the accepted standards of the global economy.

Either/or Reasoning

Throughout this text, you will notice that we notate measurements in both SI and inch terminology. Some people would demand that we use just one system of measurement in the text, in manufacturing, and throughout the world—*either* English *or* SI.

In addition, some people demand that we completely convert all of our notation of fractions to decimal. But people are accustomed to speaking in fractions—a Frenchman will order a half-liter of wine, not 500 cc—and the use of fractions does not hinder our understanding of the measurement. In computing, decimals provide the highest precision; in communication, inch-based fractions create the clearest picture of the measurement for most people.

It is the conversion of measurements between systems that can cause confusion: Thomas A. Edison (1847–1931) created a 1 3/8 in. (34.925 mm) standard for film—a standard that was misnamed 35 mm. As long as metrologists have a thorough understanding of conversion methods and their implications to precision and quality assurance, there is no reason why we cannot accommodate the use of both systems of measurement.

2–12 THE BEST SYSTEM

At this point, we can safely say that the "best" system of measurement depends on what is being measured, what use the measurement has, whether scientific, commercial, or cultural, and the audience who must understand the results of the measurement process. We must use the measurement system that helps other people understand the goals we are trying to accomplish—the goals that created the need to measure in the first place.

2–13 PRACTICAL CRITERIA

Every step in the measurement process is potentially a source of error. To achieve the most precise and reliable measurement possible, you must choose the measurement system that requires the fewest steps, from instrument selection to the final computations made with your results. To determine the best system of measurement, we use three factors:

1. Metrological factor—which act of measurement will yield usable results
2. Computational factor—which system yields figures that we can use mathematically
3. Communicative factor—which system makes it easiest for us to share the measurement with other people

In turn, each of these factors must be evaluated by four subcriteria (see Figure 2–17), whether the systems provide:

1. Maximum measurement potential
2. Minimum time required
3. Minimum error potential
4. Minimum cost incurred

Metrologically, both the metric and inch systems can handle extremely large measurements and very fine measurements. We use similar instruments in both systems; they require about the same time to operate; and they are subject to the same errors. However, as a cost consideration, the inch system requires two sets of scaled instruments: one for fractional measurements and one for decimal-inch measurements.

In *computational*, the metric system's increments are uniform, whereas the inch system must be converted among different terms (inch, foot, yard, rod, and mile) that are not systematically related. Clearly, converting among distances in the same measurement system is easier in metric terminology.

In *terms of communication*, the inch system is easier for most Americans to immediately comprehend than the metric system. Years of experience with inches, feet, yards, and so forth, give us the mental references we need to easily understand a measurement.

Metric and Inch-Pound Systems Compared

Separate factors to be compared:	Basis for comparing each factor:
1. Metrological	1. Maximum measurement potential
2. Computational	2. Minimum time required
3. Communicative	3. Minimum error potential
	4. Minimum cost incurred

Factors	Metric	Inch-Pound
Metrological		
Science	Excellent	Poor
Industry	Fair	Good
Domestic	Poor	Fair
Computational		
Science	Excellent	Poor
Industry	Good	Poor
Domestic	Good	Poor
Communicative		
Science	Good	Poor
Industry	Good	Poor

FIGURE 2–17 When the two systems are critically analyzed, neither is all good nor all bad.

We can learn the same mental references for metric, but only through the consistent, practical application of this measurement system in daily life.

Again, your choice of measurement system comes down to what you are measuring, what you are going to do with the measurements, and who you are measuring for. It is easier for scientists to express the vast distances between stars or the minute space between atoms in metric terms. But scientists also still use the term *horsepower,* which is based on the English inch system. Similarly, most screw threads are still stated in inch or "soft metric" terms, even if they are sold as "metric" screws; and, in 1961, Japanese officials had to pass a law forbidding the calibration of speedometers in miles per hour, even though the metric system had been enforced for 30 years.

2–14 THE DECIMAL-INCH SYSTEM

In order to try to eliminate some of the computational problems with the inch system, the decimal-inch system was created. It is not such a new idea because the decimal foot was used in surveying in the United States before 1856. In fact, until the Civil War, 1/64 in. was the smallest standard measurement used in practical work; thousandths of an inch were nothing more than theories.

The decimal-inch system made its first major breakthrough into popular use when, in 1930, the Ford Motor Company adopted it, followed quickly by the aircraft industry. The Society of Automotive Engineers (SAE) published a complete decimal-inch dimensioning manual in 1946. Thirteen years later, the American Standards Association (ASA) and the Society of Manufacturing Engineers (SME) chose to jointly urge greater use of the decimal-inch, proposing an American standard for its definition and use.

2–15 METROLOGICAL CONSIDERATIONS

As we said earlier in this chapter, your eye can see light through a crack as small as 2.54 μm (0.0001 in.) However, when light goes through a crack, it creates a contrast between the dark sides and the light

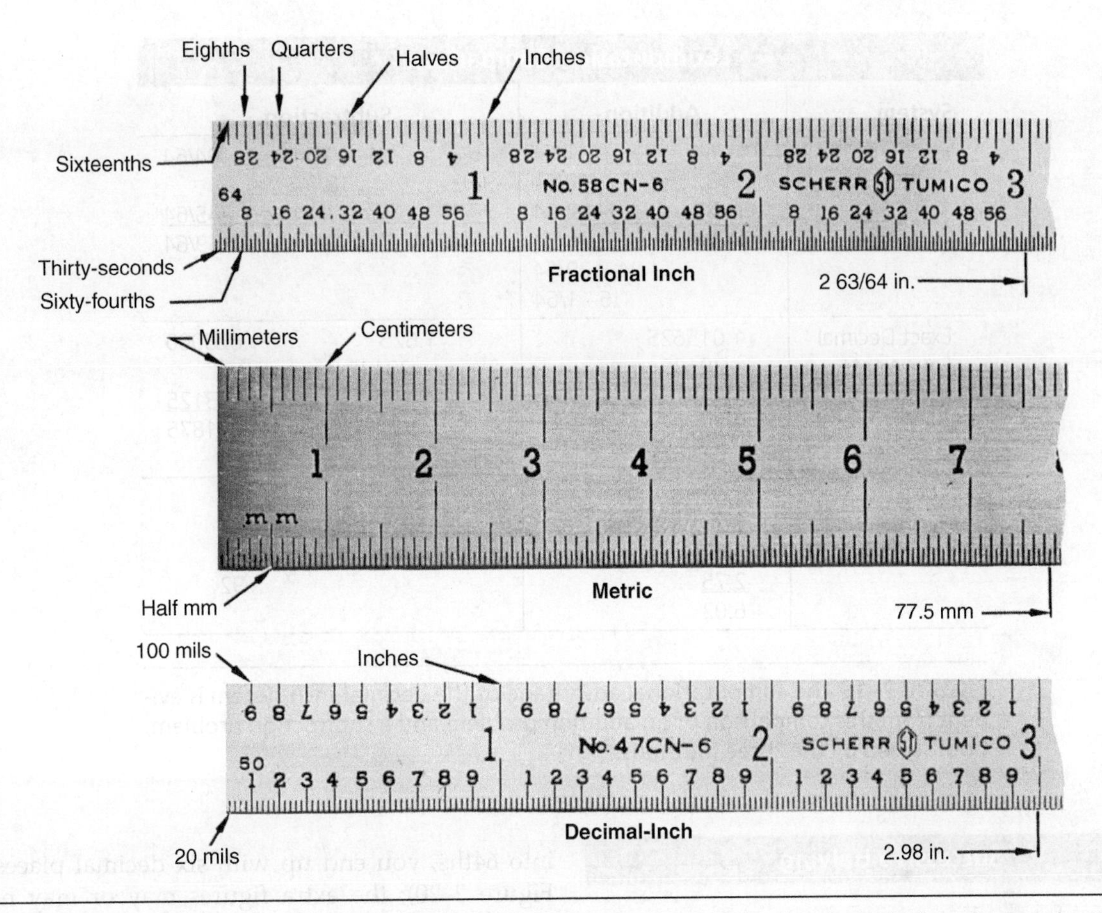

FIGURE 2–18 The decimal-inch readings provide nearly the precision of the fractional-inch scale and the simplicity of the metric scale. Steel rules with both scales are available.

coming through. Your eye would be less accurate if you were trying to read the same division—a small black mark—on a metal ruler.

In Figure 2–18, you can see the relationship among inch, metric, and decimal-inch rulers. You can also see how difficult it might be to accurately read a measurement, especially if the distance fell somewhere between marks. So, if you are trying to measure with a ruler, the traditional inch ruler is hard to read because of the variety of fractional marks; the metric ruler's divisions are either too large to be useful or too small to be read easily by the unaided eye. Users have a tendency to round to a mark too often with the decimal-inch ruler. Obviously, to make precise, fine measurements, you need a different kind of instrument.

2–16 COMPUTATIONAL CONSIDERATIONS

One advantage of the metric and decimal-inch system is obvious: there are no fractions to combine when you are making calculations. The measurements in this system add, subtract, multiply, and divide easily. During computation, you can add decimal places as needed to report the measurement to its last significant figure.

But any skilled metrologist will also be familiar with the use of fractions. You might spend up to five times as long calculating measurements in fractions because many times you first have to compute a common denominator (see Figure 2–19), then reduce the result back down.

Computation Comparison

System	Addition		Subtraction		
Fractional System	1–1/64	1–1/64	1–5/8	1–40/64	104/64
	9/32	18/64	–9/32	–18/64	
	1–31/32	1–62/64	–27/64	–27/64	–45/64
	2–3/4	2–48/64			59/64
		4–129/64			
		6– 1/64			
Exact Decimal Equivalents	1.015625			1.625	1.625000
	.28125			–. 28125	
	1.96875			– .421875	–.703125
	2.75				.921875
	6.015625				
Two-Place Decimal System	1.02			1.62	1.62
	0.28			–.28	
	1.97			–.42	–.70
	2.75				.92
	6.02				

FIGURE 2–19 The computational advantage of the decimal-inch system is evident from this comparison of an addition problem and a subtraction problem, both solved by the three methods.

Successive Halving

1	1	
1/2	0.5	1 place
1/4	0.25	2 places
1/8	0.125	3 places
1/16	0.0625	4 places
1/32	0.03125	5 places
1/64	0.015625	6 places

FIGURE 2–20 Successive halving of one (1) in order to form the common fractions results in the 64ths having six decimal places.

Of course, using decimals can create problems, too. If a measurement is originally made using the inch system, you will have to convert all fractions to decimal first. Conversion can lead to problems with rounding off. When the number "1" is divided into 64ths, you end up with six decimal places (see Figure 2–20); the extra figures may or may not be significant, but they can add up to create errors of 1/64 very quickly. Also, at times, you will need to convert the decimal result back to a fraction after all your calculations.

2–17 ROUNDING OFF NUMERICAL VALUES

When we *round off*, we eliminate unnecessary figures in any computation. However, you must know both the correct method of rounding off and the number of significant figures needed in order to round off properly.

For example, we need to know the volume of a block measuring 15.2 × 11.1 × 8.3 in.: 1,400.376 cubic in. Our original measurements were reported only to one decimal place—the tenths—but our calculations give us a volume with three decimal places. Presumably, the total volume reported with three decimal places is more accurate than the original measurements on which our calculations were based.

If we measured more precisely, we find that the block is 15.20 × 11.10 × 8.29 in.—a change in one value of only 0.01 in. But this more precise measurement changes the total volume to 1,398.6888 cu. in., which is 1.6872 cu. in. less volume than originally reported. If the other measurements were also incorrect, the reduction in volume might have been greater, the total volume might have actually increased, or the corrected measurements might have canceled out any increase or decrease in volume. For practical purposes, the volume probably would be rounded off to 1,400 cu. in., because this *area of uncertainty* exists.

We also often use the rule: the answer to any calculation can be no more accurate than the *least* accurate measurement reported. Using this rule, the calculation using only tenths should be reported in tenths: 1,400.4 cu. in. Similarly, the calculation using two decimal places should be reported as 1,398.7 in.^{3}, because only one dimension was "accurately" reported with two decimal places. We must assume that the "zero" in the last place in the other two dimensions is just a place holder until the accuracy of those measurements is verified.

In rounding off, we traditionally leave the "rounded to" digit the same if the last number is less than 5, and we increase the digit if the last number is more than 5. For example:

1,400.376	and	1,398.689
to 1,400.38		to 1,398.69
to 1,400.4		to 1,398.7
to 1,400		to 1,399
		to 1,400

By rounding, we reach the same result in both cases. But what happens if your last figure is neither more nor less than 5—it equals 5? It probably does not matter as long as you are consistent (see Figure 2–21).

Consistency is one of the most important components of reliable measurement. Of course, we must consistently choose the best measurement technique for our reporting and consistently use the same method of rounding off. Then, even when we make errors, the errors have a chance to cancel each other out or to be caught more easily, because they will not agree with our other, consistent results.

To demonstrate the importance of consistency, let us determine the total volume of five blocks by finding each block's individual volume and then adding those results for a grand total (see Figure 2–22). In the example, three values ended in 5, and the "Rounded Up" total was 0.03 in. larger than the "Rounded Down" total. If the number of blocks increases, the difference between the rounded up and rounded down totals increases, creating a corresponding increase in costs for both production and shipping.

Another option for rounding would be to go up half the time and down the other half; but then you have to keep track. Fortunately, an incorporated standard, ASA Z25.1, from the American Standards Association, has been created. This standard is: *raise (round up) the remaining last digit if it is odd and leave it the same (round down) if it is even.* If your last digit is 5, the rounded value will always be even: 1398.65 becomes 1398.6, and 1398.75 becomes 1398.8. This method of rounding off was used to create the basic 0.02 division of the decimal-inch system, when 1/64 division was rounded: 0.015625 to 0.01562 to 0.0156 to 0.016 to 0.02. (See the complete table for conversion of fractions to decimal-inch in Figure 2–23 and Appendix C.)

Figure 2–24 shows that the decimal-inch system can safely be converted to the fractional-inch system, as long as the method of rounding results in truly average values. When the true values and the rounded-off values from 1/64 through 1 are added and averaged, the sum of the rounded values, regardless of the places retained, is equal to the sum of the values given to six places. Neither the sums nor the averages have increased or decreased by rounding off.

Like many rules, this one has an exception: *the final rounded value should be created from the most precise*

Rounding Up and Rounding Down								
Value Places	Rounded up Places				Rounded Down Places			
5	4	3	2	1	4	3	2	1
2.07550	2.0755	2.076	2.08	2.1	2.0755	2.075	2.07	2.1
2.07551	2.0755	2.076	2.08	2.1	2.0755	2.075	2.07	2.1
2.07552	2.0755	2.076	2.08	2.1	2.0755	2.075	2.07	2.1
2.07553	2.0755	2.076	2.08	2.1	2.0755	2.075	2.07	2.1
2.07554	2.0755	2.076	2.08	2.1	2.0755	2.075	2.07	2.1
2.07555	2.0756	2.076	2.08	2.1	2.0755	2.075	2.07	2.1
2.07556	2.0756	2.076	2.08	2.1	2.0756	2.076	2.08	2.1
2.07557	2.0756	2.076	2.08	2.1	2.0756	2.076	2.08	2.1
2.07558	2.0756	2.076	2.08	2.1	2.0756	2.076	2.08	2.1
2.07559	2.0756	2.076	2.08	2.1	2.0756	2.076	2.08	2.1

FIGURE 2–21 Rounding up requires the digits retained to be raised when the digit eliminated to the right is a 5. Rounding down requires that it be kept the same. In the example shown, the shaded areas are the values that require the up and down decisions.

Rounding Up vs. Rounding Down			
	Calculated Volume	Rounded Up	Rounded Down
	1400.375	1400.38	1400.37
	1400.376	**1400.38**	**1400.38**
	1399.995	1400.00	1399.99
	1400.395	1400.40	1400.39
	1399.991	**1399.99**	**1399.99**
Totals	7001.132	7001.15	7001.12
Differences		over by 0.02	under by 0.01
Average T/5	1400.226 (true)	1400.230 (high)	1400.224 (low)

FIGURE 2–22 In this group of five values, the Rounded Up column yielded high values, the Rounded Down column, low values. The more items involved, the greater the total error and average error.

Fraction to Decimal Conversion Chart															
4THS	8THS	6THS	2NDS	64THS	TO 4 PLACES	TO 3 PLACES	TO 2 PLACES	4THS	8THS	16THS	32NDS	64THS	TO 4 PLACES	TO 3 PLACES	TO 2 PLACES
				1/64	0.0156	0.016	0.02					33/64	0.5156	0.516	0.52
			1/32		0.0312	0.031	0.03				17/32		0.5312	0.531	0.53
				3/64	0.0469	0.047	0.05					35/64	0.5469	0.547	0.55
		1/16			0.0625	0.062	0.06			9/16			0.5625	0.562	0.56
				5/64	0.0781	0.078	0.08					37/64	0.5781	0.578	0.58
			3/32		0.0938	0.094	0.09				19/32		0.5938	0.594	0.59
				7/64	0.1094	0.109	0.11					39/64	0.6094	0.609	0.61
	1/8				0.1250	0.125	0.12		5/8				0.6250	0.625	0.62
				9/64	0.1406	0.141	0.14					41/64	0.6406	0.641	0.64
			5/32		0.1562	0.156	0.16				21/32		0.6562	0.656	0.66
				11/64	0.1719	0.172	0.17					43/64	0.6719	0.672	0.67
		3/16			0.1875	0.188	0.19			11/16			0.6875	0.688	0.69
				13/64	0.2031	0.203	0.20					45/64	0.7031	0.703	0.70
			7/32		0.2188	0.219	0.22				23/32		0.7188	0.719	0.72
				15/64	0.2344	0.234	0.23					47/64	0.7344	0.734	0.73
1/4					0.2500	0.250	0.25	3/4					0.7500	0.750	0.75
				17/64	0.2656	0.266	0.27					49/64	0.7656	0.766	0.77
			9/32		0.2812	0.281	0.28				25/32		0.7812	0.781	0.78
				19/64	0.2969	0.297	0.30					51/64	0.7969	0.797	0.80
		5/16		21/64	0.3125	0.312	0.31			13/16		53/64	0.8125	0.812	0.81
					0.3281	0.328	0.33						0.8281	0.828	0.83
			11/32		0.3438	0.344	0.34				27/32		0.8438	0.844	0.84
				23/64	0.3594	0.359	0.36					55/64	0.8594	0.859	0.86
	3/8				0.3750	0.375	0.38		7/8				0.8750	0.875	0.88
				25/64	0.3906	0.391	0.39					57/64	0.8906	0.891	0.89
			13/32		0.4062	0.406	0.41				29/32		0.9062	0.906	0.91
				27/64	0.4219	0.422	0.42					59/64	0.9219	0.922	0.92
		7/16			0.4375	0.438	0.44			15/16			0.9375	0.938	0.94
				29/64	0.4531	0.453	0.45					61/64	0.9531	0.953	0.95
			15/32		0.4688	0.469	0.47				31/32		0.9688	0.969	0.97
				31/64	0.4844	0.484	0.48					63/64	0.9844	0.984	0.98
1/2					0.5000	0.500	0.50	1					1.0000	1.000	1.00

FIGURE 2–23 Generally, when converting fractional inches to decimal-inch, it is only necessary to convert to two places. If the fractional dimension is the basic size of a final part and a gage is being made to check it, then there is reason to use more than two decimals.

1 6 PLACES	2 5 PLACES	3 4 PLACES	4 3 PLACES	5 2 PLACES	6 1 PLACE
0.015625	0.01562	0.0156	0.016	0.02	0.0
.031250	.03125	.0312	.031	.03	.0
.046875	.04688	.0469	.047	.05	.0
.062500	.06250	.0625	.062	.06	.1
.078125	.07812	.0781	.078	.08	.1
.093750	.09375	.0938	.094	.09	.1
.109375	.10938	.1094	.109	.11	.1
.125000	.12500	.1250	.125	.12	.1
.140625	.14062	.1406	.141	.14	.1
.156250	.15625	.1562	.156	.16	.2
.171875	.17188	.1719	.172	.17	.2
.187500	.18750	.1875	.188	.19	.2
.203125	.20312	.2031	.203	.20	.2
.218750	.21875	.2188	.219	.22	.2
.234375	.23438	.2344	.234	.23	.2
.250000	.25000	.2500	.250	.25	.2
.265625	.26562	.2656	.266	.27	.3
.281250	.28125	.2812	.281	.28	.3
.296875	.29688	.2969	.297	.30	.3
.312500	.31250	.3125	.312	.31	.3
.328125	.32812	.3281	.328	.33	.3
.343750	.34375	.3438	.344	.34	.3
.359375	.35938	.3594	.359	.36	.4
.375000	.37500	.3750	.375	.38	.4
.390625	.39062	.3906	.391	.39	.4
.406250	.40625	.4062	.406	.41	.4
.421875	.42188	.4219	.422	.42	.4
.437500	.43750	.4375	.438	.44	.4
.453125	.45312	.4531	.453	.45	.5
.468750	.46875	.4688	.469	.47	.5
.484375	.48438	.4844	.484	.48	.5
.500000	.50000	.5000	.500	.50	.5
.515625	.51562	.5156	.516	.52	.5
.531250	.53125	.5312	.531	.53	.5
.546875	.54688	.5469	.547	.55	.5
.562500	.56250	.5625	.562	.56	.6
.578125	.57812	.5781	.578	.58	.6
.593750	.59375	.5938	.594	.59	.6
.609375	.60938	.6094	.609	.61	.6
.625000	.62500	.6250	.625	.62	.6
.640625	.64062	.6406	.641	.64	.6
.656250	.65625	.6562	.656	.66	.7
.671875	.67188	.6719	.672	.67	.7
.687500	.68750	.6875	.688	.69	.7
.703125	.70312	.7031	.703	.70	.7
.718750	.71875	.7188	.719	.72	.7
.734375	.73438	.7344	.734	.73	.7
.750000	.75000	.7500	.750	.75	.8
.765625	.76562	.7656	.766	.77	.8
.781250	.78125	.7812	.781	.78	.8
.796875	.79688	.7969	.797	.80	.8
.812500	.81250	.8125	.812	.81	.8
.828125	.82812	.8281	.828	.83	.8
.843750	.84375	.8438	.844	.84	.8
.859374	.85938	.8594	.859	.86	.9
.875000	.87500	.8750	.875	.88	.9
.890625	.89062	.8906	.891	.89	.9
.906250	.90625	.9062	.906	.91	.9
.921875	.92188	.9219	.922	.92	.9
.937500	.93750	.9375	.938	.94	.9
.953125	.95312	.9531	.953	.95	1.0
.968750	.96875	.9688	.969	.97	1.0
.984375	.98438	.9844	.984	.98	1.0
1.000000	1.00000	1.0000	1.000	1.00	1.0
32.500000 0.5078125	32.50000 0.50781	32.5000 0.5078	32.500 0.508	32.50 0.51	32.5 (SUM) 0.5 (AVERAGE)

FIGURE 2–24 The value of the recommended method for rounding off is demonstrated in this table. Six-place decimal equivalents of all 64 of the 64ths of an inch are successively rounded off; the successive results are the same as the averages. This shows that the rounding off did not introduce an error.

value obtainable—not from a series of roundings. For example, 0.5499 should be rounded off successively to 0.550, 0.55, and 0.5. Note that the last rounding is to 0.5, not 0.6, which would be the result if you followed the rule: the most precise value is less than 0.55, therefore, the final rounding is to 0.5. Similarly, 0.5501 is rounded off to 0.550, 0.55, and 0.6, because the most precise value obtainable is greater than 0.55.

This exception occurs only twice in the entire 64 conversions from fractions to decimal-inch:

29/64 is
0.453125 to 0.45312 to 0.4531 to 0.453 to 0.45 to 0.5

35/64 is
0.546875 to 0.54688 to 0.5469 to 0.547 to 0.55 to 0.5

If the two-place values had been exact and not the result of successive roundings, the one-place values would have been 0.4 and 0.6, respectively.

To ensure that your roundings are as precise as possible, carry the calculation two places beyond the values needed, then round off. Other rules are summarized in Figure 2–25.

What Is the Real Issue?

With the pocket calculator, we now can convert any system of measurement with ease; the computer has destroyed any remaining separation among systems of measurement. Why then are systems of measurement still an issue? The simplest, strongest reason is *world trade.*

For example, the governments of France and Japan *require* that imports be in metric dimensions. In contrast, the U.S. government places no such restrictions on other countries' products. Most manufacturers in the United States have used soft metrication to circumvent this problem; for example, a .223 caliber M16 rifle is referred to as a 5.56 mm in a metric-based country.

International standards and trade specifications sometimes result in problems for exporters. For example, if skids are standardized only in metric, they are mismatched for goods manufactured in inch-pounds. Soft metrication does not help; so our exporters must pay the penalty.

General Rules for Rounding Off

When a value is to be reduced in the number of decimal places, one of the following three rules is followed:

1. When the digit to be dropped is less than 5, there is no change in the preceding figures.
   Examples:
   0.280423 to 0.28042 to 0.2804 to 0.280 to 0.28
2. When the digit to be dropped is greater than 5, the preceding digit is increased by 1.
   Examples:
   0.046857 to 0.04686 to 0.0469 to 0.047 to 0.05
3. **When the digit to be dropped is exactly 5, round off to the nearest even number.**
   Examples:
   0.09375 to 0.0938 but 0.09385 to 0.0938

FIGURE 2–25 Whenever possible, carry the calculation two places beyond the desired value, then round off the last two significant figures.

The metric versus inch-pound controversy is finally winding down throughout the world. The advantages of operating in the global economy—an economy based almost solely in SI—have encouraged the majority of manufacturers to adopt the system and adapt to it. In some manufacturing, the fractional-inch and decimal-inch systems of measurement are still used. Our purpose in this chapter was to familiarize you with the measurement systems in use, their advantages and disadvantages, and their histories and futures. That way, you will be fully equipped to make that all-important choice of which measurement system to use.

SUMMARY

- Because it is necessary to understand each other, some terms must have specific meanings. Linear measurement expresses distance between points. It permits distances to be reproduced. A measurement consists of a unit of length and a multiplier. Each measurement begins at a reference point and terminates at a measured point. It lies along a line of measurement, which must have a known relationship to the feature being measured.
- A feature is a measurable characteristic. It is bounded by edges, usually but not always, formed by the intersection of planes. The dimension of the feature is the designer's concept of perfection. Features of actual parts are not perfect.
- Measurement shows the deviation from perfection. The measured conformity to the dimension is the accuracy. The refinement with which this can be known is the precision. The effect of accuracy and precision on attaining the desired results is reliability. Precision is essential for reliability but alone cannot produce it. Increased reliability requires increased accuracy, and that requires increased precision. Therefore, the general term used to denote progress in measurement is *precision*.
- There is no insurmountable difficulty involved in a total change from the inch-pound system to the metric system, nor is there any convincing proof that such a change is needed or desirable.
- The metric system is unquestionably superior in ease of computation. Popular use of the metric system in science clearly advocates its continued use. The units of the metric system were selected theoretically. It is not surprising that they bear little natural relation to most things in the real world, including the resolving power of the human eye.
- The inch-pound system is handicapped by an apparently disorderly assortment of units. This slows computation and sometimes clogs communication. However, its basic units are of such convenience that it finds wide use even in places in which the metric system is the legal system. While design engineers often use metric dimensions, the inch system is still the primary measurement system used in tool rooms within the United States.
- Each system has merit and has roles in which it is fully accepted. It is even possible to borrow good points from one system to enhance the other; an example of this is the decimal-inch. Furthermore, in this computer age the system of annotation is relatively unimportant and should not be an issue. Our efforts are needed far more for the application of sound principles and the perfection of our reference standards.

END-OF-CHAPTER QUESTIONS

1. Ambiguity is the enemy of dimensional metrology. Which of the following best describes a figure with four sides, each of which is 2½ in. in length?
   a. Polyhedron
   b. Parallelepiped
   c. Square
   d. Parallelogram
2. The principal dimensional measurement is length. Which of the following is the best definition of length?
   a. A dimension
   b. The shortest distance between two points
   c. The number of units of measurement that separates two points
   d. The opposite of width

3. Which of the following terms is not essential in defining a length measurement?
   a. Reference point
   b. Unit of measurement
   c. Tolerance
   d. Measured point
4. In which one or more of the following characteristics does a dimension differ from a length?
   a. A dimension is an intended size, whereas a length is the measured size.
   b. A length may be a decimal, but a dimension is a whole number.
   c. A dimension must be determined by measurement, whereas a length is stated on the drawing of the part.
   d. Lengths have tolerances, whereas dimensions do not.
5. The primary purpose of dimensional measurement is to:
   a. show how precise a part is
   b. show how close the manufacturer can produce parts
   c. justify the price
   d. communicate the designer's intent
6. Which of the following is an international standards body?
   a. ISO
   b. ABCD
   c. GD & T
   d. A2Z
7. When measurements are repeatedly taken on the same unit, the extent to which an instrument replicates its measurements is:
   a. accuracy
   b. range
   c. precision
   d. discrimination
8. Which of the following best describes the reference point?
   a. Point at which any measurement begins
   b. Position identified on a drawing as the end of a measurement
   c. Furthest point from the measured point
   d. Highest point above the reference plane
9. Which of the following is an American Dimensioning and Tolerancing Standard?
   a. ASME Y14.5 – 2009
   b. MSNBC
   c. ISO 2001
   d. American Society of Training and Development (ASTD)
10. Part features designate the portion of a part that may be considered separately from the part itself.
    a. True
    b. False
11. How many .001 inches are there in one inch?
    a. 1
    b. 100
    c. 1,000
    d. 10,000
12. Which of the following best defines an edge?
    a. The sharp corners of the part
    b. The intersection of a plane with another plane
    c. The cutting portion of an instrument
    d. Conical section
13. Which of the following best describes male features?
    a. Total lengths, heights, and widths
    b. Round rods or tubular parts
    c. Features bounded by outside dimensions
    d. Features larger than female features
14. If 1 mm equals 0.0394 inches, 700 millimeters equals how many inches?
    a. 35.53
    b. 275.8
    c. 27.58
    d. 0.003
15. The ability of two or more inspectors to obtain consistent results repeatedly when measuring the same set of parts and the same measuring instruments is called:
    a. accuracy
    b. precision
    c. reproducibility
    d. error

16. One shooter fires 10 shots in a group 5 in. in diameter but all are off the target. Then, a second shooter has a group 10 in. in diameter but three shots are through the X-ring. Which of the following statements are the most likely to be correct?
   a. The first shooter is precise but not as accurate as the second shooter.
   b. The second shooter is the luckiest.
   c. The second shooter is both precise and accurate.
   d. The first shooter is precise but inaccurate, whereas the second shooter is more precise and more accurate.

17. Accuracy may be defined several ways. Which of the following is correct?
   a. Accuracy is the measure of precision.
   b. Accuracy is the measure of conformity.
   c. Accuracy is the measure of reliability.
   d. Accuracy is the best synonym for precision.

18. Which of the following does reliability depend on?
   a. High-grade measuring instruments
   b. Both accuracy and precision
   c. Lots of luck
   d. Limited production

19. Can either accuracy or precision be improved without respect for the other value? Select the correct statement.
   a. Neither can be changed without affecting the other.
   b. Either can be increased without affecting the other.
   c. Precision can be increased without an increase in accuracy.
   d. Accuracy can have only a limited increase without requiring an increase in precision.

20. Tolerances are specified for machined parts for which of the following reasons?
   a. Manufacturing equipment has inherent variation.
   b. To ensure proper fit between mating parts.
   c. The designer wasn't sure of the exact measurement to specify.
   d. Both a and b.

21. The earliest recorded length standard was:
   a. The ulna
   b. The length of the monarch's foot
   c. The cubit
   d. The Egyptian pyramids

22. What was the primary purpose of the earliest recorded standard?
   a. To ensure the fair payment of taxes
   b. To practice early medicine
   c. To ensure that parts mated in building construction
   d. To mix cement in building the pyramids

23. What is the chief similarity between the use of this first recorded standard and present-day practices?
   a. The high cost of reliable measurement
   b. The use of metal standards
   c. The need for inches and fractions of inches
   d. The requirement that working standards agree with the master standard

24. The English system and the metric system differ in the origin of their standards. Which of the following best describes those origins?
   a. The metric standards derived from the standards in use by the crafts guilds, whereas the English standards were those used by the peasants and small merchants.
   b. Both were by decree of the existing governments.
   c. The metric standards were derived from the diameter of the earth, whereas the English standards were selected by the House of Parliament.
   d. The metric standards were placed in force by the Revolutionary Committee, whereas the English standards were already in use.

25. Why have light waves replaced metal bars as the master standards of length?
   a. Platinum-iridium bars are very expensive, but light is free.
   b. They are easier to use in the dark.
   c. Light wave standards can be more easily reproduced, and to greater accuracy, than metal bars.
   d. The use of light waves avoids political considerations.

26. There are many ways that measurement systems can be compared. Which of the following are the three best ways?
    a. Metrological superiority, computational advantage, and communications ability
    b. Finest discrimination, greatest accuracy, and easiest convertibility
    c. Agreement among nations, historical precedent, and general acceptability
    d. Lowest cost added to the manufacture of products, most easily available gages and measuring instruments, and easy-to-understand quality control procedures

27. Round off 25.7853901 to two significant decimals.
    a. 25.786
    b. 25.78
    c. 25.79
    d. 26

28. Each of the following lists is fully correct. However, one of them comes closest for defining the four most essential qualities of a measurement system. Which is it?
    a. Greatest measuring ability; least errors; least time to conduct measurements; least destruction to workpieces while being measured
    b. Lowest cost; widest range of direct measurement; adaptability to angles, surface finishes, and radii as well as lengths; minimum errors
    c. Easiest computations; self-canceling errors; clearly visible minimum scale divisions; same meaning in all languages
    d. Optimum measuring ability; minimum errors; minimum time required to measure; minimum cost

29. *Le Système International d'Unités* is abbreviated as:
    a. LSIU
    b. SI
    c. IU
    d. mm

30. Which of the following are good reasons to use the decimal-inch system?
    a. Convertibility in and out of the metric system is easy.
    b. Rounding off is easier than it is with the English system using fractions.
    c. The finest division is close to the minimum resolving power of the human eye.
    d. It does not need to be calibrated to the international meter.

31. Which of the following correctly represents a centimeter and a kilometer?
    a. 10^{22}, 10^{3}
    b. 10^{1}, 10^{21}
    c. 10^{3}, 10^{2}
    d. 10^{22}, 10^{4}

32. Which of the following best explains the reasons for rounding off?
    a. Elimination of numbers over 5
    b. Averaging the computational error
    c. Elimination of meaningless digits
    d. Simplifying computations

33. What is the exception to the rule for rounding off?
    a. Values over 9 should be left undisturbed.
    b. Calculations involving values of different decimal places should not be rounded off.
    c. The final rounding should be obtained from the most precise value.
    d. Values beyond four decimal places should not be rounded off.

Discussion Question

34. Visit the ANSI web site at www.ANSI.org. What is the mission of ANSI? Identify the current ISO quality standard.

CHAPTER THREE

MEASUREMENT AND TOLERANCES

LEARNING OBJECTIVES

- Define tolerance as it applies to metrology.
- Recognize various basic symbols used in geometric dimensioning and tolerancing.
- Define maximum material condition (MMC) and least material condition (LMC).
- Explain the advantages and disadvantages of GD&T.
- List the five types of geometric characteristics.

OVERVIEW

This chapter has been included to show the relationship between the "engineering design," "product," and "part inspection." The engineering design is what is to be manufactured. The product is what has been manufactured. The part inspection is how we compare the product with the engineering design. The common factor that binds these three areas together is geometric dimensioning and tolerancing. When properly applied, geometric dimensioning and tolerancing will ensure the most economical and effective production of the parts.

3–1 THE MEANING OF TOLERANCE

Measuring the features of any part or assembly to such precision that no two would be identical is possible. For their intended purposes, all parts can vary within limits and still be adequate. If an excessively precise instrument is used, the meaningless decimal places are rounded off. The excessive costs of such measurements are real but obscured and costly. This occurs, for example, in toolmaking. In two circumstances the measurements affect costs to such an extent that they cannot be ignored. The first case is in research. If, for example, a particular hybrid gain is being studied, insufficient precision (whether dimensions, weight, or any other

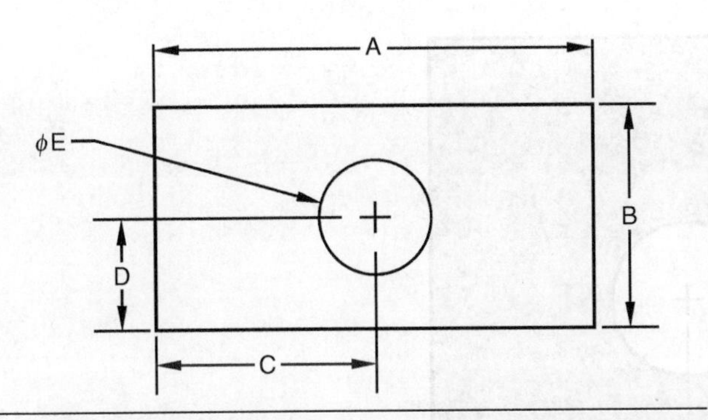

FIGURE 3–1 Tolerances apply to both dimensions of size (A, B, and E) and dimensions of location (C and D).

feature) could invalidate the effort. In contrast, excessive precision could price the research out of consideration. The second case that is even more common is in the production of goods. Too little precision produces bad products. Too much precision causes the goods to be priced out of the market. Either of these situations is made worse by the number of individuals involved and particularly by their separation, often at distant facilities. An unambiguous expression of "good enough" is clearly needed. This expression is called *tolerance.*

Tolerance is the total amount that a specific dimension is permitted to vary. In dimensional metrology, tolerances are applied to both position (where) and size (how big) dimensions. Both types of dimensions must have tolerances for economical manufacture (see Figure 3–1).

3-2 GEOMETRIC DIMENSIONING AND TOLERANCING

The authoritative document governing the use of geometric dimensioning and tolerancing (GD&T) in the United States is ASME Y14.5M, "Dimensioning and Tolerancing." Outside the United States, a series of standards under the general heading of ISO (International Standards Organization) standards governs geometric dimensioning and tolerancing. There are many standards that address specific requirements such as screw threads, drills, gears, and so forth. The United States is a member of the ISO and has contributed greatly to the documents. At this time, there is a 90 to 95% agreement between the ASME Y14.5 and the ISO standards (see Figure 1–4).

What Is Geometric Dimensioning and Tolerancing?

GD&T is a means of dimensioning and tolerancing a drawing with respect to the actual function or relationship of part features that can be most economically produced. The key words here are function and relationship. This type of dimensioning and tolerancing should be used when:

1. Features are critical to the functionality or interchange ability of the part.
2. Datum references are desirable to ensure consistency between design, manufacturing, and inspection.
3. Computerization techniques in design and manufacturing are being used or are desirable.
4. Standard interpretation or tolerance is not already implied.

The Benefits of GD&T

GD&T adds clarity to the conventional coordinate dimensioning. The universal symbols can be used to convey design intent to remote manufacturing or assembly sites, provide a common standard for dimensioning practices, enhance repeatability of part orientation, and increase interchangeability of parts.

Conceivably, the biggest advantage of using geometric tolerancing lies in increased production tolerancing. The traditional Cartesian coordinate system defines every point with x and y coordinates, creating square tolerance zones. Suppose a hole is to be located 20 mm from both the x and the y axis, as shown in Figure 3–2. Since the hole does not have to be perfectly located to work with its mating part, the dimension has a tolerance of ± 1 mm. Our permissible machining limits are shown in Figure 3–3 (greatly exaggerated).

The location dimensions of ± 1 mm establish a total tolerance zone of 2 mm, thus the location of the

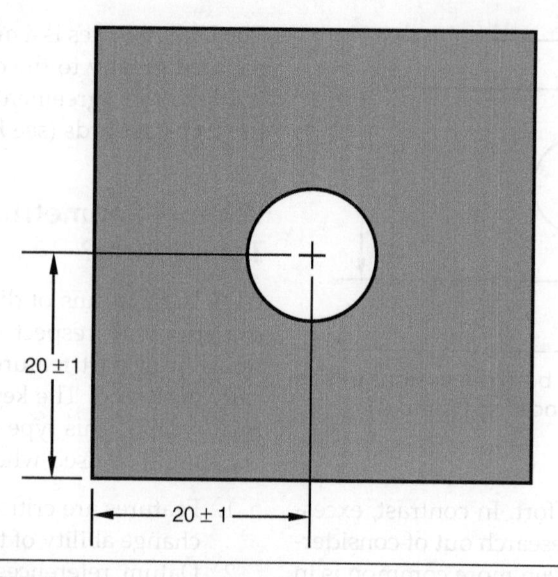

FIGURE 3–2 A square block with a hole drilled in the center, using conventional tolerancing.

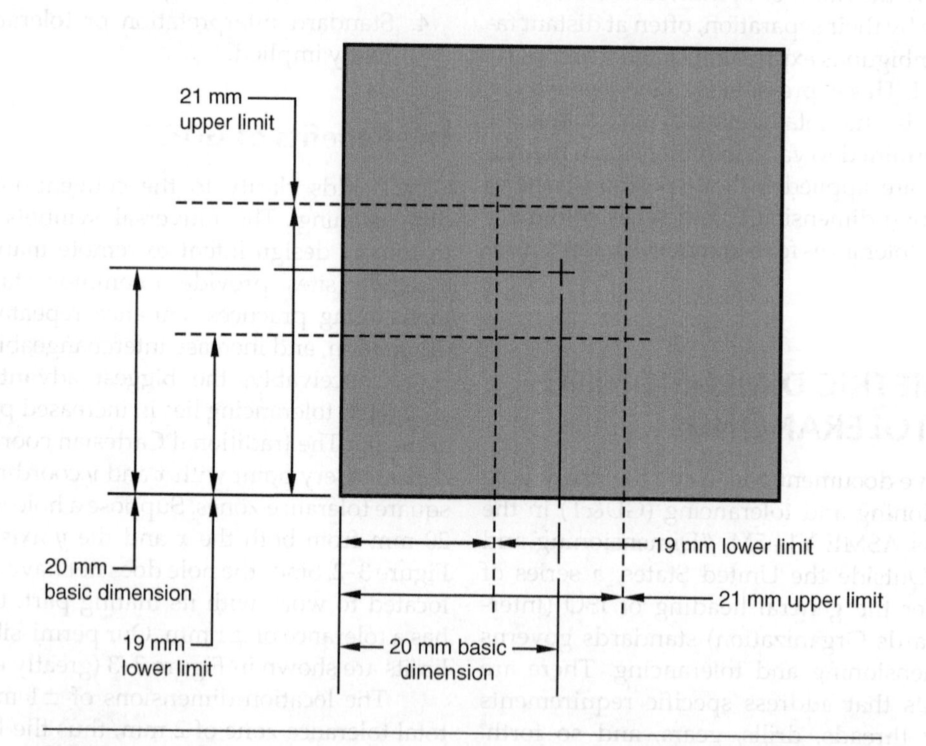

FIGURE 3–3 The conventional method prescribes a square (or rectangular) tolerance zone.

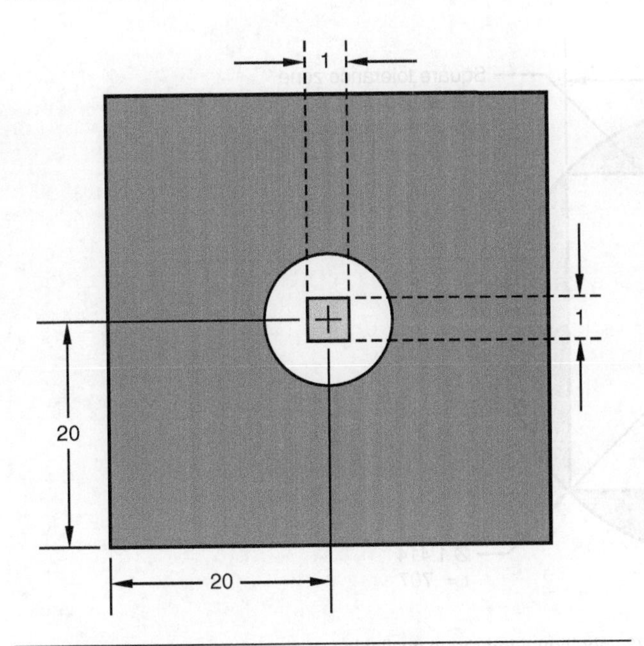

FIGURE 3–4 Coordinate location tolerance zone. The shaded area shows where the location of the hole could vary.

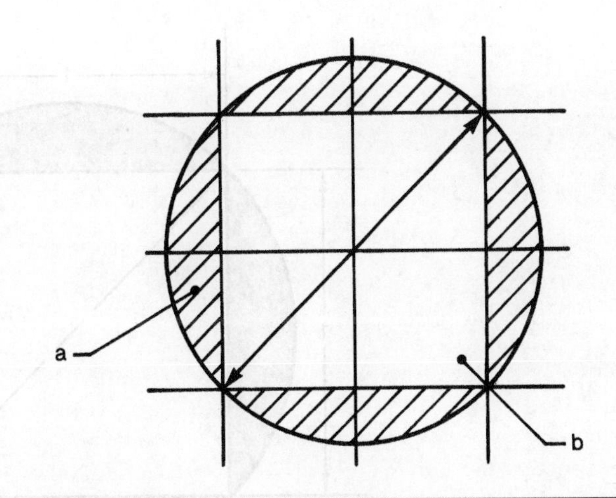

FIGURE 3–6 Point b is technically within true tolerance, whereas a is out of tolerance. Obviously, a will produce equally acceptable parts.

hole can vary by 2 mm, as shown in Figure 3–4. The center of the hole can fall anywhere within the square (shaded) area and still be an acceptable part.

Let us look more closely at the permissible center for drilling (see Figure 3–5). Only the crosshatched space is within tolerance. However, the shaded areas in Figure 3–6 are also within the same distance. A part drilled at point A is within the acceptable tolerance range, as is a part drilled at point B. If this is only one of a string of dimensions, the unnecessary restriction becomes much greater. Consider what happens when two holes are involved instead of one hole (see Figure 3–7).

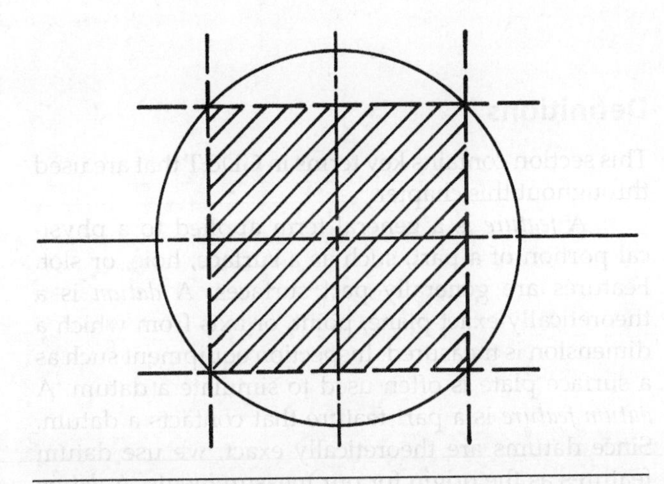

FIGURE 3–5 The conventional method requires that the drill center be within the shaded area, yet the additional spaces prescribed by the circle are also within the true tolerance required by the desired fit but not allowed.

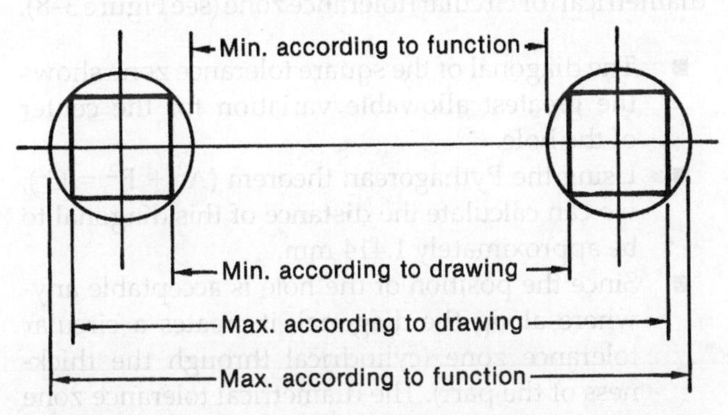

FIGURE 3–7 Geometric dimensioning and tolerancing allows us to reduce costs by not scrapping "good parts."

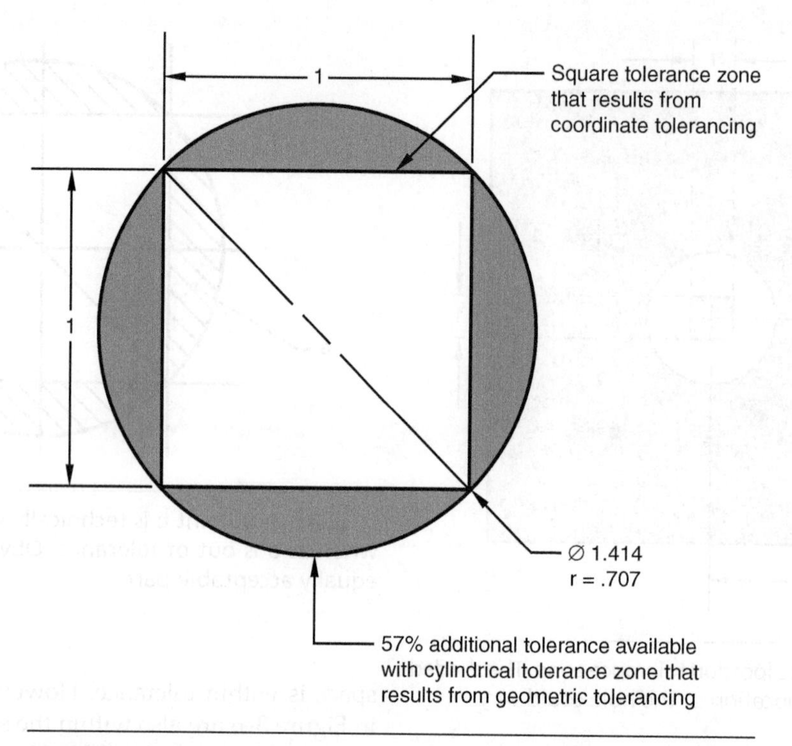

FIGURE 3–8 Area of a circle = $\pi r^2 = 3.1416 * (.707)^2 = 1.57$; area of a square = $s^2 = 1.0^2 = 1.0$. Geometric tolerancing allows a diametrical (circular) tolerance zone, resulting in a 57% increase in the allowable tolerance.

In contrast, geometric dimensioning provides a diametrical (or circular) tolerance zone (see Figure 3–8).

- The diagonal of the square tolerance zone shows the greatest allowable variation for the center of the hole.
- Using the Pythagorean theorem ($A^2 + B^2 = C^2$), we can calculate the distance of this diagonal to be approximately 1.414 mm.
- Since the position of the hole is acceptable anywhere along the diagonal, it creates a circular tolerance zone (cylindrical through the thickness of the part). The diametrical tolerance zone results in a 57% increase in the permissible area for the location of the hole.

Definitions

This section contains key terms in GD&T that are used throughout this chapter.

A *feature* is a general term applied to a physical portion of a part, such as a surface, hole, or slot. Features are generally part surfaces. A *datum* is a theoretically exact plane, point, or axis from which a dimension is measured. Inspection equipment such as a surface plate is often used to simulate a datum. A *datum feature* is a part feature that contacts a datum. Since datums are theoretically exact, we use datum features as the origin for our measurements. A *datum reference frame* is a set of three mutually perpendicular datum planes (see Figure 3–9). Metrologists create this

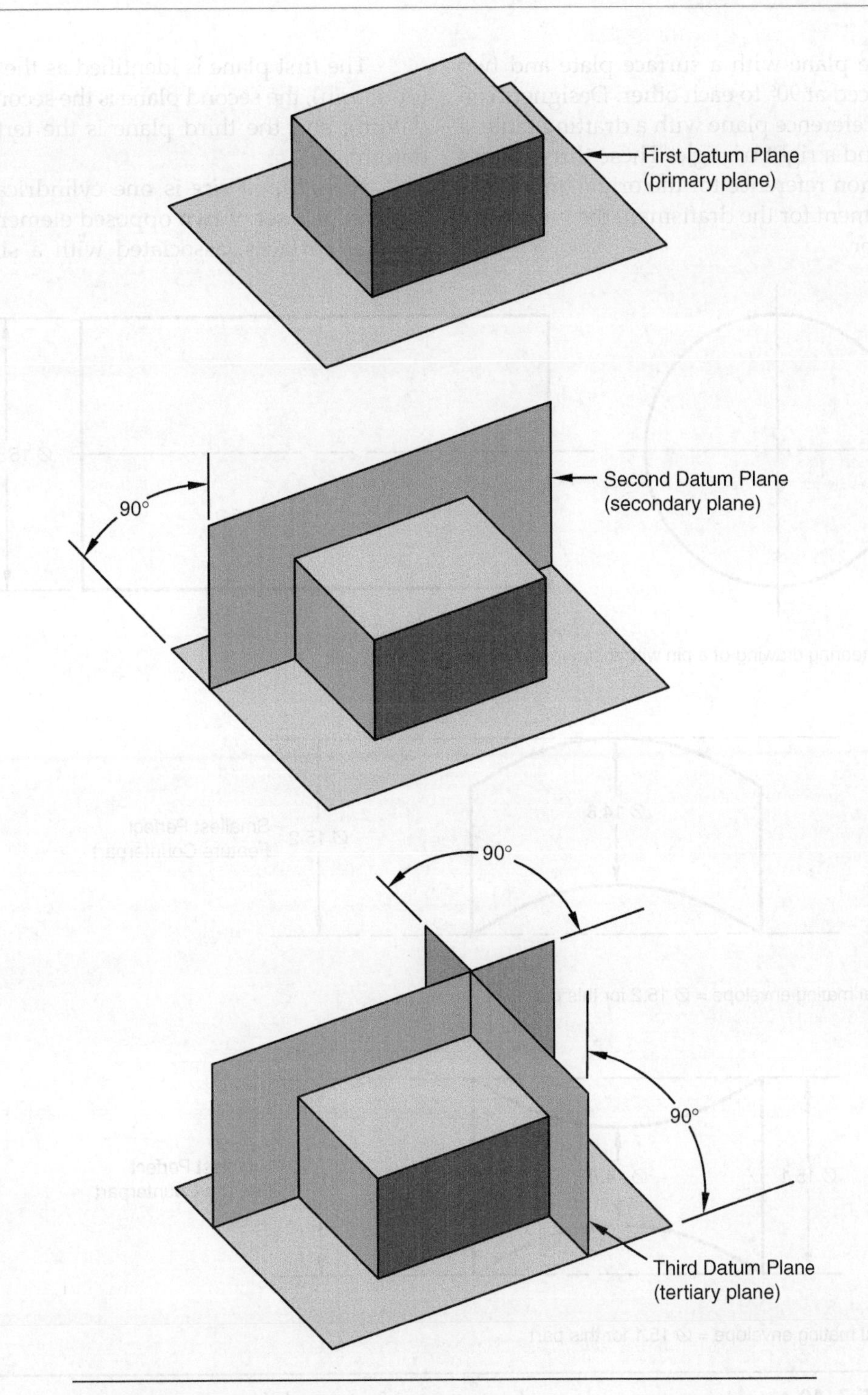

FIGURE 3–9 Datum reference frame.

datum reference plane with a surface plate and two angle plates placed at 90° to each other. Designers can create a datum reference plane with a drafting table, a straight edge, and a right triangle. These three planes provide a common reference for the origin and direction of measurement for the draftsman, the machinist, and the inspector.

The first plane is identified as the primary plane (or datum), the second plane is the secondary plane (or datum), and the third plane is the tertiary plane (or datum).

A *feature of size* is one cylindrical or spherical surface, or a set of two opposed elements or opposed parallel surfaces, associated with a size dimension.

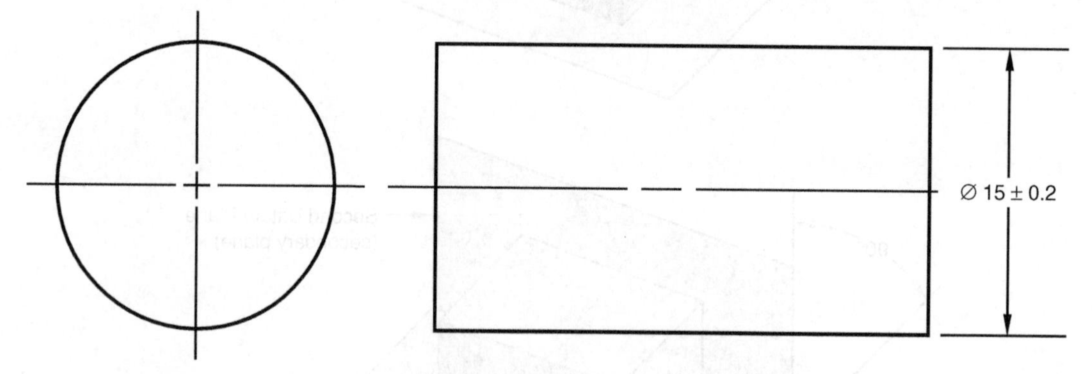

a. Engineering drawing of a pin with a dimension of 15 ± 0.2

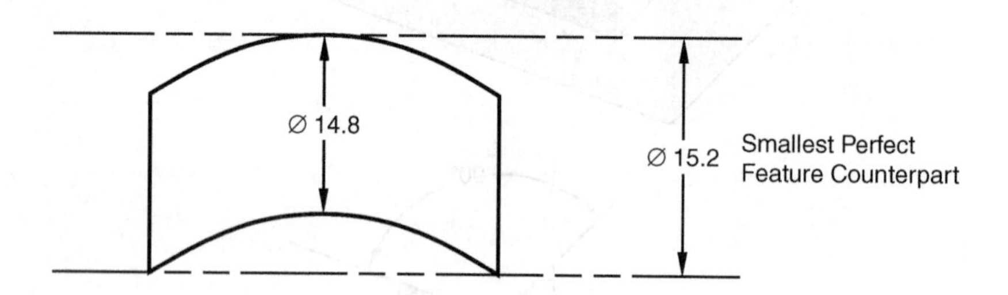

b. Actual mating envelope = Ø 15.2 for this part

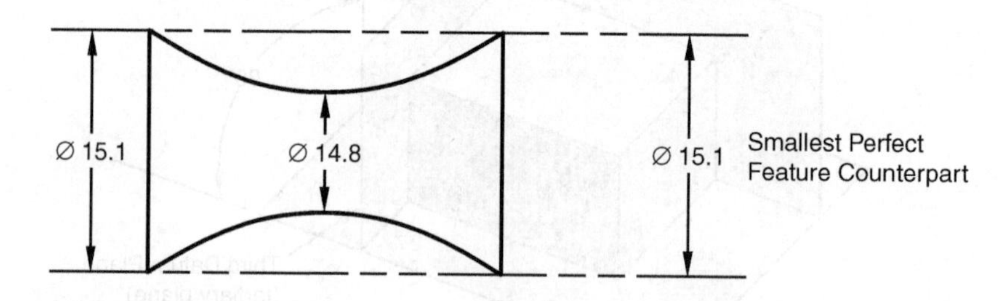

c. Actual mating envelope = Ø 15.1 for this part

FIGURE 3–10 Actual mating envelope of an external feature of size.

Note that the definition states that the surfaces or elements must be opposed. An axis, median plane, or centerpoint can be derived from a feature of size. In Figure 3–1, dimensions A, B, and E reference features of size. Dimensions C and D refer to location dimensions.

Features of size can be external or internal. External features of size are the width or length of a block or a shaft diameter. Internal features of size are the diameter of a hole or the width of a slot.

An *actual mating envelope* is defined according to whether it is an external feature, or an internal one—a pin or a hole, for instance. The actual mating envelope is a variable that is derived from the actual part.

For an external feature, the actual mating envelope is the smallest perfect feature counterpart, which can be circumscribed about the points of the surface (see Figure 3–10). It is the smallest perfect envelope that can contain the feature of size.

For an internal feature of size, the actual mating envelope is a similar counterpart of the largest size, which can be inscribed within the feature. For example, it is the largest cylindrical pin with perfect form that will fit into a hole (see Figure 3–11).

Symbols and Modifiers

The language of geometric tolerancing is a set of symbols (see Figure 3–12). These symbols are divided into five types of dimensioning control: form, profile, orientation, location, and runout.

1. **Form tolerance.** States how far an actual surface or feature is permitted to vary from the desired form implied by the drawing.
2. **Profile tolerance.** States how far an actual surface or feature is permitted to vary from the desired form on the drawing and/or vary relative to a datum.

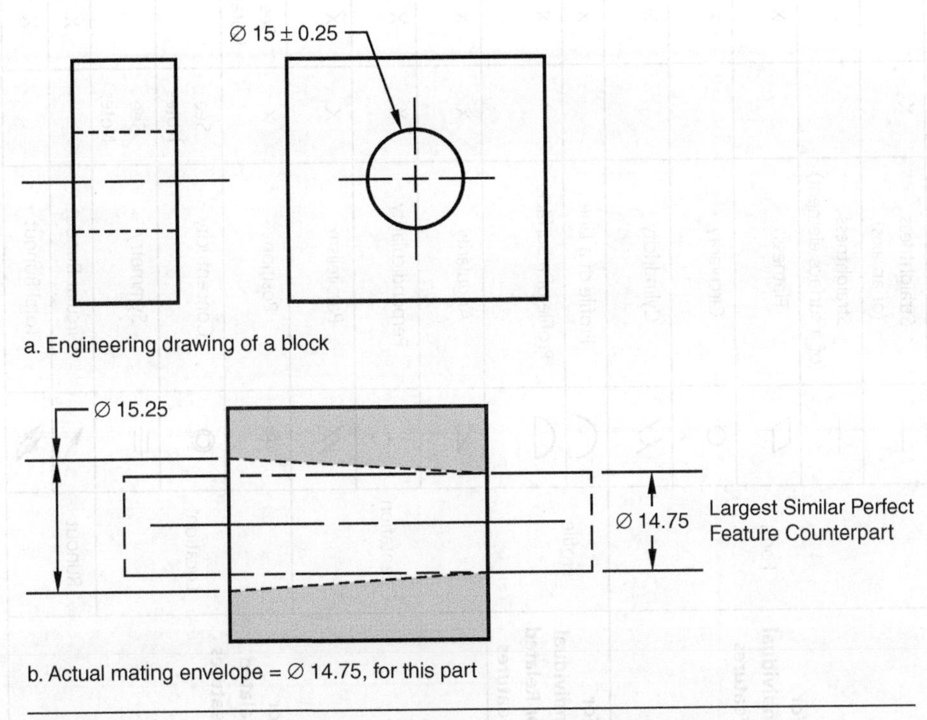

FIGURE 3–11 Actual mating envelope of an internal feature of size.

Geometric Dimensioning and Tolerancing Chart (per ASME Y14.5–1994)

	Types of Tolerance	Symbol	Characteristic	Controls: Axis	Controls: Surface	Feature Modifiers	Uses a Datum Reference	Datum Modifiers	Notes
For Individual Features	Form	—	Straightness (of an axis)	X		Yes	NEVER	N/A	Rule #1 Overridden
		—	Straightness (of a surface element)		X	No		N/A	Rule #1 applies
		⏥	Flatness		X	No		N/A	Tolerance must be less than size
		○	Circularity		X	No		N/A	Tolerance must be less than size
		⌭	Cylindricity		X	No		N/A	Tolerance must be less than size
For Individual or Related Features	Profile	⌒	Profile of a Line		X	No	SOMETIMES	Yes	Rule #1 applies
		⌓	Profile of a Surface		X	No		Yes	Rule #1 applies
For Related Features	Orientation	∠	Angularity	X	X	Yes	ALWAYS	Yes	Also controls flatness of surface
		⊥	Perpendicularity	X	X	Yes		Yes	Feature modifiers OK if diameter is used
		//	Parallelism	X	X	Yes		Yes	Also controls flatness of surface
	Location	⌖	Position	X	See Notes	Yes		Yes	Can control surface boundary
		◎	Concentricity	See Notes		No		No	Controls opposing median points
		⌯	Symmetry	See Notes		No		No	Controls opposing median points
	Runout	↗	Circular Runout		X	No		No	Rule #1 applies
		⌰	Total Runout		X	No		No	Rule #1 applies

Rule #1 summarized–Tolerance limits control the shape of an individual feature as well as the size. (See ASME Y14.5M–1994, section 2.7.1)

FIGURE 3–12 GD&T symbols.

3. **Orientation tolerance.** States how far an actual surface or feature is permitted to vary relative to a datum.
4. **Location tolerance.** States how far an actual size feature is permitted to vary from the perfect location implied by the drawing as related to a datum or other feature.
5. **Runout tolerance.** States how far an actual surface or feature is permitted to vary from the desired form implied by the drawing during a full 360° rotation of the part on a datum axis.

Geometric controls of form never use a datum reference. Form controls (straightness, flatness, circularity, or cylindricity) are always relative to themselves and not other features. Some geometric controls (orientation, location, or runout) must have a datum reference. When you think of parallelism, for instance, the feature must be parallel relative to another feature.

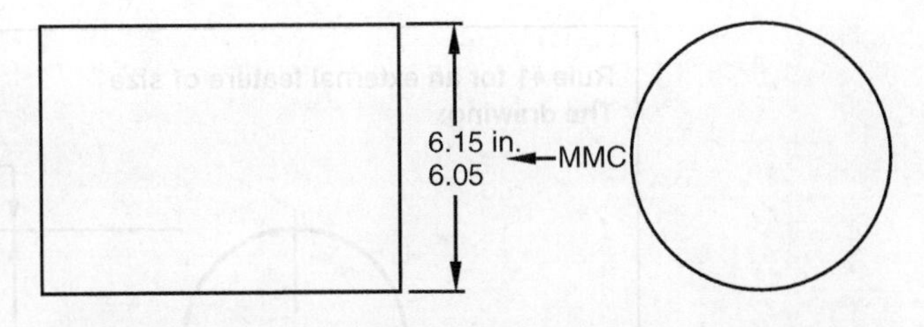

FIGURE 3–14 Maximum material condition (MMC) of an external feature like this pin equals the largest allowable diameter of 6.15 in.

Material Conditions

One of the benefits of GD&T is the ability to specify tolerances at various part sizes, or *material conditions.* A geometric tolerance can be applied to the largest size, the smallest size, or the actual size of a part feature. These symbols are called modifiers (see Figure 3–13).

Maximum Material Condition. is the condition in which a feature of size contains the material within its stated tolerance limits. An external feature of size (a pin, shaft, or block) is at its *maximum material condition* (MMC) when it is at its largest size (see Figure 3–14). An internal feature of size (a hole or a slot), is at its maximum material condition when it is at its smallest permissible size. A hole diameter retains the most material at its smallest size (see Figure 3–15). In this instance, it is easy to see that more material exists in the pin when it is at its maximum permissible size. However, the same principle exists in both hole and pin MMC conditions. Relating mating part features in this manner ensures their functional relationship and

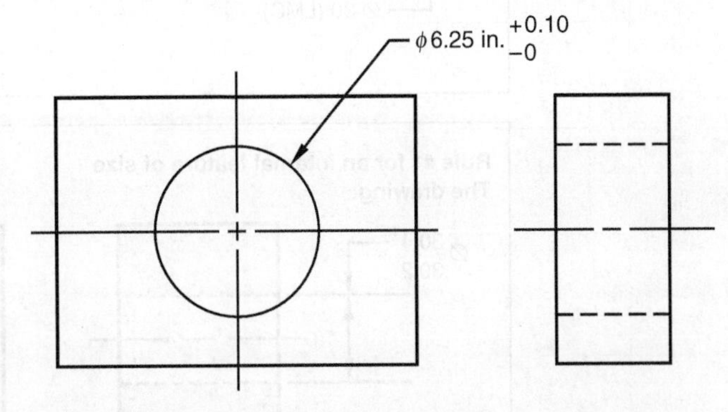

FIGURE 3–15 A hole at maximum material condition is the smallest allowable size, retaining the "most material." Here, the maximum material condition is 6.25 in.

MODIFIER	ABBREVIATION	SYMBOL
Maximum Material Condition	MMC	Ⓜ
Least Material Condition	LMC	Ⓛ
Regardless of Feature Size	RFS	None

FIGURE 3–13 Material condition modifiers.

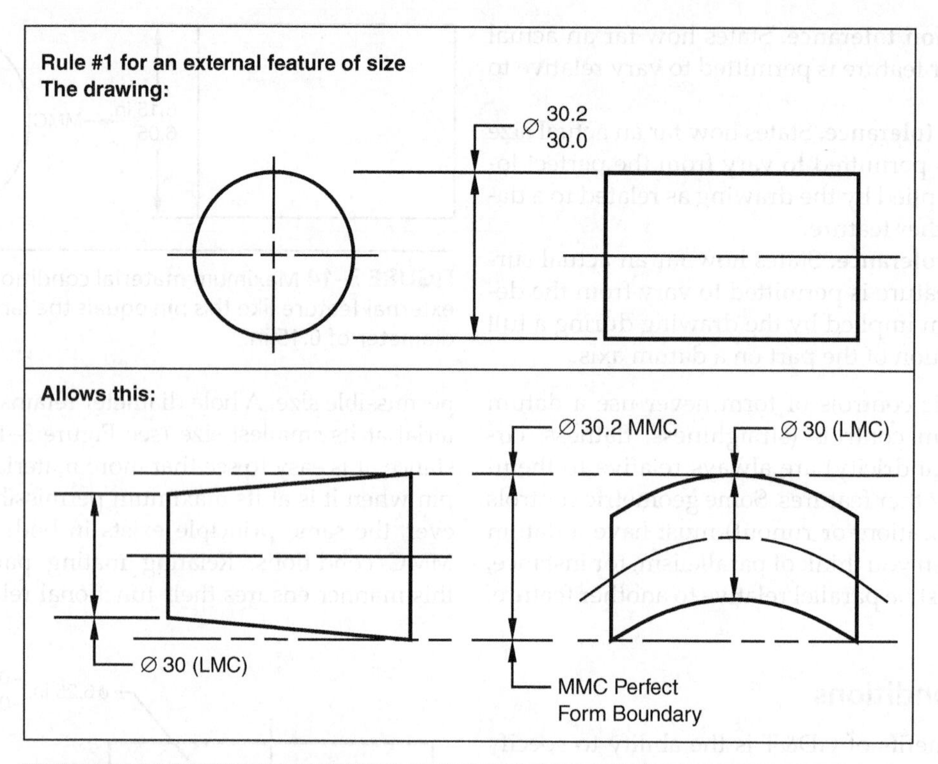

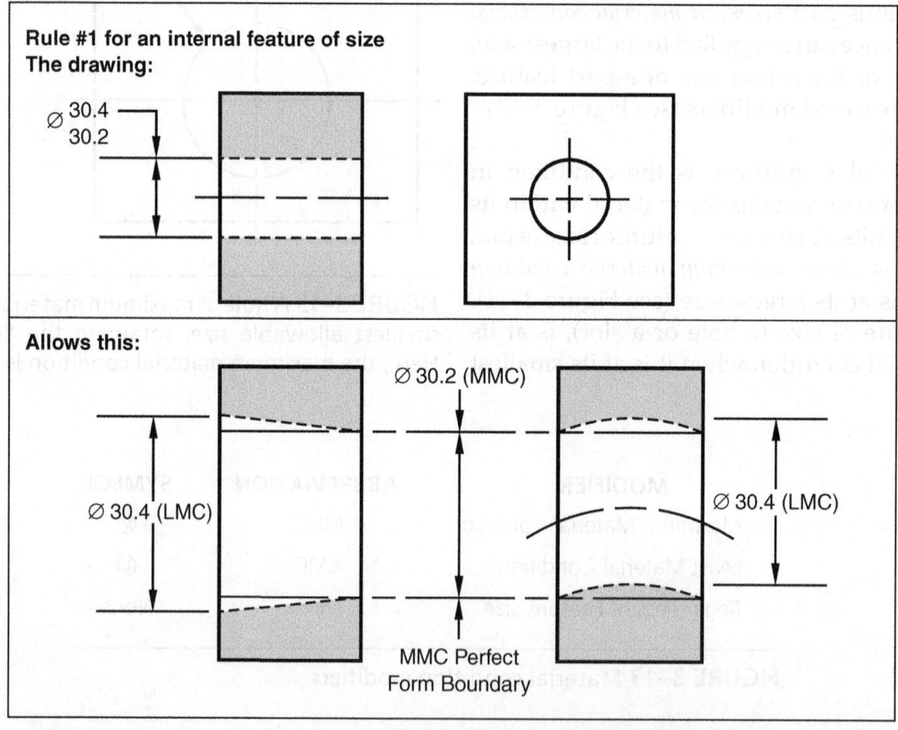

FIGURE 3–16 Rule #1—Variations in form as well as size are allowed.

establishes the criteria for determining necessary form, orientation, and position tolerances. The ø6.25 MMC hole further means that the hole must be at perfect form when the pin is at ø6.25 (Rule #1, ASME Y14.5M-1994). The pin must be at perfect form at MMC of ø 6.15. The symbol for maximum material condition is "Ⓜ".

Least Material Condition. is the condition in which a feature of size contains the least amount of material within its permissible limits—for example, the minimum shaft diameter or the maximum hole diameter. In Figure 3–14, the *least material condition* (LMC) of the pin would be ø6.05. In Figure 3–15, the least material condition of the hole would be ø6.35. The symbol for LMC is "Ⓛ".

Regardless of Feature Size. indicates a geometric tolerance that applies at any increment of the size of the feature within its permissible limits. The regardless of feature size (RFS) is implied on all geometric tolerances, unless indicated by the presence of a modifier. There is no symbol for RFS because it is the implied condition.

Rules

The Geometric Dimensioning and Tolerancing standard, like most other standards, contains specific rules. These rules provide a common foundation to apply and interpret GD&T.

Rule #1. (Individual Feature of Size Rule, ASME Y14. 5M-1994, section 2.7.1) applies to all features controlled with only plus/minus tolerances. The rule states "where only a tolerance of size is specified, the limits of size of an individual feature prescribe the extent to which variations in its geometric form, as well as the size are allowed." The result of this rule is that when only a tolerance of size is specified, the tolerance controls both the size and the form.

Also known as, "The Envelope Principle" or "Perfect Form at MMC," Rule #1 requires that at any given cross section, the combination of both form and size must be within the size limits (see Figure 3–16).

When a pin is at MMC, the pin must have perfect form. It must be perfectly straight and round. This allows the pin to fit through a boundary equal to its maximum size. If the pin departs or varies from MMC, its form is allowed to vary from perfect.

Rule #1 does not control the relationship between features; it only applies to an individual feature of size.

There are two exceptions to Rule #1. It does not apply to a feature of size on a part subject to free-state variation in the unrestrained condition. In other words, Rule #1 does not apply to flexible parts that are not restrained. The second exception is for stock sizes. Rule #1 does not apply to stock sizes such as bar stock, sheet metal, and tubing.

Rule #2. (ASME Y14.5M-194, section 2.8) states, "RFS applies, with respect to the individual tolerance, datum reference or both, where no modifying symbol is specified. MMC or LMC must be specified on the drawing where it is required." This is the reason that there is no symbol for RFS—it is always implied, unless otherwise indicated. Some geometric tolerances can only be applied at RFS. Circular runout, total runout, concentricity, and symmetry are applicable only on an RFS basis. They cannot be modified based on MMC or LMC.

Rule #3. applies to all screw threads, gears, and splines. It states, "Each tolerance of orientation or position and datum reference specified for a screw thread applies to the axis of the thread derived from the pitch cylinder." A designer must designate the tolerances from the axis of the pitch diameter. If an exception is necessary, MAJOR DIA, PITCH DIA, or MINOR DIA can be stated beneath the feature control frame.

Feature Control Frames

Geometric tolerances are specified on a drawing with a *feature control frame* (see Figure 3–17).

A feature control frame is a rectangular box that contains the geometric symbols, modifiers, and datum references. The first compartment contains the geometric characteristic symbol. The second frame references the tolerance information. If the tolerance value is preceded by a diameter symbol (ø), the shape of the tolerance zone is cylindrical. If the diameter symbol is not shown, the shape of the tolerance zone can be

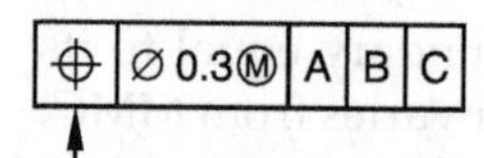

The first compartment always contains the geometric symbol.

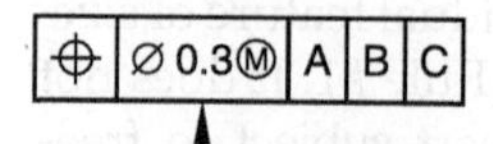

The second compartment contains the tolerance information, including the shape of the tolerance zone (diametrical = ∅), the tolerance value (0.3), and any modifiers that describe the tolerance condition (Ⓜ).

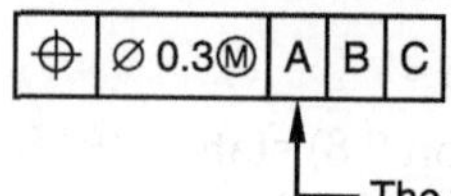

The third, fourth, and fifth compartments (when used) specify the related datums, if necessary.

FIGURE 3–17 Feature control frames.

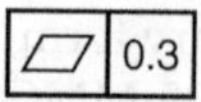

a. This feature control frame specifies a flatness tolerance of 0.3. Note that the units are generally specified on the engineering drawing.

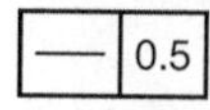

b. This feature control frame specifies a straightness tolerance of 0.5.

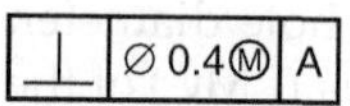

c. This feature control frame specifies perpendicular, with a cylindrical tolerance zone of 0.4, at maximum material condition to datum A.

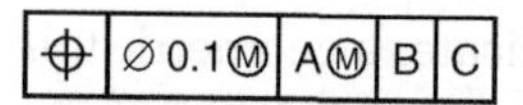

d. This feature control frame specifies true position, with a cylindrical tolerance zone of 0.1 at maximum material condition to datum A at maximum material condition, datum B, and datum C.

FIGURE 3–18 Examples of feature control frames.

parallel lines or planes, or a uniform boundary of profile. The tolerance value given is always a total tolerance value.

When a datum reference is required, it is specified in the third, and if necessary, fourth and fifth frames. These compartments are known as the *datum reference* portion of the feature control frame. The third compartment is known as the primary datum, the fourth compartment is the secondary datum, and the fifth compartment is the tertiary datum. It is not always necessary to use any or all datums, depending on the type of geometric control (see Figure 3–18).

3–3 THE APPLICATION OF GEOMETRIC TOLERANCING

When tolerances of size and location do not provide adequate control, part features and relationships can be further defined with geometric tolerancing. Form and profile tolerances control each feature's size and shape, while orientation, location, and runout tolerances can define the relationship of individual features of a part.

Form Tolerances

Form tolerances state how far an actual surface or feature is permitted to vary from the desired form. Straightness, flatness, circularity, and cylindricity are most frequently applied to single features or portions of a feature. These controls are specified without a datum reference because the features are not controlled in relation to another feature.

Straightness

Straightness is the condition where one line element of an axis, or a surface, is in a straight line. Straightness tolerances can be applied to an axis or to a surface (see Figure 3–19). When applied to an axis, straightness is specified in the view where the axis is a straight line (see Figure 3–20). The tolerance zone is a space between two parallel straight lines. For axis control, the tolerance is implied RFS unless a modifier is specified.

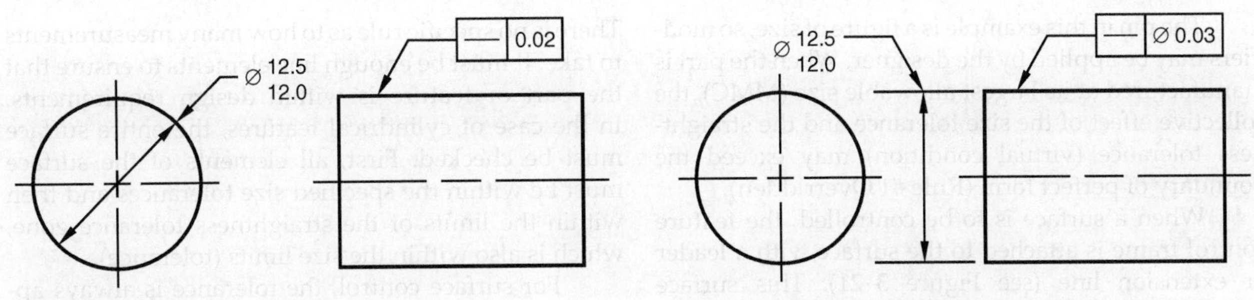

Types of Tolerance	Symbol	Characteristic	Controls		Feature Modifiers	Uses a Datum Reference	Datum Modifiers	Notes
			Axis	Surface				
Form	—	Straightness (of an axis)	X		Yes	NEVER	N/A	Rule #1 Overridden
	—	Straightness (of a surface element)		X	No		N/A	Rule #1 applies

Note: The part must also be within the size limits.

FIGURE 3–19 Straightness tolerance.

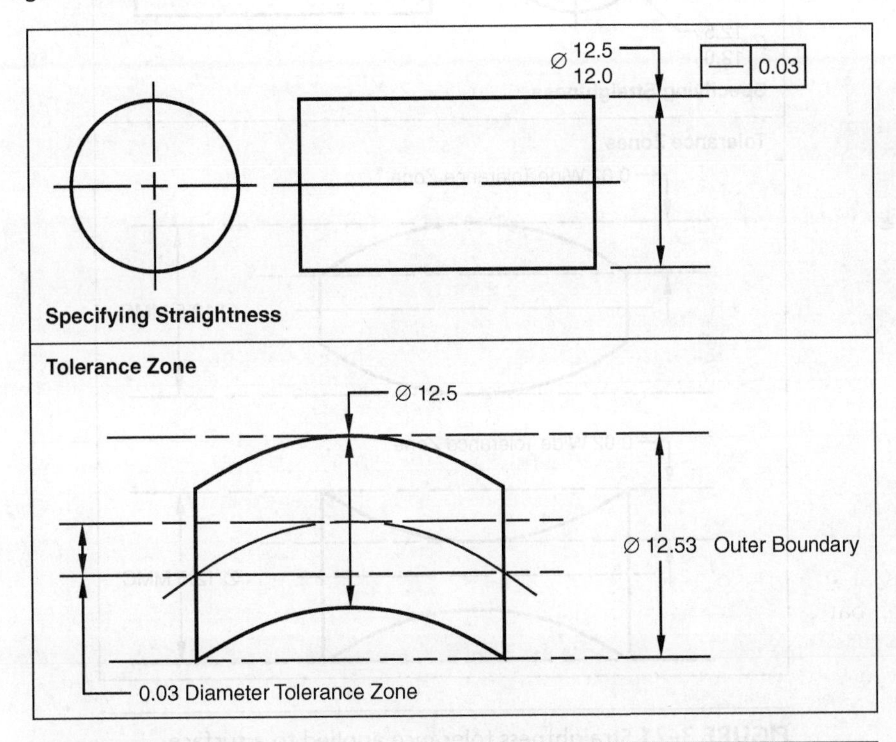

FIGURE 3–20 Straightness tolerance applied to an axis.

The pin in this example is a figure of size, so modifiers may be applied by the designer. When the part is manufactured at its largest allowable size (MMC), the collective effect of the size tolerance and the straightness tolerance (virtual condition) may exceed the boundary of perfect form (Rule #1 Overridden).

When a surface is to be controlled, the feature control frame is attached to the surface with a leader or extension line (see Figure 3–21). This surface may be measured or verified with a dial indicator. There is no specific rule as to how many measurements to take. It must be enough line elements to ensure that the part or feature is within design requirements. In the case of cylindrical features, the entire surface must be checked. First, all elements of the surface must be within the specified size tolerances and then within the limits of the straightness tolerance zone, which is also within the size limits (tolerance).

For surface control, the tolerance is always applied regardless of feature size because the straightness

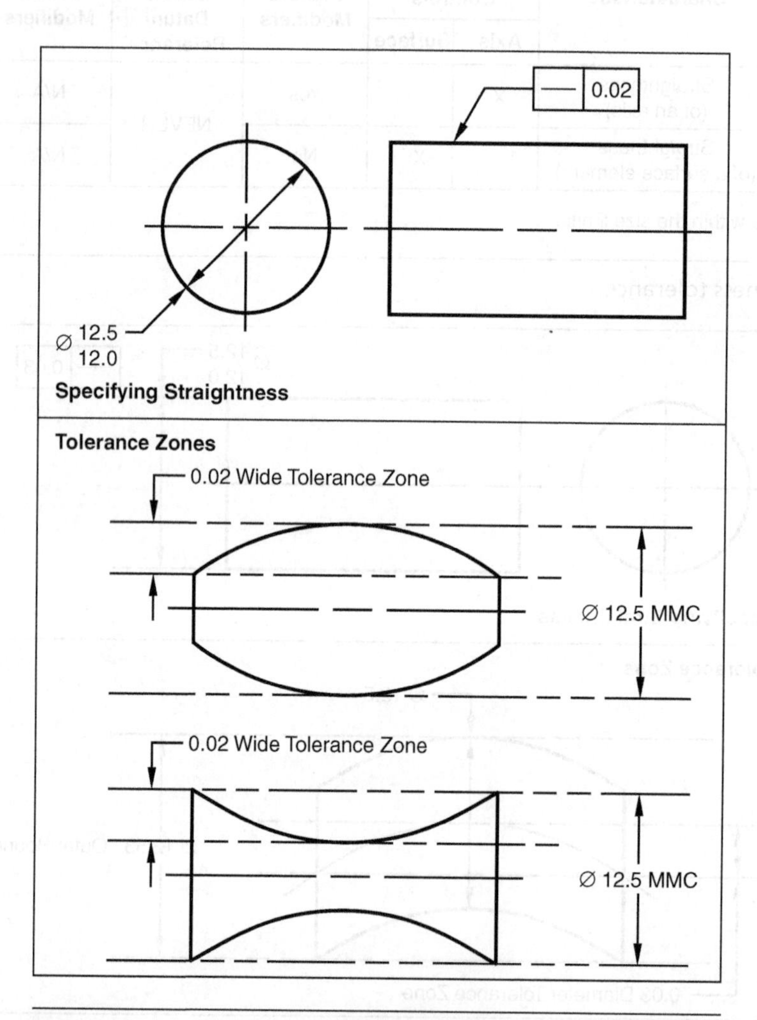

FIGURE 3–21 Straightness tolerance applied to a surface.

tolerance controls line elements that have no size. The feature must be within the stated size limits at each cross-sectional measurement.

Flatness

Flatness is the condition of a surface having all elements in one plane. When flatness is specified, the feature control frame is attached directly to the surface or to an extension line of the surface.

The flatness tolerance zone is defined by two parallel planes. All points of the surface must be within the limits of the tolerance zone defined by these two planes (see Figure 3–22). The smaller the tolerance zone, the flatter the surface. The flatness tolerance must be less than the size tolerance.

Flatness does not have a datum reference (a surface is flat relative to itself and not other features). It is always applied regardless of feature size—no feature modifiers such as MMC or LMC are allowed.

Circularity (Roundness)

Circularity is identified with any given cross section taken perpendicular to the axis of a cylinder or cone or through the center of a sphere. The feature control frame is connected with a leader (see Figure 3–23).

The tolerance zone is bounded by two concentric circles. Each circular element of the surface must be contained within these concentric circles. The circularity tolerance must be less than the size tolerance, except for parts subject to free state variation.

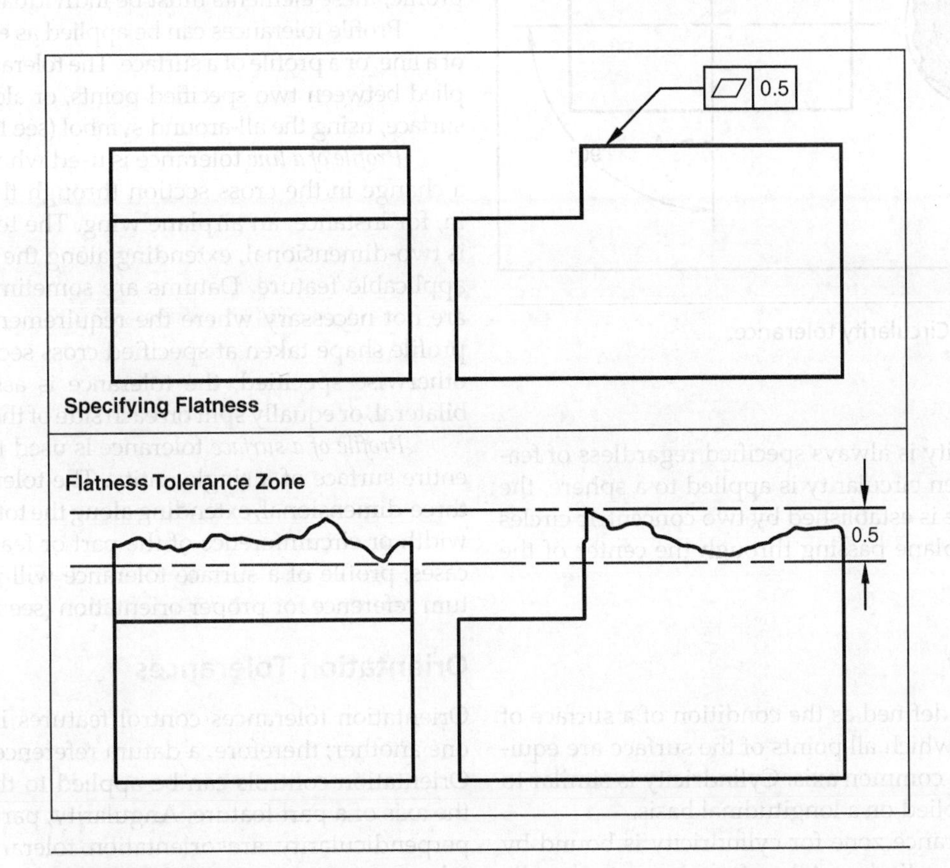

FIGURE 3–22 Flatness tolerance.

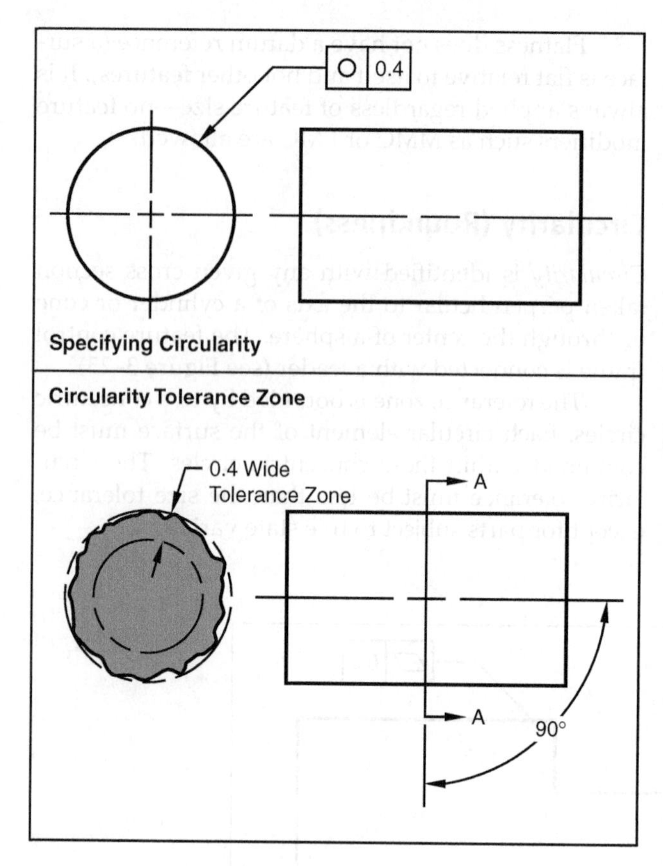

FIGURE 3–23 Circularity tolerance.

Circularity is always specified regardless of feature size. When circularity is applied to a sphere, the tolerance zone is established by two concentric circles created by a plane passing through the center of the sphere.

Cylindricity

Cylindricity is defined as the condition of a surface of revolution in which all points of the surface are equidistant from a common axis. Cylindricity is similar to circularity applied on a longitudinal basis.

The tolerance zone for cylindricity is bound by two concentric cylinders. All surface elements must lie within these concentric circles. The tolerance applies simultaneously to both the circular and the longitudinal elements of the surface (see Figure 3–24).

The cylindricity tolerance must be less than the size tolerance, and is always applied regardless of feature size.

Profile Tolerances

Profile tolerancing allows us to control the form or shape of a surface. A profile is the outline of an object represented by a cross section through the part or by an end view of the part. Basic dimensions are generally used to define a profile. These can include basic radii, basic angular dimensions, basic coordinate dimensions, basic size dimensions, or formulas. If the drawing specifies individual tolerances for elements or points of a profile, these elements must be individually verified.

Profile tolerances can be applied as either a profile of a line, or a profile of a surface. The tolerance can be applied between two specified points, or along the entire surface, using the all-around symbol (see Figure 3–25).

Profile of a line tolerance is used where parts have a change in the cross section through the length—as in, for instance, an airplane wing. The tolerance zone is two-dimensional, extending along the length of the applicable feature. Datums are sometimes used, but are not necessary where the requirement is only the profile shape taken at specified cross sections. Unless otherwise specified, the tolerance is assumed to be bilateral, or equally split on each side of the true profile.

Profile of a surface tolerance is used to control the entire surface of a single entity. The tolerance zone is three-dimensional, extending along the total length and width or circumference of the part or feature. In most cases, profile of a surface tolerance will require a datum reference for proper orientation (see Figure 3–26).

Orientation Tolerances

Orientation tolerances control features in relation to one another; therefore, a datum reference is required. Orientation controls can be applied to the surface or the axis of a part feature. Angularity, parallelism, and perpendicularity are orientation tolerances. (Profile tolerances can sometimes be orientation tolerances as

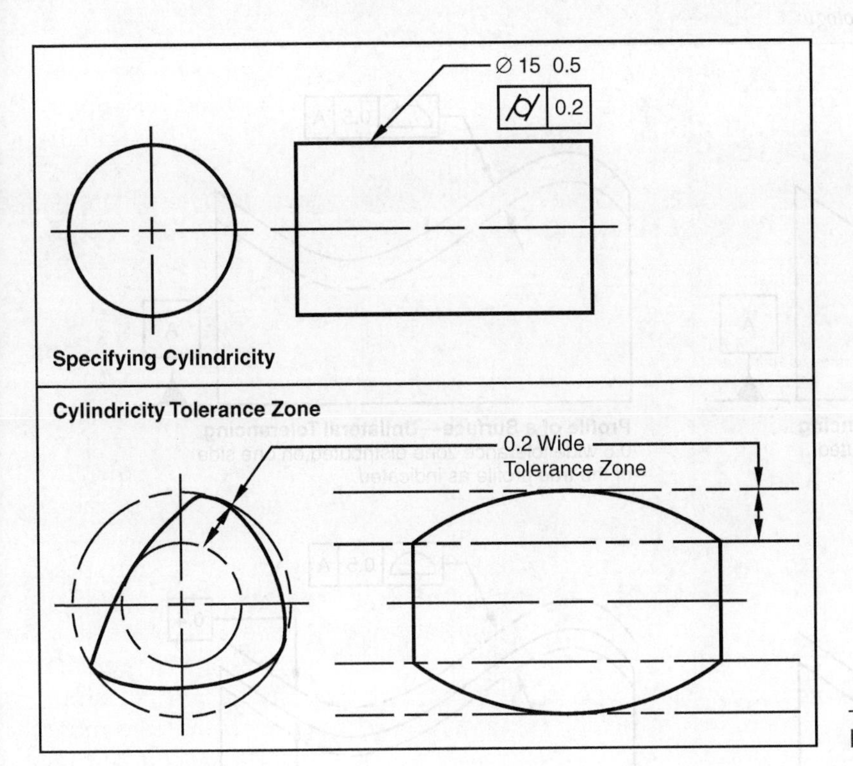

FIGURE 3–24 Cylindricity tolerance.

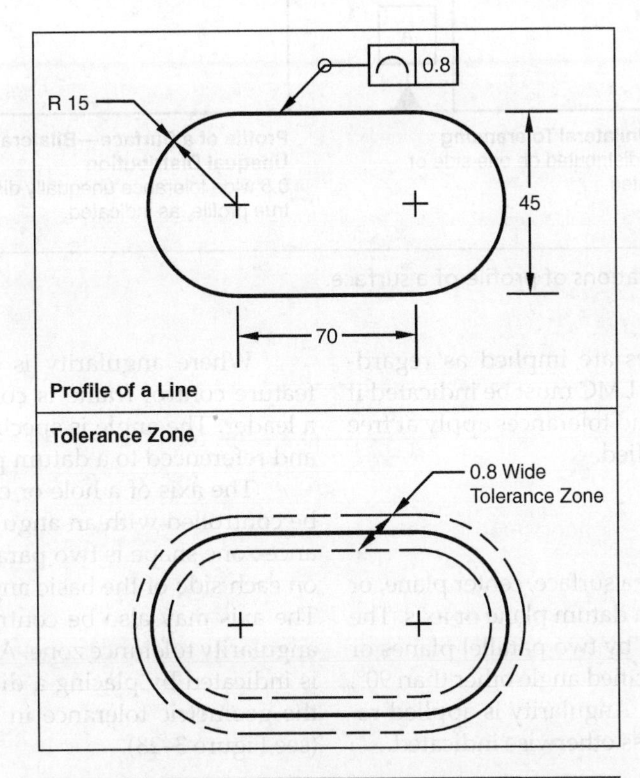

FIGURE 3–25 Profile of a line with all-around symbol.

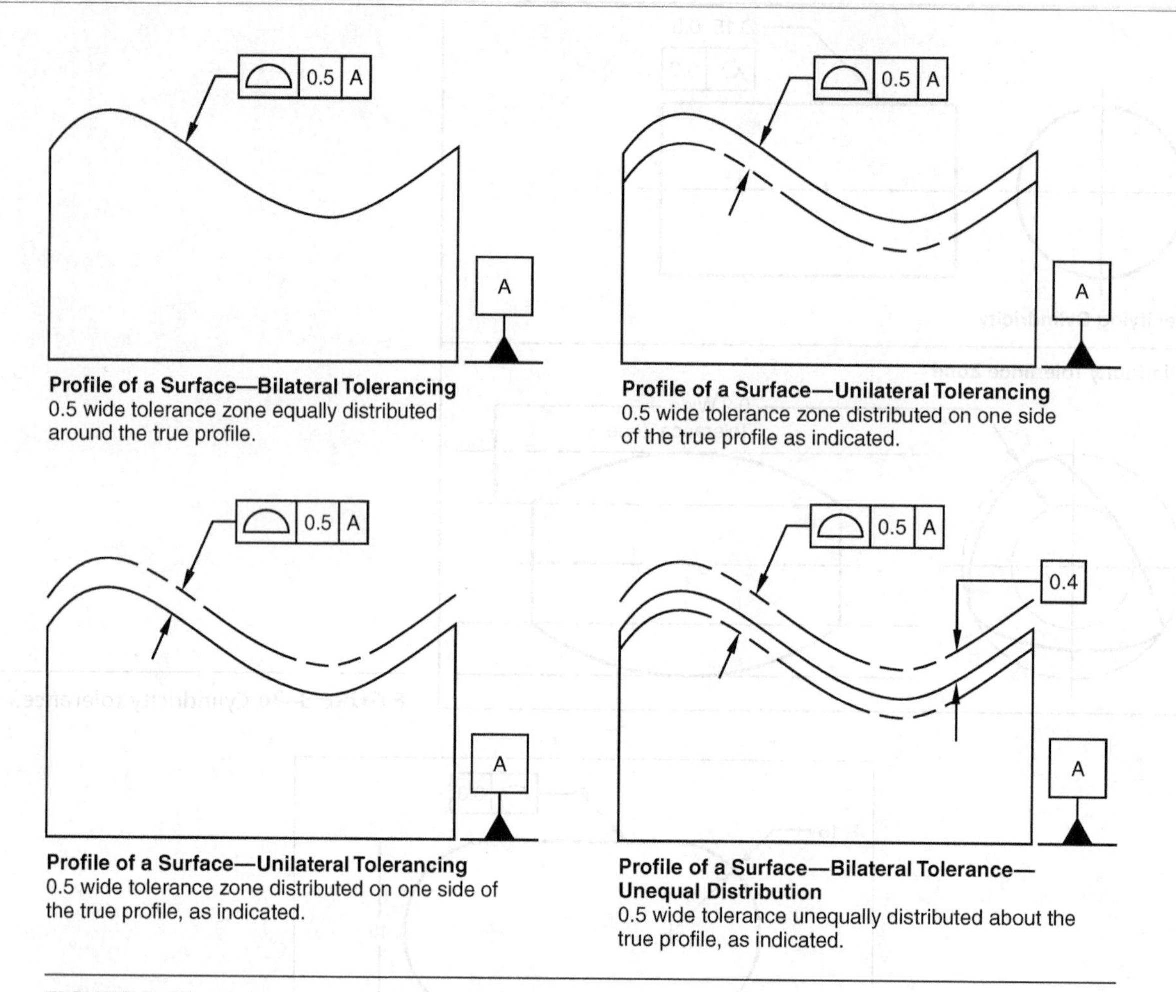

FIGURE 3–26 Applications of profile of a surface.

well.) Orientation tolerances are implied as regardless of feature size; MMC or LMC must be indicated if necessary. All dimensions and tolerances apply at free state unless otherwise specified.

Angularity

Angularity is the condition of a surface, center plane, or axis at a specific angle from a datum plane or axis. The tolerance zone is established by two parallel planes or a cylindrical zone at any specified angle other than 90°, to a datum plane or an axis. Angularity is applied regardless of feature size, unless otherwise indicated.

Where angularity is applied to a surface, the feature control frame is connected to the surface by a leader. The angle is specified by a basic dimension, and referenced to a datum plane (see Figure 3–27).

The axis of a hole or cylindrical feature can also be controlled with an angularity tolerance. The tolerance zone shape is two parallel planes equally spaced on each side of the basic angle from the datum plane. The axis may also be controlled within a cylindrical angularity tolerance zone. A cylindrical tolerance zone is indicated by placing a diameter symbol in front of the geometric tolerance in the feature control frame (see Figure 3–28).

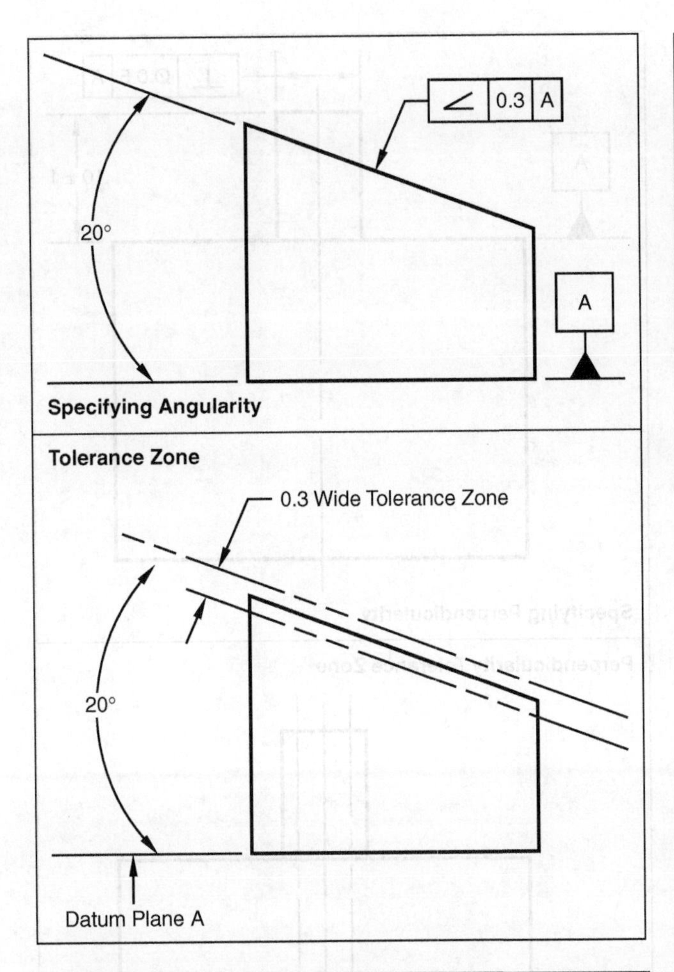

FIGURE 3–27 Angularity tolerance applied to a surface.

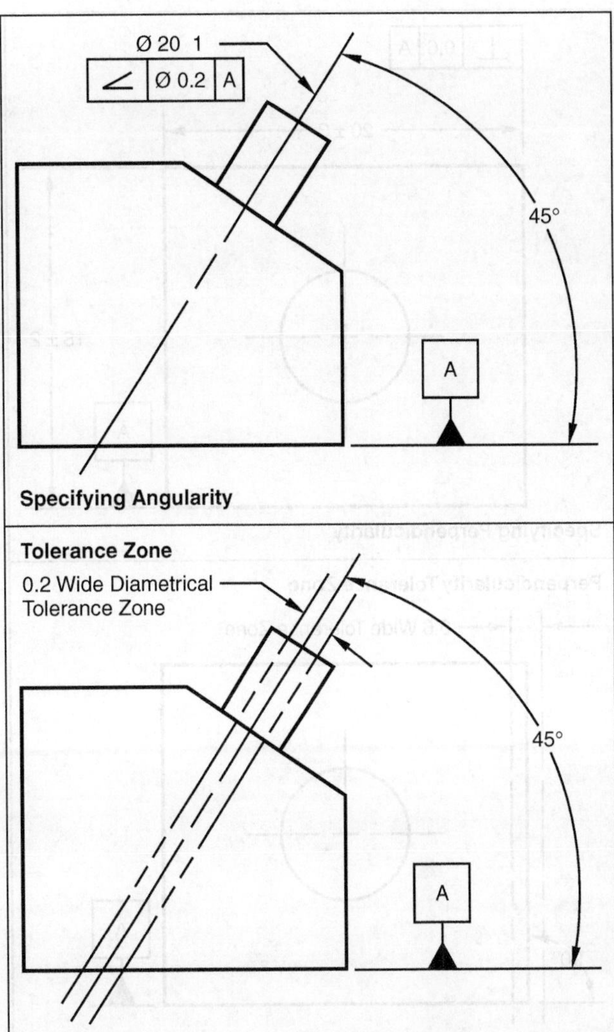

FIGURE 3–28 Angularity tolerance applied to an axis.

Perpendicularity

Perpendicularity is the condition when a surface, center plane, or axis is at exactly 90° to a datum. The tolerance zone for perpendicularity is established by two parallel planes or cylindrical zones that are 90° to a specified datum plane or axis. Perpendicularity is applied regardless of feature size, unless otherwise indicated.

Perpendicularity can be applied to a surface or an axis. When applied to a surface, the shape of the tolerance zone is two parallel planes that are at a 90° angle to a datum plane. The tolerance value identifies the size of the tolerance zone. All elements of the surface must lie within this tolerance zone (see Figure 3–29). When applied to a surface, perpendicularity also limits flatness.

A perpendicular tolerance may be applied to the axis of a shaft or a hole (see Figure 3–30). The diameter symbol in the feature control frame indicates a cylindrical tolerance zone. Regardless of feature size, the axis of the pin must lie within a cylindrical tolerance

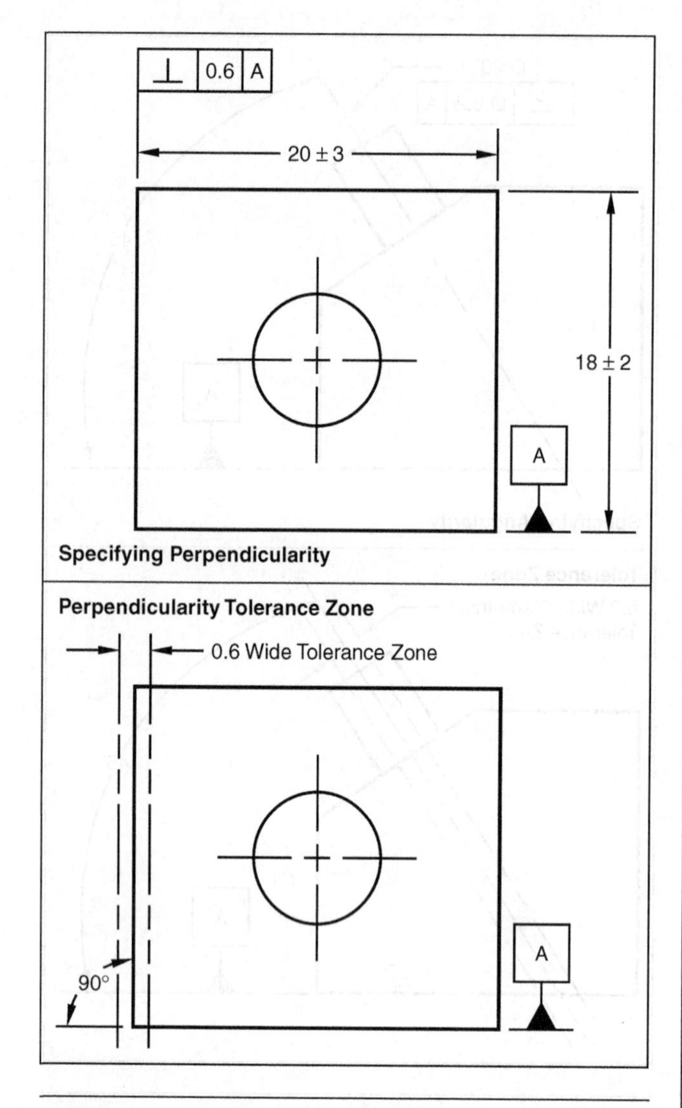

FIGURE 3–29 Perpendicularity tolerance applied to a surface.

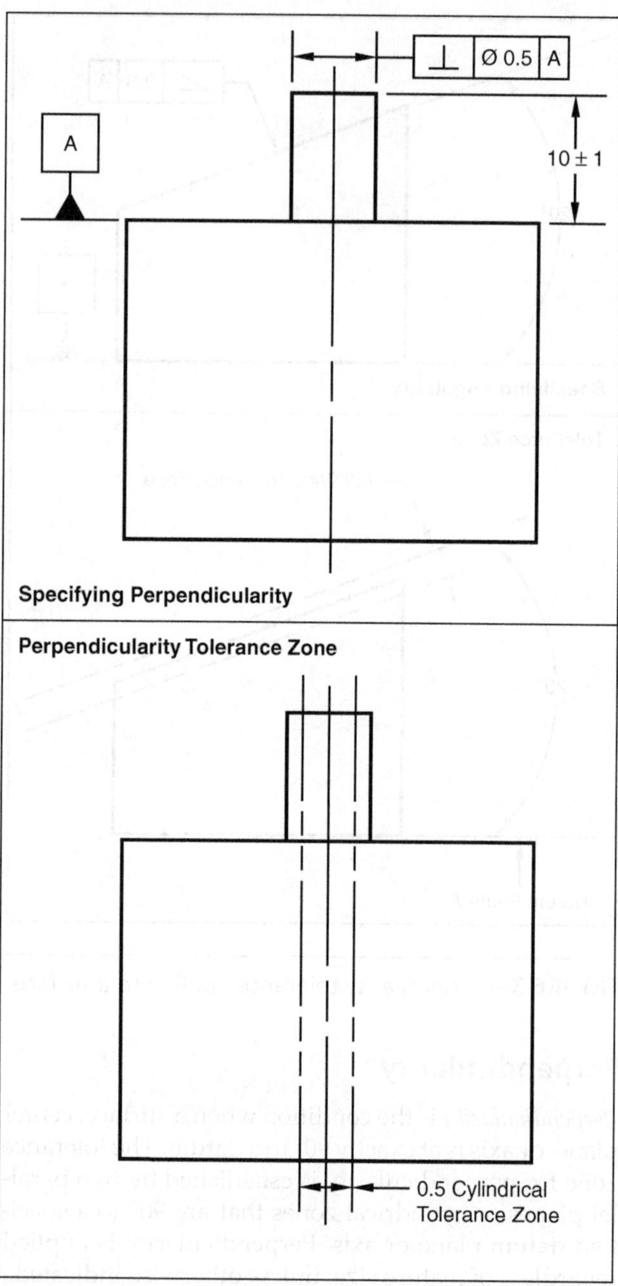

FIGURE 3–30 Perpendicularity tolerance applied to an axis.

zone of 0.5 diameter. The tolerance zone extends for the entire height of the pin, perpendicular to datum A.

Parallelism

Parallelism is the condition where a surface, center plane, or axis is exactly parallel to a datum. Parallelism may be applied to a surface, resulting in a tolerance zone of

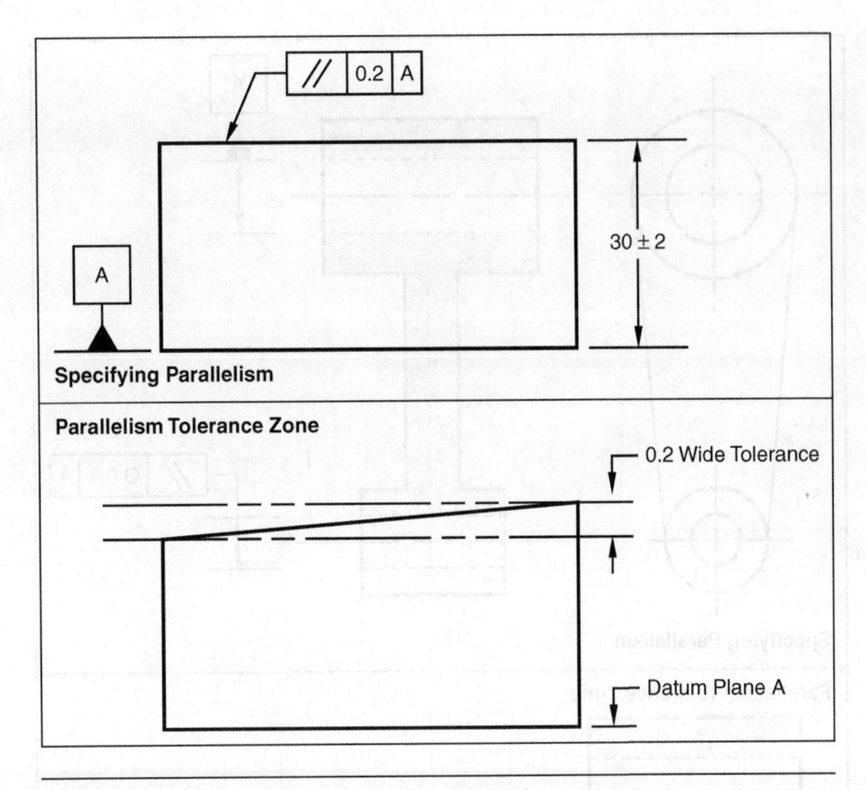

FIGURE 3–31 Parallelism tolerance applied to a surface.

two parallel planes, or applied to an axis, resulting in a cylindrical tolerance zone. Like perpendicularity, the tolerance zone limits the flatness of the specified feature.

When a surface is controlled parallel to a datum, all elements of that surface must lie within two parallel planes, parallel to the datum (see Figure 3–31).

Parallelism may be applied to the axis of two or more features where a parallel relationship is required (see Figure 3–32). In this case, the feature axis must lie within a 0.4 cylindrical zone parallel to datum axis A. The controlled axis must also be within the specified tolerance of location.

Location Tolerances

Location tolerances are used to locate or position features from datums. Location tolerances include position, concentricity, and symmetry. Positional tolerancing provides the maximum benefit of GD&T, allowing an increase in tolerance of the feature by 57%, increasing the interchangeability of parts, and reducing scrap. Location tolerances can be used to control position, symmetry, and coaxiality.

Position

Positional tolerance defines a condition where the center, axis, or center plane of a feature of size is allowed to vary from true position. *True position* is the theoretically exact location of a feature. The location of each feature is given by a basic dimension, and the location tolerance is indicated by the position symbol, a tolerance value, applicable modifiers, and datum references (see Figure 3–33). The tolerance zone is generally a cylindrical tolerance zone where the centerline of the feature is located. Where a feature other than a cylindrical shape is indicated, the tolerance zone is established by two parallel straight lines or planes. The tolerance is assumed to be regardless of feature size, unless otherwise indicated.

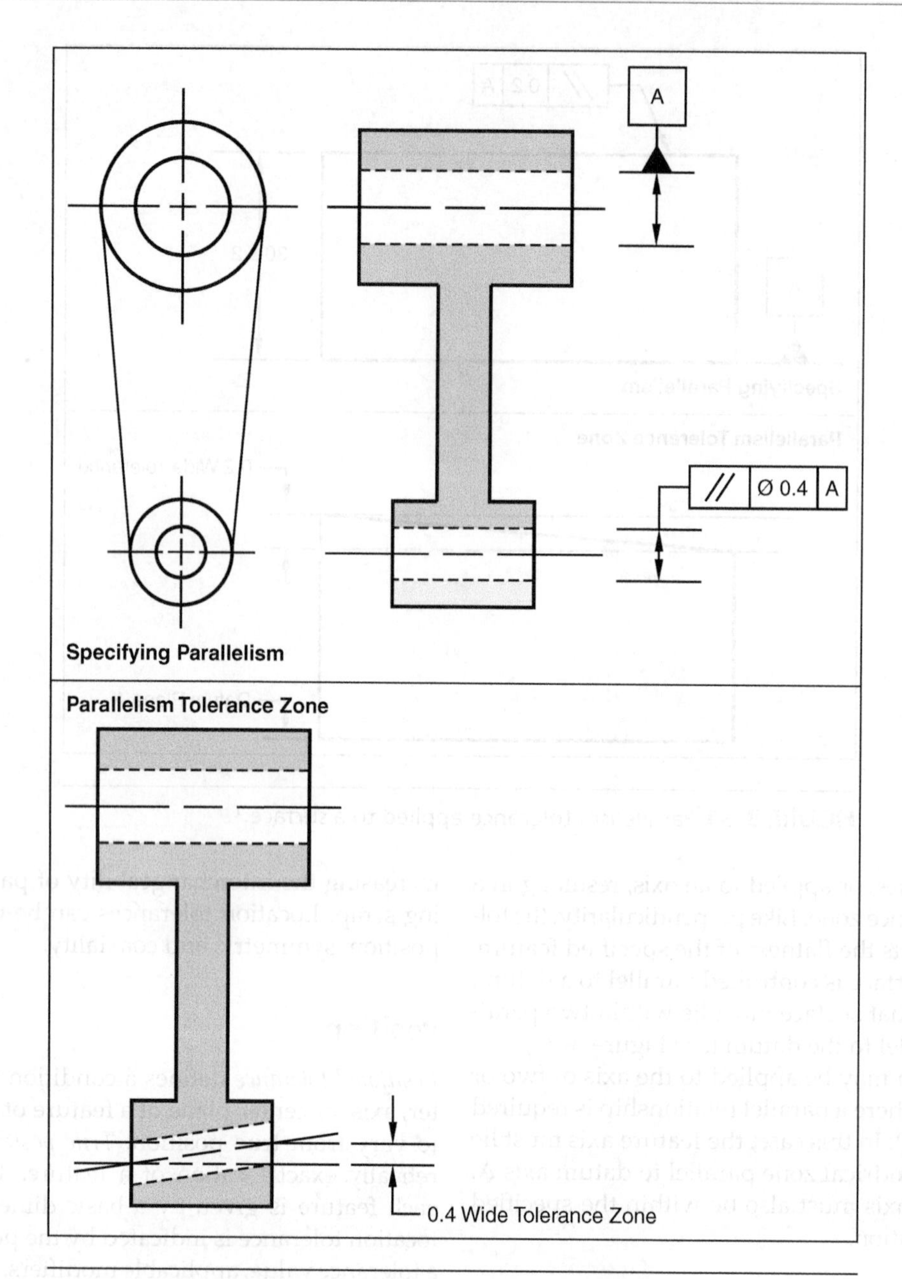

FIGURE 3–32 Parallelism tolerance applied to an axis.

Positional Tolerance at MMC. The application of MMC allows the tolerance zone to increase in diameter as the feature size departs from maximum material condition, increasing production flexibility while maintaining part functionality. Maximum material condition means that the actual size of the feature contains the most material. For a pin or a shaft, this is its largest permissible size. A hole containing maximum material would be its smallest permissible size.

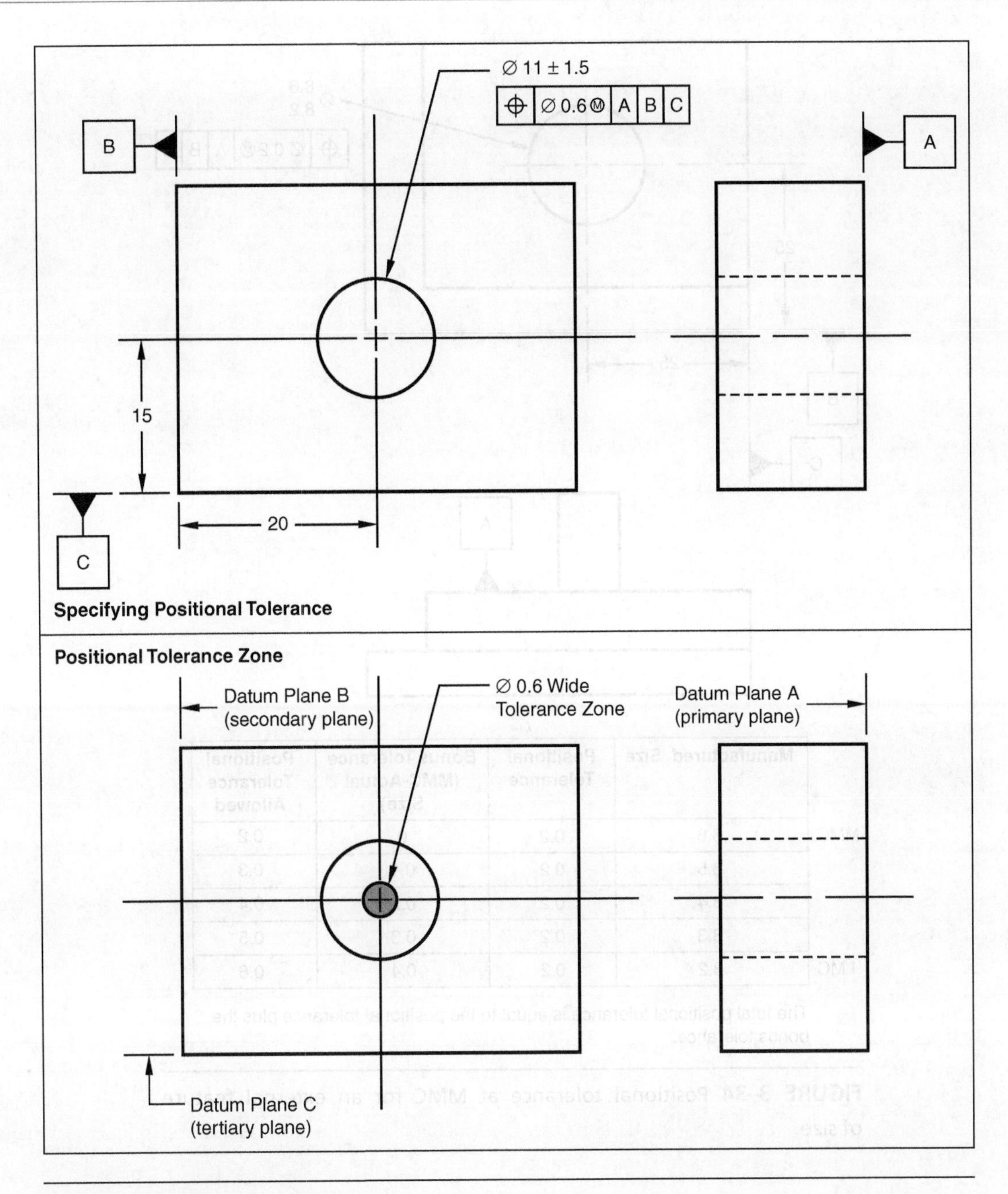

FIGURE 3–33 Position tolerance.

A positional tolerance at MMC means that the tolerance applies when the feature is at MMC condition. As the feature departs from MMC, additional (or "bonus") tolerance is allowed. The positional tolerance is allowed to increase an amount equal to the amount of difference from MMC. The maximum amount of tolerance is when the feature is manufactured at LMC.

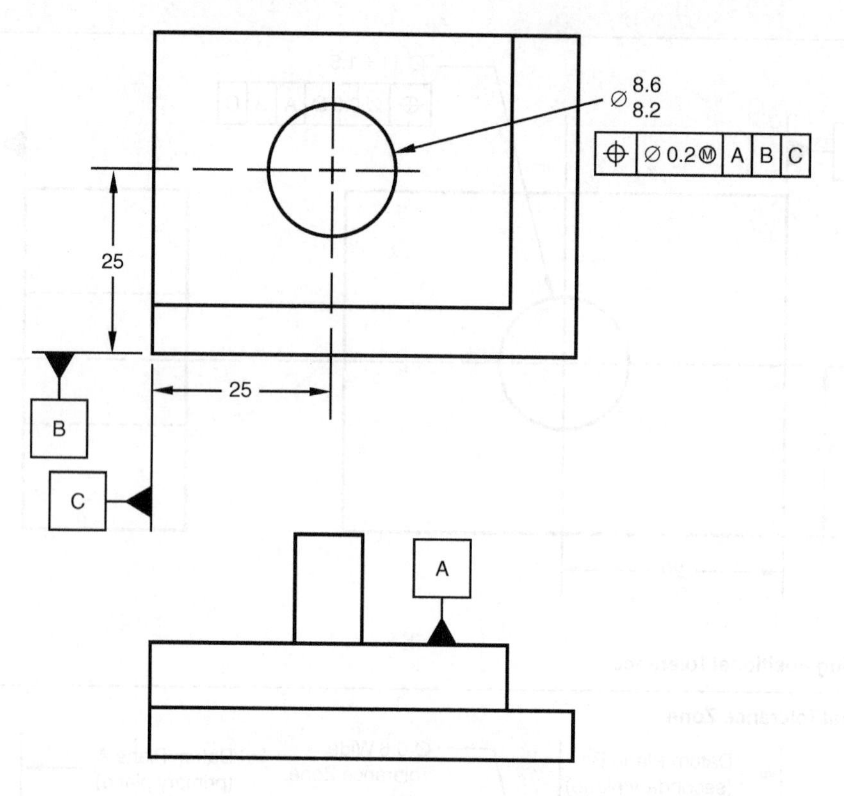

	Manufactured Size	Positional Tolerance	Bonus Tolerance (MMC-Actual Size)	Positional Tolerance Allowed
MMC	8.6	0.2	0	0.2
	8.5	0.2	0.1	0.3
	8.4	0.2	0.2	0.4
	8.3	0.2	0.3	0.5
LMC	8.2	0.2	0.4	0.6

The total positional tolerance is equal to the positional tolerance plus the bonus tolerance.

FIGURE 3–34 Positional tolerance at MMC for an external feature of size.

For an external feature of size (see Figure 3–34), the bonus tolerance is equal to the amount the actual size departs from the maximum material condition. (Bonus tolerance = MMC − Manufactured size.) The total positional tolerance is the allowed positional tolerance plus the bonus tolerance. (Total positional tolerance = Positional tolerance + Bonus tolerance.)

For an internal feature of size, such as a hole, the bonus tolerance is equal to the amount the actual

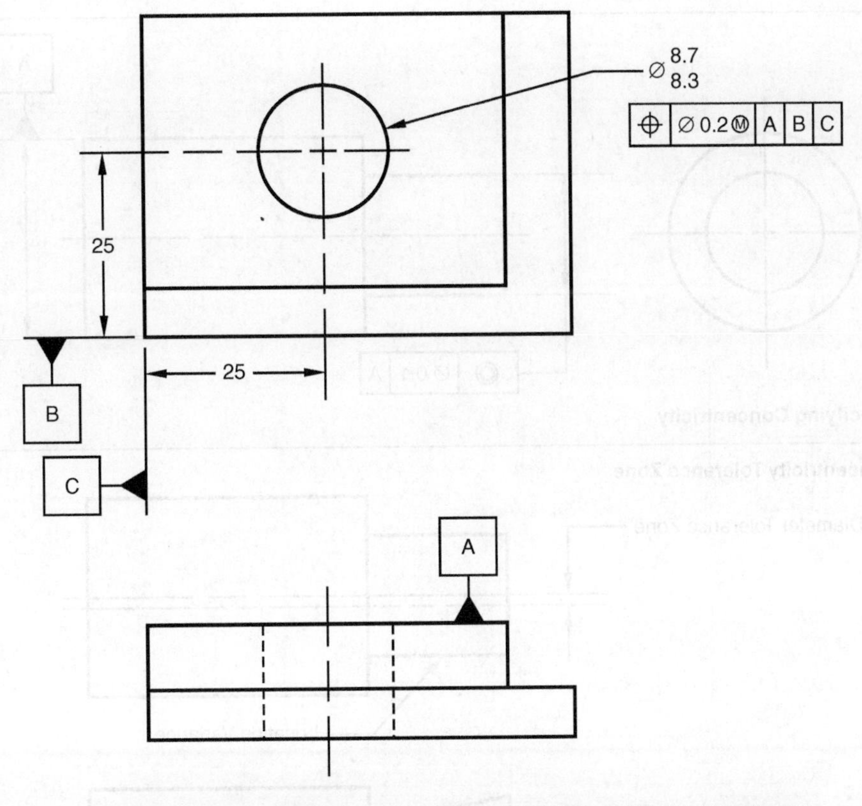

	Manufactured Size	Positional Tolerance	Bonus Tolerance (Actual Size- MMC)	Positional Tolerance Allowed
MMC	8.3	0.2	0	0.2
	8.4	0.2	0.1	0.3
	8.5	0.2	0.2	0.4
	8.6	0.2	0.3	0.5
LMC	8.7	0.2	0.4	0.6

The total positional tolerance is equal to the positional tolerance plus the bonus tolerance.

FIGURE 3–35 Positional tolerance at MMC for an internal feature of size.

size departs from maximum material condition (see Figure 3–35). (Bonus tolerance = Actual size − MMC.) The total allowed tolerance is the bonus tolerance plus the positional tolerance (Total positional tolerance = Bonus tolerance + Positional tolerance.)

Concentricity

Concentricity is used to control the relationship between the axes of two or more cylindrical features. When the axes of each cylinder fall on the same centerline, they are concentric (see Figure 3–36).

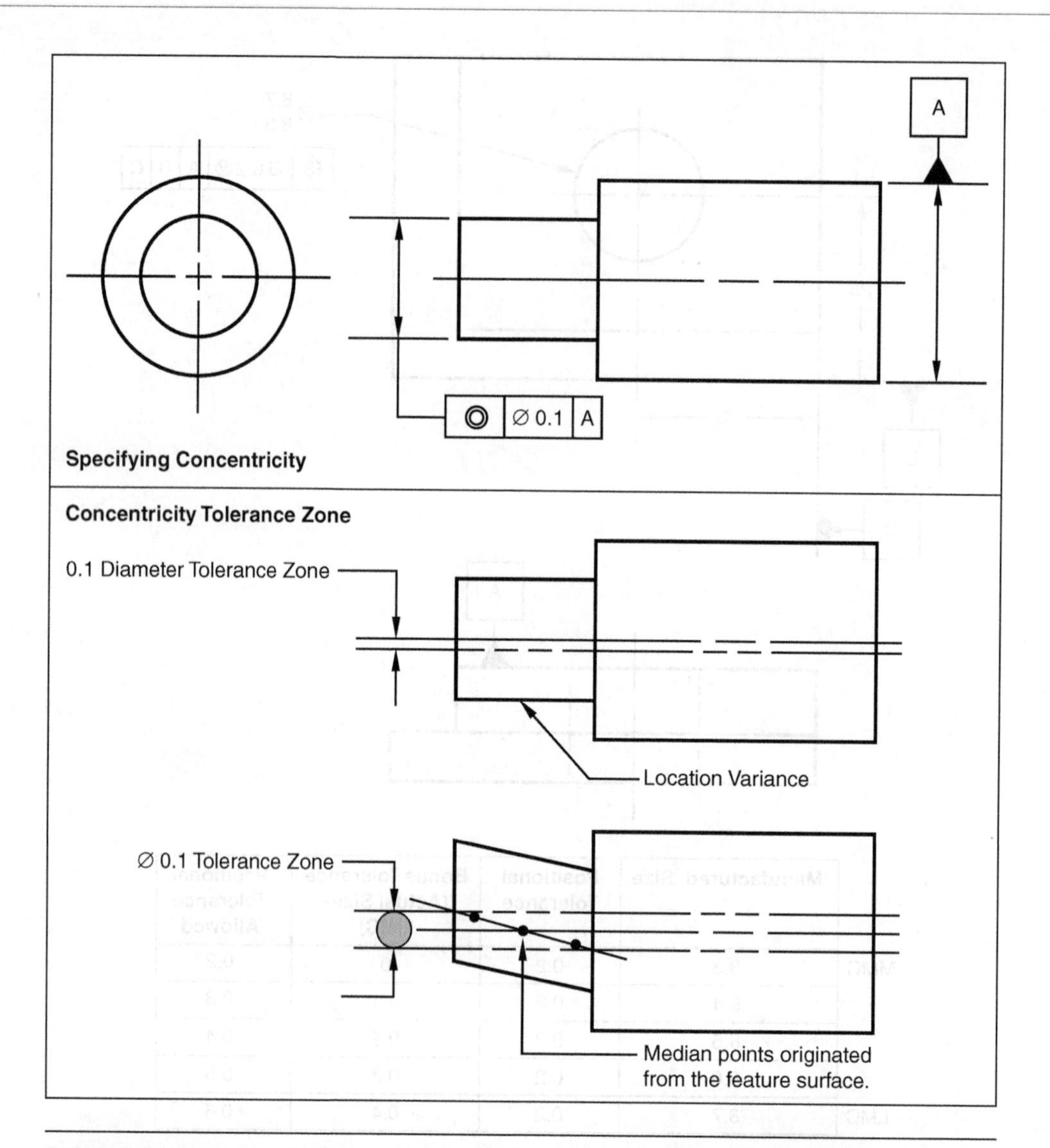

FIGURE 3–36 Concentricity tolerance. Concentricity is defined as the condition where all median points of diametrically opposed elements of a cylinder are congruent with the axis of a datum feature.

The tolerance zone for concentricity is a cylindrical tolerance zone whose axis coincides with the axis of the datum feature. Concentricity is a relative measurement, so it requires a datum specification. The tolerance can only be applied on an RFS basis.

Symmetry

Symmetry is a positional tolerance where the median points of all opposed elements of two or more features are congruent with the axis or center plane of a datum

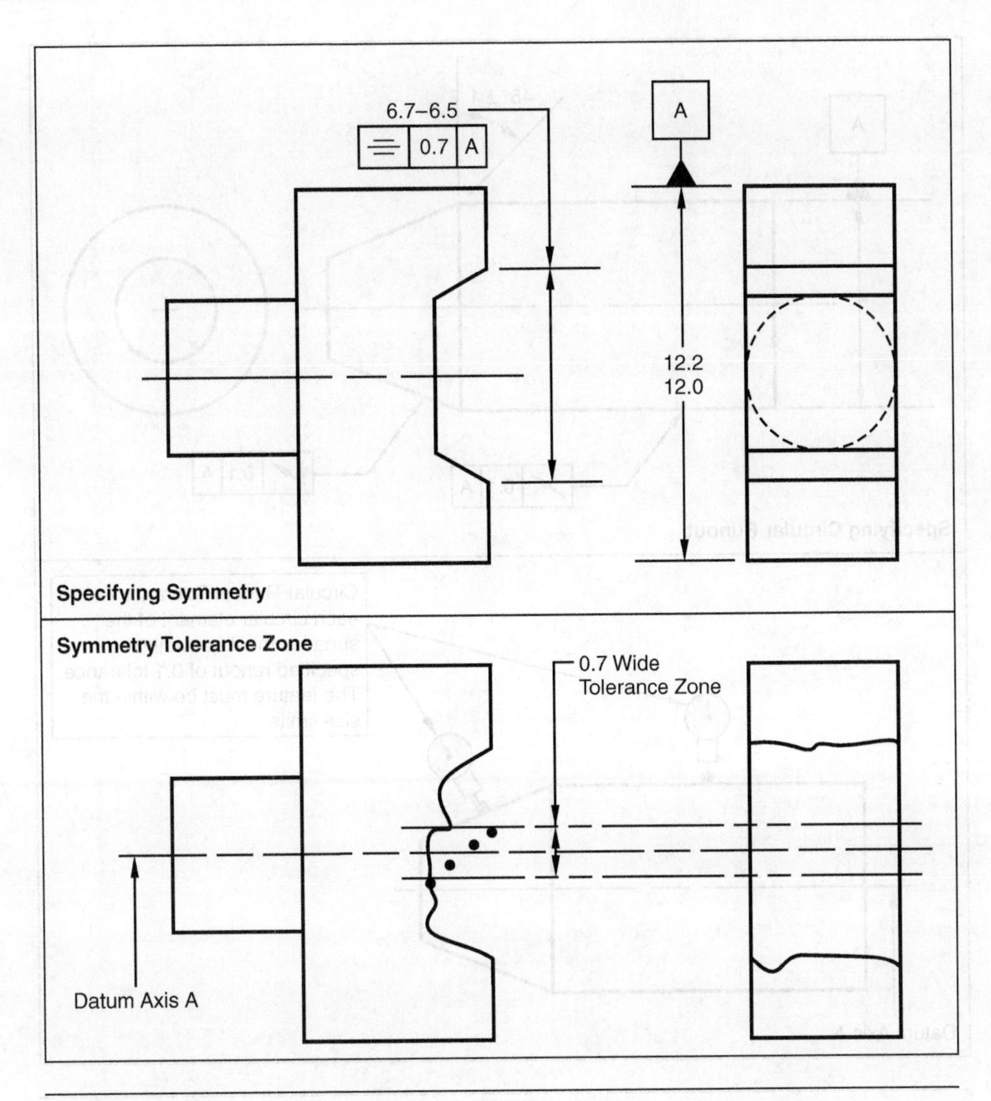

FIGURE 3–37 Symmetry tolerance.

feature. Concentricity and symmetry are similar concepts. The difference is that they are applied to different geometric configurations—concentricity applies to cylindrical features while symmetry is applied to planar features. The tolerance zone is centered about the center plane of the datum (see Figure 3–37). Symmetry is always used with a datum reference, and applied on an RFS basis.

Runout Tolerances

Runout tolerances are a combination of tolerances used to control the relationship of one or more features of a part to a datum axis. The features can be surfaces perpendicular or surfaces around the datum axis.

There are two types of runout control: circular runout and total runout. Circular runout is less

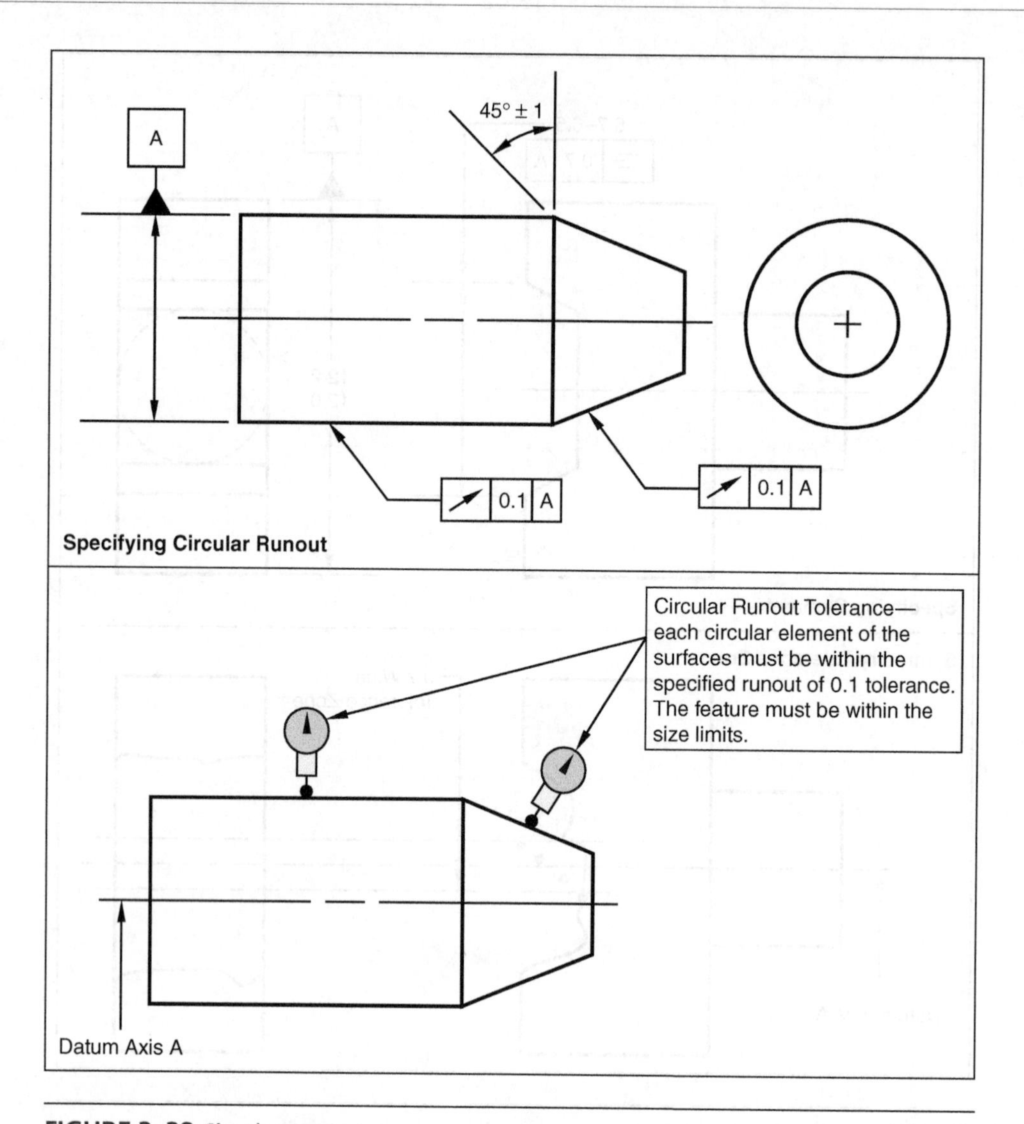

FIGURE 3–38 Circular runout.

complex than total runout. Both circular runout and total runout require a datum reference. Runout is always applied regardless of feature size.

Circular Runout

Circular runout controls circularity and coaxiality (the condition where two or more features share a common axis). The tolerance is measured by full indicator movement of a dial indicator placed at several locations while the part is rotated 360°. Circular runout is measured as a single circular element at each measured location. The full indicator movement displays total tolerance (see Figure 3–38).

Total Runout

Total runout is used to provide total composite control of all surface elements (see Figure 3–39). The tolerance is applied to both the circular elements and the

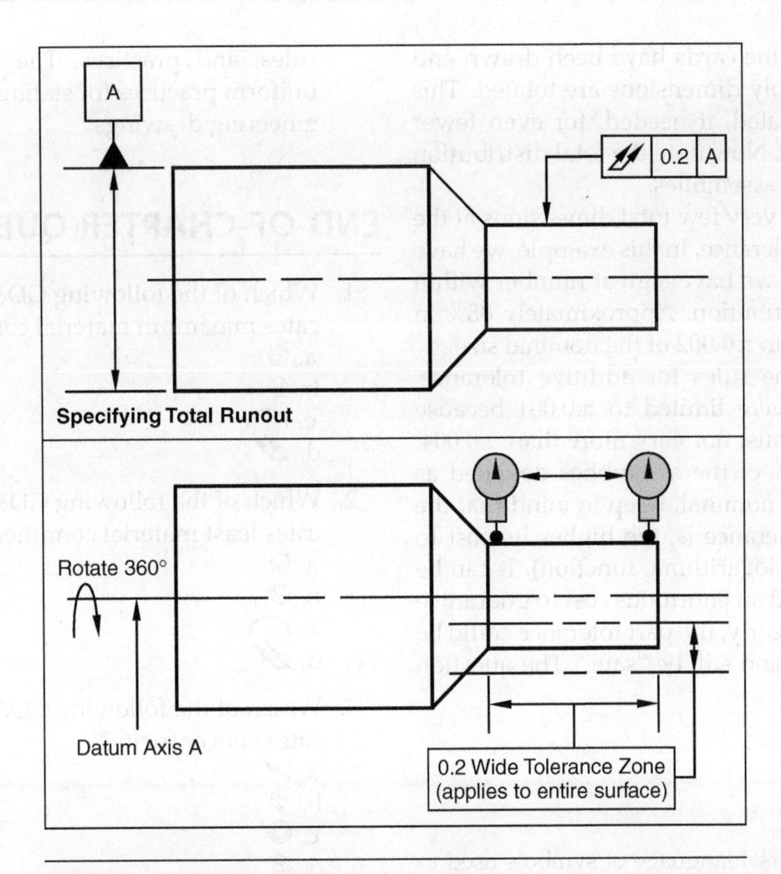

FIGURE 3–39 Total runout.

profile. When applied to the surfaces around and at right angles to a datum axis, total runout may be used to control a combination of circularity, straightness, angularity, taper, and profile. It is also used to control wobble, concavity, or convexity when applied perpendicular to a datum axis.

Statistical Tolerancing

In the next chapter, we will discuss statistics and statistical process control (SPC). In that discussion, we will, for the most part, discuss how statistics and SPC are applied. The economics of Just in Time (JIT) manufacturing, short cycle manufacturing, and complete interchangeability have further driven the needs for SPC.

The traditional method for interchangeable manufacturing is making sure the tolerances "add up," producing the term *additive tolerances*. In the beginning of the chapter, we saw the wasteful effects of this type of tolerance. We can show this by using cards instead of actual parts.

Assume that an assembly consists of four parts and that their tolerances are additive to form an assembly specification of 2.00 ± 0.004. Assume that each component has a tolerance of ±0.001. Let us round off our assumption so there are only three sizes of each part coming from the manufacturing operation: 0.449, 0.500, and 0.501. To simulate production, prepare four sets of three cards each. Label these cards as 20.001, 0.000, and 10.001. Then, select one card from each set randomly. Add the sizes and tally the total.

Continue until all of the cards have been drawn and the imaginary assembly dimensions are totaled. This process may be repeated, if needed, for even fewer "chance" occurrences. Now note the total distribution of dimensions for the assemblies.

We clearly have very few total dimensions at the extreme ends of the tolerance. In this example, we have only two. In contrast, we have a great number within the center of the distribution. Approximately 88% in this example are within ±0.002 of the nominal size.

By following the rules for additive tolerance, the part tolerances were limited to ±0.001 because the total assemblies must not vary more than ±0.004. In practice, only 2.5% of the assemblies deviated as much as ±0.004 from nominal. Keep in mind that the narrower the part tolerance is, the higher its cost to manufacture (often a logarithmic function). It can be seen that we have paid an enormous cost to guarantee interchangeability. Clearly, the part tolerance could be opened considerably and still be "safe." The question is "How much?"

SUMMARY

- GD&T is a universal language of symbols used to convey design intent from the design stage, through manufacturing, to inspection of the final product.
- Machine drawings were conventionally dimensioned by Cartesian coordinates—vertical and horizontal dimensions. Cartesian coordinates prescribe a square tolerance zone. Geometric tolerancing allows a diametrical or cylindrical tolerance zone, increasing the size of the tolerance zone by approximately 57%.
- Geometric tolerancing provides manufacturing with increased tolerancing for mating parts, reducing scrap, and lowering manufacturing costs.
- While this chapter provides a brief summary of GD&T concepts, the ASME Y14.5M-1994 Dimensioning and Tolerancing Standard should be made available to designers and inspectors for a complete set of definitions, fundamental rules, and practices. The standard establishes uniform practices for stating and interpreting engineering drawings.

END-OF-CHAPTER QUESTIONS

1. Which of the following GD&T symbols indicates maximum material condition?
   a. Ⓛ
   b. Ⓜ
   c. ⌒
   d. ⌰
2. Which of the following GD&T symbols indicates least material condition?
   a. Ⓛ
   b. Ⓜ
   c. ⌒
   d. ⌰
3. Which of the following GD&T symbols indicates concentricity?
   a. ↗
   b. ⌖
   c. ◎
   d. ⌭
4. Which of the following GD&T symbols indicates flatness?
   a. ⌖
   b. ⏤
   c. ⊥
   d. ⏥
5. Which of the following GD&T symbols indicates cylindricity?
   a. ⌖
   b. ⌭
   c. ○
   d. ⌓
6. Which of the following feature control frames is incorrect?
   a. | ⏥ | 2.0 | A |
   b. | ⏥ | 0.5 |

c. [⌯ | 0.4 | A]
d. [⌖ | ∅ 0.8 Ⓜ | A Ⓜ]

7. Which of the following geometric characteristic symbols are used to control an individual feature's form?
   a. ∠ ⊥ //
   b. Ⓜ Ⓛ Ⓢ
   c. — ⏥ ○
   d. ⌖ ◎ ⌭

8. Tolerances of location include:
   a. position, concentricity, and symmetry
   b. position, perpendicularity, and flatness
   c. symmetry, concentricity, and runout
   d. circularity, cylindricity, and runout

9. Orientation tolerances include:
   a. angularity, profile of a line, and runout
   b. profile of a line, profile of a surface, and total runout
   c. angularity, perpendicularity, and parallelism
   d. perpendicularity, parallelism, and symmetry

10. The effect of Rule #1 is that it controls:
    a. the size and location of an individual feature of size
    b. the size and orientation of an individual feature of size
    c. the size and form of an individual feature of size
    d. the orientation and form of an individual feature of size

11. When applied to a surface, angularity also controls:
    a. size
    b. location
    c. position
    d. flatness

12. The difference between concentricity and symmetry is:
    a. the use of modifiers
    b. the datum references required
    c. the geometric shape of the tolerance zone
    d. concentricity is a less stringent requirement

13. Which of the following feature control frames is incorrect?
    a. [∠ | 0.7]
    b. [⌖ | ∅ 0.4 Ⓜ | A | B]
    c. [⌖ | ∅ 0.4 Ⓛ | A | B]
    d. [∠ | 0.7 | B]

14. Which of the following GD&T symbols indicates total runout?
    a. ⌰
    b. ⌯
    c. ↗
    d. ⌭

15. Which of the following GD&T symbols indicates circular runout?
    a. ⌰
    b. ○
    c. ↗
    d. ⌭

16. Which of the following feature control frames is incorrect?
    a. [⌓ | 0.9 | A]
    b. [○ | 0.7 | C]
    c. [⊥ | ∅ 1.0 | B]
    d. [○ | 0.4]

4

CHAPTER FOUR

STATISTICS AND METROLOGY

The long range contribution of statistics depends not so much upon getting a lot of highly trained statisticians into industry as it does in creating a statistically minded generation of physicists, chemists, engineers, technicians, and others who will in any way have a hand in developing and directing the production processes of tomorrow.

W. A. Shewhart (1891–1967) and W. E. Deming (1902–1994)

LEARNING OBJECTIVES

- State the relationship between measurements and statistics.
- Explain why statistical control is based on measurements that have already been taken.
- Explain how statistical control allows you to predict what the future measurements will be.

OVERVIEW

If we adhere strictly to the fundamentals of metrology, we find only one important connection between statistics and measurement: The values used for statistical analysis are determined by measurement, whether linear, mass, temperature, and so on. Having said that, we have eliminated the need for any further reference to statistics.

What follows has nothing to do with metrology as such. It is included because the ability to use metrology is every bit as important as the ability to make the measurements. Even then, it could not be justified simply by improving our abilities as line inspectors, research microbiologists, or in some other specialized field. A discussion of statistics is included because it improves our ability to understand and perform in every field in which data must be processed. It is difficult, if not impossible, to name one profession that does not require this ability.

When this subject of *statistics* or statistical methods is mentioned, its value is often waved aside because of the incorrect perception of it. The principles that follow are applicable to almost all works in metrology. In fact, when we step back and take a look, they apply to almost everything in life.

In earlier times, *quality control* (QC) was a simple matter, and the term *quality assurance* (QA) was not used. The assembly department expected to get parts that would fit. If a part was too large, the shop was advised and the cutting tool was advanced slightly. When an order was placed with a supplier, the material received was expected to have some defective parts. Today, that is far too expensive. The high speeds of modern mechanisms call for closer tolerances, and high cost does not allow for large inventories of materials. In the past, these required far more control of the production machines, and thus far more inspection. This is now costly and unnecessary, thanks to *statistical process control* (SPC).

Although statistics as a branch of mathematics was well understood at a much earlier date, it was the efforts of Dr. Walter A. Shewhart (1891–1967) of the Bell Telephone Laboratories around 1924 that established the techniques for applying statistics to the control of industrial processes. The control charts that are fundamental to SPC today still bear his name: *Shewhart Control Charts*. Despite the proven advantages of SPC, it was not used extensively until World War II when unprecedented production demands required improved methods. Dr. Shewhart died shortly thereafter.

The concept of *statistical quality control* (SQC) became widely used in the late 1970s and 1980s and is the driving force behind most quality programs today. SPC deals specifically with the use of statistics to control a process, typically some type of control chart or precontrol. The term *statistical quality control*, or SQC, deals with the wider use of statistics to control the product from the design stages to the final product, which is shipped to the customer. SQC includes such activities as:

- Process capability studies
- SPC
- Statistical-based sample inspection
- Statistical design of experiments

4-1 BASIC STATISTICS SECOND LINE

When a problem is approached from the standpoint of statistics, we go through a process of gathering information or data. These data are then compiled in the form of a table or graph. This is known as *descriptive statistics*. Descriptive statistics gives us information regarding the data we gather. Frequently, we need to develop a simple statement based on the sample about the population. To do this, we need to have an understanding of *probability*. Then, to finalize the procedure, we need to apply the theories of statistics to draw valid conclusions about the data. This procedure is called *inferential statistics*. Each of these topics is discussed briefly in this chapter. If more in-depth knowledge is needed, please refer to the references in Appendix E.

Descriptive Statistics

Descriptive statistics consists of four general areas: central tendency or what is the predominate value of the data; dispersion or the spread of the data; position of the data with respect to other data or data points; and graphic presentation.

Central Tendency. The four measures of central tendency are mean, median, mode, and midrange. We frequently call this value the "average," "mean," or "typical" value. The terms *average* and *mean* have very specific statistical definitions.

Mean. The *arithmetic mean* or *mean*, $\bar{x}$ (read as "x bar"), which we commonly call the average, is the summation of all the values divided by the total number of values.

$$\bar{x} = \frac{\sum_{i=1}^{n} x_1}{n} \qquad \textit{Equation 4–1}$$

This is the value you receive when you ask for your average grade or the average high or low temperature for a given month in the city or town you live in.

ILLUSTRATION 4–1

A set of five measurements was taken from a sample of five shifts. The values in millimeters are 25.06, 25.03, 25.08, 25.05, and 25.03. Find the mean.

SOLUTION
Using Equation 4–1, we find

$$\bar{x} = \frac{25.06 + 25.03 + 25.08 + 25.05 + 25.03}{5}$$
$$= \frac{125.25}{5}$$
$$= 25.05$$

Median. The *median*, M_d (read M sub d), is the value of the data that occupies the middle of the data when arranged in ascending order. When the data set contains an odd number of data points, then the value is one of the data points. When the data set contains an even number of data points, then the value is the "middle" point between the two numbers. Note that the median "breaks" the data set into two subsets. The depth (the number of positions from either end), or position, of the median is determined by the formula:

$$d(M_d) = \frac{n+1}{2} \qquad \textit{Equation 4–2}$$

where 1 is the position of the smallest value, and n is the number of pieces of data.

ILLUSTRATION 4–2

Using the data set from Illustration 4–1, find the median.

SOLUTION (the odd number of data points in the data set)

STEP 1. Rank the data in descending order.

25.08, 25.06, 25.05, 25.03, 25.03

STEP 2. Calculate the depth of the median.

$$d(M_d) = \frac{5+1}{2} = 3$$

STEP 3. Determine the third position. This is the median.

25.05

SOLUTION (the even number of data points in the data set). Assume the following data set: 25.08, 25.06, 25.03, 25.03.

STEP 1. Rank the data in descending order.

25.08, 25.06, 25.03, 25.03

STEP 2. Calculate the depth of the median.

$$d(M_d) = \frac{4+1}{2} = 2.5$$

STEP 3. Determine the position halfway between data points 2 and 3. This is the median. (Note the number was 25.045 and rounded to 25.04.)

25.04

Mode. The *mode*, M_o (read "M sub o"), is the value that occurs with the greatest frequency, the most common value. A mode may not exist or it may not be unique.

ILLUSTRATION 4–3

Using the data set from Illustration 4–1, find the mode.

SOLUTION

STEP 1. Rank the data in descending order.

25.08, 25.06, 25.05, 25.03, 25.03

STEP 2. Determine the frequency of each value.

Value	Frequency
25.08	1
25.06	1
25.05	1
25.03	2

STEP 3. Determine the mode(s).

25.03

Midrange. The *midrange* is midway between the highest and lowest values. It is found by averaging the high and low values:

$$\text{midrange} = \frac{H + L}{2} \qquad \textit{Equation 4–3}$$

The midrange is the numerical value halfway between the two extreme values, the high H and the low L.

ILLUSTRATION 4–4

Using the data set from Illustration 4–1, find the midrange.

SOLUTION

STEP 1. Rank the data in ascending order.

25.08, 25.06, 25.05, 25.03, 25.03

STEP 2. Select the highest value, H, and the lowest value, L.

$H = 25.08$
$L = 25.03$

STEP 3. Determine the midrange.

$$\text{midrange} = \frac{25.08 + 25.03}{2}$$

$$\text{midrange} = \frac{50.11}{2}$$

$$\text{midrange} = 25.05$$

The four measures of central tendency represent four different methods of describing the middle value. The four values may be the same, but more likely the calculation will result in different values. In the preceding illustrations, the mean was 25.05, the median was 25.05, the mode was 25.03, and the midrange was 25.05.

Dispersion

The three measures of dispersion are range, variance, and standard deviation. These terms assign a numerical value to the amount of spread of the data set. The variance and standard deviation are deviations from the mean.

Range. The *range* is the simplest measure of dispersion. It is the difference between the largest (highest) valued data (H) and the smallest (lowest) valued data (L) in the data set.

$$\text{Range} = H - L \qquad \textit{Equation 4–4}$$

ILLUSTRATION 4–5

Using the data set from Illustration 4–1, find the range.

SOLUTION

STEP 1. Rank the data in ascending order.

25.08, 25.06, 25.05, 25.03, 25.03

STEP 2. Determine the highest value (H).

$H = 25.08$

STEP 3. Determine the lowest value (L).

$L = 25.03$

STEP 4. Determine the range (difference) between the highest and the lowest values.

$\text{Range} = 25.08 - 25.03$
$\text{Range} = 0.05$

Variance. The variance, s^2, is calculated by using the following formula:

$$s^2 = \frac{\sum_{i=1}^{i=n} (x - \bar{x})^2}{n - 1} = 1 \qquad \textit{Equation 4–5}$$

where x is the individual value, $\bar{x}$ is the mean, and n is the sample size. A simpler formula to use is:

$$s^2 = \frac{\sum_{i=1}^{i=n} x^2 - \frac{\left[\sum_{i=1}^{i=n} x\right]^2}{n}}{n-1} \qquad \textit{Equation 4–6}$$

ILLUSTRATION 4–6

Given a set of dimensions from a machined block (25.6 mm, 25.3 mm, 25.8 mm, 25.8 mm, 25.5 mm, 25.5 mm, and 25.3 mm), find the variance.

STEP 1. List the data using the following format: where n = number, x = value, x^2 = value squared.

n	x	x^2
1	25.6	655.36
2	25.3	640.09
3	25.8	665.64
4	25.8	665.64
5	25.5	650.25
6	25.5	650.25
7	25.3	640.09

STEP 2. Find the sum of columns x and x^2.

178.8 4567.32

These are the sum of x (Σ x) and the sum of x^2 (Σx^2) in Equation 4–6.

STEP 3. Calculate the variance using Equation 4–6.

$$s^2 = \frac{\sum(x^2) - \frac{\sum(x^2)}{n}}{n-1}$$

$$= \frac{4567.32 - \frac{31969.44}{7}}{7-1}$$

$$= \frac{4567.32 - 4567.06}{6}$$

$$= \frac{0.26}{6}$$

$$= 0.043$$

Standard Deviation. The *standard deviation* of a sample is the positive square root of the variance.

$$s = \sqrt{s^2} = \sqrt{\frac{\sum_{i=1}^{i=n}(x-\bar{x})^2}{n-1}} \qquad \textit{Equation 4–7}$$

$$s = \sqrt{s^2} = \sqrt{\frac{\sum_{i=1}^{i=n} x^2 - \frac{\left[\sum_{i=1}^{i=n} x\right]^2}{n}}{n-1}} \qquad \textit{Equation 4–8}$$

ILLUSTRATION 4–7

Using the data in Illustration 4–6, calculate the standard deviation.

STEP 1. Use Equation 4–8.

$$s = \sqrt{s^2}$$

$$= \sqrt{0.043}$$

$$= 0.21$$

Position

Measures of position allow us to compare one piece of data to the set of data. The two most frequently used forms are *quartiles* and *percentiles.* Quartiles split our data into *four* groups of approximately the same number of data Q1, is a number such that, one-fourth of the data is smaller in value than Q1 (see Figure 4–1).

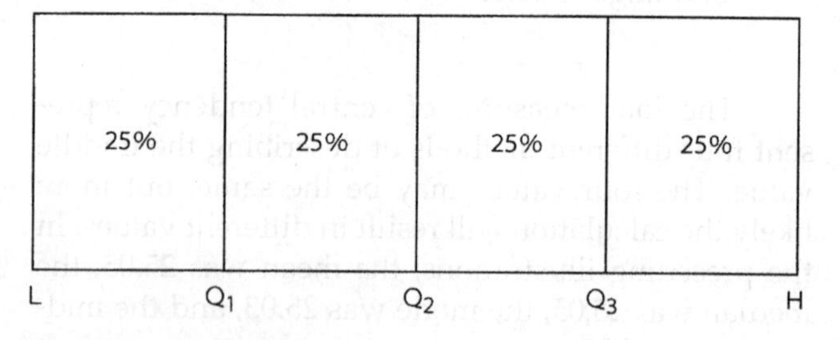

FIGURE 4–1 Quartiles

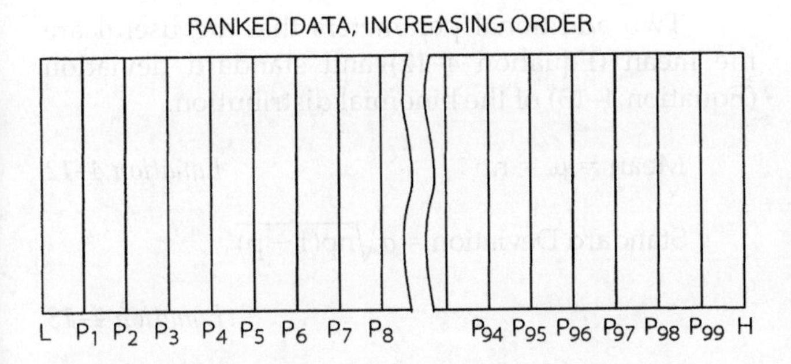

FIGURE 4–2 Percentiles

Q2, is the median. Approximately three-quarters of the data lie below the value of Q3.

Percentiles are determined in the same manner as the quartiles except the data are ranked into 100 parts. See Figure 4–2 for an example.

4–2 PROBABILITY

Nature of Probability

Let us consider an experiment in which we toss two coins simultaneously and record the number of tails that occur in each trial. The only outcomes are 0T (no tails), 1T (one tail), and 2T (two tails). See the summary of the results in Table 4–1 if we tossed the coins 10 times and recorded our outcome after each trial.

The results are shown in Table 4–2 if we repeated the experiment 10 times for a total of 100 coin tosses.

Because we did the experiment 10 times for a total of 100 trials, we obtained 24% with no tails, 54%

TABLE 4–1 Coin Toss

Outcome	Frequency
0 Tails	1
1 Tail	5
2 Tails	4

TABLE 4–2 Results of 100 Trials in a Coin Toss

Outcome	1	2	3	4	5	6	7	8	9	10	Total
0 Tails	1	2	1	4	2	2	5	3	0	4	24
1 Tail	5	5	6	5	5	7	5	4	8	4	54
2 Tails	4	3	3	1	3	1	0	3	2	2	22

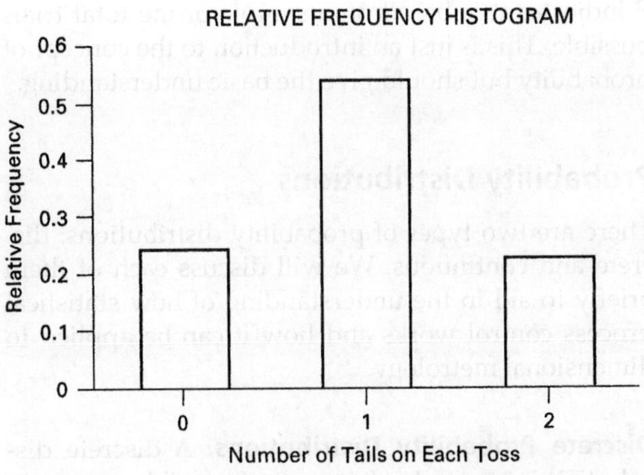

FIGURE 4–3 Relative frequency histogram of a coin toss

with one tail, and 22% with two tails. The results are shown in the *histograms* in Figure 4–3. What would happen if we repeated this experiment an additional 100 times? Would the frequencies change, and if so, how much? If we repeated this experiment several hundred times, we would approach the ratio of 1:2:1 or 25% with no tails, 50% with one tail, and 25% with two tails. These relative frequencies accurately reflect the idea and concept of probability.

Probability of Events. From the previous example, we can now define what is meant by probability, and specifically what is meant by the "probability of an event." The probability of an event is the relative frequency with which an event will occur. Using the previous example, the probability of tossing the coins and getting zero tails is one in four, or 25%. Notice that the total for all of the events is 100%. In all cases, the total of all events is 100%. Equation 4–9 shows the way the value of probability is expressed.

$$P'(\text{event}) = \frac{n(\text{event})}{n_{\text{total}}} \qquad \textit{Equation 4–9}$$

This is read as Prime (P′), the probability of the event, is equal to n(event), the number of times the event occurred, divided by n_{total}, the total times the experiment was attempted. The "prime" symbol (′) after the

P indicates it is based on a sample of the total trials possible. This is just an introduction to the concept of probability but should give the basic understanding.

Probability Distributions

There are two types of probability distributions: discrete and continuous. We will discuss each of them briefly to aid in the understanding of how statistical process control works and how it can be applied to dimensional metrology.

Discrete Probability Distributions. A discrete distribution is a sample of any set of possible outcomes that have a unique result. The previous example of the coin toss is an example of a discrete probability distribution. The results of the coin toss can only be one of three outcomes. The outcomes are all "discrete" or separate results. The outcome is always one of two discrete results: head or tail. There cannot be any other result; if there is, then the experiment or trial is invalid. The rolling of dice is another example of a discrete distribution because there are six possible outcomes from this distribution. In metrology, the results of a go/no-go gage would represent a discrete distribution.

Binomial Probability Distribution. The binomial distribution is shown in Equation 4–10.

$$P(D = x) = \left[\frac{n}{x}\right] p^x (1 - p)^{n-x} \quad x = 0,1,2,\ldots,n$$

Equation 4–10

In Equation 4–10, p is the fraction of the population that contains a *defect*. Because we are working with a sample, we use the term *p′* (p prime) to differentiate the difference between a sample and population. To calculate p′, we use Equation 4–11.

$$P' = \frac{D}{n}$$ *Equation 4–11*

where D is the number of units with a defect and n is the total sample from which D was taken.

Two additional parameters that are useful are the mean (Equation 4–12) and standard deviation (Equation 4–13) of the binomial distribution.

$$\text{Mean} = \mu = np$$ *Equation 4–12*

$$\text{Standard Deviation} = \sigma\sqrt{np(1-p)}$$

Equation 4–13

Poisson Distribution. The Poisson distribution, named after French mathematician S. D. Poisson (1781–1840), is shown in Equation 4–14.

$$f(x) = \frac{e^{-\lambda}\lambda^x}{x!}$$ *Equation 4–14*

In Equation 4-14, λ (the Greek letter lambda) is the term for the mean or average number of defects per unit. The unique item regarding the Poisson distribution is that the mean and the variance are always the same value. The calculation of the sample mean number of defects is shown in Equation 4–15.

$$\lambda' = \frac{D}{n}$$ *Equation 4–15*

In Equation 4–15, λ is the total number of defects found in our sample of n items. The calculation of the mean and standard deviation are shown in Equations 4–16 and 4–17, respectively.

$$\text{Mean} = \mu = \lambda$$ *Equation 4–16*

$$\text{Standard Deviation} = \sigma = \sqrt{\lambda}$$ *Equation 4–17*

Continuous Probability Distributions. In metrology, we normally work with continuous distributions. A continuous distribution is only limited by our ability to discern between two different measurements. The continuous probability distribution that is used the most in SPC and in inspection is the normal probability distribution.

Normal Probability Distribution. The Gaussian distribution or *normal distribution* dates back to eighteenth-century Germany where the mathematician, Gauss, found that repeated measurements of the same

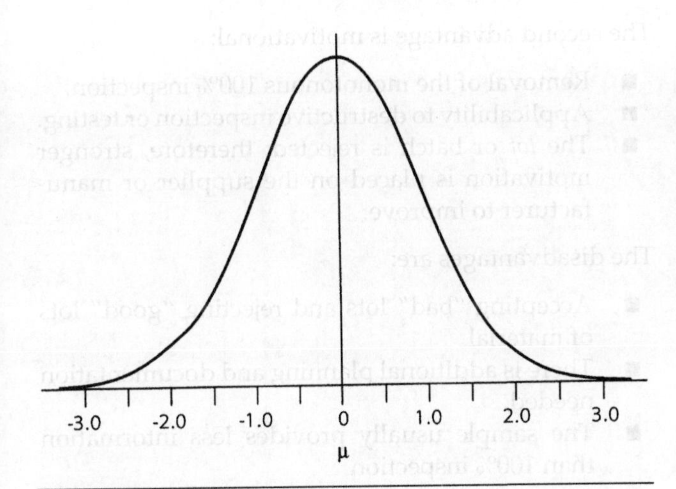

FIGURE 4–4 Graph of Gaussian or normal distribution

astronomical quantity produced a pattern similar to the continuous curve in Figure 4–4. The formula for the Gaussian or normal curve is shown in Equation 4–18.

$$f(x) = \frac{1}{\sqrt{2\pi}\sigma} e^{-\frac{1}{2}\left[\frac{(x-\mu)^2}{\sigma}\right]} \quad \text{for } -\infty < x > \infty$$

Equation 4–18

It is shown not to frighten the reader, but rather to point out the two parameters that are present in the equation: μ and σ. The mean, μ, and standard deviation, σ, are contained in the equation. Notice when we calculated the mean and standard deviation we used the terms *x* and *s* to describe the central tendency and dispersion. At that time, we were discussing it in terms of a sample. A probability distribution is stated in terms of the population. The population is the set of all possible outcomes. The sample is a subset of the population.

The calculation of the population mean is the same as the calculation of the sample mean, but that does not mean they are always equal to each other. The sample mean will approach the population mean when the sample size approaches the population. Thanks to the central limit theorem in statistics, a sample of 30 or more will give a good approximation for the *population*. There are also ways to estimate the confidence of the sample mean, but they are beyond the scope of this chapter and text.

The calculation of the population standard deviation is slightly different from the calculation of the sample standard deviation (see Equations 4–7 and 4–8). Equation 4–19 is used for calculating the population standard deviation.

$$s^2 = \frac{\sum_{i=1}^{i=n} (x - \bar{x})^2}{n}$$

Equation 4–19

For computational purposes, use Equation 4–20.

$$s = \sqrt{s^2} = \sqrt{\frac{\sum_{i=1}^{i=n} x^2 - \frac{\left[\sum_{i=1}^{i-n} x\right]^2}{n}}{n}}$$

Equation 4–20

The properties of normal distribution are key to descriptive statistics and inferential statistics in general and *control charts* and acceptance sampling systems specifically. The mean, central tendency, of the normal distribution is such that 50% of the points lie on each side of the mean. The standard deviation, the measurement of the dispersion, is such that ±1 standard deviation from the mean will contain 68.3% of the population. When you take ±2 standard deviations from the mean, it will contain 94.5% of the data, and ±3 standard deviations from the mean will contain 99.7% of the data. This can be carried out beyond the three standard deviations because the distribution goes to infinity. The position of the standard deviations and their percentages are shown graphically in Figure 4–5. This information is key in the development and interpretation of control charts.

Sampling and Sample Variability. Because we normally select a sample of the parts we are manufacturing, we need to understand a little more about sampling and sampling variability. When we did the initial coin toss, we obtained the results shown in Table 4–1. When we did the coin toss a second time, we did not get the exact same results of the first. In fact, when we did the

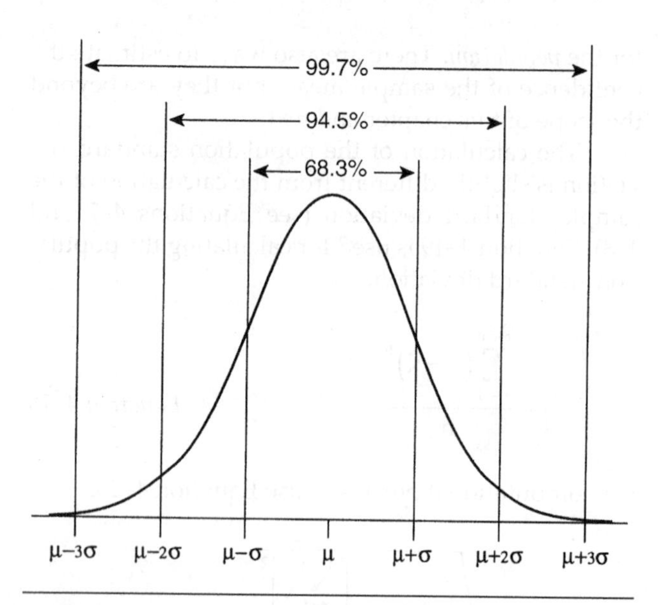

FIGURE 4–5 Percentages for Gaussian or normal distribution

coin toss a total of 10 times (see Table 4–2), we obtained the same result only on outcomes 2 and 5. In other words, 9 of the 10 outcomes were different. The difference is due to sample variability. This is why it is so important to obtain a random or representative sample of the work being performed. Again, if more information is needed about sampling and sampling variability, please see Appendix E.

4-3 ACCEPTANCE SAMPLING

Acceptance sampling is used to determine if the material produced is acceptable for use. Because this process is conducted after the material is produced, any unacceptable material would have to be reworked or scrapped. There is a major difference when comparing acceptance sampling with statistical process control. SPC allows the process to be controlled so all material is acceptable.

The advantages of acceptance sampling are primarily economical because of the following:

- Only part of the material needs to be inspected.
- There is less handling damage during inspection.
- Fewer inspectors are needed.

The second advantage is motivational:

- Removal of the monotonous 100% inspection.
- Applicability to destructive inspection or testing.
- The *lot* or batch is rejected; therefore, stronger motivation is placed on the supplier or manufacturer to improve.

The disadvantages are:

- Accepting "bad" lots and rejecting "good" lots of material.
- There is additional planning and documentation needed.
- The sample usually provides less information than 100% inspection.

When a sampling plan is being developed, the first consideration is to determine if the sampling plan is to accept or reject the lot of material immediately being evaluated, or to determine which of the lot of material produced is acceptable. The first inspection is called the *Type A sampling plan,* whereas the second is called the *Type B sampling plan.* This affects the probability distribution used to determine the sampling plan. In addition to the reason for the sampling plan, the type of inspection must be determined.

Type A sampling plans specify the quality level for each lot in terms of percent defective at a given risk of being accepted by the customer. They are known as lot tolerance percent defective plans, or LTPD plans. Type B sampling plans specify the quality level in terms of the percentage of accepting the submitted lots at a given quality level. They are known as *acceptable quality levels* or AQL plans.

Both LTPD and AQL plans provide lot-by-lot protection. A second type of protection is the limiting average percentage of defective items. These types of plans are known as the *average outgoing quality limit* or AOQL plans.

Key Definitions in Sampling Plans

Producer's Risk. The producer's risk is the probability that a "good" lot will be rejected by the sampling plan. The producer's risk traditionally will vary from 0.001 to 0.10, depending on the individual sampling plan. The risk is stated in conjunction with the maximum

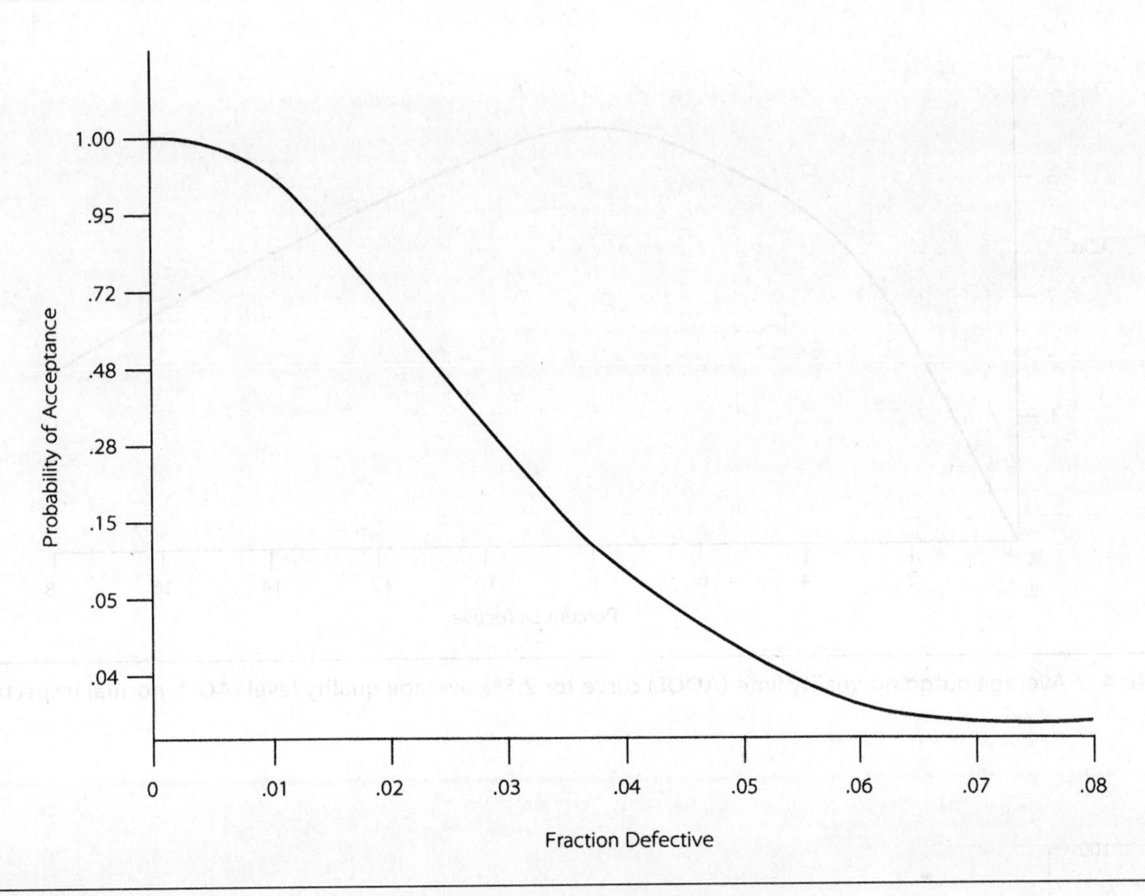

FIGURE 4–6 Operating characteristic (OC) curve for sampling inspection plan with n = 100, c = 2

quality level that the plan will accept. This is normally called the AQL.

Consumer's Risk. The consumer's risk is the probability that a "bad" lot will be accepted by the sampling plan. The risk is stated in conjunction with the definition of rejectable quality, such as lot tolerance percent defective. Traditionally, only 0.10 is used for the consumer's risk.

Operating Characteristic Curve. The *operating characteristic (OC) curve* is a graph of the lot fraction defective versus the probability of accepting the lot. Figure 4–6 shows an example of an OC curve.

Average Outgoing Quality Limit. We stated that the AQL and LTPD are two common quality indices for sampling plans. A third quality index is the AOQL, which is the worst case of quality level a consumer would receive when an inspection plan is in effect and all rejected lots have been 100% inspected with the removal of defective material. Figure 4–7 shows the relationship of the AOQL to the lot percent defective.

Average Total Inspection. The average total inspection (ATI) is the average inspection required when 100% inspection of the rejected lots is included. Figure 4–8 is an example of an ATI curve.

Types of Sampling Plans

There are many types of sampling plans. Table 4–3 shows the types of sampling plans and their typical uses. The sampling plans discussed here will be limited to two: ANSI/ASQC Z1.4 (*Sampling Procedures*

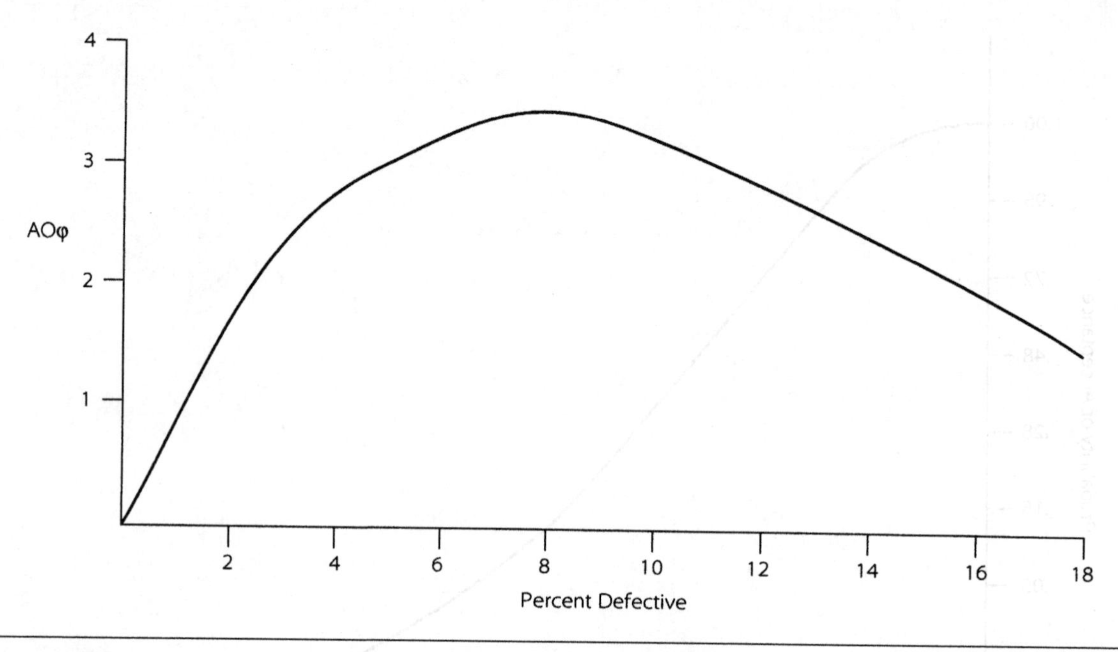

FIGURE 4–7 Average outgoing quality limit (AOQL) curve for 2.5% average quality level (AQL), normal inspection

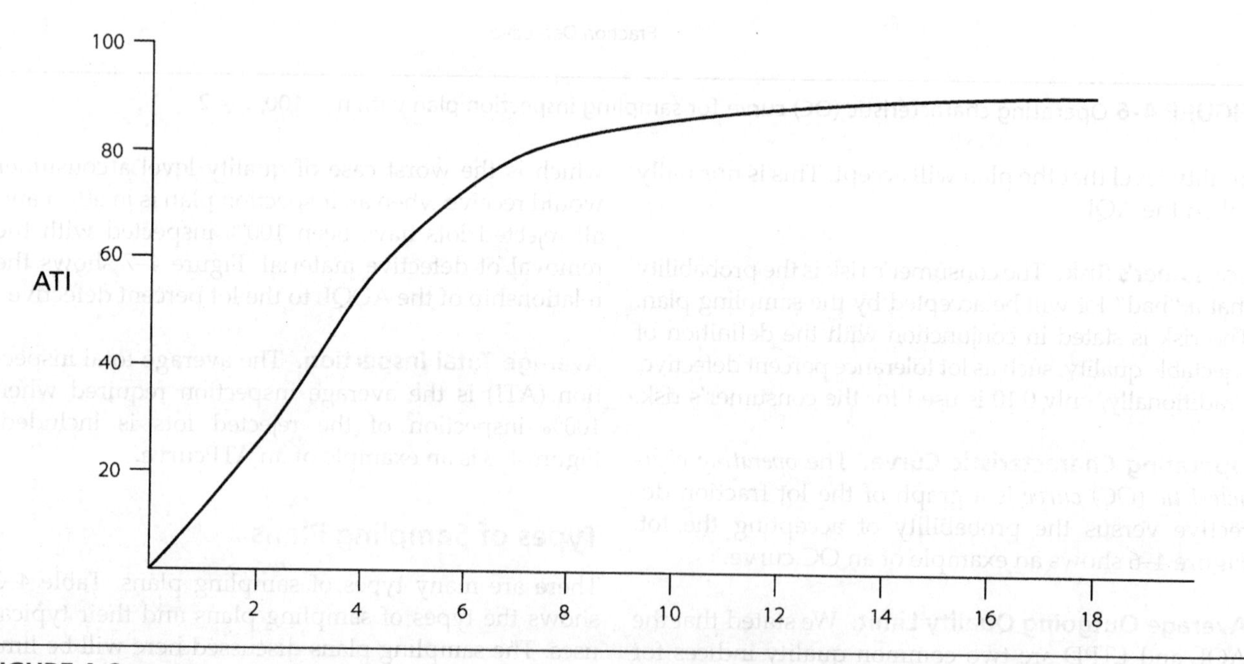

FIGURE 4–8 Average total inspection (ATI) for 2.5% average quality level (AQL), normal inspection

TABLE 4–3 Sampling Plans and Uses

Type of Sampling Plan	Typical Use of Sampling Plan
Single	General use
Double	General use with known good quality
Multiple	General use with known excellent quality
Sequential	General use with optimal sample size
Bulk	Bulk material sampling
Rectification	No lot can exceed a given percent defective
Continuous	Used when no natural lot occurs
Compliance	Maximizes consumer protection
Reliability	Controls reliability of product

and *Tables for Inspection by Attributes*) and ANSI/ASQC Z1.9 (*Sampling Procedures and Tables for Inspection by Variables for Percent Nonconforming*). These sampling systems are multiuse plans and include single, double, and multiple sampling plans. These sampling plans are equivalent to *MIL-STD-105* and MIL-STD-414.

ANSI/ASQC Z1.4. Sampling Procedures and Tables for Inspection by Attributes, Z1.4 (MIL-STD-105), provides items to consider.

- A choice of 26 AQL values ranging from 0.010 to 1000.0 (values of 10 or less can be considered as percent defective or defects per hundred units; values of 10 or greater must be considered as defects per hundred units).
- The probability of accepting at AQL quality varies from 89 to 99.5%.
- Seven inspection levels, three general and four special levels.
- Defect classification as critical, major, or minor.
- The customer may specify separate AQLs for each class or specify an AQL for each kind of defect that a product may show.

To choose a plan, the customer must determine the AQL, the lot size, the type of sampling (single, double, or multiple), and the inspection level. Once

TABLE 4–4 Sample Size Codes

	Special Inspection Level				General Inspection		
Lot Size 2	**S1**	**S2**	**S3**	**S4**	**I**	**II**	**III**
1–8	A	A	A	A	A	A	B
9–15	A	A	A	A	A	B	C
16–25	A	A	B	B	B	C	D
26–50	A	B	B	C	C	D	E
51–90	B	B	C	C	C	F	F
91–150	B	B	C	D	D	F	G
151–280	B	C	D	E	E	G	H
281–500	B	C	D	E	F	H	J
501–1200	C	C	E	F	G	J	K
1201–3200	C	D	E	G	H	K	L
3201–10000	C	D	F	G	J	L	M
10001–35000	C	D	F	H	K	M	N
35001–150000	D	E	G	J	L	N	P
150001–500000	D	E	G	J	M	P	Q
500001–OVER	D	E	H	K	N	Q	R

this information is known, a sample size can be determined and what the accept-reject criteria are. Table 4–4 shows the lot size to the inspection level. The normal inspection is level II. Once the correct inspection code is determined, the sample size can be determined by using the inspection code and Table 4–5.

ILLUSTRATION 4–8

You receive 450 pieces of a new machined part. What is the correct sample size to determine the acceptance of the part using a go/no-go snap gage?

STEP 1. Using Table 4–4, determine the sample code.

281 to 500—general inspection level II is H

STEP 2. Using Table 4–5, determine the sampling size.

H = 50

Therefore, the sample size would be 50 units for attribute inspection.

Assume an engineer determines the product can have an AQL of 0.65% and the material is received in

TABLE 4–5 In the table below, Ac is the acceptable number and Re is the rejection number. When the sample size (by code letter, see Table 4-4) and the desired acceptable quality level (AQL) are known, the table may be used to find the sample size and acceptable number. When the arrow is pointing downward, the first sampling plan (size and acceptance number) below the arrow is used. When pointed upward, the first sampling plan above the arrow is used.

Single Sampling Plans for Normal Inspection (Master Table)

Sample Size Code Letter	Sample Size	Acceptable Quality Levels (Normal Inspection)																									
		0.010	0.015	0.025	0.040	0.065	0.10	0.15	0.25	0.40	0.65	1.0	1.5	2.5	4.0	6.5	10	15	25	40	65	100	150	250	400	650	1000
		Ac Re	Ac Re	Ac Re	Ac Re	Ac Re	Ac Re	Ac Re	Ac Re	Ac Re	Ac Re	Ac Re	Ac Re	Ac Re	Ac Re	Ac Re	Ac Re	Ac Re	Ac Re	Ac Re	Ac Re	Ac Re	Ac Re	Ac Re	Ac Re	Ac Re	Ac Re
A	2	↓	↓	↓	↓	↓	↓	↓	↓	↓	↓	↓	↓	↓	↓	0 1	↓	↓	1 2	2 3	3 4	5 6	7 8	10 11	14 15	21 22	30 31
B	3	↓	↓	↓	↓	↓	↓	↓	↓	↓	↓	↓	↓	↓	0 1	↑	↓	1 2	2 3	3 4	5 6	7 8	10 11	14 15	21 22	30 31	44 45
C	5	↓	↓	↓	↓	↓	↓	↓	↓	↓	↓	↓	↓	0 1	↑	↓	1 2	2 3	3 4	5 6	7 8	10 11	14 15	21 22	30 31	44 45	↑
D	8	↓	↓	↓	↓	↓	↓	↓	↓	↓	↓	↓	0 1	↑	↓	1 2	2 3	3 4	5 6	7 8	10 11	14 15	21 22	30 31	44 45	↑	↑
E	13	↓	↓	↓	↓	↓	↓	↓	↓	↓	↓	0 1	↑	↓	1 2	2 3	3 4	5 6	1 2	10 11	14 15	21 22	30 31	44 45	↑	↑	↑
F	20	↓	↓	↓	↓	↓	↓	↓	↓	↓	0 1	↑	↓	1 2	2 3	3 4	5 6	7 8	10 11	14 15	21 22	↑	↑	↑	↑	↑	↑
G	32	↓	↓	↓	↓	↓	↓	↓	↓	0 1	↑	↓	1 2	2 3	3 4	5 6	7 8	10 11	14 15	21 22	↑	↑	↑	↑	↑	↑	↑
H	50	↓	↓	↓	↓	↓	↓	↓	0 1	↑	↓	1 2	2 3	3 4	5 6	7 8	10 11	14 15	21 22	↑	↑	↑	↑	↑	↑	↑	↑
J	80	↓	↓	↓	↓	↓	↓	0 1	↑	↓	1 2	2 3	3 4	5 6	7 8	10 11	14 15	21 22	↑	↑	↑	↑	↑	↑	↑	↑	↑
K	125	↓	↓	↓	↓	↓	0 1	↑	↓	1 2	2 3	3 4	5 6	7 8	10 11	14 15	21 22	↑	↑	↑	↑	↑	↑	↑	↑	↑	↑
L	200	↓	↓	↓	↓	0 1	↑	↓	1 2	2 3	3 4	5 6	7 8	10 11	14 15	21 22	↑	↑	↑	↑	↑	↑	↑	↑	↑	↑	↑
M	315	↓	↓	↓	0 1	↑	↓	1 2	2 3	3 4	5 6	7 8	10 11	14 15	21 22	↑	↑	↑	↑	↑	↑	↑	↑	↑	↑	↑	↑
N	500	↓	↓	0 1	↑	↓	1 2	2 3	3 4	5 6	7 8	10 11	14 15	21 22	↑	↑	↑	↑	↑	↑	↑	↑	↑	↑	↑	↑	↑
P	800	↓	0 1	↑	↓	1 2	2 3	3 4	5 6	7 8	10 11	14 15	21 22	↑	↑	↑	↑	↑	↑	↑	↑	↑	↑	↑	↑	↑	↑
Q	1250	0 1	↑	↓	1 2	2 3	3 4	5 6	7 8	10 11	14 15	21 22	↑	↑	↑	↑	↑	↑	↑	↑	↑	↑	↑	↑	↑	↑	↑
R	2000	↑	↑	1 2	2 3	3 4	5 6	7 8	10 11	14 15	21 22	↑	↑	↑	↑	↑	↑	↑	↑	↑	↑	↑	↑	↑	↑	↑	↑

= Use first sampling plan below arrow. If sample size equals or exceeds lot or batch size, do 100 percent inspection.
= Use first sampling plan above arrow.
= Acceptance number.
= Rejection number.

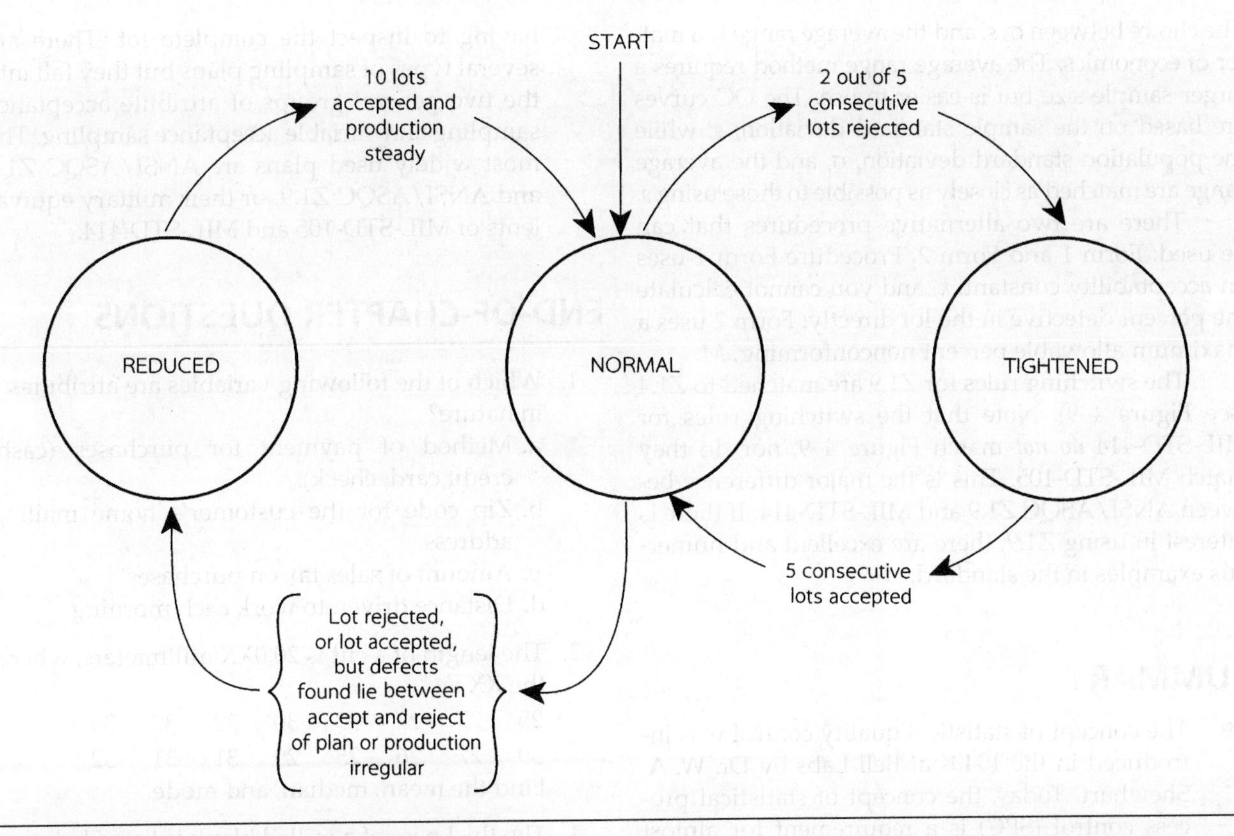

FIGURE 4–9 Switching rules for ANSI/ASQC Z1.4

1,500-piece lots. What would the inspection sample size and accept-reject level be?

By using Table 4–4, we would determine that the correct sample code would be "K," because the lot size is between 1,201 and 3,200 units for level II inspection. Then, going to Table 4–5, we would find sample size code letter "K" and determine that the sample size would be 125 units. Then, going to the 0.65 column, we would find the accept-reject values of accept on 2 and reject on 3.

The inspector would obtain a 125-piece *random sample* of material and do an attribute inspection of all pieces. After determining the number of rejectable units, and if the number of rejectable units is two or less, then the lot would be accepted.

The preceding illustration addresses a single inspection. The overall sampling system is designed for a continuous flow of material and will only guarantee material at a given acceptable quality level over a period of time. Figure 4–9 identifies the switching rules for Z1.4.

ANSI/ASQC Z1.9. *Sampling Procedures and Tables for Inspection by Variables for Percent Nonconforming,* ANSI/ASQC Z1.9 (MIL-STD-414), is an AQL type of sampling plan that assumes the individual measurements being taken are normally distributed. Series errors can be introduced if this is not the case. The sampling plan allows for the use of three different measures of variability:

1. Known population standard deviation, σ
2. Sample standard deviation, s
3. The average range

The choice between σ, s, and the average range is a matter of economics. The average range method requires a larger sample size but is easier to use. The OC curves are based on the sample standard deviation, *s*, while the population standard deviation, σ, and the average range are matched as closely as possible to those using *s*.

There are two alternative procedures that can be used: Form 1 and Form 2. Procedure Form 1 uses an acceptability constant, *k*, and you cannot calculate the percent defective in the lot directly. Form 2 uses a maximum allowable percent nonconforming, *M*.

The switching rules for Z1.9 are matched to Z1.4 (see Figure 4–9). Note that the switching rules for MIL-STD-414 *do not* match Figure 4–9, nor do they match MIL-STD-105. This is the major difference between ANSI/ASQC Z1.9 and MIL-STD-414. If there is interest in using Z1.9, there are excellent and numerous examples in the standard.

SUMMARY

- The concept of statistical quality control was introduced in the 1940s at Bell Labs by Dr. W. A. Shewhart. Today, the concept of statistical process control (SPC) is a requirement for almost any business to be successful.
- One main area of statistics is descriptive statistics. Here we are describing the data we have. In describing data, we need two key components: the central tendency and the dispersion. The mean, median, and mode are all terms for the central tendency. The variance, standard deviation, and range are names for the dispersion.
- Probability plays a key part in SPC. It allows us to predict what a population will look like. Probability is a key component of sampling.
- By tying the concepts of probability and descriptive statistics together, we can develop probability distributions. These are mathematical models of the distributions. There are two types of probability distributions: attribute and variable distribution. Both play key roles in SPC.
- Acceptance sampling plans are used to predict the acceptability of lots of material without having to inspect the complete lot. There are several types of sampling plans but they fall into the two general groups of attribute acceptance sampling and variable acceptance sampling. The most widely used plans are ANSI/ASQC Z1.4 and ANSI/ASQC Z1.9, or their military equivalents of MIL-STD-105 and MIL-STD-414.

END-OF-CHAPTER QUESTIONS

1. Which of the following variables are attributes in nature?
   a. Method of payment for purchases (cash, credit card, check)
   b. Zip code for the customer's home mailing address
   c. Amount of sales tax on purchases
   d. Distance driven to work each morning
2. The length of a cut is 20.0XX millimeters, where the XX is
   25 27 27 33 30 32 30 34
   30 27 26 25 29 31 31 32
   Find the mean, median, and mode.
3. The thickness of a milled blank is listed below in millimeters.
   15.02 15.04 15.07 15.08 15.09
   Find the range, sample standard deviation, and sample variance.
4. A complex aluminum part has been manufactured and the machine operator wants to determine if the parts are acceptable. What is the sample size needed if a total of 150 parts were manufactured and they were going to check them on a go/no-go gage. Assume the AQL is 0.65%?
5. A complex aluminum part has been manufactured and the machine operator wants to determine if the parts are acceptable. What is the sample size that would be needed if a total of 150 parts were manufactured and they were going to check them on a coordinate measuring machine? Assume the AQL is 0.065%.

CHAPTER FIVE

MEASUREMENT WITH GRADUATED SCALES AND SCALED INSTRUMENTS

Man's industrial progress has been linked to his ability to divide and measure the inch and meter finer and finer.

Louis Polk, Sheffield Corporation

LEARNING OBJECTIVES

- Know the distinction between rules and scales and understand their role.
- Be able to use graduated scales within the recognized limitations.
- Recognize the primary sources of error for measurement with graduated scales.
- Understand the role of the basic measurement instruments that are still the foundation of toolroom and prototype work.
- Recognize some of the fundamental principles of metrology from the basic instruments.
- Understand the relationship between scale divisions and discrimination.
- Describe the significance of feel in the use of precision instruments.
- Explain the pervasive role of error in repeated measurements.

OVERVIEW

If you rounded up all the rulers, yardsticks, machinist's rules, and measuring tapes nearby, laid them on a flat surface, and matched their graduations, you would end up with a picture like Figure 5–1. No two rules are alike. Even those that, at a glance, appear to agree probably will not—if you look at the graduations with a magnifying glass.

Rulers, or *graduated scales,* are most commonly used in layout and inspection, the two *opposite* ends of the production process. Each depends on the

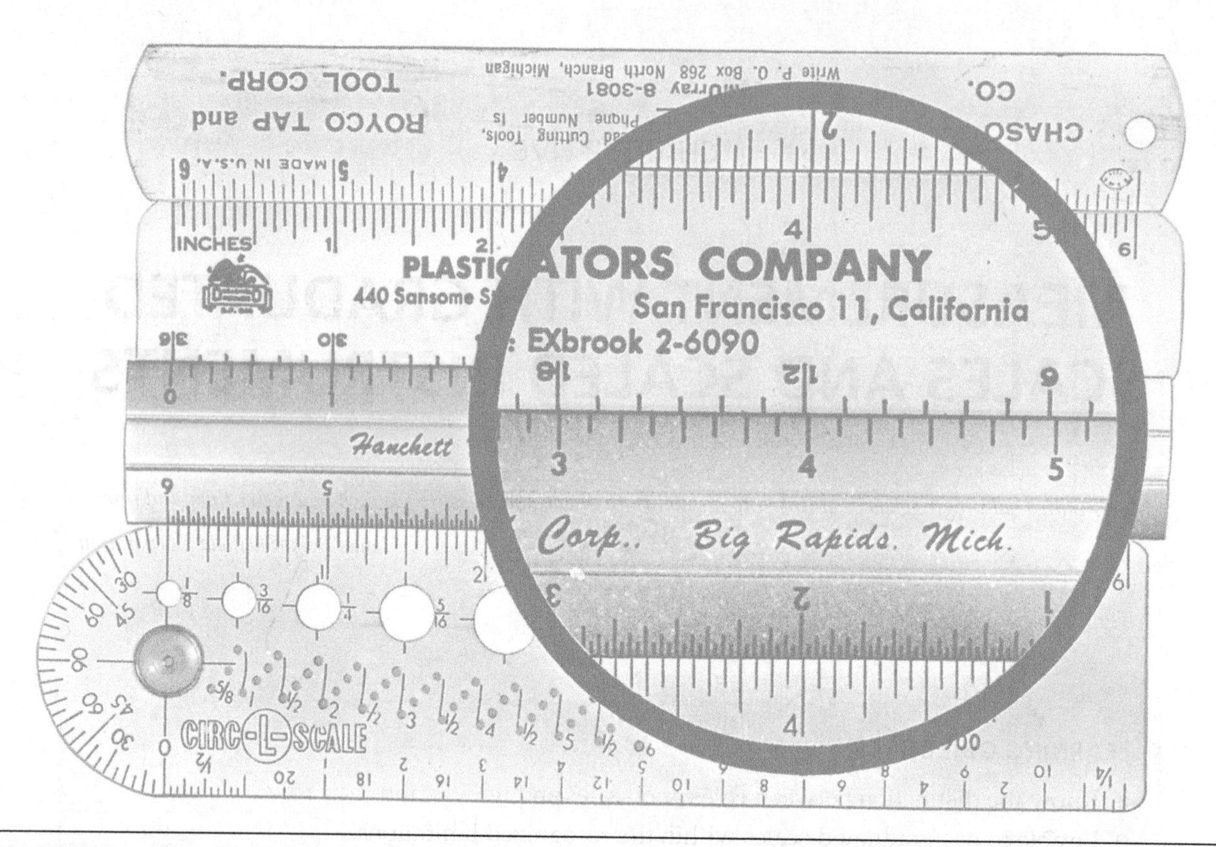

FIGURE 5–1 A simple comparison of the scales and rules emphasizes the principal problem in measurement—error. Every measurement contains some error. For reliability, it must be recognized.

other—you cannot inspect for the quality of production without knowing what the original design dimensions were—and both depend on measurement. In the inspection process, you must determine if material or work performed is according to specifications (see Figure 5–2). In the layout process, we prepare the specifications, such as scribing lines to show the limits for band machining and the location of the centers of diameters to be drilled or bored to size, that our production processes must meet (see Figure 5–3).

Using *machining centers* and *coordinate measurement machines*, manufacturers have mechanized both layout and inspection. Machining centers vary in sophistication and can machine parts to preprogrammed dimensions along multiple axes. Coordinate measurement machines are similar, but they inspect the results of the machining (see Chapter 18). They may apply advanced technologies, but their basic principles lie in the techniques used when you measure with any graduated scale.

For rough parts, the *layout* provides the features to which the part is machined, limiting the precision you can achieve. Of course, no matter how carefully your layout is created, the part may still be ruined by bad machining; similarly, no matter how carefully a part is machined, *if the layout is in error, the part must be in error, too.* The layout helps us machine parts sufficiently close to specifications while allowing for errors that will occur in machining.

For precision parts, the layout furnishes a guide, and the precision of the machine tool determines the completed dimensions. The closer the layout is to

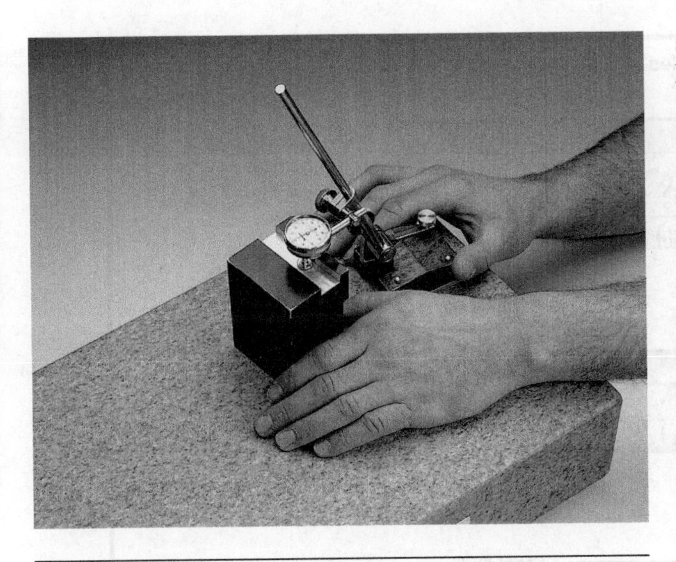

FIGURE 5–2 Surface plate work is measurement taken from an auxiliary reference surface called a surface plate. (*Courtesy of the L.S. Starrett Company*)

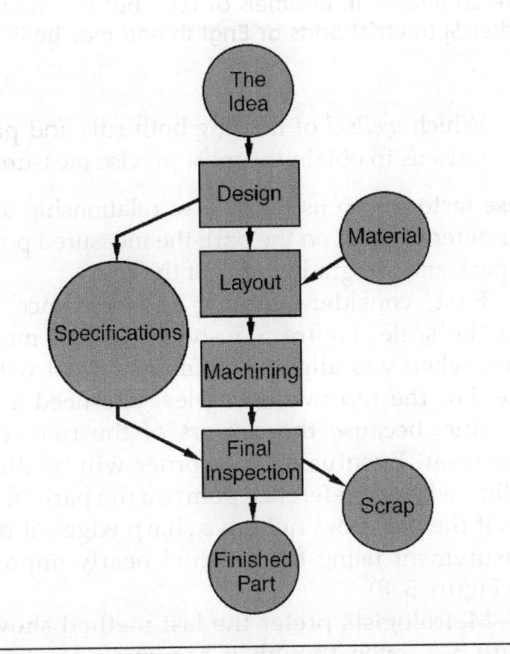

FIGURE 5–3 Measurement is used at each step. The inspection step is the use of measurement for verification.

specifications, the less work production workers will have to do in machining operations. Rapid, accurate layouts will increase speed and reduce costs.

For example, a part is machined to rough size, heat treated, then surface ground to specified dimensions. The least expensive machining method—band machining—removes material in blocks, while grinding removes material in chips. Grinding is more precise than band machining, but it is also more expensive. If we carefully band machine to a tighter limit, such as 0.8 mm (1/32 in.) instead of 1.6 mm (1/16 in.), we can reduce both the time and expense of grinding.

Layout and inspection both require high levels of precision and the correct precision instrument for each measurement. With the correct precision instrument, we ensure that specifications will be met and have been met. And the *precision instruments* we use start with the steel rule.

5–1 THE STEEL RULE

There is a difference between a scale and a rule: a scale is graduated in *proportion* to a unit of length; a *rule* is the unit of length, its divisions, and its multiples. One unit on a scale may represent one mile in the "real world"; and a scale allows engineers and draftspeople to depict small or large items at a convenient size, not actual size (see Figure 5–4). A 6-inch rule, on the other hand, represents 6 one-inch divisions (multiples of one inch), 12 half-inch divisions, 24 quarter-inch divisions, and so forth.

Although they vary in some features, steel rules are all narrow steel strips with one set or more of graduated marks. These marks are referred to as a *scale.* Steel rules represent lengths from a fraction of an inch to several feet, but the most popular size of rule still seems to be one that you can fit in your pocket.

The number of subdivisions of a unit of length on a rule is called its *discrimination* (see Figure 5–5), and the discrimination of rules is as varied as rules themselves (see Figure 5–6).

When you measure with a rule, you use the interchange method of measurement because you

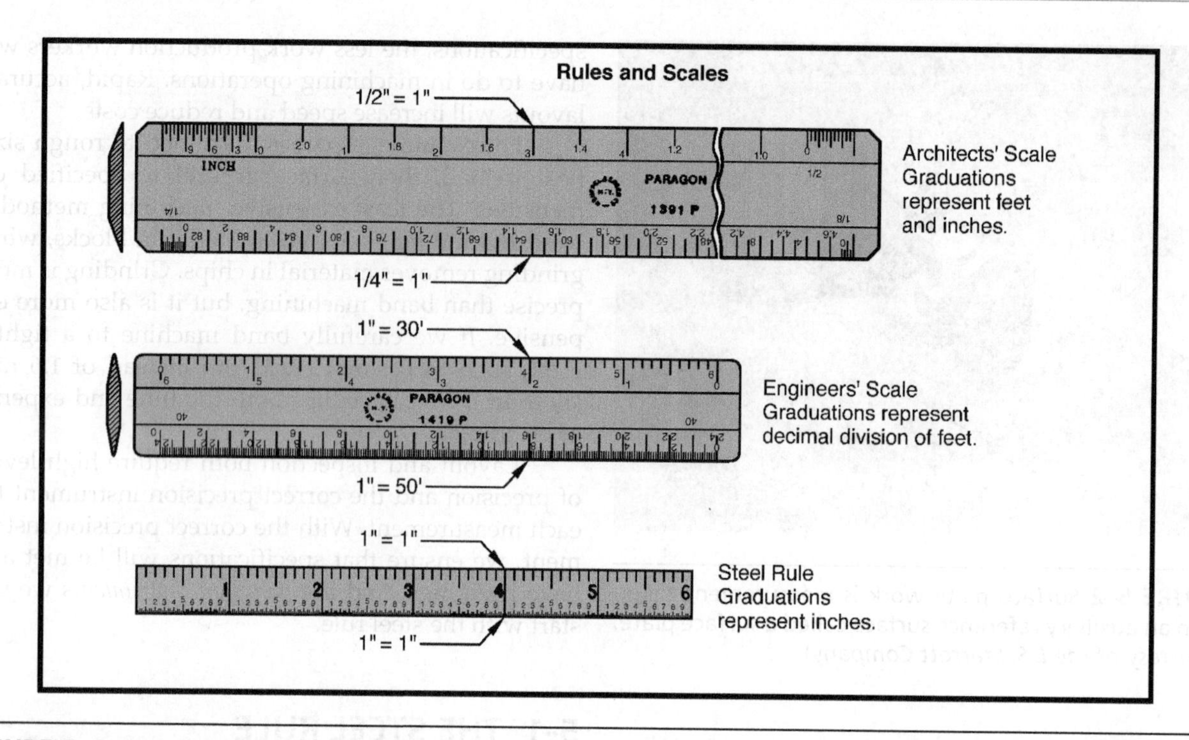

FIGURE 5–4 The architects' scale is read in feet and inches, the engineers' in decimals of feet, but the machinists' steel rule reads directly in units of length. These may be in either SI (metric) units or English and may be in either decimals or fractions and decimals for English.

observe both ends of the part feature at the same time (see Chapter 2, Figure 2–6). Of course, some people argue that you are using the displacement method because your eye "displaces" from end to end. Whichever method of measurement is employed, the rule you are using is the standard for measurement. You also read the measurement directly from the rule, so measuring with a graduated standard or rule is commonly called *direct measurement*.

The Reference Point

You must consider three factors when using a steel rule:

1. Which *style* of rule will do the best job
2. Which measurement divisions (scale) should be used
3. Which *method* of holding both rule and part allows us to obtain the most precise measurement

These factors help us to the best relationship among the reference point on the part, the measured point on the part, and the graduations of the rule.

First, consider aligning the reference point with the scale. Figure 5–7 shows several methods to use when you align the reference point with the scale. For the top two examples, you need a fairly new rule, because the corners of the rule receive more wear. Eventually, the corner will be difficult to align with the reference point on the part. In addition, if the part does not have sharp edges, it makes measurement using this method nearly impossible (see Figure 5–8).

Metrologists prefer the last method shown in Figure 5–7, even though it requires an extra step of very careful alignment. The two lines (the edge

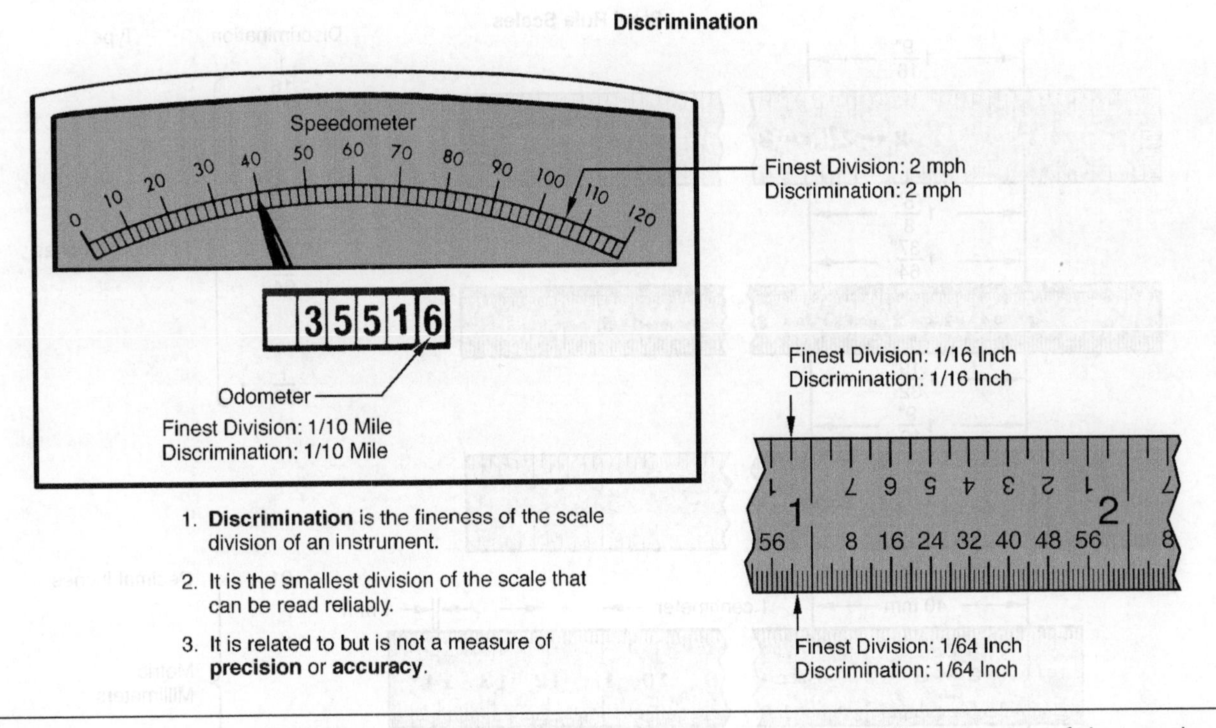

FIGURE 5–5 Anyone who has argued with a police officer knows that the assumed discrimination of the speedometer is not considered proof of its accuracy.

appears as a line) can be aligned very precisely and checked with your eye. If you use a *knee*, make sure it is as free from chips, burrs, and poor surface finish as possible because these irregularities could result in inaccurate measurements. Similarly, the hook on a rule should be checked often. *Every attachment used in measurement can contribute to the possibility of error.*

Another important factor is the size of the rule. If you choose a rule longer than necessary, it may be difficult for you to position it; if the rule is too short, you may need to add multiple measurements together, decreasing the precision of your results. To measure a dimension in a recess, you need a short rule; for a slot, you need a narrow rule. Another special measurement problem is shown in Figure 5–9.

The Measured Point

You should read the scale on any rule or other measurement instrument to the nearest graduation for two reasons: one, when you interpolate between graduations, you compromise the reliability of your measurement; and two, there usually is a rule nearby with sufficient discrimination for the job. As you choose increasingly finer scales, one of three things will always happen:

1. The measured point will coincide with a graduation when the reference point and the measured point are carefully aligned.
2. As fine a division as is needed will be reached.
3. The finest scale will be reached (see Figure 5–10).

You probably will not measure a dimension to 1/64 in. if 1/16 in. is all you need, unless you need the

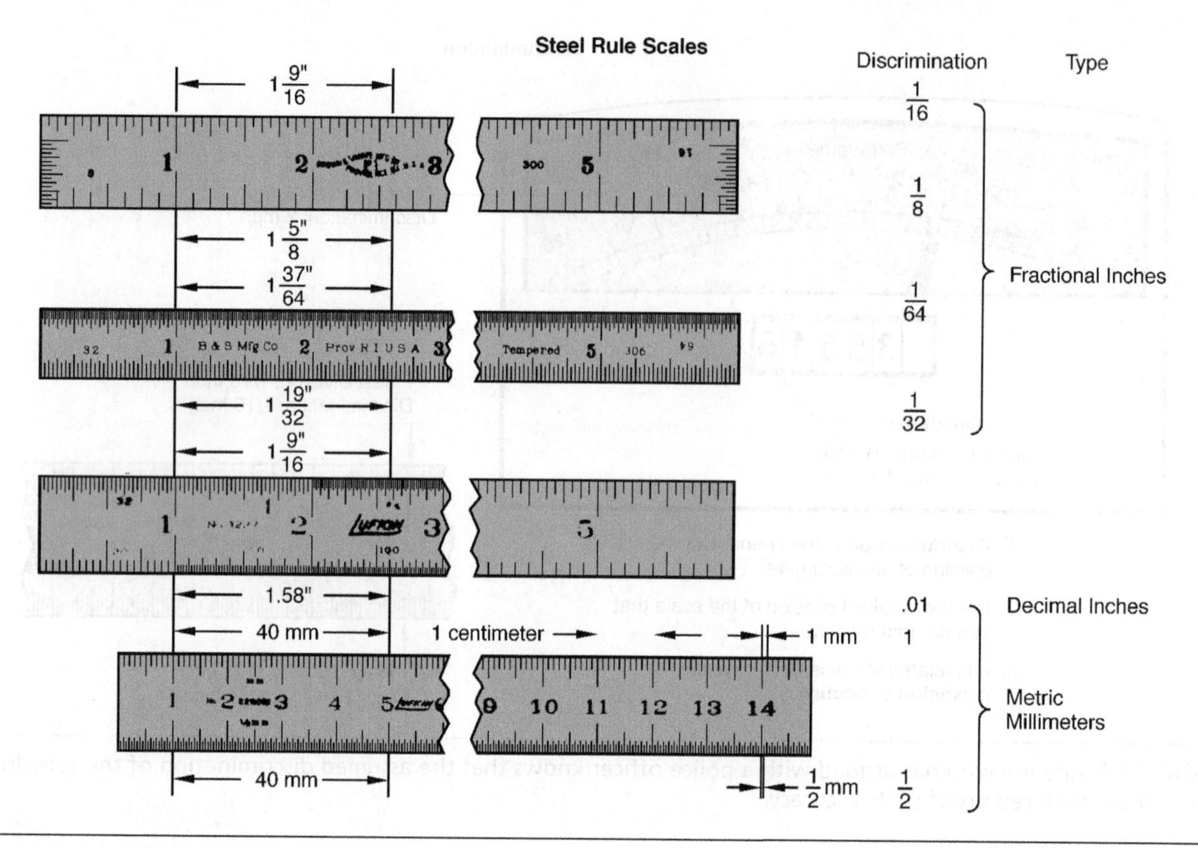

FIGURE 5–6 Many scales are available. The four scales shown above are the most popular in the United States. The decimal-inch is standard in many plants.

additional reliability that the 1/32 in. discrimination will provide. Similarly, a measurement made with 0.5 mm discrimination will be more reliable than a measurement made at 1 mm.

The only acceptable time to interpolate is when you cannot find an instrument discriminating to provide an accurate measurement; then you should be very cautious to minimize any contributing sources of error. When you are using a 1/64 in. scale and the measured point falls between two graduations, we conventionally take the closest graduation as the measurement. If the space seems equally divided, use the measurement that provides the best *margin of safety,* keeping in mind the bias, which is explained later in this chapter. We provided guidelines for rounding off using a decimal-inch or metric graduated rule in Chapter 2.

5–2 THE ROLE OF ERROR

Measurement errors with steel rules come from the following: an inherent instrument error or *tool;* an observational error or *eye;* a *manipulative error* or *hand;* and bias. You can eliminate instrument error by choosing a quality steel rule. A quality steel rule will inherently have an error factor of point three mil (0.0003 in.) per inch of length—one-tenth the width of a usual graduation. Inherent error, therefore, affects only very long measurements.

The 1/64 in. divisions on a fractional-inch rule create a maximum error of about three mil (0.003 in.) or half that measurement (0.0015 in.) on either side of the scale mark. This potential error of one-tenth of the discrimination is significant in your selection of a measurement instrument and will be discussed later.

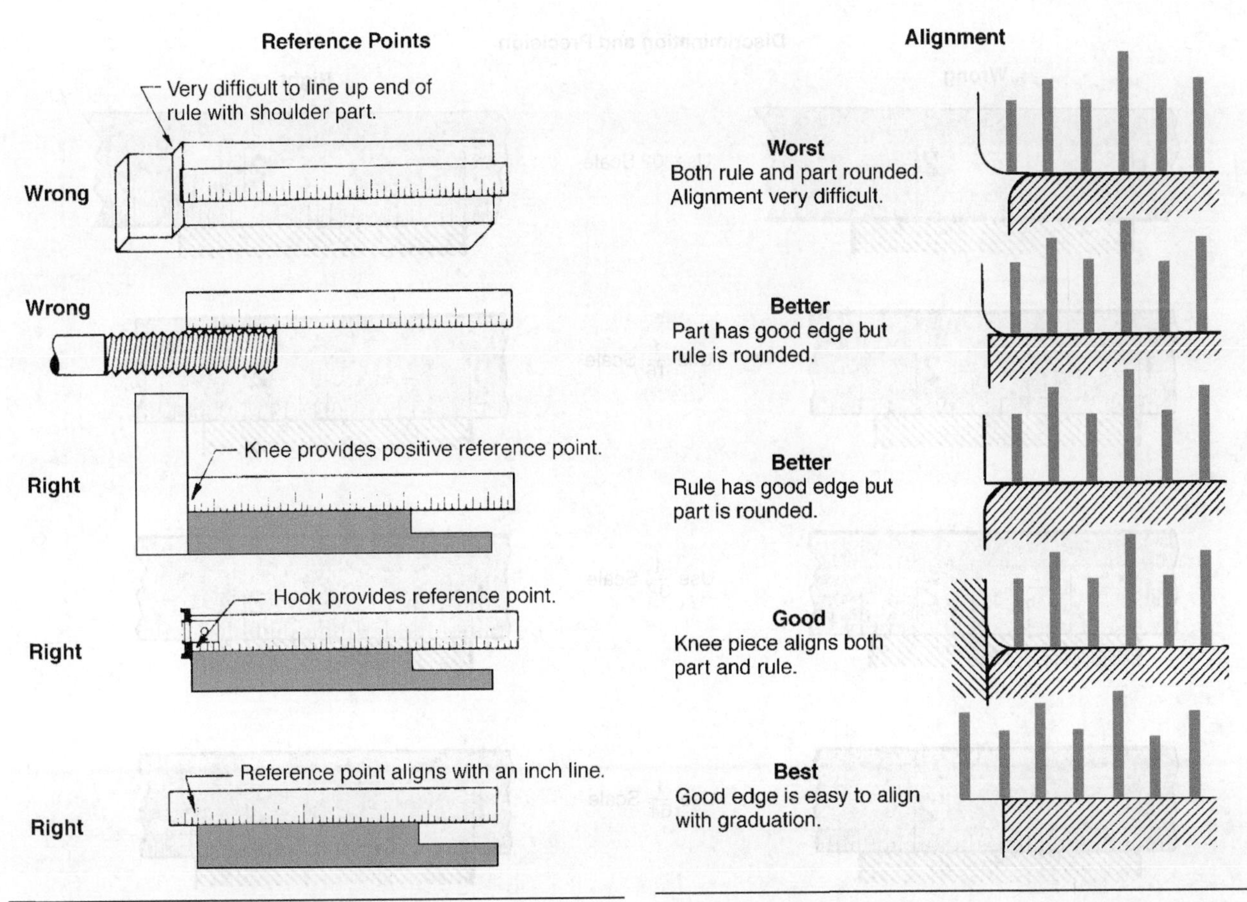

FIGURE 5–7 The right way to use a rule is usually the easiest, fastest, and most reliable.

FIGURE 5–8 Use of the end of a rule invites error. Examination with a magnifying glass will usually disclose wear.

The smallest division on a decimal-inch rule is approximately 28% larger than the smallest division on a fractional-inch rule. Therefore, decimal-inch rules are less discriminating, but they are also easier to read. Because the width of the graduations is the same in either case, the observational error is decreased with the decimal-inch scale. ASA Z 75.1 for decimal-inch scales limits the overall length tolerance on a 6-in. steel rule to plus 0.004 in. minus 0.002 in. For a 12 or 18 in. scale, the tolerance is extended to plus 0.005 minus 0.002 in.

The only time that the full width of one line might alter the measurement is when both the reference point and measured point are aligned with graduations (see Figure 5–11). Quality rules are

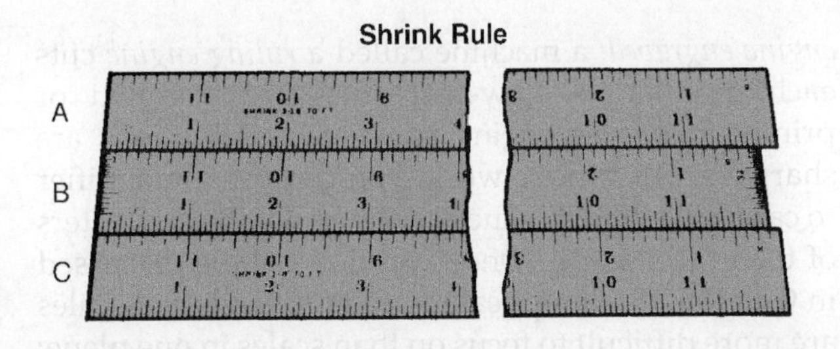

FIGURE 5–9 Molten metal shrinks as it cools and solidifies. To compensate for the shrinkage, pattern makers use shrink rules that automatically compensate. A is a rule for brass, B is a standard rule, and C is for cast iron. Care must be taken not to use a shrink rule accidentally for other than its intended purpose.

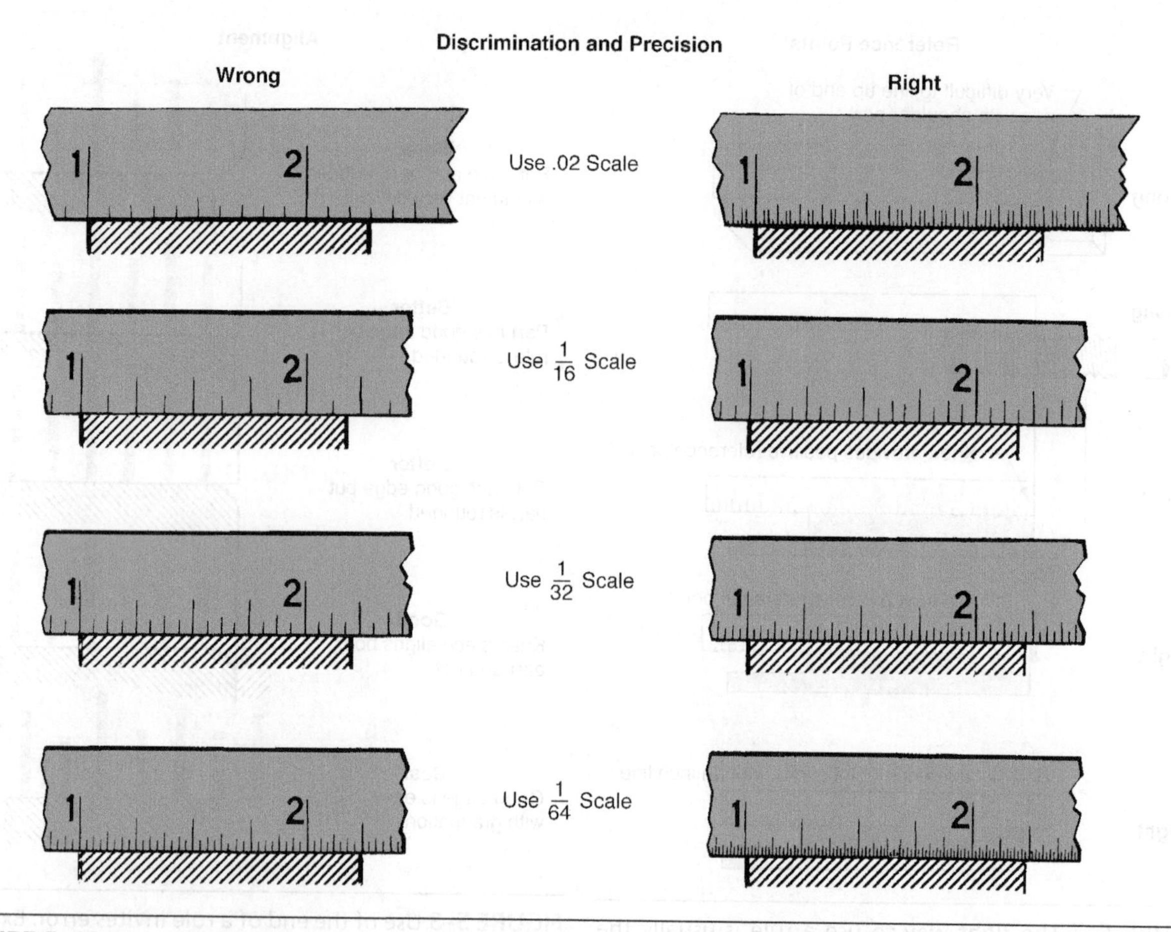

FIGURE 5–10 The only excuse for not using an instrument of the proper discrimination is not having one. Unfortunately, this happens often and carefully considered compromises must be made.

engine engraved: a machine called a *ruling engine* cuts each graduation. Low-cost rules are stamped or printed, whereas engine engraved graduations are sharp Vs. For precise work, you can use a magnifier to carefully align the lines on the part with the centers of the graduations. (One exception will be discussed in Chapter 18.) In optical metrology, engraved scales are more difficult to focus on than scales in one plane; therefore, we use high-quality, printed scales.

Observational Error

Parallax is an important form of observational error in which an object appears to shift when the observer changes his or her position. For example, the speedometer in your car appears to read slower to the passenger in your front seat. Parallax does not affect measurement when the scale edge of the rule is directly on the line of measurement.

Figure 5–12 is exaggerated to show parallax. In the top drawing, the observer at B would correctly measure X as 16 divisions, while A would measure 15, and C would measure 17. When the scale is placed directly in contact with the line of measurement (lower drawing), all three observers correctly measure X as 16 divisions.

To combat parallax: always place the scale edge of the rule as close to the line of measurement as possible (that may be why thin steel rules are popular),

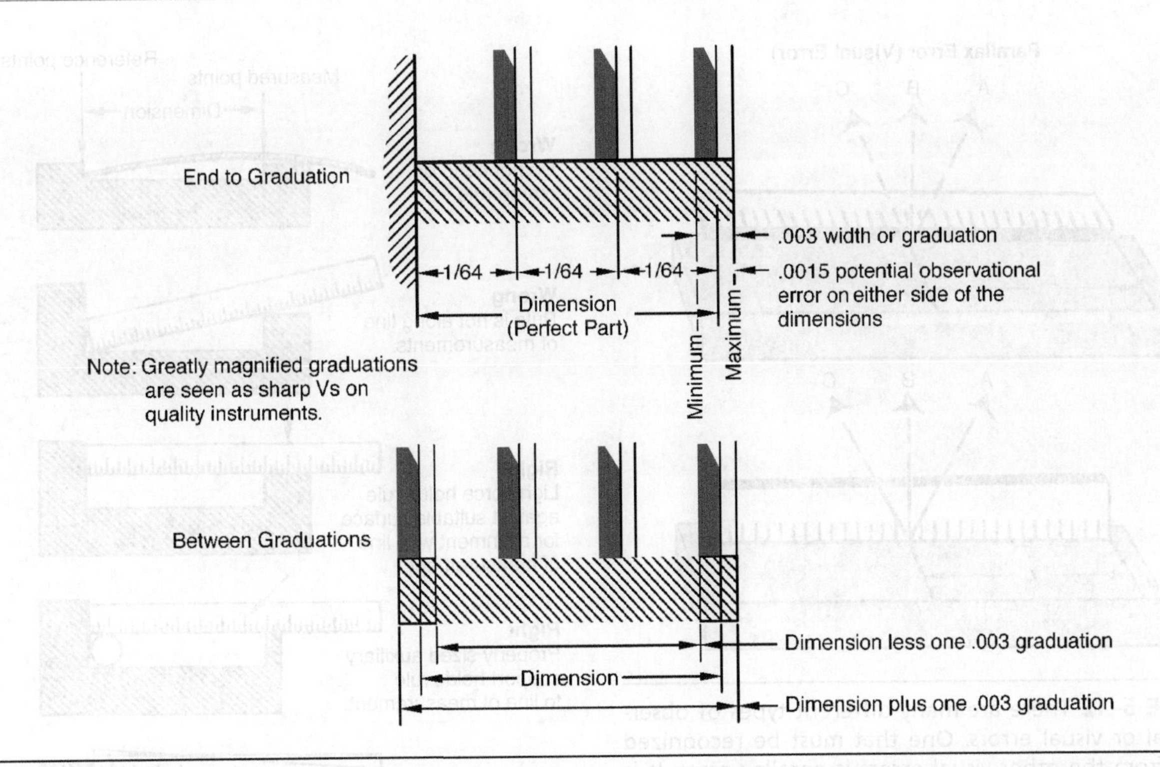

FIGURE 5–11 The maximum observational error that can be caused by the graduation widths is one-half graduation for an end setting, and one graduation for a setting between two graduations. With decimal-inch scales, the margin of safety is greater because 0.02 in. is approximately 28% wider than 1/64 in.

and align your head so that your line of sight is perpendicular to the measured point on the line of measurement, as B in Figure 5–12. As long as you recognize parallax, you can almost always correct it.

Manipulative Error

It is not easy to hold both the part and the rule and measure to 0.5 mm (0.02 in.), 1/64 in., or even closer. When you try to measure any closer, you need to correct for parallax. Decimal-inch or metric rules help, but the very act of moving your head perpendicular to the measured point will probably cause a shift between the rule and the part, generating error.

Some of the common manipulative errors are shown in Figure 5–13. In some cases, the part or the scale, or both, may have to be held in your hand. Even when they are lightweight, it is easy for one or more to slip. When the parts are heavy, it is difficult to prevent them from slipping. Many common manipulative errors are caused by "cramping"—the use of excessive force. When you squeeze a rule or other instrument tightly, you may be forcing it against the part. For reliable measurements, always use a light touch.

The Problem of Bias

Bias means that we unconsciously influence each measurement we make. When you consciously recognize your biases, you can combat them. Even then, some bias usually remains. The exercises in the accompanying lab manual can help you identify bias and will give you practice in technical report writing.

Assume that a dimension needs to be 14 cm (5 1/2 in.). Because that graduation is easier to read than 13.5 cm (5 31/64 in.) or 14.5 cm (5 33/64 in.), you might be biased to read 14 cm instead of the accurate measurement (see Figure 5–14). Bias will also affect reading

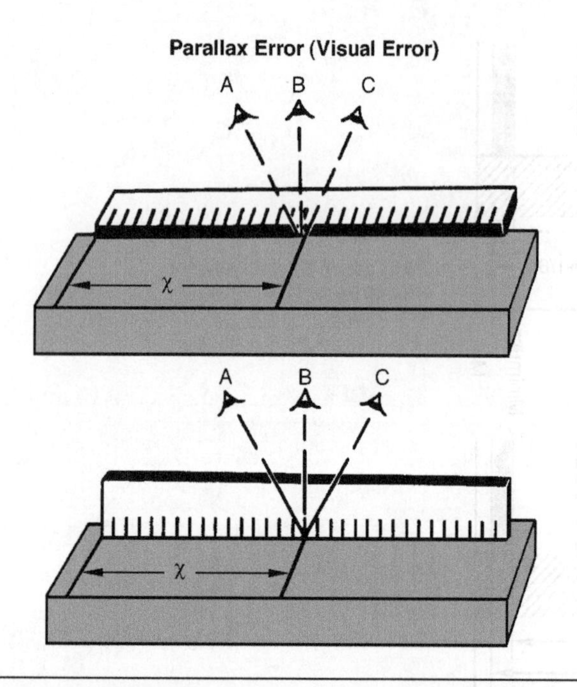

FIGURE 5–12 There are many different types of observational or visual errors. One that must be recognized apart from the other visual errors is parallax error. It is minimized by having the line of measurement of the rule as close as possible to the feature being measured.

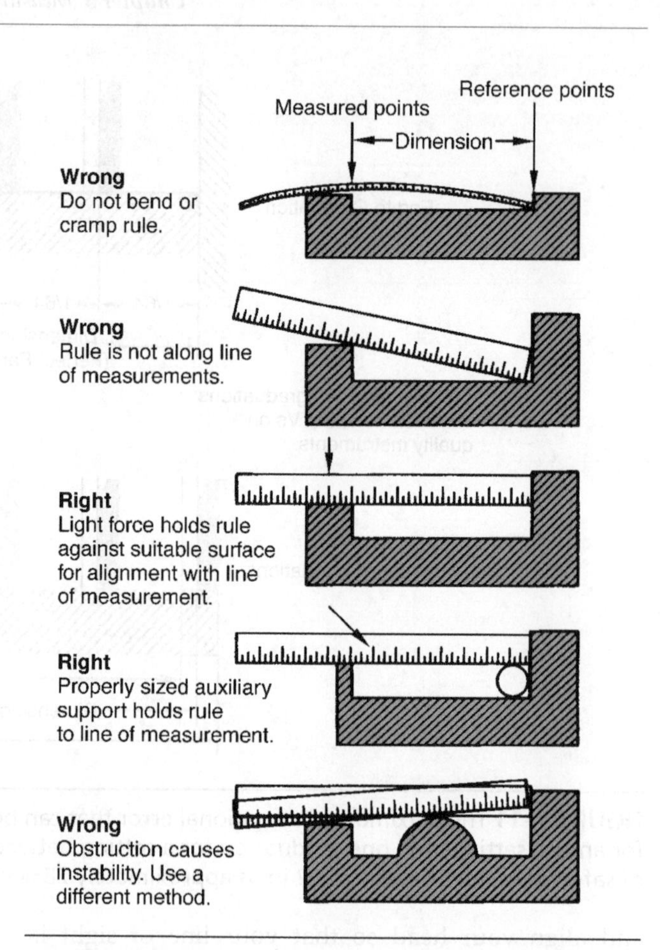

FIGURE 5–13 These potential errors are exaggerated. In practice, they may be much less obvious.

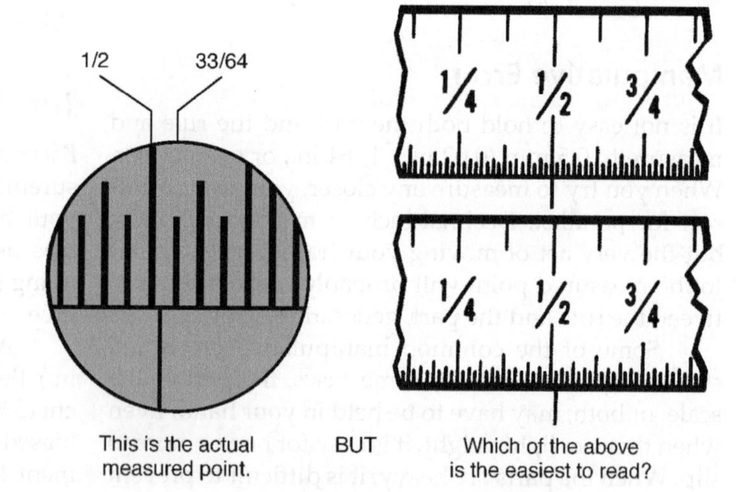

FIGURE 5–14 It is a normal tendency to take the easiest route, especially if it produces hoped-for results. The finer the scale, the more serious this problem becomes.

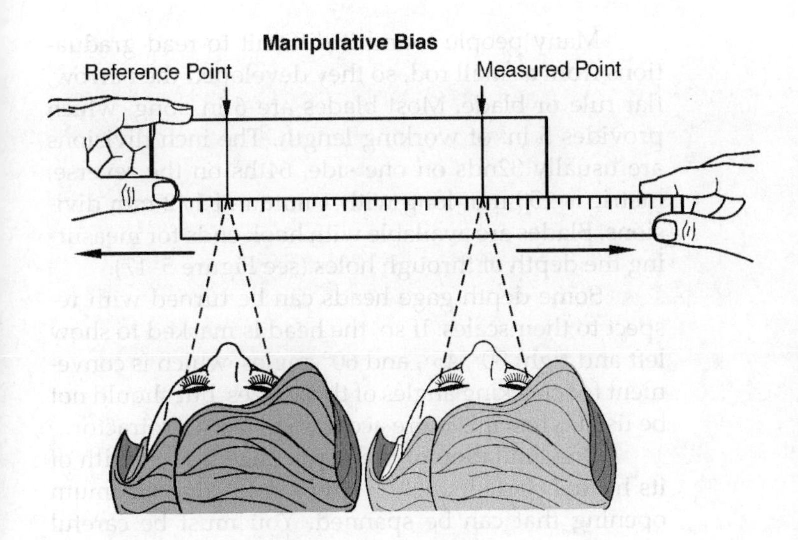

FIGURE 5–15 When the line of sight is moved from reference point to measured point, the natural tendency is for the hands to move slightly apart. This is exacerbated by bias. The larger the measurement, the more serious this becomes.

a measured point between the 13.97 cm (5 1/2 in.) and 14.01 cm (5 33/64 in.) graduations unless you consciously fight this tendency.

The act of moving your line of sight to prevent parallax will usually cause slippage, but these errors statistically fall on (are biased toward) the side of the desired results (see Figure 5–15). If the operation is tedious or repetitious, your progressive measurements will be less reliable, because you become biased toward the results you have most consistently been receiving.

In most cases, precise measurements require several readings that are averaged. "Wrong" readings are thrown out; however, your bias may cause you to believe that a set of results is wrong when they are correct. *Never reject any reading unless it is definitely known to be fallacious.* One exception is in the control of some processes, the highest and lowest readings are discarded and the remaining readings are averaged.

5-3 SCALED INSTRUMENTS

We can divide layout and inspection instruments into three groups: First, rules with mechanical refinements to make them more useful, such as the combination square. Second, related devices that are used with rules, but are not rules themselves, like calipers. And third, vernier instruments, like the vernier height gage, which are the highly precise refinements of the steel rule (see Chapter 6).

These instruments use the basic principles we have been discussing. Mechanical refinements make it more convenient for you to use the instrument; related devices allow you to use a rule for a wider variety of applications. Verniers are the only instruments mentioned so far that allow us to extend our ability to measure beyond what we can do with a steel rule.

The Depth Gage

A popular modification of the steel rule, the depth gage (see Figure 5–16) measures into holes, grooves, and recesses transferring inaccessible reference and measurement points to an instrument where these points are accessible. In its simplest form (on the right), a small rod slides through a T-head or *stock.* You span the shoulders of a recess with the head and push the rod into the recess until it touches the bottom. You then lock the rod into place with a screw clamp, withdraw the gage, and read the measurement on the length of rod protruding from the head.

Rods are available in both plain and graduated styles and are still used in depth gages, particularly for small holes. Because every transfer of measurement reduces reliability, you will find that graduated rods are more convenient. They eliminate the transfer of measurement.

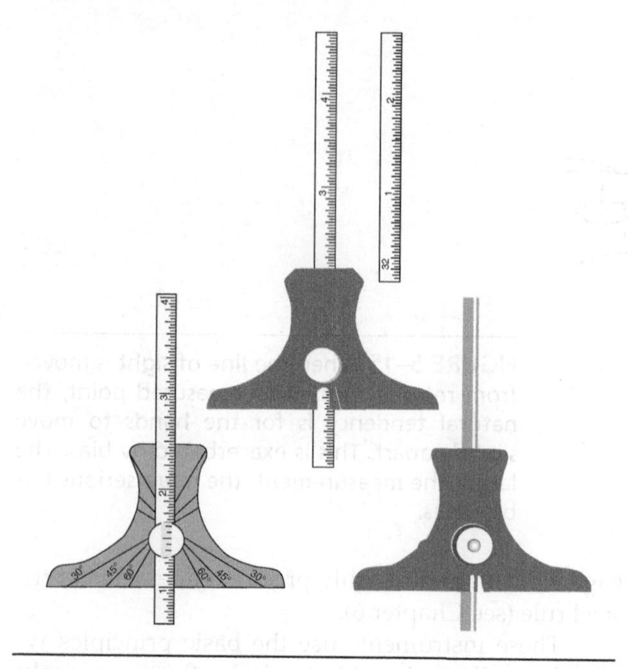

FIGURE 5–16 Types of depth gages.

Many people found it difficult to read graduations from a small rod, so they developed the narrow, flat rule or blade. Most blades are 6 in. long, which provides 5 in. of working length. The inch divisions are usually 32nds on one side, 64ths on the reverse; metric is 150 mm long with 1 mm and 1/2 mm divisions. Blades are available with hook ends for measuring the depth of through holes (see Figure 5–17).

Some depth gage heads can be turned with respect to their scales. If so, the head is marked to show left and right 30°, 45°, and 60° angles, which is convenient for checking angles of those sizes, but should not be used when it is more accurate to use a protractor.

One limitation of the depth gage is the width of its head: approximately 5 cm (2 in.) is the maximum opening that can be spanned. You must be careful when using a depth gage to keep the base of the head perpendicular to the line of measurement. In addition, you must hold the end of the blade against the desired

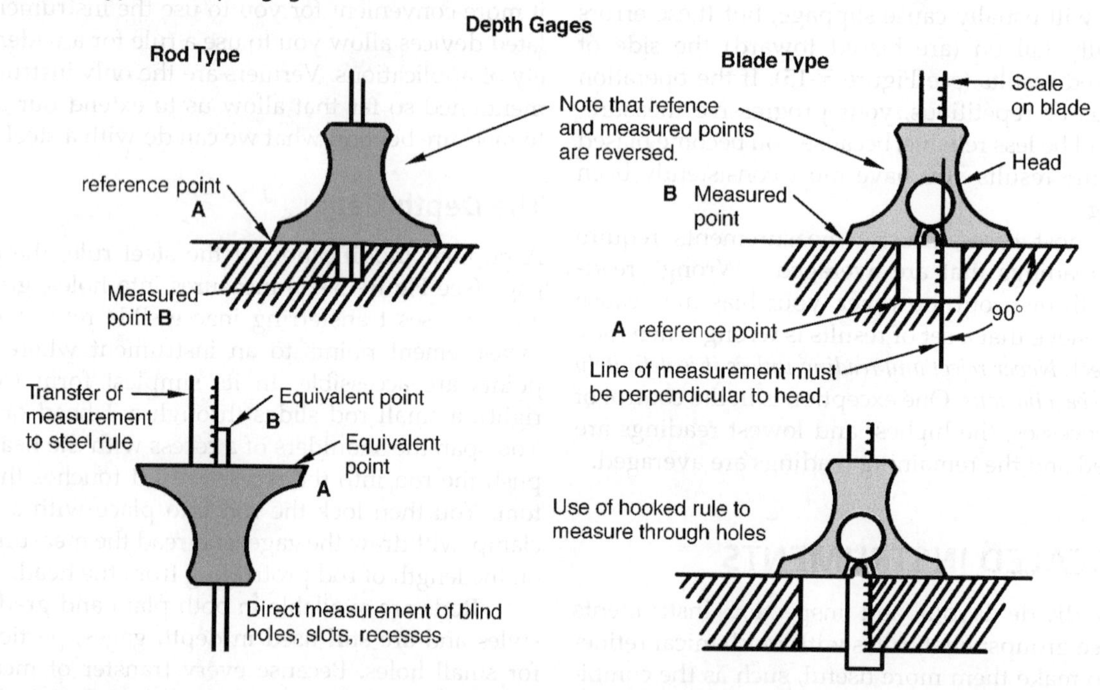

FIGURE 5–17 The depth gage is frequently used to check the progress when machining holes and recesses. When close to the finished size, a more precise instrument is substituted.

reference, which may be difficult when the reference is in the bottom of a blind hole.

For most depth gage measurements, the end of the blade actually touches the part, causing the inevitable wear and limiting the reliability of the instrument. Check your instrument frequently and replace it when it is worn by the width of one graduation line.

The Combination Square

The combination square is one of the most useful variations of the steel rule, and it is used extensively in tool and die making, pattern making, model and prototype work, and for machine setup. The combination square was invented in 1887 by Laroy S. Starrett, who later founded the L. S. Starrett Company to market this product.

The more difficult a dimension is to measure, the less reliable the measurement is. But a combination square makes it more convenient to measure as compared to a steel rule, so it increases the reliability of our measurements. We could then say that the combination square is more precise (see Figure 5–18).

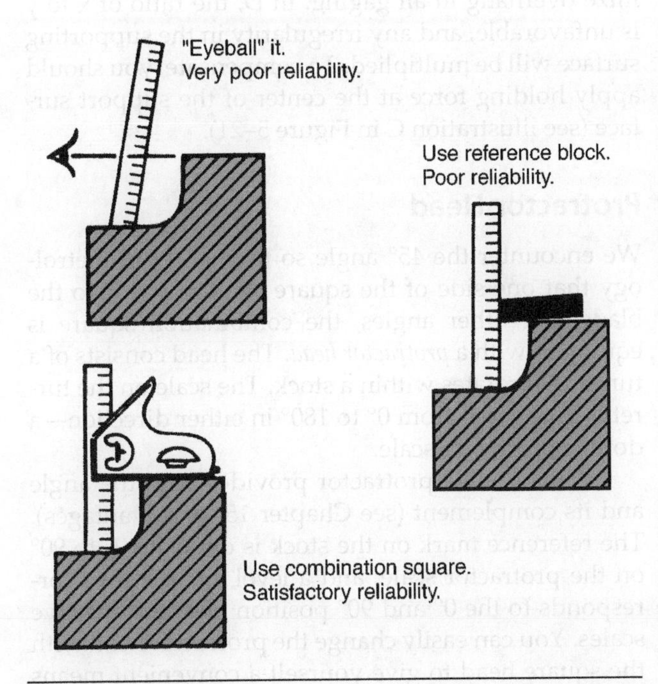

FIGURE 5–18 A combination square improves reliability.

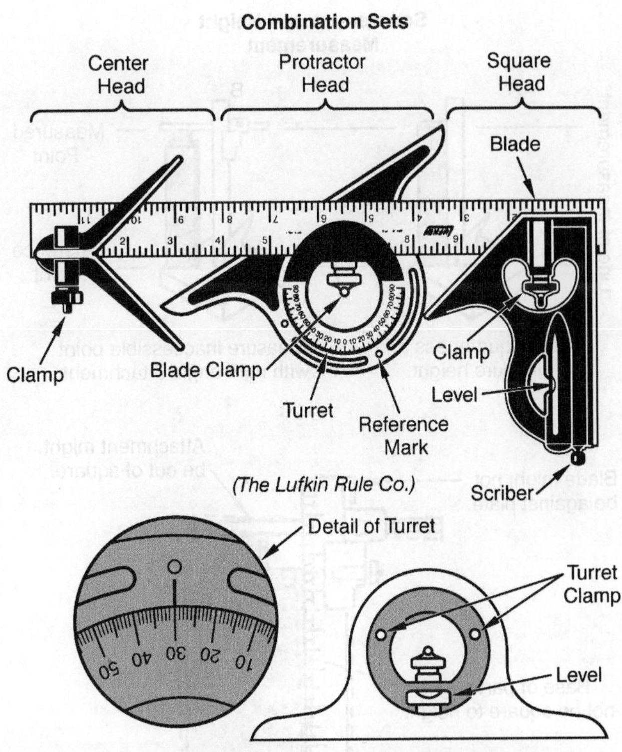

FIGURE 5–19 The steel rule and square head are called a combination square. Adding the center head and protractor head changes the name to combination set.

It does not, however, increase the accuracy of the measurement, because accuracy is related to the scale that is machined on the blade. As with any measurement instrument, you should use the combination square that has the correct discrimination for the job.

A combination square consists of a *blade* and a *square head* (see Figure 5–19). Size is determined by the length of the head. The most popular of the three sizes is 10 cm (4 in.) long. A combination square's head may be either plain or hardened and has a clamp that locks the blade into any position, a spirit (nonfreezing) level, and a scriber.

Square Head

A square head adds two things to the steel rule: a right angle reference and a means to transfer either

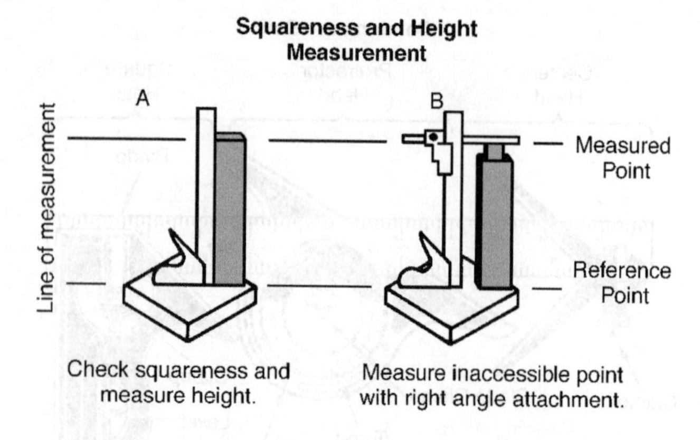

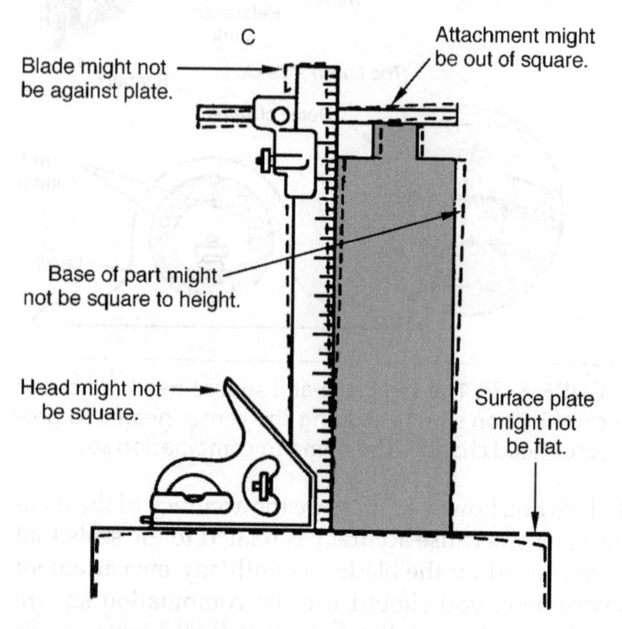

FIGURE 5–20 Combining squareness and length measurement in one instrument may combine errors or eliminate errors. It is up to the skill of the measurer.

the measured point or the reference point from the work to the rule.

In Figure 5–20, illustration A shows that the combination square can be used from a flat reference, such as a surface plate—a precisely flat surface from which measurements are made. You can measure the squareness of a part height to its base and/or the part's height. Adapters can be used to hold a steel rule at right angles to the blade, illustration B, which correctly establishes the reference point and measured point on the work with the blade and improves reliability. As with any measurement, you can make some basic mistakes, as in illustration C. Even if all four of these errors—each one no greater than the thickness of a human hair—were made, the total result could be greater than the 0.4 mm (1/64 in.) sensitivity you might expect.

For depth measurements (see Figure 5–21), the combination square offers a greater measurement range and better reliability than the depth gage. Note that the support surface must be square to the line of measurement, and that an error could occur if a feature of the part is confused with the real line measurement. Misplaced holding force (in B) has the same effect as out-of-square support.

You can use attachments to extend the range of depth measurements (see illustrations C and D), if you remember that every attachment is a new source of potential errors. For example, the overhang in the illustration is of concern because you should minimize overhang in all gaging. In D, the ratio of x to y is unfavorable, and any irregularity in the supporting surface will be multiplied. To compensate, you should apply holding force at the center of the support surface (see illustration C in Figure 5–21).

Protractor Head

We encounter the 45° angle so frequently in metrology that one side of the square head is at 45° to the blade. For other angles, the combination square is equipped with a *protractor head.* The head consists of a turret that rotates within a stock. The scale on the turret is graduated from 0° to 180° in either direction—a double protractor scale.

This double protractor provides both the angle and its complement (see Chapter 15 for advantages). The reference mark on the stock is opposite 0° to 90° on the protractor scale, and a level in the turret corresponds to the 0° and 90° position on the respective scales. You can easily change the protractor head with the square head to give yourself a convenient means for checking angles no closer than 1° (see Figure 5–22).

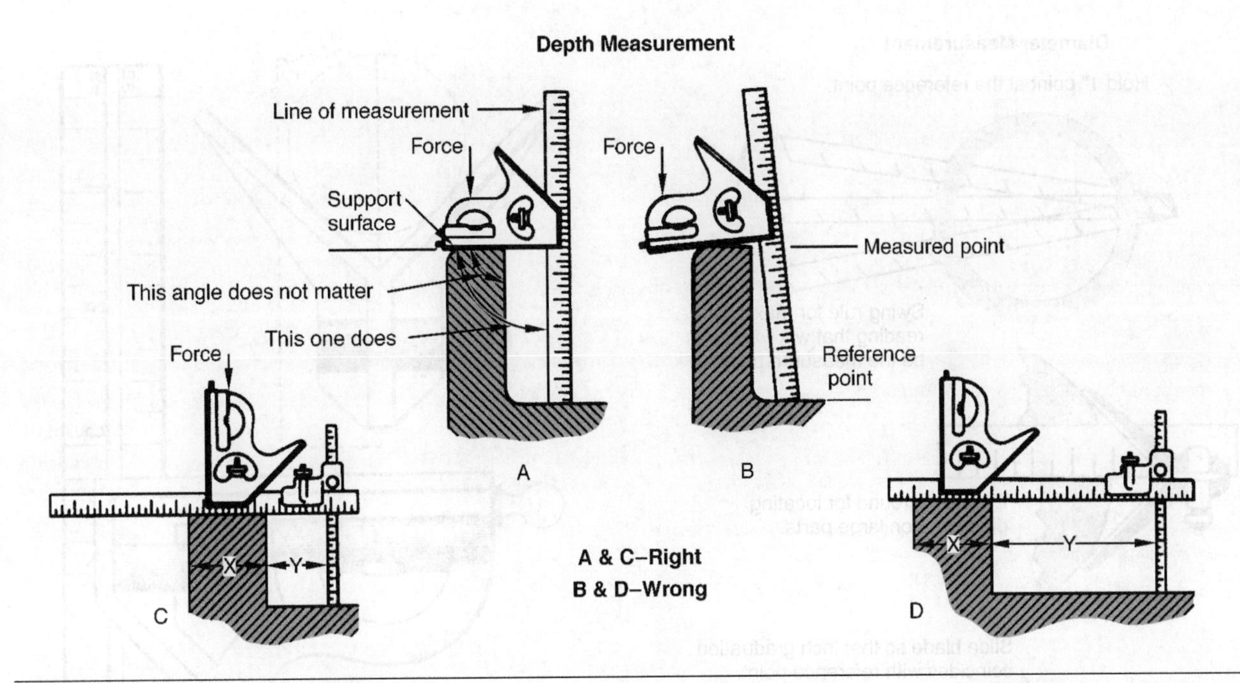

FIGURE 5–21 The combination square can be used like a depth gage. Being larger, it magnifies its range and the errors. Overhang is always a problem.

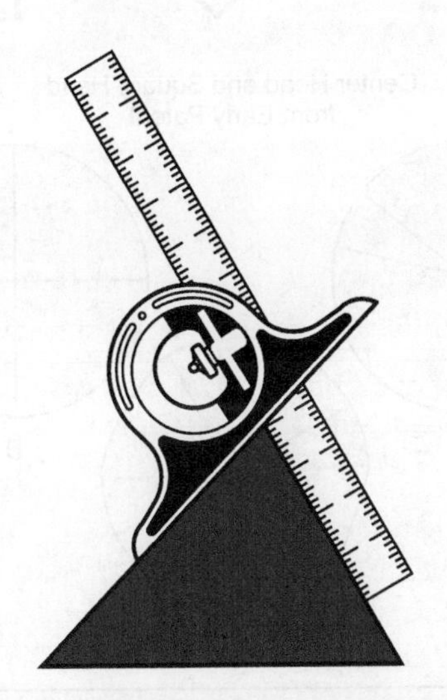

FIGURE 5–22 The protractor head with sliding blade forms a versatile instrument for the measurement of angles.

Center Head

When you place a blade in the *center head,* one edge of the blade bisects the V-angle of the head so the blade edge lies along the centerline of any circle placed against the faces of the V. It is a useful placement for both measurement and inspection of a wide range of parts.

The center head method of measuring diameters (see Figure 5–23) is far more reliable than the steel rule method. The center head eliminates the need to "juggle" the rule and the part while you try to make a reading. The major limitation to the center head method is parallax: both a reference point setting and a measured point reading must be made, but the blade is very thick. Unless you view the points properly, the error could be 0.08 mm (1/32 in.) or more.

For layout, the center head provides one of the best ways to find the center of a shaft. Theoretically, the intersection of any two diameters (Figure 5–24A) is at the center of a circle; the reliability of the intersection occurring in the center increases as the diameters

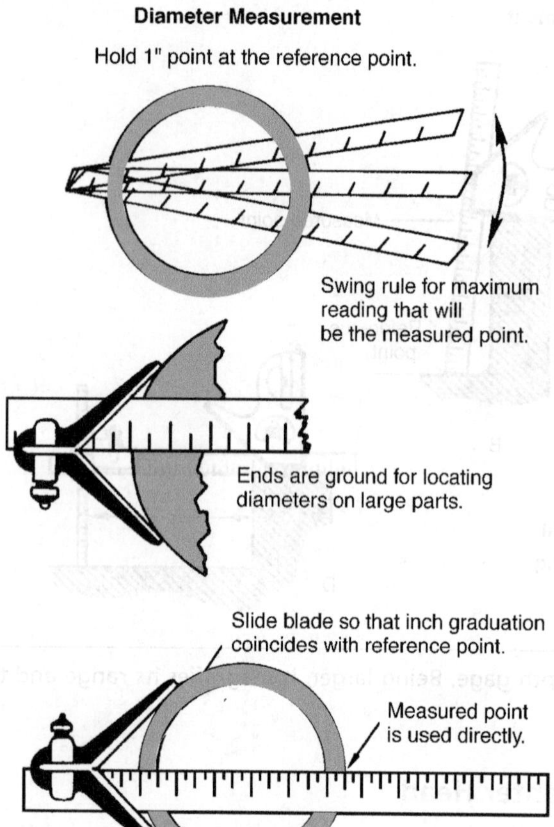

Diameter Measurement
Hold 1" point at the reference point.
Swing rule for maximum reading that will be the measured point.
Ends are ground for locating diameters on large parts.
Slide blade so that inch graduation coincides with reference point.
Measured point is used directly.

FIGURE 5–23 The center head speeds the measurement of diameters and improves reliability.

Center Head and Square Head from Early Patent
A
B
C

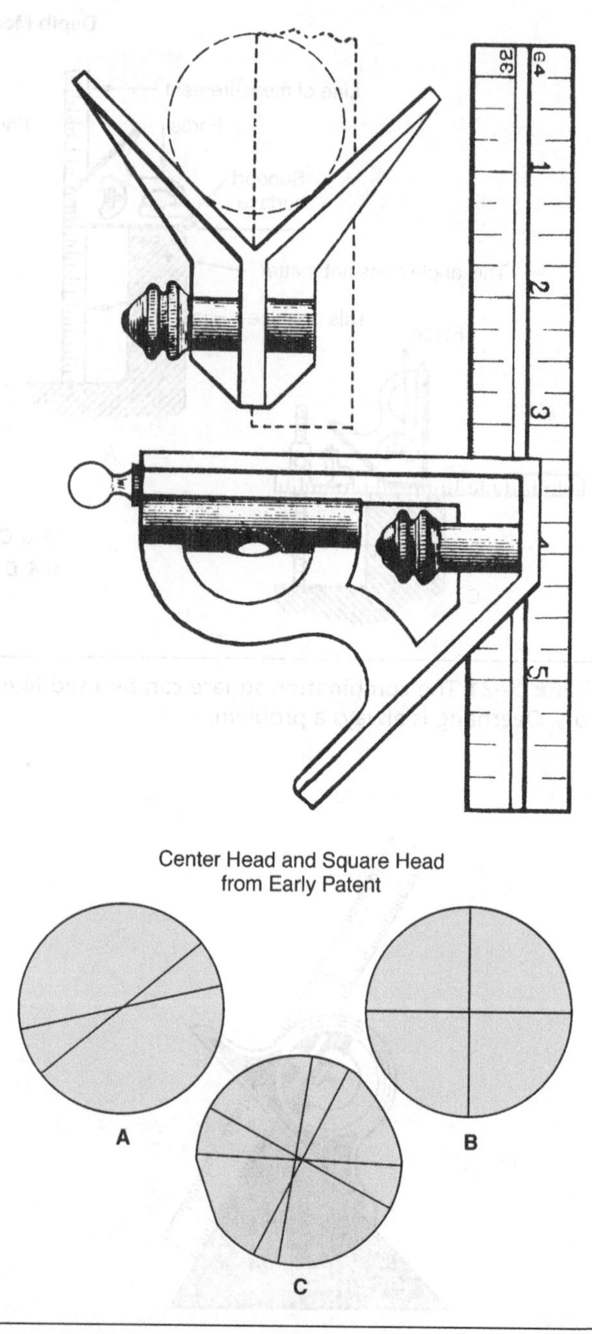

FIGURE 5–24 The center head was considered early in the development of the combination square by L. S. Starrett as the patent drawing shows. In the circles, B would be a more reliable center than A, whereas C is an average for an irregular shaft.

approach right angles (illustration B). If the shaft is slightly irregular, you should measure several diameters; then choose the center of the resulting polygon as the practical center of the shaft.

All the components of a combination square are replaceable—but they are also subject to wear. Check them frequently, and, if you use your combination square often, you might invest in hardened heads. Hardened heads are more expensive, but in the long run they are more economical.

Blades

The most popular combination square (10 cm head [4 in.]) and larger combination squares all use blades

that are 50 mm (1 in.) wide by 2.38 mm (3/32 in.) deep. Blades are available in the English system from 4 in. to 24 in. in length and are machined with a variety of scales. Blade lengths for metric instruments range from 150 mm to 600 mm. Blades have four scales:

1. 8ths, 16ths, 32nds, 64ths or 4R
2. 32nds, 64ths, 50ths, 100ths or 16R
3. 32nds, 1/2 mm, 64ths, 1 mm
4. 1 mm, 1/2 mm on both sides

Because graduations vary among manufacturers, you must be careful when you select a combination square. You also need to be careful when you read the measurement on the blade because the head covers part of the scale, thus hiding the divisions. You may be forced to interpolate the measurement if your measured point falls within the area that is covered (see Figure 5–25).

Each blade has four scales on it, but the blade can have only two positions with respect to the head. Therefore, you must use the blade that places the scale

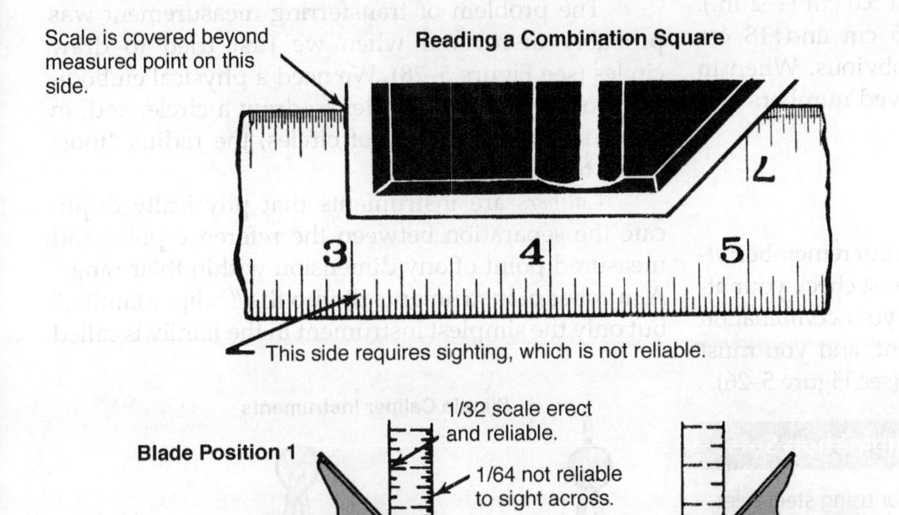

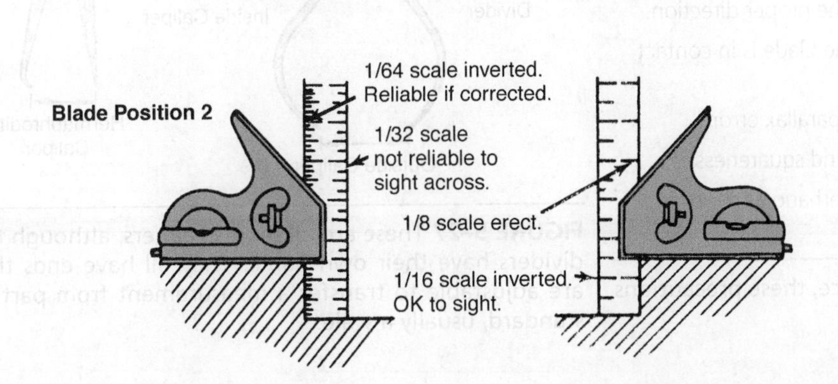

FIGURE 5–25 When a blade has only two positions by four scales, a little judgment improves reliability. When reading near the center of the scale, particular care is required.

you are using in the best position for making the measurement. In addition, you can safely sight from the head to the scale for 1/8 or 1/16 in., but you must have the scale you need on the head side if you are measuring with the finer 1/32 or 1/64 in. scales. Because the dimensions uncovered by the head decrease in size—go from larger to smaller—you have to adjust your measuring style to compensate.

Near the ends of the blade, the larger-to-smaller scale is less confusing. On a 30 cm (12 in.) blade, if the reading is between 25 cm and 28 cm (10–11 in.), you will be able to tell easily whether or not the measurement should be between 2.5 cm and 5.0 cm (1–2 in.). When your reading is between 15 cm and 18 cm (6–7 in.), the measurement is less obvious. When in doubt, you should ignore the engraved numbers and count divisions from the end.

Accessories

Accessories are fairly self-explanatory, but remember: *attachments add error in every case.* You must check your attachments as frequently as you check your combination square to ensure reliable measurement, and you must take care in the measurement process (see Figure 5–26).

Reliable Measurement with Combination Sets

1. Observe all applicable precautions for using steel rules.
2. Ensure that the heads have not worn excessively.
3. Be sure that the line of measurement is parallel to the blade.
4. Apply holding force to ensure stability during measurement.
5. Be sure that the scale is read in the proper direction.
6. For work from a plate, be sure the blade is in contact with the plate.
7. Take particular care to minimize parallax error.
8. Check all attachments for wear and squareness.
9. If attachments are used, keep overhang as short as possible.

FIGURE 5–26 With very little practice, these precautions become automatic.

5–4 CALIPERS: THE ORIGINAL TRANSFER INSTRUMENTS

Later, we will discuss ways to reliably measure to within millionths of an inch by transfer of measurement. The caliper (see Figure 5–27) is the forerunner of that measurement system.

Calipers and more sophisticated instruments may not look the same, but they share the same basic measurement principles. Calipers, like the combination square, are rarely used in production inspection, but they are still widely used for toolroom and related work.

The problem of transferring measurement was probably discovered when we first tried to draw circles (see Figure 5–28). We need a physical embodiment of the radius in order to draw a circle, and, in order to draw a variety of circles, the radius "tool" needs to be adjustable.

Calipers are instruments that physically duplicate the separation between the reference point and measured point of any dimension within their range. A number of instruments fit into the "caliper family," but only the simplest instrument in the family is called

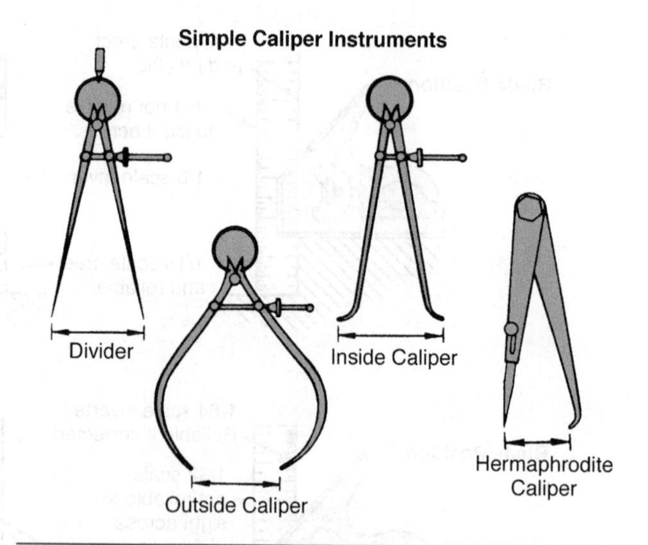

FIGURE 5–27 These are all simple calipers, although the dividers have their own name. They all have ends that are adjustable to transfer a measurement from part to standard, usually a scale.

Caliper Defined

Distance Measured

Reference Point

Measured Points

A circle can be **approximated** with a graduated scale if sufficient measured points are plotted.

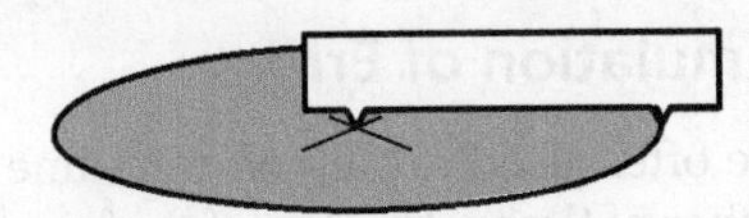

If the distance is made into a radius bar, an infinite number of measured points can be swung around any reference point.

If, instead of a solid radius bar, the points are adjustable, then a range of radii can be selected.

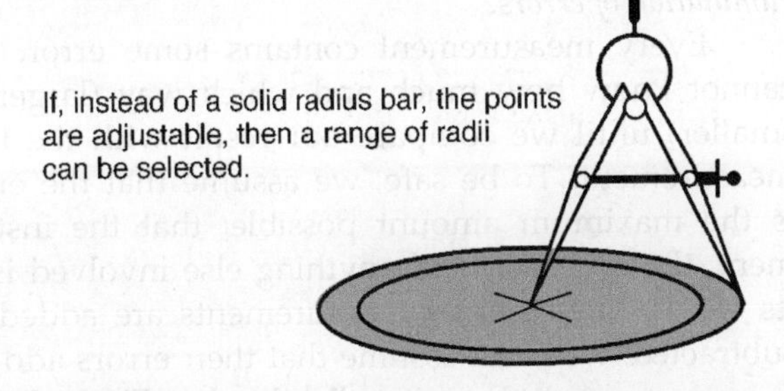

FIGURE 5–28 A caliper is an instrument for mechanically duplicating a measurement. Unless otherwise specified, it is an adjustable instrument. Not all calipers are termed that; dividers are an example.

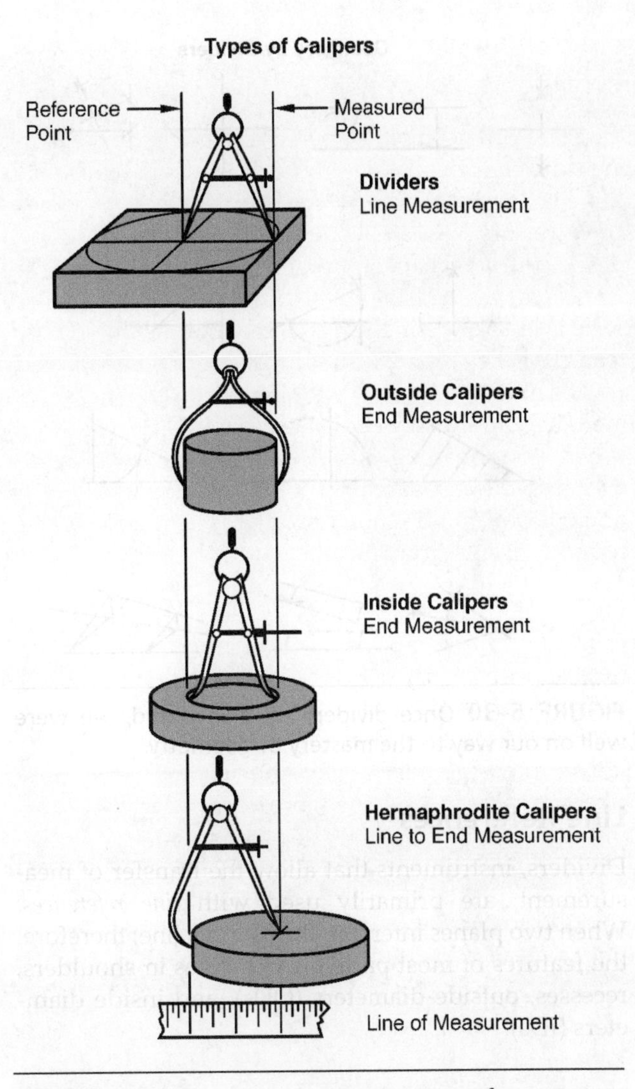

FIGURE 5–29 All caliper instruments transfer measurements. These are the basic types. Other versions bear little resemblance, except in principle of operation.

"calipers" (see Figure 5–29). We use calipers for transfer of measurement only. More sophisticated calipers incorporate their own scales and standards of length (see Chapter 6).

Dividers for Line Measurement

Dividers, a style of calipers, originally helped us draw circles. They led to the study of geometry and enabled premeasurement craftspeople to control the position and size of what was produced. With dividers, any straight line or arc could be "divided" into convenient spaces (see Figure 5–30). Each item then had its own standard length—a size that was specific to that item and its duplicates, something which did not work well for mass production.

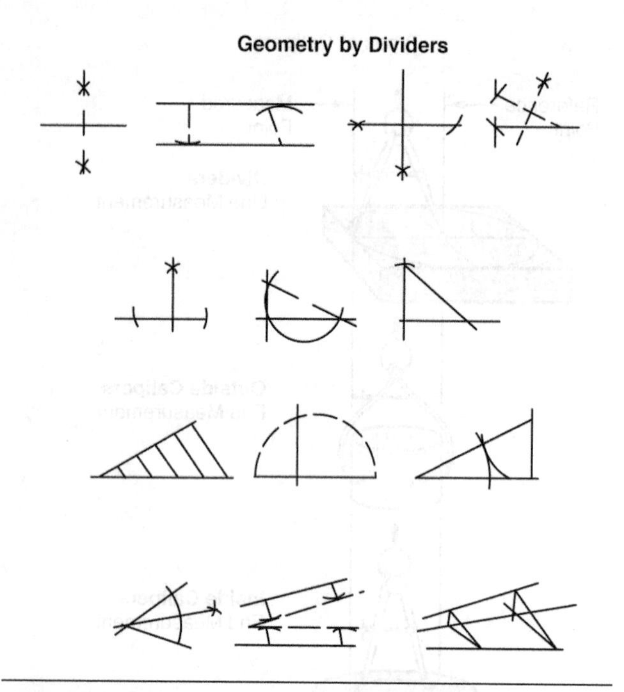

FIGURE 5–30 Once dividers were invented, we were well on our way to the mastery of geometry.

Line References

Dividers, instruments that allow the transfer of measurement, are primarily used with *line references.* When two planes intersect, they form a line; therefore, the features of most parts provide lines in shoulders, recesses, outside diameters (ODs), and inside diameters (IDs).

Transfer Measurement

Today, the primary use for dividers is still scribing arcs. Other important uses are: transferring a dimension of a part to a scale for measurement; transferring a dimension from one part to another; and transferring a dimension from a scale to a layout of a part.

The "lines" to which the dividers must be set (see Figure 5–31A) determine the reliability of the measurement obtained by the dividers (see Figure 5–31). In every transfer of measurement, you actually do two separate operations: from the unknown to the gage (divider) and from the gage to the known (scale, standard, or a part to be duplicated). *The reliability of each operation influences the reliability of the entire measurement.*

When you set a divider to a dimension, you place one leg on a rule graduation (metric or inch) for the reference point. You then adjust the divider until the other leg touches the desired graduation for the measured point. Good graduations are V-shaped, so a divider with sharp points can be adjusted fairly easily and accurately. Because a dull point can cause considerable inaccuracy, points can be honed with a stone or returned to the factory for repair.

The Accumulation of Errors

Dividers are often used to step off the same measurement a number of times. This transfer of measurement style is useful for laying out bolt circles or dividing the circumference of a circle, but it can also lead to the *accumulation of errors.*

Every measurement contains some error. We cannot know how much and which way (larger or smaller) until we compare our result with the true measurement. To be safe, we assume that the error is the maximum amount possible: that the instrument, the observer, and anything else involved is at its worst. But whether measurements are added or subtracted, we must assume that their errors add, or our measurement loses its reliability (see Figure 5–32).

Although this assumption adds to the safety margin for our measurements, it is like assuming that your opponent in a poker game has a royal flush after each deal. It is statistically unlikely and makes modern statistical dimensioning even more important to measurement.

We know that the number of distinct operations in a measurement decreases reliability because errors are potentially added in each operation. So there can be only two reasons to add an operation to a measurement process: one, when a measurement cannot be made without the added operation); and two, when the errors added by the new operation actually help decrease the number of errors in the total measurement process (see Figure 5–33). A number of measurement operations are known as serial measurements.

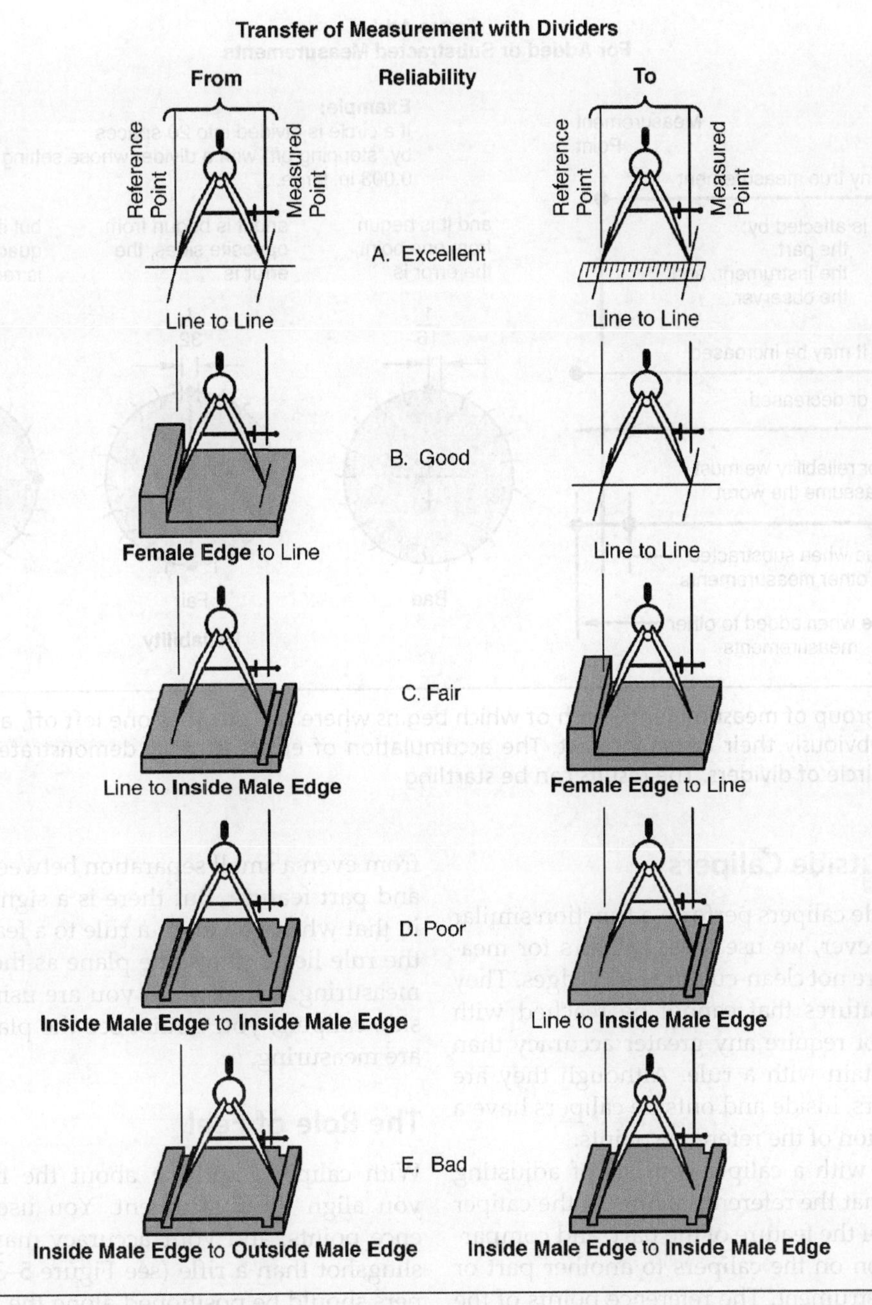

FIGURE 5–31 The important thing to remember in the transfer of measurements is that two, not one, measurements are involved. Each contributes errors.

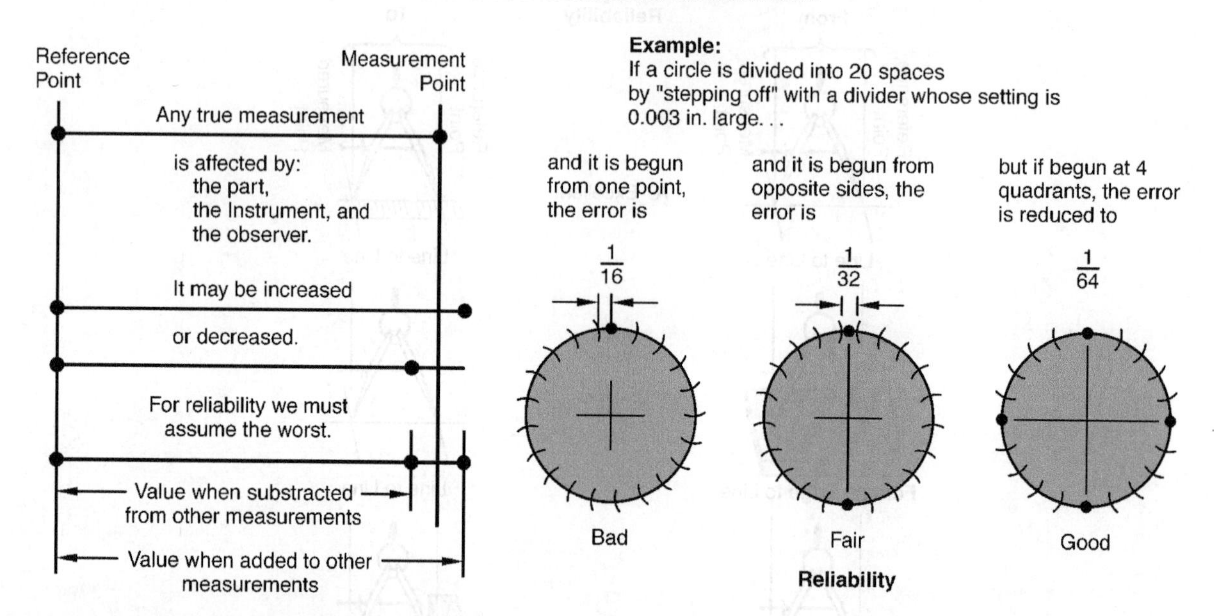

FIGURE 5–32 A group of measurements, each of which begins where the previous one left off, are known as serial measurements. Obviously their errors interact. The accumulation of errors is easily demonstrated by stepping off spaces around a circle of dividers. The results can be startling.

Inside and Outside Calipers

Inside and outside calipers perform a function similar to dividers; however, we use these calipers for measurements that are not clean-cut lines and edges. They also measure features that cannot be reached with a rule and do not require any greater accuracy than what we can obtain with a rule. Although they are similar to dividers, inside and outside calipers have a different separation of the reference points.

Measuring with a caliper consists of adjusting the opening so that the reference points of the caliper duplicate those of the feature of the part, and comparing the separation on the calipers to another part or measurement instrument. The reference points of the caliper must lie along the same line as the feature of the part, the line of measurement.

We use this measurement process for steel rules and combination squares, and parallax can result from even a small separation between the instrument and part feature. But there is a significant difference in that when you align a rule to a feature, the edge of the rule lies in the same plane as the feature you are measuring. Often when you are using inside or outside calipers, you cannot see the plane in which you are measuring.

The Role of Feel

With calipers, nothing about the instrument helps you align the instrument. You use only the reference points, and your accuracy may be more like a slingshot than a rifle (see Figure 5–34). Outside calipers should be positioned along the line of minimum separation, the distances between surfaces that bind the feature being measured; inside calipers should be positioned along the line of maximum distance (see Figure 5–35).

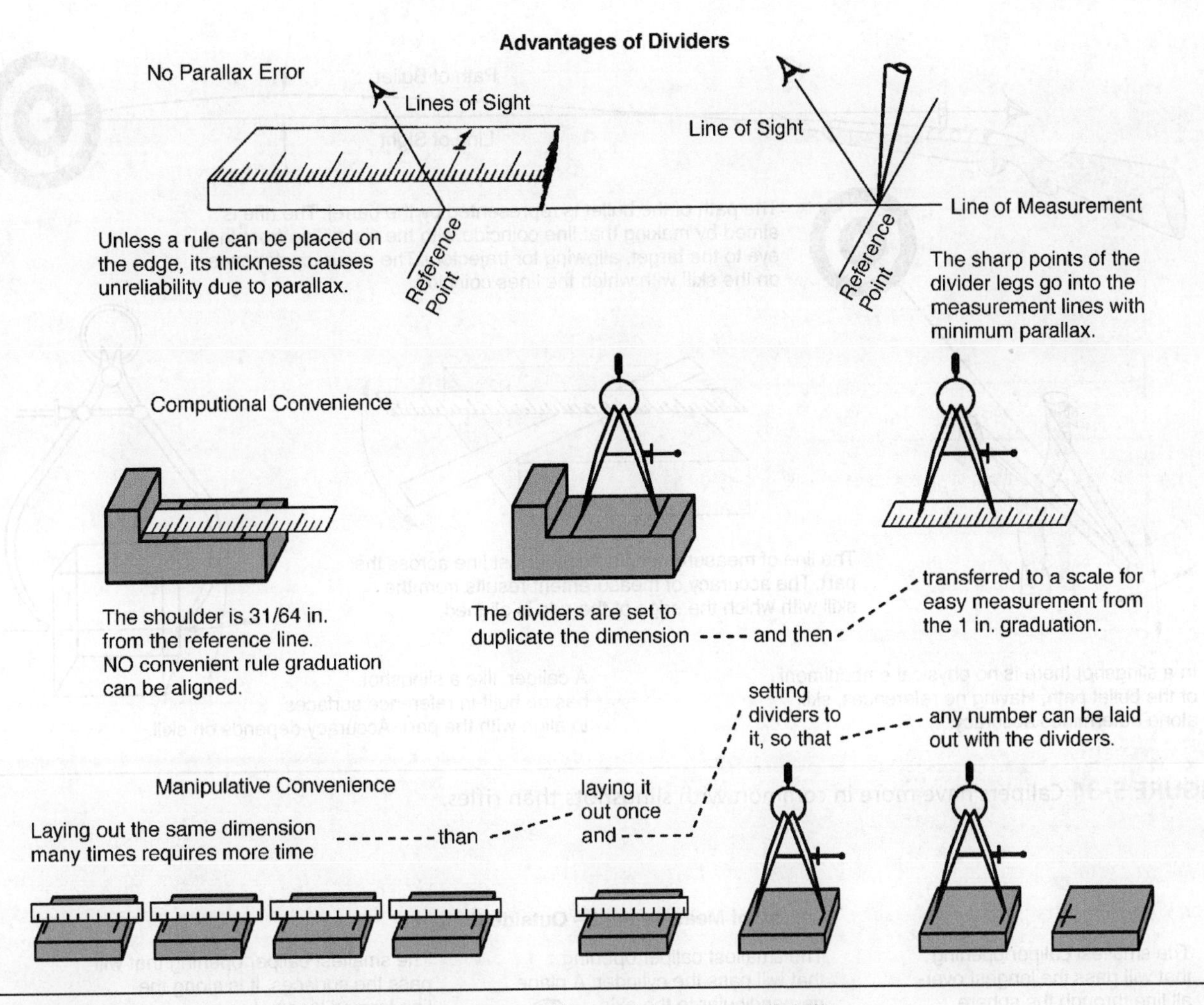

FIGURE 5–33 Properly used, dividers can save time without decreasing reliability.

So how do you find the measurement lines? Mostly by *feel*, which is difficult because surfaces vary. For example, when you measure a sphere, you have nothing but feel to guide the measurement. With a cylinder, you use a right-angle plane for the diameter and often a shoulder or a groove provides that plane automatically (see Figure 5–36). In these cases, the outside caliper is a relatively precise instrument.

It is more difficult to measure rectangles with calipers because it can be hard to determine the actual shortest distance between the faces. In everyday work, outside calipers are used with shoulders to keep them from shifting too far (see Figure 5–36). You use much the same accessories with inside calipers (see Figure 5–37). Inside calipers are slightly more reliable because the instrument actually helps locate the inside surfaces that you need to measure.

The Use of Calipers

When you use calipers, they are set to the feature and the distance is transferred to a measuring instrument, or they are set to the dimension and used as a fixed gage.

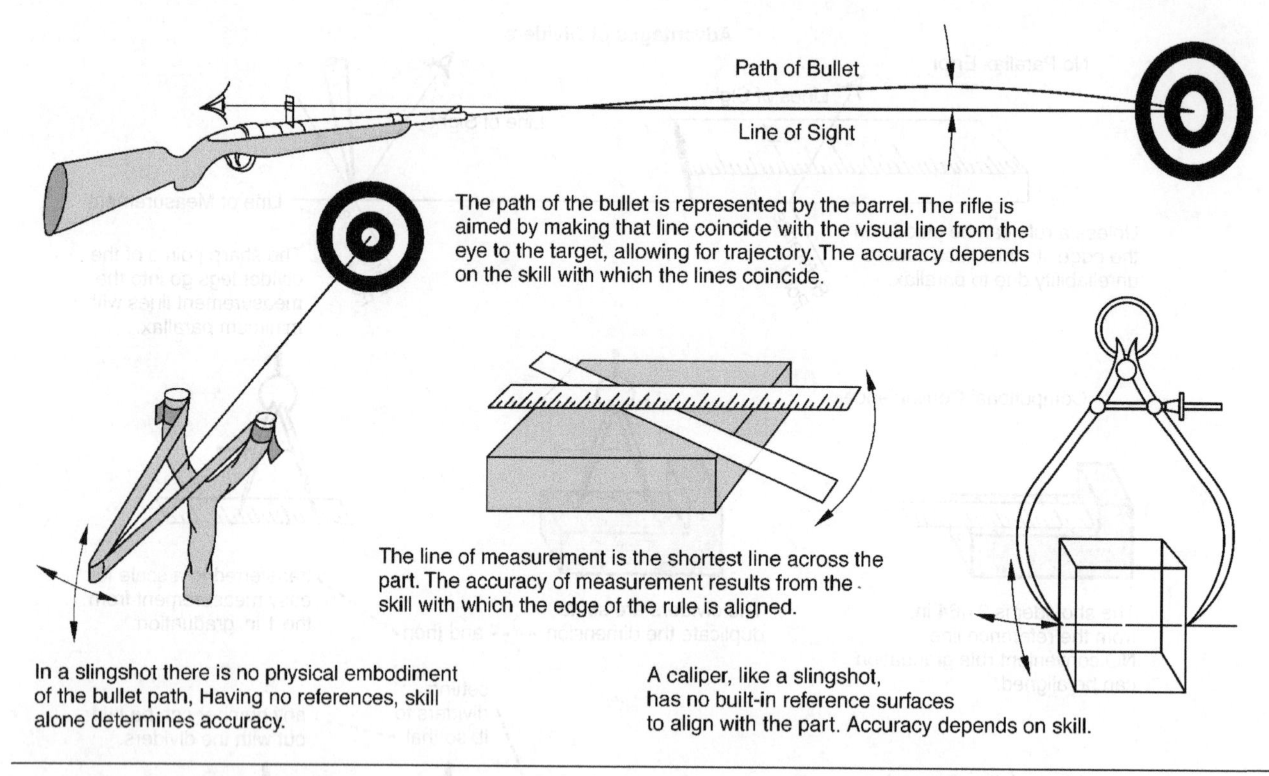

FIGURE 5–34 Calipers have more in common with slingshots than rifles.

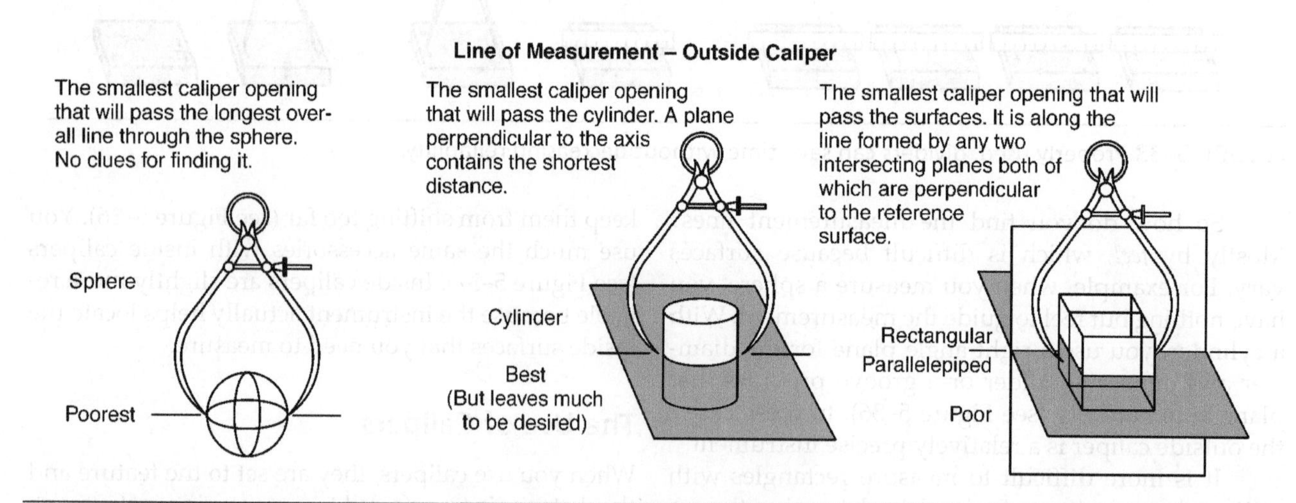

FIGURE 5–35 Seek the smallest caliper opening that will pass the part.

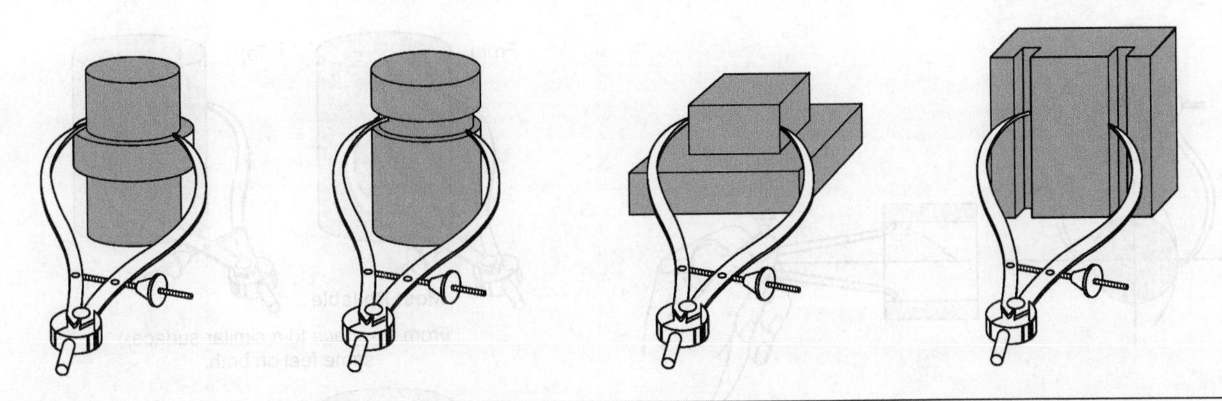

FIGURE 5–36 Greatest reliability is achieved when a shoulder or recess on the part restricts the line of measurement to a plane perpendicular to the part.

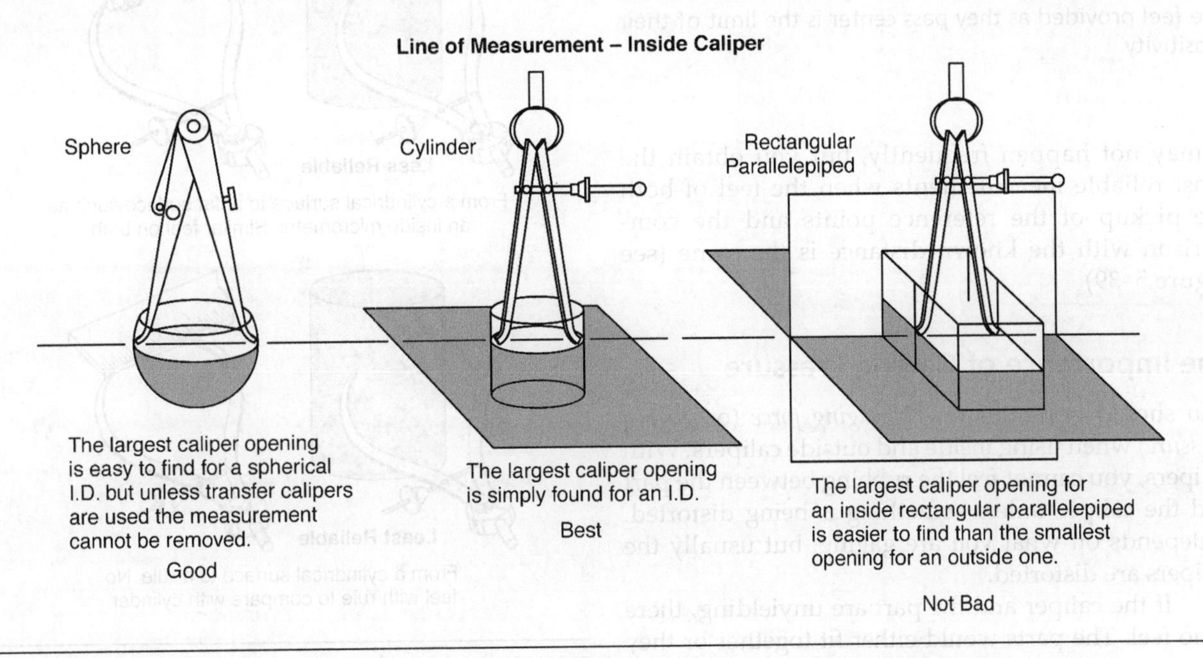

FIGURE 5–37 With calipers, inside measurements are somewhat easier to take than outside.

Outside calipers set to the desired dimension show whether the feature is or is not too large, but they do not show if the dimension is too small. Similarly, an inside caliper set to the desired dimension shows whether or not the feature is too small. Under controlled conditions, we can gain valuable knowledge from this "sorting" process. That is why this procedure is still used by industry today, although not very often with calipers.

To measure precisely with calipers, you need to feel the instrument rubbing lightly against the part as you rock it across center (see Figure 5–38).

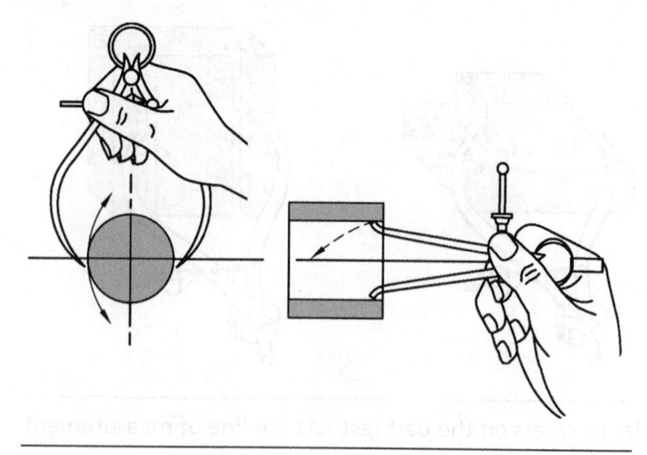

FIGURE 5–38 Calipers are carefully rocked over center. The feel provided as they pass center is the limit of their sensitivity.

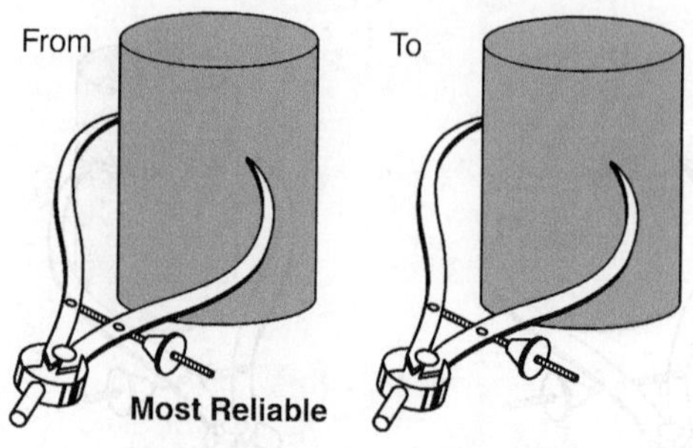

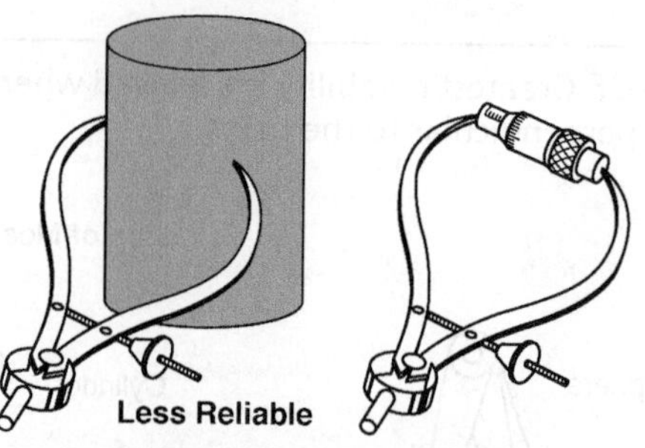

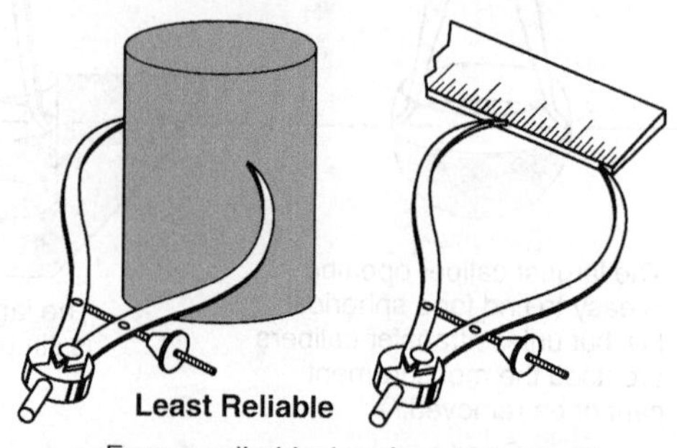

FIGURE 5–39 For greatest reliability with outside calipers, the feel on the work should duplicate the feel on the instrument.

It may not happen frequently, but you obtain the most reliable measurements when the feel of both the pickup of the reference points and the comparison with the known distance is the same (see Figure 5–39).

The Importance of Gaging Pressure

You should be familiar with *gaging force* (or *gaging pressure*) when using inside and outside calipers. With calipers, you cannot feel the rubbing between the part and the caliper unless something is being distorted. It depends on what you are gaging, but usually the calipers are distorted.

If the caliper and the part are unyielding, there is no feel. The parts would either fit together or they would not. In every gaging situation, one of the elements will yield, creating distortion—an error. When you are measuring to a millionth of an inch, even gravity can cause distortion.

The lighter the feel, the more reliable your measurements are. Of course, the part and the instrument must also be extremely clean: a coating of dust-laden oil or grease can drastically alter the feel. You can learn to use a light feel by practicing and by being careful in your measurements. You also need to have a comfortable measuring position, and, as parts become larger, it can be more difficult to get comfortable.

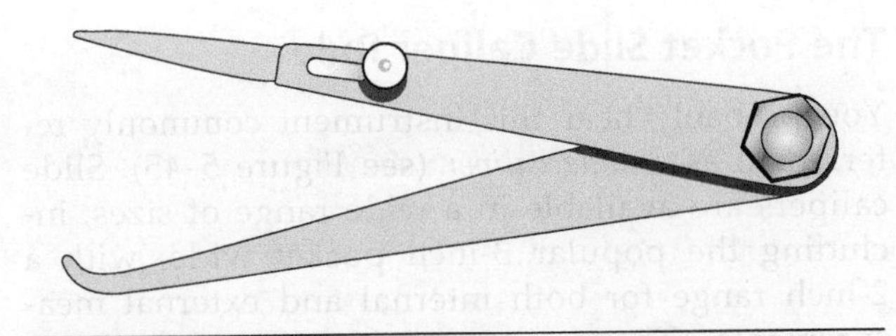

FIGURE 5–40 The hermaphrodite caliper transfer measurements bounded by a line and an edge. Its use is limited.

Layout Instruments

We use several scribing instrument rules: the beam trammel, the hermaphrodite caliper, the surface gage, and the scriber.

Hermaphrodite calipers, sometimes called "odd legs" calipers have one curved caliper leg and one divider leg (see Figure 5–40). The caliper leg is reversible so that you can make contact with either inside or outside reference surfaces. They also make it possible for you to pick up a measurement bounded by a line and an edge.

A popular shop instrument, hermaphrodite calipers are used widely for scribing lines parallel with an edge, although they are not the most reliable way to do so (see Figure 5–41). A more practical and reliable use for these calipers is finding the centers of shafts for which the center head cannot be used (see Figure 5–42).

The *surface gage* transfers height measurements and is used most often on a surface plate (also discussed in Chapter 6). See Figure 5–43. You can obtain a more reliable measurement using a surface plate than you can holding both the instrument and the part in your hand. However, the surface gage is not used often for other measurements.

Unlike hermaphrodite calipers and other hand-held instruments, the surface gage (Figure 5–44) eliminates many alignment problems; however, you still have the potential to create errors because of the power of your eye. You can minimize error by using a magnifier. For inspections, you can substitute a dial indicator for the scriber in order to minimize errors (Chapter 10).

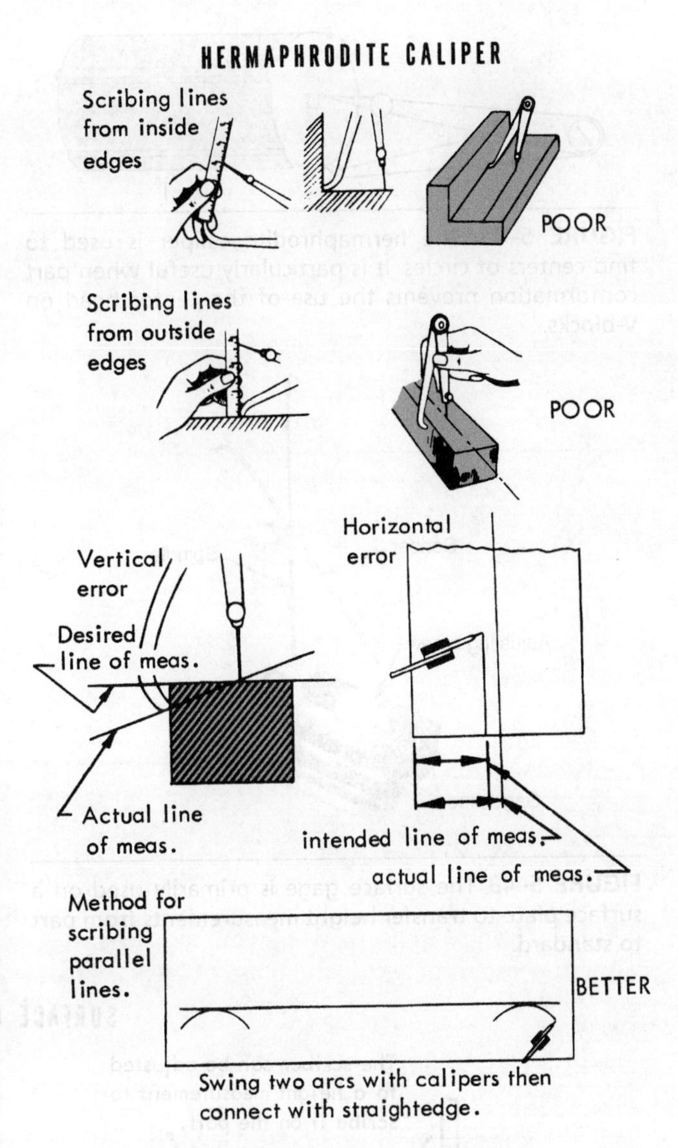

FIGURE 5–41 If you must use this instrument you should be aware of the changes for positional misalignment and their consequences.

The *scriber,* although not a measuring instrument, is frequently required to transfer measurements. Use your common sense when handling a scribe and care for it like you care for your other instrument accessories.

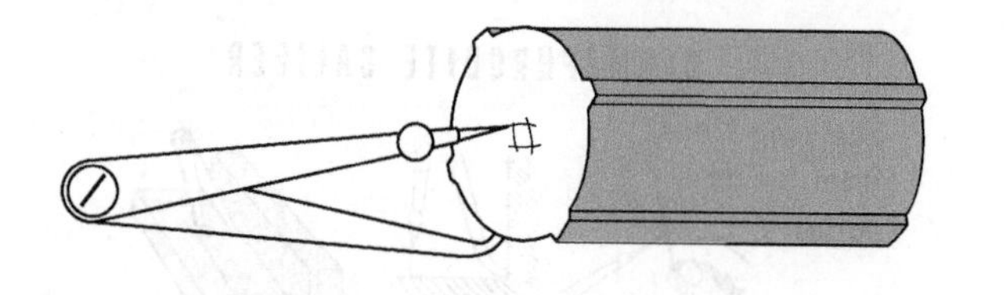

FIGURE 5–42 The hermaphrodite caliper is used to find centers of circles. It is particularly useful when part conformation prevents the use of the center head on V-blocks.

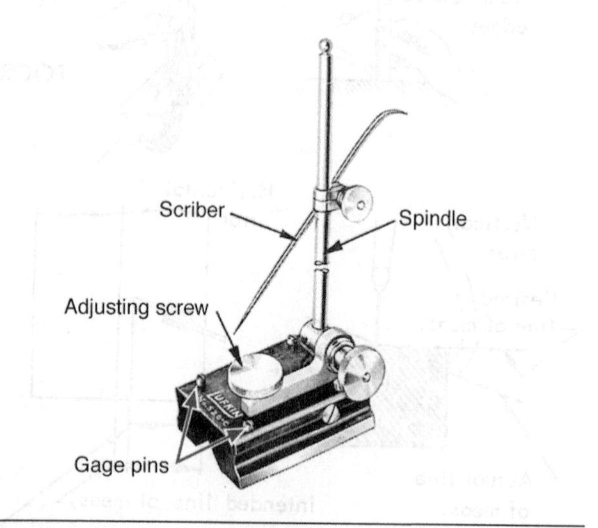

FIGURE 5–43 The surface gage is primarily used on a surface plate to transfer height measurements from part to standard.

The Pocket Slide Caliper Rule

You probably hear this instrument commonly referred to as a *slide caliper* (see Figure 5–45). Slide calipers are available in a wide range of sizes, including the popular 3-inch pocket wide, with a 2-inch range for both internal and external measurements. They are available in sizes up to 80 inches with fractional-inch, decimal-inch, and metric scales. The *stock* usually has a coarser scale, whereas the slide's scale is finer.

You can fix the slide caliper into any position with the lock, which can help you when you are rough turning a long shaft. Simply lock the calipers after each measurement, and you will not have to remember the diameter between passes—the slide calipers act as a memory instrument, remembering for you. You can then check the measurement, unlock the calipers, and recheck the measurement.

You might think that this locking mechanism would make the slide caliper into a handy gage. *The slide caliper is not intended to be a snap gage* because you cannot adjust the instrument for excessive wear, as snap gaging requires (see Figure 5–46).

There are three advantages to slide calipers:

1. They provide positive contact between the instrument and the reference and measured

SURFACE GAGES

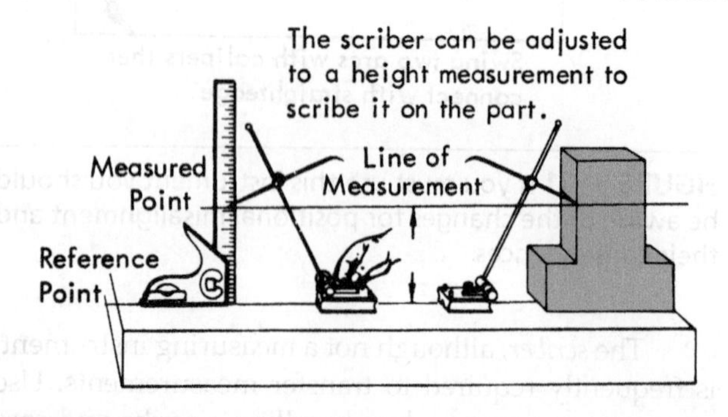

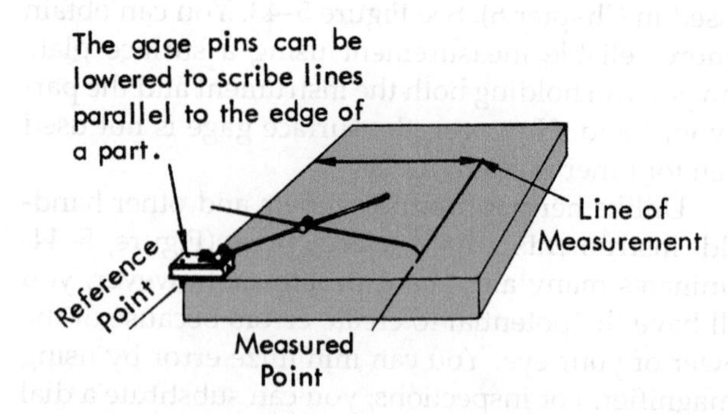

FIGURE 5–44 With a surface gage, there is stable contact between the part and the instrument along the reference point line. (*Courtesy of Apex Tool Group*)

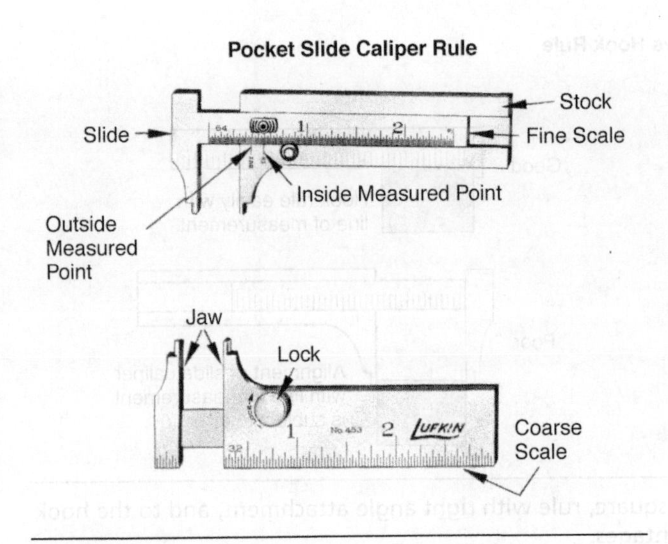

FIGURE 5–45 This handy shop instrument is usually called a slide caliper.

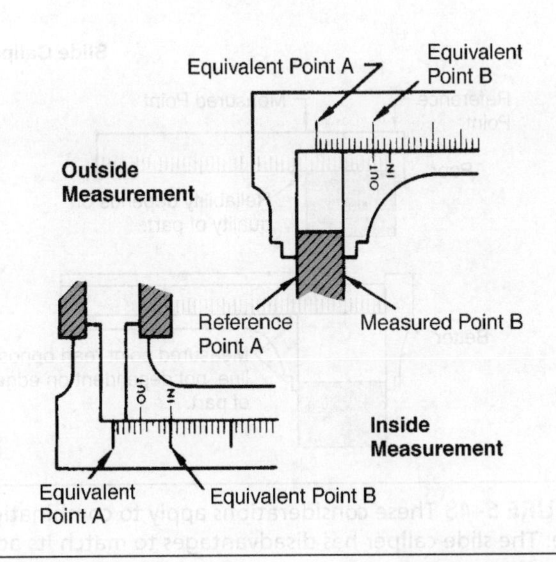

FIGURE 5–47 Care must be taken to use the correct equivalent point. (*Courtesy of Flexbar Machine Corp.*)

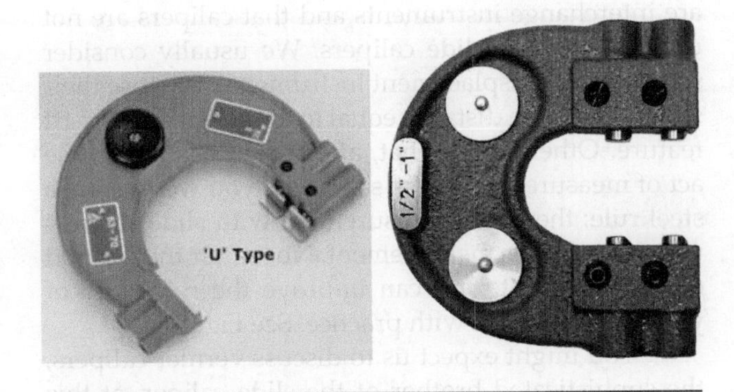

FIGURE 5–46 Calipers should never be used as snap gages. Caliper instruments are used for measuring, whereas snap gages are used for sorting. In this typical example, the part must pass through the first (right) opening but not through the second (left) opening. This is referenced to as a go/no-go gage.

points of the feature for both internal and external measurements (see Figure 5–47), which even the hook rule cannot provide (see Figure 5–48).

2. They substitute a line on the instrument for a feature to use as a measured point, so slightly rounded corners do not affect the precision of setting or reading.
3. You have the built-in memory provided by the slide lock.

Two major disadvantages of slide calipers are:

1. Like all caliper instruments, they can produce errors if positioned incorrectly.
2. You cannot adjust the instrument to compensate for wear.

You must recheck slide calipers frequently for wear along both the internal and external jaw surfaces. If you regularly work to 0.4 mm (1/64 in.), you should replace slide calipers when wear reaches beyond the width of one graduation.

Minor disadvantages are the slide calipers' limited discrimination, 0.01 mm (1/64 in.), and that they can only measure between two inside or two outside references. Slide calipers cannot be used

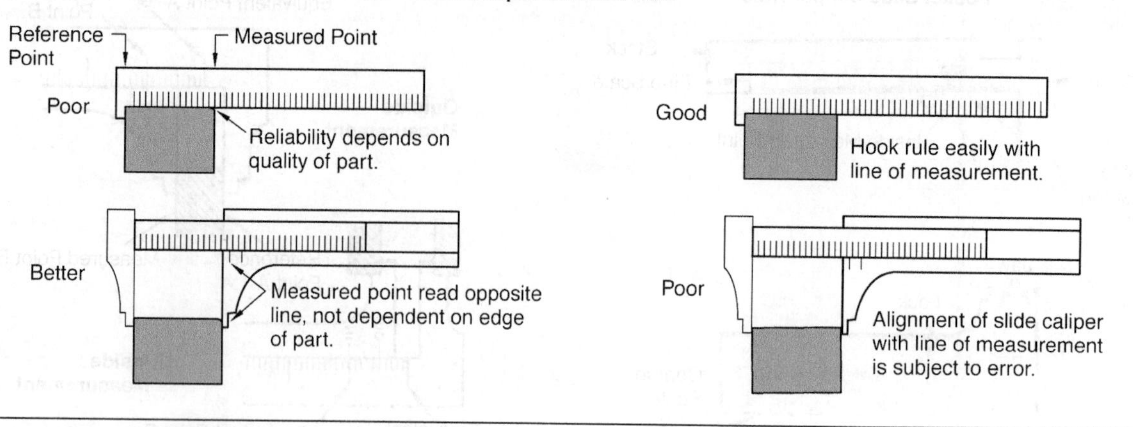

FIGURE 5–48 These considerations apply to combination square, rule with right angle attachment, and to the hook rule. The slide caliper has disadvantages to match its advantages.

Slide Caliper	
Advantages	1. Combines rule, inside, and outside calipers in one instrument. 2. Provides positive contact with reference and measured points. 3. Substitutes line-to-line for line-to-edge readings. 4. Has built-in memory.
Disadvantages	1. No wear adjustment. 2. Subject to misalignment. 3. Limited discrimination. 4. Cannot caliper inside to outside part features.

FIGURE 5–49 Although slide calipers are handy, careful measurement practices must be applied.

to measure from an outside to an inside reference (see Figure 5–49).

Most people usually feel that scaled instruments are interchange instruments and that calipers are not until we discuss slide calipers. We usually consider slide calipers displacement instruments, because their slides displace a distance equal to the length of the part feature. Others argue that, after contact is made, the act of measurement is the same as if you were using a steel rule; therefore, measurement with slide calipers uses interchange measurement. No matter the method of measurement, you can improve the reliability of your measurement with practice. See Figure 5–50.

You might expect us to discuss vernier calipers, the sophisticated brother of the slide caliper, at this point. But vernier calipers increase the precision of measurement from millimeters to micrometers (64ths to thousandths). So we will cover vernier calipers in their own chapter.

Metrological Data for Scaled Instruments								Reliability	
Instrument	Type of Measurement	Normal Range	Designated Precision	Discrimination	Sensitivity	Linearity	Practical Tolerance for Skilled Measurement	Practical Manufacturing Tolerance
Depth gage:								
metric	direct	150 mm	0.5 mm	0.5 mm	0.5 mm	0.005/mm	±0.5 mm	±0.5 mm
decimal-inch	direct	6 in.	0.02 in.	0.02 in.	0.02 in.	0.0003/in.	±0.02 in.	±002 in.
fractional-inch	direct	6 in.	1/64 in.	1/64 in.	1/64 in.	0.0003/in.	±1/64 in.	±1/64 in.
Combination sets:								
metric	direct	150 mm	0.5 mm	0.5 mm	0.5 mm	0.005/mm	±0.5 mm	±0.5 mm
decimal-inch	direct	4 in.	0.01 in.	0.01 in.	0.01 in.	0.0003/in.	±0.01 in.	±0.02 in.
fractional-inch	direct	6 in.	1/64 in.	1/64 in.	1/64 in.	0.0003/in.	±1/64 in.	±1.32 in.
Calipers:								
metric	transfer	150 mm	none	none	0.25 mm	none	±0.5 mm	±1 mm
decimal-inch	transfer	3 in.	none	none	0.005 in.	none	±0.02 in.	±0.08 in.
fractional-inch	transfer	3 in.	none	none	0.005 in.	none	±0.02 in.	±0.08 in.
Slide Calipers:								
metric	direct	130 mm	0.5 mm	0.5 mm	0.5 mm	0.005/mm	±0.5 mm	±0.5 mm
decimal-inch	direct	5 in.	0.01 in.	0.01 in.	0.01 in.	0.0003/in.	±0.01 in.	±0.02 in.
fractional-inch	direct	5 in.	1/64 in.	1/64 in.	1/64 in.	0.0003/in.	±1/64 in.	±1.32/in.

FIGURE 5–50 Note that although the caliper instruments have relatively good sensitivity, their reliability is low for a skilled operator and very low for manufacturing applications because two transfers are required for each measurement.

SUMMARY

- There is an oft-repeated claim that a rule can be used for measurement to a mil. It can, and frequently is, but not reliably (see Figure 5–51). When measurement is not reliable, manufacturing costs rise, more hand fitting is required, pieces are scrapped, and assemblies wear out faster and are harder to repair.
- When building parts whose measurements are to be within 0.4 mm (1/64 in.), you can reliably use a rule. But if you are instructing other people to make parts that must fit together, the situation is different. If the measurements required for proper functioning are no finer than 0.8 mm (1/32), you may require that they make them to 0.4 mm (1/64). Even that would not provide adequate control, as you will realize when we discuss the rule of ten-to-one in Chapter 14.
- Choose a scale that is graduated the same as the part dimensioning if you are working to a drawing. Use metric scales for metric parts, decimal-inch scales for decimal-inch parts, and so forth (see Figure 5–52).

Reliability Check List for Scaled Instruments	
Inspection of Instrument:	1. Set up periodic system for inspection, depending on use. 2. Inspect contact surfaces with magnifier for wear or abuse. 3. Remove burrs from sliding and contact surfaces. 4. Compare readings against an instrument of higher precision, greater accuracy, and with known calibration. 5. Check all mechanical actions for proper functioning. 6. Clean and lubricate internal parts. 7. Check alignment against square of known calibration.
Use:	1. Never use a measuring instrument for a hand tool (scraper, chip digger, burring tool, mallet, screwdriver or sledge-hammer). 2. Never use beyond intended size range (do not force open). 3. Never use beyond discrimination or recommended precision. 4. Keep contact force to a minimum. 5. Avoid excessive movements causing wear. 6. Clean both part and instrument before using. 7. Substitute mechanical support for hand support whenever possible. 8. Guard against parallax when reading. 9. Have entire setup rigidly supported. 10. Do not overtighten anything.
Care: 1.	1. Lubricate instruments before replacing in case. 2. Keep away from moisture. 3. Do not pile instruments together or with other objects. 4. Do not mark tools in any way that interferes with use. 5. Do not hesitate to throw away a worn or defective tool.

FIGURE 5–51 Some of these suggestions seem self-evident. Carelessness can cause an expensive part to be scrapped or a careful experiment to yield incorrect results.

- Steel rules are a simple type of scaled instrument. They are quick, inexpensive, and easy to use. Accurate measurement with a steel rule will increase with practice and the use of good measurement procedures (see Figure 5–53).
- As with the higher precision instruments, their accuracy is traceable to the National Institute of Standards and Technology (see Figure 5–54). Figures 5–55 and 5–56 illustrate various types of rules and tapes.
- Most of the basic measurement procedures and measurement instruments derive from the simple steel rule. These instruments were required to lay out parts before machining as well as to check parts afterward. Among these are the depth gage, combination square, caliper, and slide caliper.
- The steel rule, combination square, and depth gage are scaled instruments. Worn scales, incorrect placement of the scale, or observation

Metrological Data for Scaled Instruments								Reliability
Instrument	Type of Measurement	Normal Precision	Designated Precision	Discrimination	Sensitivity	Linearity	Practical Tolerance for Skilled Measurement	Practical Manufacturing Tolerance
Ordinary rulers	direct	12 in.	1/16 in.	1/16 in.	1/16 in.	1/16 in.	1/16 in.	never
Steel rules:								
Decimal-inch	direct	6, 12, 18, 24 in.	0.02 in.	0.02 in.	0.02 in.	0.0003/in.	±0.02 in.	±0.04 in.
Fractional-inch	direct	6, 12, 18, 24 in.	1/64 in.	1/64 in.	1/64 in.	0.0003/in.	±1/64 in.	±1/32 in.
Steel tapes:								
Decimal-inch	direct	100 ft.	0.10 in.	0.10 in.	0.10 in.	0.01 in.	±0.10 in.	±0.30 in.
Fractional-inch	direct	100 ft.	1/8 in.	1/8 in.	1/8 in.	0.01 in.	±1/8 in.	±3/8 in.

FIGURE 5–52 For scaled instruments, some of these columns are relative. They will have important meaning, however, for higher-precision instruments.

Reliable Measurements with Steel Rules

1. Choose the proper rule and the proper scale for the measurement.
2. Have both the part and the rule clean.
3. If at all possible have either the part or the rule resting on a stable surface.
4. Align the scale edge of the rule and the line of measurement of the part as closely as possible.
5. Align the reference point to the rule so that unsharp edges on either will not interface.
6. Read the measured point from a point directly opposite, use magnifier for greatest precision, or to avoid fatigue from repeated related measurements.
7. Repeat with higher discrimination scales until a graduation aligns with the measured point, desired discrimination is reached, or the finest scale is reached.
8. Remind oneself of bias and repeat operations.
9. Repeat sufficient times for the needed reliability based on your skill.

FIGURE 5–53 These seemingly obvious steps can quadruple the chances that your steel rule measurements will be accurate to the full discrimination of the rule.

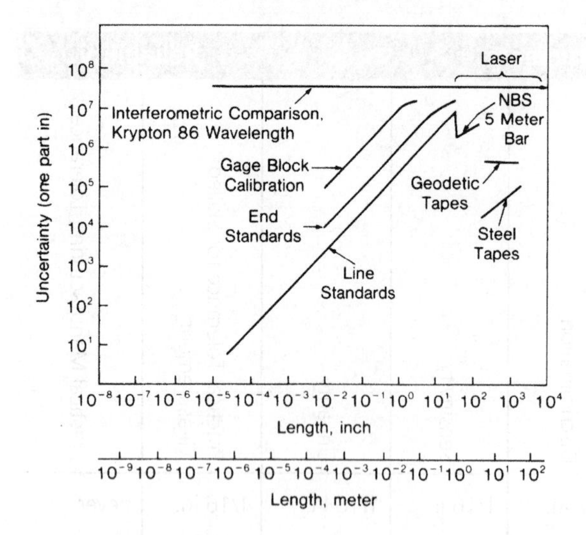

FIGURE 5–54 This graph from the National Institute of Standards and Technology shows the relative accuracy of the major types of measurement instruments. It shows the accuracy in terms of "uncertainty" (vertical) plotted against the length of measurement (horizontal). As expected, in most cases the uncertainty increases as the length increases. Although steel tapes are among the lowest precision instruments, NIST is capable of calibrating them precisely.

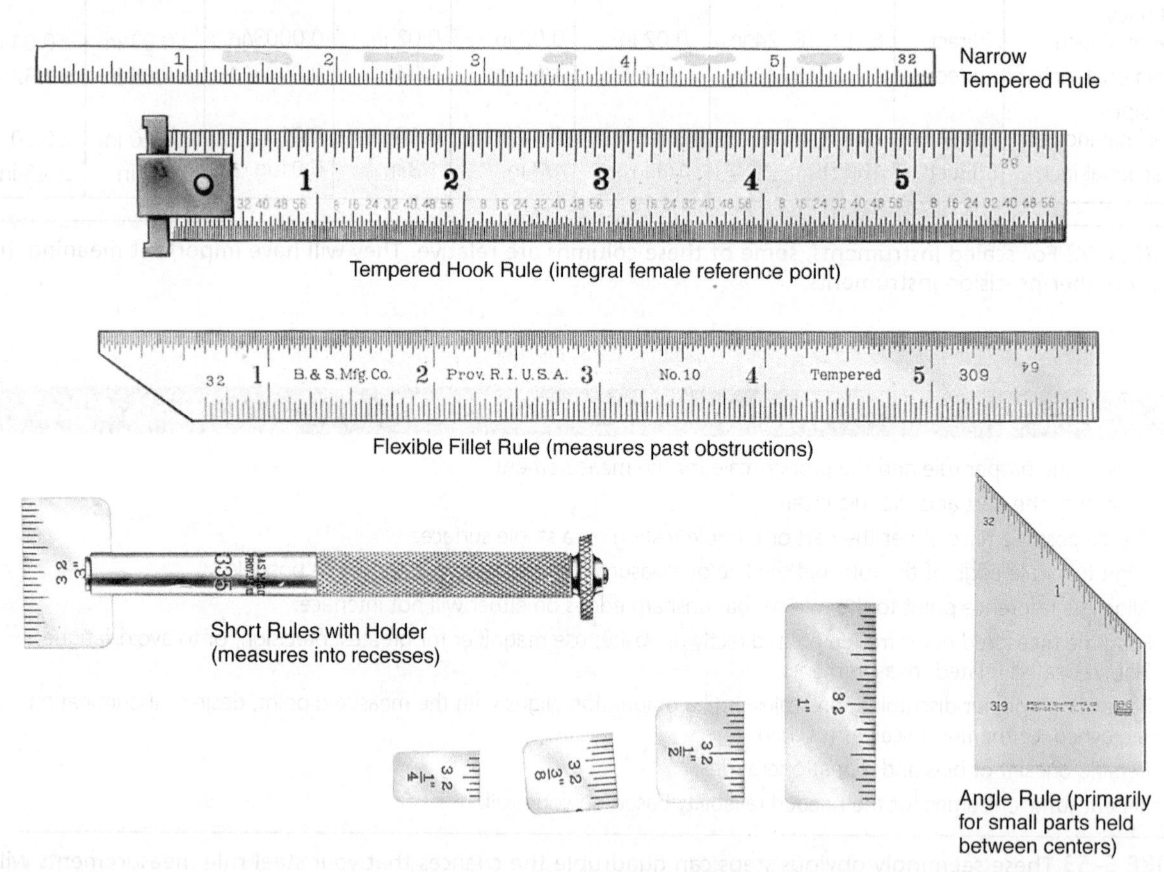

FIGURE 5–55 This is a small sampling of many available types of steel rules. *(Courtesy of Hexagon Metrology Inc. TESA)*

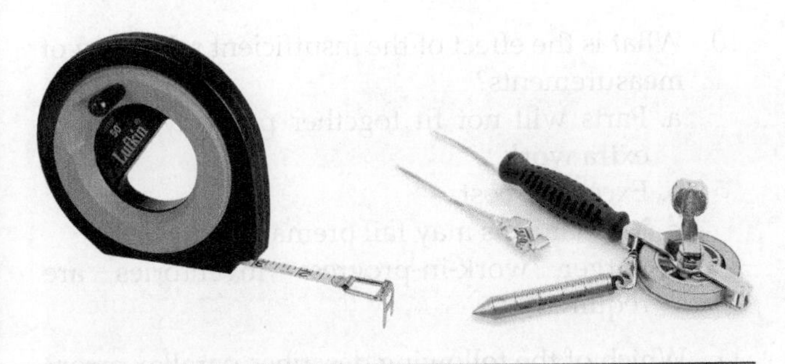

FIGURE 5–56 Steel tapes are extensions of steel rules and obey the same principles. For very precise measurement, temperature expansion and contraction must be considered. *(Courtesy of Apex Tool Group)*

error (parallax error) can introduce measurement error. Calipers are transfer instruments—they physically duplicate the distance between the reference point and the measurement point, which is then transferred to a scale. Transfer instruments add an extra step, introducing additional opportunity for measurement error.

- Because we need reliability, all steps must be assumed to add their maximum error, although some may cancel out. With sufficient information, this can be alleviated.
- Instruments that require contact, such as calipers, depend on the feel of the user for their precision. Cleanliness will promote reliability, but only skill will provide accuracy. This is a fundamental consideration.
- Any instrument incorporating a scale that does not provide a means for compensating for wear must always be suspect. This applies to the simple steel rule, the slide caliper, and to more sophisticated instruments.

END-OF-CHAPTER QUESTIONS

1. Scales and rules differ. Which of the following statements is correct?
   a. A scale is divided in proportion to a unit of length, whereas the divisions of a rule are the units of length.
   b. Rules are used by children in school, whereas scales are required for toolmakers and machinists.
   c. Advertising giveaways are rules, whereas expensive, precision-made equivalents are scales.
   d. Rules have divisions to 1/16 in., whereas scales have divisions as fine as 1/64 in.
2. What is the advantage of the scale?
   a. It fits all drafting machines.
   b. It is computer compatible.
   c. It enables large articles to be drawn on small sheets of paper and small articles to be magnified to fit large sheets of paper.
   d. It provides dimensions in feet.
3. Clarify the term scale by selecting the best of the following descriptions.
   a. Scales read in decimal-inches and metric units, whereas rules only read in fractional inches.
   b. The term scale must never be used for anything other than measuring instruments that are divided in proportion to their units of length.
   c. A scale may either be a measuring instrument or the measuring edge of a scaled instrument.
   d. A scale is the theoretical measuring instrument, whereas a rule is its practical embodiment.
4. Which of the following pairs relate most closely?
   a. Accuracy - discrimination
   b. Precision - accuracy
   c. Scale divisions - accuracy
   d. Scale subdivisions - discrimination
5. A large variety of attachments is available for use with scaled instruments. Which of the following statements about them is most essential?
   a. They simplify measurement but are hard to find when you need them.
   b. They are a waste of money once measuring skills have been developed.
   c. They facilitate measurement but add errors.
   d. Some are essential and these are built into the instruments.

6. Interpolation refers to judging the next decimal point when the instrument is between divisions. When should this be used?
   a. Always, then round off
   b. Whenever the scale has divisions 10 times the size of the last decimal point of the dimension
   c. Whenever an instrument with finer divisions is not available
   d. When temperatures in the measuring environment fluctuate
7. Which of the following best describes the sources of measurement error when using a steel rule?
   a. Inherent instrument error, manipulative error, temperature error, and observational error
   b. Attachment error, manipulative error, temperature error, and observational error
   c. Attachment error, manipulative error, bias, and inherent instrument error
   d. Inherent instrument error, manipulative error, bias, and observational error
8. Bias seriously reduces the reliability of measurement with graduated scales. Why?
   a. Because we instinctively guess the correct measurement based on our experience.
   b. We have almost always seen or heard something that affected our reading of the scale.
   c. The 64th divisions on scales are harder to read than the 32nds; the 32nds harder than the 16ths, and so forth. Hence, we pick the larger divisions.
   d. Too coarse a scale is chosen; this requires interpolation, an act inducing bias.
9. Steel rules are not recommended for control of high production. Which of the following is the best explanation for this?
   a. They do not have sufficient discrimination for most production.
   b. They are more subject to manipulative error than most instruments.
   c. They are not sufficiently reliable.
   d. They are more subject to observational error than most instruments.
10. What is the effect of the insufficient reliability of measurements?
    a. Parts will not fit together properly without extra work.
    b. Excessive cost.
    c. Mechanisms may fail prematurely.
    d. Larger work-in-progress inventories are required.
11. Which of the following describes parallax error?
    a. It is another name for observational error.
    b. It is the apparent shifting of an object when the position of the observer is changed.
    c. It is error caused by the distance between the scale and the feature being measured.
    d. It is the difference in measurement that occurs when the object is viewed from the left instead of the right or vice versa.
12. Which of the following methods is the best way to eliminate parallax error?
    a. Close one eye when taking the measurement.
    b. Align the rule so that the scale being read is along the line of measurement.
    c. Center the scale along the feature being measured.
    d. Take the measurements first from one side, then from the other side, and then average them.
13. When selecting a steel rule for a given measurement, which of the following combinations is preferable?
    a. Engine-engraved scales, shortest length that contains the measurement yet can readily be handled, a scale that discriminates to the same precision that is required in the measurement
    b. Made of the same material as the workpiece, shortest length that contains the measurement yet can be readily handled, a scale that discriminates to the same precision that is required for the measurement
    c. Engine-engraved scales, length at least 1½ times the required measurement, a scale that discriminates to the same precision that is required in the measurement

d. Engine-engraved scales, same material as the workpiece, at the same temperature as the workpiece, a scale that discriminates to the same precision that is required in the measurement

14. One important use for measurement instruments is layout. Which of the following is the best explanation of this?
   a. Layout is required for all mass production.
   b. All cylindrical and spherical parts require layout in order to be manufactured.
   c. Layout is another form of inspection.
   d. It is a preliminary operation that facilitates final machining.

15. When the steel rule is used for layout, various attachments are often used with it. Why?
   a. They enable interpolation to be used for the measurement readings.
   b. They prevent the unskilled from taking measurements that might be unreliable.
   c. Convenience.
   d. They prevent confusion between the reference end and the measured end.

16. The depth gage is the instrument of choice for the measurement of slots, recesses, and blind holes. Which of the following is the best explanation for this?
   a. Its blade is narrow but the graduations are easily read.
   b. Its graduations are easily read and most depth gages can be set to 30°, 45°, and 60° angles.
   c. Its blade is narrow enough to enter most openings and its head provides a right angle reference to the part.
   d. Its head rests on the two sides of the opening, which makes it easy to handle.

17. Which of the following best explains the importance of the combination square set?
   a. It is a universal tool suitable for both production inspection and individual measurements in toolmaking, and so forth.
   b. Only the combination square has blades with four separate scales.
   c. It facilitates a number of different measurement techniques and exemplifies the basic principles of dimensional metrology.
   d. It is more useful for locating centers of round parts than any of the other scaled instruments.

18. The combination square is sometimes used instead of a depth gage. Which of the following comments about this practice is most important?
   a. It provides greater range and is often more convenient. However, it also has greater potential for error.
   b. Under no condition should it be tolerated unless readings to 1/16 in. or greater are acceptable.
   c. It is more convenient because of its blades, which have four scales.
   d. It provides greater range together with freedom from parallax errors.

19. What is the chief limitation in the use of the center head?
   a. It is limited to round bars.
   b. You cannot measure from the end of a scale but must select a division and set the blade to it.
   c. You can only use one of the four scales of the blade at one time.
   d. Parallax error.

20. What is the chief reason for studying calipers?
   a. They are essential production inspection instruments.
   b. They are one of the oldest known instruments.
   c. They are one of the few instruments used both for inspection of production and skilled work such as tool, die, pattern, and model making.
   d. They demonstrate the principle of transfer measurement.

21. Calipers transfer measurement from a workpiece to a scale. Which of the following statements relates most closely to that requirement?
   a. All measurements are transferred from a workpiece to a scale.

b. Two potential errors are added. One is the error from the feel of the engagement of the caliper on the workpiece. The other is the fact that every added step adds a new potential for error.
c. Caliper measurements are vulnerable to parallax errors.
d. The caliper method depends more on the inherent accuracy of the scale than do other types of measurements with scales.

22. Some calipers have locks so that they may be locked to a fixed opening. Which of the following best describes the proper use of this feature?
a. It keeps them from opening up when not in use.
b. It provides a memory of the last measurement.
c. It enables them to be used as snap gages.
d. It prevents the measurement from being lost if the caliper is dropped.

23. Why was a discussion about snap gages included?
a. Because calipers can properly be used as snap gages.
b. Because calipers and snap gages are actually the same instruments but snap gages are built heavier.
c. Because calipers are called snap gages when used for production inspection.
d. Because we needed to distinguish between gaging and measurement.

24. Why are calipers important for both layout and machining?
a. Both require transfer measurements.
b. Both involve production inspection.
c. For accuracy, the machining should be inspected with the same instruments used for its layout.
d. Calipers are the only instrument almost certain to be on hand for both activities.

25. Which of the following is the most important relationship between dividers and calipers?
a. Calipers have locks, whereas dividers do not.
b. Dividers are used for layout, whereas calipers are used for machining.
c. Dividers are calipers intended to be used with line references.
d. Calipers are used to 64ths, whereas dividers may only be used to 32nds.

26. Why are measurements with an inside caliper considered to be more reliable than those taken with an outside caliper?
a. Because you cannot get an outside caliper into a recess
b. Because the concave surfaces help center the caliper
c. Because inside measurements are smaller than outside measurements
d. Because inside surfaces are usually machined better than outside surfaces

27. When using calipers, which of the several variables must be most carefully controlled?
a. The selection of the proper scale
b. The size of the selected caliper
c. The feel between the legs and the workpiece
d. The slippage of a caliper joint

28. Select the group of terms in which all items are compatible.
a. Dividers, steel rule, scriber, surface gage
b. Dividers, steel rule, surface gage, combination square
c. Dividers, beam trammel, surface gage, combination square
d. Beam trammel, surface gage, hermaphrodite caliper, common scriber

29. Which of the following is commonly used with the surface gage?
a. Steel rule
b. Combination square
c. Common scriber
d. Surface plate

30. Which of the following is not a desirable feature of the slide caliper?
a. It provides positive contact between the instrument and the reference.
b. You can adjust the instrument for wear.
c. It substitutes a line for a part feature to use as a measured point.
d. It is quick and easy to use.

CHAPTER SIX

VERNIER INSTRUMENTS

LEARNING OBJECTIVES

- Briefly describe the vernier family of instruments.
- Read vernier instruments.
- Recognize the differences between precision, accuracy, and reliability for these instruments.
- Identify the positional problems inherent in measurement.

OVERVIEW

Today, we make our most important measurements in millimeters and thousandths of an inch. These measurements may not put space vehicles in orbit, but they do provide justification for funding such projects. They keep the modern, mass-production economy alive and progressing.

We obtain these measurements today with digital readout instruments and dial calipers, but these modern instruments are based on the basic metrological principles pioneered by the use of vernier instruments in either the English inch or metric system. If you are able to measure reliably with vernier instruments, you are unlikely to make the common errors associated with their modern, digital versions. In addition, you probably will not need special instructions to reliably use these modern instruments, because you understand the basic principles of measurements with vernier calipers.

Vernier instruments are used most often in toolrooms, model making, and laboratory work, but rarely for modern production inspection. They are referred to as one of the "nonprecision" instruments, not for their lack of precision, but because they lack the amplification of other instruments.

In the previous chapter, we saw how our own senses—sight and feel—limited reliability of the measurements we obtained. We can, however, increase the accuracy of these senses with mechanical, electronic, and optical enhancements. One of the simplest, the *vernier*, is a system of scales invented by Pierre Vernier (1580–1637) in 1631. Vernier scales offer a fine discrimination, which we use by lining up two marks on the instrument.

6-1 VERNIER INSTRUMENTS

Vernier instruments today also include height gages (see Figure 6–1), depth gages, gear tooth instruments, and protractors. These simple scale instruments all use verniers to increase their amplification, which, in turn, increases their discrimination. Adding verniers to instruments that use other means of amplification, such as micrometers (see Chapter 7), increases their precision.

When a vernier scale is attached to an instrument, it slides parallel to the line of measurement (see Figure 6–2), so that the main scale also slides parallel to the line of measurement. Scales are mounted so you can make readings from one scale to the other with a minimum of parallax error.

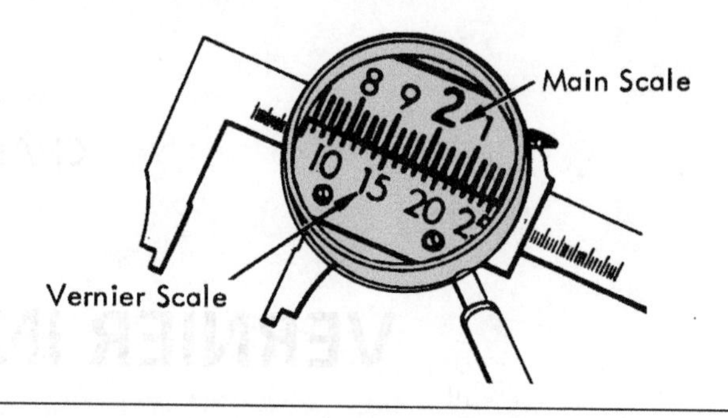

FIGURE 6–2 The simple addition of a vernier scale adds amplification to the scale reading.

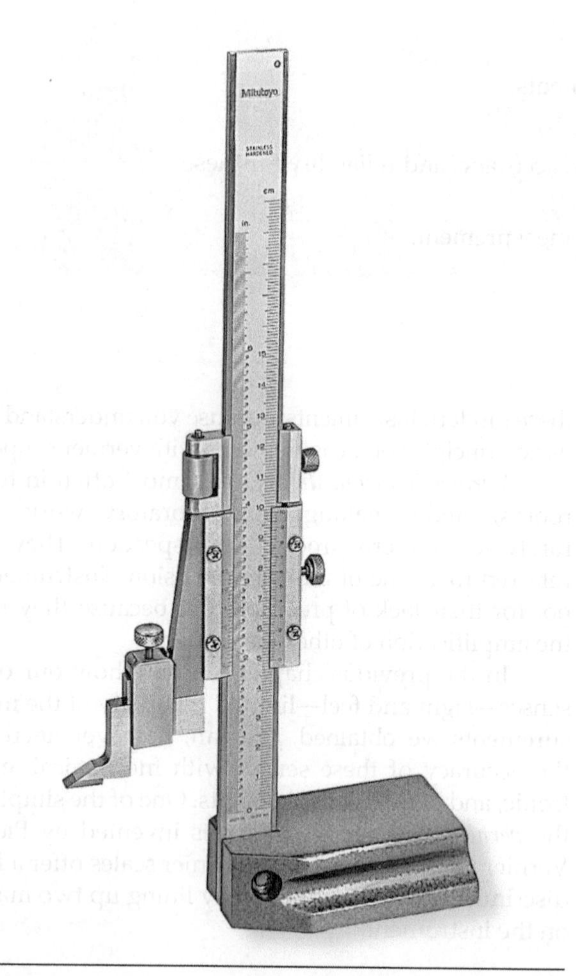

FIGURE 6–1 Vernier height gages are among the most popular measurement instruments. *(Courtesy of Mitutoyo American Corporation)*

For example, we can use a vernier scale and a main scale in decimal-inches—the vernier has been a decimal instrument from its inception. Each inch is divided into 10 parts, and each division is subdivided into quarters (see Figure 6–3), making the smallest division 1/40 in., or 0.025 in., read as 25 mil.

The vernier scale on a caliper is attached to the sliding jaw and moves along the main scale. The inch vernier scale has 25 divisions in the same length that the main scale has 24 divisions. The difference between a main scale division and a vernier division is 1/25 of a main scale division. 1/25 of 0.025 in. equals 0.001 in. or 1 mil, which is the discrimination of the instrument or the *least count*. You read a vernier instrument or the controls on machines by adding the total readings from the main scale and the thousandths readings on the *vernier scale* (see Figures 6–3 and 6–4). Remember, only when the vernier is at zero do two lines, 0 and 25, of the vernier scale coincide with the main scale.

Take these steps to read a vernier instrument:

1. Read the number of whole divisions on the main scale that appear *to the left of zero* (0) on the vernier.
2. Read the largest numbered graduation on the main scale that lies *to the right of the index* (0) on the vernier scale. Read them as even 100 mil (0.100, 0.200, and so forth). Add to the whole reading from step 1 (1.100, 2.100, and so forth).

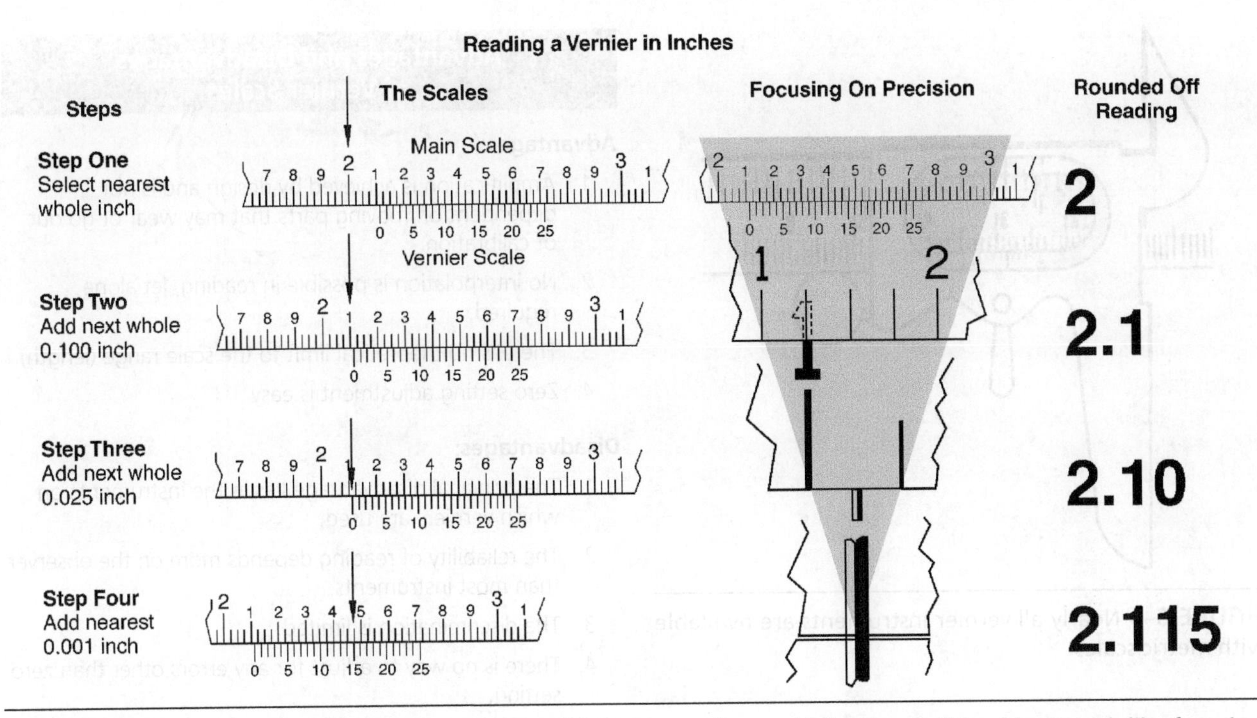

FIGURE 6–3 Each of the steps for reading a vernier raises the results to a higher level of precision, much like focusing a microscope.

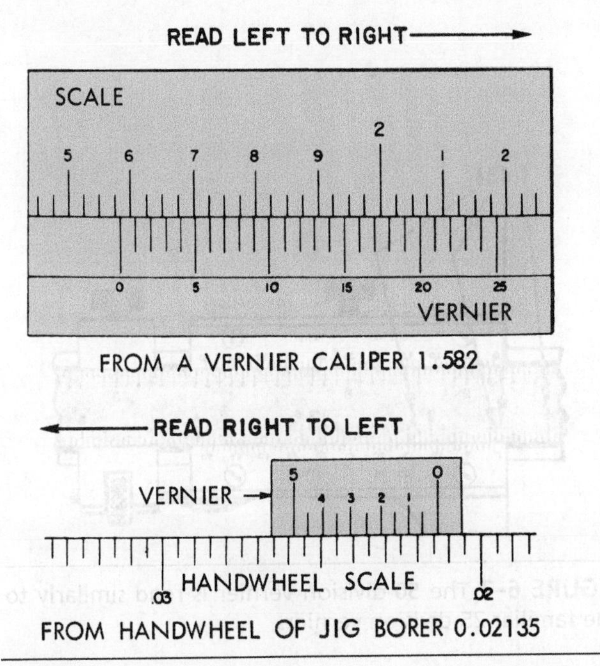

FIGURE 6–4 Practice is required for reliable reading of vernier scales.

3. Read the largest whole minor division to the right of the index. Read these graduations in increments of 25 mil (0.000, 0.025, 0.050, and 0.075 in.). Add to the sum of steps 1 and 2 (1.125, 1.150 in., and so forth).
4. Find the vernier graduation that most exactly coincides with any graduation on the main scale. This measurement is the nearest mil and may be any whole number from zero to 25 (0.000, 0.001. . . 0.024, 0.025). Add the result to the sum of the previous three steps.

For metric verniers, each graduation (see Figure 6–5) on the main bar is 1.0 mm, and every tenth graduation is numbered 10 mm, 20 mm, and so forth. The vernier plate is divided into 50 graduations, each representing 0.02 mm. Every fifth line is marked in sequence from 0.10 mm, 0.20 mm, and so forth, to 0.50 mm. With these graduations, you can read directly in hundredths of a millimeter.

In Figure 6–5, read the number of millimeters between the 0 line on the main stationary bar and the 0 line on the vernier plate, 27 mm. Find the graduation

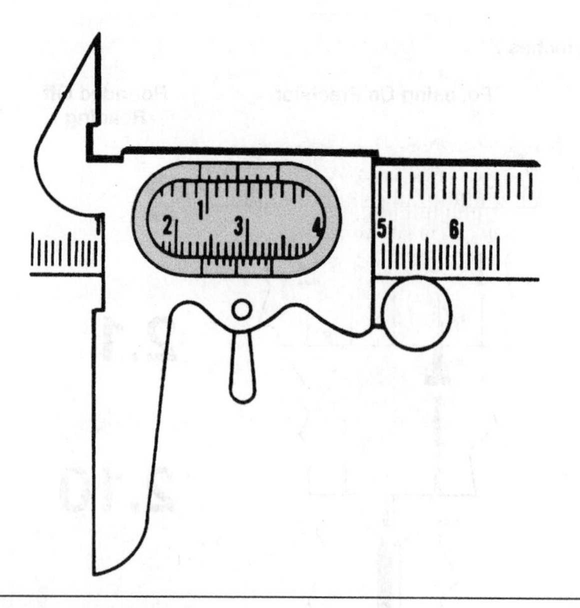

FIGURE 6–5 Nearly all vernier instruments are available with metric scales.

on the vernier plate that coincides exactly with a line on the stationary bar, 0.42 mm, which is the hundredths of a millimeter reading. Add the two values to achieve the total measurement: 27.00 mm on the bar + 0.42 mm from the vernier plate = 27.42 mm total measurement.

We have listed the advantages and disadvantages for vernier instruments in Figure 6–6. Essentially, vernier scales make the instruments they are used on more reliable. You might wonder why discrimination of vernier instruments is limited. Again, our senses limit our ability to measure. A vernier designed to read to 0.0001 from an inch scale would have lines grouped so closely together (lines so nearly in coincidence) that they would be indistinguishable. Many modern vernier instruments do have expanded scales; they are meant to achieve greater reliability through improved readability.

The 50-division vernier (see Figure 6–7) makes use of the amazing ability of the eye to bring lines into coincidence. You would need vision to be twice as sharp as what you would need for a 25-division

Advantages and Disadvantages of Vernier Scales

Advantages:

1. Amplification is achieved by design and is not dependent on moving parts that may wear or go out of calibration.
2. No interpolation is possible in reading, let alone required.
3. There is no theoretical limit to the scale range (length).
4. Zero setting adjustment is easy.

Disadvantages:

1. The principal disadvantage lies in the instruments on which verniers are used.
2. The reliability of reading depends more on the observer than most instruments.
3. The discrimination is limited.
4. There is no way to adjust for any errors other than zero setting.

FIGURE 6–6 Important note: These advantages and disadvantages apply to vernier scales, not to complete vernier instruments.

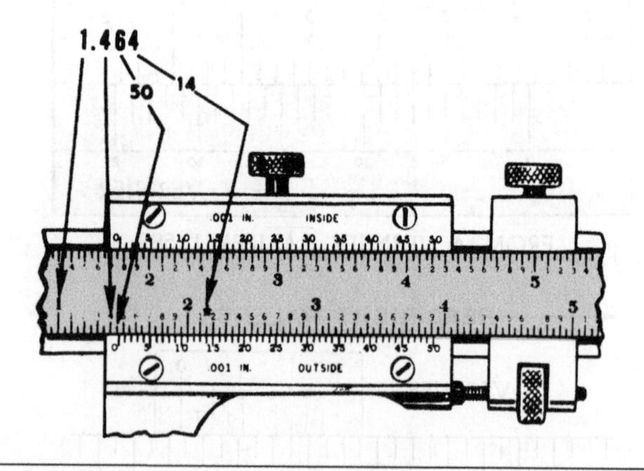

FIGURE 6–7 The 50-division vernier is read similarly to the familiar 25-division verniers.

vernier, but the 50-division is easy to use. It does not, however, increase the accuracy of measurement (which is a function of instrument construction) or the discrimination, but it does increase the reliability. Another adaptation of the Vernier principle is to protractor scales (see Chapter 12).

6–2 VERNIER CALIPER

Although they are the simplest of vernier instruments, vernier calipers are important in toolroom, die-making, model-making, and similar applications. They provide long measurement ranges (6 to 80 in.) and are economical. You can find vernier calipers in use in virtually all metrology labs. Understanding how they work provides a foundation for measurement by other instruments: height gages, vernier micrometers, dial and digital calipers. One pair of vernier calipers can *substitute* for many outside and inside micrometers, but they cannot *replace* them.

Vernier calipers are slide calipers with a vernier scale attached. Because the vernier increases its discrimination to 0.02 mm (1 mil, 0.001 in.), we need a fine adjustment and a means for a zero setting (see Figure 6–8). For convenience, some instruments are graduated on one side for inside measurement and on the other side for outside measurement. The jaws of most vernier calipers are reduced to thin nibs for small inside measurement.

You can make reliable measurements with the vernier caliper, but it is not easy. The reason why is a fundamental of all highly reliable measurement and is the most important point of this entire chapter: It has to do with alignment. Vernier calipers demonstrate clearly how many problems bad alignment can create.

Abbe's Law

Abbe's Law was one of the first fruits of the Industrial Revolution. It came as a result of the work of Ernest Abbe, a remarkable physicist and leading industrialist.

After graduating from the university at Jena, Abbe stayed on as an instructor. In 1870, he became a professor of physics, and, in 1878, was named the director of observatories. At Jena, Abbe met Carl Zeiss, founder of the famous Zeiss works. When Zeiss died in 1888, he left his entire business to Professor Abbe, which Abbe reorganized so that a portion of the profits went to both the employees and the university. With his own money, Abbe founded the Carl Zeiss-Stiftung, an organization that continues to promote

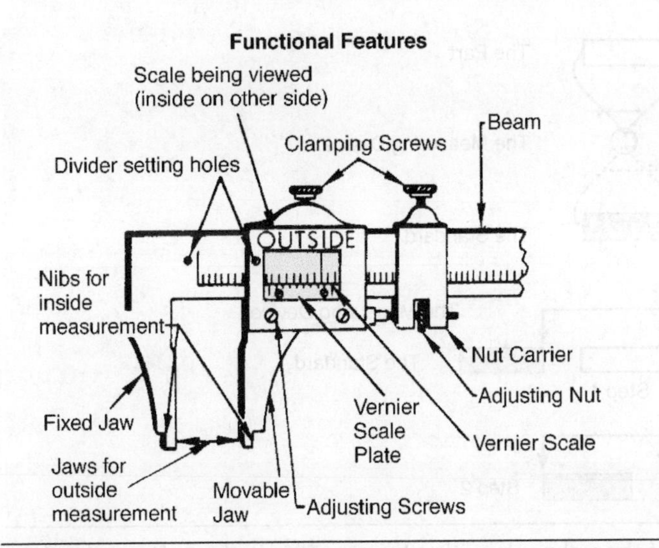

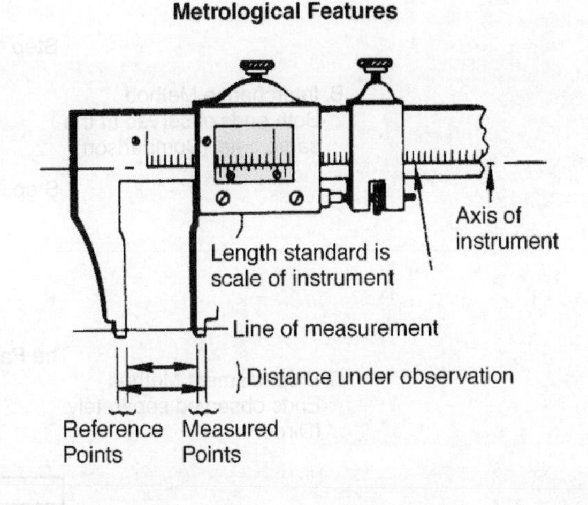

FIGURE 6–8 The vernier caliper is a slide caliper with trimmings.

scientific research and social betterment. Abbe still found time for his scientific work, primarily working in optics.

"Abbe's Law," "Abbe's Principle," or more often, "The Comparator Principle," is the result of some of Ernest Abbe's work. The law can be paraphrased as follows: *maximum accuracy may be obtained only when the standard is in line with the axis of the part being measured.*

To understand the implications of Abbe's Law, consider the fundamental premise of measurement: *If two different quantities each equal a third, they must be equal to each other (A = B = C).* In all measurement, you must have a standard, a feature to measure, and a measurement device to use to compare the two (see Figures 6–7 and 6–9A). With scaled instruments, such as a vernier caliper, there is little distinction between the standard and the device, so the reliability of the result depends on how you relate the standard to the part—either with the *interchange method* or the *displacement method.*

The Interchange Method

Until now, we have used the interchange method most often. With it, you observe both ends of the measured length at the same time, which is exactly how we use caliper instruments. The caliper observes both ends of

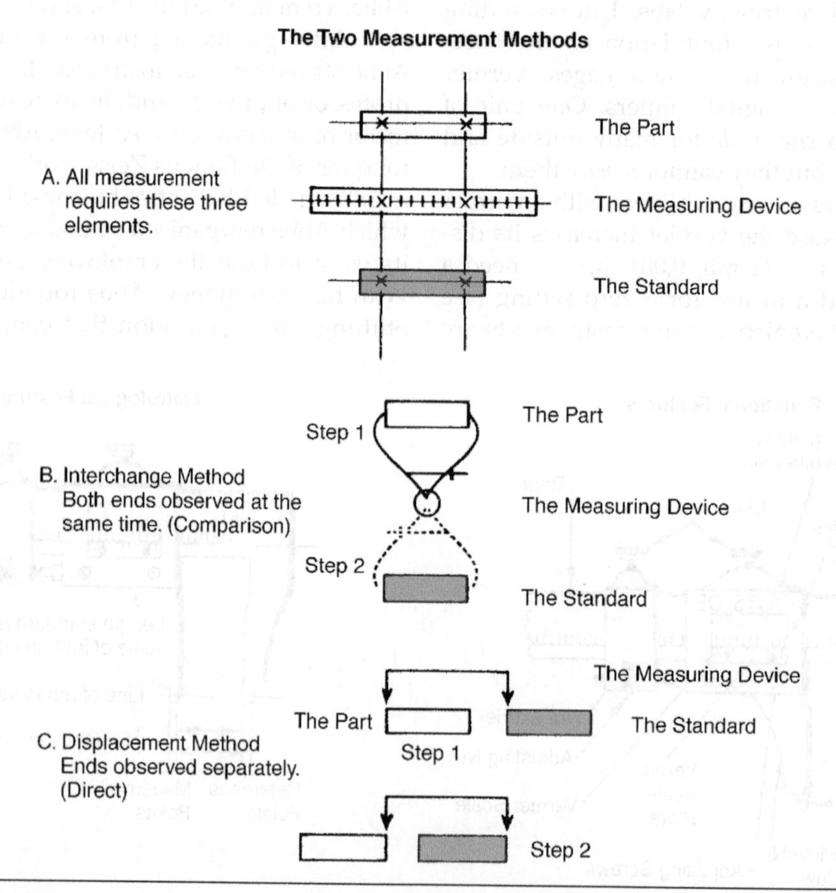

FIGURE 6–9 All measurement consists of comparison of the unknown with a known. The methods for comparison vary but all fall into either B or C. Vernier calipers are of the displacement type.

the part (see Figures 6–6A and 6–9B) and then compares the separation with both ends of the standard simultaneously. This measurement method is the basis of *transverse-comparators,* whether mechanical, optical, pneumatic, or electronic (see Chapters 10 and 14).

The Displacement Method

With the displacement method, we use the same longitudinal (lengthwise) movement on both the part and the standard and relate these two movements to each other. The displacement method is different from the interchange method in that you only observe one end of the measured length at one time. The part and the standard are both lined up with an index mark, then they are moved together. You then see if another set of their reference points lines up with the same index mark (see Figures 6–6B and 6–9C), and if they do, their lengths are the same. We use the displacement method primarily for positioning in machine tools and measuring machines.

The vernier caliper shows us simply how the displacement method is used and how difficult it is to follow Abbe's law. You need to take the steps shown in Figure 6–10 in order to make a displacement measurement with the vernier caliper. Notice the similarity to Figure 6–9C, except that the *measurement device* is the instrument, and the instrument contains both the standard and the means for relating the standard to the part.

To position the standard and the part (step 1), you simply close the jaws and zero the setting because the vernier caliper is a contact instrument. Any movement of the caliper jaw is a longitudinal displacement. The part is not physically being measured, but it is still, in some way, being observed. In step 2, you observe the other ends. When you observe the part, you need to make physical contact between the part and movable jaw. For the standard, we accept the vernier reading, and, in practice, we conclude that the reading is the desired length. However, there are always errors.

For the reading you obtain to be the exact length, the beam of the caliper would have to be perfectly straight, and the jaw would need to be absolutely at 90° to the beam. However, it is impossible to make a perfectly straight beam (or scale standard). In addition,

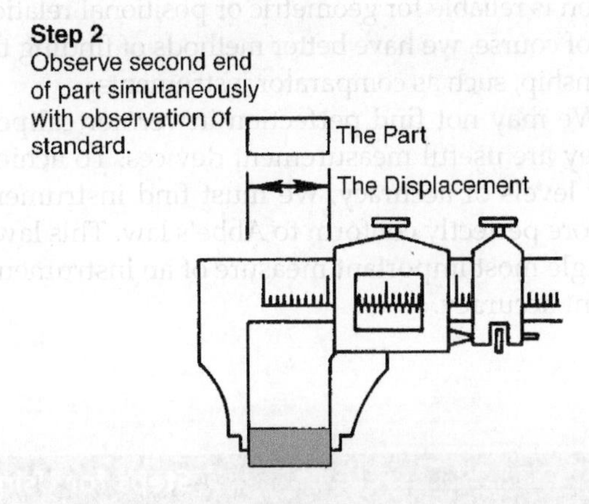

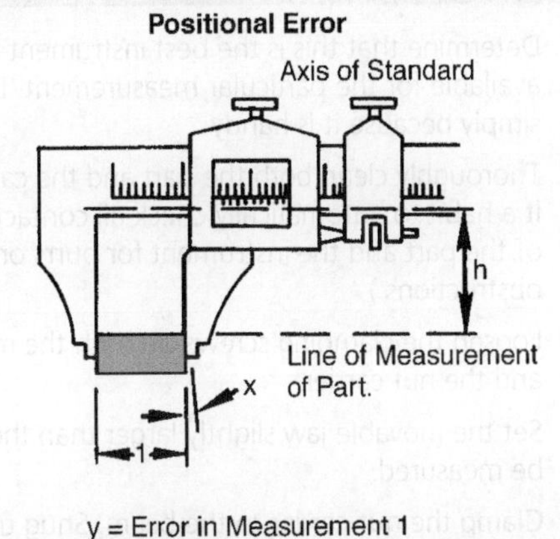

FIGURE 6–10 The vernier caliper is shown to be a displacement instrument, but one that does not conform to Abbe's law. X can never be zero and neither can h.

even if the jaw was nearly perfectly square, the jaw must "slide" along the beam; therefore, there must be some play between the beam and jaw, which means that, at any given moment, the jaw may be tilted slightly.

Obviously, perfection is not the goal of vernier calipers, but we do hope to obtain an instrument reading that contains zero error. We need two conditions to make zero errors possible: the angularity error (x) must equal zero, and the distance (h) must equal zero. The first condition cannot happen, so we can only manipulate the second condition. When h equals zero, the standard is in line with the line of measurement. The measurement situation is reliable for geometric or positional relationships; of course, we have better methods of finding this relationship, such as comparator instruments.

We may not find perfection in vernier calipers, but they are useful measurement devices. To achieve higher levels of accuracy, we must find instruments that more perfectly conform to Abbe's law. This law is the single most important measure of an instrument's inherent accuracy.

The Importance of the Clamping Screw

The manufacturers of vernier calipers have tried to minimize x error by adding a clamping screw that locks the movable jaw to the beam. The addition of a clamping screw to vernier calipers causes many people to think of them as a memory instrument—a means to retain a reading. Still others believe that the clamping screw simply prevents a reading from moving between the time the caliper is set to a part feature and the time the part is removed.

Measurement with Vernier Calipers

You can make reliable measurements with English or metric vernier calipers. You should take certain precautions, however, which apply to vernier calipers and instruments of even higher amplification (see Figure 6–11). Steps 1, 2, 10, 11, 12, and 14 recur throughout measurement and are among the most important guidelines you will learn in this book.

Steps for Using the Vernier Caliper

1. Determine that this is the best instrument you have available for the particular measurement. Do not use simply because it is handy.
2. Thoroughly clean both the part and the caliper. (Make it a habit to automatically check all contact surfaces of the part and the instrument for burrs or other obstructions.)
3. Loosen the clamping screws on both the movable jaw and the nut carrier.
4. Set the movable jaw slightly larger than the feature to be measured.
5. Clamp the nut carrier to the beam. Snug up but do not lock the clamping screw on the movable jaw.
6. Place the fixed jaw in contact with the reference point of the part feature (see Figure 6–15).
7. Align the beam of the caliper to be as nearly parallel to the line of measurement as possible—in both planes.
8. Turn the adjusting nut so that the movable jaw just touches the part. Tighten the clamp screw on the movable jaw without disturbing the feel between the caliper and the part.
9. Read in place without disturbing part of the caliper, if possible. If not, remove the caliper.
10. Record the reading on scratch paper, chalk on the part, or on part drawing. Do not trust memory.
11. Repeat the measurement steps a sufficient number of times to rule out any obviously incorrect readings and average the others for the desired measurement.
12. Loosen both clamps, slide the movable jaw open, and remove the work if not already done.
13. Clean, lubricate, and replace the instrument in its box.
14. Ask yourself what errors may remain in your measurement.

FIGURE 6–11 These steps apply to most vernier caliper measurements. Do not slight those that appear obvious.

These steps help you save time and apply to both English and metric vernier devices.

You make inside and outside measurements using similar steps, but it is important that you remember to read the proper side of the caliper for the measurement you are taking.

Alignment Consideration

You must be very careful of the alignment between a vernier caliper and the part (see Figure 6–12) because the geometry is vital to reliable measurement (see Figure 6–13). Of course, when we measure to such fine discrimination, like 1 mil, it is almost impossible to locate a misalignment with our eyes. In most cases, you have to rely on the *feel*. That is why it is so important for you to understand gaging feel and to develop your own sensitivity to it. The accompanying lab manual has practice exercises for measuring with a vernier caliper that will help you develop that sensitivity.

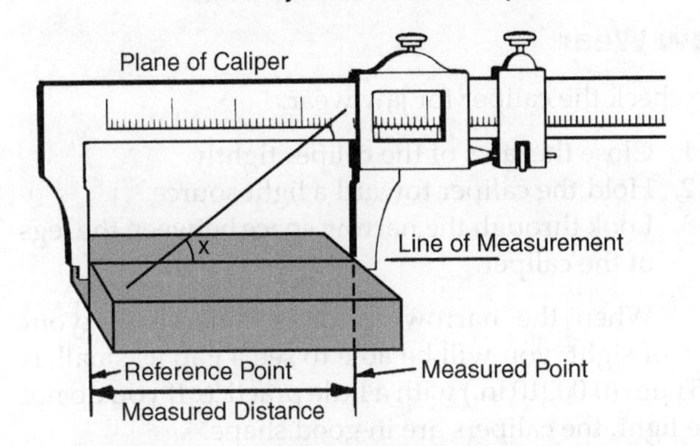

FIGURE 6–12 The four essential points.

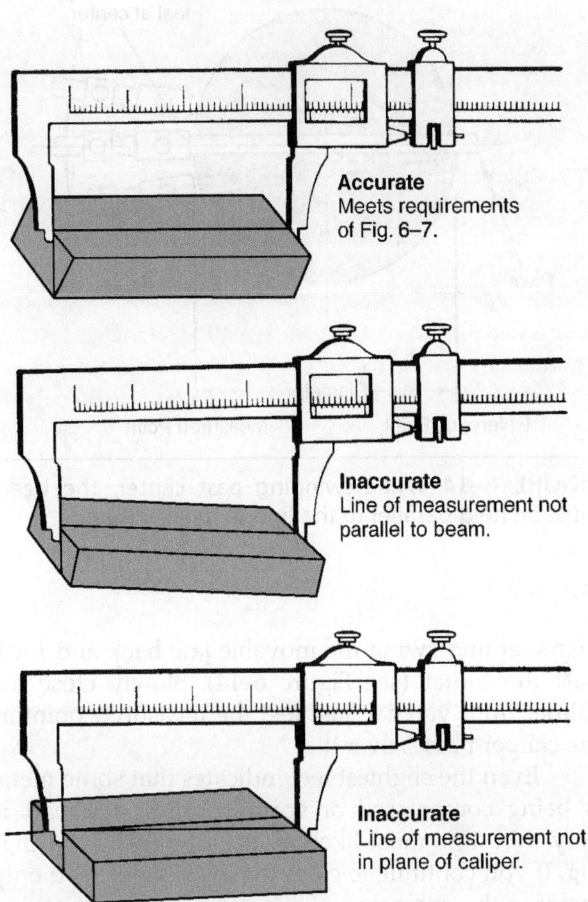

FIGURE 6–13 Because it does not comply with Abbe's law, positional considerations are of great importance when using the vernier caliper.

Outside Diameter

You will probably find it easier to measure a diameter than the distance between flat surfaces because the diameter is the greatest distance between the reference point and the measured point. You might have a parallelism misalignment, which could result in a reading longer than the true value (a plus error), but most misalignment errors will be shorter (a minus error). To measure an outside diameter, you must hold the caliper so it is balanced and does not tend to upend. Hold the fixed jaw against the edge of the

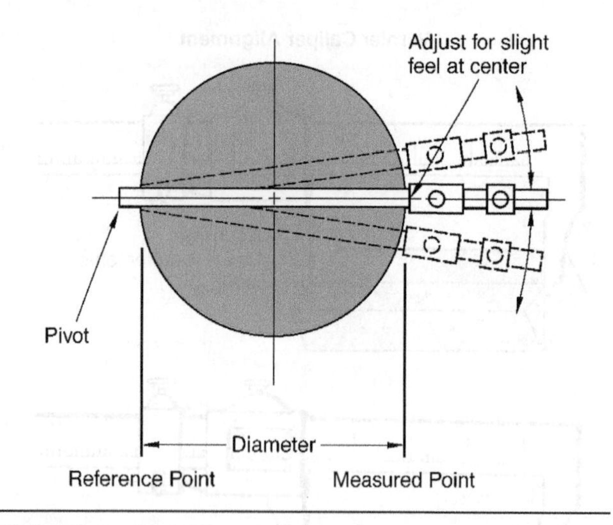

FIGURE 6–14 While swinging past center, the beam must be held parallel to the line of measurement.

diameter and swing the movable jaw back and forth past the center (see Figure 6–14). Slowly close the caliper until you can just feel the measured point as the caliper passes over it.

Even the slightest feel indicates that some metal is being compressed or sprung; unless the part is very thin, it is most likely that the caliper is springing. If you continue to close the instrument, you only increase the springing of the caliper. Heavy gaging pressure causes the jaws to wear more rapidly, burnishes the part, or even damages the caliper.

Inside Diameter

When you measure an inside diameter, you need to use a more careful touch, but the measurement principles are the same. When you first feel the work, a slight misalignment could cause a plus error, so you should carefully move the jaw in a small circle in all directions. When you are closer to the line of measurement, the movement of the instrument will feel more free again; then open the adjustment until you pick up the feel again. Move the jaw again—this time the circle will be smaller—and, when you approach the line of measurement, make another adjustment. Continue this process until the jaws are squarely on the measured and reference points of the part feature. This technique is called centralizing, and the best way to learn it is to practice.

Flat Surfaces

As we pointed out during our discussion of outside calipers, measuring between flat surfaces is much more difficult because large vernier calipers apply tremendous leverage to compress or spring both to the part and to the caliper itself. For example, Figure 6–15 shows that 0.002 in. (50 μm) curvature in 10 in. (25.4 cm) shortens the reading by 0.0016 in. (406 μm).

Accuracy Checks for Vernier Calipers' Manipulative Error

The most important cause of unreliable vernier caliper readings is manipulation during measurement (see Figure 6–16). You should check, or calibrate, your instruments frequently, to ensure that instrument defects do not lead to additional errors.

Jaw Wear

To check the caliper for jaw wear:

1. Close the jaws of the caliper tightly.
2. Hold the caliper toward a light source.
3. Look through the narrow space between the legs of the caliper.

When the narrow space is squarely in your line of sight, you will be able to see a gap as small as 2.54 μm (0.00010 in.) with a little practice. If you do not see light, the calipers are in good shape.

How much light can you see between the jaws of a caliper before they should be repaired? Books recommend that when the wear exceeds 5 μm (0.0002 in.), you should return the caliper to the manufacturer for repair. But how do you identify a gap that is 5 μm (0.0002 in.) wide? Fortunately, there are other ways to measure wear: optical flats, for example, provide a quick and easy check. To be practical, you should check the light gap periodically, and, if you suspect

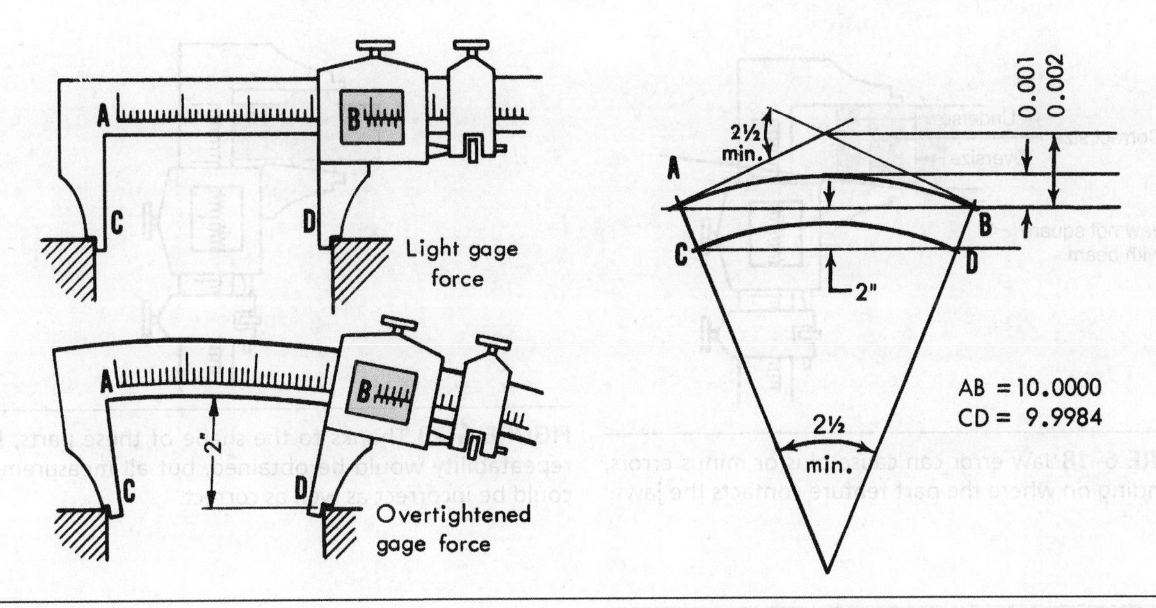

FIGURE 6–15 Overtightening causes beams to bow. Only 0.002 in. bow in 10 in. will shorten the reading by 0.016.

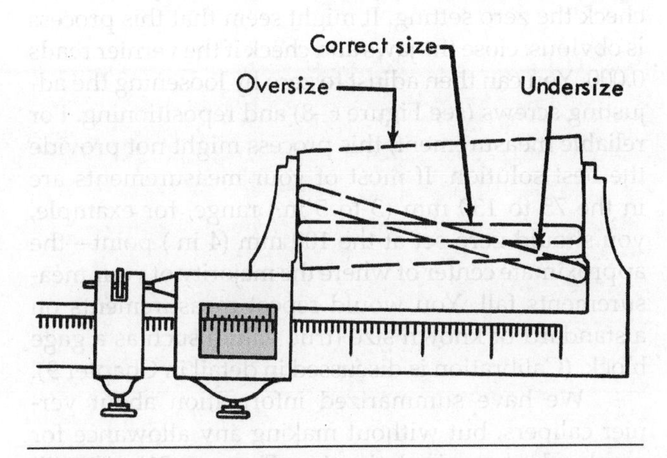

FIGURE 6–16 Manipulation of the part and instrument is by far the greatest cause of errors with vernier calipers.

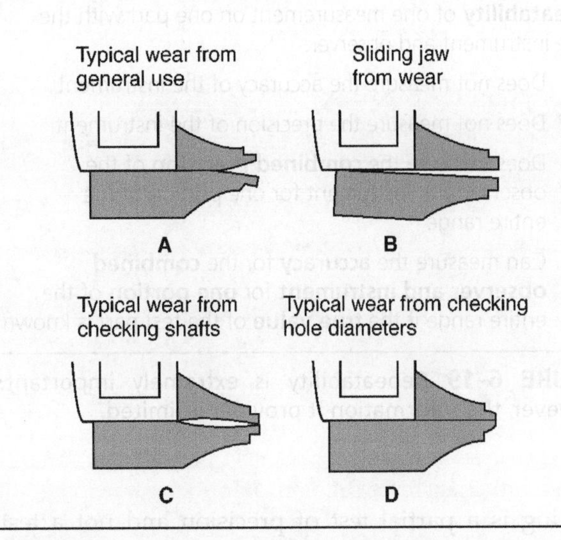

FIGURE 6–17 Wear may take very different forms, as shown in these exaggerated examples.

wear, consult with an experienced gage maker. The effect of wear varies with the use of the calipers; typical wear patterns (exaggerated so that they are easier to see) are shown in Figure 6–17. Figure 6–18 shows the effect of either type A or B jaws.

Repeatability

We frequently use repeated measurements to check for accuracy. Although they are a valuable means for checking instruments, simple repetition of the same

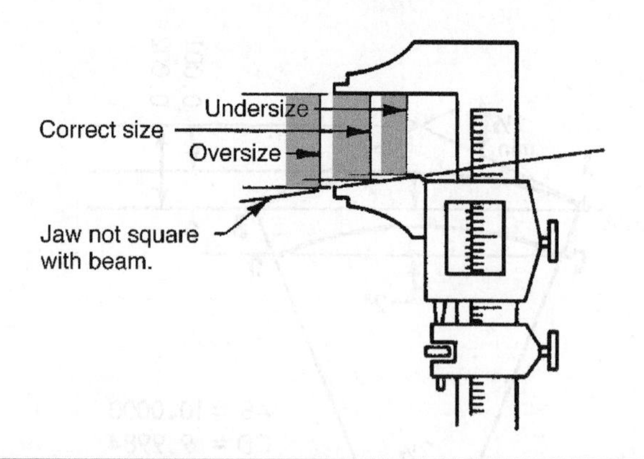

FIGURE 6–18 Jaw error can cause plus or minus errors, depending on where the part feature contacts the jaws.

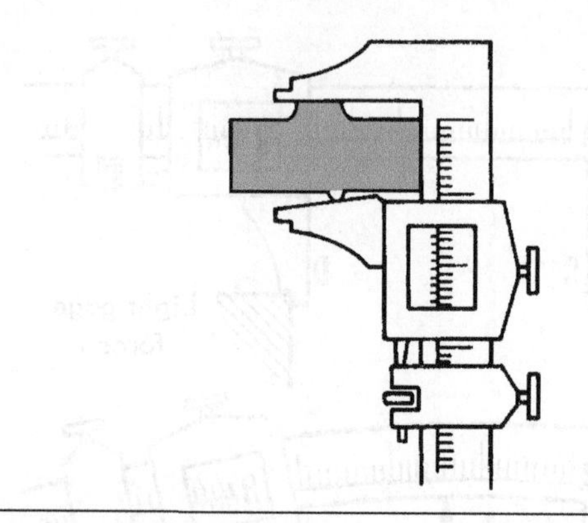

FIGURE 6–20 Thanks to the shape of these parts, high repeatability would be obtained, but all measurements could be incorrect as well as correct.

Repeatability, Accuracy, and Precision

Repeatability of one measurement on one part with the same instrument and observer:

1. Does not measure the accuracy of the instrument
2. Does not measure the precision of the instrument
3. Does measure the **combined precision** of the observer and instrument for one portion of the entire range
4. Can measure the **accuracy** for the **combined observer and instrument** for **one portion** of the entire range if the **true value** of the test part is known

FIGURE 6–19 Repeatability is extremely important. However, the information it provides is limited.

reading is a partial test of precision and not a test of accuracy at all. If the true value of the test part is known, repeated measurements only check the accuracy on that one measurement (see Figure 6–19).

You can repeat incorrect measurements very reliably under certain conditions (see Figure 6–20). Therefore, you should perform repeatability tests with standards (gage blocks or master cylinders) of known value.

Zero Setting

After checking the caliper jaws for wear, you should check the zero setting. It might seem that this process is obvious: close the jaws and check if the vernier reads 0.000. You can then adjust to zero by loosening the adjusting screws (see Figure 6–8) and repositioning. For reliable measurement, this process might not provide the best solution. If most of your measurements are in the 75 to 130 mm (3 to 5 in.) range, for example, you should zero set at the 100 mm (4 in.) point—the approximate center of where the majority of your measurements fall. You would repeat measurements on a standard of known size (true value) such as a gage block. (Calibration is discussed in detail in Chapter 9).

We have summarized information about vernier calipers, but without making any allowance for good or bad manipulation (see Figure 6–21). We will include a table like this one for each important instrument, because the table will help you select the best instrument for the job, whether in the physics laboratory or on the shop floor.

Dial Calipers

The vernier caliper, of course, is an extension of the slide caliper. By adding vernier scales, we increase the

Metrological Data for Vernier Instruments							Reliability	
Instrument	**Type of Measurement**	**Normal Range**	**Designated Precision**	**Discrimination**	**Sensitivity**	**Linearity**	**Practical Tolerance for Skilled Measurement**	**Practical Manufacturing Tolerance**
Vernier caliper								
Metric	direct	150 mm	0.02 mm	0.02 mm	0.02 mm	0.005/mm	±0.04 mm	±0.05 mm
Inch	direct	6 in.	0.001 in.	0.001 in.	0.001 in.	0.0003/in.	±0.002 in.	±0.010 in.
Vernier depth gage								
Metric	direct	150 mm	0.02 mm	0.02mm	0.02 mm	0.005/mm	±0.04 mm	±0.05 mm
Inch	direct	6 in.	0.001 in.	0.001 in.	0.001 in.	0.0003/in.	±0.002 in.	±0.010 in.
Vernier height gage								
Metric	direct	300 mm	0.02 mm	0.02 mm	0.02 mm	0.005/mm	±0.04 mm	±0.05 mm
Inch	direct	6 in.	0.001 in.	0.001 in.	0.001 in.	0.0003/in.	±0.002 in.	±0.010 in.

FIGURE 6–21 There is virtually nothing on record to establish the reliability of vernier instruments. These data are considered conservative but practical.

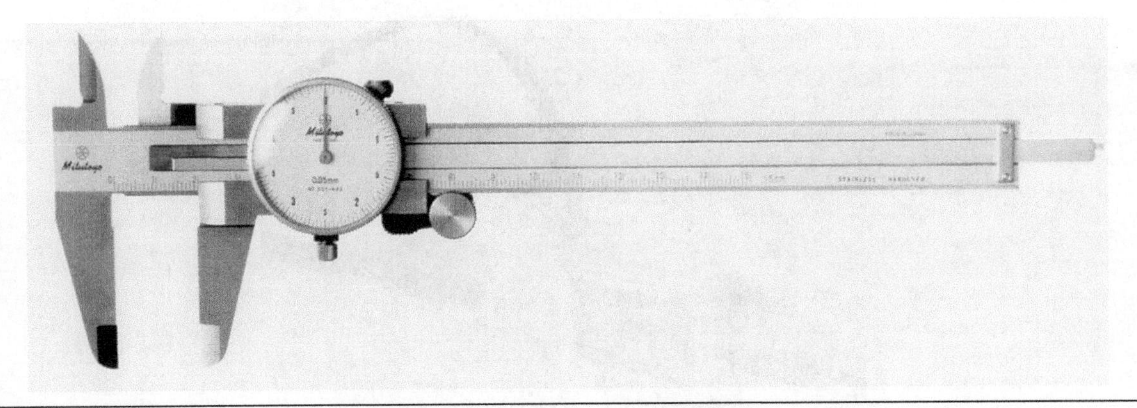

FIGURE 6–22 Dial calipers have mechanical amplification of the travel of the movable jaw, thereby eliminating the need for the vernier for fine readings. The dial reads directly to 0.001 in. (0.05 mm for the metric version) and covers 0.100 in. of jaw travel. (*Courtesy of Mitutoyo American Corporation*)

precision by an order of magnitude. The vernier scales by themselves do not improve the accuracy of the instrument: if the primary scale of a vernier caliper is inaccurate, we could make very fine measurements of the wrong dimension. Primary scales, however, can be produced to much finer precision and accuracy than the eye can resolve.

The vernier scale is not the only way to achieve greater perfection or "expand the scale." *Dial calipers* (Figure 6–22) use a vernier scale with a dial readout.

Dial calipers are also commonly found with an electronic readout (Figure 6–23).

The dial mechanism will malfunction more often than the simple vernier scale. Normally, dial calipers are accurate to 20 μm per 150 mm (0.001 in. per 6 in.) of travel. In contrast, a vernier reading has the same level of accuracy any place along the scale, but you must always, at the same time, consider the accuracy of the main scale.

Electronic Digital Calipers

Another extension of scale reading is used in the *electronic digital caliper* (Figure 6–23). These battery-operated devices not only count the distance traveled by the movable jaw and display the count on a digital readout, but they also provide ports for computer cables, so that the data can easily be used in statistical process control (SPC).

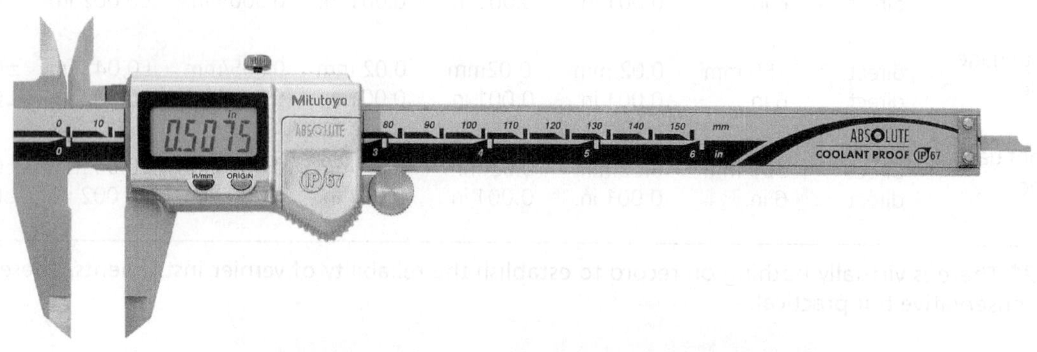

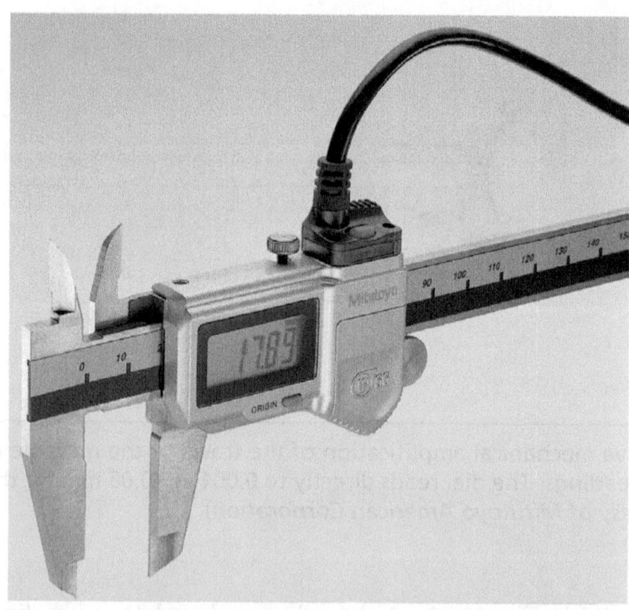

FIGURE 6–23 An electronic digital readout allows the caliper to calculate as well as measure. (*Courtesy of Mitutoyo American Corporation*)

Because the digital display makes the instrument even easier to read, electronic digital calipers are very useful for less-experienced users. Their greatest benefit, however, is the computer connection. You can set the measurement system to either English or SI and convert freely from one system to the other.

Electronic digital calipers use a floating zero, which allows you to make any point within the scale range the reference, setting it to zero. The digital readout then displays either minus or plus deviations in the jaw movement, so you do not have to do calculations, which inherently add error.

Also, electronic digital calipers provide calibration, including the user's gaging force for each measurement. In the *limit mode*, the instrument automatically warns when any measurement is above or below the tolerance; in the *peak mode*, the instrument recalls from memory the maximum and minimum measurements in a series.

Finally, electronic digital calipers allow you to directly connect with other electronic devices. You can directly print a record of measurements if you are connected to a printer, or you can interface with a computer for SPC data collection.

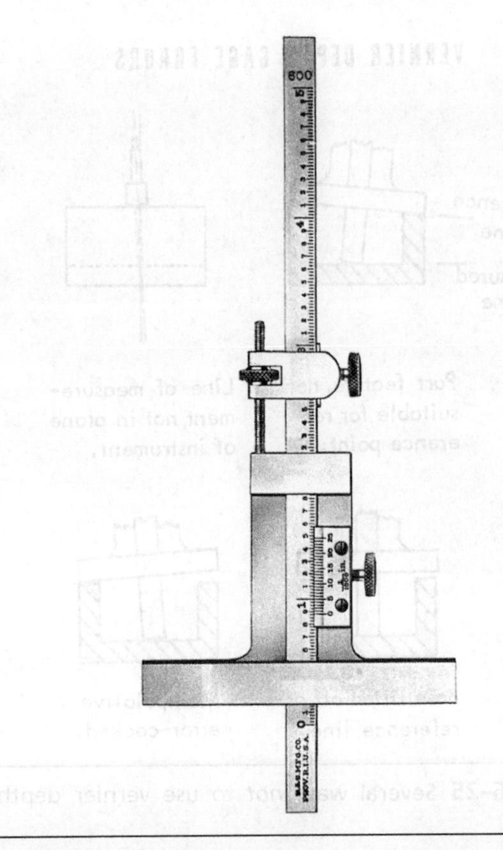

FIGURE 6–24 A vernier depth gage deserves the same care given the vernier caliper.

6–3 VERNIER DEPTH GAGE

If you take a depth gage, as discussed in the last chapter, and add a vernier, you get a vernier depth gage (see Figure 6–24).

We can justify using a fairly unreliable instrument, such as a vernier caliper, because it provides a wide range inexpensively. However, depth micrometers beat vernier depth gages in terms of centimeters per dollar and reliability per dollar. The important point we need to learn about vernier depth gages, as well as depth micrometers, is that they conform with Abbe's law.

We do use a vernier depth gage, rather than a depth micrometer, when the slot we are measuring is too narrow for the 2 mm (0.100 in.) rod of the micrometer but wide enough for the 0.8 mm (1/32 in.) blade to enter. It is easier to make manipulative errors with the vernier depth gage than any other measuring instrument (see Figure 6–25). We read vernier depth gages the same way we read vernier calipers, and the vernier depth gage requires the same maintenance to ensure its reliability.

We can also use dial versions of the depth gage (see Figure 6–26). Like dial calipers, they read to 0.02 mm (0.001 in.) with a total travel of 2.5 mm (0.100 in.), and they have the same advantages and limitations as the dial calipers.

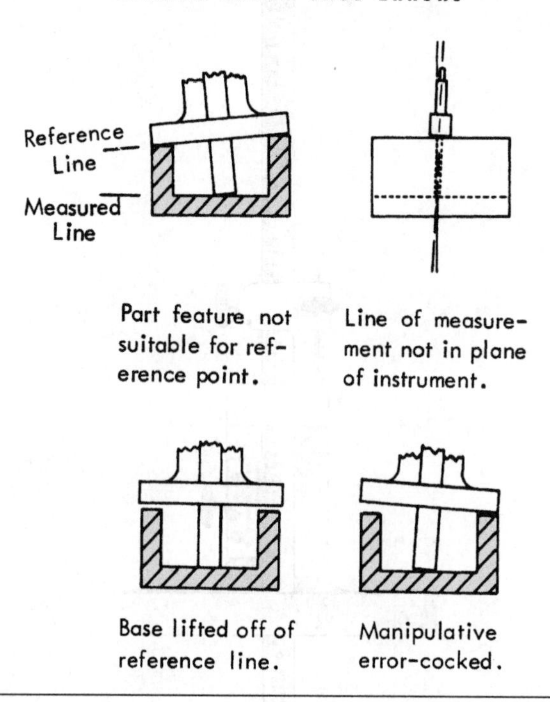

FIGURE 6–25 Several ways *not* to use vernier depth gages.

6-4 THE VERNIER HEIGHT GAGE

Because of this instrument's unfortunate ratio of base length to height, it has been used to make more erroneous measurements than any other precision measuring instrument.

As an example, we offer this quote from *Military Standard, Gage Inspection,* MIL-STD-120, Article 8: "After a moderate amount of use, the accuracy of this gage is generally slightly less than that of ordinary calipers. The only advantage in using a vernier height gage in this way (direct height measurement) is that it is very fast."

We can do so much with the vernier height gage, but maybe that is what causes all the trouble. The height gage (see Figure 6–27) is essentially a vernier caliper with an entire surface plate as its fixed jaw. (Base attachments are available to convert vernier calipers to height gage use.) The surface plate is not part of the height gage, but the height gage makes efficient use of the surface plate, because the gage sits right on the plate. We often use the height gage directly on a reference surface of a large part, and we need to keep in mind the guidelines for surface plate use for these measurements.

Because height gages have so many refinements, they are often considered very versatile. They are available in sizes (based on maximum measuring

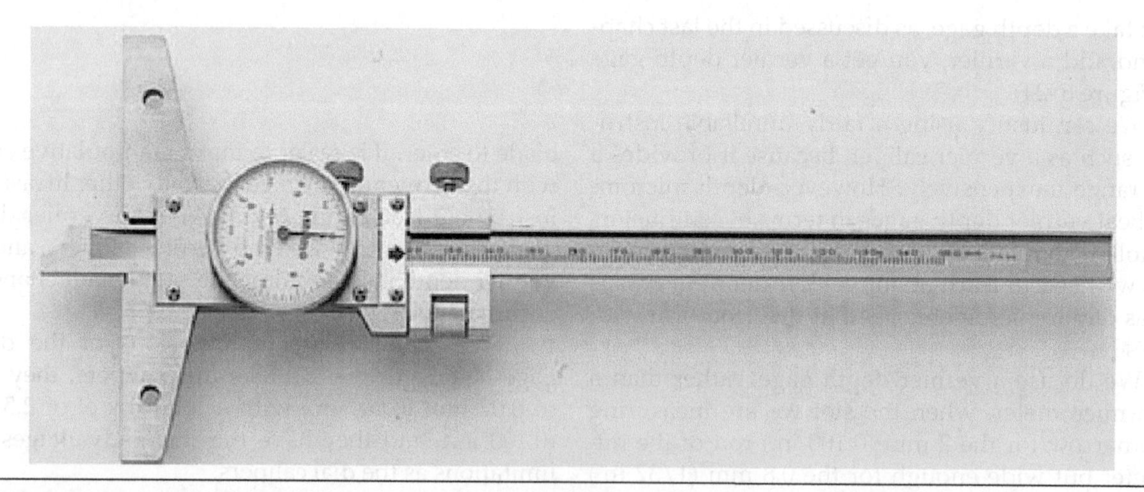

FIGURE 6–26 For the dial depth gage, a mechanical dial replaces the vernier scale. It reads to 0.001 in. (0.05 mm for the metric version) and has travel of 0.100 in. (*Courtesy of Mitutoyo American Corporation*)

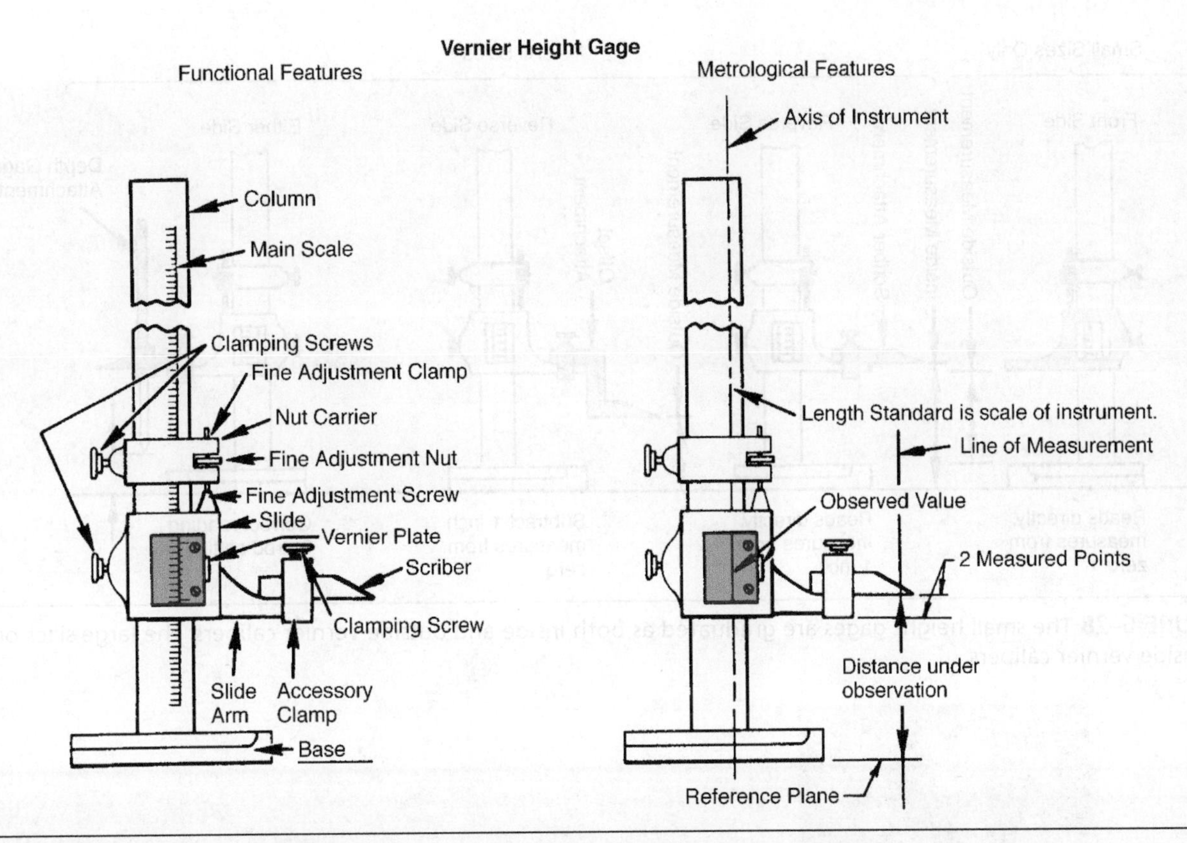

FIGURE 6–27 The versatile height gage is associated with surface plate inspection.

height) from 250 to 1500 mm (10 to 60 in.). You can use these sizes either as outside or inside vernier calipers, but only the small sizes have a second scale so you can directly read outside dimensions (see Figure 6–28).

However, like the vernier caliper, the vernier height gage violates Abbe's law. You can expect accuracy of 1 mil (0.001 in.) with a height gage, and you cannot reliably use it to control the manufacture of parts with tolerances closer than 5 mil (0.005 in.). This level of accuracy is not usually guaranteed to the upper ends of height gages 18 inches or higher (see Figure 6–21 and the summary at the end of the chapter). Three of the four popular attachments for vernier height gages (see Figure 6–28) are the scriber, the depth gage attachment, and the indicator holder.

Although we do not recommend this use, you can align the scriber with an edge of a part resting on the surface plate and measure the height of that edge above the reference base. You should use an indicator with the height gage for height measurement, except when you measure horizontal lines on the part. You must use a magnifying glass to accurately align the edge of the scriber to the line. We commonly use the scribe for layout (see Figure 6–29). Although it is used in similar applications, the offset scriber also can be used all the way down to the level of the surface plate (see Figure 6–30).

With a depth gage attachment (see Figure 6–31), we can convert the instrument to a depth gage with a very large range, which allows us to measure relative height differences in inaccessible places.

The most widely used attachment today is the indicator holder (see Figures 6–31 and 6–32). The indicator magnifies the movement of a probe so that the need for feel is minimized. For greatest reliability, you should not use the indicator for measuring.

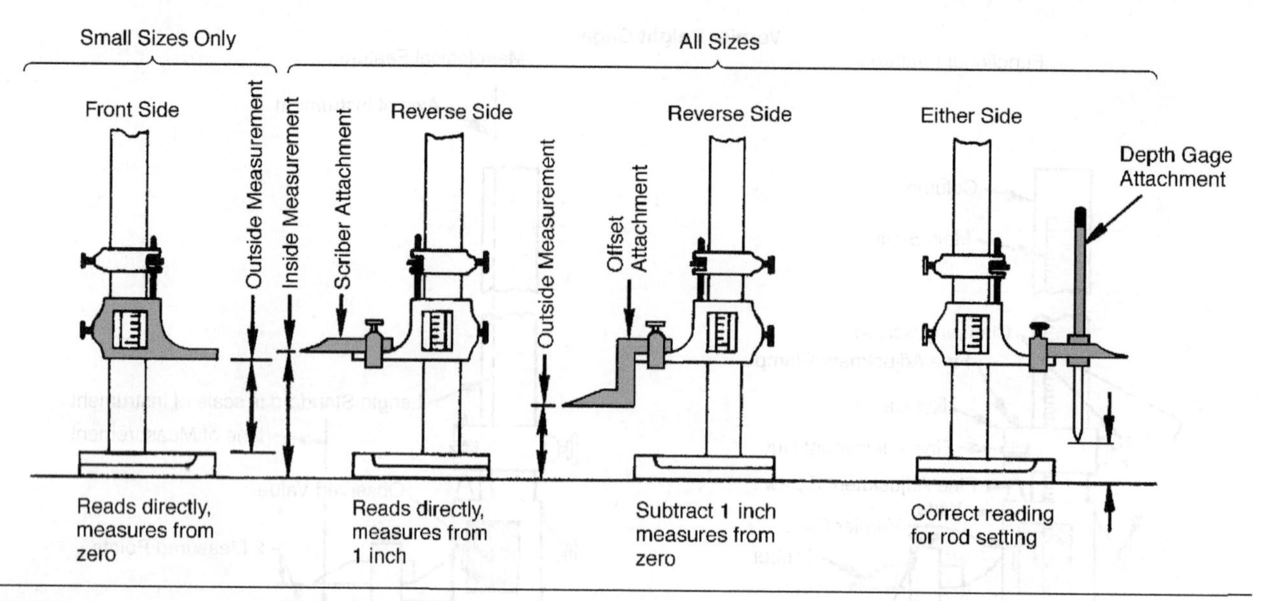

FIGURE 6–28 The small height gages are graduated as both inside and outside vernier calipers; the large sizes only as inside vernier calipers.

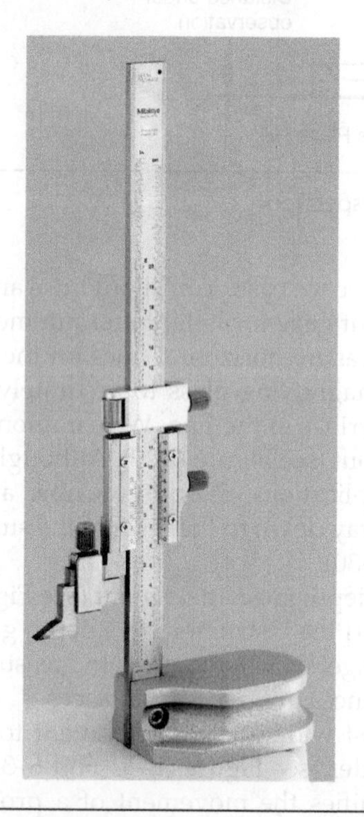

FIGURE 6–29 The most important use of the scriber is for layout rather than measurement. (*Courtesy of Mitutoyo American Corporation*)

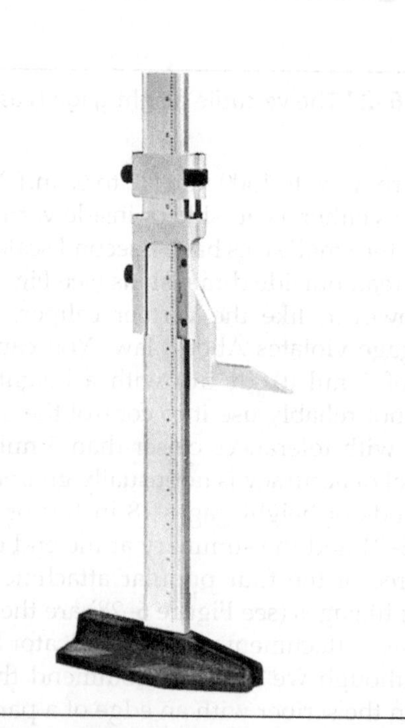

FIGURE 6–30 The offset scriber can be used from the surface of the plate.

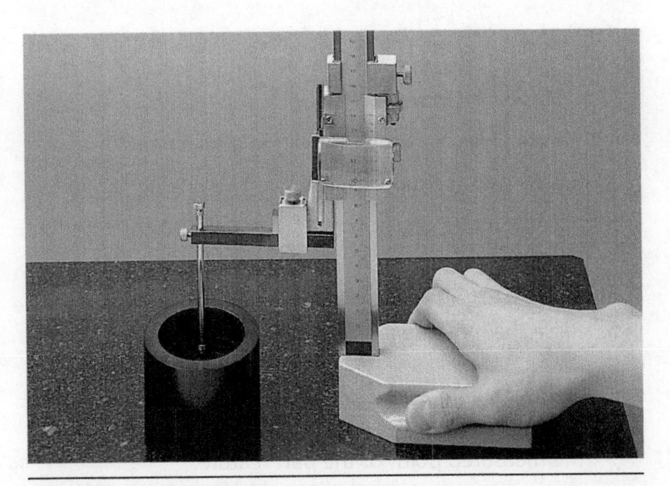

FIGURE 6–31 The depth gage attachment gets into otherwise inaccessible places. (*Courtesy of Mitutoyo American Corporation*)

You should use it only for zero setting. You should then take the measurements from the vernier scale of the height gage after you have moved the slide arm up or down to return the indicator reading to zero. The indicator does not convert the instrument into a dial height gage.

Dial and Electronic Height Gages

We use dial and electronic readouts on height gages for the same reason we use them on vernier calipers (see Figure 6–33).

The *dial height gage*, available in English and SI, has heights up to 600 mm (24 in.). As with dial calipers, the height gage dial reads to 20 μm (0.001 in.) and has a range of 0.100 in. It also has an accuracy of ±20 μm (0.001 in.) for the full range and repeatability of 12 μm (0.0005 in.).

The *electronic height gage* is even more precise. It reads to 0.0001 in. (10 μm) and claims repeatability of ±0.0001 in. (10 μm). You should note that accuracy is omitted from the specifications for these instruments because electronic amplification means that there is theoretically no limit to their precision. However, precision does not ensure accuracy, and when you try to make readings beyond the capability of the instrument, your results are meaningless.

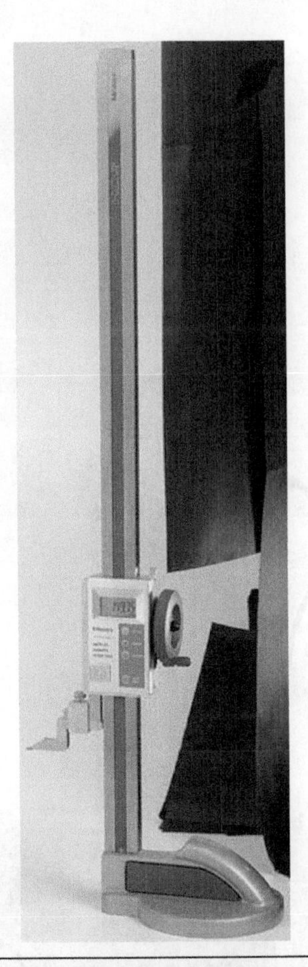

FIGURE 6–32 An indicator holder is an important attachment for a height gage. It substitutes movement of the indicator for the feel of metal-to-metal contact, which takes considerable skill. (*Courtesy of Mitutoyo American Corporation*)

Whereas the dial height gage and electronic height gage make reading measurements more convenient, they do not improve the metrological characteristics (see Figure 6–21) or add to the fundamental principles necessary for reliable measurement with vernier height gages.

Using the Height Gage

You use basically the same steps when measuring with the height gage as you do with vernier calipers. Step 6 in Figure 6–11 is simplified because the instrument

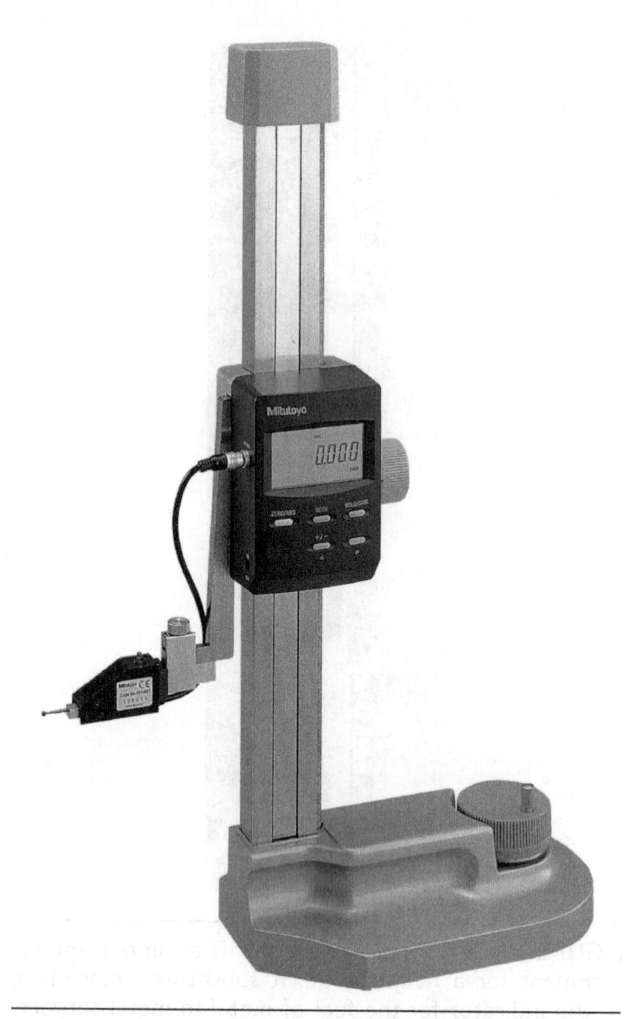

FIGURE 6–33 An electronic height gage adds convenience in measurement. The electronic version can also input directly into statistical quality control systems. (*Courtesy of Mitutoyo American Corporation*)

rests on a surface plate. Most often when you measure with a height gage, you will use an indicator so that you *centralize* using your eye rather than using feel.

When you use the depth gage attachment, you make depth or height measurements past obstructions (see Figure 6–31). Follow general steps 1, 2, and 3 for the vernier caliper, then proceed as shown in Figure 6–34.

Steps for Using Depth Gage with Vernier Height Gage

1. Thoroughly clean and inspect the attachment.
2. Fasten it to the movable jaw.
3. Set the depth gage rod with its end in contact with the surface place.
4. Write down the reading. Do not rely on memory.
5. Raise the movable jaw to clear the obstruction on the part.
6. Place the height gage in position and lower the movable jaw until the rod of the depth gage contacts the measured point of the part feature.
7. Note the reading. The difference between this reading and the reading in step 4 is the height of the measured point above the reference surface.

FIGURE 6–34 In general, the steps for using the vernier height gage resemble those for using the vernier caliper. The depth gage attachment is the exception. Proceed as shown in this checklist.

When you need to make several depth measurements, you might find it easier to work from a whole centimeter or inch than from a decimal. To do so, hold the rod against the surface plate and adjust the movable jaw to the nearest centimeter or inch. After clamping the rod, recheck your measurement and carefully adjust as required. See Figure 6–35.

Checking the Height Gage

You should check height gages like you check vernier calipers—with the addition of checking the base-to-column squareness. Base-to-column squareness must be checked from a surface plate, so surface plate errors must also be checked. Therefore, height gage inspection and calibration should be done by the gage inspection department.

You can perform several tests on your own, however, that show whether the instrument needs to be calibrated. Regular visual inspection can reveal burrs, which can be removed, and nicks that may indicate

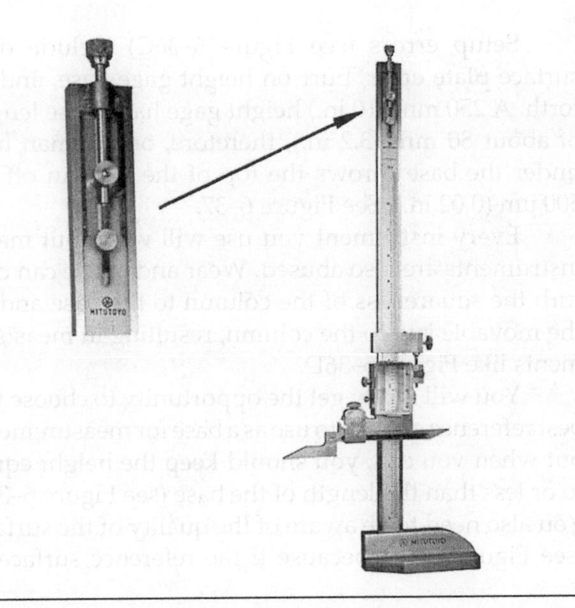

FIGURE 6–35 A height gage with an adjustable main scale can be set for any reference height. This cuts down the chance for observational and computational errors. (*Courtesy of Mitutoyo American Corporation*)

severe abuse or damage. In addition, the base should not rock on a flat surface, which you can determine by applying pressure at various places. A truly flat base will "wring" to a fine surface plate and you will feel drag or suction when you slide the instrument.

The best way to test accuracy and precision is to make repeated readings of gage block heights at various points throughout the range. Again, it is not enough to repeat the measurements. If, after you have corrected the zero setting by adjusting the vernier plate, the precision (dispersion of repeated readings) or the accuracy (repeating readings of the gage block height) is more than the *discrimination* of the instrument (20 µm / 0.001 in.), you should return the instrument to the manufacturer for repair. You should always inspect your accessories as carefully as you inspect the instrument itself.

The Problem with Height Gages

The problem with height gages is height. The height of the instrument in relation to the small base creates a lever action, which adds to errors. The higher the measured plane is, the greater the effect of many of these errors. In addition, the height gage adds errors from setup to those it generates. Figure 6–36A shows that a height gage is inherently unstable: the tall, thin column sways freely. You probably cannot see a wobble of 20 µm (0.001 in.) or even 0.8 mm (1/32 in.), but that much sway can destroy the reliability of your measurements.

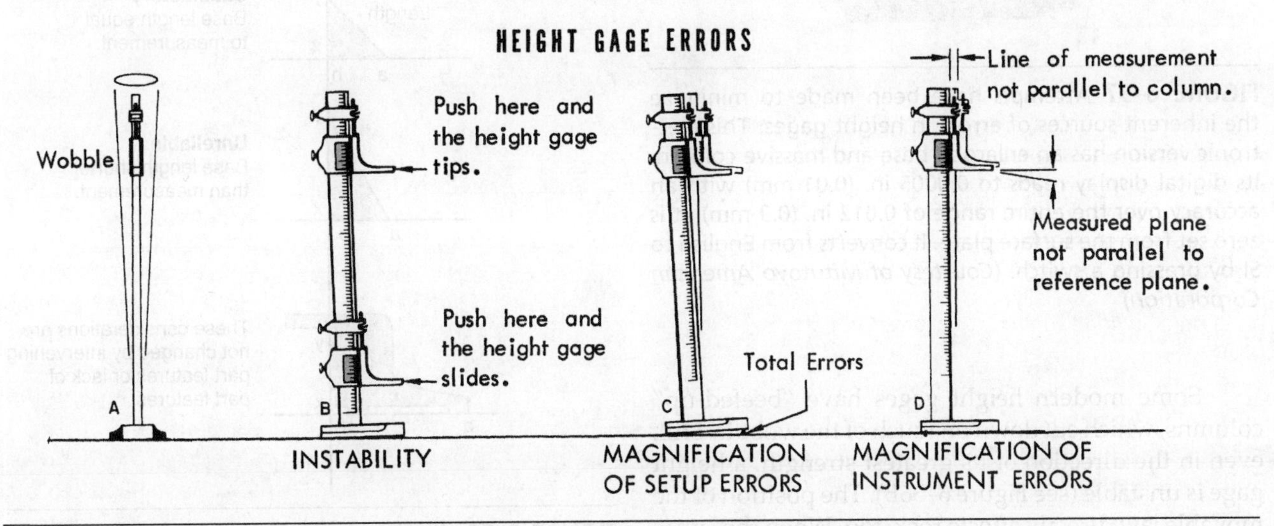

FIGURE 6–36 The very height of a height gage creates its measurement problems.

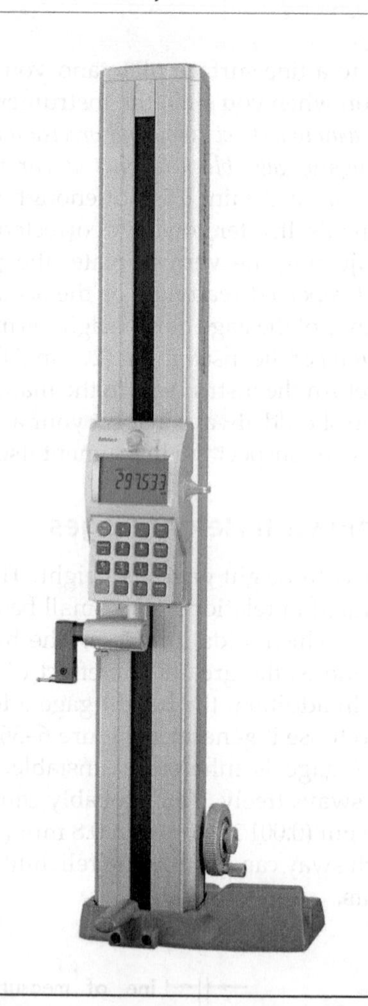

FIGURE 6–37 Attempts have been made to minimize the inherent sources of errors in height gages. This electronic version has an enlarged base and massive column. Its digital display reads to 0.0005 in. (0.01 mm) with an accuracy over the entire range of 0.012 in. (0.3 mm). It is zero set from the surface plate. It converts from English to SI by pressing a switch. (*Courtesy of Mitutoyo American Corporation*)

Some modern height gages have "beefed-up" columns, which cut down on much of the wobble. Still, even in the direction of its greatest strength, a height gage is unstable (see Figure 6–36B). The position of the movable jaw directly affects rock, too. When the jaw is low, rock is at a minimum; when the jaw is high, the column may rock more.

Setup errors (see Figure 6–36C) include dirt, surface plate error, burr on height gage base, and so forth. A 250 mm (10 in.) height gage has a base length of about 80 mm (3.2 in.); therefore, one human hair under the base throws the top of the column off by 500 μm (0.02 in.). See Figure 6–37.

Every instrument you use will wear, but many instruments are also abused. Wear and abuse can disturb the squareness of the column to the base and of the movable jaw to the column, resulting in measurements like Figure 6–36D.

You will rarely get the opportunity to choose the best reference surface to use as a base for measurement, but when you can, you should keep the height equal to or less than the length of the base (see Figure 6–38). You also need to be aware of the quality of the surface (see Figure 6–39), because if the reference surface is

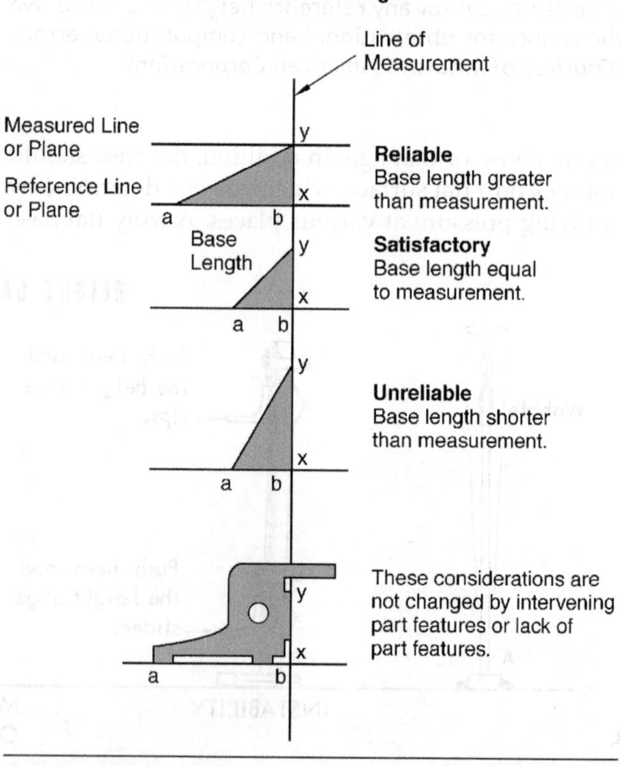

FIGURE 6–38 You cannot always select the best surface for a reference. When you can, the above relationships should be remembered.

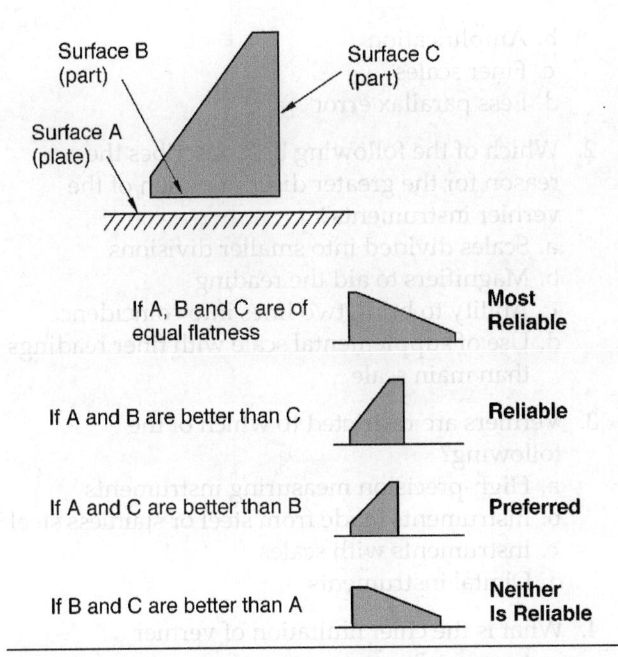

FIGURE 6–39 When you choose a support surface, consider its condition as well as the ratio of base to height.

Suggestions for Reliable Height Gage Measurement

1. Do not use the height gage for measurements requiring greater discrimination (Figure 6–28).
2. Know that the instrument is in calibration.
3. Observe scrupulous cleanliness.
4. Move the height gage by the base, not the column.
5. Allow the height gage, surface plate, and part to stabilize (reach the same temperature).
6. Prefer low setups to high ones.
7. Repeat each measurement.
8. Question the alignment of instrument, reference surface, and part for each use.
9. Know that the surface plate is sufficiently flat for the desired measurement.

FIGURE 6–40 If these suggestions are followed carefully, they will force you to explore better means for height measurement.

FIGURE 6–41 The gear tooth vernier caliper measures both tooth thickness and height. (*Courtesy of Hexagon Metrology, Inc., TESA*)

unknown or poor, all hope of reliable measurement with a height gage is lost. You might find following the suggestions in Figure 6–40 helpful.

The other important vernier instruments are vernier protractors (see Chapter 12) and gear tooth vernier calipers. The *gear tooth vernier caliper* is actually two vernier instruments combined (see Figure 6–41). One part of the instrument acts as a vernier caliper to measure the tooth thickness at the pitch line. The other part of the instrument is a *tongue* that travels between the jaws to measure the distance from the top of the tooth to the pitch line of the tooth.

6–5 THREE ELEMENTS OF MEASUREMENT

As discussed, every measurement requires three elements: the object to measure, the standard of length to compare it with, and the means for comparison. Every instrument discussed has met these three requirements; however, the instruments discussed in this chapter have the standard built into the means for comparison. Without these three elements of measurement, you cannot take measurements, and no measurement can be more accurate than the positional relationship of these elements.

SUMMARY

- The familiar vernier instruments are the vernier caliper and vernier height gage. They discriminate in mils, the area in which most practical measurement takes place. Essentially, they are steel rules with refinements. The vernier scale provides amplification so that the resolving power of the eye is increased many times.
- The advantages of vernier instruments are their long measuring range and convenience. Their chief disadvantages are reliance on the feel of the user and their susceptibility to misalignment of all kinds. Vernier instruments have been largely supplanted by higher amplification instruments wherever reliability is essential.
- The study of vernier instruments is important because they vividly demonstrate the importance of geometry and positional relationships in measurement. Furthermore, although the vernier instruments have been supplanted for the control of production, they remain important tools of the skilled craftsman. It will be a long time before the vernier caliper disappears from the toolbox of the master toolmaker or the vernier height gage from the surface plate of the model maker. For the latter, Figure 6–21 will assist with instrument selection.
- Dial versions of these instruments add convenience in reading. Electronic versions provide digital readouts, which are even easier to read. However, neither of these improves the metrological capabilities nor decreases the problems in making reliable measurements.

END-OF-CHAPTER QUESTIONS

1. Which of the following is the primary advantage of the vernier instruments over those discussed previously?
   a. Longer range
   b. Amplification
   c. Finer scales
   d. Less parallax error
2. Which of the following best describes the reason for the greater discrimination of the vernier instruments?
   a. Scales divided into smaller divisions
   b. Magnifiers to aid the reading
   c. Ability to bring two lines into coincidence
   d. Use of supplemental scale with finer readings than main scale
3. Verniers are restricted to which of the following?
   a. High-precision measuring instruments
   b. Instruments made from steel or stainless steel
   c. Instruments with scales
   d. Digital instruments
4. What is the chief limitation of vernier instruments?
   a. Hard to read
   b. High cost
   c. Lack of multiple scales
   d. Discrimination may exceed inherent accuracy
5. Of the many advantages of vernier instruments, which of the following pairs is the most important?
   a. Long measuring range, great stability
   b. Great stability, measurements to 20 μm (0.0008 in.)
   c. Measurements to 20 μm (0.0008 in.)
   d. Accuracy to 3 μm (0.0001 in.), precision to 20 μm (0.0008 in.)
6. Which of the following difficulties apply to vernier instrument use?
   a. Takes excessive time to measure
   b. Parallax error
   c. Limited discrimination
   d. Alignment
7. Which of the following part features is easiest to measure with a vernier caliper?
   a. Large distances between outside planes
   b. Heights from a surface plate
   c. Cylindrical features
   d. Concave features

8. Which of the following measurement situations could best be fulfilled by the vernier depth gage?
   a. Depth over 150 mm (6 in.) with required discrimination of 800 µm (1/32 in.)
   b. Depth measurements of inside measurements at angles to the reference surfaces
   c. Depth measurements of small recesses with required discrimination closer than 300 µm (1/64 in.)
   d. Depth measurements of medium size production runs
9. Which of the following is the greatest limitation of vernier height gages?
   a. Inherent inaccuracy of the column scale
   b. Low amplification of vernier instruments
   c. Wear of base
   d. Lack of stability
10. Which of the following best explains extensive use of the vernier height gage?
    a. Most measurements require the long range.
    b. Convenience and versatility
    c. Recognized requirement for stability
    d. Discrimination to 20 µm (0.0008 in.)
11. Several attachments extend the use of the height gage. Which of the following are the most important?
    a. Surface plate
    b. Scriber, offset scriber, dial indicator, holder, and depth gage
    c. Auxiliary vernier scale
    d. Scriber, hook rule, dial indicator, holder, and depth gage
12. Surface plates are the most common reference surfaces for use with height gages. Which of the following describes the way they interact?
    a. There is no relationship.
    b. The surface plate supports the height gage.
    c. Plus errors of the surface plate reverse their sign when combined with the height gage readings.
    d. Any flatness error in the surface plate is multiplied by the height gage.
13. Which of the following statements is valid for a height gage?
    a. In testing a vernier instrument, zero setting repeatability is the most important test.
    b. Accuracy may be checked at any point in the range by the repeatability of successive measurements at that point.
    c. Repeated readings of gage block heights at random points in the range confirm accuracy.
    d. Accuracy may be checked by comparing jaw movement against dial comparator readings.
14. Which of the following statements is most nearly correct for vernier instruments?
    a. Discrimination of 20 µm (0.0008 in.) can be expected throughout their range.
    b. Accuracy of 20 µm (0.0008 in.) can be expected throughout their range.
    c. They cannot be considered reliable for measurement closer than 400 µm (1/64 in.).
    d. Accuracy decreases as the length of the measurement increases.
15. Dial height gages have one of the following relationships with vernier height gages. Select the correct answer.
    a. Ensured discrimination to 20 µm (0.0008 in.), together with ensured accuracy to 20 µm (0.0008 in.) for full scale reading
    b. Greater convenience with no decrease of discrimination or accuracy
    c. Greater convenience but with reduced accuracy
    d. Compensation for instability of instrument
16. Electronic digital height gages provide only two advantages over vernier height gages. Indicate which of the following, if any, applies.
    a. Reading convenience
    b. Input into statistical quality control systems
    c. Reduced inherent errors
    d. Instability problem cancels out

7

CHAPTER SEVEN

MICROMETER INSTRUMENTS

As a rule, inventions—which are truly solutions—are not arrived at quickly. They seem to appear suddenly but the groundwork has usually been long in preparing. It is the essence of this philosophy that man's needs are balanced by his powers. That as his needs increase, the powers increase. . .

Louis H. Sullivan (1856–1924), *(architect who pioneered the skyscraper)*

LEARNING OBJECTIVES

- Explain why increased amplification results in better readability.
- Describe how Abbe's law contributes to micrometer measurement.
- Explain the way the screw adapts to the measurement of linear displacements.
- Explain the importance of cleanliness and alignment increases as the amplification increases.
- Explain why the basic micrometer principle has developed into a family of diverse instruments.

OVERVIEW

Vernier instruments differ from rules in that they have amplification. This gives them greater precision. Micrometers are instruments that achieve still higher amplification by the use of screw threads. They would not be able to make full use of this amplification if they had the same geometry as the vernier instruments. Fortunately, micrometers obey Abbe's law. The line of measurement is in line with the axis of the instrument.

Background

In the hands of a skilled metrologist, a micrometer is among the most accurate of hand tools. The micrometer technically falls into the category of "caliper," in

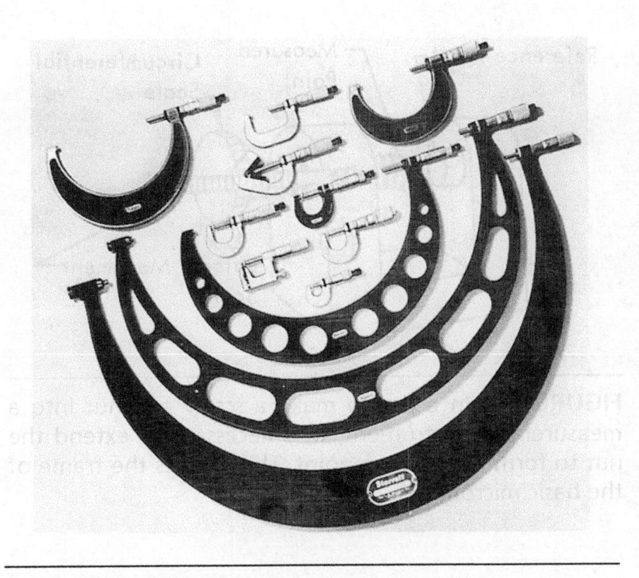

FIGURE 7–1 A wide range of types and sizes of micrometers are used for measurement. *(Courtesy of the L.S. Starrett Company)*

that it is two opposing ends joined by a frame. It is sometimes referred to as a "micrometer calipers," or "mics." An object is measured with micrometers by placing the part between the anvil and the spindle (a very accurately machined screw). Micrometers come in a variety of sizes and types (see Figure 7–1).

7–1 MICROMETERS

The designated micrometer size is its largest opening, not its range. A quick check of catalogs shows that micrometers are available from a 0 to 25 mm range, to a 375 to 400 mm range, to even larger (e.g., from a 0 to 0.5 in. range to a 19 to 20 in. range). Larger sizes can be obtained in both metric and inch. Regardless of the overall size, the range is usually limited to 25 mm (1 in.). Digital micrometers are available in 25, 50, 75, and 100 mm sizes, with readouts to 0.002 mm (1, 2, 3, and 4 inches with 0.0002 in.). Metric micrometers with vernier scales for discrimination of 0.002 mm are available up to 50 mm. (Up to 12 inches in 1 inch ranges with discrimination of 0.0001 in. are also available.) Other useful features are the ratchet stop and clamping ring.

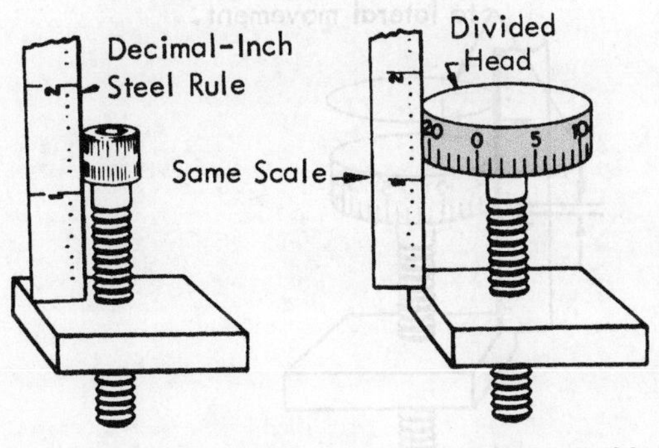

FIGURE 7–2 If we use a rule to measure the advance of the thread, discrimination is limited by the rule. If we calibrate the rotation of the screw, then the screw itself becomes the element that limits discrimination.

All micrometers are based on the relation of a screw's circular movement to its axial movement. Consider a screw threaded through a plate (see Figure 7–2). Each turn of the screw would move it a distance that could be measured with a steel rule. The accuracy with which we could measure the changes is limited by all of the variables associated with rules. The discrimination is that of the rule, or 0.02 inches for a decimal-inch rule of a scaled instrument.

Instead of the rule, we could place a special scale alongside the head, and this could be calibrated in whole turns. The discrimination is then one turn. If we try to increase this discrimination by subdividing the whole-turn spaces, we again run into the resolving power of the eye as the limit. However, if we divide the head into spaces, we have effectively increased the discrimination many times over. Figure 7–3 shows that the amount of amplification increases with an increase of the circumference and a decrease in the screw thread (the lateral travel of the screw in one revolution).

A screw by itself cannot do much measuring. Even with a nut as in Figures 7–2 and 7–3, it would be difficult. The screw furnishes just one end of the

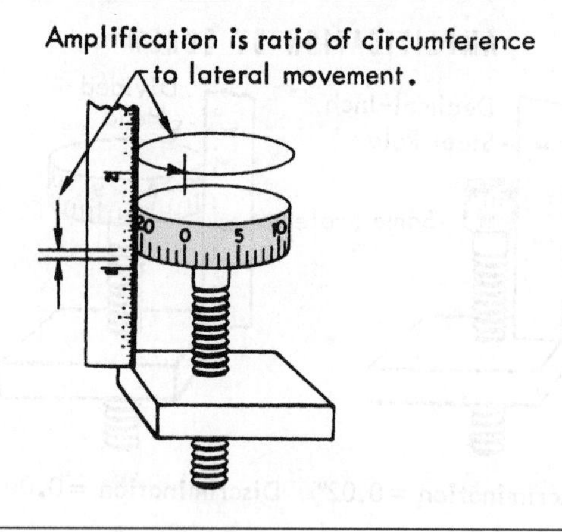

FIGURE 7–3 The greater the ratio between the circumference and the lead of the screw, the higher the possible amplification.

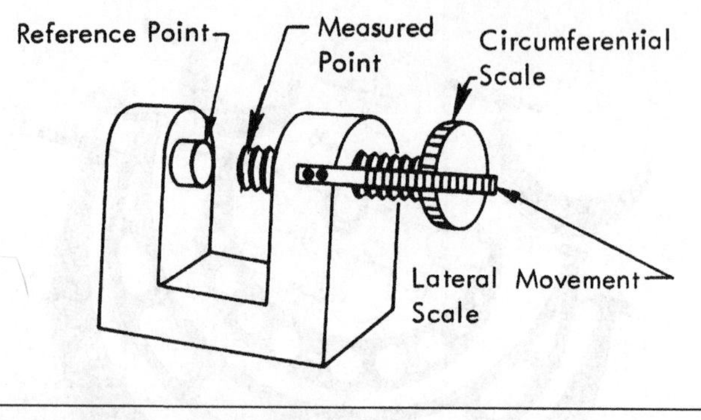

FIGURE 7–4 In order to make a screw and nut into a measurement instrument, it is necessary to extend the nut to form a reference point. This creates the frame of the basic micrometer.

measurement partnership. It is the measured point. Still needed is the reference point, which is obtained by extending the nut to form a frame (see Figure 7–4).

For a metric micrometer, the pitch of the spindle screw is set at one-half millimeter (0.5 mm); one revolution of the thimble advances the spindle toward or away from the anvil the same 0.5 mm distance. The thimble is divided into 50 divisions; thus, each graduation is equal to 0.01 mm. For an inch micrometer, the screw has 40 threads per inch. Each revolution moves the screw 1/40 inch or 0.025 inch. The thimble is divided into 25 spaces. Each division thus represents 0.001 inch. The addition of the vernier increases the discrimination of the metric micrometer from 0.1 mm to 0.002 mm. The inch micrometer increases the discrimination from 0.001 to 0.0001 inch.

Micrometer Readings

Figure 7–5 names the most-often used functional parts of a micrometer. The barrel of the micrometer is divided into 0.01 mm (0.025 in.) graduations. Each fifth one is elongated and numbered for easy reading (it's the fourth one on an inch micrometer). These large divisions are 0.5 mm (0.100 in. micrometer) each. As mentioned, the thimble is divided into 50 graduations (25 for inch) of 0.01 mm (0.001 in.) each. One important element of reliability of an instrument is its *readability* (see Figure 7–6). The smallest division on a metric rule is 0.5 mm, and on the micrometer, the smallest division is 0.002 mm—an increase in discrimination of 25 times. The micrometer is clearly more readable than a rule or the caliper. All other things being equal, the micrometer should be more reliable than a vernier.

There are three steps for reading a micrometer (see Figure 7–7).

1. Note the highest figure on the barrel that is uncovered by the thimble. That is the one-hundredth millimeter part of the reading.
2. Note the whole number of graduations between the figure and the thimble. It may be 5, 10, or 15. Each of these represents 5 mm; the divisions between are normally 0.5 mm or one revolution.
3. Read the thimble opposite the index on the barrel. This will give the 0.01 mm reading from 0.00 to 0.49 inch (zero and 0.50 inch on the thimble coincide).

There are only two precautions to observe when reading a micrometer. The thimble must be read in the right direction. For example, in Figure 7–7 the thimble might be carelessly read as 0.14 instead of 0.16 mm.

The other precaution concerns the zero position on the thimble. When passing the index line on

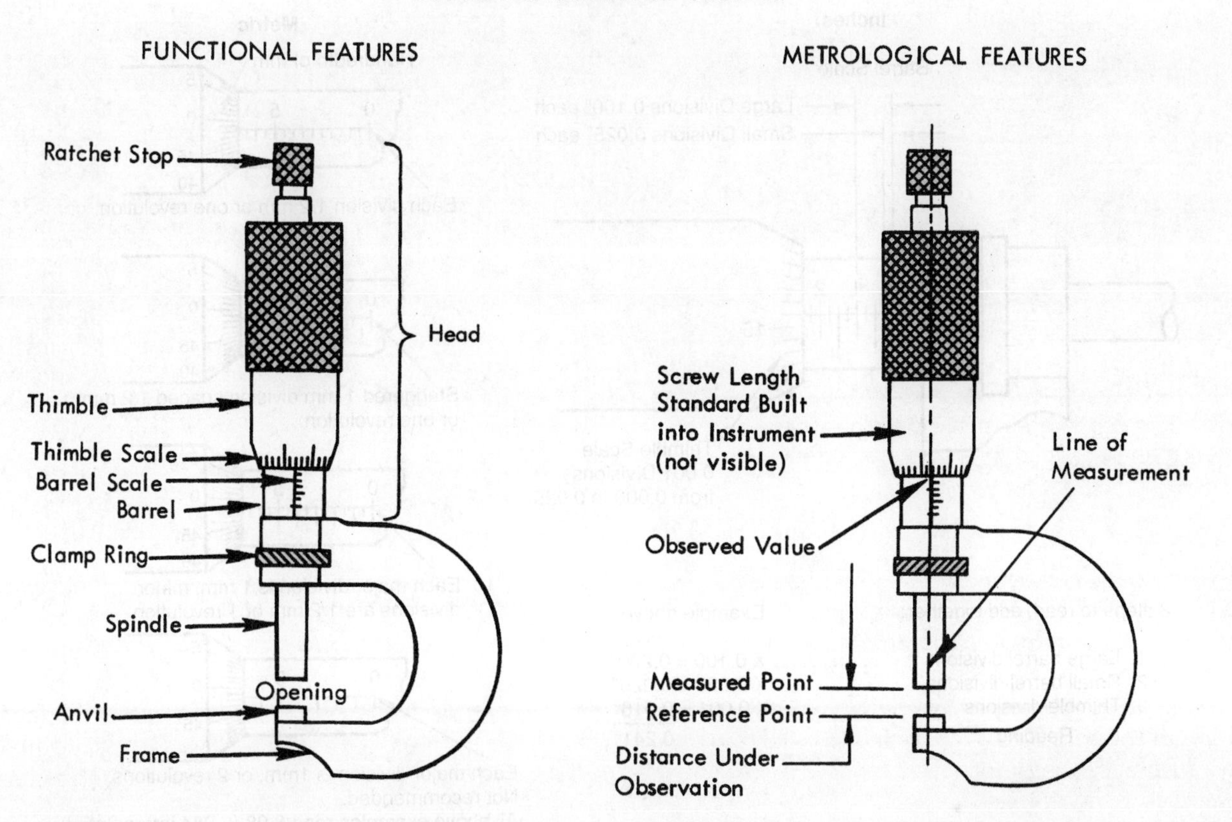

FIGURE 7–5 The functional features make the metrological ones usable for practical measurements.

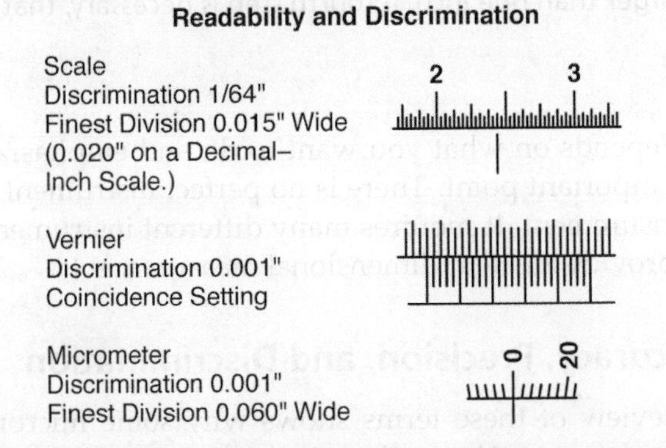

FIGURE 7–6 The smallest division on the micrometer reads to 0.001 but is four times as wide as a 1/64 in. graduation on a scale. Although the micrometer and vernier have the same discrimination, the micrometer is the most readable.

the barrel, there is a chance to read an extra 0.50 mm (0.025 in.). This is caused by the fact that the next barrel graduation has begun to show but has not yet fully appeared (see Figure 7–8). This is avoided by being careful to read only full divisions on the barrel. This error is particularly dangerous because, once started, it is usually continued without detection. Some errors are progressive and if continued will eventually call attention to themselves. Stepping off intervals with dividers (see Figure 5–33) is an example. Frequently, a machine operator has read the micrometer 0.50 mm incorrectly on the first reading of the day and continued to produce scrap the balance of the day. Certainly if the operator did not suspect this first reading, it would seem that he or she would nevertheless have

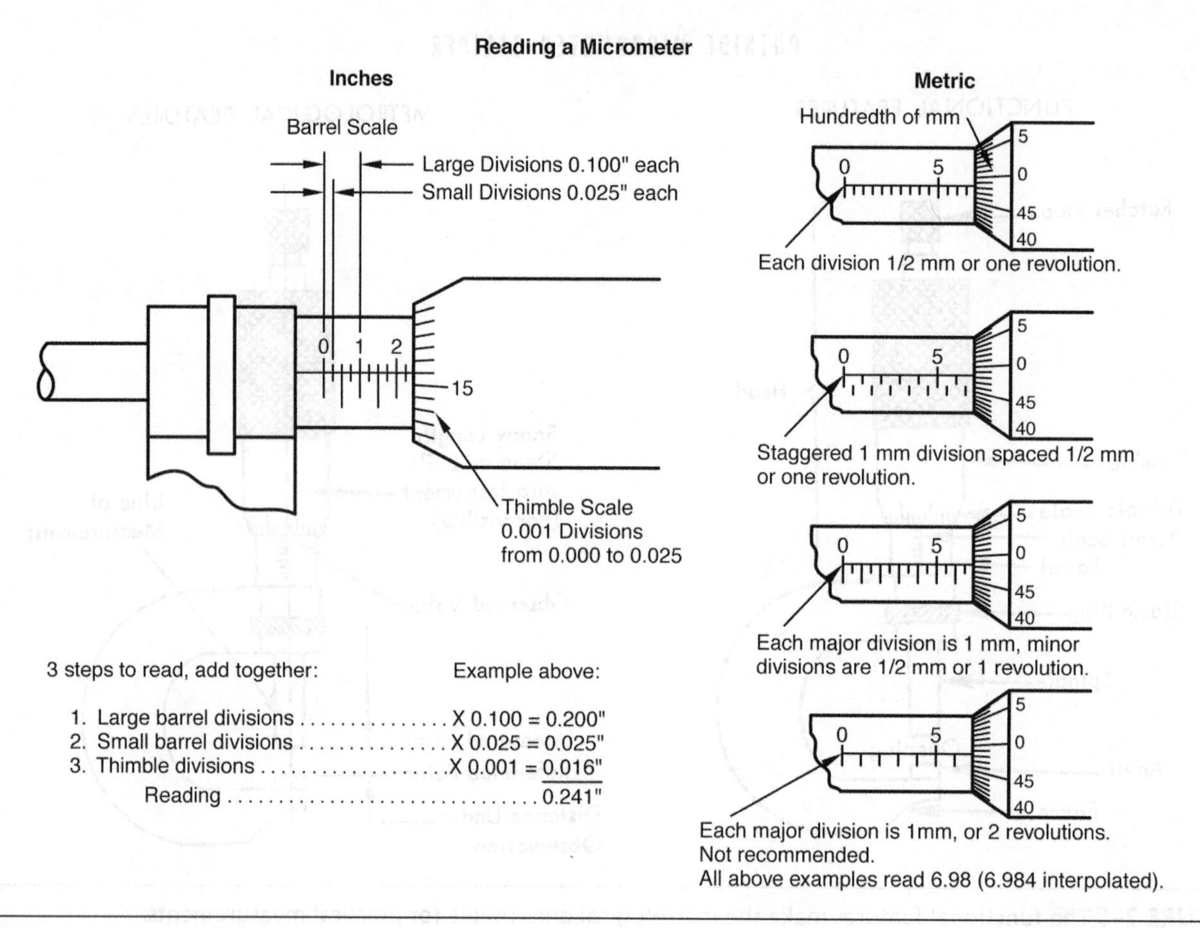

FIGURE 7–7 A micrometer is read by simply totaling the number of whole divisions on the barrel scale and adding the thousandths from the thimble scale. Note: If the frame is larger than one inch, a fourth step is necessary, that of adding the frame size to the total.

reason to suspect the 137th reading. Suspect every measurement.

One manufacturer has found a novel method to eliminate this observational error. The barrel graduations are at an angle (see Figure 7–9). There is no way a graduation significant to the reading can be completely covered. This problem is also eliminated by the digital micrometers.

The popularity of the micrometer is due to other factors in addition to its improved reliability. These are summarized in Figure 7–10. Note that *end measurement* is listed as both an advantage and a disadvantage. It depends on what you want to do and emphasizes an important point. There is no perfect instrument in measurement. It requires many different instruments to provide efficient dimensional measurement.

Accuracy, Precision, and Discrimination

A review of these terms shows why some micrometers have vernier scales. Accuracy, precision, and discrimination are related. Accuracy is the relation between the readings (observed values) and the true values. Precision is the "fineness" of the instrument

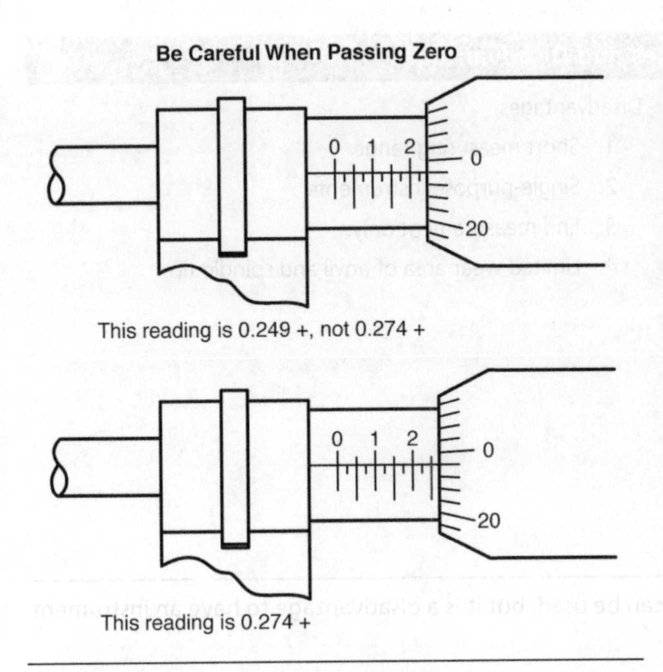

FIGURE 7–8 One of the few errors that can occur when passing zero on the thimble.

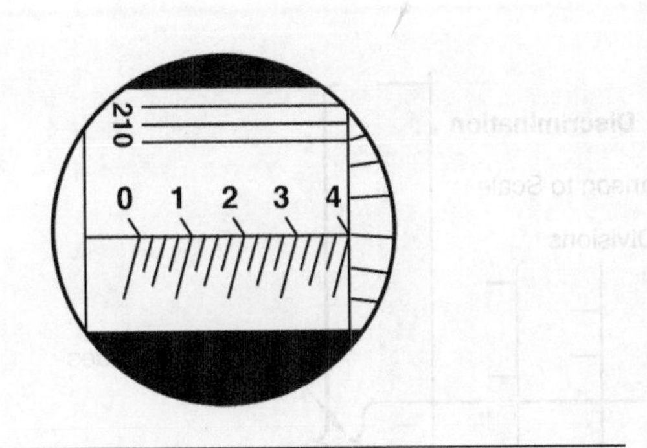

FIGURE 7–9 The slanting graduations on the barrel of this micrometer reduce the chances of observational errors such as the one shown in Figure 7–8.

or the dispersion of repeated readings. Discrimination is the smallest readable division. The discrimination given to an instrument is arbitrarily selected by the manufacturer. It is based on practical considerations and is a measure of only one thing, the integrity of the manufacturer.

Figure 7–11 analyzes an imaginary instrument. Assume that several readings are taken of a standard of known length. All readings deviate from the true value. They are fanned out in the illustration to show this dispersion. This is the precision, a measurable amount. Note that all of the readings are greater than the true value. Now consider four possible scales for the imaginary instrument: A, B, C, and D. The smallest scale divisions are shown at A. This scale would cause serious errors. Although the readings are fine, they are virtually meaningless. The total dispersion is greater than the smallest graduation. In fact, as drawn, the minimum error is nearly one graduation. These graduations measure the instrument instead of the part. The situation would be better with scale B, which is about as good as flipping a coin. A single reading could be 5 as easily as 6. In either case, its value would be reduced by at least the minimum error.

In C, the discrimination is sufficient to include all of the possible readings and allow for the error. But that is not saying much. When you measure to 5, you do not expect that the true value might actually be 6.75 inches. It does show that extending the size of graduations with respect to the precision (or increasing the precision with respect to the size of the graduations) increases the reliability of correct measurement. An increase to 10 times the width would give the scale at D. Even at the extremes of dispersion and instrument error, the readings will be correct 90% of the time. This would be considered a reliable scale for the particular instrument in our time and in this example.

If that ratio of scale division to precision (and accuracy) is practical, consider what happens if the graduations are twice as wide as in D. On one hand, the reliability of the instrument is increased. Note: The reliability of the instrument is increased, not the reliability of the measurement. It is the latter that we ultimately want. On the other hand, the wide graduations invite interpolation.

Interpolation is to instrument reading what rounding off is to computation. Like rounding off, the problem occurs when the reading is nearly halfway between two graduations. Which one should be chosen? Unfortunately, interpolation adds human error and should be chosen only as a last resort. If the

Micrometer Advantages and Disadvantages

Advantages

1. More accurate than rules.
2. Greater precision than calipers.
3. Better readability than rules or verniers.
4. No parallax error.
5. Small, portable, and easy to handle.
6. Relatively inexpensive.
7. Retains accuracy better than verniers.
8. Has wear adjustment.*
9. End measurement.

* For screw only. Anvil and spindle tip wear remain a potential source of error.

Disadvantages

1. Short measuring range.
2. Single-purpose instruments.
3. End measurement only.
4. Limited wear area of anvil and spindle tip.

FIGURE 7–10 End measurement is an advantage where it can be used, but it is a disadvantage to have an instrument that can only be used for end measurement.

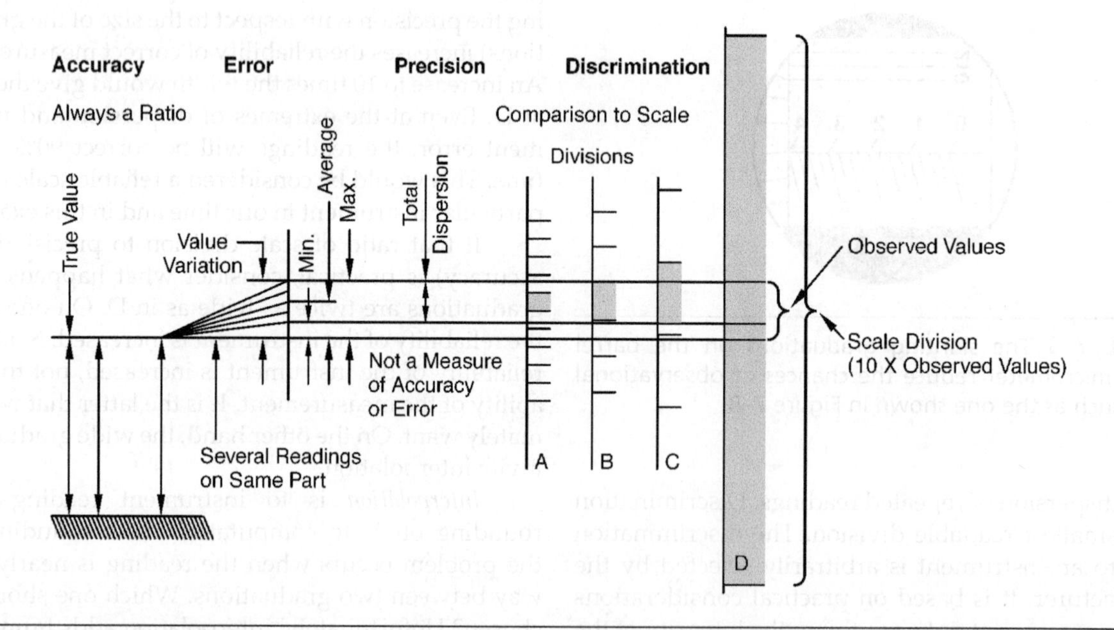

FIGURE 7–11 Discrimination *is* related to instrument precision, and *should be* related to accuracy.

scale can be divided into smaller graduations (and the same ratio retained), there will be less need to interpolate. Furthermore, the full practical capability of the instrument will be used, which is of importance whenever cost of measurement is considered. This is the reason that instruments are provided with scales with as high a discrimination as possible. It is understandable that sometimes the manufacturer may be a bit too optimistic.

Micrometer Construction

Micrometers are very similar in construction, although many manufacturers have developed special features, which are described in their literature.

Frame. Figure 7–12 shows the typical construction. The frame may be of a number of constructions. Full finished solid frames are popular for the 25, 50, 75, and 100 mm sizes (1 to 4 in.). Drop-forged frames, both plated and black finished, are common, as well as tubular frames that are lightweight and have favorable heat conduction characteristics.

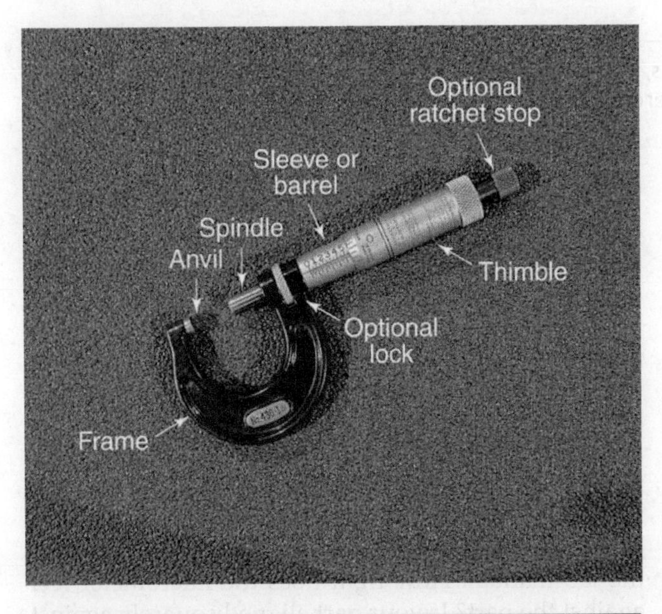

FIGURE 7–12 Parts of the outside micrometer.

Anvil. The hardened anvil is pressed or screwed into the frame. It is replaceable in a factory overhaul. The anvil was once adjustable in micrometers, which permitted some wear adjustment. The wear is seldom uniform across the anvil; therefore, this adjustment seldom completely eliminated the effect of error. The anvil is no longer adjustable; however, micrometers with interchangeable anvils are available (see Figure 7–13). Carbide anvils as well as carbide spindle tips are available, providing a far better answer to wear.

Barrel. The barrel is fastened to the frame. It carries both the barrel graduations and the nut in which the screw turns. The nut is short compared with the screw, resulting in much more rapid wear. To compensate for this, the nut is slotted and has a tapered thread on its outside. The adjusting nut draws the slotted nut closed as it is tightened, thereby removing the unwanted clearance (play). The manuals supplied with micrometers show how this is done.

Thimble. The thimble is attached to the screw. Not the least of the purposes of the thimble is to protect the screw and nut from abrasive dust. It also carries the 0.01 mm (0.001 in.) scale.

Both zero settings and manufacturing problems are solved neatly by making the screw and thimble two separate parts. Zero setting is accomplished by loosening the parts and resetting the thimble scale. The typical method is shown in Figure 7–14.

Two other important construction features are the ratchet stop and the clamp ring. Both are taken up when the use of the micrometer is discussed.

Vernier Micrometers

The design and manufacture of the measuring portion of micrometers are intended to give them an inherent accuracy of 0.002 mm (0.0001 in.). We have repeatedly observed that accuracy is a ratio. Therefore, this means 0.002 mm (0.0001 in.) to the entire range of 25 mm (1 in.). Placing a vernier scale on the micrometer permits us to make readings to the limit of the instrument accuracy.

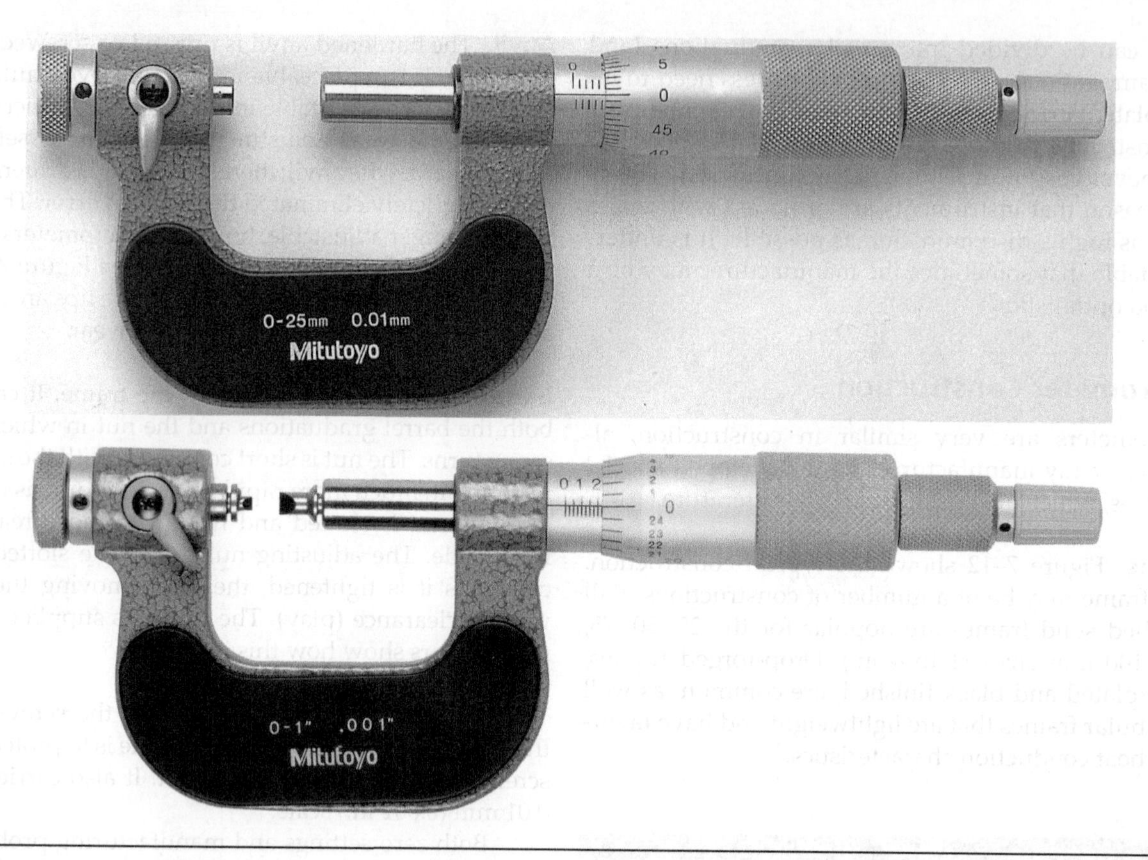

FIGURE 7–13 To increase the applications of micrometers, versions are available with interchangeable anvils. Note that when the anvil is changed, the instrument must be zero checked. *(Courtesy of Mitutoyo American Corporation)*

The point to be emphasized is that the 0.002 mm (0.0001 in.) readings approximately correspond to scale B in Figure 7–11. They are not reliable measurements for 0.002 mm (0.0001 in.) increments. However, they are extremely valuable because they provide highly reliable measurements to 0.01 mm increments and increased reliability of smaller measurements, such as 0.006 mm increments.

The principle of measurement with the vernier micrometer is exactly the same as for vernier calipers. The vernier scale is on the barrel (see Figure 7–12). It consists of 11 equally spaced lines. The first 10 are numbered zero to 9. The last one is marked with a second zero. The 10 divisions thus created span nine divisions on the thimble (see Figure 7–15). One vernier division, therefore, is $1/10 \times 9/1000$, which equals 9/10000. The difference between a vernier and a thimble division is 10/10000 less 9/10000, or 1/10000. That is a reading of 0.002 mm (0.0001 in.) and is found by matching a vernier graduation to a thimble graduation. The steps for reading a vernier micrometer are shown in Figures 7–15 and 7–16. They are identical to the steps for reading a plain micrometer except that the 0.01 mm (0.0001 in.) is added. Note that when the micrometer is exactly on the 0.01 mm (0.001 in.), readings of both zero graduations on the vernier scale coincide with lines on the thimble. Practice measuring a gage block with vernier micrometers, as described in the lab manual. How tight does the micrometer feel against the part? Is your part aligned squarely against the micrometer blades? This practice allows you to make accurate, repeatable measurements.

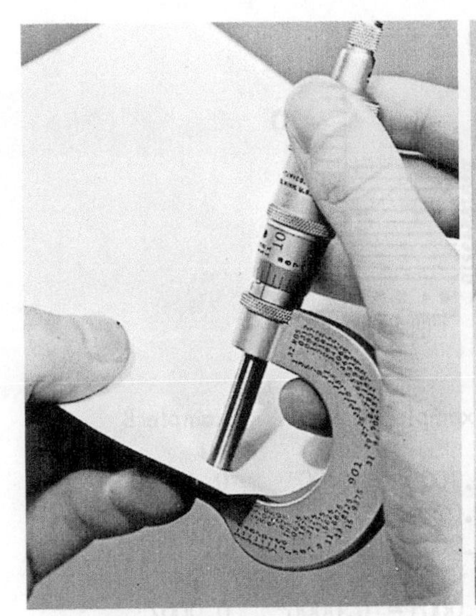

1. Carefully clean the measuring surfaces by pulling a piece of soft paper between the surfaces while they are in light contact with the paper. Do not use a hard paper.

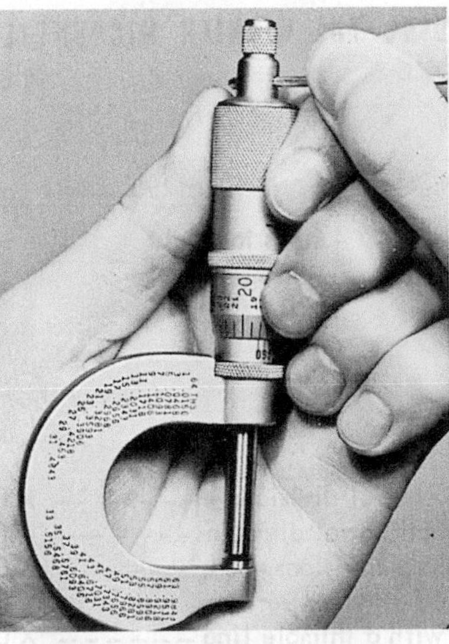

2. With the anvil and spindle apart, unlock cap with spanner wrench, then tighten cap lightly with fingers to bring slight tension between thimble and spindle.

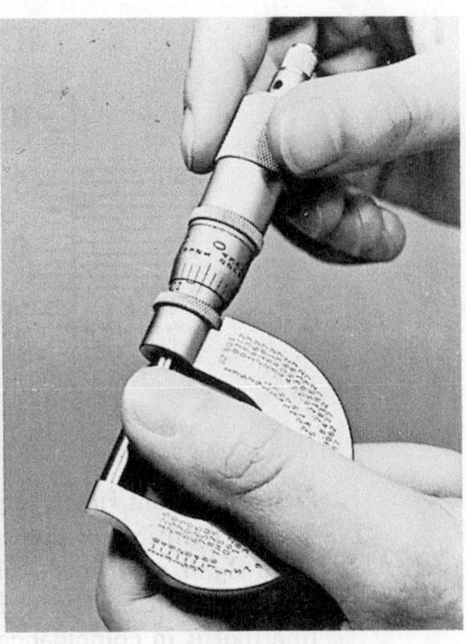

3. Bring anvil and spindle together by turning spindle and set zero line on thimble to coincide with line on barrel.

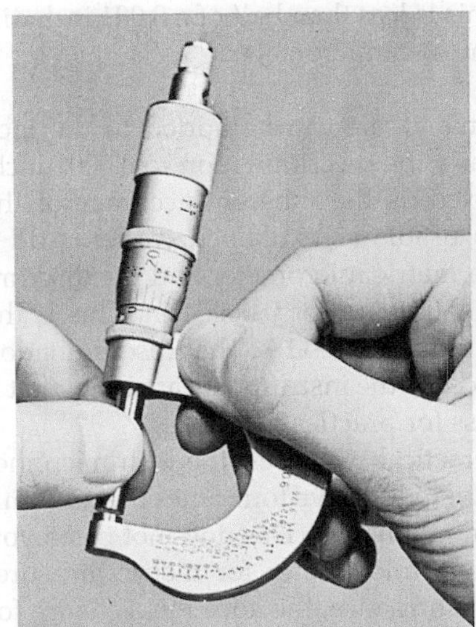

4. Move spindle away from anvil by turning spindle, not by turning thimble.

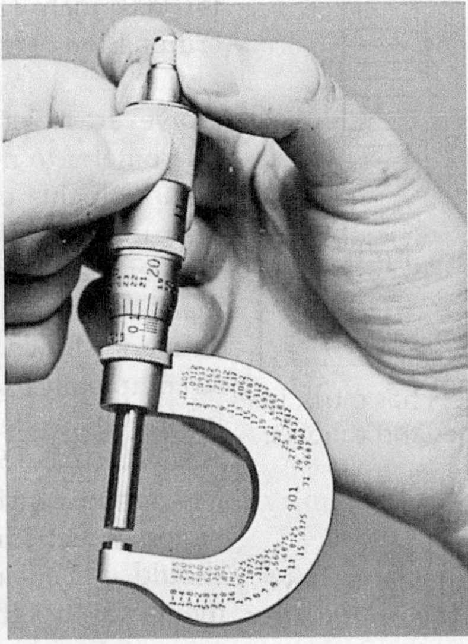

5. Holding thimble only, tighten cap with fingers. Be careful not to touch frame.

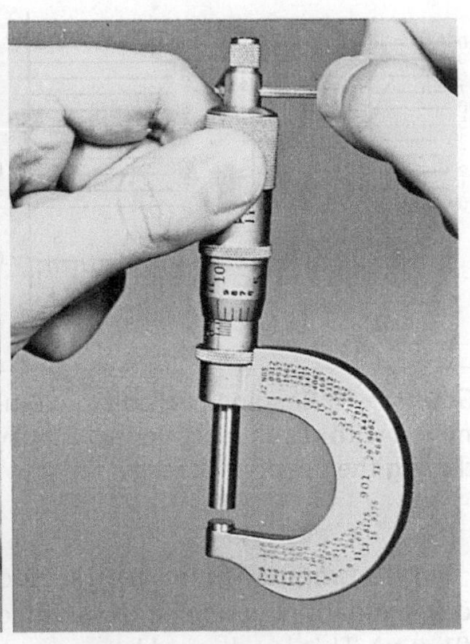

6. Lock cap with wrench still holding thimble only, and adjustment is complete.

FIGURE 7–14 This is a typical procedure to zero set a micrometer. Be sure to grip the micrometer as shown. *(Courtesy of Hexagon Metrology, Inc., TESA)*

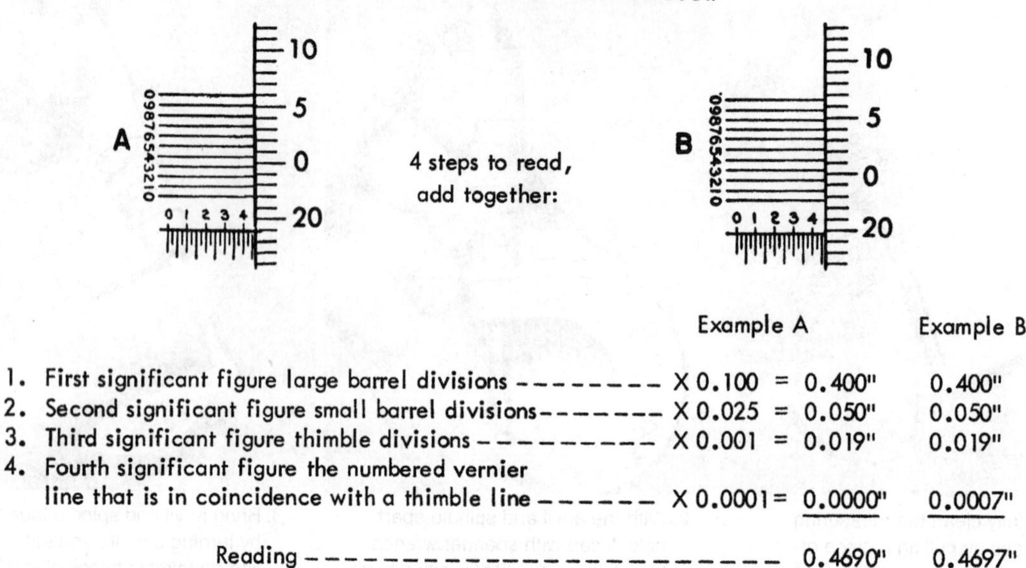

FIGURE 7–15 The vernier micrometer is read the same as a standard micrometer.

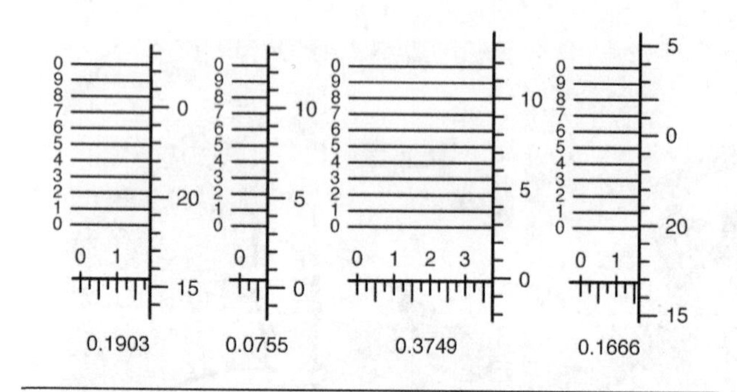

FIGURE 7–16 Reading micrometer verniers has much in common with reading vernier height gages and calipers, including the need for practice.

SI Measurements. SI or metric micrometers have a discrimination of one hundredth of a millimeter (0.01 mm). This is achieved by a screw with a one-half millimeter pitch. Here is another instance in which the inch-pound system happens to fit practical situations better than the metric system. A plain metric micrometer provides better discrimination than does a plain inch micrometer. One hundredth of a millimeter is 0.0004 inch. This is less than half of a 0.001 inch discrimination of the inch micrometer.

However, a vernier scale can be added to the inch micrometer to raise the discrimination to 0.0001 inch. Although this exceeds the inherent accuracy of the instrument, it is usable as explained earlier. Add a vernier scale to a metric micrometer and the discrimination becomes 0.001 mm or 0.00004 in. This is the situation shown in Figure 7–11A. That discrimination so completely exceeds the instrument capability that it is virtually useless for practical purposes.

Thus, for practical purposes, the inch micrometers may be considered to have four times the discrimination of metric micrometers. This does not mean you should use an inch micrometer for metric measurements. Besides the nuisance, the additional chance for computational errors would eliminate any advantages.

Digital Micrometers

In Chapter 6, it was shown how both mechanical and electronic amplification have been applied to vernier

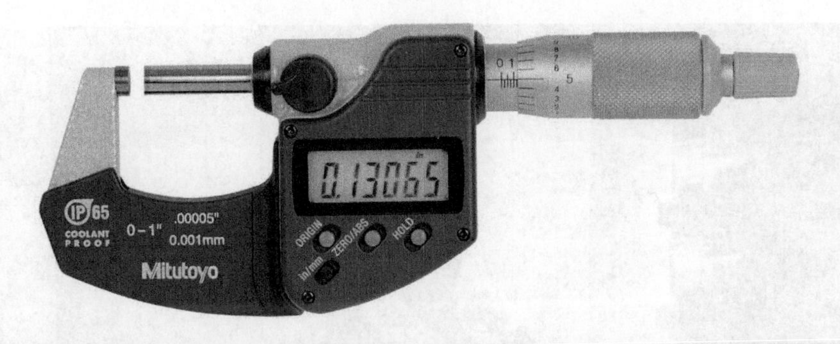

FIGURE 7–17 This micrometer has a digital readout that can be converted from inches to millimeters. *(Courtesy of Mitutoyo American Corporation)*

instruments in order to simplify their readings. This has also been done for *micrometer instruments*. In both cases, they provide digital readouts.

Figure 7–17 shows a micrometer with a digital readout. The resolution on this micrometer is to 0.00005 inch (0.001 mm). With this discrimination, the last digit displayed would read a zero or a five when measuring in inches.

The obvious advantage of this micrometer is the ease of reading, although there is another advantage. In the earlier section about micrometer readings, there was a warning that the thimble may accidentally be read in the wrong direction and that there is a chance to read an extra 0.50 mm (0.025 in.). Both of these problems are eliminated by the digital readout.

One of the several electronic versions is shown in Figure 7–18. It is shown in a stand, which allows the metrologist to use them as a bench micrometer. The term multifunction type is coming into use for these micrometers because of the ease with which their readings may be processed. A push of a button converts from the standard SI readings to decimal-inch. Similarly, they may be set to read workpiece size with respect to their preset tolerance.

Considerable statistical information can be obtained directly from the measurements. This information includes standard deviation, maximum, minimum, and mean readings of a lot, and total number of measurements taken. As Figure 7–18 shows, electronic digital micrometers may be connected to a "dedicated" printer. By the same means, two or more instruments can input the same computer.

Some manufacturers make several models for extended ranges, as they do for their conventional micrometers. All have built-in batteries with remarkably long life.

While care must be taken to keep most micrometers from being damaged by dirt and oil, some micrometers are manufactured with the shop floor in mind (Figure 7–20). This micrometer is rated as an IP65. The Ingress Protection rating is a two-digit number established by the International Electro Technical Commission, used to rate a piece of electronic equipment. The two digits represent different forms of environmental influence: the first digit represents protection against the ingress of solid objects; the second digit represents protection against the ingress of liquids. The larger the value of each digit, the greater the protection. For example, a product rated IP58 would be better protected against environmental factors than another similar product rated as IP34.

Consider the advantages of using the micrometers pictured in Figures 7–18 and 7–19 compared to

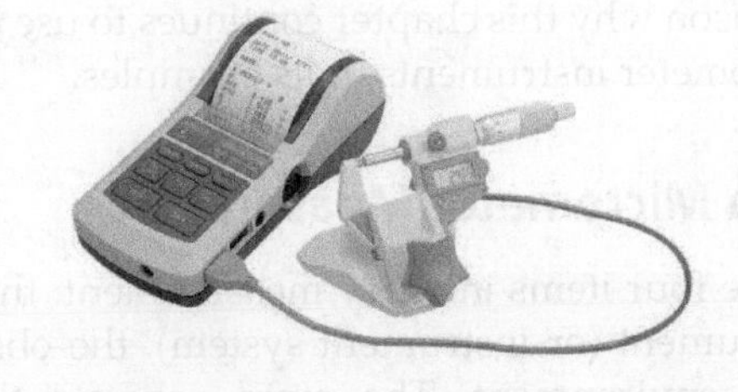

FIGURE 7–18 Digital micrometers with stand and electronic printout. *(Courtesy of Mitutoyo American Corporation)*

FIGURE 7–19 Coolant proof micrometers are designed for environments where cutting oil or fluids are used. *(Courtesy of Mitutoyo American Corporation)*

the early modified version in Figure 7–20. The digital micrometers are without question a convenience. The electronic versions are also capable of inputting to printers and computer systems. However, two points about them must be emphasized. First, they are not a revolutionary breakthrough in *metrology*. Second, they in no way change the basic considerations. That is the reason why this chapter continues to use the basic micrometer instruments in its examples.

What a Micrometer Measures

There are four items in every measurement: the part, the instrument (or instrument system), the observer, and the environment. The more accurate the desired measurement, the more control there must be over these items. Although the total control required increases, the emphasis may change, as shown in Figure 7–21. Environment is a constant in these limited examples and has been omitted.

The control of part conditions requires flatter surfaces, better surface finish, and sharper edges. Obviously, a measurement cannot be made to greater precision than permitted by these part conditions. Control of inherent instrument accuracy does not imply that greater care is taken in the manufacture of micrometers than of vernier instruments. Satisfactory manufacturing standards are presumed for both. This is primarily a measure of design features. Control of the observer includes all steps, from care of the instrument to any computation required to put the measurement in usable form, and there are many.

FIGURE 7–20 Nearly three-quarters of a century ago, Carl Johansson, while developing gage block set combinations, modified a micrometer to give better control. He did not use it to measure gage blocks as has often been stated. Gage blocks are used to measure micrometers, not vice versa. He used it to control manufacturing operations up to the final sizing.

In the case of the micrometer, the manufacturer has built some of the control that the user would have to furnish with the less precise vernier caliper.

The basic measurement act with a micrometer is the same as with all previous instruments. The instrument is set to duplicate in itself the separation between the reference point and the measured point of the part feature. In previous instruments, this required parallel paths for the line of measurement and the axis of the instrument. Because these can never be perfectly parallel, some serious errors were expected. With the micrometer, the axis of the instrument lies along the line of measurement. As predicted by Abbe's law, this minimizes error. Any variation from this caused by faulty use will show up as an error in the micrometer reading.

The spindle tip and anvil of the micrometer are areas that provide both the advantage and the problem of micrometer measurements. Accurate measurements are made only between external points, external lines, certain external planes, and combinations of these (see Figure 7–22). The certain planes referred to are parallel planes that can be brought into coincidence with the planes of the micrometer's contact surfaces. Although these are limitations caused by the

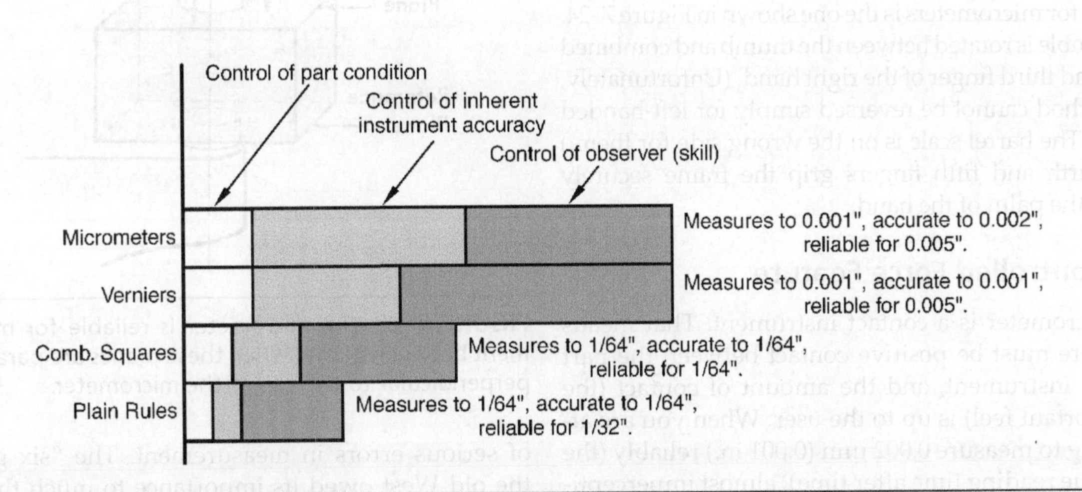

FIGURE 7–21 The total control increases as the requirements become more stringent. The emphasis shifts from the user to instrument maker.

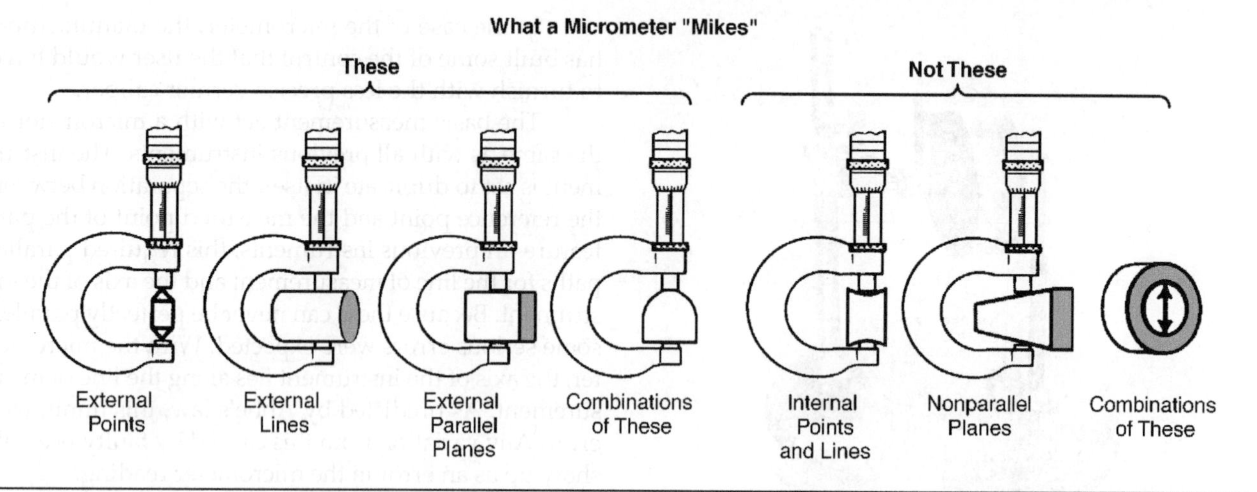

FIGURE 7–22 Although versatile, a micrometer is completely inflexible in use. It only produces reliable measurement in the first four cases. Unfortunately, the conditions shown by the next two often cannot be detected by the eye.

area of contact, these planes of the micrometer help it locate squarely on the reference surfaces of the part (see Figure 7–23).

7–2 USING THE MICROMETER

There is no correct way to hold a micrometer that is suitable for all measurements. By far the most useful method for micrometers is the one shown in Figure 7–24. The thimble is rotated between the thumb and combined index and third finger of the right hand. (Unfortunately, this method cannot be reversed simply for left-handed people. The barrel scale is on the wrong side for them.) The fourth and fifth fingers grip the frame securely against the palm of the hand.

The Controlled Force Feature

The micrometer is a contact instrument. That means that there must be positive contact between the part and the instrument, and the amount of contact (the all-important feel) is up to the user. When you are attempting to measure 0.002 mm (0.001 in.) reliably (the same true reading time after time), almost imperceptible differences in gaging force can be very important. Because human beings vary so widely, this is a source

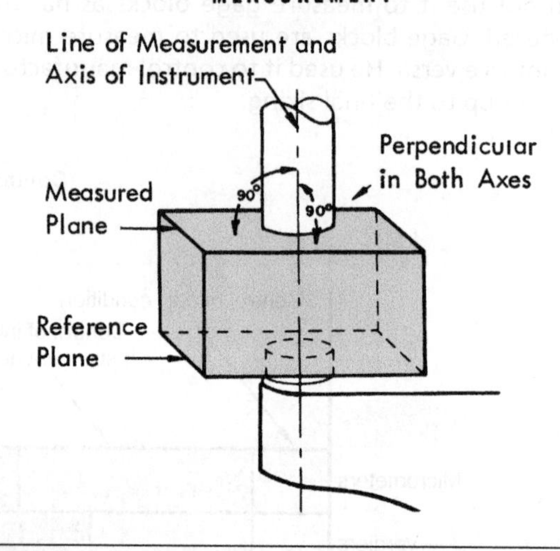

FIGURE 7–23 The micrometer is reliable for measurement between planes when these planes are parallel and perpendicular to the axis of the micrometer.

of serious errors in measurement. The "six gun" of the old West owed its importance to much the same reason. It was known as the "great equalizer" because the big bully and the undersized whelp were equally

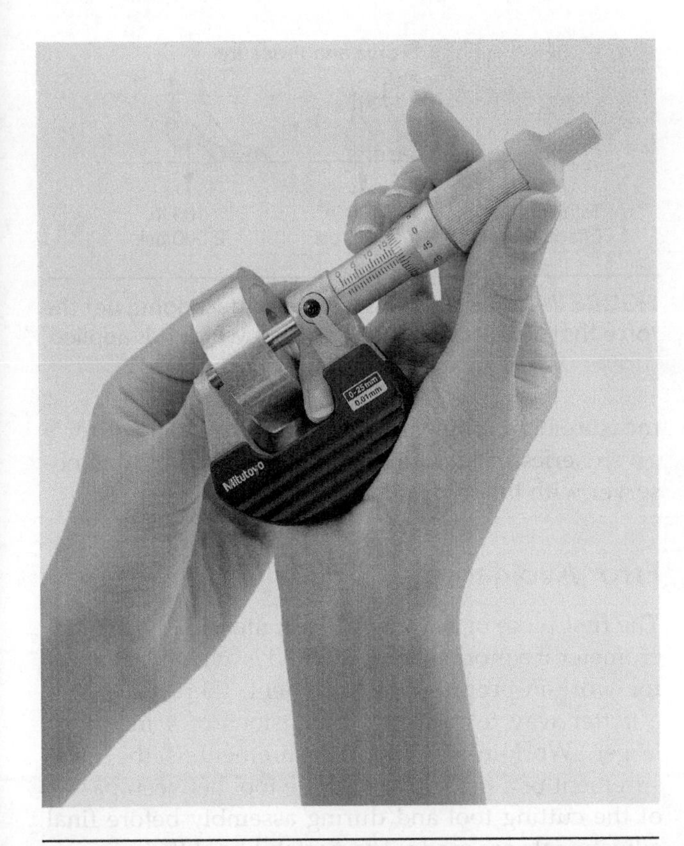

FIGURE 7–24 Practice will help you to judge the correct gaging force for reliable measurement. *(Courtesy of Mitutoyo American Corporation)*

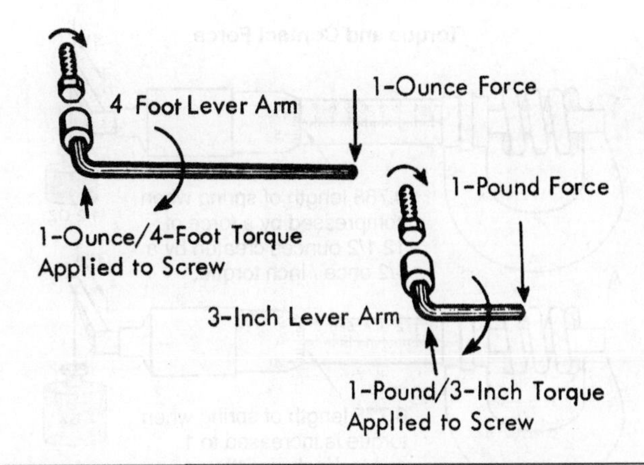

FIGURE 7–25 Both of these wrenches are applying the same torque to the screw: 1 pound/4-ft. torque = 16 oz./48-in. torque = 1 oz./3-in. torque.

persuasive with its aid. So it is with the controlled force feature on micrometers. There are two mechanical types: friction thimbles and ratchet stops.

Basically, the ratchet stop is an overriding clutch that "kicks out" at a predetermined torque. This holds the gaging force to the same amount for each measurement regardless of the firmness in human physiques.

Torque is the amount of twist. It is based on the principle of the lever (see Figure 7–25). When torque is applied to the rotation of a screw thread, force results. Force is the forward push of the thread as it is turned by the torque. In a constant instrument, it is the force that develops the *feel* of the contact. This is shown in Figure 7–26. All material is somewhat elastic. Whenever there is the feel of contact, the instrument and/or the part are being distorted. This is symbolized by the spring in the illustration. Instead of fingers turning the ratchet stop, a lever arm has been attached. Weights hanging from the arm create the torque, which generates the force, which in turn squeezes the spring.

With the 1-inch lever arm and 0.5 ounce weight in Figure 7–26A, you should expect a relatively large force because of the amplification of the screw thread. Threads, however, are a very inefficient means for the power transmission. Figuring only 100% efficiency, the force against the spring is squeezed down to a length of 0.783 inch as read on the micrometer. As the weight is increased, the increased force closes the micrometer tighter, as shown in Figure 7–26B. At 1 ounce, the force is 25 ounces, and the spring is compressed to 0.775 inch. If 1 ounce is the limit at which the ratchet stop is set to disengage, any further addition of weight does not change the reading. The clutch simply "kicks out" before the spring is compressed any further. The 1-pound weight in Figure 7–26C results in the same 0.775 inch reading that the 1-ounce weight produced.

Note that the term gaging pressure is often used to mean *gaging force.* Force is the total push in units of weight. Pressure is force acting on a specified area, such as pounds per square inch (psi). It is pressure that we are concerned with in measurement, not force. It

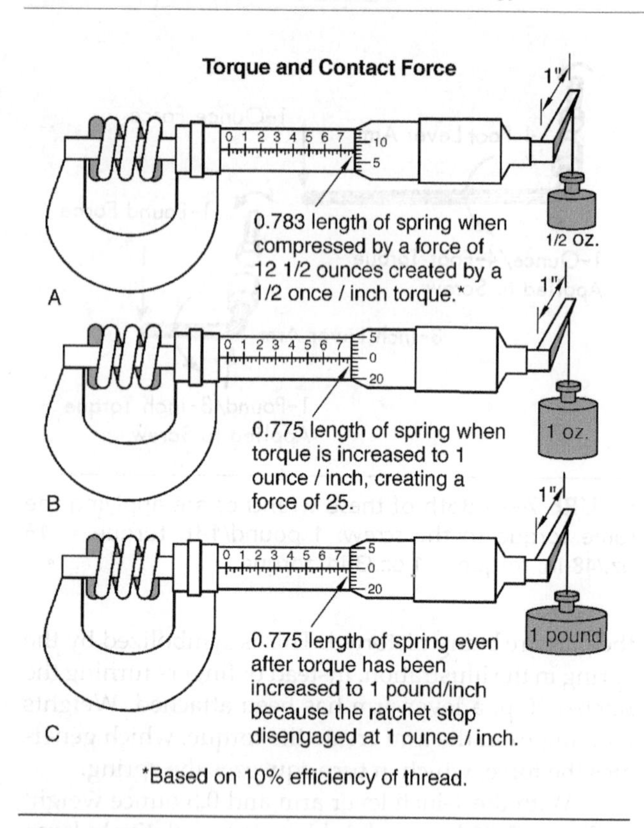

FIGURE 7–26 The amount that the spring is compressed depends on the thimble torque but is limited by the ratchet stop.

can be deceiving, as shown in Figure 7–27. The heel of the right shoe exerts many times the pressure against the floor than that of the left one.

When a micrometer is used, the frame is sprung open and the part is compressed. The ratchet stop on a micrometer is not intended to eliminate the compression of the work and expansion of the micrometer frame. It is intended to equalize this effect for all measurements. Unfortunately, it does not succeed.

The reasons that the ratchet stop does not ensure completely repeatable readings are varied. Of course, there are bound to be some mechanical variations within the ratchet stops.

The friction of the screw in the nut is another important one. A tight adjustment will use more of the torque to overcome friction than to apply to the

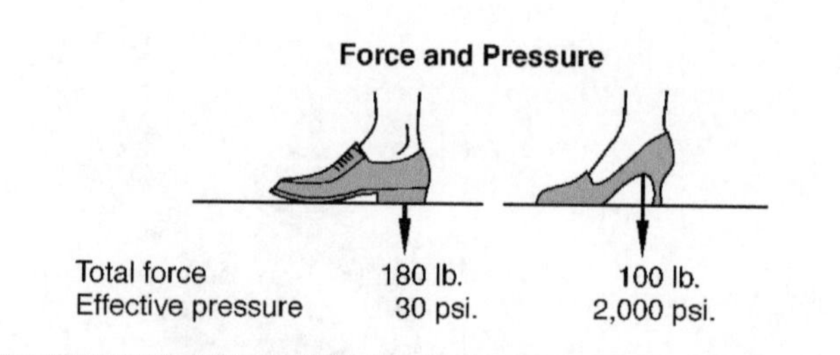

FIGURE 7–27 It is the pressure of a contact point, not the force that indents the surface against which it is applied.

measurement. This variable should be constant for a given series of measurements performed by one observer with the same micrometer.

Error Avoidance

The first piece of advice is do not measure with a micrometer if a more reliable method is available. Except for work-in-progress measurement, there usually is a better way to measure dimensions of 4 inches or larger. Work-in-progress measurement is the measurement of a part on a machine tool between passes of the cutting tool and during assembly before final adjustments are made. The portability of the micrometer makes it suitable for these uses. Although the height gage was thoroughly castigated in Chapter 6, its use in conjunction with electronic indicators and gage blocks is usually more reliable for large dimensions than micrometers. The balance of the advice on alignment errors requires that we distinguish between flat and cylindrical measurement.

Measuring Flat Parts

The proper way to approach the part size is shown in Figure 7–28. Before placing the micrometer on the part, bring it to nearly the desired opening. Do this by rolling the thimble along the hand or arm—not by twirling. When placing the micrometer onto the reference plane of the part, hold it firmly in place with one hand. Use the feel of stability (no rock) to show when the axis of the micrometer is perpendicular to the reference plane. Rapidly close the micrometer using the

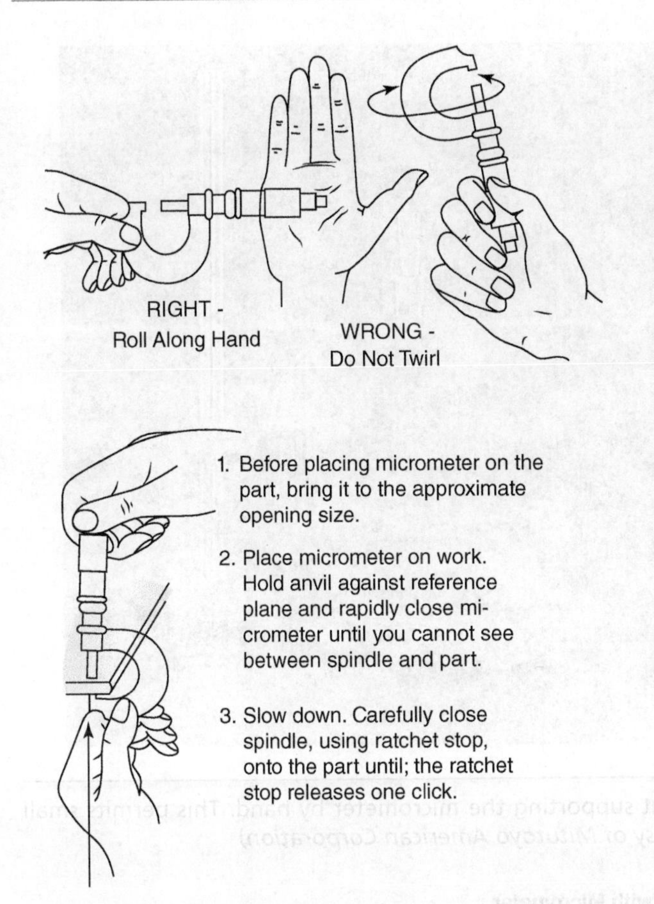

FIGURE 7–28 Often the fastest way to measure reliably is to slow down.

ratchet until the spindle is nearly on the measured plane of the part. This usually can be determined visually. If the part is hit before expected, back off slightly and then slowly and gently close the spindle until the ratchet stop disengages one click.

Note that this procedure requires two hands. If the micrometer is handled with only one hand, the ratchet stop cannot be reached, and reliability will suffer. This works for any part that is large enough to support itself without moving around during measurement. It is awkward for small parts. Tool stands are available to hold the micrometer so that one hand may be used on the ratchet stop while the other hand is free to handle the part (see Figure 7–29). A tool stand is easily improvised if one is not available.

When the part is large, the same procedure is followed, but greater care must be taken in alignment. The feel of the anvil "homing" against the reference surface is very deceptive. With a large micrometer, the lever arm is so great that an almost imperceptible misalignment can indent the anvil into the part or spring the micrometer frame. A simplified and exaggerated example is shown in Figure 7–30.

Measuring Cylindrical Parts

Books on measurement and manufacturers' literature abound with photos of shafts and other diameters being measured with the ratchet stop. As much as the ratchet stop is to be recommended, it is only suitable for diameters that are very small or very large.

Accurate measurement of cylinders requires that the line of measurement be a diameter. When the diameter of the part is no larger than the diameter of the anvil and spindle, the alignment problem is simple, as shown in Figure 7–31A. If the micrometer is closed and the diameter overhangs the anvil and spindle, the part pops out. Otherwise, the micrometer is closed to the desired feel or until the ratchet stop disengages.

It would be difficult to have the condition shown in Figure 7–31B except for thin wall tubing. Compare this with an odd-shaped part, such as the one in Figure 7–31C. It is entirely possible to have severe misalignment and still produce the same feel as in Figure 7–31A.

As expected, the problems in cylinder measurement become greater as the diameter increases. In Figure 7–32, A represents the case for very small diameters. As the diameter approaches 1 inch, the danger of condition B becomes serious. It is easy to measure a 25 mm diameter standard (micrometer checking disk) undersized by 0.01 mm. Its fine finish does not give as much of a feel as a part with a rougher surface. A relatively small cylinder as in B usually can be aligned by eye to the axis of the micrometer. When the part is quite large, say 4 inches or greater, the eye is not reliable for alignment as at C.

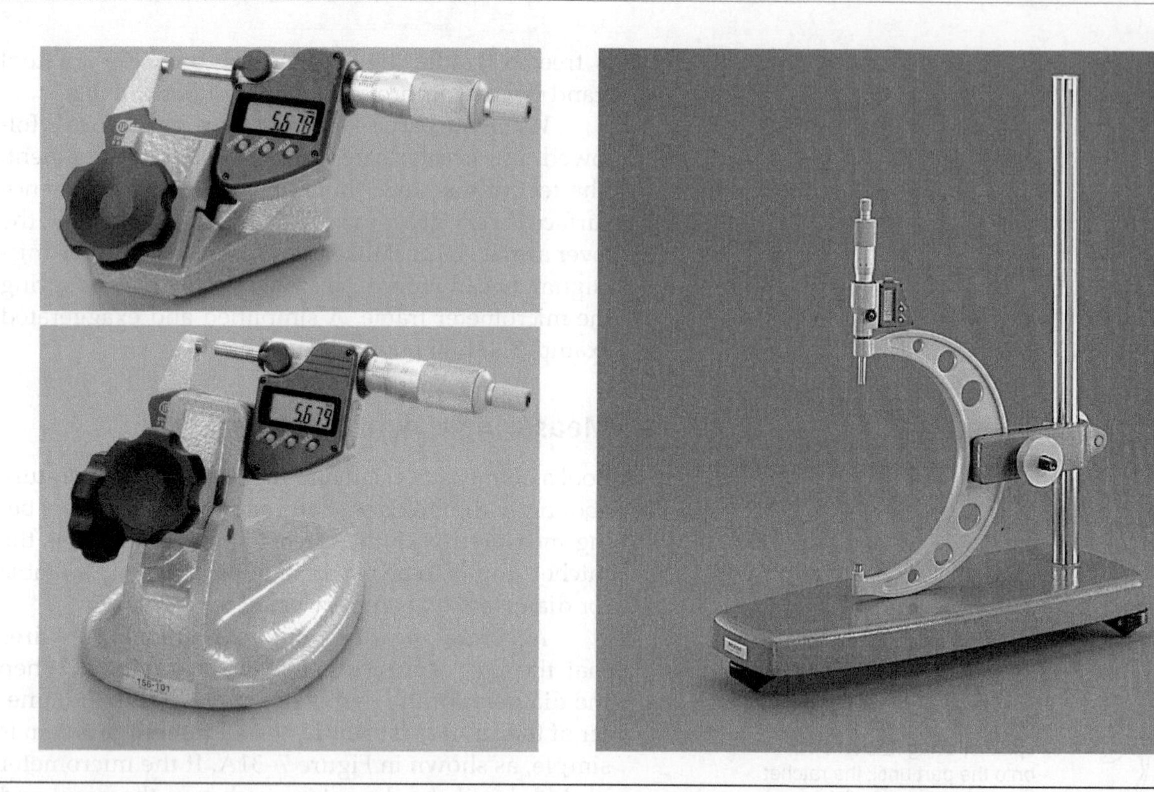

FIGURE 7–29 A tool stand permits measurement without supporting the micrometer by hand. This permits small parts to be measured using the ratchet stop (left). *(Courtesy of Mitutoyo American Corporation)*

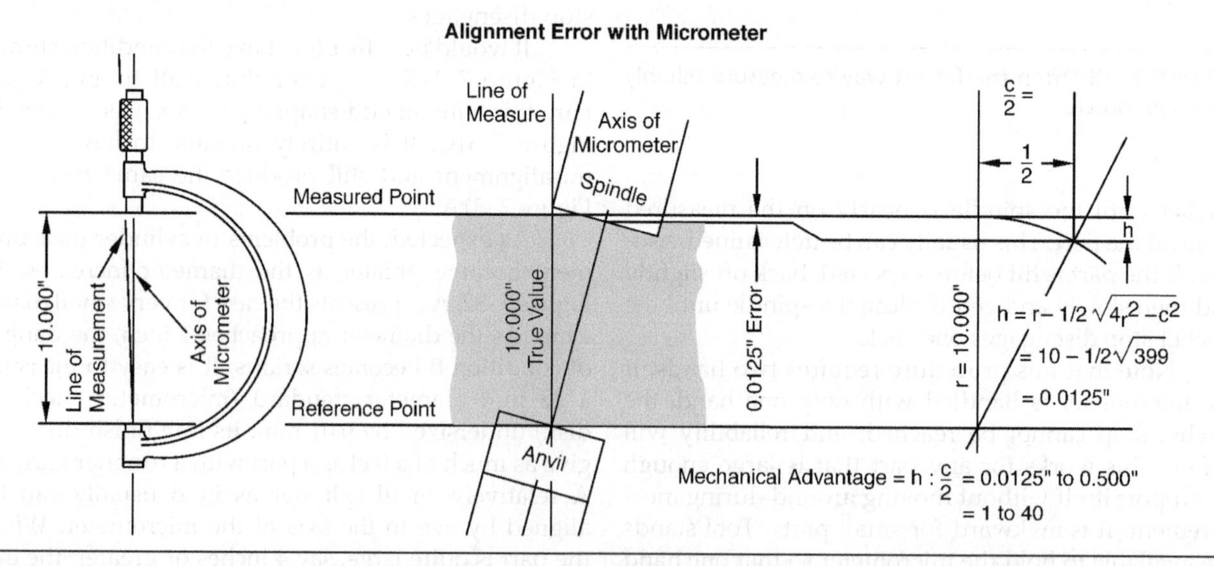

FIGURE 7–30 These oversimplified drawings show that a ½-in. misalignment when measuring 10 in. has produced an error of 12½ mil (0.0125 in.), and the 40-to-1 mechanical advantage easily forces the spindle and anvil into the part being measured.

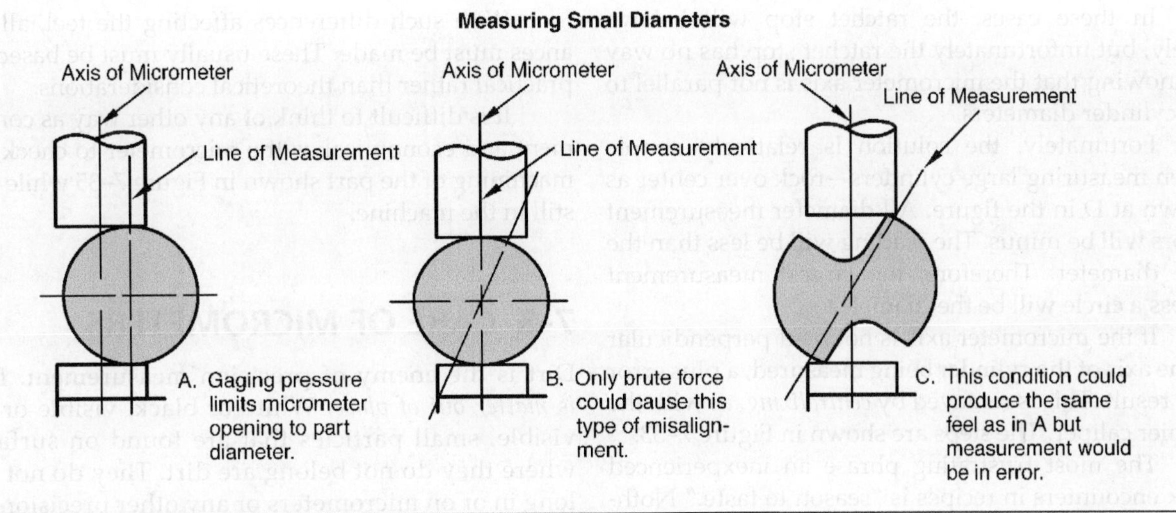

FIGURE 7–31 Circles generally provide the same diameter any place through the center. That eliminates alignment problems for very small cylinders.

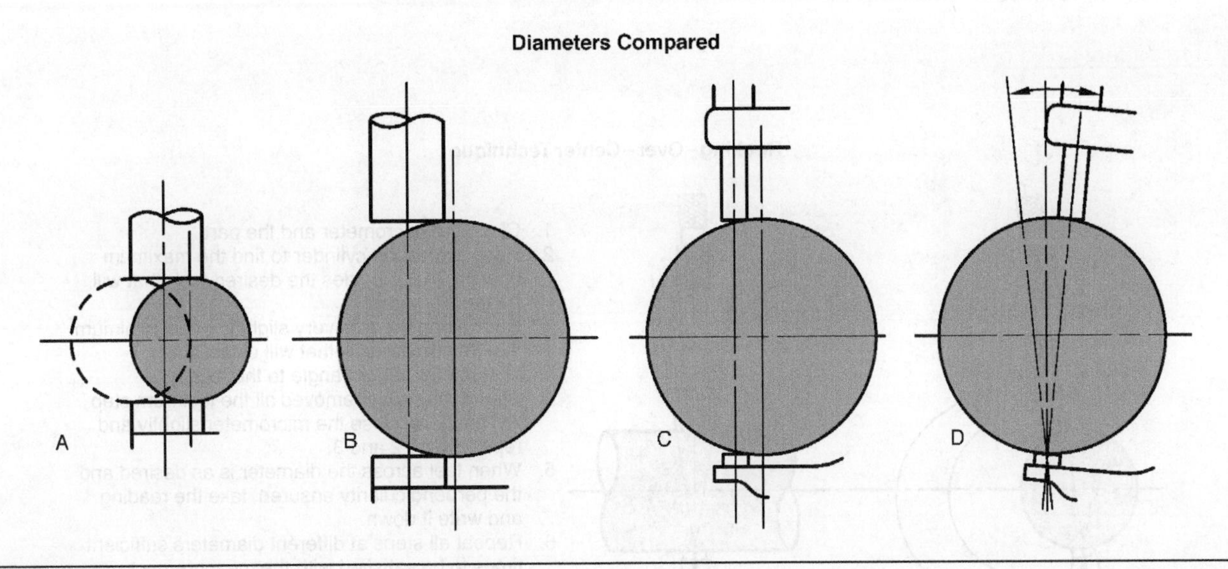

FIGURE 7–32 The ratchet stop can be used for very small cylinders, but rocking over center is required for large cylinders.

In these cases, the ratchet stop will behave nicely, but unfortunately the ratchet stop has no way of knowing that the micrometer axis is not parallel to the cylinder diameters.

Fortunately, the solution is relatively simple when measuring large cylinders—rock over center as shown at D in the figure. All diameter measurement errors will be minus. The reading will be less than the true diameter. Therefore, the largest measurement across a circle will be the diameter.

If the micrometer axis is not held perpendicular to the axis of the cylinder being measured, a plus error will result. This is removed by *centralizing*, as with the vernier caliper. The steps are shown in Figure 7–33.

The most frustrating phrase an inexperienced cook encounters in recipes is "season to taste." Nothing will save the dish at that point except experience or luck. That is the way it is with the feel required in step 5. Only experience will teach how much feel is the right amount so that the micrometer reading is consistently close to the true value. Some of the factors involved are itemized in Figure 7–34.

With such differences affecting the feel, allowances must be made. These usually must be based on practical rather than theoretical considerations.

It is difficult to think of any other way as convenient and economical as the micrometer to check the machining of the part shown in Figure 7–35 while it is still in the machine.

7-3 CARE OF MICROMETERS

Dirt is the enemy of precision measurement. *Dirt is matter out of place.* White or black, visible or invisible, small particles that are found on surfaces where they do not belong are dirt. They do not belong in or on micrometers or any other precision instrument. Although there are hand tools designed to use in cutting oils and fluids, they should be cleaned at the end of the day and stored in a protected location.

After each use, the micrometer should be wiped clean and put in a safe place, preferably in its own

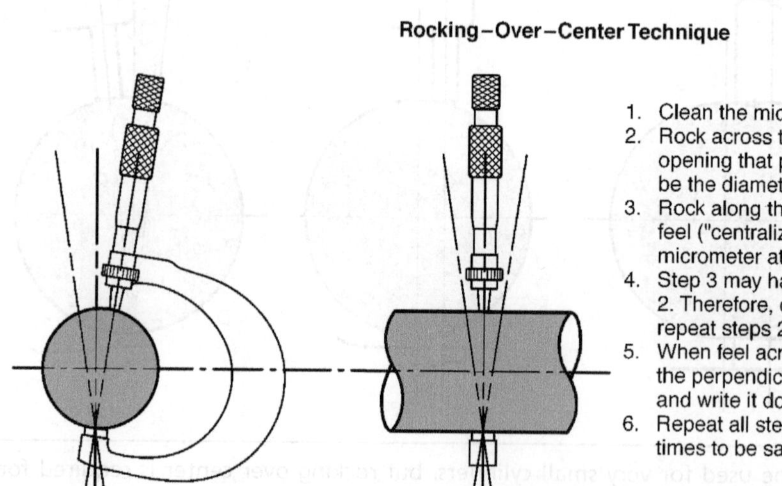

FIGURE 7–33 These steps are automatic for skilled metrologists.

External Micrometer Measurement Summarized

Flat Parts

1. Clean the contact surfaces of the part and of the micrometer.
2. Open slightly larger than part feature.
3. Seat anvil squarely against reference surface of part.
4. Using ratchet, slowly close the micrometer until the ratchet clicks once.
5. Record reading.
6. Repeat entire procedure several times and average readings.

Cylindrical Parts

1. Clean the contact surfaces of the part and of the micrometer.
2. Open slightly larger than part feature.
3. Seat anvil squarely against reference surface of part.
4. Rock back and forth across diameter, closing micrometer by small steps.
5. When first contact is felt, rock sideways to find position over center.
6. Repeat steps 4 and 5 until perpendicular position is found and spindle just contacts the measured point as it passes over center.

Feel in Measurement

Influencing Factors:

1. Size of the part. If the micrometer is very large, it will be awkward and may be heavy. The person using it must be more intent on support than on feel.
2. Closely related to the first is the position of the measurement. Even a 25-millimeter micrometer may provide inadequate feel if used at arm's length through a recess in a large machine.
3. Shape of part. If the feel is checked against gage blocks and then duplicated on a cylindrical part of the same size, the reading will vary 0.02 mm or more.
4. Surface finish affects the feel. A coarse finish will produce a more pronounced feel than a fine finish.

Micrometer Do's and Don'ts

Clamp Ring:

1. Do use it as a memory device to preserve a reading until repeated.
2. Don't use it to make the micrometer into a snap gage.

Ratchet Stop:

1. Do use it for every measurement between flat surfaces.

Don't Expect It to Guarantee Reliable Measurement if:

1. the micrometer is dirty.
2. the micrometer is poorly lubricated.
3. the micrometer is poorly adjusted.
4. the micrometer is closed too rapidly.

Don't Use It for Measuring Diameters Unless They Are Very Large or Very Small.

FIGURE 7–34 The rules for reliable measurement with micrometer instruments may be stated concisely, but only experience will develop the necessary skills.

FIGURE 7–35 Micrometers are a convenient tool to check work in process. *(Courtesy of Mitutoyo American Corporation)*

case. Similarly, at the end of each day of use, it should be wiped clean, visually inspected, oiled, and replaced in its case, to await its next use.

This is a good place to mention oil on precision instruments. It is highly essential on steel surfaces for lubrication and to protect against both rust and corrosion. Perspiration and cutting fluids can be highly corrosive. The actual rust or corrosion will not be visible unless the instrument has been grievously neglected. That does not mean it does not exist. Invisible corrosion is rubbed off when the micrometer is used. This needless loss accumulates and virtually destroys accuracy.

A thin covering of oil has virtually no effect on the measurement until 0.002 mm or less is measured. Unfortunately, the air is full of dust, and dust collects on oiled surfaces. Although the oil film may be of negligible thickness, the buildup of dust can create errors of 0.002 mm or more. Worst of all, most dust is composed of silicon particles. These are highly abrasive and can cause rapid wear of fine surfaces.

Never leave a micrometer in the closed position when not in use or tightened on its standard. Finely finished steel surfaces corrode rapidly when left in close contact.

Do not immerse the micrometer in solvent if it sticks. There are three reasons for this: first, it washes away needed lubrication; second, the solvent may carry the foreign matter further in between the sliding surfaces, possibly causing abrasive particles to become embedded and difficult to remove; and third, immersing the micrometer in solvent may camouflage the real reason for sticking, which could be more serious.

If a micrometer sticks, first inspect it carefully for evidence of damage. Then, disassemble it, clean all the parts in *clean* solvent, lubricate them, and reassemble. If the sticking persists, loosen the adjusting nut slightly. If that does not correct it, have the micrometer repaired by a qualified person.

A handy method to clean the contact surfaces is to close the micrometer lightly on a piece of soft paper, and then withdraw. The paper will probably leave fuzz and lint on the surfaces. Wiping the micrometers with lint-free paper is the best practice. *Never use compressed air to clean any precision instrument.* The high velocity forces abrasive particles into the mechanism as well as away from it.

Micrometers are very rugged for precision instruments, but they can be sprung or bent. Never "cramp" or force a micrometer when aligning it on a part. There is a human tendency to make things fit. This is the bias mentioned in Chapter 5. For reliable measurement, you must recognize that you have bias in your actions and consciously correct for it. Every time a micrometer is cramped, there is danger of a small permanent set that could eventually destroy the reliability yet escape detection.

Overtightening is perhaps the worst of all sins of micrometer use. There are three reasons for it: ignorance, bias, and carelessness. Many people using micrometers daily are not aware of the errors caused by overtightening. Some are not sufficiently sensitive to feel for correct use. Certainly these people should use only ratchet stop–equipped micrometers. Unfortunately, many of these people operate boring mills and lathes in which nearly all the measurements are diameters, and ratchet stops do not help. That is one reason why indicator-type instruments are often preferred to micrometers.

Bias has been discussed as a cause of cramping. It is also the largest cause for overtightening. A person should be completely impersonal when measuring. Forget everything about the dimension desired. Seek only one thing—the correct measurement. After you are certain you have measured reliably, and only then, compare the measurement with the desired dimension.

Carelessness causes errors whenever it creeps into measurement. For accuracy, distractions must be avoided. The greatest chance for serious overtightening occurs when measuring diameters. Do not have the opening too large when beginning the measurement because there is a tendency to close it in large steps; you will find there is no feel when the micrometer is open too far. Closing the micrometer too quickly forces it past the center point, which causes the instrument to make contact with the part where the diameter is not at its largest point. The measurement is read undersized before realizing that the micrometers are overtightened, resulting in an incorrect reading.

Inspection of the Micrometer

Calibration of an instrument refers to an important inspection process. The instrument is compared with standards of known accuracy to determine its condition. *Out of calibration* means two things. Most frequently it means that calibration has shown that the instrument no longer conforms with the specified standards that apply to that particular class of instruments. It also means that the stipulated period of time between calibrations has been exceeded and the instrument can no longer be considered reliable.

In a manufacturing operation, the quality control department establishes the period for calibration of the instruments. The period is found by applying statistics to past experience records. In the laboratory or small shop, calibration is equally important but frequently overlooked until something goes wrong. Protect all measurement data. Calibrate all instruments before undertaking critical work. The closer the measurement requirements are to the specified accuracy of the instrument, the more frequently calibration is needed.

Gage blocks and optical flats are now in general use wherever precision measurement is employed. They provide reliable means for calibrating micrometers. The primary features that require calibration are the condition of the contact surfaces and the actual accuracy of the measurement.

The contact surfaces should be checked for parallelism and flatness before the accuracy check. There are several easy methods that can be applied in the shop. A ball, of course, presents nearly a point contact. It therefore provides a convenient means for checking flatness and parallelism of the contact surfaces. The ball "explores" the surfaces as the micrometer is opened and closed on it. Great care must be taken to apply uniform gaging pressure each time and to read the micrometer carefully.

Note: Tests such as this show when something is wrong and requires further attention. *They do not show that nothing is wrong*. They should not be substituted for thorough, periodic calibration. A thorough check of accuracy requires gage blocks. However, a partial but useful check of a 25 mm micrometer can be made at zero and 25 mm in the shop.

After the micrometer is checked for surface defects and burrs, which are then removed, it is closed tightly using the ratchet stop. If the micrometer does not have a ratchet stop, close it with the same gaging pressure you use when measuring. The micrometer should read zero exactly. If not, reset it according to the manufacturer's instructions, as shown earlier in the chapter (see Figure 7–14).

During the zero setting, check the adjustment nut for the desired degree of friction as has been discussed. Also check for excessive looseness that would warn of a worn-out instrument. Be sure that the final zero setting is made after any adjustment. These precautions will certainly improve reliability.

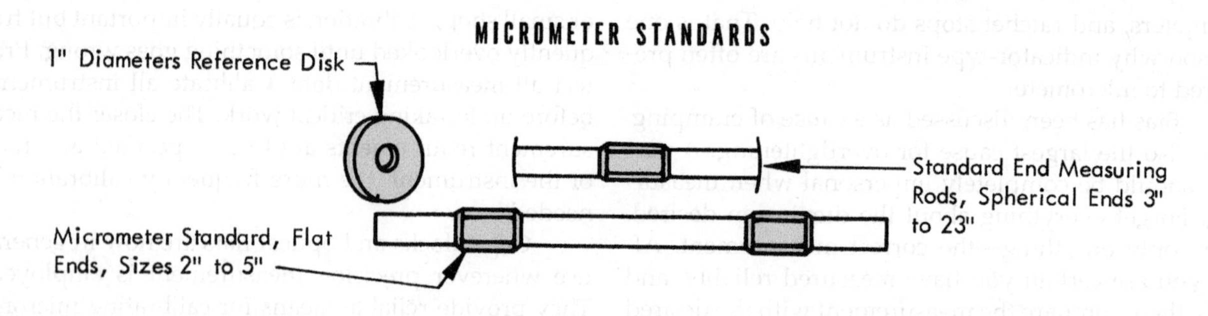

FIGURE 7–36 Standards are used to check the calibration of micrometers. The end surfaces of the standard end measuring rods are portions of spheres. If flat parts are being measured, the flat-ended micrometer standard is preferred for calibration because the feel is nearly duplicated. Similarly, the spherical ends more nearly duplicate the feel of measurements made on cylindrical parts. Gage blocks are frequently used in place of these standards.

Setting Standards

Accurate masters, called *setting standards*, are frequently purchased with micrometers and are usually furnished in sets of micrometers (see Figure 7–36). For 25 and 50 mm micrometers, these are cylindrical-shaped and are called checking disks. For large sizes, they are rods. After the closed setting of the micrometer is checked (against the anvil with a 25 mm size; against a setting standard for larger sizes), the open setting should be checked against a setting standard. If the reading is more than 0.002 mm, the micrometer should receive expert inspection and repair if needed.

Whether a micrometer is sent in for service because of a 0.002 mm error depends on two factors. First, unless you can reliably repeat measurements 0.002 mm (0.001 in.) without helping them along, there is no reason to suspect the micrometer. Second, if the measurements you are making are ±0.01 mm or coarser, it is desirable to have a 0.002 mm accuracy but 0.001 mm should be sufficient. A better solution in cases such as these is to select an instrument of higher precision. The use of gage blocks and comparator instruments is often the answer, as shown in later chapters.

In all maintenance and calibration of precision instruments, it is important to refer to the manuals furnished with them.

7-4 VARIATIONS OF MICROMETERS

Micrometers are a popular and versatile tool, and there are many special application micrometers. Some are included here so you can become familiar with the many types and varieties of micrometers.

The screw thread micrometer (see Figure 7–37) has a 55° or 60° anvil and conical spindle for measuring pitch diameter. The disk micrometer and blade micrometer (see Figures 7–38 and 7–39), permit micrometer measurements of part features that otherwise cannot be reached. The point micrometer (see Figure 7–40) comes with 15° or 30° measuring points. The V-anvil micrometer (see Figure 7–41) measures the outside diameter of cutting tools (taps, reamers, end mills) with an odd number of flutes.

A limit micrometer can be set with upper and lower limits, and thus used as a go/no-go gage (see Figure 7–42). Can seam micrometers (Figures 7–43) come in three options: aluminum, steel, and sprayer cans.

The hub micrometer is primarily used for in-process measurement (see Figure 7–44). Measuring wall thickness of tubing is frequently required and, not surprisingly, there are several ways to "mike" it (see Figure 7–45). The instrument that bridges the gap between shop instruments and gage room instruments is the bench micrometer, shown in Figure 7–46. When not available, the next best thing is the tool stand shown in Figure 7–29.

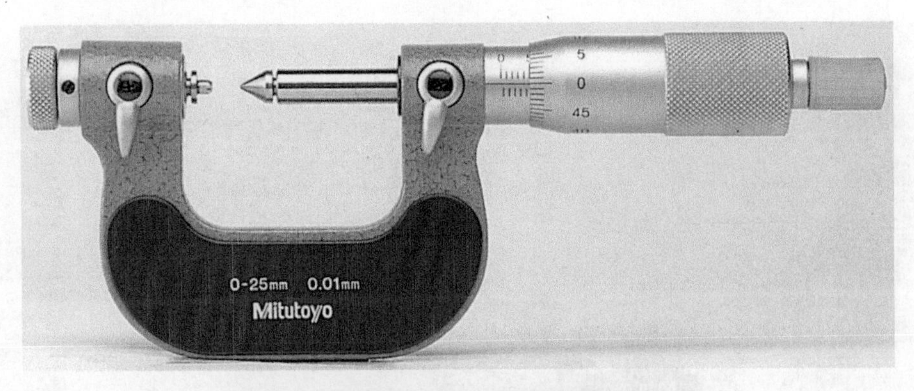

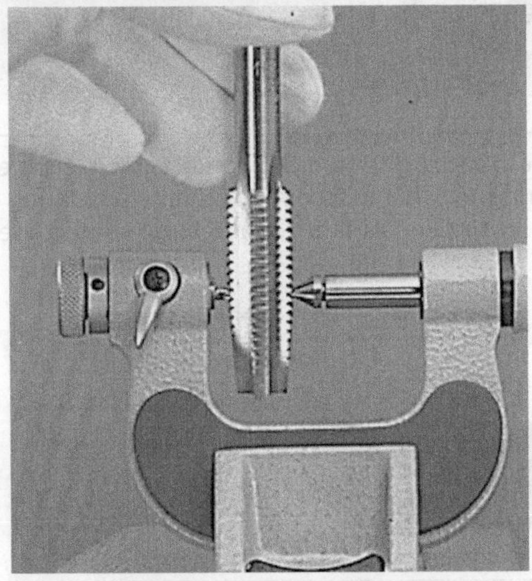

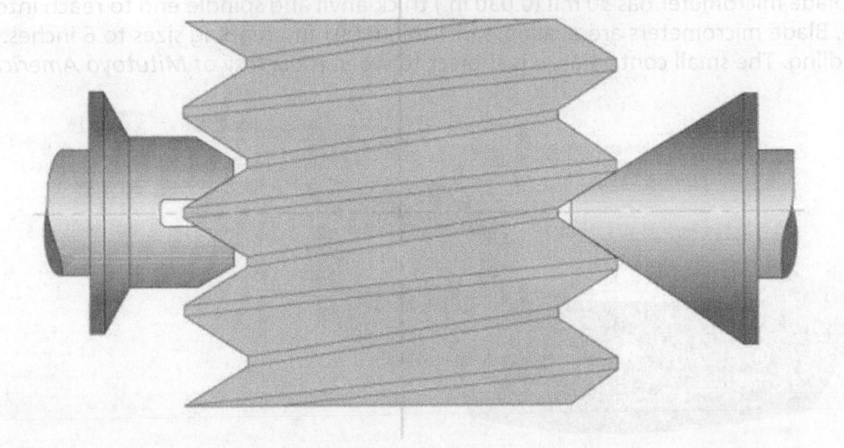

FIGURE 7–37 Screw thread micrometers. *(Courtesy of Mitutoyo American Corporation)*

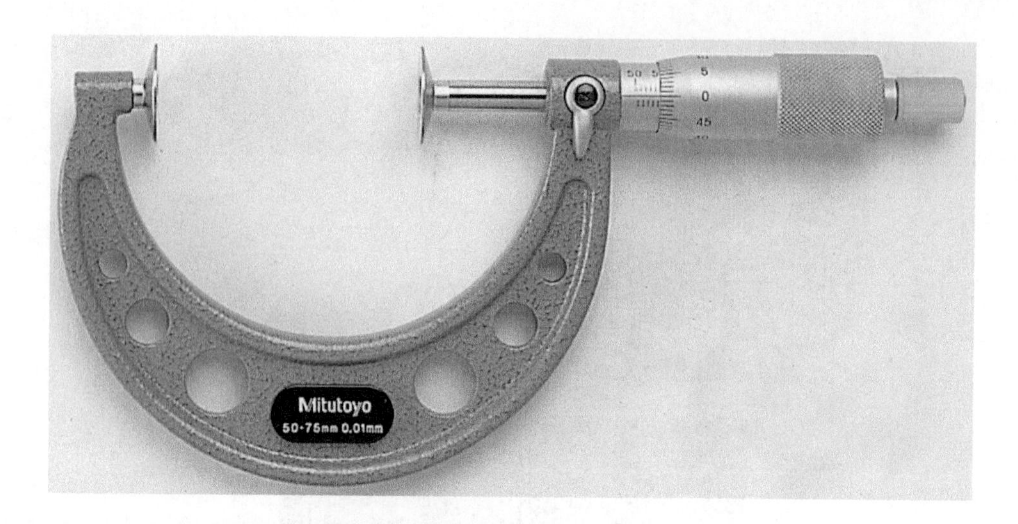

FIGURE 7–38 Disc micrometers have narrow discs for the contacts. This enables them to enter slots as narrow as 1/32 inch. They are useful for measuring to 1 mil (0.001 in.). A similar appearing micrometer but with thicker discs is used for measuring paper and other soft materials that require large contact areas. *(Courtesy of Mitutoyo American Corporation)*

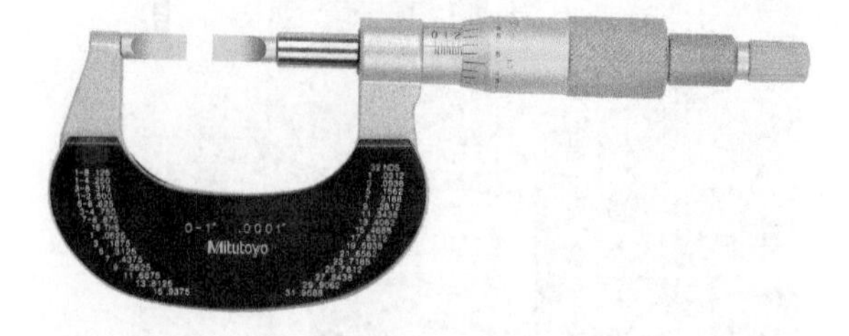

FIGURE 7–39 The blade micrometer has 30 mil (0.030 in.) thick anvil and spindle end to reach into recesses. The spindle does not rotate. Blade micrometers are available in 1 mil (0.001 in.) reading sizes to 6 inches. Blade micrometers require careful handling. The small contact area is subject to wear. *(Courtesy of Mitutoyo American Corporation)*

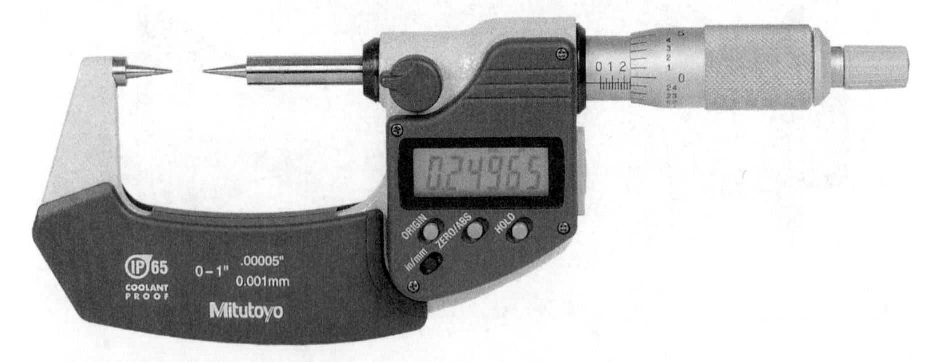

FIGURE 7–40 Point micrometers are made with a pointed anvil and spindle to measure the web thickness of drills, small keyways, and grooves. *(Courtesy of Mitutoyo American Corporation)*

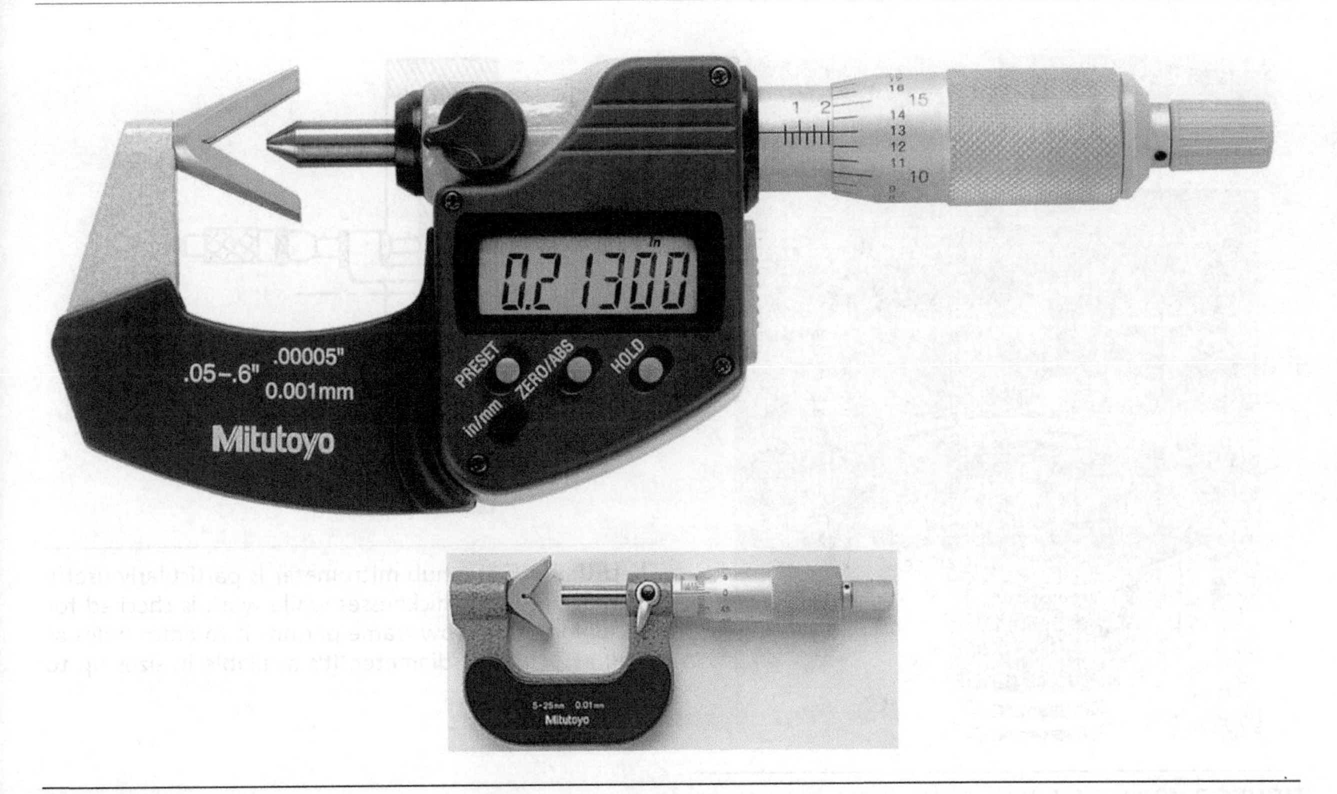

FIGURE 7–41 V-anvil micrometers are used for measuring cutting tools such as taps, reamers, and end mills. *(Courtesy of Mitutoyo American Corporation)*

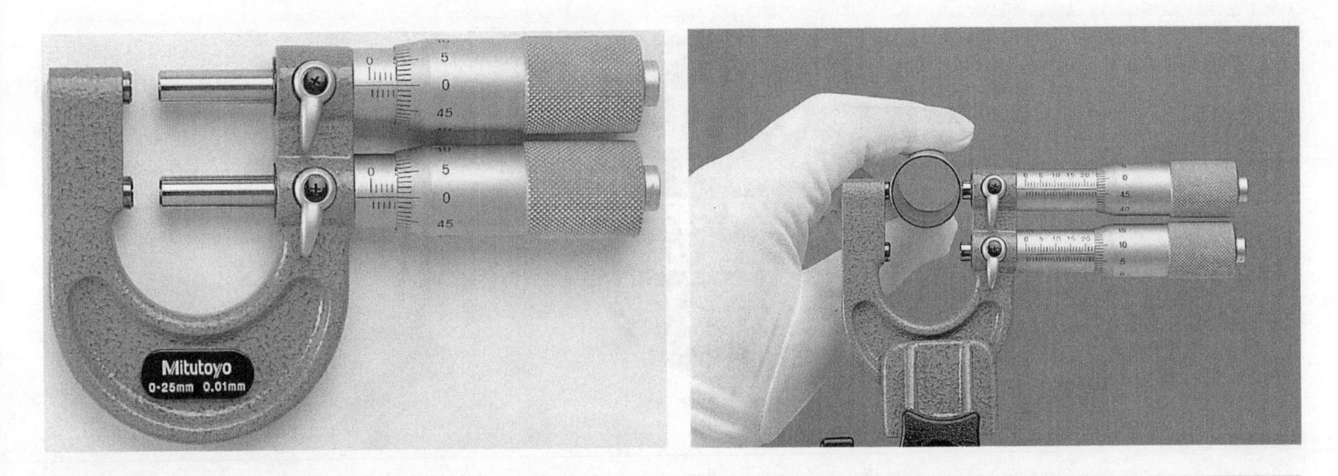

FIGURE 7–42 A limit micrometer can be used as a go/no-go gage by setting the upper and lower limits. *(Courtesy of Mitutoyo American Corporation)*

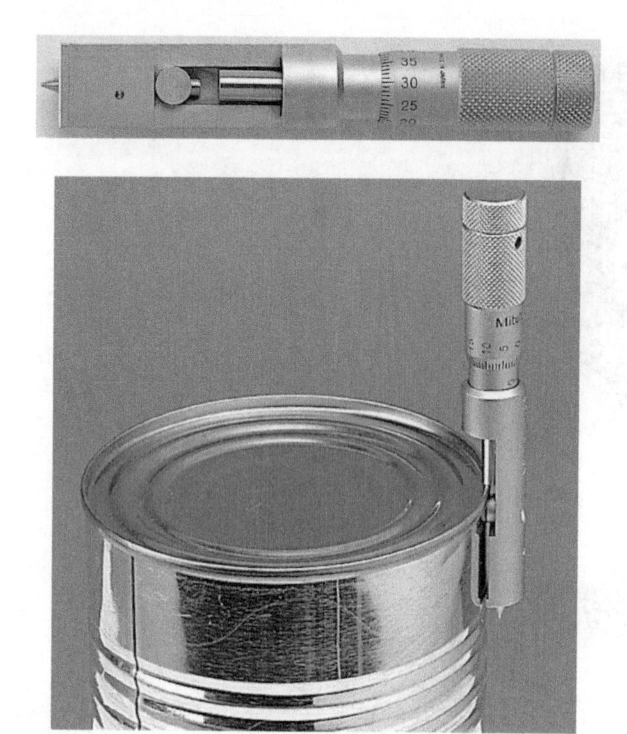

FIGURE 7–43 Many of the canned goods in your home have been inspected with a can seam micrometer. It measures the width, height, and depth of can seams. *(Courtesy of Mitutoyo American Corporation)*

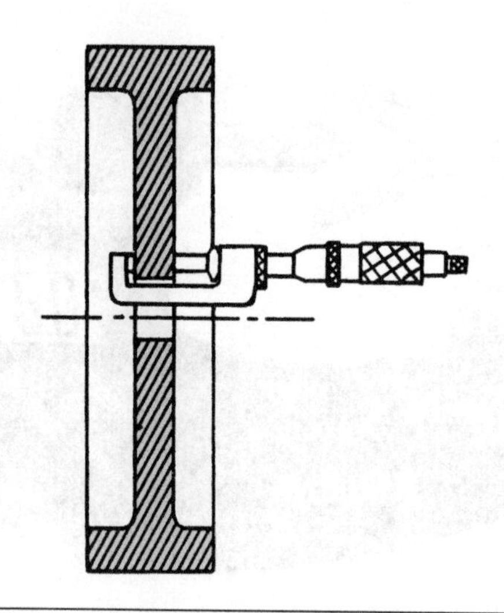

FIGURE 7–44 The hub micrometer is particularly useful for checking web thicknesses while work is checked for machining. Its narrow frame permits it to enter holes as small as ¾ inch in diameter. It's available in sizes up to 5 inches.

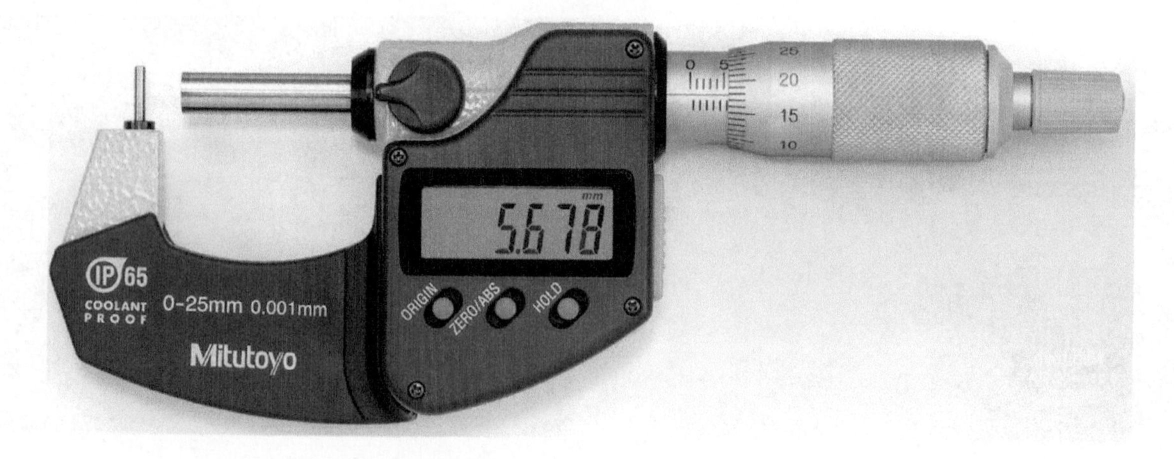

FIGURE 7–45 Several micrometers are specially adapted for measuring tubing. The one shown here has a pin anvil. *(Courtesy of Mitutoyo American Corporation)*

FIGURE 7–46 The bench micrometer is by far the most reliable micrometer for measuring to tenths of a mil (0.0001 in.). Its discrimination is directly in tenths mil. No vernier is required. It owes greater reliability to its superior stability as well as the large diameter of the head. *(Courtesy of L. S. Starrett Company)*

SUMMARY

- Knowledge of micrometers is important because they are among the most frequently encountered instruments.
- Relatively simple and inexpensive instruments can provide high accuracy when they conform with Abbe's law as well as other good design principles.
- Whenever you gain something, you must give up something. Micrometers provide better reliability than do vernier instruments, but at a sacrifice in range and versatility. Think of the range you will be restricted to when you use instruments that measure to 2 μm or 0.0001 in.
- Micrometers show the relationship between discrimination and accuracy. At the same discrimination as vernier instruments, they provide greater accuracy (less error). However, if a vernier scale is added to a micrometer, its discrimination then exceeds its accuracy. The vernier micrometer provides high reliability for measurements in the plain micrometer range but is unreliable at its maximum discrimination.
- Micrometers are contact instruments. This means that there must be some distortion, however minute, for every measurement. This must be controlled either by skill (the feel of the contact) or by mechanical limiting of the force of contact (the ratchet stop).
- The proper geometry for accurate measurement can only be achieved by adjusting the method to the configuration of the part. For nearly all cylindrical and spherical part features, contact instruments must rely on feel. One technique is centralizing, but even that ultimately depends on skill. For parallel plane part features, the ratchet stop may substitute for most of the skill of feel. However, no matter how well designed the instrument, careless handling will result in measurement errors.
- Much measurement with micrometers involves the transfer of measurement through an intermediate step. This doubles the measurement problem and greatly reduces the reliability. If a more suitable instrument is not obtainable, the only course open is to reduce the level of accuracy expected in the result.

END-OF-CHAPTER QUESTIONS

1. Which of the following is the correct way to open and close micrometers?
   a. Run the thimble across a hard surface.
   b. Hold the adjusting nut and turn the ratchet screw.
   c. Run the thimble across your hand or arm.
   d. Hold it by the thimble and twirl it.
2. A standard inch micrometer will have how many pitch threads per inch?
   a. 25
   b. 40
   c. 50
   d. 100
3. What is the chief advantage of the micrometer over the vernier caliper?
   a. Easier to carry
   b. Greater precision
   c. Can be used as a gage
   d. Easier to read
4. Which of the following is the most nearly correct?
   a. The direct measurement range of micrometers is 1 inch, but that of vernier calipers is 6 inches.
   b. It requires several micrometers to span the range of one vernier caliper.
   c. Vernier instruments have greater range but micrometer instruments have finer discrimination.
   d. There are micrometers that open as widely as the largest vernier calipers. Therefore, vernier instruments can no longer be considered state of the art.
5. If a micrometer is considered to be more accurate than a vernier instrument, which of the following would be the best explanation?
   a. The micrometer is easier to handle.
   b. The standard of a micrometer is a steel screw, whereas the standard of a vernier is only a scale.
   c. The screw of the micrometer provides more reliable discrimination than the reading of a vernier scale.
   d. It is easier to read a circular scale on a micrometer barrel than a straight scale on a vernier beam.
6. Which of the following are limitations of micrometers?
   a. Suitable for end measurement only
   b. Require oiling
   c. Temperature control of frame
   d. Measuring range
7. When measuring to 2 μm, why is a micrometer more reliable than a vernier instrument?
   a. It is not easy to find the correct place on the vernier scale.
   b. It is quicker. If you are not rushed, there is less chance for error.
   c. There is less chance for parallax error.
   d. A 2 μm can be read directly on the micrometer.
8. Which of the following are comparisons of related parts?
   a. Vernier beam—micrometer thimble
   b. Vernier scale of vernier instrument—thimble scale of micrometer
   c. Vernier movable jaw—micrometer frame
   d. Fixed jaw of vernier—spindle of micrometer
9. How is the precision of a micrometer related to its accuracy?
   a. By the inverse square law.
   b. Same as for all other instruments.
   c. The precision must be 10 times the accuracy.
   d. Precision is the dispersion of repeated readings, whereas accuracy is their adherence to a standard.
10. Which of the following is most nearly a correct statement about the maximum inherent accuracy of a micrometer?
    a. Depends on the size of the micrometer
    b. 2 μm for metric and 0.0001 in. for inch
    c. 0.01 mm for metric and 0.001 in. for inch
    d. 0.001 mm less zero setting error
11. Which of the following is the most appropriate statement about the use of a vernier on a micrometer to read to 0.1 mil (0.0001 in.)?
    a. It allows the micrometer to be used for measurements to 0.1 mil.
    b. It does not change the inherent reliability of micrometer measurements.

c. It eliminates the parallax error from the reading of the thimble scale.
d. It increases the reliability of 1 mil (0.001 in.) measurements.

12. As a general statement, allowing for exceptions in specific measurement situations, which of the following is the best way to improve micrometer reliability?
    a. Use with right hand only
    b. Measure quickly
    c. Apply tension with clamp ring
    d. Close thimble with ratchet stop

13. What is the function of the clamp ring?
    a. To lock a micrometer when put in storage
    b. To prevent temperature changes from changing a measurement
    c. To adjust for differences in measurement feel
    d. To lock a micrometer so that a reading may be remembered

14. If a micrometer does not have a ratchet stop or friction thimble, which of the following should be followed?
    a. Wear gloves when using.
    b. Apply some tension with the clamp ring.
    c. Overtighten until there is strong resistance to further tightening, then back off 0.002 in. (0.004 in. for work over 6 in.).
    d. Take several measurements and average them.

15. As diameters and lengths increase, which of the following considerations becomes most important?
    a. The selection of a micrometer with sufficient size
    b. The selection of the best micrometer attachments to achieve a desired range
    c. The use of long travel micrometers
    d. Particular care for the best positional relationship

16. Which of the following combination of activities is adequate for micrometer care?
    a. Keep clean, lubricate, do not leave in tightened position, do not overtighten, clean frequently with compressed air or by soaking in bath of solvent.
    b. Handle with right hand only, read in good light, do not overtighten, do not warm the micrometer.
    c. Keep clean, lubricate, take multiple readings, do not overtighten, clean frequently with compressed air or by soaking in bath of clean solvent.
    d. Keep clean, lubricate, do not leave in tightened position, do not overtighten, do not use compressed air to clean, do not soak in solvent solution.

17. Why is it more difficult to calibrate an inside micrometer than an outside micrometer?
    a. Inside micrometers must be calibrated to 0.00005 in., whereas outside micrometers are calibrated to 0.0001 in.
    b. Inside micrometers are more difficult to hold.
    c. Inside micrometers do not obey Abbe's law, whereas outside micrometers do.
    d. The outside micrometer is compared directly with the standard, whereas both the inside micrometer and its standard must be compared with a comparator.

18. What additional precautions are necessary for using the inside micrometer as compared with the outside micrometer?
    a. The workpiece must be oriented so that the scale may be read.
    b. It must be calibrated before each use.
    c. Check the shoulders of the rods to be sure they are tight and free of dirt.
    d. A calculator is needed to convert the results.

19. How do small hole gages differ from telescope gages?
    a. One is for inside dimensions, the other for outside.
    b. Small hole gages are for spherical measurements, whereas telescope gages are for cylindrical measurement.
    c. Telescope gages are transfer instruments, whereas small hole gages measure directly.
    d. Telescope gages are used for larger dimensions than small hole gages.

20. What alternative methods are there to the depth micrometer for the measurement of recesses to 1 mil (0.001 in.)?
    a. Direct reading of depth gage scale
    b. Vernier depth gage
    c. Inside micrometer
    d. Telescope gage

21. What is the most important precaution to take when reading the depth micrometer?
    a. Ascertain the alignment of the base to the line of measurement.
    b. Do not use ratchet stops for depths over 1 inch.
    c. Apply reversed scales.
    d. Be aware of parallax error.

22. For reliable measurement with a depth micrometer, certain considerations must be carefully observed. Which of the following does not apply?
    a. Frequent inspection of the end of the rod for wear
    b. Clamping of base to workpiece
    c. Shoulders of workpiece should be perpendicular to the line of measurement
    d. The use of a ratchet stop

23. Using a 0–25 mm range micrometer, what are the readings shown next?
    a. ______________mm

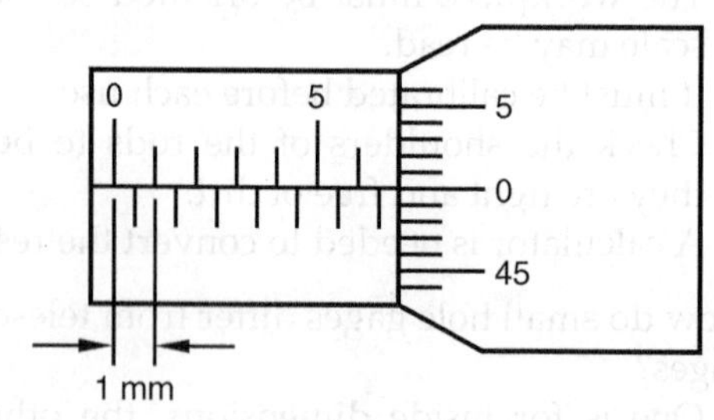

b. ______________mm

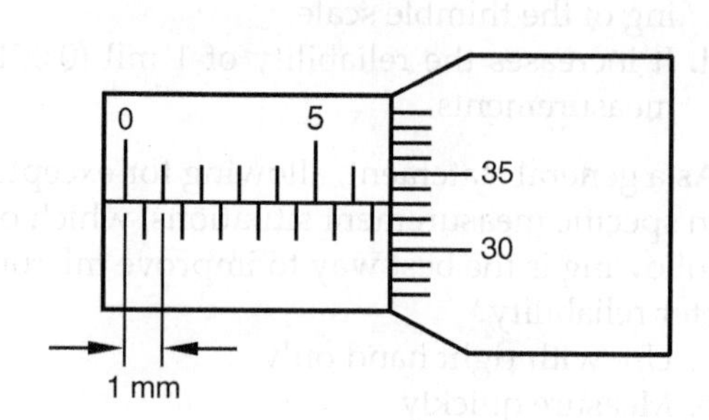

24. Using a 0–1 in. micrometer, what are the readings shown next?
    a. ______________inches

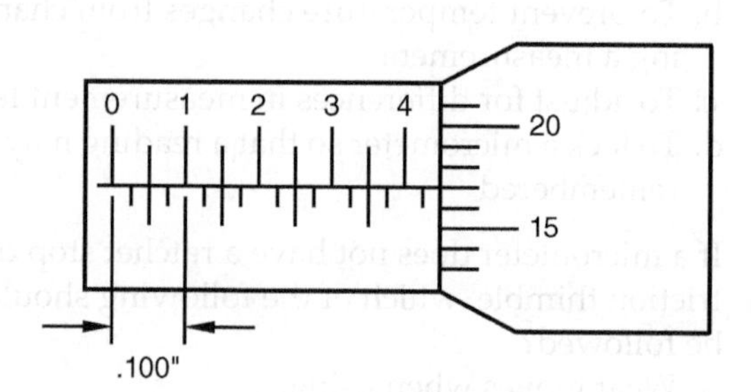

b. ______________inches

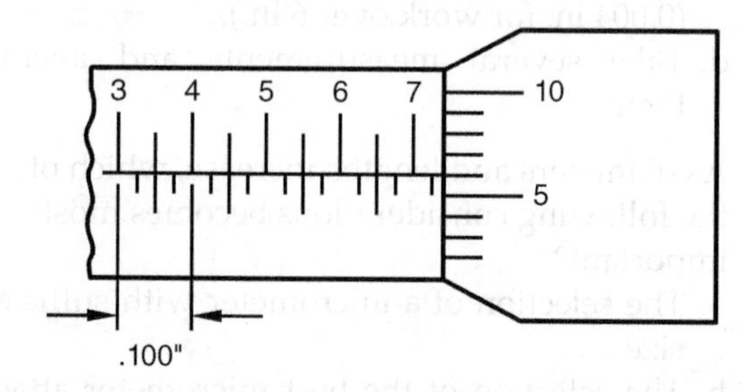

CHAPTER EIGHT

DEVELOPMENT AND USE OF GAGE BLOCKS

LEARNING OBJECTIVES

- State reasons for the need for practical length standards.
- Explain why traceability is required from the lowest to the highest standards.
- State the relationship between light waves and standards.
- Describe the mathematical basis for a gage block series.
- Explain the importance of gage block calibrations.
- Explain why gage blocks should be used whenever possible.
- List recommended methods for using gage blocks.
- Demonstrate how to wring blocks together effectively.
- Combine blocks for any desired dimension.

8-1 DEVELOPMENT OF GAGE BLOCKS

By now, you should understand that *precision*—the ability to divide into small increments—means little unless you can make the divisions with *accuracy*. Accuracy means adherence to a *standard*: all parts made to that one standard so that the parts can be used in predictable relationships to each other. We also need both precision and accuracy so that scientists can communicate with each other the quantitative measurement knowledge necessary for research.

Before the Industrial Revolution, we used vernier instruments to increase the precision of a craftsperson's sight and feel. The craftsperson, however, was still concerned with the production of a complete, functioning unit, not the manufacture of multiples of the same part for mass production. Machine tools made it easier for parts to be made; however, the parts were useless unless they could be assembled into a functioning unit. We needed a method to transfer the measurement skill from the craftsperson to the manufacturing operation.

Eli Whitney (1765–1825) is credited with the introduction of interchangeable manufacture, which is manufacturing that requires all parts within one operation to be made to one standard of length. At this time, the standard was often a handmade model of the device, *not* a measurement system. We believe that Henry Ford (1863–1947) still used this method for producing the Model T, and you might have a hard time convincing anyone who has ever worked with a car that this "handmade model" method of standardizing is not used today.

Amazingly enough, all the principles of measurement we still use today were available by 1900, but they were not consistently used in manufacturing. It was not until World War I, when war-time production meant life and death to thousands, that manufacturers began to understand that they needed both accuracy and precision for reliable production.

As products are used and produced on a more global scale, we have a greater need for universally accepted standards to ensure uniform production. The standard is the definition of the ultimate in accuracy—the absolute master from which all others are judged. Any deviation from the standard is error. A standard is different: It is any replica of the standard or even a replica of a replica.

We need standards today more than ever, because of increases in:

- Speeds of mechanisms
- Numbers of components
- Costs of manufacturing
- International trade
- The complexity of systems

Whenever one of these variables increases, standardization must improve, or the reliability of production and products decreases. For example, missiles and space exploration involve tremendous increases in all three variables; and the production of these technologies has also demanded the most from our measurement standards. You might think that international standards would develop hand in hand with precision manufacturing processes, but, unfortunately, the two have not progressed together. Although we had made great progress in measuring the international standard and correlating the lengths of the various national standards, this progress was only for line standards.

Line standards did not help people who were trying to make things that would run faster and do it for less money. They were working with *end standards.* In the beginning of the eighteenth century, "gage sticks" were used in Sweden. About 1890, go/no-go gages were included in rifle manufacturing, but these gages had to be made to control the size of each part. Because gages wear, manufacturers had to create separate sets of setting gages for each working gage—setting gages which, because they also were subject to wear, were often too expensive to make to the exacting standards that would ensure reliability in production.

Although *Carl E. Johansson* is popularly credited with the invention of gage blocks, they were actually used long before his time. Johansson's major contribution is in the area of standardization of gage blocks.

Around 1890, Hjalmar Ellstrom, chief mechanic in a Swedish arms factory, manufactured parallel-surfaced steel measuring blocks, which could be used together in different measuring combinations. The only problem was that they could not maintain enough of them for all of the factory's measuring requirements. Johansson worked with Ellstrom with one theory in mind: Make a set of master gages that could be combined for all sizes within its range.

Johansson not only recognized the need for gage blocks and designed a system of block sizes that would satisfy a variety of measuring needs, but he also saw that these blocks, if they were going to be used on a worldwide basis, must be calibrated to the international standard. For example, Johansson proposed 20°C as the standard measurement for temperature, because it was the average temperature in European machine shops, and it converted to a whole number in the Fahrenheit scale, 68°F. Through the efforts of Johansson, Ellstrom, and others, end standard replicas of the international standards became available to the major metrology laboratories of the world.

Gage blocks are still used today because of the efforts of William E. Hoke of the United States Bureau of Standards. He developed mechanical methods for lapping (final finishing) that made mass production of *gage blocks* possible.

Johansson's Original Standard

How do you create a gage block set from scratch? Where do you get the standard from which to start? We will take you through the method Johansson used, because there are many practical applications of the principle he used.

First, Johansson made a 100 mm gage block as close to that dimension as he could, developing some of his own instruments (see Figure 7–20) along the way. He sent the block to the International Bureau of Standards at Sevres, France. Instead of asking them to confirm its measurement, he asked the laboratory to determine the exact temperature at which the block would be 100 mm long. Eventually, Johansson received his answer from a rather unhappy laboratories staff: The block was exactly 100 mm long at 20.63°C.

Next, Johansson made a block that measured 100 mm at 20°C. Using the coefficient of expansion, he added 0.7 μm to the length of the block.

Finally, he used the single master to calibrate every block in the set. Johansson used subdivision based on simultaneous equations: If both the sum and difference of two or more quantities are known, the individual quantities may be found (see Figure 8–1).

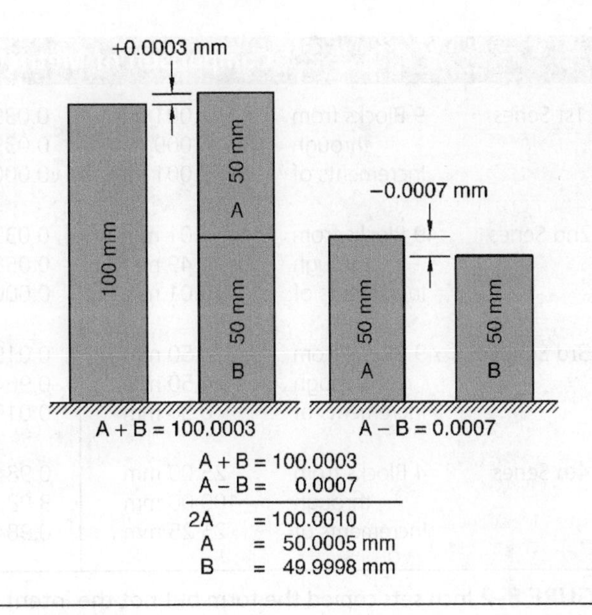

FIGURE 8–1 Starting with one known block, Johansson used simultaneous equations to derive all of the other blocks in the set.

First Inch Sets Introduced

English-based metrologists, however, did not derive much benefit from gage blocks. Inch sets imitate the form of Johansson's sets, but they were not made with the same intention behind them (see Figure 8–2). For example, whereas the smallest block in the metric set (1.001 mm) equals 0.0394 in., the smallest block in the inch set is 0.1001 in.—*2½ times larger than the smallest metric block.* The discrimination of the metric set equals 0.001 mm (0.0000394 in.) and the inch set equals 0.0001 in., again 2½ times larger.

Johansson was concerned about the practical range of his gage blocks, as well as their precision; therefore, the inch blocks that were created do not meet his standards for practicality. If they were practical, block manufacturers would not have had to put so many different sets on the market to compensate for the inadequate, basic inch set.

Originally, Johansson made an inch set in 1906 as part of a program to reconcile official differences between the length standard of the English and metric systems. For reconciliation purposes, his English block sets are just as accurate as his metric sets, but they do lack the small sizes and the fine discrimination of the metric sets.

Johansson spent much of his later life in administration and as a part of the international standards community. He probably was never approached with the idea of creating the best possible combination of inch block sets.

Comparison of Metric and Inch Gage Block Sets

1st Series	9 Blocks from	1.001 mm	0.039439 in.	1st Series	9 Blocks from	0.1001 in.
	through	1.009 mm	0.039755 in.		through	0.1009 in.
	Increments of	0.001 mm	0.000039 in.		Increments of	0.0001 in.
2nd Series	49 Blocks from	1.01 mm	0.039794 in.	2nd Series	49 Blocks from	0.101 in.
	through	1.49 mm	0.058661 in.		through	0.149 in.
	Increments of	0.01 mm	0.000394 in.		Increments of	0.001 in.
3rd Series	49 Blocks from	0.50 mm	0.0197 in.	3rd Series	19 Blocks from	0.050 in.
	through	24.50 mm	0.9646 in.		through	0.950 in.
	Increments of	0.50 mm	0.0197 in.		Increments of	0.050 in.
4th Series	4 Blocks from	25.00 mm	0.9842 in.	4th Series	4 Blocks from	1.000 in.
	through	100.00 mm	3.9370 in.		through	4.000 in.
	Increments of	25.25 mm	0.9843 in.		Increments of	1.0000 in.

FIGURE 8–2 Inch sets copied the form but not the intent of Johansson's standard set. Note the sameness of the digits but the difference in the actual sizes.

From Light Waves to End Standards

The metric system was a result of the French Revolution, before which there was no systematic system of measurements. Taxes and rent payments were levied based on the highest possible measurement; goods were sold based on the smallest possible measurement. At that same time, independent thought was sweeping the world; this French "meter" was a measurement on which all independent nations could agree.

Unfortunately, the international meter was a line standard, making it difficult to replicate and verify for other countries. Nearly 25 years passed from the time the United States adopted the metric system as its standard and the time we received our official replica. Even this official standard inherently contains error.

Whenever you do work that requires a subdivision of the standard that is smaller than the known error in the standard, the work you do is not in conformance with the standard and cannot be considered accurate. Metrologists have worked for decades to reduce the error, but the only way to eliminate error is to base the standard on an unchanging physical phenomenon. This phenomenon should also have units that are smaller than any conceivable error. The wavelength of light was the obvious choice for this phenomenon.

Light had been suggested as a standard as early as 1827. Later, Albert A. Michelson (1852–1931) and Edward W. Morley (1838–1923) constructed an *interferometer* to test the theory of ether drift; and, in 1892–1893, Michelson used the same principles to learn the wavelength of the red cadmium spectral line (see Figure 8–3).

Measurement is relative: One thing is always measured in terms of another. Therefore, in measuring the wavelength of a particular color of light, Michelson also defined the unit of length of the standard that he was using in relation to that particular color of light's wavelength.

As we discussed in Chapter 2, a *unit of length* (meter) is an arbitrary way to describe lengths as either fractions or multiples. A *standard of length* (the International Prototype Meter) is the physical thing that defines the unit of length (see Figure 8–4). Before the use of light waves, the unit of length and standard of length were two separate things; with light, they are the same. Also, with the right laboratory equipment, we can reproduce the standard anywhere and anytime and know that measurement will agree with

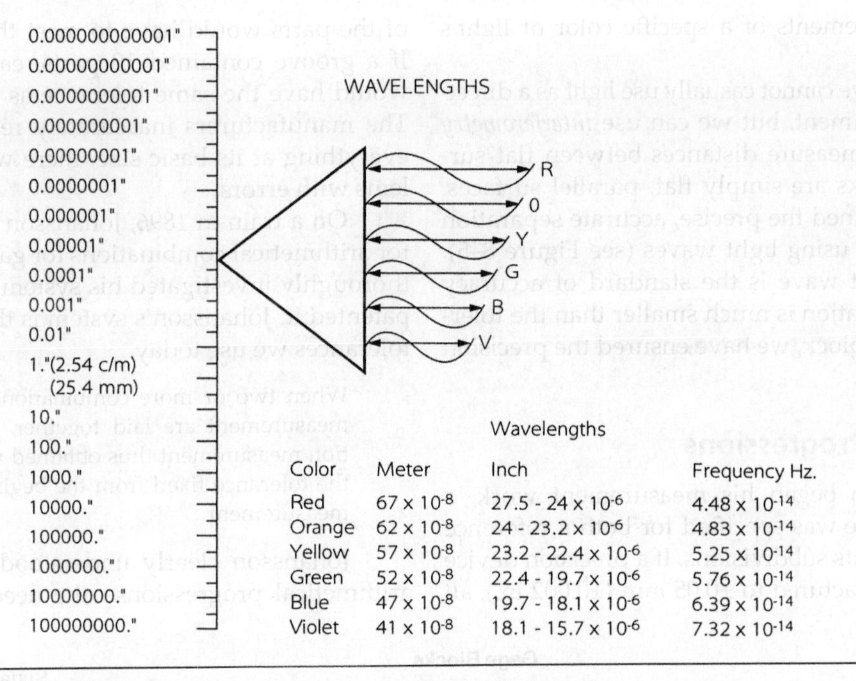

Color	Meter	Wavelengths Inch	Frequency Hz.
Red	67×10^{-8}	$27.5 - 24 \times 10^{-6}$	4.48×10^{-14}
Orange	62×10^{-8}	$24 - 23.2 \times 10^{-6}$	4.83×10^{-14}
Yellow	57×10^{-8}	$23.2 - 22.4 \times 10^{-6}$	5.25×10^{-14}
Green	52×10^{-8}	$22.4 - 19.7 \times 10^{-6}$	5.76×10^{-14}
Blue	47×10^{-8}	$19.7 - 18.1 \times 10^{-6}$	6.39×10^{-14}
Violet	41×10^{-8}	$18.1 - 15.7 \times 10^{-6}$	7.32×10^{-14}

FIGURE 8–3 When Michelson determined that the number of cadmium red wavelengths corresponded to the length of the meter, he simultaneously established light as a basis for length standards.

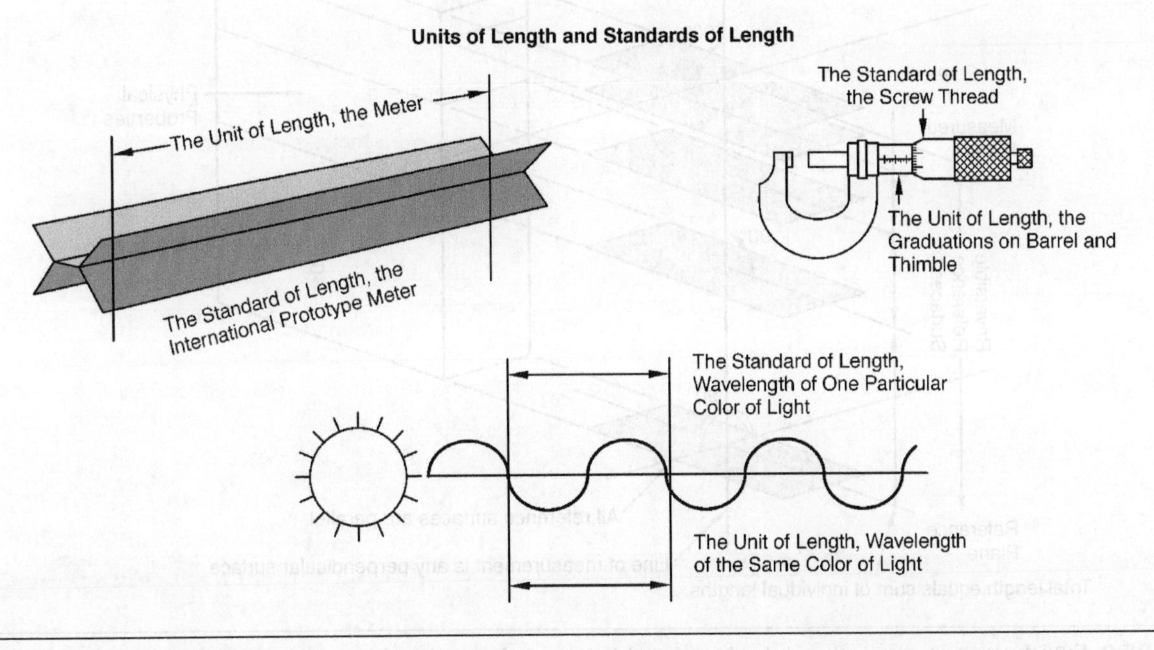

FIGURE 8–4 Only for light does the unit of length and standard of length become one. Therefore, light waves are the standard for both accuracy and precision.

all other measurements of a specific color of light's wavelength.

Of course, we cannot casually use light as a direct-measuring instrument, but we can use *interferometry* (Chapter 16) to measure distances between flat surfaces. Gage blocks are simply flat, parallel surfaces. We have determined the precise, accurate separation of these surfaces using light waves (see Figure 8–5). Because the light wave is the standard of accuracy and its discrimination is much smaller than the tolerance of the gage block, we have ensured the precision of the gage block.

Arithmetical Progressions

Before Johansson began his measurement work, a constant tolerance was specified for both a reference tolerance and all its subdivisions. If a precision device was being manufactured to ±0.05 mm (±0.002 in.), all of the parts would have to meet the same tolerance. If a groove contained 10 parts, each of its 10 parts would have the same tolerance as the entire groove. The manufacturers inaccurately reasoned that, with everything at its basic size, there would be no problems with errors.

On a train in 1896, Johansson recorded the idea for arithmetical combinations for gage blocks. He also thoroughly investigated his system of tolerances and patented it. Johansson's system is the basic system of tolerances we use today.

> When two or more combinations of a certain measurement are laid together, the combination measurement thus obtained will lie within the tolerance fixed from the beginning for this measurement.

Johansson clearly understood the need to use arithmetical progressions—the need to combine the

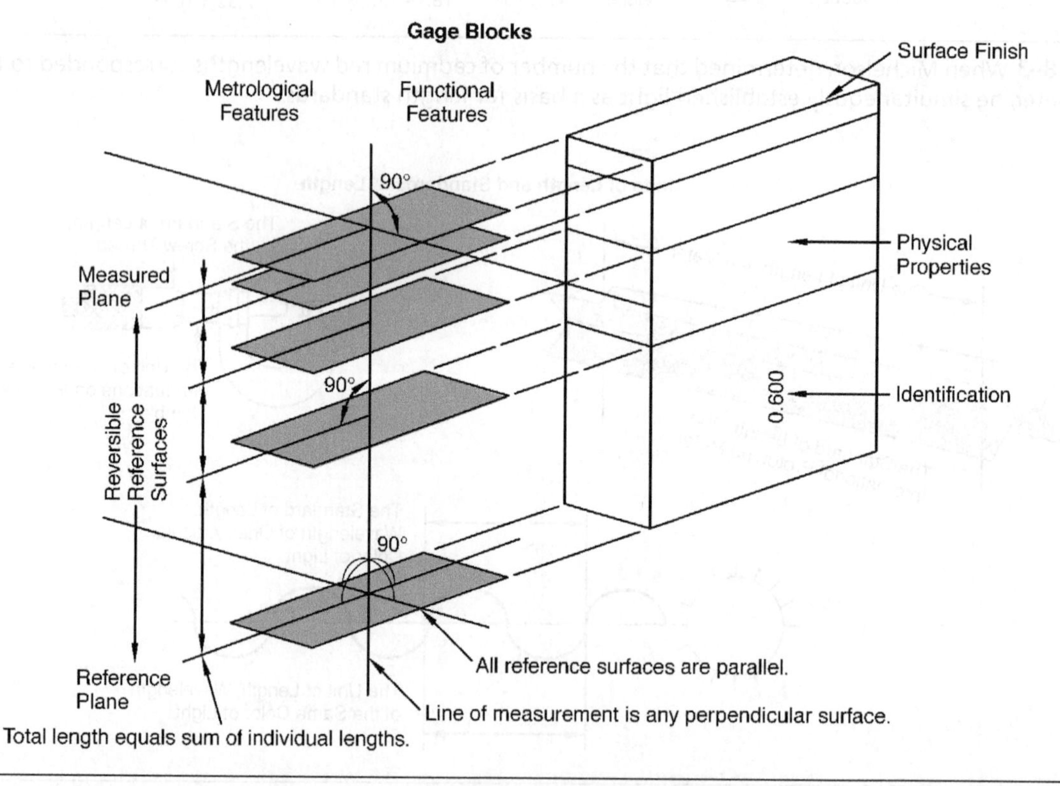

FIGURE 8–5 Both the metrological and the functional features of gage blocks suit them for exacting measurements.

individual lengths and have the result fall within a standard tolerance, rather than have each piece built to an individual tolerance. Unfortunately, he also understood that it would take time and untold billions of dollars wasted in scrap losses for the industries of the world to accept his version of tolerancing.

Manufacturers needed gage blocks to establish standards for accuracy that were as close to what could be attained by the actual manufacturing process. In Johansson's own words:

> So it was mainly in France that Johansson's standard gage block sets were acquired, even though they often ended up in a glass case in the laboratory instead of in the workshop where they belonged.

Clearly, Johansson meant for his gage block to be used—an attitude he reinforced when he modified the original set so it would cover the full range required for practical shop use (see Figure 8–6). Still wanting to see his sets used in the shop, Johansson also designed the end standards and holders required for practical use of the blocks, which are covered later in this chapter.

8–2 MODERN GAGE BLOCKS

Although we will not cover the manufacturing of gage blocks, you should understand that while we talk about millionths of an inch, it is quite another thing to embody those millionths in a block of metal.

Gage Block Materials

The practical, functional characteristics of gage blocks are:

- They must be made from something that can be accurately sized and finished.
- They must be stable and not change size of their own accord.
- They must withstand considerable wear.
- They must be practical, affordable, and wear and corrosion resistant.

The majority of gage blocks are made from hardened alloy steel. Hardening is required to allow the blocks to resist wear, but it also traps stresses in the material, which are slowly released as the block ages. As the stresses are released, the blocks also lose

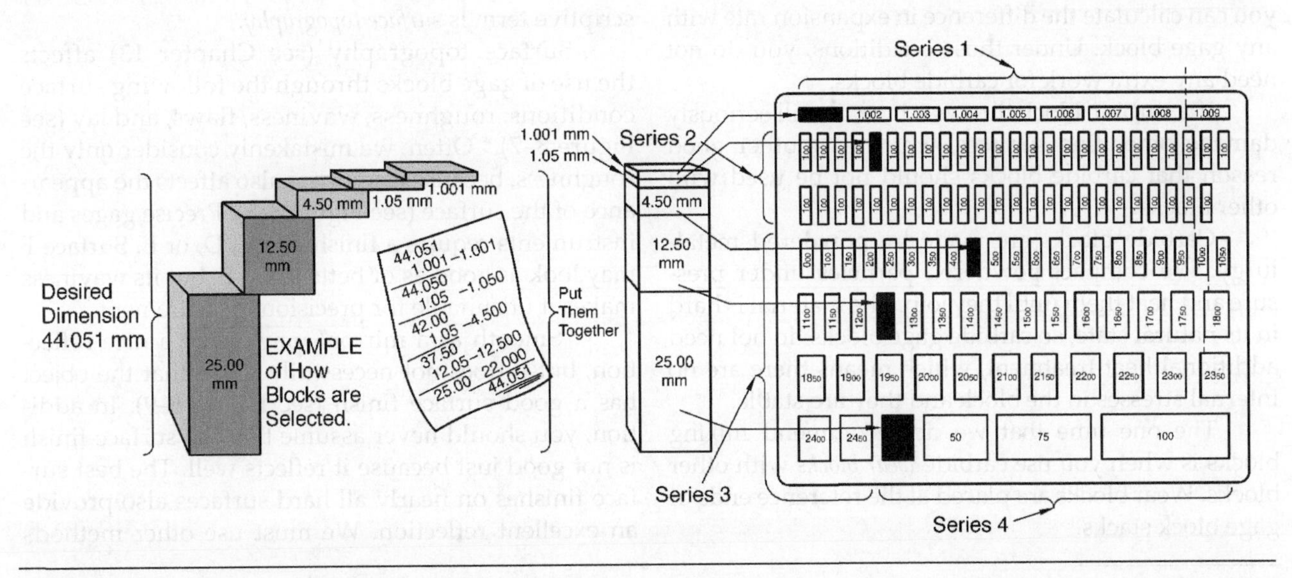

FIGURE 8–6 In Johansson's patent, 11 blocks in 4 series combine to form any dimension from 2 mm to 202 mm in steps of 0.001 mm, for a total of 200,000 combinations.

their calibrated size and become unstable. We lessen the effects of hardening stress with an elaborate heat-treatment process that cools the blocks to –200°F or lower.

You might think that it would be better to just harden the surface of the gage block and leave the interior free from stress. But in order to withstand the pounding of comparator points and normal use, we need a core that is harder than soft steel provides—a core as strong as the casing. Steel gage blocks also have very nearly the same coefficient of temperature expansion as most of the steels with which they are used, and they are low in cost.

At the opposite extreme, carbide blocks provide the longest service life, are extremely wear resistant but very expensive, and have very different coefficients of expansion than most materials. If they are used in standard temperature laboratories, their low coefficient of expansion does not cause problems; however, if they are used at ordinary temperatures with steel parts, we must correct for the expansion rate. Their expansion rate is one reason why carbide blocks should never be mixed with other blocks. One exception is if you are measuring aluminum parts at ordinary temperatures to 2 μm (0.0001 in.) or closer, you can calculate the difference in expansion rate with any gage block. Under those conditions, you do not need any extra work for carbide blocks.

If a carbide block throws a burr, it will seriously damage other fine surfaces, which is another good reason that carbide blocks should not be used with other blocks.

Carbide blocks are made by powdered metallurgy: We compact powdered particles under pressure and heat them until they join. The material is hard in its natural state, so carbide gage blocks do not need additional heat treatment, which means there are no internal stresses in the block and they are stable.

The one time that we do recommend mixing blocks is when you use carbide *wear blocks* with other blocks. Wear blocks are placed at the reference ends of gage block stacks.

For wear resistance and price, all other blocks fall between steel and carbide. Among these blocks are chromium-plated blocks and stainless steel blocks, which are both surface hardened. Unlike ordinary steel blocks, the non-heat-treated interior on these blocks is hard and tough enough to adequately support the outer case. Carbide, chromium-plated, ceramic, and stainless steel blocks are also resistant to corrosion.

Macrogeometry and Microgeometry

"Macro" means large, and "micro" means small. *Macrogeometry* generally refers to the general shape of a part, and the features can usually be easily measured with instruments like the ones we have discussed. *Microgeometry* refers to the minute analysis of shape. We cannot measure to very high precision or accuracy without the following microgeometric features: flatness, parallelism, straightness, roundness, and surface finish.

When we use end standards, our length measurements depend on the condition of the contacting surfaces, commonly but inaccurately called the *surface finish* or *surface texture.* The proper, completely descriptive term is *surface topography.*

Surface topography (see Chapter 13) affects the use of gage blocks through the following surface conditions: roughness, waviness, flaws, and lay (see Figure 8–7).* Often, we mistakenly consider only the roughness, because roughness also affects the appearance of the surface (see Figure 8–8). Precise gages and instruments require a finish like C, D, or E. Surface F may look as good as or better than E, but its waviness makes it unsuitable for precision measurement.

"Smooth as a mirror" may create a nice reflection, but it does not necessarily mean that the object has a good surface finish (see Figure 8–9). In addition, you should never assume that the surface finish is not good just because it reflects well. The best surface finishes on nearly all hard surfaces also provide an excellent reflection. We must use other methods

*"Surface Roughness, Waviness and Lay," ASA Bulletin B46.1, New York, American Standards Association, 1947.

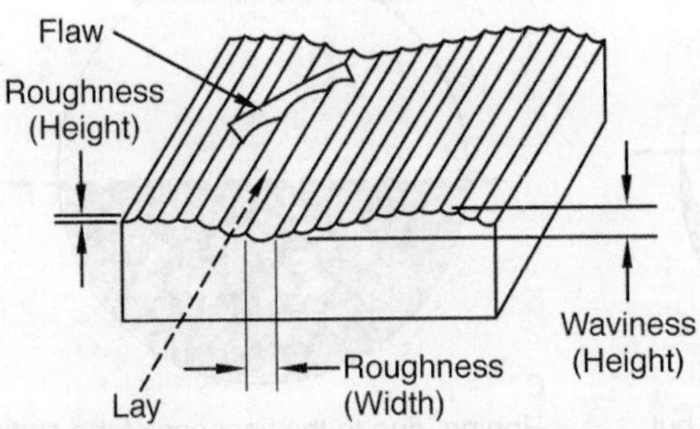

Roughness is finely spaced surface irregularities, in a consistent pattern, produced by machining or processing.

Waviness is an irregular surface condition of greater spacing than roughness. Usually caused by deflections or vibrations, not by the cutting edge.

Flaws are irregularities that do not appear in a consistent pattern.

Lay is the predominant direction of surface pattern.

FIGURE 8–7 These terms are usually used to describe surface conditions.

besides our vision to determine the quality of the surface finish.

In the past, we used a method called the root mean square (*RMS*) to specify surface finish, but this system was inadequate (see Figure 8–10). Today, we specify surface finish with an arithmetical mean or average—the distance between the mean line and the crests or peaks.

Generally, wear is proportional to surface finish: the larger the bearing area, the better the load is distributed, and the lower the pressure on the surface of the block (see Figure 7–27). Each minute scratch on the surface reduces the amount of area available to take the pressure of contact (see Figure 8–11). Also, any irregularities trap abrasive dust, which wears the surface of the block and the part when the two are "wrung."

Fine surface finish can be deceptive. One manufacturer developed an improved finish that was applied to all gage blocks. During the final quality check, the inspectors discovered that they occasionally had to reject a block because it had visible scratches. The actual scratches were often more shallow and narrower than the finishing marks on other blocks, but the improved finish made the scratches visible, thus unacceptable for the consumer.

Gage Block Grades

Precision gage blocks are fundamental to dimensional quality control. The major characteristics for choosing quality gage blocks include accuracy, wear resistance, surface finish, and dimensional stability, or thermal conductivity.

The previous specification for gage blocks, Federal Specification GGG-G-15C, has been withdrawn, although some manufacturers still refer to it. The prevailing standard for gage blocks in the United States is ASME (American Society of Mechanical Engineers) B89.1.9. The size tolerances of the new standard apply to all points on the gage surface, not just to the reference points.

There are five gage block grades under the ASME B89 standard (see Figure 8–12). Grade K blocks were introduced as reference masters for calibration where compensations to the deviations of the masters are applied (due to larger length tolerances). The geometry of the blocks (flatness and parallelism) is tightly controlled to minimize error.

The Grade 00 gage block has the same properties as the Grade K, but has a tighter length tolerance, so that Grade 00 can be used as Grade K, but Grade K cannot be used as Grade 00.

Grade 0 blocks are a replacement for the GGG-G-15C Grade 2 gage blocks.

Grades AS-1 and AS-2 are intended for use in applications where extreme accuracy is not required, as on the shop floor. The accompanying lab manual exercise has you record the grade and tolerance level of a gage block set.

Sizes and Shapes

There has been a variety of gage block shapes and sizes. The two most commonly used are rectangular

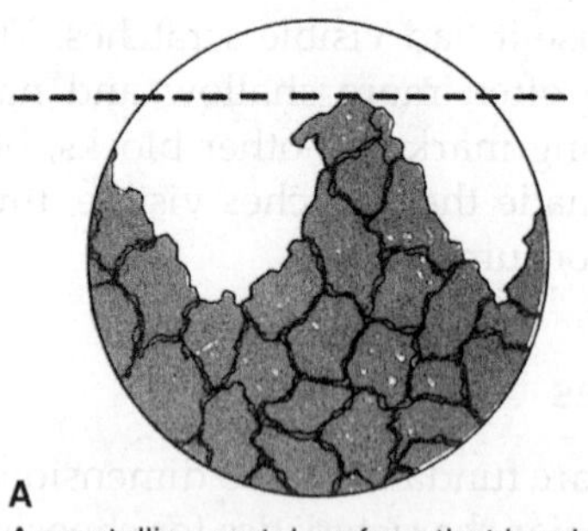

A
A crystalline metal surface that has been machined on a lathe may be visualized as having high peaks and deep valleys.

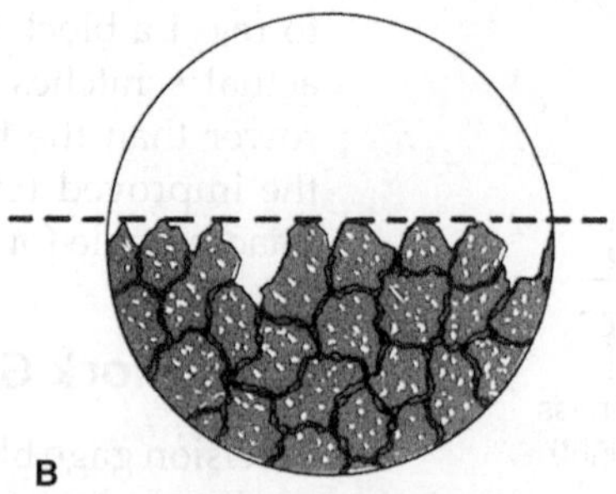

B
Grinding reduces the high peaks but produces new and smaller peaks and valleys, the dimensions depending on the grit size.

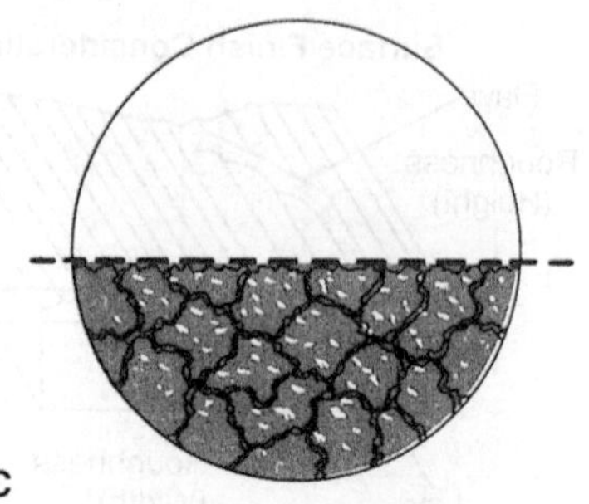

C
Honing, due to the fineness of the cutting grits and irregularity of cutting motion, reduces a surface to one of fine scratches.

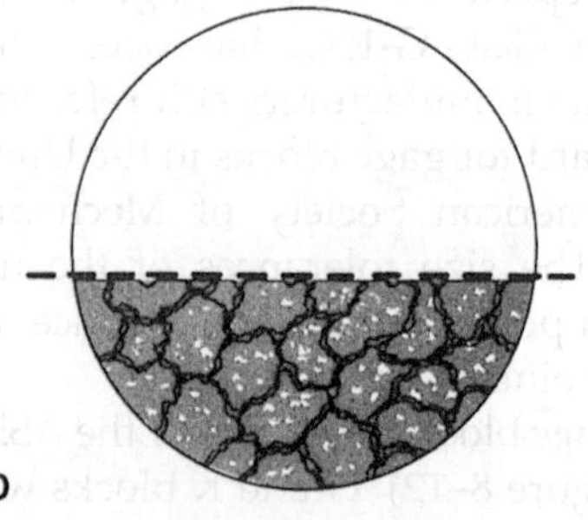

D
Lapping normally leaves a finish of minute scratches. The scratch pitch is finer than that left by a hone.

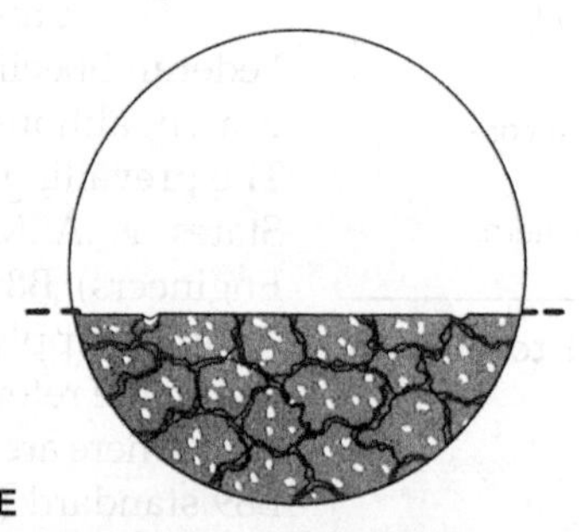

E
Gage block surfaces have a high degree of flatness as well as surface finish. Only occasional extremely shallow scratches are left.

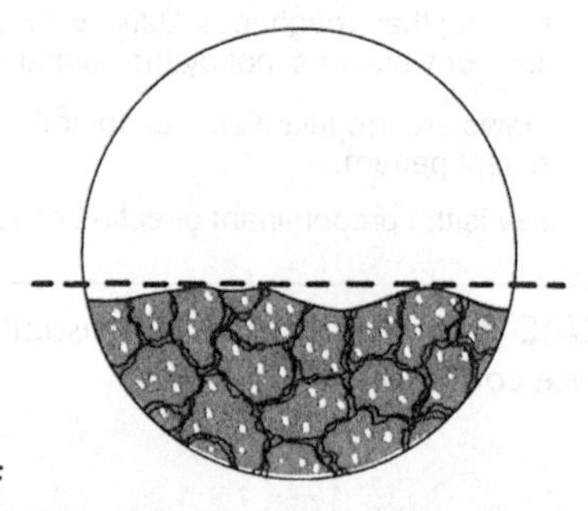

F
Buffing produces sufficient heat to bring about "plastic flow" and while aiding reflectability will reduce the quality of surface flatness.

FIGURE 8–8 Comparison of surface finish shows the exacting requirements for gage blocks.

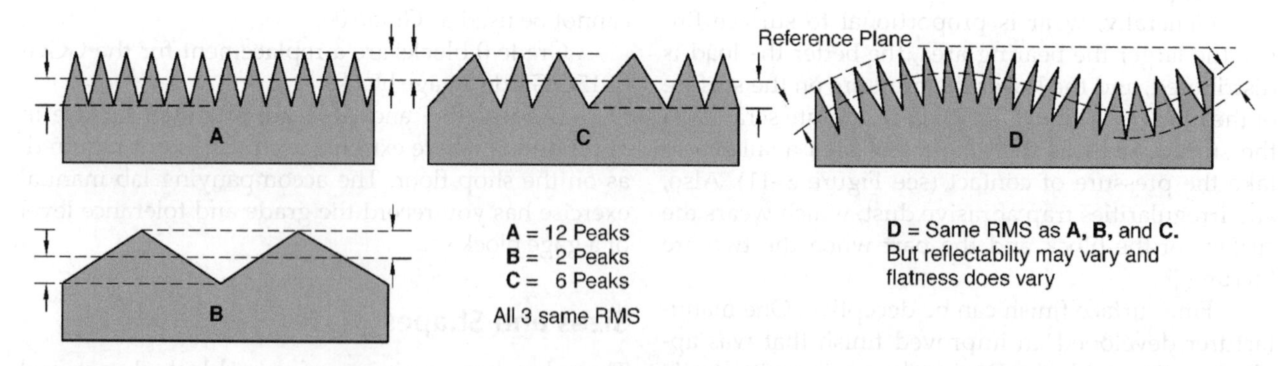

FIGURE 8–9 The top drawings show that an irregular surface or poor surface may have better reflectivity than a good surface finish. The bottom drawing shows the reason for this.

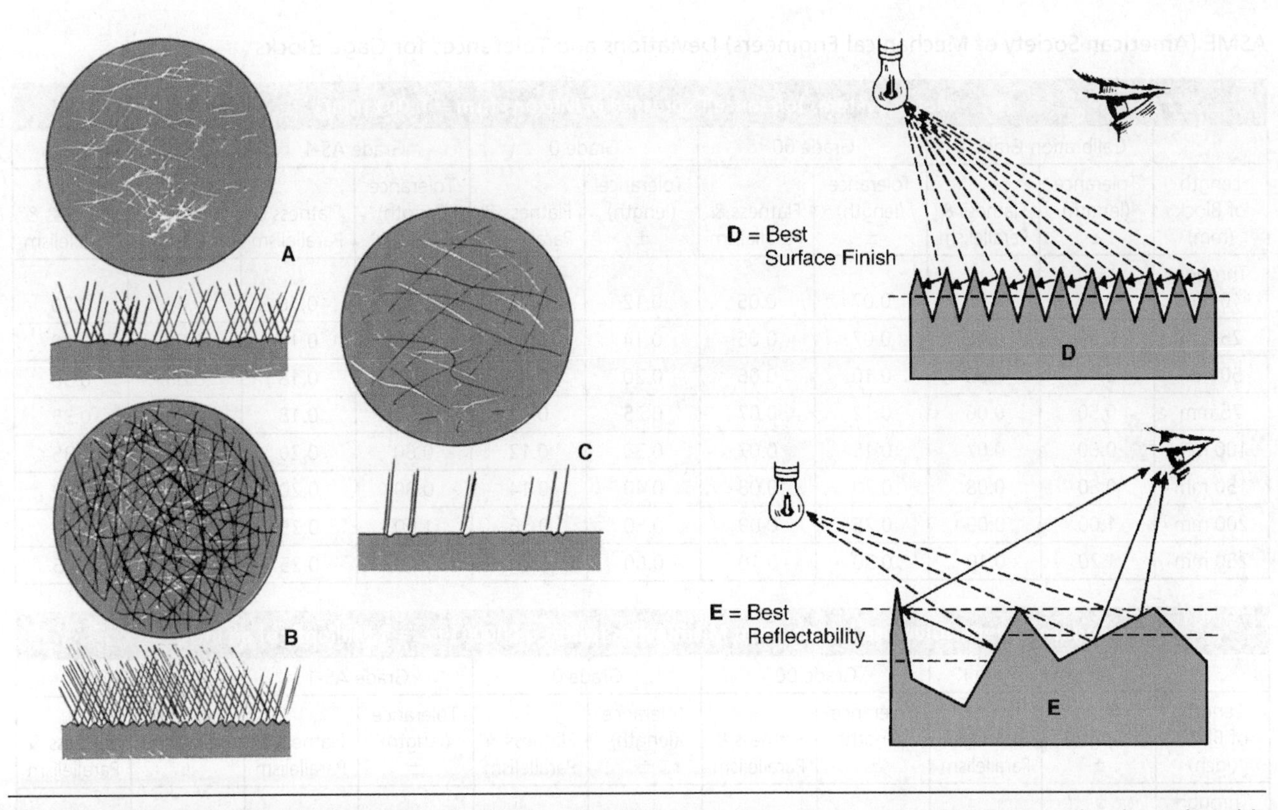

FIGURE 8–10 Surface finish is measured by the height of the crests from the average. These examples show that this does not fully describe the surface.

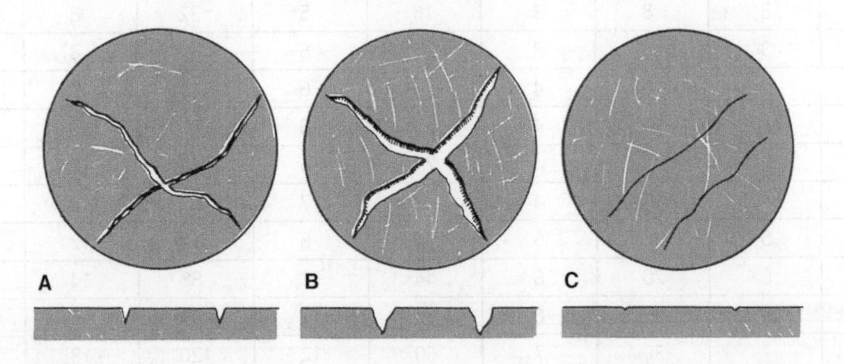

FIGURE 8–11 The top width of scratches reduces the load supporting area and thus reduces the wear life.

ASME (American Society of Mechanical Engineers) Deviations and Tolerances for Gage Blocks

Metric System: Tolerance Expressed in Microns (μm = 0.001 mm)										
	Calibration Grade K		Grade 00		Grade 0		Grade AS-1		Grade AS-2	
Length of Block (mm)	Tolerance (length) ±	Flatness & Parallelism	Tolerance (length) ±	Flatness & Parallelism	Tolerance (length) ±	Flatness & Parallelism	Tolerance (length) ±	Flatness & Parallelism	Tolerance (length) ±	Flatness & Parallelism
Through 10 mm	0.20	0.05	0.07	0.05	0.12	0.10	0.20	0.16	0.45	0.30
25 mm	0.30	0.05	0.07	0.05	0.14	0.10	0.30	0.16	0.60	0.30
50 mm	0.40	0.06	0.10	0.06	0.20	0.10	0.40	0.18	0.80	0.30
75 mm	0.50	0.06	0.12	0.07	0.25	0.12	0.50	0.18	1.00	0.35
100 mm	0.60	0.07	0.15	0.07	0.30	0.12	0.60	0.20	1.20	0.35
150 mm	0.80	0.08	0.20	0.08	0.40	0.14	0.80	0.20	1.60	0.40
200 mm	1.00	0.09	0.25	0.09	0.50	0.16	1.00	0.25	2.00	0.40
250 mm	1.20	0.10	0.30	0.10	0.60	0.16	1.20	0.25	2.40	0.45

English System: Tolerance Expressed in Microinches (μin = 0.000001 in. = 1 Millionth of an Inch)										
	Calibration Grade K		Grade 00		Grade 0		Grade AS-1		Grade AS-2	
Length of Block (inch)	Tolerance (length) ±	Flatness & Parallelism	Tolerance (length) ±	Flatness & Parallelism	Tolerance (length) ±	Flatness & Parallelism	Tolerance (length) ±	Flatness & Parallelism	Tolerance (length) ±	Flatness & Parallelism
Through 0.05″	12	2	4	2	6	4	12	6	24	12
0.4″	10	2	3	2	5	4	8	6	18	12
1″	12	2	3	2	6	4	12	6	24	12
2″	16	2	4	2	8	4	16	6	32	12
3″	20	2	5	3	10	4	20	6	40	14
4″	24	3	6	3	12	5	24	8	48	14
5″	32	3	8	3	16	5	32	8	64	16
6″	32	3	8	3	16	5	32	8	64	16
7″	40	4	10	4	20	6	40	10	80	16
8″	40	4	10	4	20	6	40	10	80	16
10″	48	4	12	4	24	6	48	10	104	18
12″	56	4	14	4	28	7	56	10	112	20
16″	72	5	18	5	36	8	72	12	144	20
20″	88	6	20	6	44	10	88	14	176	24
24″	104	6	25	6	52	10	104	16	200	28
28″	120	7	30	7	60	12	120	18	240	28

FIGURE 8–12 While the gage blocks of individual manufacturers may vary, all meet or exceed some required specification, including the old Federal Specification GGG–G–15C, ANSI/ASME B 89.1.9, or ISO 3650.

and square blocks. The square blocks may or may not have a center hole. Both shapes are covered by the ASME B89.1.9 standard. Both are available in thicknesses from 254 μm to 508 mm (0.010 in. to 20 in.) in several series. A series is a group of blocks all of one nominal size in which one digit changes consecutively through the series (3 mm to 508 mm in 0.001 mm steps). We usually purchase blocks in sets of one or more series.

The number of blocks in a set varies among manufacturers, but sets are typically made up of 112, 88, and 45 blocks for metric (121, 86, 56, 38, and 35 for English). Manufacturers also produce sets to extend the range—either larger or smaller—of these basic sets. Long block sets have blocks up to 450 mm (20.000 in.); thin block sets start at 254 μm (0.010 in.) and increase in increments of 0.005 mm (0.010 in.) to 3 mm (0.090 in.). You can also purchase holders, end standards, and accessories for both types of blocks.

Whether you use square or rectangular blocks is usually a matter of personal preference; however, there are some technical considerations. Square blocks have greater surface area, last longer, and are less likely to be toppled over because of their greater mass. However, they are much more expensive than rectangular blocks. You need to balance three factors when choosing a set of blocks: the range of the set, the number of combinations that require more than three blocks to form, and the number of people who will use it each day.

Range, of course, is determined by the manufacturer of the set. It is just a matter of choosing the appropriate range for your applications.

A set of blocks will give you a number of options to form a given length; however, the combination in which you use the fewest blocks is always fastest and eliminates the chance for additional errors. With a large set of blocks, it is easier for you to make more lengths with fewer combined blocks. So, although a 38-piece set might have a sufficient range for most applications, it may not provide the number of blocks you actually need to minimize blocks used for a measurement.

Finally, if many people will be using the set, opt for the larger block set. With a number of blocks available, you can minimize the amount of time people spend waiting for the blocks they need. The larger the number of users, the larger the set should be.

Wear Allowance

From time to time, gage block manufacturers market gage block sets with a "built-in wear allowance." They create this allowance by tolerancing the block on the plus side: the usual 10.00003 mm tolerance of a grade 0.5 block becomes 10.00006 mm and 20.0 mm. Using this tolerancing, the expected wear only brings the block closer into its basic size. But these blocks fail to use the true meaning of tolerance, effectively *doubling* the amount that the block can deviate from its intended size.

A dimension consists of the basic size and the tolerance. The tolerance is *not* part of the size: it is *a rule about that size.* A part feature 1.000 mm, +0.0001, −0.00005 mm does not mean that the part feature measures 1.0001 in. That measurement places the part on the borderline *beyond which* parts are rejected. Creating these borderlines is the function of tolerance, so tolerance varies with the size limit, not the absolute size. A 1.00000 mm gage block with a tolerance of ±0.00003, would allow measurements between 0.99997 mm (the lower allowable size) and 1.00003 mm (the uppermost allowable size).

When we use gage blocks, we frequently use them in combination. Because there are bilateral tolerances (±) in each block, these uncertainties have a chance to cancel each other out—they may not, but they could. Blocks with built-in wear allowance are manufactured with unilateral tolerances, so that the uncertainties cannot cancel and must be accumulated.

If manufacturers could consistently produce blocks exactly at the basic size, we would not need tolerances. But, of course, manufacturers cannot guarantee such consistency. The tolerance allows a needed "leeway." Manufacturers of built-in wear blocks can miss the mark by twice as much as other manufacturers will, and they still will not have to scrap the block. Fewer scrapped blocks means less waste, and reducing waste may have more to do with creating a "wear allowance" in a block than reducing the customer's wear errors does.

Further, when built-in wear blocks are returned to the laboratory for calibration, very few of them are rejected because of wear. Most defective blocks are rejected because of nicks, scratches, and corrosion, not because of the day-in, day-out wear caused by wringing of clean blocks and setting comparators.

Referring to Figure 8–12, you might wonder why gage block tolerances are not split equally plus and minus except in the Grade 1, but standard tolerances need to reflect both practical considerations and theoretical ones. They also need to take into consideration the capabilities of manufacturing and the intention of the product. Tolerances should be adjusted honestly as the theories, capabilities, and intentions of manufacturing change.

8–3 CALIBRATION OF GAGE BLOCKS

Just because we use gage blocks to calibrate other instruments does not mean that gage blocks do not need to be calibrated, too. A change in one standard can add uncertainty to several of the gages in that set, which in turn can lead to increased scrap losses, assembly costs, and field service costs. In contrast, if the master set of gage blocks goes out of tolerance, the entire plant's performance suffers and the potential losses can include millions of dollars.

You should maintain all precision instruments with a systematic, periodic inspection program. The length of time between inspections depends on how often, how long, and how carefully the instruments are used. Calibration is expensive, so unnecessary calibration should be avoided. ASME recommends at least annual calibration for grades 00 and K, and monthly to semiannually for grades 0, AS-1, and AS-2. Most NIST customers send master blocks for recalibration every two years. Since master blocks are used in a clean, dry environment, and not extensively, this schedule may be adequate. Despite common misconceptions, NIST has no regulatory power over instrument calibrations. You can use statistics to help set up a calibration program to maintain the desired level of assurance without excessive costs.

Progressive manufacturers and research laboratories often use gage blocks directly for measurement. To ensure reliability, blocks should be sent to the *metrology laboratory* for calibration when any of these conditions exist:

- Visual inspection shows abnormal wear when compared to past calibrations.
- Wringing becomes difficult.
- You suspect the reliability of the block(s) when there is an increase in rejects.

Typically, a plant using many sets of gage blocks will calibrate its working blocks against its own master sets. Conservative manufacturers will always have at least two sets of master blocks—with one on hand and the other at the metrology laboratory being calibrated. Smaller users send their working sets out for calibration.

Working blocks can be safely calibrated within an organization, but master blocks must be calibrated against *grand masters,* whose lengths can be directly traced to the national standard of length. For many years, the NIST calibrated master gage block sets, but this service is not properly that organization's function. It is required by law to maintain our *national* standards.

All manufacturers of gage blocks provide calibration services and sometimes replacement of worn blocks, not only for their own blocks but also for the blocks of other manufacturers. Manufacturers also seem to be evenhanded in the evaluation of blocks, whether their own or their competitors'. Independent metrology laboratories are an alternative if you suspect the calibration services that manufacturers offer.

Calibration does more than show whether the blocks are in or out of tolerance; it provides you with the deviation from the nominal lengths. For the most precise measurements, you can use this information to adjust your results for the deviation.

Many plants progressively replace gage block sets. As a set wears, it is downgraded on the Federal Accuracy Grades. If a newly purchased set is Grade 1, it is moved down to a lower precision level, if possible, or the set is discarded when it has worn past the tolerances of that grade. This process continues until the set cannot be reliably used for any application in

Gage Block Calibration

What is it?
Verification that the block is within the tolerance specified.

How is it done?
By comparison with master standards of known calibration and know traceability to the national length standards.

When is it done?
1. Periodically, according to a systematic program based on statistical quality control.
2. Not less often than annually for grade AA, quarterly for grade A, and semiannually for grade B.
3. When required by visual inspection, wringability decline, or slipping quality control level.

Who calibrates working sets?
The user's gage department or any of the facilities for calibrating master sets.

Who calibrates master sets?
1. Independent metrology laboratories
2. Manufacturers of gage blocks
3. NIST

What is done with worn blocks?
They are replaced.

What is done with worn sets?
They are regarded for a lower level of precision.

FIGURE 8–13 Considering the critical use of gage blocks, their calibration is of crucial importance.

the manufacturing plant. You can use this process for gage blocks only if all of the sets are under control and well identified, as with a color dot or band on the side surfaces. Blocks should never be etched or defaced on the contact surfaces.

Although each situation is different, modern statistical quality control and the guidelines in Figure 8–13 can help you develop a calibration plan for each gage block set and each application in any manufacturing plant.

8-4 GAGE BLOCK APPLICATIONS

At the lowest level of precision, we seldom use gage blocks in the place of rules. If you invest in good rules from the start, all you have to do is use them until the ends wear.

As we step up our demand for precision, we sometimes prefer to use gage blocks instead of vernier instruments, even if the specified tolerance is within the discrimination of the vernier instrument. We prefer to use gage blocks instead of vernier calipers when the part is more than 5 inches in length and instead of vernier height gages when the part is more than 10 inches in height. Also, if the person making the measurement is not a highly skilled professional, you should recommend that he or she use gage blocks.

Micrometers have the same discrimination but higher reliability than vernier instruments (see Chapter 7); therefore, we prefer to use gage blocks under the same circumstances for micrometers as for vernier instruments. Because micrometers require little skill to produce comparable results, you should use them only for measurements requiring no finer than 0.01 mm discrimination up to 150 mm. Beyond those limits and even at these low-precision requirements, you may prefer to use gage blocks. For measurements of more than 200 mm, you should use gage blocks, but you must make the decision of which instrument to use based on practical considerations, too.

For precision to 0.002 mm (0.0001 in.), you cannot use vernier instruments, but you *can* use 25 mm and 50 mm micrometers. They are convenient for these measurements, but you must ensure that there is a margin of safety, which must be the width of the tolerance. You will not encounter many problems at ±0.01 mm, but even with a small change in tolerance—to ±0.002 mm, for example—you need to exercise greater care. At any closer tolerance, you might as well make a gage block setup, rather than waste as much time as it would require to make the measurement accurately with a vernier micrometer.

Gage blocks primarily use the comparison method of measurement. In order to take advantage of this method, you need:

- Knowledge of gage block handling and care
- The ability to figure gage block combinations
- The ability to wring blocks reliably
- An understanding of the principles of comparison measurement

Figure 8–14 summarizes general facts about gage block use, while Figure 8–15 summarizes gage block terminology.

Care of Gage Blocks

If you follow a few rules, you can extend the life of gage blocks, reduce the time required to wring combinations, and, most importantly, enhance the reliability of gage block measurement (see Figure 8–16). All these rules are important, but we need to highlight a few.

The Place for Gage Block Measurement

Consider gage blocks when:

1. Precision increases
2. Length increases
3. Importance of reliability increases
4. Skill of measurer decreases

The General Rule:

Use gage blocks for every measurement unless adequate reliability can be more economically obtained by another method.

Rules by precision required:

For 0.05 mm or finer, use only high amplification comparison instruments set to gage blocks. Check the working gage block set with a master gage block set (0.002 inch or finer).

For 0.025 to 0.005 mm: (0.001 to 0.0002 in.)

1. Use only high amplification instruments set to gage blocks for all lengths.
2. Over 50 mm, use only comparison instruments set to gage blocks.
3. Under 50 mm, use vernier micrometers checked with gage blocks.

For 0.050 to 0.025 mm: (0.002 to 0.001 in.)

1. Use comparison instruments (such as dial indicators) set to gage blocks for all lengths.
2. Use micrometers to approximately 50 mm. Check with gage blocks.

For 0.125 to 0.050 mm: (0.005 to 0.002 in.)

1. Use comparison instruments (such as dial indicators) set to gage blocks for all lengths.
2. Use micrometer instruments to approximately 125 mm. Check with gage blocks.
3. Use vernier instruments to approximately 125 mm. Check with gage blocks.

For 0.5 to 0.125 mm: (0.015 to 0.005 in.)

Use micrometers or vernier instruments. Check with gage blocks.

For precision less than 0.5 mm: (0.015 in.)

Gage blocks are rarely preferred.

FIGURE 8–14 Gage blocks are needed at many levels. These rules are based on average measurements, equipment, and skill. The availability of a measuring microscope or an interferometer obviously alters them, and they cannot be applied to extremely large or small measurements.

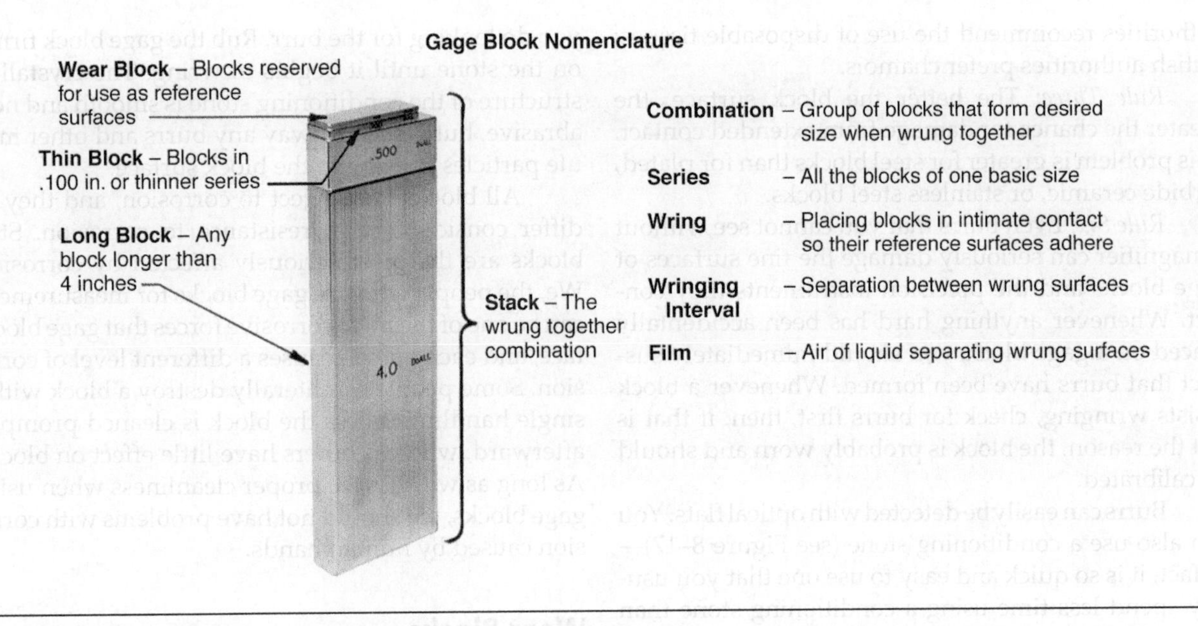

FIGURE 8–15 These terms are used when working with gage blocks.

Rules for Gage Block Care

1. Never attempt to wring or otherwise use gage blocks that have been in contact with chips, dust- or dirt-laden cutting fluids.
2. Before using, clean blocks with a high-grade solvent or commercial gage block cleaner. Wipe dry with a lint-free tissue.
3. Do not allow blocks to remain wrung together for long periods. Separate daily.
4. When not in use, place blocks in a safe place where they will not be damaged, preferably in their case.
5. Before putting blocks away, clean the blocks and cover with a noncorrosive oil or grease or commercial preservative.
6. Be on constant guard for burrs. If anything has been placed on a block or if it does not wring readily, use a conditioning stone immediately.
7. Thoroughly clean the gage block case periodically.

FIGURE 8–16 Obeying these rules will improve reliability, speed measurement, and lower the cost of measurement.

Rule Two: Evaporation of solvents cools gage blocks severely, causing them to shrink. When working to 0.05 μm or finer, always try to keep the blocks clean. However, do not soak blocks in solvent, do not clean them frequently with aerosol cleaners, and do not use compressed air to remove excessive solvent.

To properly clean gage blocks, simply apply solvent and quickly wipe dry with a lint-free tissue. It is important that you remove the solvent as quickly as possible so you minimize cooling. Repeat these cleaning steps as often as necessary to remove any visual signs of dirt, stains, or oil. In the United States, most authorities recommend the use of disposable tissues; British authorities prefer chamois.

Rule Three: The better the block surface, the greater the chance for damage from extended contact. This problem is greater for steel blocks than for plated, carbide ceramic, or stainless steel blocks.

Rule Six: Even burrs that you cannot see without a magnifier can seriously damage the fine surfaces of gage blocks and the precision instruments they contact. Whenever anything hard has been accidentally placed on a gage block, you should immediately suspect that burrs have been formed. Whenever a block resists wringing, check for burrs first, then, if that is not the reason, the block is probably worn and should be calibrated.

Burrs can easily be detected with optical flats. You can also use a conditioning stone (see Figure 8–17)—in fact, it is so quick and easy to use one that you usually spend less time using a conditioning stone than you do looking for the burr. Rub the gage block firmly on the stone until it begins to wring. The crystalline structure of the conditioning stone is smooth and nonabrasive, but it shears away any burrs and other minute particles sticking to the block surface.

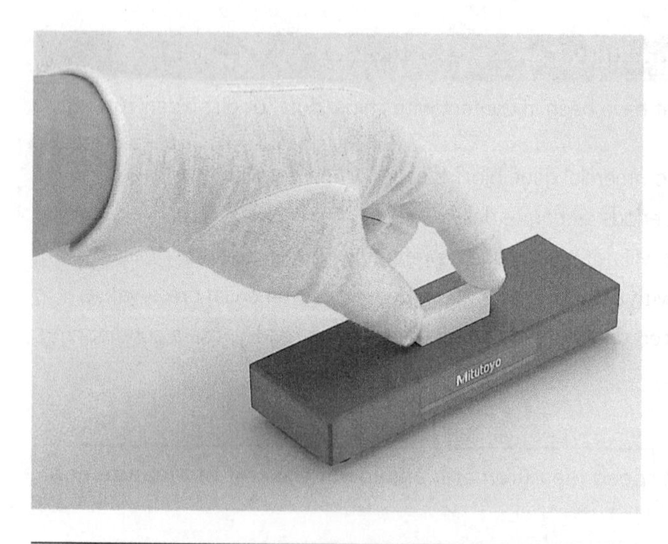

FIGURE 8–17 The conditioning stone is one of your best allies to ensure reliable measurement and long life from gage blocks. The blocks are firmly wrung along its smooth, crystalline surface to remove burrs and nicks without abrading the surfaces. *(Courtesy of Mitutoya American Corporation)*

All blocks are subject to corrosion, and they all differ considerably in resistance to corrosion. Steel blocks are the most seriously affected by corrosion. We, the people who use gage blocks for measurement, can be one of the most corrosive forces that gage blocks face, and each person causes a different level of corrosion. Some people can literally destroy a block with a single handling unless the block is cleaned promptly afterward, whereas others have little effect on blocks. As long as we observe proper cleanliness when using gage blocks, we should not have problems with corrosion caused by human hands.

Wear Blocks

You can use wear blocks to increase the life of a set of gage blocks. Wear blocks are special blocks specifically used as reference surfaces at the ends of gage block stacks. You should use them when blocks are used for direct comparison. You do not need to use them when you use a gage block holder, because the holder provides end standards for the reference surfaces.

When we use wear blocks, the total wear of the stack, which would usually be distributed over all the blocks, is concentrated on the two end blocks. This concentration of wear extends the life of the entire set of gage blocks. Wear blocks that have worn out can be replaced economically. You should always use the same face of a wear block as the reference surface; for example, if it is etched on an end surface, always have that surface showing.

Wear blocks are available in both steel and carbide. Carbide blocks are more expensive, but they provide longer useful life. The difference in the coefficient of expansion is not a serious problem with carbide wear blocks, because the blocks themselves are thin. Finally, remember the damage burrs can cause to carbide blocks. You probably should pass carbide blocks over a conditioning stone after almost every use.

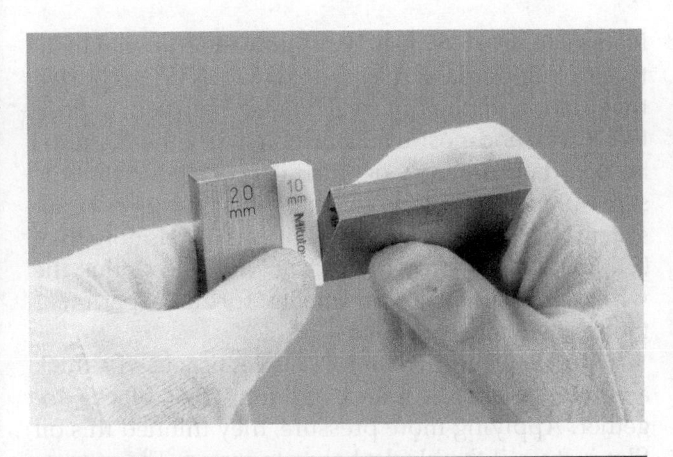

FIGURE 8–18 These blocks show how tenaciously finely finished surfaces will adhere to each other. *(Courtesy of Mitutoya American Corporation)*

Wringing Gage Blocks

If two sufficiently flat, smooth surfaces come into direct contact, they will adhere or *wring* (see Figure 8–18). Wringing makes it convenient for us to handle a stack of blocks as one unit and more.

The Theory of Wringing

Whenever two surfaces are brought together, *some space* still exists between them. Peaks in a surface finish usually push the surfaces apart (see Figure 8–19), or, for better finished surfaces, oil or grease films occupy the space. If the film is cleaned away, a minute air gap will still separate the surfaces. Sometimes, when the surfaces are almost perfect, the gap becomes so small it acts like a liquid film. We refer to this gap as an *air film.* You can try to force this film out, too, but some separation always remains.

We call this remaining separation the *wringing interval.* If this interval was missing, you would have one solid piece, not two parts.

You can reduce the wringing interval to 0.0254 μm (1 μin.), although we do not often achieve this small interval. Under normal measuring conditions, you can reduce the interval to between 0.025 μm and 0.050 μm (1 μin. and 2 μin.). If you wring the blocks properly, you can ignore the wringing interval for most practical measurements.

So why do surfaces adhere? Some say it is a magnetic attraction, but we cannot demonstrate this

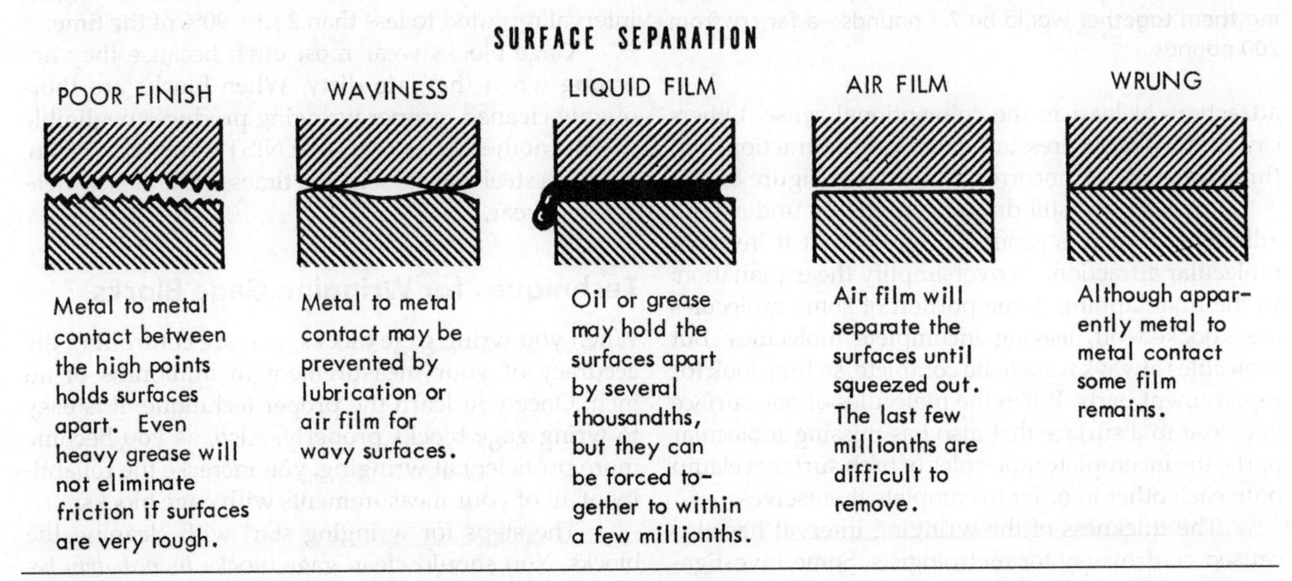

FIGURE 8–19 Air, oil, grease, and surface irregularities hold surfaces apart.

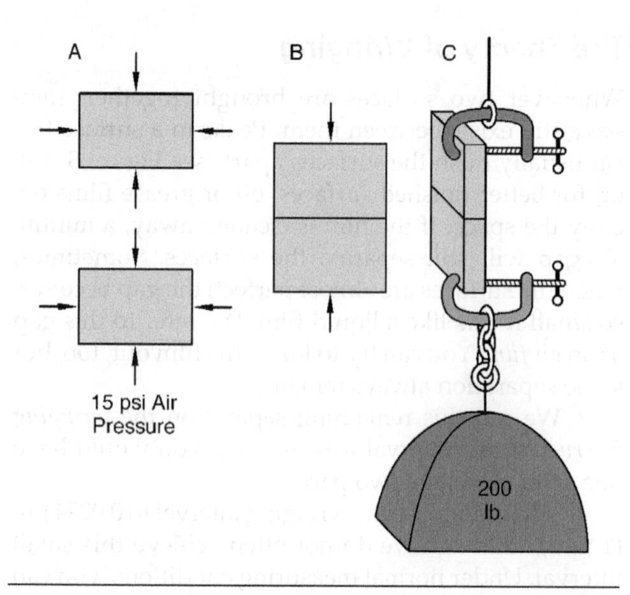

FIGURE 8–20 When the gage blocks are not in contact, as in A, the 15 psi atmospheric pressure acts on all sides and balances. When placed together as in B, the effects on the sides balance but the pressure is now closed off between the blocks. This leaves air pressure on both ends holding the blocks together. But C shows the blocks can resist a 200-pound pull. Based on 15 psi atmospheric pressure and a block area of 0.525 in., the force holding them together would be 7.7 pounds—a far cry from 200 pounds.

attraction, at least in the conventional sense. Others say atmospheric pressure causes the attraction, but this explanation is incorrect as well (see Figure 8–20).

In short, we still do not completely understand adherence, but it is generally agreed that it involves molecular attraction. To oversimplify the explanation: In the final lapping, some portion of some molecules are knocked off, leaving incomplete molecules. But molecules always want to be complete, so they look for replacement parts. When the molecules of one surface get close to a surface that also has missing molecular parts, the incomplete molecules of both surfaces clamp onto each other in order to complete themselves.

The thickness of the wringing interval has also caused controversy for metrologists. Some investigators have claimed that it is negative—the surfaces actually intertwine, so that the space ends up being less than the sum of the parts. In 1922, the United States National Bureau of Standards (now the National Institute of Standards and Technology or NIST) reported that the wringing interval varied between a positive 1 µin. and positive 2.8 µin. In 1956, the National Standards Laboratory of Australia concluded that the separation between two wrung surfaces was not consistently positive or negative but was never more than 0.4 µin. Although recent work at NIST shows that the wringing interval is even smaller, further experimentation is obviously needed.

In an experiment at NIST, scientists used a thick application of oil, almost "gluing" two blocks together. Applying more pressure, they thinned this oil film just until the blocks began to wring. The separation at this point was approximately 7 µin. more than the final wrung separation. If the scientists stopped the mechanical manipulation after the wringing began, the surfaces pulled together by themselves within 20 minutes to 2 hours, depending on the film material.

If the surface topography (finish and flatness) of the blocks is fine, the blocks will be easier to wring together, and the wringing interval will be smaller. However, after a certain point, further improvement in finish does not increase the uniformity of the wringing interval. Tests at NIST demonstrated that the wringing interval repeated to less than 2 µin. 90% of the time.

Gage blocks wear most often because they are wrung when they are dirty. When blocks are thoroughly cleaned, proper wringing produces negligible wear. Another series of tests at NIST wrung two clean, stainless steel gage blocks 100 times without any measurable wear.

Techniques for Wringing Gage Blocks

When you wring gage blocks, you are controlling the accuracy of your measurement to millionths of an inch. Once you learn the proper technique, it is easy to wring gage blocks properly. Also, as you become more proficient at wringing, you increase the reliability of all of your measurements with gage blocks.

The steps for wringing start with cleaning the blocks. You should clean gage blocks *immediately before* wringing. Sometimes, someone will methodically clean the gage blocks, then just as methodically get them dirty again before wringing. Timing is vital to

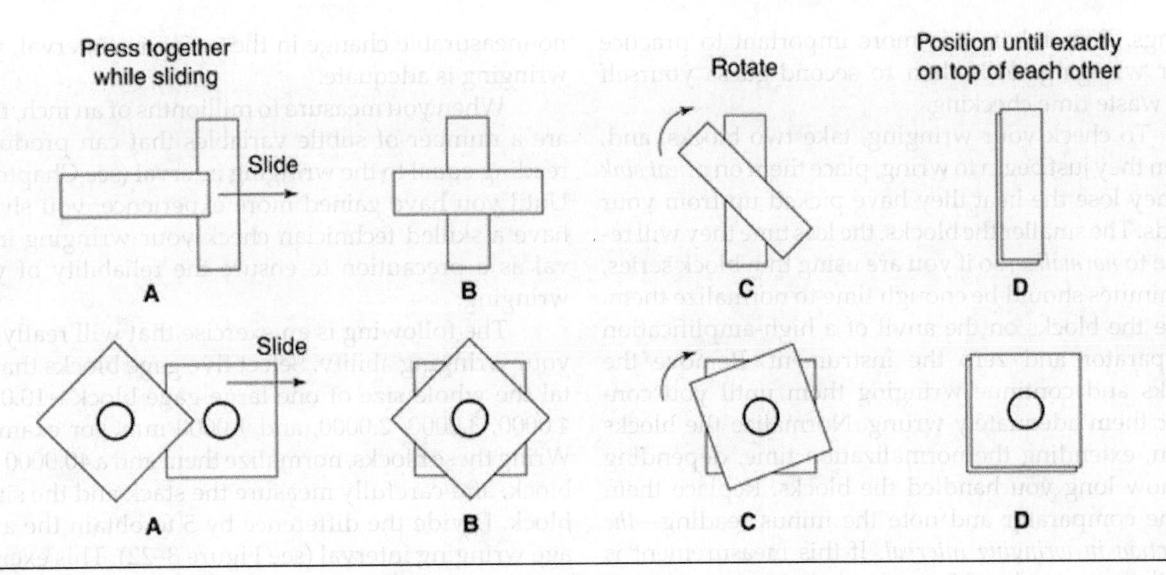

FIGURE 8–21 These steps need to be practiced to develop the feel for wringing.

proper wringing. If you delay even one minute between cleaning the blocks and wringing, dust will settle on them, but you can use a camel's hair brush to remove the dust.

To wring blocks, follow these steps

1. Be sure the gaging surfaces are clean.
2. To start, overlap the gaging surfaces about 1/8 inch.
3. While pressing blocks lightly together, slip one over the other. (Blocks should start to adhere.)
4. Rotate blocks smoothly until gaging surfaces are fully mated.

Practicing these steps for wringing gage blocks before working on them will help you with Lab 6, Exercise 2 from your lab manual.

As blocks wear, they become more difficult to wring unless you use an oil film. You may use a small amount of a noncorrosive oil that does not carry microscopic dust particles onto the blocks, applied most often with a dropper. The oil spreads itself over the block surfaces during the wringing.

If you follow the instructions for wringing, you can also minimize the effect of dust: Sliding the blocks together (step A through step C) (Figure 8–21) should push away dust from the contact surfaces rather than trap it between them. The more worn the blocks are, the more pressure you will have to apply to ensure that the blocks adhere. When one pass is not enough to start the wring, move back from step C to step B (not all the way back to the beginning) and start again.

When you feel resistance, the wringing action has begun. As you continue to slide the blocks, you will feel the wringing get stronger; but *blocks are not fully wrung just because they adhere.* You should continue to slide the blocks until further movement does not increase the adhesion. Be careful until the blocks are wrung, because a circular sliding motion can cause serious wear or even damage from dust trapped between the surfaces. After the blocks are wrung, you can wring the blocks the rest of the way by moving one block through a small arc.

It is important that you learn to recognize when gage blocks have been wrung properly; generally, we can repeat the wringing interval reliably to within 0.025 μm. This measurement is sufficiently small that we can consider the uncertainty from wringing negligible; however, poor wringing can result in errors of 0.250 μm (10 μin.) or more. Thus, in a stack of five blocks, you would end up with a 1 μm (40 μin.) total interval.

While you are still learning, we recommend that you periodically check your progress with a high-amplification comparator between several practice wrings. Remember, it is more important to practice your wringing skills than to second-guess yourself and waste time checking.

To check your wringing, take two blocks, and, when they just begin to wring, place them on a *heat sink* so they lose the heat they have picked up from your hands. The smaller the blocks, the less time they will require to *normalize*, so if you are using thin block series, 15 minutes should be enough time to normalize them. Place the blocks on the anvil of a high-amplification comparator and zero the instrument. Remove the blocks and continue wringing them until you consider them adequately wrung. Normalize the blocks again, extending the normalization time, depending on how long you handled the blocks. Replace them in the comparator and note the minus reading—*the reduction in wringing interval.* If this measurement is less than or equal to 0.2 μm (6 μin.), you have probably wrung the blocks as much as possible. Remove the blocks, wring them some more, normalize the blocks again, and recheck your wringing. If there has been no measurable change in the wringing interval, your wringing is adequate.

When you measure to millionths of an inch, there are a number of subtle variables that can produce a reading equal to the wringing interval (see Chapter 9). Until you have gained more experience, you should have a skilled technician check your wringing interval as a precaution to ensure the reliability of your wringing.

The following is an exercise that will really test your wringing ability. Select five gage blocks that total the whole size of one large gage block—10.0000, 4.0000, 3.0000, 2.0000, and 1.0000 mm, for example. Wring these blocks, normalize them and a 40.0000 mm block, and carefully measure the stack and the single block. Divide the difference by 5 to obtain the average wringing interval (see Figure 8–22). This exercise will not give you a true measure of your average wringing because the sample is too small, but you can use the result to guide your continuing wringing efforts.

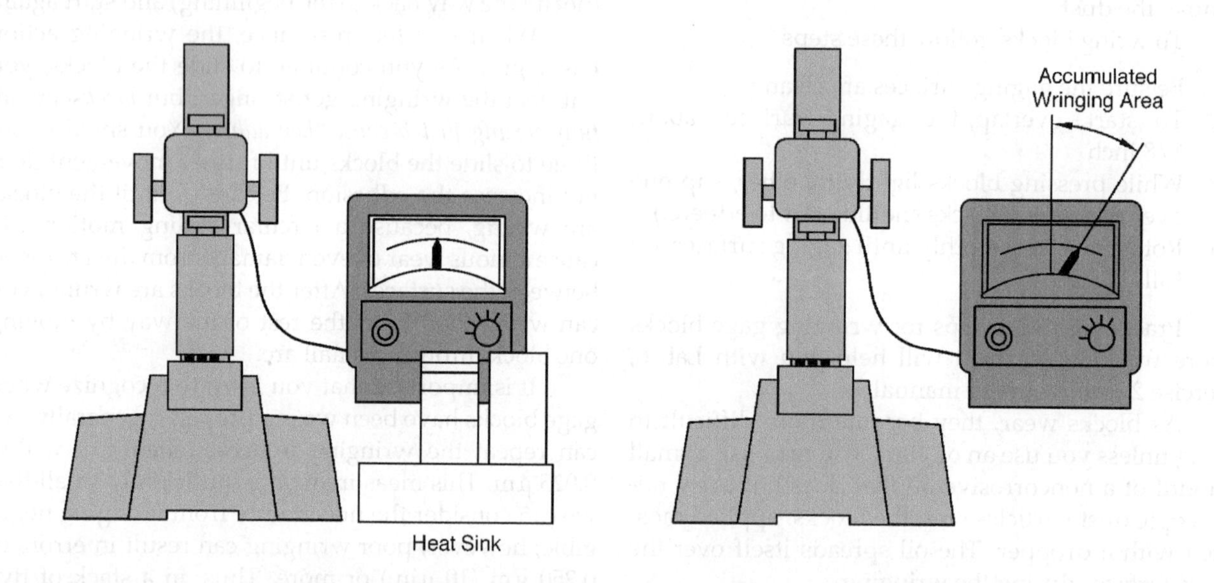

FIGURE 8–22 Patience and a high amplification comparator are all that are needed to test wringing ability. The stack of several wrung blocks is normalized with a single block of the same nominal length. The comparator is zeroed on the single block. That block is replaced with the stack. The plus reading will be your total accumulated wringing interval.

8–5 COMBINING GAGE BLOCKS

Gage blocks are produced in several series, and Johansson's system allows a large number of dimensional combinations. However, we do need a system to decide which blocks are needed to make up a given dimension. That system should save time, reduce the chance of error, use the smallest number of blocks, and leave the most blocks in the set so you can duplicate the dimension.

Combining for a Given Dimension

When you use this system, you need to consider the figures to the right of the decimal point first, starting with the figure farthest to the right, and continue to eliminate figures from right to left.

In Figure 8–23, how can we set the dimension up with the fewest blocks? Typically, gage block sets are tabulated in an ascending magnitude, so you work with the blocks from the top down, that is from the smallest increment first.

For our example dimension of 24.817 mm, start with the 7 and eliminate it from further consideration by choosing a 2.007 mm block. Subtract this measurement from the total dimension and move on to the final figure in the result of your subtraction. Eliminate the 1 with a block designated 2.31 mm, and subtract. Then, eliminate the 5 with the 20.5 mm block and subtract. Simply wring the blocks together, and the unit dimension will be 24.817 mm.

Suppose you need a duplicate combination of blocks for the 24.817 measurement. Because you have already used the only block in the 1.001 mm series

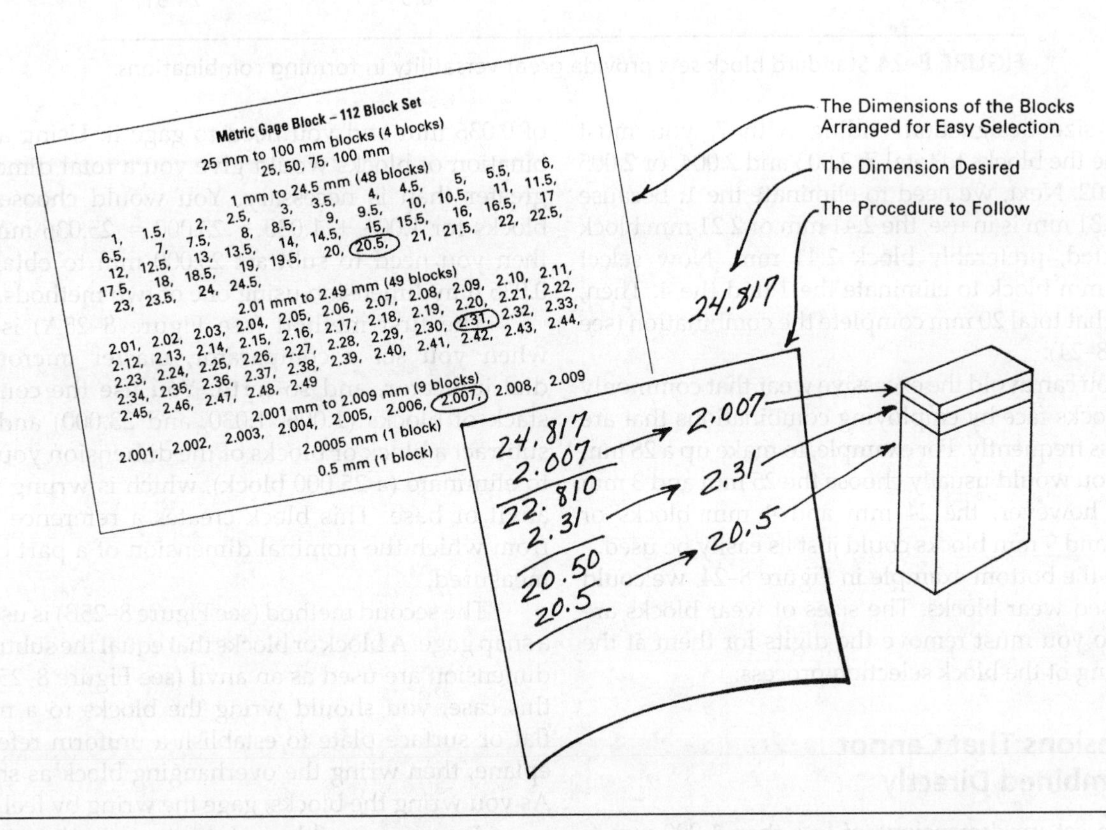

FIGURE 8–23 Elimination of digits from the right selects the blocks required for any combination.

Alternate Combinations

Finding a second combination choice:

Dimension sought	24.817 mm	Proof
To eliminate the 7, add 1.003 and 1.004	2.007	2.007
Result	22.810	
To eliminate the 1	2.41	2.41
Result	20.4	
To eliminate the 4	2.4	2.4
Result	18.0	
To eliminate the 8	8	8
Result	10	
To eliminate the 10	10	10
Result	0.0	24.817

FIGURE 8–24 Standard block sets provide great versatility in forming combinations.

with a size designation ending with 7, you must combine the blocks to total 7: 2.003 and 2.004, or 2.005 and 2.002. Next, we need to eliminate the 1. Because block 2.31 mm is in use, the 2.41 mm or 2.21 mm block is selected, preferably block 2.41 mm. Now select the 1.4 mm block to eliminate the 1 and the 4. Then, blocks that total 20 mm complete the combination (see Figure 8–24).

You can avoid the excessive wear that commonly used blocks face by employing combinations that are used less frequently. For example, to make up a 28 mm stack, you would usually choose the 25 mm and 3 mm blocks; however, the 24 mm and 4 mm blocks or 19 mm and 9 mm blocks could just as easily be used.

In the bottom example in Figure 8–24, we could have used wear blocks. The sizes of wear blocks are fixed, so you must remove the digits for them at the beginning of the block selection process.

Dimensions That Cannot Be Combined Directly

You can set up dimensions of less than 1.000 mm—for example, let's say a part has a nominal dimension of 0.036 mm and you need to gage it. Using a combination of blocks would give you a total dimension greater than is necessary. You would choose gage blocks for 1.006 + 1.030 + 23.000 = 25.036 mm. But then you need to subtract 25.000 mm to obtain the 0.036 mm dimension using one of two methods.

The first method (see Figure 8–25A) is used when you set a comparator, master micrometer, dial indicator, and so forth. You use the complete stack of blocks (1.006, 1.030, and 23.000) and then subtract a block or blocks of the dimension you need to eliminate (a 25.000 block), which is wrung to the anvil or base. This block creates a reference plane from which the nominal dimension of a part can be measured.

The second method (see Figure 8–25B) is used for a snap gage. A block or blocks that equal the subtracted dimension are used as an anvil (see Figure 8–25B). In this case, you should wring the blocks to a master flat or surface plate to establish a uniform reference eplane, then wring the overhanging block as shown. As you wring the blocks, gage the wring by feel.

As you wring blocks to an oversized part, typically the wring will be broken. Of course, today,

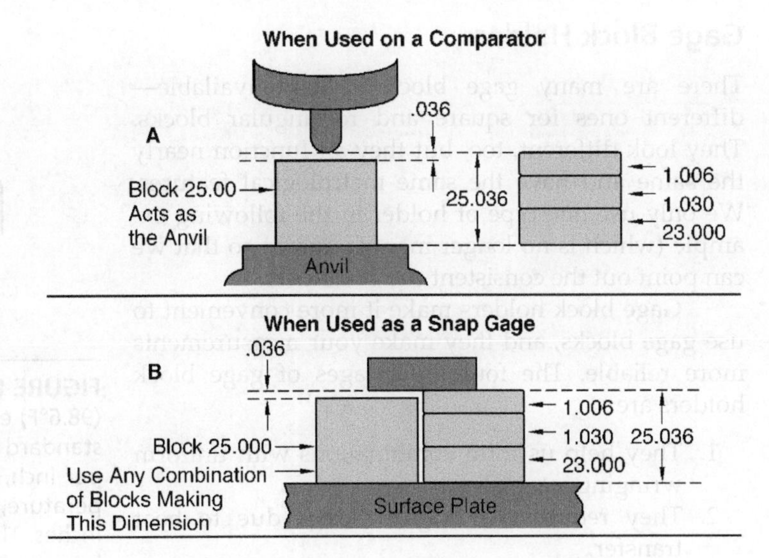

FIGURE 8–25 Dimensions too small to be combined with the available set may be formed as the difference between stacks of blocks.

electronic calculators and computer programs are available to help you select gage block combinations.

Combining to Find a Dimension

Not only can we use gage blocks to create stacks to a specific length, but we can also use them to measure slots. For a slot, maybe a snap gage opening or an opening on a part, you should first use a scale to approximate the opening, which in our example is approximately 12 mm. Usually, you would build a stack of blocks and try them in the opening, but your block stack will either be too large or too small. Any effort to correct the size would only waste time.

We use a technique called *bridging* to speed up our selection process (see Chapter 9). *Bridging* consists of choosing successive values on both sides of the desired unknown value until, with each selection, you reduce the spread and zero in on the correct dimension. To do this with gage blocks, you must create a trial stack with blocks from more than one series so you can make changes in an orderly fashion in either large or small increments.

To begin, build up a stack smaller than the opening—for example, 1.1 mm smaller. Add a 1.220 mm block, or, if this block is too small, change it to a 1.270 mm block. If the stack is too large when you check it, you know that the final dimension is between 12.220 and 12.270 mm. Next, replace the 1.270 mm block with a 1.250 mm block, and determine if the stack is too large or too small (for our example, we will assume it is too large). Replace the 1.250 mm block with a 1.230 mm block, and determine if the stack is too large or small. Continue this process until you find the correct 0.01 mm block, and then repeat the process for the 1.001 mm series of blocks until you find the actual size.

You will need to practice in order to learn the correct feel. It should be tight enough that there is no perceptible wobble, but the blocks should slide quite easily through the entire opening. Also, you should be careful of openings with a good finish. The stack might wring to one surface or another and appear to fit snugly. Make sure you repeat the process to check the fit.

During this process, you must also remember the effect of heat transfer from your hands. When you believe you have a stack that duplicates the size of the opening, lay both the part and the stack aside to normalize, preferably on a heat sink. After an hour or more, try the fit again; there is a chance the fit will have changed by a small amount.

Gage Block Holders

There are many gage block holders available—different ones for square and rectangular blocks. They look different, too, but they all function nearly the same and have the same metrological features. We only use one type of holder in the following example (which is no longer manufactured) so that we can point out the consistent use of holders.

Gage block holders make it more convenient to use gage blocks, and they make your measurements more reliable. The four advantages of gage block holders are:

1. They help us form combinations with uniform wringing intervals.
2. They reduce the length change due to heat transfer.
3. They make it more convenient to handle stacks of blocks.
4. They help prevent costly damage to blocks.

In addition, some of our most valuable gage block techniques are impossible without gage block holders.

All gage block holders provide a frame into which the blocks are constrained. The constraining force is commonly applied by screw thread, which gives us the advantage of carefully controlling the compression force. Compression never takes the place of good wringing, but it does add to the reliability of gage block measurement. Many holders provide a way for you to control the clamping force.

As we have mentioned, heat is a problem in high-precision measurement. Whenever you handle a gage block stack, it absorbs heat from your hands. Holders for rectangular blocks stand between the stack and the hands, which slows, but does not eliminate, heat transfer.

When you measure to a precision of 25 μm or finer and for all lengths 150 mm or greater, you should always wear insulating gloves or use plastic insulated tweezers (see Figure 8–26). Remember, as a professional metrologist, it is up to you to determine all the tools you need to ensure the precision and reliability of your measurements.

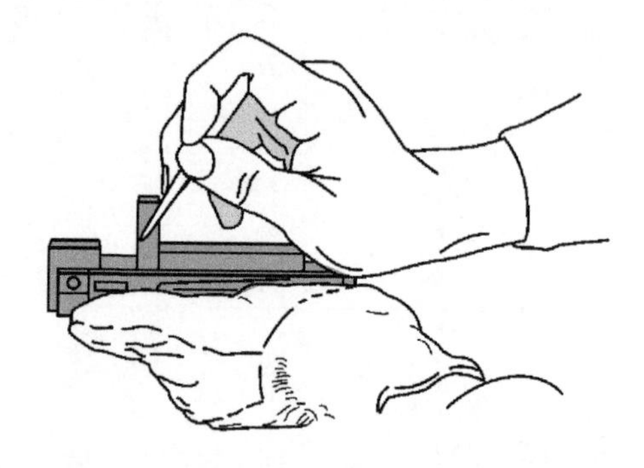

FIGURE 8–26 Remember that body temperature is 37°C (98.6°F) even on a cold day. The expansion coefficient of standard gage block steel is 6.4 microinches per degree per inch, which means that for a 1 degree rise in temperature, a 4-inch stack of blocks will expand 25.6 microinches. Therefore, insulating gloves and tweezers should be used for all precision work.

The final way in which square and rectangular gage block holders reduce heat transfer is that they enable you to assemble stacks in easier-to-use, self-supporting assemblies. These assemblies require less handling than blocks without holders, and they help minimize the negative effects temperature expansion can have on measurement (see Figure 8–27).

Of course, as important as metrological reasons for using gage block holders are, most of us use them for practical reasons, and the most important reason may be protection if a block stack falls. When a high stack of blocks tips over, you cannot consider those blocks reliable again until they are recalibrated. In addition, the precision surface onto which the blocks fell cannot be reliably used until it is recalibrated, too.

Square Gage Block Holders

Although square blocks are less likely to fall, they, too, are at risk. Gage block holders for square or rectangular blocks decrease the likelihood of them toppling over, and they also decrease the damage that happens when they do topple. Holders prevent or minimize damage to the blocks themselves, to the part surface, and to precision reference surfaces with which the blocks are used.

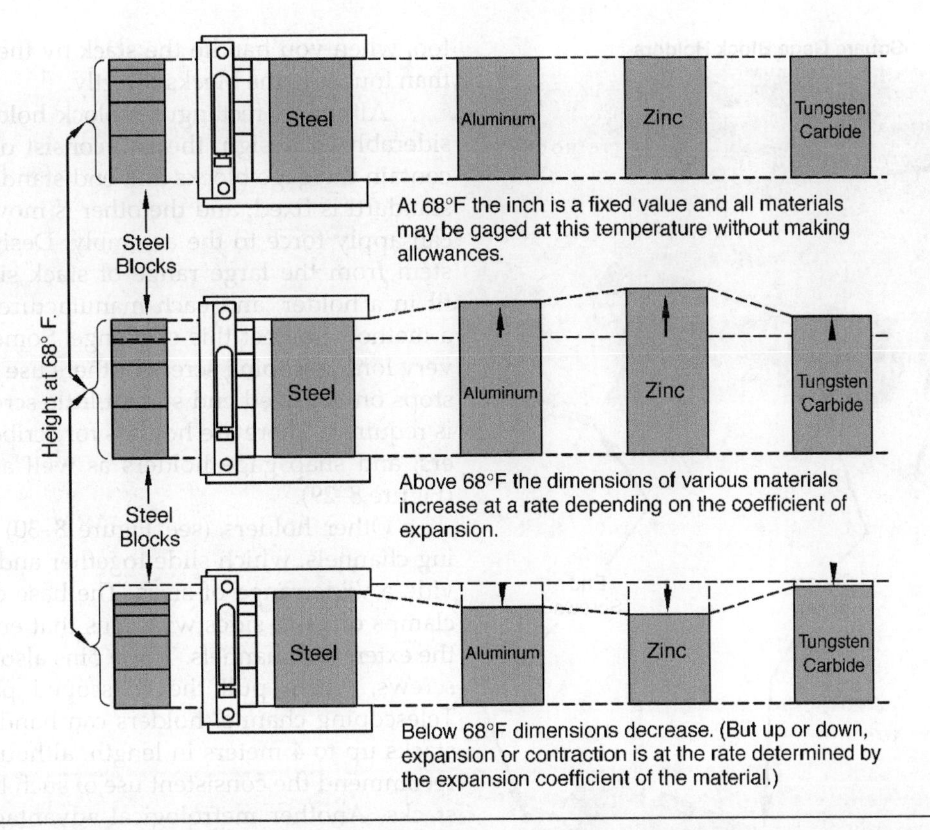

FIGURE 8–27 These exaggerated drawings show the effect of temperature changes on measurement.

The holders for square blocks are very similar (see Figure 8–28) consisting of expandable rods that pass through the center holes of the blocks. These rods have threaded ends so that the holder can be secured against the end standards. To reliably use square gage block holders, you should fully wring the blocks with the end standards, assemble the stack with the tie rods, and firmly draw up the end screw with a torque screwdriver. When you follow this process, you are creating an assembly with a repeatable, uniform total wringing interval and sufficient rigidity for practical measurement.

When you use a torque screwdriver, you will notice that you can make sufficiently reliable assemblies for most measurements without wringing the blocks. You can simply thread the blocks onto the tie rods and clamp with the end screws; however, you can only use this method satisfactorily if the gage blocks are clean. The only way to ensure there is no dust between surfaces is to use the wringing procedure described earlier. A compromise procedure is to wring the blocks lightly and then clamp the blocks, relying on the clamping force to squeeze out excess film and form uniform wringing intervals.

Rectangular Gage Block Holders

Because rectangular blocks are lower in price than square blocks, they are used for the majority of measurements. We can compensate for their instability and other inadequacies by using well-designed holders.

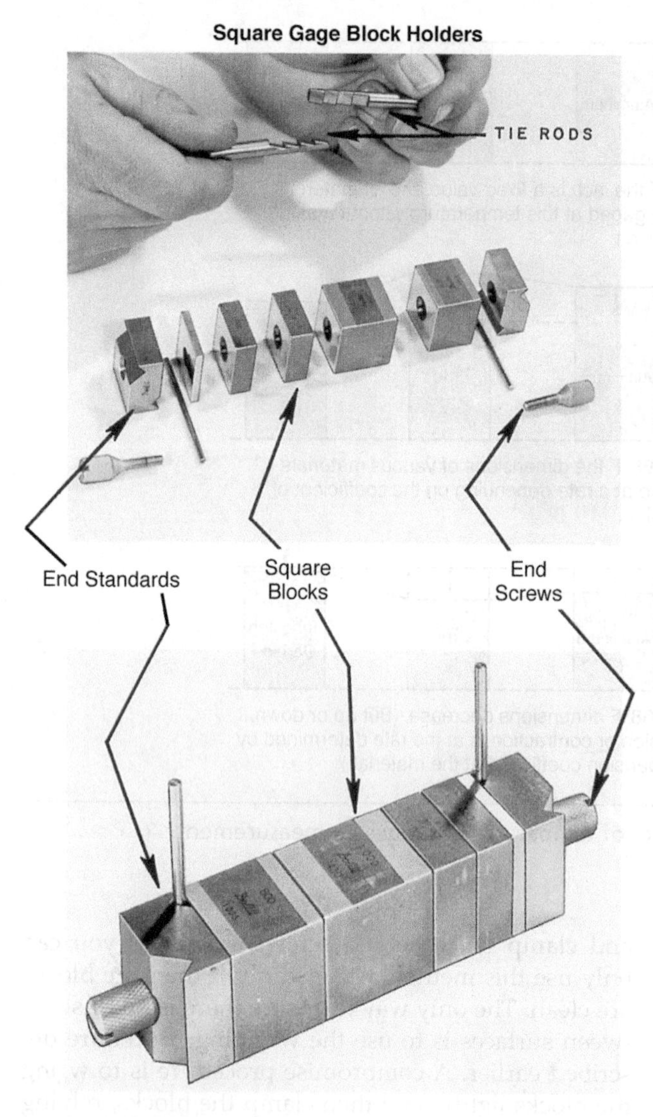

FIGURE 8–28 This pin gage assembly shows the use of square gage block holders.

Rectangular blocks are narrow and very unstable when stacked vertically. When held to a base by holders, they are as stable as square blocks. Also, their large, flat sides and small mass make them subject to heat expansion. Block holders help with this problem, too, when you handle the stack by the holder rather than touching the blocks directly.

Although rectangular block holders vary considerably in design, they all consist of a channel to contain the gage blocks and end standards. One end standard is fixed, and the other is movable so that it can apply force to the assembly. Design differences stem from the large range of stack sizes that must fit in a holder, and each manufacturer has devised a method to meet this challenge. Some versions use very long clamping screws; others use channels with stops on the fixed end so that little screw movement is required. There are holders for scribers and dividers, and snap gage holders as well as gage blocks (Figure 8–29).

Other holders (see Figure 8–30) use telescoping channels, which slide together and apart to provide a wide range of sizes. The base channels have clamps on both sides with pins that engage holes in the extension channels. These pins also have knurled screws, which pull the telescoped parts together. Telescoping channel holders can handle gage block stacks up to 4 meters in length, although we do not recommend the consistent use of such tall gage block stacks. Another metrological advantage is that the force that controls the wringing interval flows along the axis of the stack of blocks; therefore, the blocks in the stack do not bend.

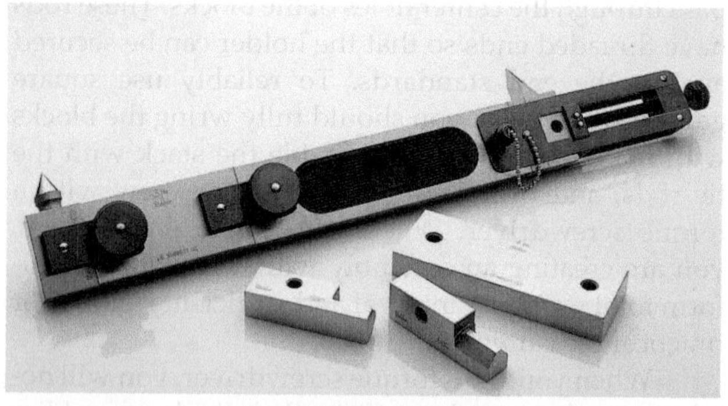

FIGURE 8–29 Scribers, dividers, and a snap gage (*Courtesy of the L. S. Starrett Company.*)

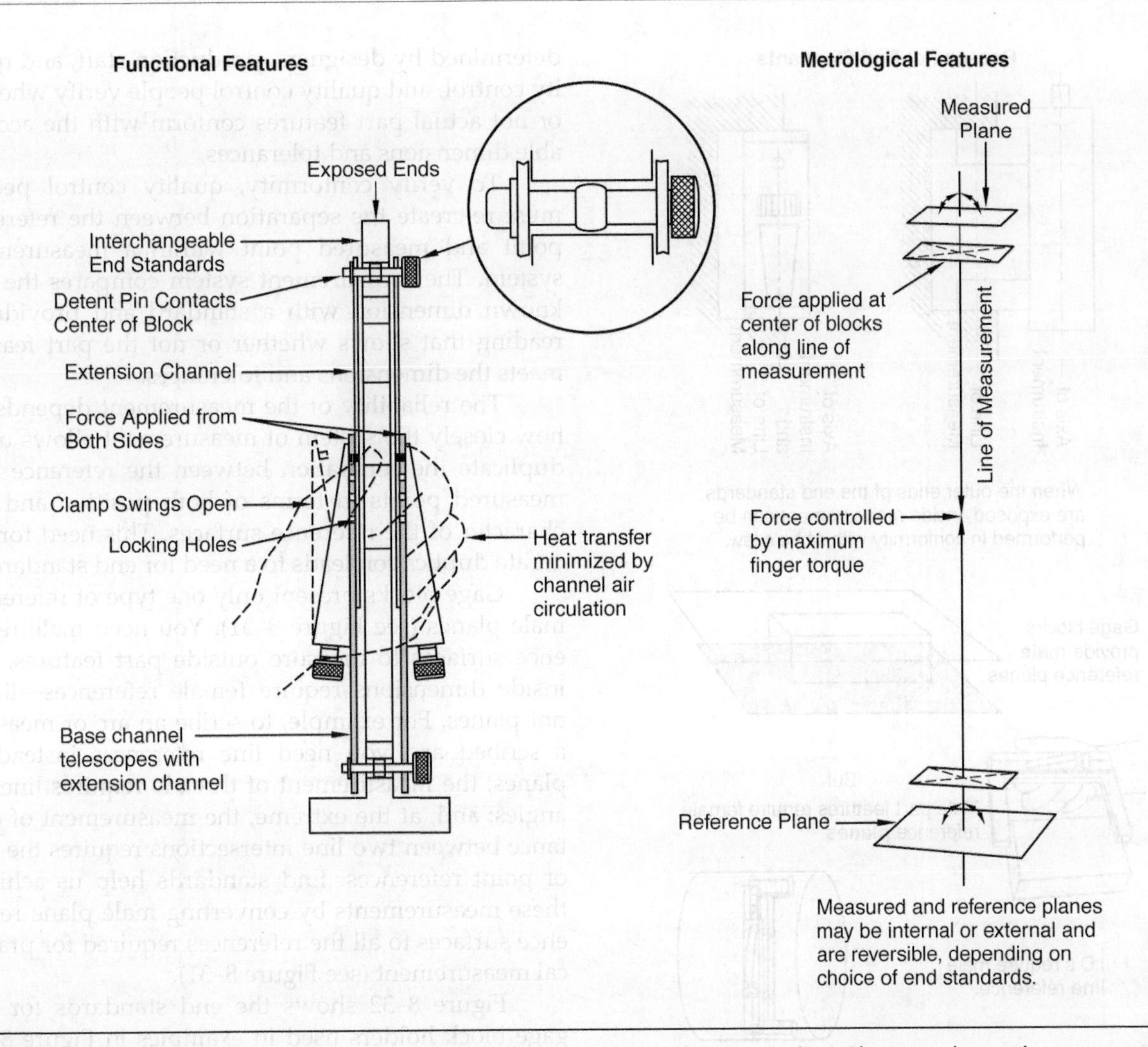

FIGURE 8–30 Gage block holders vary considerably in mechanical construction. The one shown is representative of good design features.

The open construction of the channels minimizes heat transfer, and their design permits the end standards to be exposed fully, which offers a distinct metrological advantage (see Figure 8–31).

End Standards

One of the most important functions of a gage block holder is to adapt gage blocks so that you can use them for more than just setting comparators. We use end standards to adapt gage blocks for other measurements.

In Chapter 1, you learned that a dimension is the size that a designer feels a part feature should be so that we can manufacture perfect parts. A dimension is a specified separation between two points that lie along a line of measurement. Tolerance is the allowable deviation from the dimension, which is

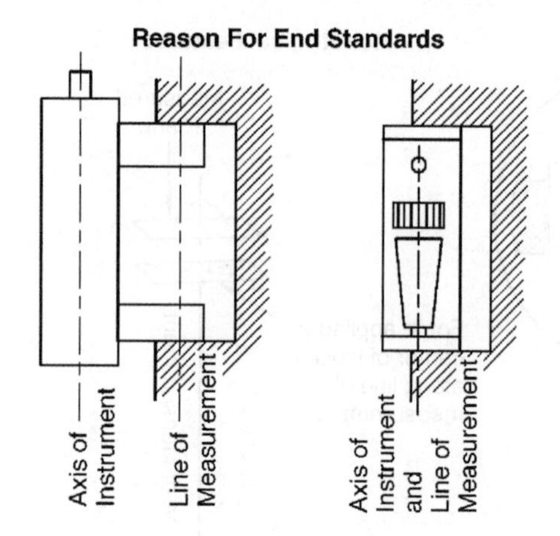

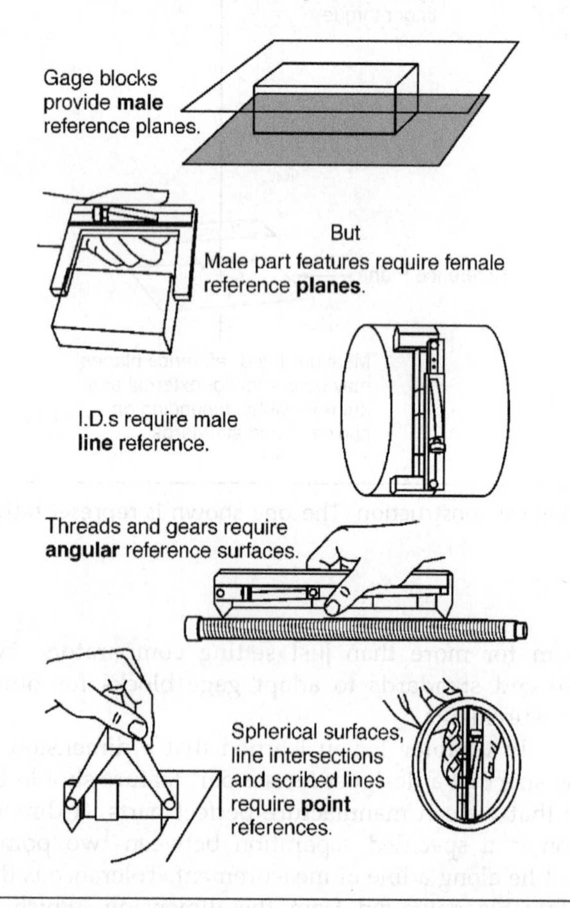

FIGURE 8–31 End standards adapt gage block reference surfaces to part features.

determined by designers, production staff, and quality control, and quality control people verify whether or not actual part features conform with the acceptable dimensions and tolerances.

To verify conformity, quality control people must re-create the separation between the reference point and measured point within a measurement system. The measurement system compares the unknown dimension with a standard and provides a reading that shows whether or not the part feature meets the dimensions and tolerances.

The reliability of the measurement depends on how closely the system of measurement allows us to duplicate the separation between the reference and measured points in terms of both position and the character of the reference surfaces. This need for accurate duplication leads to a need for end standards.

Gage blocks present only one type of reference: male planes (see Figure 8–31). You need male reference surfaces to measure outside part features, but inside dimensions require female references—lines, not planes. For example, to scribe an arc or measure a scribed arc, you need line references instead of planes; the measurement of threads requires lines at angles; and, at the extreme, the measurement of distance between two line intersections requires the use of point references. End standards help us achieve these measurements by converting male plane reference surfaces to all the references required for practical measurement (see Figure 8–31).

Figure 8–32 shows the end standards for the gage block holders used in examples in Figure 8–30 and 8–31. Similar end standards are available for other styles of gage block holders, all providing the same function even though they differ in appearance.

Gage blocks with and without holders and end standards are used for a variety of purposes (see Figure 8–33) and for applications that are pertinent to metrology (numbers 1 and 2) and those that are not. Always, the basic objective is reliable measurement. These measurements are performed to different levels of precision and under widely divergent circumstances, but all applications require end standards and efficient holders. The principles behind these accessories and their advantages are as valuable for gaging attributes as they are for precision comparator measurements.

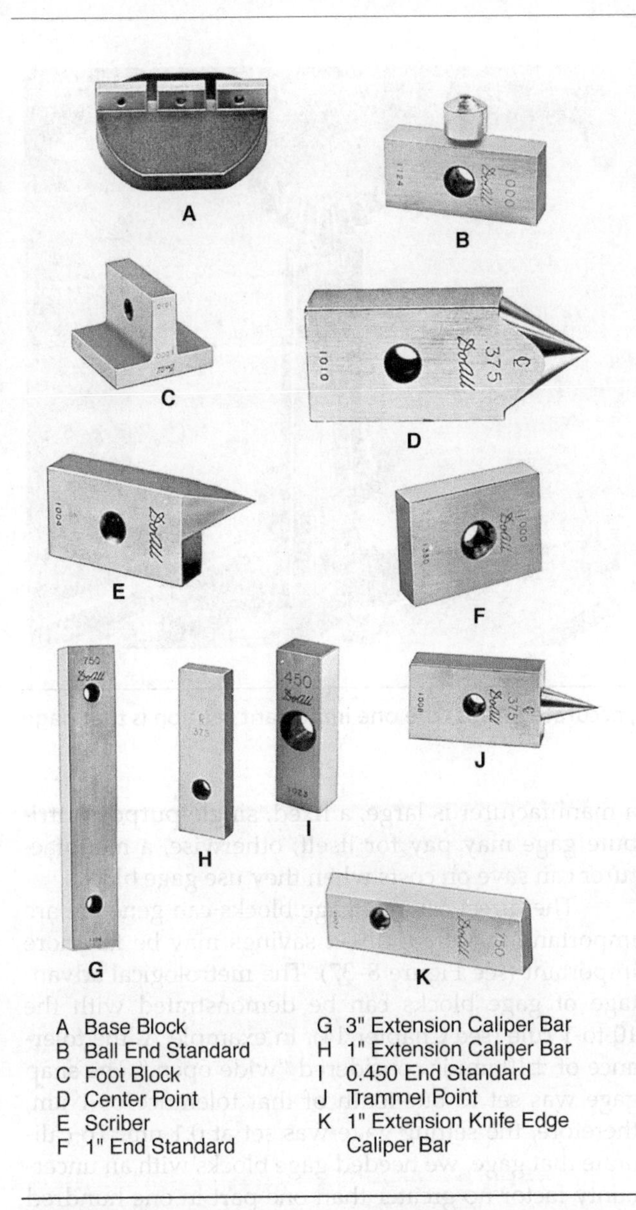

FIGURE 8–32 The design of end standards varies according to the types of gage block holders with which they are used. These types are typical of the references provided by end standards.

Principal Uses for Gage Blocks

1. Calibration of other instruments and lesser standards
2. Setting of comparators and indicator-type instruments
3. Attribute gaging
4. Machine setup and precision assembly
5. Layout

FIGURE 8–33 These are the five principal uses for gage blocks.

Setup, Layout, and Assembly

In some cases, it takes longer to make a measurement with gage blocks, but you gain in precision, accuracy, and reliability when the blocks are used. One exception is machine setups. Even when high precision is a low priority, gage blocks may be the fastest and most convenient method of measurement.

You can quickly check spacing relationships before operating a machine with gage blocks (see Figure 8–34). In the past, you would make an approximate setup, but measuring between two milling cutters was not easy because of the compound cutting angles. You would make a cut, measure it, and alter the spacing. Continuing this process, you would finally achieve the tolerance of the part. The flat reference planes of gage blocks are ideal for pinpointing the location of these cuts. Note: you must use wear blocks for this type of work. Direct applications of gage blocks are shown in Figure 8–35.

For layout applications, we use special gage block end standards to precisely scribe lines. These lines show the part features for future machining and represent the measured ends of dimensions. The reference ends may be points such as punch marks, lines such as other scribed lines, intersections of scribed lines, or plane surfaces.

Special end standards have been developed, including the center point, trammel point, and scriber, and typical examples are shown in Figure 8–32. We primarily use the scriber when the reference end of the dimension is a plane—for example, see Figure 8–29, where the surface plate is the reference. The reference could as easily have been a part feature, and the measurement can be made horizontally or vertically.

Attribute Gaging with Gage Blocks

Attribute gaging is a sorting operation, which is not truly the function of a metrologist. Once we attribute

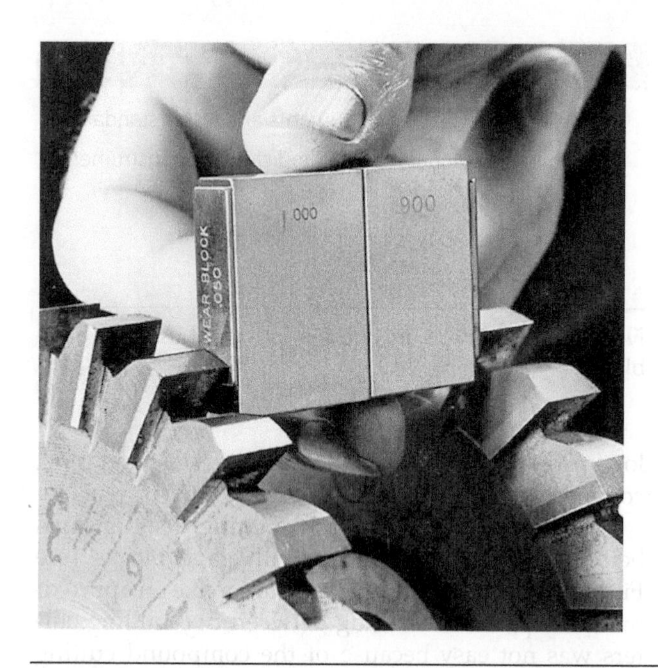

FIGURE 8–34 Gage blocks are highly recommended for fast, accurate setups. The one important caution is that gage blocks are never used in this manner without wear blocks.

gaging in the production process, the decision is merely whether the part is good or bad, whether it is a go or no-go, and so forth. If you continue into manufacturing, you will get to know attribute gaging well, along with inspection and quality control. We do not widely use gage blocks in attribute gaging, but there are sound metrological and functional reasons for using them (see Figure 8–36), such as for go and no-go gages.

If you use gage blocks as an attribute gage for an inside measurement, the flat top surface of the extension bar will limit your measurement. You could use the blocks to measure to the bottom of flat surfaces or outside diameters, but you could not use them to measure to an inside diameter unless you also use a knife-edge caliper bar with the knife edge up (see Figure 8–31).

Why are gage blocks so useful as gages? Gage blocks provide discrimination in 1 μm or 100 μin. One set of gage blocks with suitable holders and end standards can generate 10,000 different sizes of fixed gages for each inch of its range. If the production of a manufacturer is large, a fixed, single-purpose attribute gage may pay for itself; otherwise, a manufacturer can save on costs when they use gage blocks.

The direct savings gage blocks can generate are important, but the indirect savings may be far more important (see Figure 8–37). The metrological advantage of gage blocks can be demonstrated with the 10-to-1 rule (see Chapter 14). In example A, the tolerance of ±10 μm is considered "wide open." The snap gage was set to one-tenth of that tolerance or 1 μm; therefore, the setting gage was set at 0.1 μm. To calibrate that gage, we needed gage blocks with an uncertainty factor no greater than one part in one hundred thousand or 0.01 μm. If we assume that the comparator used to calibrate the working set to the master blocks is perfect (which it is not), the master blocks would have to be accurate to one part in one billion or 1 nanometer.

In example B, the tolerance is closed down to ±1 μm, but then the demand for discrimination exceeds the capability of existing equipment, as shown by the shaded areas. Fortunately, the demand for this discrimination happens in the final calibration stage, where the

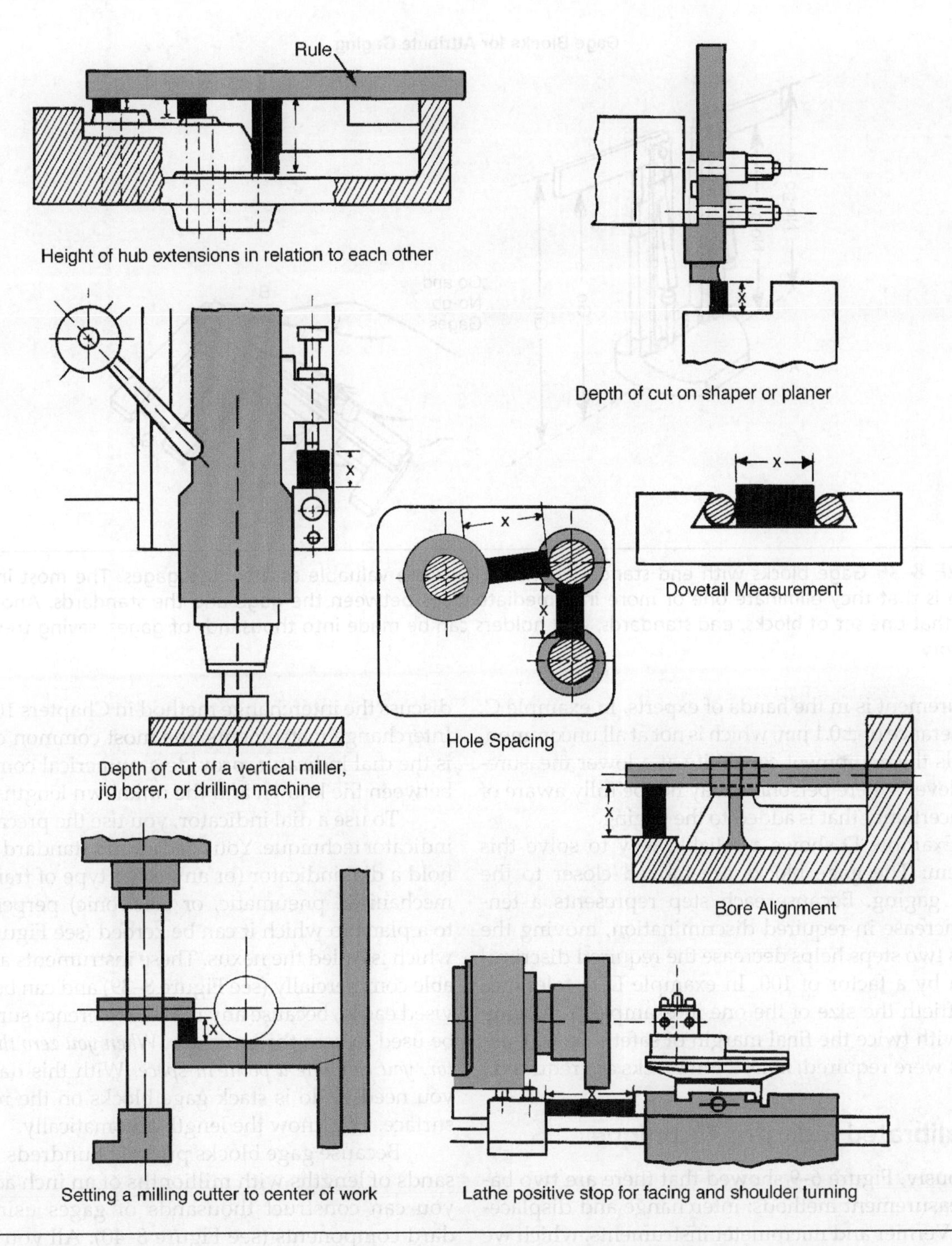

FIGURE 8–35 Direct use of gage blocks to set desired dimensions (x) eliminates errors from transfer of measurement. Gage blocks are widely used with sine bars and sine plates for angle measurements as discussed in Chapter 12.

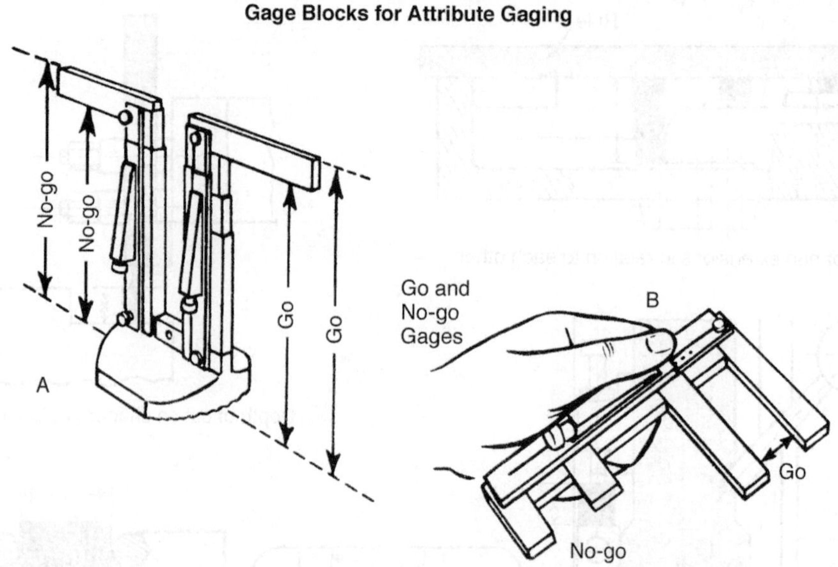

Gage Blocks for Attribute Gaging

FIGURE 8–36 Gage blocks with end standards and holders are valuable as attribute gages. The most important reason is that they eliminate one or more intermediate steps between the gage and the standards. Another reason is that one set of blocks, end standards, and holders can be made into thousands of gages, saving tremendous inventory.

measurement is in the hands of experts. In example C, the tolerance of ±0.1 μm, which is not at all uncommon, exceeds the equipment capability at a lower measurement level, where personnel may not be fully aware of the uncertainty that is added to the gaging.

Example D shows a reliable way to solve this problem: The gage blocks are moved closer to the actual gaging. Because each step represents a tenfold increase in required discrimination, moving the blocks two steps helps decrease the required discrimination by a factor of 100. In example D, a tolerance one-fiftieth the size of the one in example A is being held with twice the final margin of safety. In A, 1 μm blocks were required; in D, 2 μm blocks are required.

Precalibrated Indicator Technique

Previously, Figure 6–9 showed that there are two basic measurement methods: interchange and displacement. Vernier and micrometer instruments, which we have discussed, use the displacement method. We will discuss the interchange method in Chapters 10 and 14. Interchange instruments, the most common of which is the dial indicator, provide a numerical comparison between the known and the unknown lengths.

To use a dial indicator, you use the precalibrated indicator technique: You need an end standard that will hold a dial indicator (or any other type of transducer, mechanical, pneumatic, or electronic) perpendicular to a plane to which it can be zeroed (see Figure 8–38), which is called the nexus. These instruments are available commercially (see Figure 8–39) and can be improvised easily, because any reliable reference surface can be used to zero the indicator. *When you zero the indicator, you calibrate a point in space.* With this datum, all you need to do is stack gage blocks on the reference surface. You know the length automatically.

Because gage blocks provide hundreds of thousands of lengths with millionths of an inch accuracy, you can construct thousands of gages using standard components (see Figure 8–40). All you need to do to recheck this gage is remove the blocks, repeat

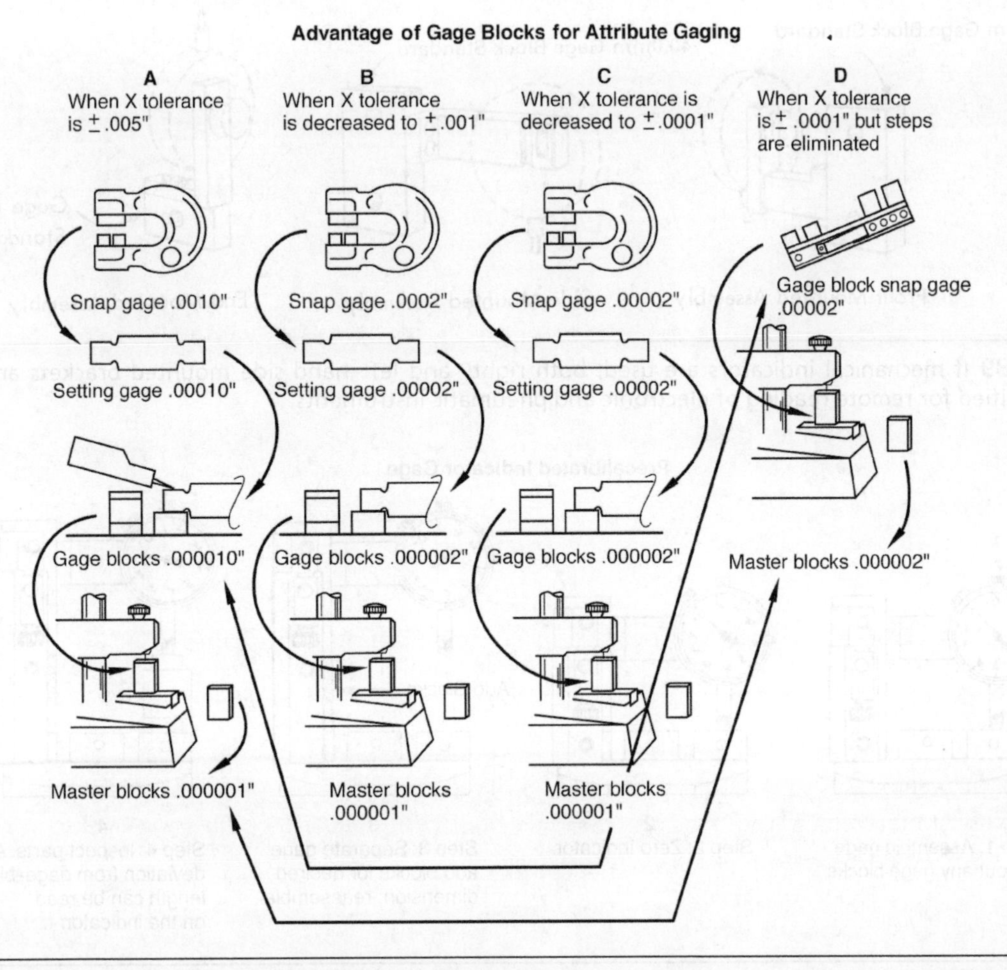

FIGURE 8–37 In example D, a tolerance, 1/50th as large as the one in example A, is being controlled with double the final margin of safety. This is achieved by the direct use of gage blocks for attribute gaging.

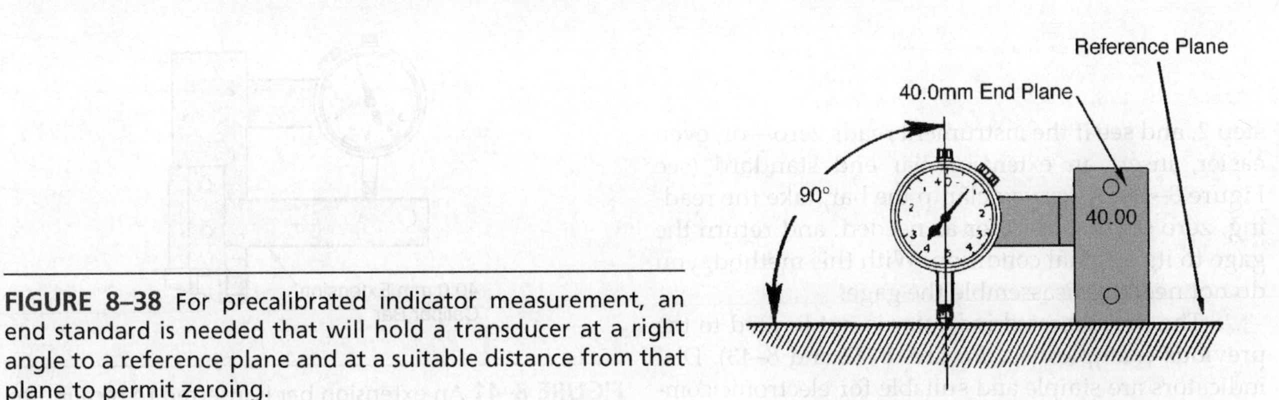

FIGURE 8–38 For precalibrated indicator measurement, an end standard is needed that will hold a transducer at a right angle to a reference plane and at a suitable distance from that plane to permit zeroing.

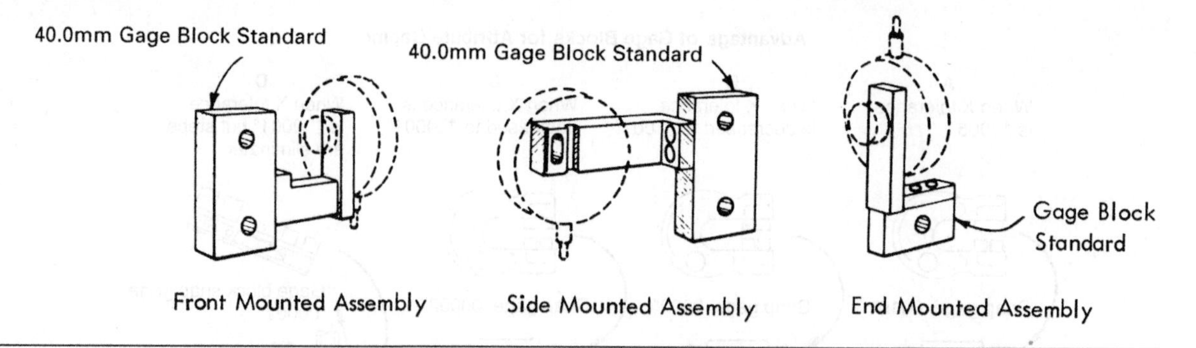

FIGURE 8–39 If mechanical indicators are used, both right- and left-hand side mounted brackets are required. This is simplified for remote reading of electronic and pneumatic instruments.

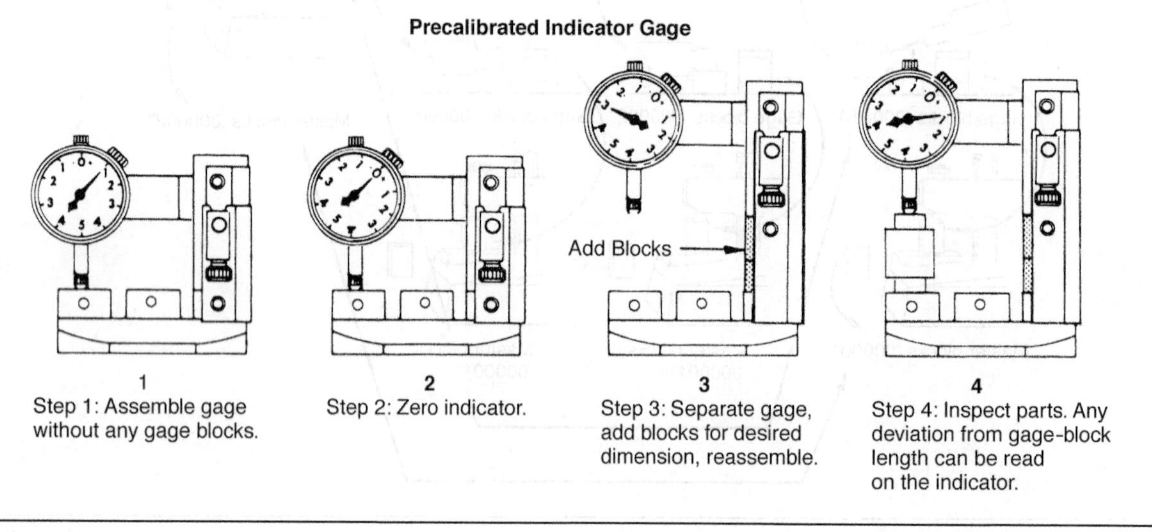

FIGURE 8–40 These four simple steps provide a nearly limitless range of gages with microinch accuracy, even though their precision is in thousandths of an inch.

step 2, and see if the instrument reads zero—or, even easier, insert an extension bar end standard (see Figure 8–41). After you clamp the bar, take the reading, zero set the indicator as needed, and return the gage to its original condition. With this method, you do not need to disassemble the gage.

The precalibrated indicator is not limited to the previous examples (see Figures 8–42 and 8–43). Dial indicators are simple and suitable for electronic comparators, too.

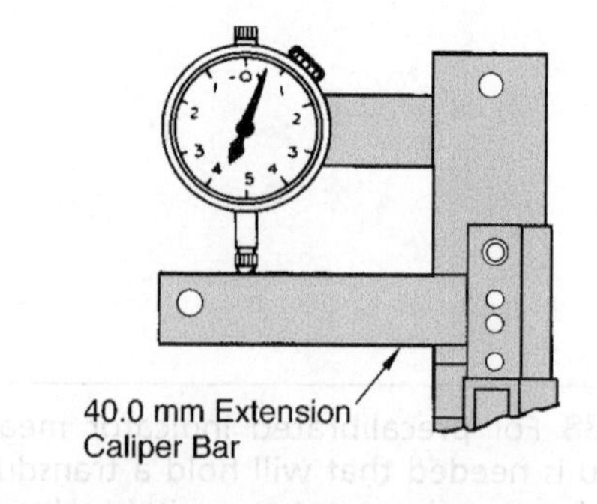

FIGURE 8–41 An extension bar is inserted to recalibrate the gage.

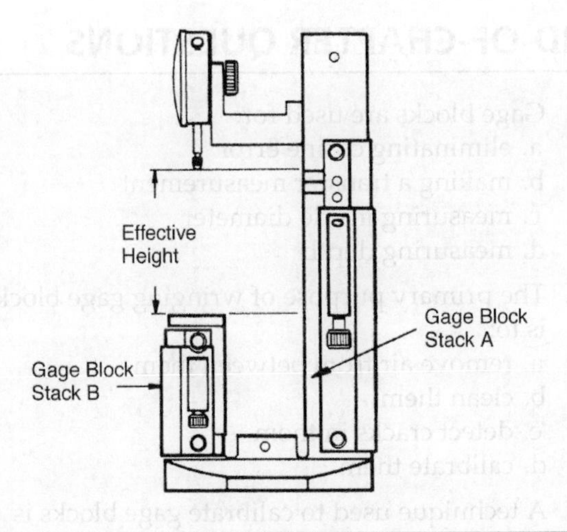

FIGURE 8–42 The use of standard holders and end standards gives great versatility to the precalibrated indicator technique. The effective height in this example is the length difference between stacks A and B. Therefore, this method can be used for dimensions that are ordinarily too small for gage block methods.

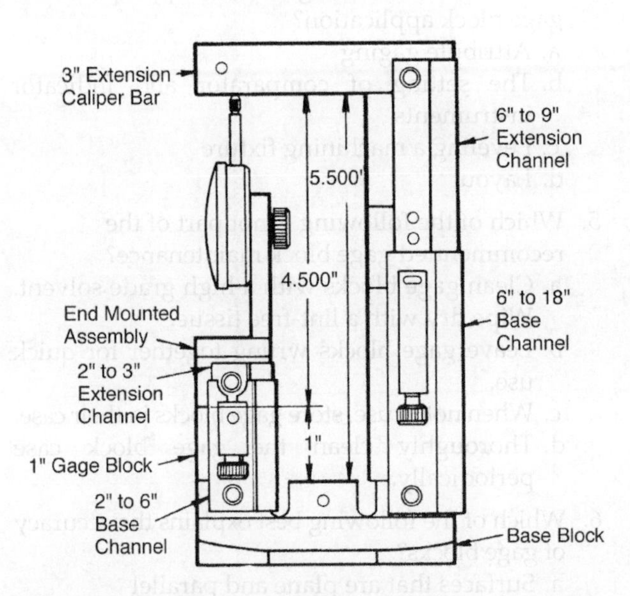

FIGURE 8–43 The precalibrated indicator technique can also be used for inside measurement, as shown in this example. It is being zeroed in the illustration. Following that, gage blocks will be inserted in one or both stacks to create the desired dimension. The extra step, of course, somewhat reduces the reliability.

SUMMARY

- To sum up this chapter, never assume that a measurement to millionths means anything more than that there are six places to the right of the decimal point. Unless based on standards, the accuracy is doubtful.
- Gage blocks provide the most practical standard for linear measurement. They are readily available. A small number forms thousands of dimensions. The errors involved with them are known and therefore predictable. Well-organized channels verify their accuracy.
- The size of a gage block set required depends not only on the smallest and largest dimensions to be formed, but also on the average number of blocks required to form combinations. A large set will require fewer blocks, therefore saving time and reducing the wringing errors.
- The material and shape of the gage blocks selected will depend on both economic and metrological factors. The cost per use will be far lower for a set of hard but expensive material, if extensively used. Cost per use will soar if expensive blocks are rarely used. The probability of wear will be reduced for any given number of combinations. This decision must be tempered by environmental conditions. Abrasive dust calls for the hardest blocks, whereas temperature expansion problems may recommend the lower priced steel blocks. Although all gage blocks follow federal specifications, care must be applied to their purchase.
- Gage blocks provide access to 0.03 µm or 10 µin. accuracy. Without gage blocks, even with 0.03 µm (10 µin.) reading capability, electronic comparators achieving 0.3 µm (10 µin.) accuracy is difficult. Gage blocks should be used for all measurements from 25 µm (0.001 µin.) discrimination and finer when this is possible. They should be obligatory for measurements to 0.25 µm (10 µin.) discrimination or finer.
- The inherent reliability from gage blocks can only be achieved by proper wringing. That requires considerable skill but can be achieved by anyone with knowledge and practice. Blocks

adhering to each other do not show that they are wrung properly. When fully wrung, the wringing interval is about 0.02 μm (1 μin.) That is negligible for most practical measurement.

- To form a stack of blocks of a given dimension, use the following procedure. Select blocks that, when their lengths are added, eliminate the last digits from the right. To form a stack of blocks to find an unknown dimension, the bridging technique is used. The test stack is successively reduced by steps of one-half until it duplicates the feature being measured. Very small dimensions are formed by the differences in lengths of two stacks of blocks.
- For many applications, end standards and holders are needed. Sometimes this is for stability and convenience. In other cases, it is to extend gage blocks from their basic role as male gages to other roles, such as line standards and female standards.
- Whenever tolerances of 2 μm (100 μin.) or smaller are being held, it becomes important that temperature compensation is considered. For large parts (more than 250 mm or 10 in.), this is important even for 0.02 mm (0.001 in.) tolerances. If the parts are steel, it is only necessary that they are at the same temperature as the gage blocks. If they are not steel and both the parts and the gage blocks are at 20°C (68°F), no compensation is needed. However, at any other temperature the difference must be calculated based on the respective coefficients of expansion.
- The basic rule is to use gage block setups unless there is a good reason not to. Therefore, they are unquestionably required for calibration of linear instruments. They should also be considered for layout, machine setup, assembly, and attribute gaging. The precalibrated indicator technique brings laboratory standards directly to the line inspection operation, thereby improving reliability.
- All of the advantages of gage blocks are nullified if the gage blocks themselves are not reliable. In fact, this may cause irreparable harm. Therefore, gage blocks must be kept in calibration.

END-OF-CHAPTER QUESTIONS

1. Gage blocks are used for:
   a. eliminating cosine error
   b. making a transfer measurement
   c. measuring inside diameter
   d. measuring depth
2. The primary purpose of wringing gage blocks is to:
   a. remove air from between them
   b. clean them
   c. detect cracks in them
   d. calibrate them
3. A technique used to calibrate gage blocks is known as:
   a. wringing
   b. magnetic resonance
   c. interferometry
   d. ultrasonic
4. Which of the following is not an appropriate gage block application?
   a. Attribute gaging
   b. The setting of comparator and indicator instruments
   c. Leveling a machining fixture
   d. Layout
5. Which of the following is not part of the recommended gage block maintenance?
   a. Clean gage blocks with a high grade solvent. Wipe dry with a lint-free tissue.
   b. Leave gage blocks wrung together for quick use.
   c. When not in use, store gage blocks in their case.
   d. Thoroughly clean the gage block case periodically.
6. Which of the following best explains the accuracy of gage blocks?
   a. Surfaces that are plane and parallel
   b. A surface finish with a 0.4 microinch arithmetical average or better
   c. Heat-treated steel that has been stress relieved
   d. Their traceability to the National Institute of Standards and Technology

7. Which of the following people is generally credited with having the greatest influence on interchangeable manufacture?
   a. Hjalmar Ellstrom
   b. Henry Ford
   c. Eli Whitney
   d. C. E. Johansson

8. Which of the following best explains how interchangeable manufacturing hastened the development of gage blocks?
   a. Improved alloys and heat treatment for steel
   b. Convenience of interplant standardization
   c. Convenience of intraplant standardization
   d. Extended the metric system of standards

9. Which of the following do we attribute to C. E. Johansson?
   a. Introduced interchangeable manufacturing
   b. Developed a process for lapping gage blocks
   c. Devised a system of gage block sizes
   d. Introduced Henry Ford to gage blocks

10. What happens when work is performed that requires a subdivision of the standard that is smaller than the error existing in the standard?
    a. It is necessary to subtract the error in the standard.
    b. The measurement is reversed and the results averaged.
    c. The result cannot be considered accurate.
    d. It's impossible. It cannot happen.

11. Each gage block is marked with its basic size. Which of the following best explains what that means?
    a. It is essential in order to put each block back into its proper place in the set.
    b. It is the minimum size of that block. Any tolerance will be a wear allowance in the plus direction.
    c. It is the exact size of that block after calibration to NIST.
    d. It is the basic size and may vary by the amount of the tolerance.

12. Which of the following affirms the accuracy of a gage block?
    a. The bill of sale that accompanies each order for gage blocks
    b. The use of a comparator that divides the gage block tolerance by a factor of 10
    c. Traceability—the calibration to the next highest class of blocks, which in turn is calibrated to a still higher one until NIST is reached
    d. The theory of reversibility. Two readings are taken from opposite directions and averaged so that their errors may cancel out.

13. Which of the following is the best reason for frequent calibration of gage blocks?
    a. To pass the responsibility for errors to a higher level
    b. To determine the amount of wear and possible instability
    c. To ensure that the calibration instruments do not drift
    d. To fulfill the requirement of MIL-Q-8958

14. How are gage blocks calibrated?
    a. By repeat measurements with a known workpiece and taking the RMS of the readings
    b. By the use of high-amplification comparators
    c. By comparison with blocks of a higher grade
    d. By carefully cleaning them, and then using them at only 20°C (68°F)

15. What should be done with gage blocks that fail the calibration for their grade?
    a. Record the readings in the master calibration record and hold the blocks in reserve.
    b. Stamp the symbol "X" with metal stamps on each reference surface.
    c. Lap both surfaces until the next smaller size is reached.
    d. Move the blocks to the next lowest tolerance for which they fulfill the requirements.

16. Which of the following is the best criterion for the selection of a set of gage blocks?
    a. Lowest overall cost
    b. Lowest cost per block

c. Sufficient range to provide one group of blocks for every dimension likely to be used
d. Best cost-effectiveness based on the number of people who may be making up similar measurements at the same time

17. Which of the following best explains the need for gage blocks in practical manufacturing?
a. Need for precision increases
b. Value of the part increases
c. Increased skill of the inspectors is available
d. Temperatures are other than 68°F

18. Which of the following best describes wringing?
a. Squeezing the oil out from between two blocks
b. Causing blocks to adhere by molecular attraction
c. The effect of atmospheric pressure to hold blocks together
d. The high-frequency sound resulting from two gage blocks struck together

19. Which of the following results in wringing error with properly prepared blocks?
a. Variations in the thickness of air or oil film between the blocks
b. Burrs
c. Corrosion from the perspiration of the users of the blocks
d. Blocks of different temperature

20. Which of the following are the best explanations for the use of wear blocks?
a. They expand in direct proportion to their reduction in length by wear over a calculated period of time.
b. They are made of carbide, which is harder than steel.
c. They protect the other gages in the set from wear resulting from direct contact with the workpiece or other gaging instruments.
d. They are available without charge from the manufacturers of gage blocks.

21. Which of the following precautions should be used before wringing gage blocks?
a. They must be absolutely clean with no trace of oil.
b. They must all be from the same series.
c. From smallest to largest, they must vary by 50% of their respective sizes.
d. Potential burrs should be removed with a deburring stone.

22. Which of the following helps produce uniform wringing intervals?
a. Light oil diluted with solvent
b. Use at 68°F
c. Uniform contact pressure from controlled pressure holders
d. The same surface finish for both surfaces

23. Dirt is obviously undesirable, but which of the following best describes the greatest problem with dirt?
a. It gets the parts dirty.
b. It causes excessive wear of the blocks.
c. It causes unreliability of the wringing interval.
d. Valuable time must be spent cleaning.

24. How can you tell that you have wrung blocks together properly?
a. Submit them to a gage lab for checking.
b. Only by extensive practice.
c. All the oil has been squeezed out.
d. Checking with a high-amplification comparator reveals no change from subsequent wringing.

25. Which of the following is one of the steps for selecting a combination of blocks to form a given dimension?
a. Selective halving
b. Writing out all combinations and selecting the closest one
c. Determining the fewest number of blocks
d. Eliminating the last digit to the right

26. Which of the following are reasons to utilize two stacks of gage blocks to form one dimension?
a. The dimensions are too small for the available set of blocks
b. As a double-check for reliability
c. For attribute gaging
d. To correct for thermal expansion

27. Excessive handling of blocks should be avoided. Why?
    a. It dirties the blocks.
    b. It causes excessive wear.
    c. It causes corrosion.
    d. It heats the blocks.

28. What explains the more extensive use of rectangular blocks over square ones?
    a. They are easier to handle.
    b. Some sets will fit in a shop coat pocket.
    c. They have less surface to clean.
    d. They cost less.

29. Which of the following best explains the development of end standards?
    a. Gage blocks have only male planes.
    b. To reach into recesses
    c. To scribe lines
    d. To form attribute gages

30. When compared with other methods of measurement, which of the following is the chief advantage of gage blocks?
    a. Gage blocks are self-calibrating.
    b. Gage block methods are relatively slow.
    c. Gage blocks provide better precision, accuracy, and reliability.
    d. Gage block methods have a longer range.

31. In addition to the improved metrological features, what are the other chief advantages of using gage blocks and their accessories?
    a. The versatility and ability to replace many other gages
    b. Speed of setup
    c. Freedom from the effects of heat
    d. Cleanliness

CHAPTER NINE

CALIBRATION

. . . let it not be feared that erroneous deductions may be made from recorded facts: the errors which arise from the absence of facts are far more numerous and durable than those which result from unsound reasoning respecting true data.

Charles Babbage (1792–1871)
(Nineteenth-century English inventor who conceived the computer as an instrument for research)

LEARNING OBJECTIVES

- Explain the role of calibration.
- Describe the significance of traceability.
- Explain the effect of errors.
- Describe general calibration techniques.
- State the limitation of the 10-to-1 rule.

OVERVIEW

Very often, a user calibrates an instrument meticulously, and then goes back to work and repeats the same errors again and again. A calibrated instrument may give many users a false sense of security—if the instrument is set correctly, the measurement must be right—but calibration may mask more serious errors. Error can cause problems, but properly applied calibration methods can reduce or even eliminate these problems.

Calibration is not a cure-all for measurement problems; it only determines adherence to standards. As we have emphasized, accuracy is adherence to a standard; therefore, calibration is the measurement of accuracy. With a fixed gage, the calibration process simply determines the amount by which the gage is larger or smaller than its nominal size. With a measuring instrument, calibration is the complex relationship between a change in input and a change in output.

When we calibrate an instrument, in the strictest sense of the term we are establishing numerical values for metrological features according to accepted standards. We can only know one of these features, the inherent accuracy, in terms of a higher standard; therefore, each calibration depends on the calibration of a higher instrument or standard, until we base the calibration on the international standard. There may be many steps (see Figure 9–1) between the working measurement that you obtain and the international standard. Each calibration step (see Figure 9–1) depends on many factors.

Any metrologist's ability to measure is affected by the environment, the functional features of the instrument, and his or her training. These factors, along with the design features of the instrument, combine in precision, and these factors combined with precision interpret accuracy. You must be able to trace the accuracy of any part feature measured directly back to the international standard. At each step, all the factors contribute error that distorts the accuracy; however, only the inherent accuracy carries the accuracy forward. Ensuring accuracy is the function of the calibration group.

As defined by the *International Vocabulary of Basic and General Terms in Metrology (VIM)* and adopted by NIST, traceability is the "property of the result of a measurement or the value of a standard whereby it can be related to stated references, usually national or international standards, through an unbroken chain of comparisons all having stated uncertainties." *Traceability* requires the establishment of an unbroken chain of comparisons to stated references.

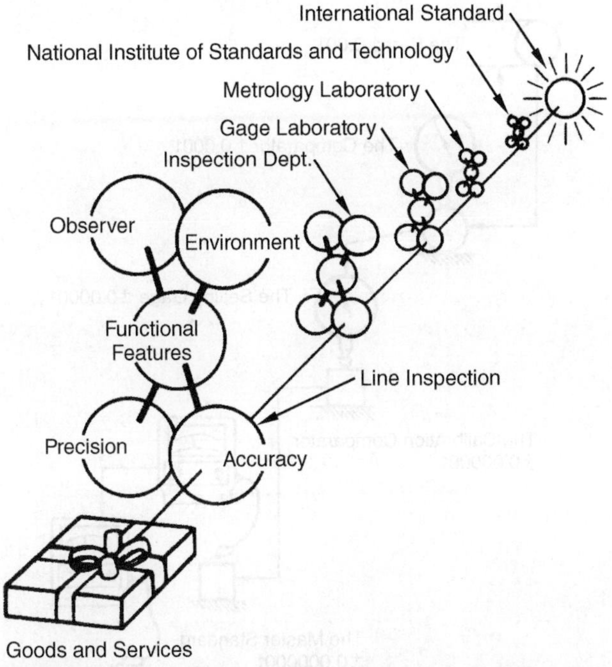

FIGURE 9–1 Calibration is the authentication of the accuracy lineage of any measurement tracing back to the absolute standard. The other metrological and functional features at each step simply help get the job done.

The Rule of 10-to-1

Wherever people are concerned with practical measurement, you will hear that the measuring instrument should be 10 times as accurate as the part. The generally accepted explanation is that, by doing so, we limit the amount of instrument error to 1%.

The *10:1 rule* is a requirement of the ISO/TS 16949 standard and applies to *inspection*. When you inspect a part, you are required to use a gauge that has a resolution of 1/10th of the total tolerance of the dimension being measured. If you are measuring a part which has a tolerance of ±0.04 mm with a total tolerance of 0.08 mm, you will need a gage that can discriminate to 0.008 mm.

The *4:1 rule* is an ANSI/NCSL-Z-540 *calibration* requirement. If you are calibrating a gage, you need to use a check standard that has an uncertainty of measurement that is less than or equal to 25% of the total gage tolerance. For example, if you are calibrating a caliper with a tolerance of ±0.04 mm, with a total tolerance of 0.8mm. You will need a Check Standard (master) that will have an uncertainty of measurement no greater than 0.02 mm.

It is important to note that the standards referenced here are guidelines for choosing the appropriate instrument for inspection and calibration. Compliance

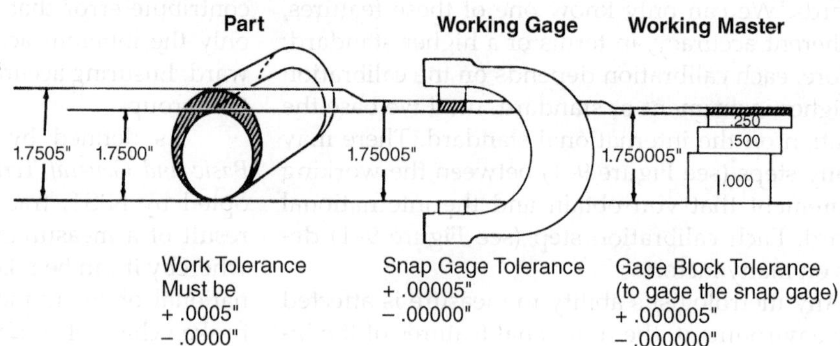

FIGURE 9–2 This is a pragmatic rule, not a law. It is based on practical results. Often, practical considerations force us to deviate from it.

to these "rules" is voluntary, unless a company or lab is seeking accreditation.

When we apply statistical process control, however, we must question even the 10-to-1 rule. We most often question the accuracy of the application of the rule and the costs it generates, but we rarely ever question whether we should use the 10-to-1 rule (see Figure 9–2). The standard may need to have 1,000 times the precision of the actual part, which can exceed the ability of present-day measurement (Figure 9–3).

The rule, as we stated, deals only with precision, not accuracy. Accuracy is derived from the standard; therefore, if the standard is not accurate, we have no hope of measuring accurately even if we follow the 10-to-1 rule (see Figure 9–4). When you understand the 10-to-1 rule, you can also understand why an instrument with more than one scale is more practical for measurement

9–1 THE ROLE OF ERROR

Because the act of measurement alters both the item being measured and the measurement system, we never know an absolutely true value. This fact is known as *Heisenberg's Uncertainty Principle,* named for Werner Karl Heisenberg (1901–1976). The error may be small in practical situations, but it is always present.

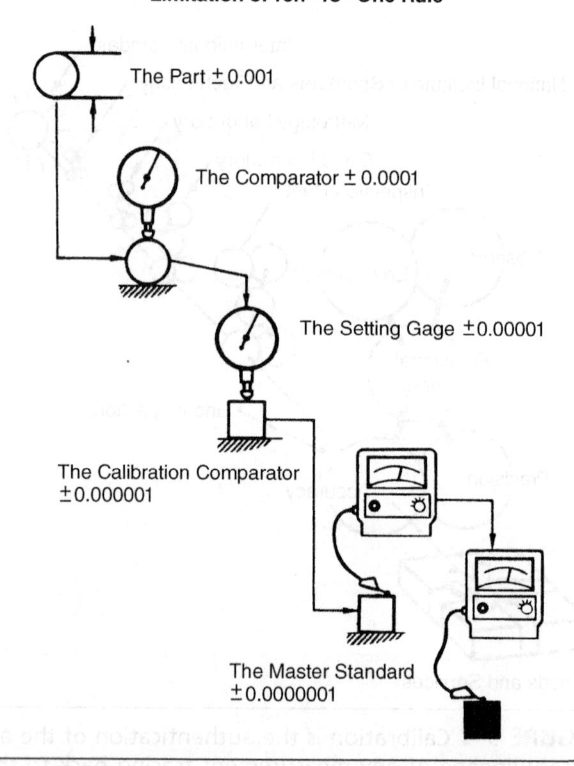

FIGURE 9–3 The demands of the 10-to-1 rule quickly exceed present measuring instrumentation. The National Institute of Standards and Technology is currently investigating 1/10 millionth measurements and finer.

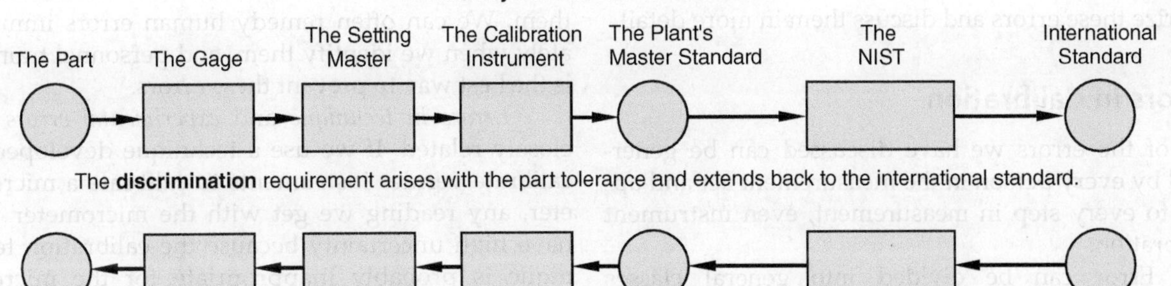

The **accuracy** arises at the international standard and progresses through each calibration step to the part.

Traceability is the procedure for determining that none of the connecting links between the part and NIST has been severed. It is required by law on government contracts and by sound economics for all other measurements.

MIL-1-45208a Inspection System Requirements, MIL-Q-9858A Quality Control Requirements, and MIL-C-45662A Calibration System Requirements are the applicable standards.

FIGURE 9–4 The 10-to-1 rule deals with the precision of the measuring instruments, but carried to its limit it reaches the international standard. By definition, this is the ultimate existing accuracy. If any link along the way is broken, the accuracy is lost; hence, the need for traceability.

In addition, we must always assume that we do not know everything we need to know about the standard—that there is some error in our knowledge. When we repeat a measurement, the amount by which the results disagree is called the *discrepancy.* The causes of discrepancy, or *random errors,* must also be the causes of differences between the readings and the true value. The smaller the random errors, the greater the precision. This agrees with a statement in Chapter 2 that precision is a measure of the dispersion of readings.

Even if all of the individual readings are the same, you may still have errors, because all the readings could vary from the uniform error or a *systematic error.* One example is a kink in a steel tape: All measurements that include the kink will be too short or will contain the same amount of minus error. The smaller the systematic errors, the higher the accuracy; but most measurements contain both random and systematic errors.

If you can evaluate an error by any logical process, it is a *determinate error;* if not, the error is *indeterminate.* You can evaluate random errors with a study of repeated readings, making the errors determinate. Some systematic errors cannot be determined by any experimentation, so we only know that they exist by inference. Fortunately, we can evaluate these errors by comparing them with a higher authority, which, in measurement, is calibration. Whether a systematic error in linear measurement is determinate or indeterminate depends on whether there is a standard available for comparison.

We correct errors in different ways for random and systematic errors. In our tape example, we could determine the systematic error caused by the kink during calibration to a standard. When we know the amount of error, we simply add the amount of the error from the readings. More typically, random error would change the length of the tape with temperature change, but these errors can be corrected with a calculation. In these cases, however, the correction would not be constant. The correction would have to be calculated for each reading at the time of that reading, resulting in a different value for each temperature.

Besides systematic and random errors, we can also encounter and prevent *illegitimate errors.* The first two types of errors exist in all measurement, but illegitimate errors should not exist at all. We now categorize these errors and discuss them in more detail.

Errors in Calibration

All of the errors we have discussed can be generated by every person in the measurement act and apply to every step in measurement, even instrument calibration.

Error can be divided into general classes (see Figure 9–5) and then subdivided into types of errors: *frequency* relates the error to some observable event, such as a change in measurement or change of observer; *location,* which, in this chart, is arbitrarily limited to instrument (including setup and reference surfaces, if separate), observer, and technique; *cause* is a general category for malfunctions; *detection* assumes that quality assurance has shown that a problem exists and directs you to general steps that may isolate the error; *remedy,* and, most importantly, *prevention.*

You can detect systematic errors, because they will recur regularly and periodically; but you might also have a tendency to overlook these errors, just because they show up so regularly. Things that are systematic also might seem to have a legitimate place in the operation, even when we really do not want them there. *Calibration errors* do not mean that the instrument is incorrectly set: They mean that there is a disparity between the input signal and the reading. These errors are caused by false elements, such as a scale spacing that does not match the amplification, or a probe that is too long. You can detect calibration errors by correctly recalibrating the instrument, especially to gage blocks; sometimes calibration errors can be corrected by adjusting the instrument. If you cannot correct these errors, you must replace the instrument. Thorough, regular evaluation of an instrument will prevent calibration errors.

Human errors can be particularly difficult to detect, especially when you are making the errors yourself. Human errors can include a tendency to read high or low or using the wrong instrument for your "handedness," as in a left-handed person being forced to use a right-handed instrument. Human errors usually change with each new observer, which can give us clues that the errors exist and how we can detect them. We can often remedy human errors immediately when we identify them, and personnel training is the best way to prevent these errors.

Errors in technique and *experimental errors* are closely related. If we use a technique developed to calibrate vernier instruments to calibrate a micrometer, any reading we get with the micrometer will have high uncertainty because the calibration technique is probably inappropriate for the micrometer. On the other hand, if you do not know what calibration technique to use, you might experiment and try one. If the unknown technique works, you then know that the technique is appropriate—at least in that one situation; however, if it fails, you have generated an experimental error. Technique and experimental errors can usually be detected if you use another method, and you can remedy these errors by changing your method. Education helps prevent these errors.

You can recognize random errors because they lack consistency. These errors of judgment are often caused by a lack of self-discipline or a lack of adequate instructions. To test for this kind of error, change observers; to prevent these errors, obtain more training.

Because measuring conditions often change, we can also often generate random errors, which can occur in the technique, observer, or instrument. When these errors occur in the instrument, they usually are due to a disturbed element—anything from improperly soldered connections to backlash, play, or temperature changes in mechanical systems. Lack of constraint often causes random error in these situations. It may be difficult to detect the cause of these errors, but, if you systematically substitute elements, you should be able to identify the culprit, which should be corrected or replaced. Environmental controls can prevent some random errors, as can adequate instrument specifications or standards.

Definition is an evaluation of consistency of the measured quantity. We can also call it an attempt to get more information out of the measurement system than we put in. If an object has sides that are not smooth and parallel, we cannot tell much about the separation of the sides no matter how perfect our measurement

Classification of Measurement Errors

Types of Errors	Frequency	Location	Cause	Detection	Remedy	Prevention
1. Systematic or Fixed	**Periodic**	**Instrument Observer or Technique**	**Recurring malfunction of one or more elements**	**Comparison or Substitution**		
a. Calibration	Changes with measurement	Instrument	False elements, design and construction errors	Comparison to superior standard	Replace instrument	Instrument evaluation
b. Human	Changes with observers	Observer	Bias, physical peculiarities, behavioral traits	Substitute observer	Train observer	Training
c. Technique	Changes with measurements	Instrument or Observer	Use of the *known* method but in a situation for which it is not satisfactory	Substitute method	Change method	Education
d. Experimental	Changes with measurement	Instrument or Observer	Use of an *unknown* method in a situation for which it is not satisfactory	Substitute method	Change method	Education
2. Random or Accidental	**Random**	**Instrument Observer Technique or Part**	**Erratic malfunction of one or more elements or part**	**Comparison or Substitution**		
a. Judgment	Changes with observers	Observer	Lack of discipline or precise instructions	Substitute observer	Tighter controls and procedures	Training
b. Conditions	Changes without regard to system	Instrument or Part controlled	Disturbed element caused by external influences such as vibration or temperature change	Systematic substitution of elements	Replace troublesome element	Environmental controls, instrument standards
c. Definition	Changes with position of measurement	Part	Attempt to get more information out of the measurement system than was put in	Analysis of part	Change requirements of part or measurement	Education
3. Illegitimate	**Random Periodic and Continuous**	**Instrument or Observer**	**Outside interference or other completely avoidable disturbances**	**Reappraisal of procedure**		
a. Mistakes	Changes with observers	Instrument or Observer	Wrong decision in choice and/or use of measurement instruments	Analysis of measurement system and technique	Replace or retain faulty element	Education
b. Computational	Random or Continuous	Observer	Environmental factors, fatigue and poor instrumentation	Substitute computation	Change environment technique or instrument	Personnel and environment evaluation
c. Chaotic	Random	Observer or Instrument	Extreme external disturbance	Self-detecting	Stop measurement until ended	Environmental analysis

FIGURE 9–5 Any table such as this one is necessarily general in nature. It is a guide rather than a rule.

system is. This uncertainty is in the measured quantity itself. You detect definition error by fully inspecting and analyzing the part, and you remedy definition error by changing procedures. Either the requirements of the part or the instructions for measurement must be changed. Prevention comes from a better understanding of measurement capabilities.

There should be no illegitimate errors, which are the result of mistakes and carelessness. Because some illegitimate errors creep into measurement, repetition is a legitimate way to detect errors. But illegitimate errors can be difficult to detect because they can be random, periodic, or continuous. Continuous errors, by their very nature, resist detection by repetition.

We can make *mistakes* at any level—from the observer to the person calibrating the standards. Many times, we choose the wrong instrument or measurement system for the job; and we may only notice those mistakes when the parts will not assemble or function properly. To remedy the faulty element, you may need to do a complete analysis of the measurement system, because better understanding may be the only practical prevention for this error.

Computational errors may be random or continuous, but once an error has started, it usually establishes itself in the computation. These errors are affected by environment, fatigue, and instrumentation, including instruments that do not zero set and those that read in the reverse direction from the desired measurement. Computational errors can often be detected simply by having another person do the computational steps, but the remedy and prevention of computational errors can be difficult because they require a deep understanding of the capabilities of observers in relation to changes of instruments, techniques, and environment.

Chaotic errors are extreme disturbances that ruin or hide the measurement results. These errors include vibration, shock, sudden glare, extreme noise, and uncommon events such as flooding. Unlike all the other errors, chaotic errors actually call attention to themselves, and the only way to remedy them is to wait until the disturbance ends and redo the measurement. A thorough analysis of environmental factors can help prevent chaotic error.

The Interaction of Errors

We cannot just detect error; we also need to decide what effect errors have on each other, if any. We can classify errors as independent, dependent, and correlated (see Figure 9–6).

For example, if we use a micrometer that has an error in it to measure the length and width of a rectangle, any calculation of area would combine the errors from both the length and width. These errors cannot cancel out or compensate for each other, so they are called *dependent errors.* Dependent errors are total errors that cannot be separated from the separate errors that make them up.

If, in another example, the micrometer is in calibration, then its total error is so small that it will not materially affect the reliability of the measurement you make with it. The total instrument error cannot exceed more than 1% of the total measurement error. If, however, the person calculating the area of the same rectangle measures the length and width carelessly, then the measurement errors could compensate or cancel completely. We might still compute a correct total area from inaccurate measurements, but the reliability of our measurements and calculations decreases. When the person making the computation has no way of knowing when the results are correct and when they are not, we call the errors *independent errors.*

Correlated errors are caused by a systematic relationship between independent and dependent errors. If we need to know about the surface finish of the rectangle we have been discussing, every value for the surface finish will be affected by the error in the area calculation. The final calculation may compensate for other measurement errors or increase them, but for any one surface finish, the error will be constant.

In the real world, your error problems will not all be so simple. You can expect that errors in real calibration situations will be far better concealed and much more mysterious about their relationships. In addition, only a small number of all the errors in measurement are metrological errors in the instrument, which again emphasizes the importance of cleanliness, temperature, condition of the reference surfaces, and geometrical alignment for reliable measurement.

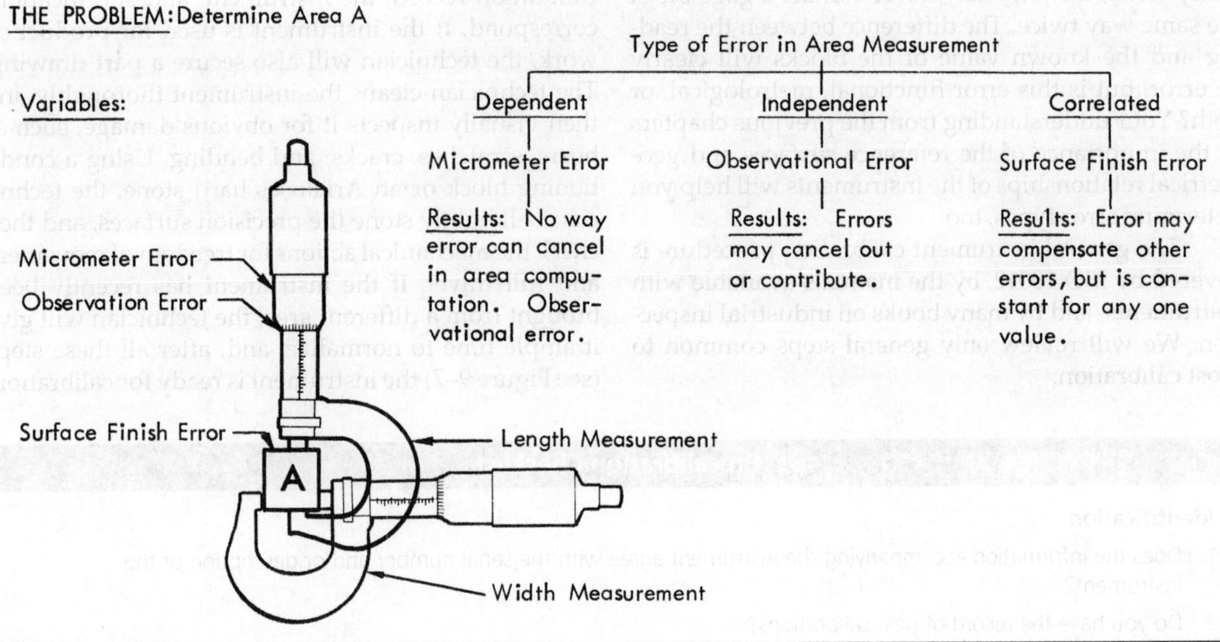

FIGURE 9–6 Even though errors can interact to produce accurate results, those results are not reliable unless the errors and their interaction are known.

We cannot use gage blocks effectively to pinpoint calibration errors until we can isolate those errors in specific variables.

Even after eliminating these potential errors, we cannot calibrate any instrument just by checking a couple of readings. We must analyze each instrument and its applications separately. Much of this information is available in manufacturers' manuals, national and international printed standards (specifications), and books on inspection.

9–2 BASIC CALIBRATION PROCEDURE

It is easier to discuss calibration if we start at the opposite end with the *uncertainty* of the instrument. An instrument's uncertainty is made up of all the composite errors from each of the instrument's separate characteristics, and each possible source of error is a *variable*. For more complicated instruments, many characteristics can generate errors.

Calibration consists of three steps on the metrological side:

1. Determine the variables that might contribute to instrument uncertainty.
2. Measure the error contributed by each variable.
3. Determine the net effect of the interaction of the variables on the instrument's measurement capability.

Gage blocks are the most precise and most accurate means of measuring the separate variables, but these variables are usually tightly intertwined with the functional characteristics of the instrument we are calibrating. We cannot measure most of the errors from functional characteristics even with gage blocks, and in many cases, these functional errors prevent gage blocks from measuring the metrological features with any degree of reliability.

For example, if the jaws of a vernier caliper are badly worn, the jaws may never contact a gage block the same way twice. The difference between the reading and the known value of the blocks will clearly be error, but is this error functional, metrological, or both? Your understanding from the previous chapters of the importance of the reference surfaces and geometrical relationships of the instruments will help you determine error types, too.

The general instrument calibration procedure is covered by ISO 10012, by the manuals available with instruments, and by many books on industrial inspection. We will review only general steps common to most calibration.

The calibration technician makes sure that the calibration record, the instrument, and specifications correspond. If the instrument is used for production work, the technician will also secure a part drawing. The technician cleans the instrument thoroughly and then visually inspects it for obvious damage, such as burrs, scratches, cracks, and bending. Using a conditioning block or an Arkansas hard stone, the technician will lightly stone the precision surfaces, and then check the mechanical actions for freedom of movement and full travel. If the instrument has recently been brought from a different area, the technician will give it ample time to normalize, and, after all these steps (see Figure 9–7) the instrument is ready for calibration.

Precalibration Checklist

A. Identification

1. Does the information accompanying the instrument agree with the serial number and/or description of the instrument?
2. Do you have the record of past calibrations?
3. Do you have the manual for the instrument?
4. Do you have all applicable federal specifications?
5. Do you have a report from the user concerning real or imaginary troubles with the instrument?
6. Do you have instructions for disposing of the instrument after calibration?

B. Requirements

1. Is the environment for the calibration suitable for the precision expected? Have the drafts, temperature change, vibration, and interference been checked and minimized?
2. Are suitable standards on hand and are they in calibration?
3. Are the necessary calibration instruments and accessories on hand and are they in calibration?
4. Is a heat sink of adequate capacity on hand?
5. Are the necessary supplies on hand?
6. Are paper and pencil on hand?
7. Are the necessary packaging materials on hand?

C. Preparation

1. Have the instrument, the standards, and the calibration instruments been normalized?
2. Has the instrument been visually inspected?
3. Have the references and contact surfaces been inspected for damage, wear, and alignment?

FIGURE 9–7 These practical considerations will speed calibration without impairing reliability.

A measurement is useless if the instrument you use checks perfectly under the calibration conditions but produces large errors under actual measurement conditions. We can minimize this problem by duplicating the actual measurement conditions as closely as possible when the instrument is calibrated. Ideally, we calibrate the instrument to a standard that is the same size, shape, and material as the part the instrument is measuring. For a production gage that only checks one dimension, we can approximate the conditions for calibration, but for a measurement instrument that is expected to perform reliably throughout its range, calibration is extremely complicated.

Because total error is the sum of many separate errors, errors can add up to one total for one reading and a completely different total for another reading. You must determine the individual variables that can produce errors before you can check whether they are responsible for errors.

Calibration of a Vernier Caliper

In order to completely calibrate an instrument, a technician must compare the instrument to the standard at every point to which the instrument can discriminate. For a 300 mm vernier caliper with a discrimination of 0.01 mm and a range of 300 mm, the technician must make 30,000 separate calibration measurements.

As a calibration technician, you would not have to go far into this process before discovering a pattern to the errors. You might find that the observed values get successively larger than the true values, or that there is a uniform fluctuation, or even a random change in magnitude of error. In all these cases except the random change, once you observe the pattern, you can skip many readings and still have a reasonably accurate calibration. This process is often used in calibration to make the number of measurements more practical.

In the case of the vernier caliper, if you decide to save time by only checking multiples of 10 mm (see Figure 9–8A), you could check the instrument and have no error, yet this selective checking could still hide serious errors. The most obvious error is that, although the inch increments are uniform, they could also be uniformly too high or too low in comparison with the true values (see Figure 9–8B). In another common error (see Figure 9–8C), the readings depart from their true values at some points and then return to the true values. This same variation (see Figure 9–8D) can happen even in smaller ranges, like the 0.1 mm shown.

Visualization by Graphing

The easiest way to visualize errors is to graph them, as shown in Figure 9–9 which we have also exaggerated for demonstration purposes. You will create a similar graph of your own in the exercises in the accompanying lab manual. In these graphs, the true values are vertical divisions; the observed readings that correspond to the true values are horizontal divisions. A perfect instrument generates a straight line, which we call a *curve* (see Figure 9–9A), so that reading down or to the left results in the same number. If we encounter a plus error, as at 40 mm, the plus reading is a point above the line, and a minus error, as at 30 mm, is a point below the line.

In Figure 9–9B, we have graphed the conditions shown in Figure 9–8B. The uniform +0.02 mm error has created a curve that is exactly parallel to the perfect instrument curve but too high at every calibration point. If we extend the curve to the left, we can see that a +0.02 mm error in zero setting the instrument is responsible for this consistent error. We can correct the error by readjusting the instrument.

When we plot the readings from Figure 9–8C, their curve (see Figure 9–9C) clearly shows a *topical error*—an error of one portion of a scale or surface, most commonly caused by wear. This error is systematic, not random, because even though its position on the scale of the instrument is not systematic, its effect on measurement is. Whenever a measurement falls into the worn area of a scale, you will be able to predict an error in the measurement.

In Figure 9–8D, the readings depart from their true values but return to them periodically as in our previous example. In contrast, however, the full cycle

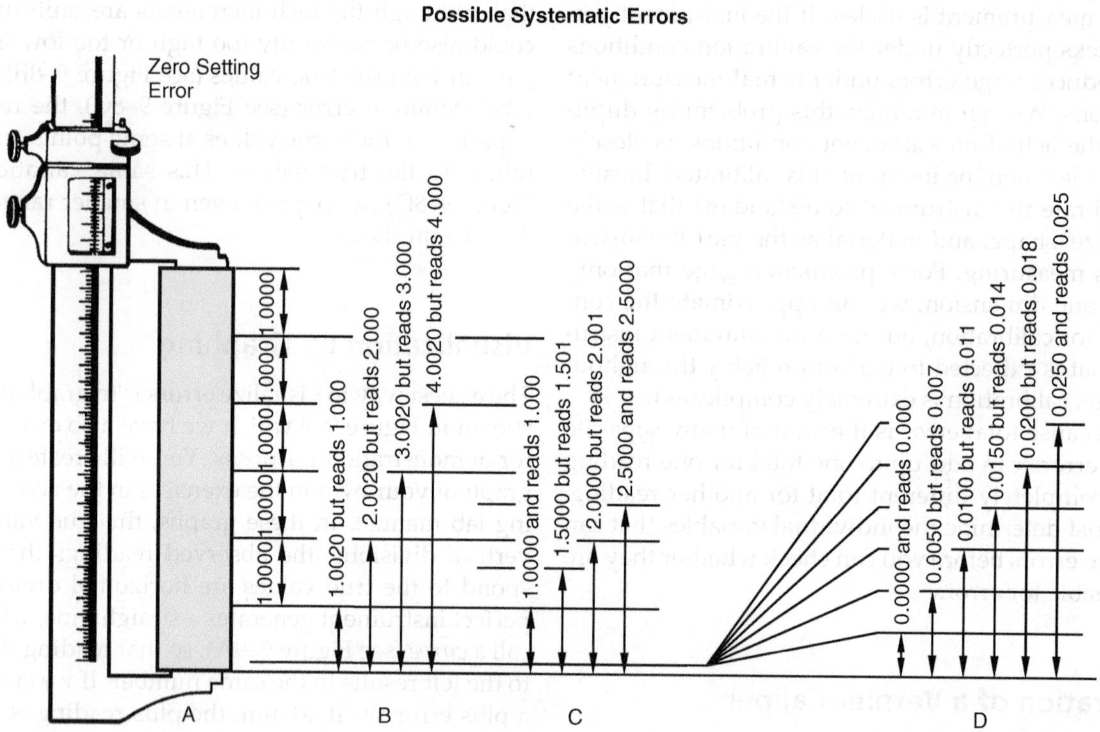

FIGURE 9–8 Inch-to-inch calibration as at A could fail to disclose the systematic errors shown at B, C, and D. More than one could exist at once.

of departure from the true values takes place between zero and 0.5 mm (0.025 in.). This error could be caused by wear of the beam of the instrument or an error in the vernier scale, which would certainly be the case if the pattern repeats each 0.5 mm, as shown in Figure 9–9D.

Variables to Be Calibrated

How many readings do we need to check all the variables that may contain errors? First, you must determine the variables involved, and then you can determine the readings needed. For vernier calipers, there are always at least four variables: two are metrological, the main scale and the vernier scale; and the other two are functional, the mechanical action and the condition of the contact faces, or jaws (see Figure 9–10). Once you recognize these errors, you should check them and then continue on to correct any metrological errors—the variables that might be caused by use of the instrument.

Continuing this example, you need to investigate three variables of the main scale, all of which are potential sources of systematic calibration errors, although one might be illegitimate. The first, manufacturing error, is unlikely except when you are using poor-quality instruments and will assume any of the characteristic curves we have graphed. The second, wear, generally creates a curve, as in Figure 9–9C. If it occurs at one or more distinct places, it is a topical error, and if this occurs uniformly throughout the range, it reduces the precision of the instrument. The third, linearity error, is a uniform increase or decrease

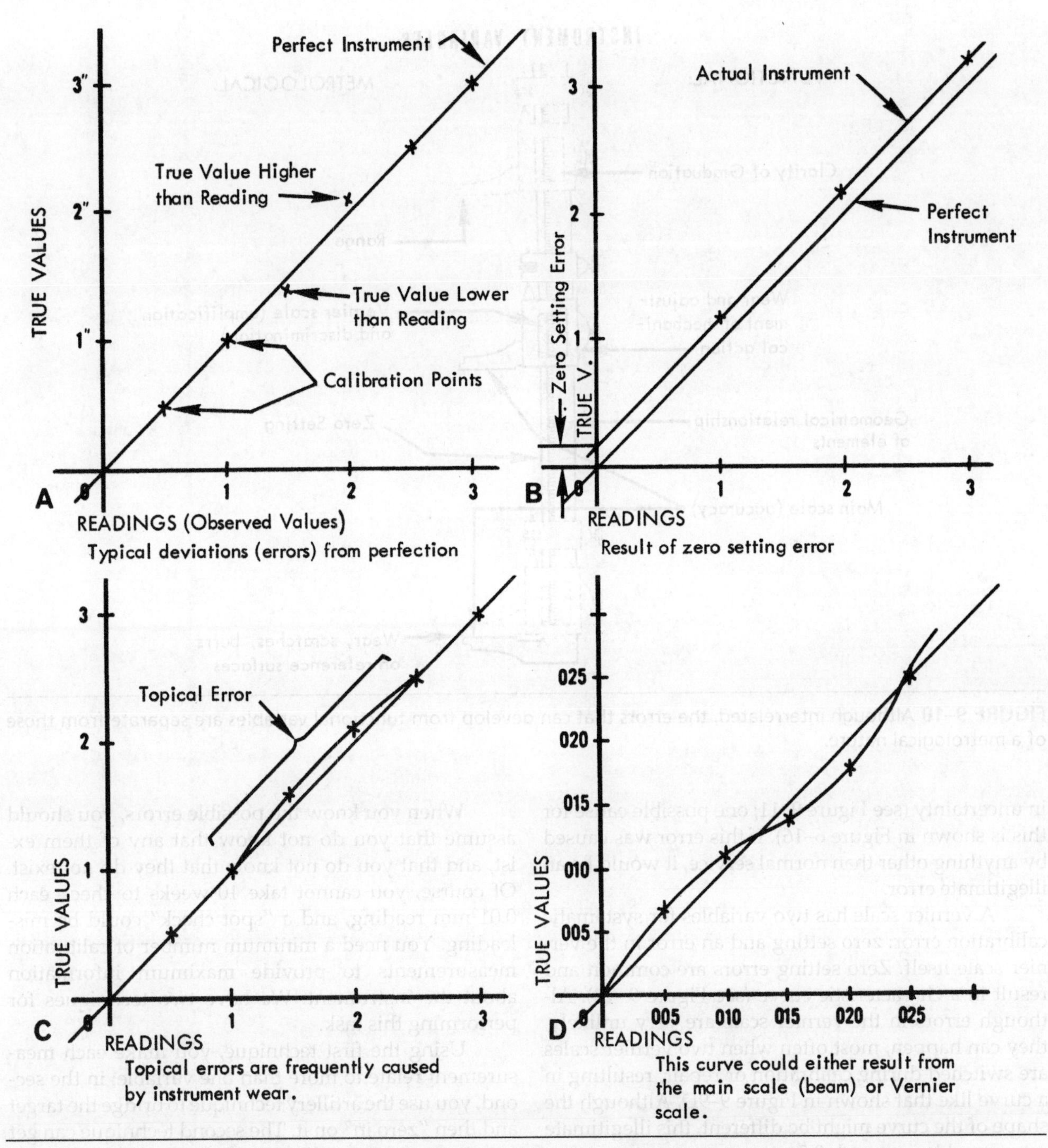

FIGURE 9–9 Graphs of errors permit visualization of the types of uncertainty that the instrument may have.

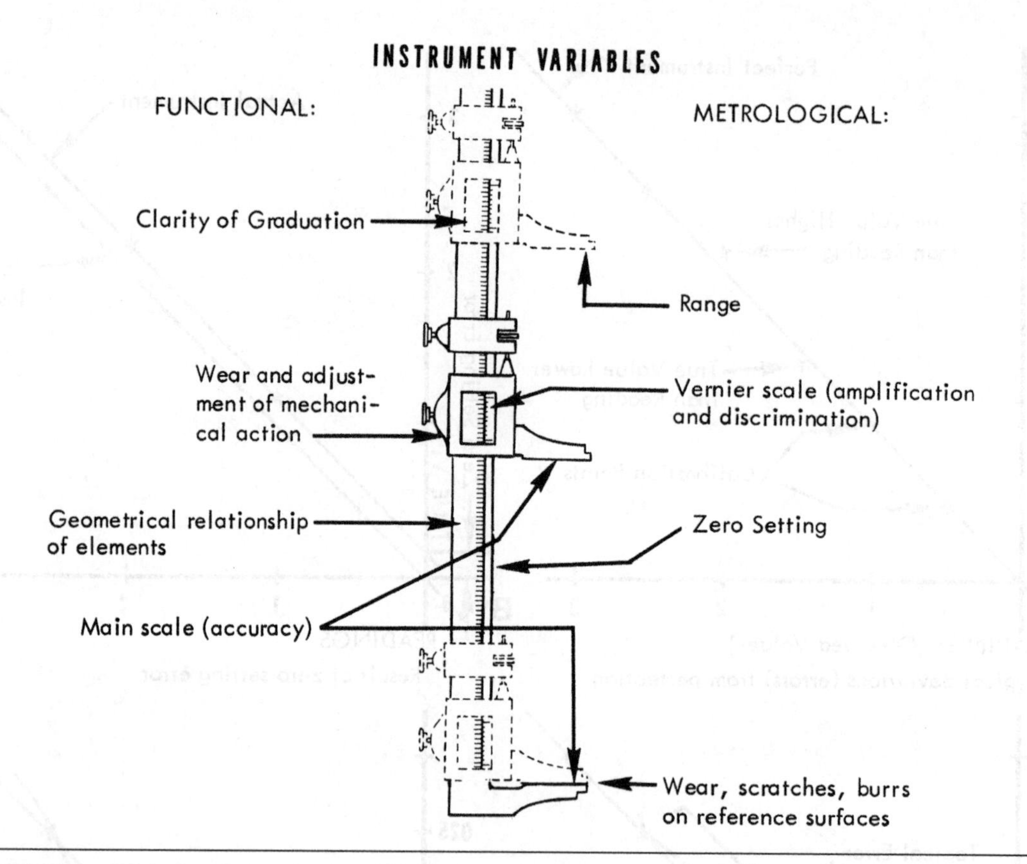

FIGURE 9–10 Although interrelated, the errors that can develop from functional variables are separate from those of a metrological nature.

in uncertainty (see Figure 9–11; one possible cause for this is shown in Figure 6–16). If this error was caused by anything other than normal service, it would be an illegitimate error.

A vernier scale has two variables for systematic calibration error: zero setting and an error in the vernier scale itself. Zero setting errors are common and result in a characteristic curve (see Figure 9–9B). Although errors in the vernier scale are very unlikely, they can happen, most often when two vernier scales are switched during calibration or repair, resulting in a curve like that shown in Figure 9–9D. Although the shape of the curve might be different, this illegitimate error would occur each 0.5 mm.

When you know the possible errors, you should assume that you do not know that any of them exist, and that you do not know that they do not exist. Of course, you cannot take 10 weeks to check each 0.01 mm reading, and a "spot check" could be misleading. You need a minimum number of calibration measurements to provide maximum information about the instrument. We have two techniques for performing this task.

Using the first technique, you make each measurement relate to more than one variable; in the second, you use the artillery technique to bridge the target and then "zero in" on it. The second technique can get you "on target" rapidly if you know something about

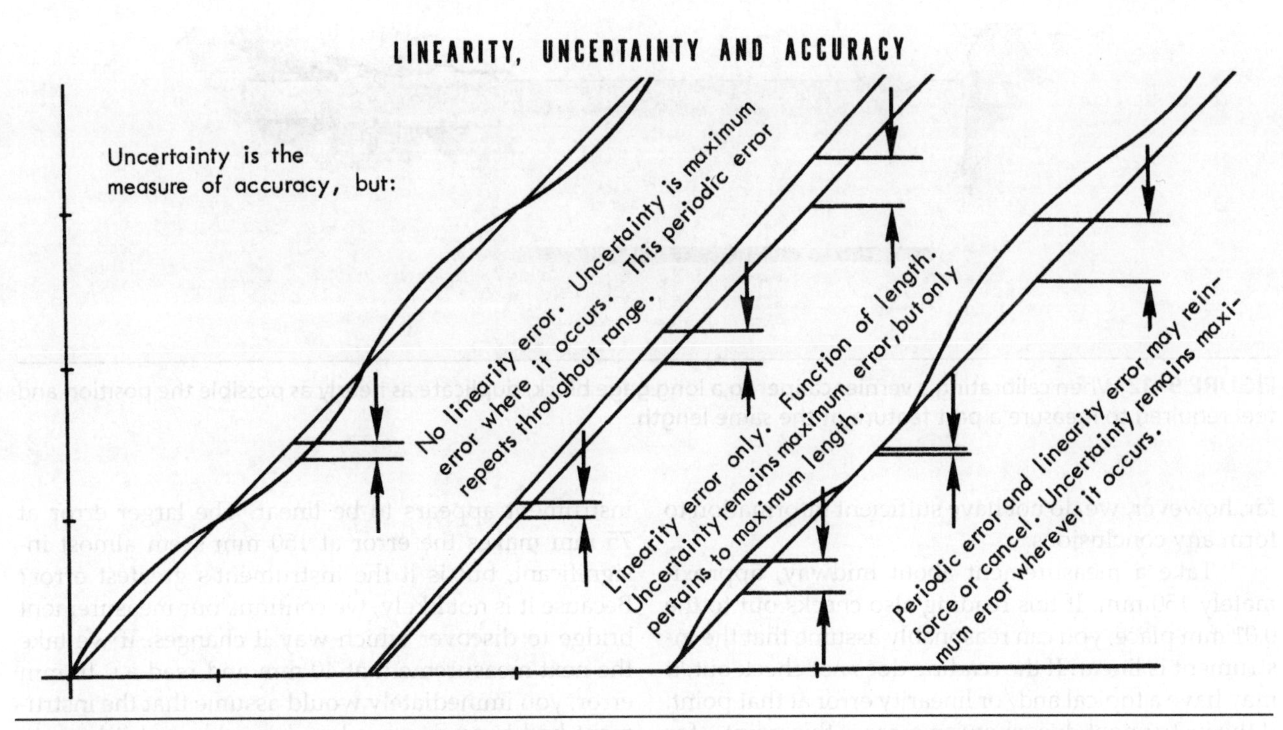

FIGURE 9–11 The uncertainty of an instrument is the maximum error that it may add (algebraically) to a reading. It is the combination of all errors, only two of which are shown here.

the previous use of an instrument. For our example, however, we will assume you do not know what errors you may find.

Because zero setting error is common to both the linearity and topical checks, you should use this check first, and, second, make a calibration measurement at nearly the extreme range of the instrument—for example, 250 mm. You make this simple second check by carefully closing the caliper on a 250 mm gage block. You should have the gage block lying flat on a supporting surface, and you should hold the caliper the same way you would if you were measuring an unknown part feature. Remember to centralize the caliper carefully (see Chapter 6) to duplicate the normal feel (see Figure 9–12). Carefully lock the movable jaw before you read the vernier and repeat this operation until the repeated measurements corroborate each other.

By repeating this process, you could also give bias a chance to ruin all your careful work, unless you guard against it. You do not perform the corroboration steps to make all of the measurements read the same; you use corroboration to warn you of any illegitimate errors that may have occurred.

For example, if you read the 250.00 mm measurement—which is as closely as you can read the vernier using a magnifying glass—the measurement does not prove that there is no linearity error in the instrument and no topical error at this point on the scale. These factors are independent errors that could cancel out, and if this reading had been in the range of the majority of use, the errors very likely could have canceled. We chose the long measurement because, in most cases, it would be beyond the range where topical error might compensate for a linearity error. So

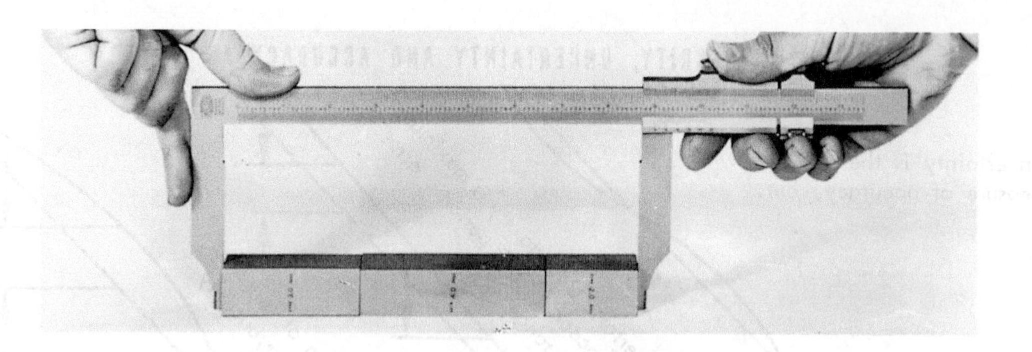

FIGURE 9–12 When calibrating a vernier caliper to a long gage block, duplicate as nearly as possible the position and feel required to measure a part feature of the same length.

far, however, we do not have sufficient information to form any conclusions.

Take a measurement about midway, approximately 150 mm. If this reading also checks out to the 0.01 mm place, you can reasonably assume that the instrument is linear. If the reading does not check out, it may have a topical and/or linearity error at that point. If the instrument does show an error at this point—for example, a 0.01 mm error—you know that you need to investigate the instrument's performance further. Rather than try to pinpoint small places of error, you should continue to get a "big picture." Measurements on either side of the last one, such as 75 mm and 200 mm, continue to give you this big picture.

As you continue checking the instrument, you should record the observed values, also called "calibration measurements" or "readings," which, in our example, are 75.03 and 200.00 mm. You can see that a pattern is emerging:

True Values	Observed Values
0.000	0.000
75.000	75.003
150.000	150.001
200.000	200.000
250.000	250.000

You do not have very much information in this table, but you can detect trends—for example, the instrument appears to be linear. The larger error at 75 mm makes the error at 150 mm seem almost insignificant, but is it the instrument's greatest error? Because it is not likely, we continue our measurement bridge to discover which way it changes. If we take the next measurement at 40 mm and read a 0.10 mm error, you immediately would assume that the instrument had been damaged and would send it back to the manufacturer for repair. If the reading came out 40.03 mm, however, you could not make a conclusion and would continue your measurement bridge at approximately 100 mm. If the 100.00 read 100.03 mm, you would conclude that a topical error with a maximum value existed some place on one side or the other of 75 mm, probably on the high side. You would reach this conclusion because there was evidence of error at 150 mm. If you check at 125 mm and find a 0.02 mm error, you would check on the other side of the scale at 4 in. If that point reveals a 0.002 in. error, you would continue checking between the values you already used until you had enough information to make a conclusion about the instrument.

How much is enough information? It depends on why you are calibrating the instrument. If you are calibrating it to determine whether or not you can continue to use it, when you get one verifiable out-of-tolerance reading, you have enough information to withdraw the instrument from use until it is repaired. If you are simply adjusting the instrument in the field, you know when you have enough information to pinpoint the variable that needs adjustment. In our caliper example, you would probably check the play of the movable jaw on the beam after the 5-inch reading, and, if you found that the jaw was loose, you would tighten it and try the same calibration measurements again. If the jaw adjustment eliminated the error, you would only continue to calibrate the instrument to ensure that it was now in calibration.

So far, our example calibrations have only determined the condition of the main scale and whether the vernier scale was zero set. Although unlikely—and no measurement we have taken so far has shown that it is true—the vernier scale could be out of calibration; therefore, you should make a measurement within the 0.5 mm range of the vernier scale. If this disclosed an error, you would use the bridging and zeroing-in technique in that 0.5 mm range, just as in the 250 mm example.

You will see from our examples that arbitrarily choosing even intervals is *not* the way to calibrate an instrument. You should make measurements at intervals that disclose the change in error that can be expected from the separate variables. Often, a small number of measurements will provide more information than a large number. Of course, you must know what the variable you are checking is in each particular case, as our next example with a micrometer shows.

Calibration of Micrometers

We discussed calibrating micrometers in Chapter 7, but we omitted using gage blocks to pinpoint the variables peculiar to screw thread standards.

The *pitch* of a thread is the distance between one point on one thread and the same point on the next thread, which is expressed in terms of a given length—for example, threads per inch. The *lead* is the amount of movement between the male and female threads for one revolution. Therefore, in a single-thread screw, the lead and the pitch are numerically the same. When we discuss screws as length standards, pitch conveniently refers to advancement by full turns, whereas lead refers to advancement up to a full turn. Pitch and lead, then, are two variables in screw thread standards: They are independent and may be either plus or minus in each case. There are other variables, but because micrometer screws are short, we will discuss them later.

Pitch and lead are roughly equivalent to the main scale and vernier scale variables of the vernier caliper. However, the same vernier scale variables repeat, whereas no two points of the screw thread are exactly the same. Any lead defect, usually created in the manufacturing process or by use, has a tendency to repeat itself. For example, if a slight eccentricity causes a thread to rub the mating thread unusually hard during a third of its revolution, this rubbing would also happen to the next thread, and so on. All the threads tend to have a wear pattern in the same general area (see Figure 9–13), which simplifies micrometer calibration. The error in the lead variable is usually periodic, but in order to calibrate a micrometer screw, you must check both pitch and lead even if other "calibrations" are traditionally used in the workplace.

Zero setting check, a lead variable, is the first step for calibration, but it provides very little additional information. You do not need to use "feel" when you

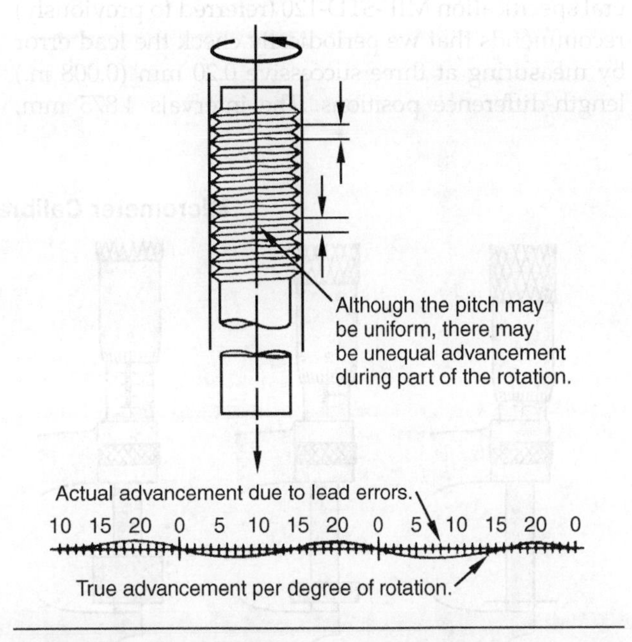

FIGURE 9–13 The axial and rotational thread errors are independent and affect micrometer accuracy. This is called a "drunken" thread.

zero set, so this process is not like any other process you encounter in actual micrometer measurement. If you add a calibration check at 25 mm, you get more information. However, this measurement alone does not really provide enough information about the condition of the micrometer. Because the lead and pitch variables are independent, they could compensate at the 25 mm position, and just one more calibration measurement will help you determine whether they are compensating. For example, if the micrometer had a +0.002 mm error at 25 mm and you measured the same error at 15 mm, you should check the zero setting first. If you measured an error of more than 0.02 mm or any minus value, it would be enough information to remove the instrument from service until it could be thoroughly calibrated.

In an attempt to be thorough, we often choose the 5 mm, 10 mm, 15 mm, and 20 mm positions for our checking, but with these positions, the thimble rotation stops in exactly the same position. When you use these four positions, you are actually doing a good check of the pitch value—you are not checking the lead at all. Federal specification MIL-STD-120 (referred to previously) recommends that we periodically check the lead error by measuring at three successive 0.20 mm (0.008 in.) length-difference positions. The intervals 4.875 mm, 9.750 mm, 14.625 mm, 19.500 mm, and 25.000 mm or a similar series are often suggested to check both variables at the same time (see Figure 9–14).

The Observational Variable in Calibration

"Feel" has such a significant effect on calibration of micrometers that we need to discuss it separately. Feel affects all measurement that even the most precise instruments in which the gaging force control is built in can be affected by the handling they receive. For instruments like micrometers or gages like snap gages, we need to be keenly aware of feel during calibration.

For example, the user of a micrometer always applies 3 pounds of gaging force, but the instrument is mistakenly calibrated for 3 ounces. The user would read a minus error in every measurement—an error as substantial as 0.002 mm or more. Logically, the instrument should be calibrated with the same feel as the user applies during measurement with the instrument; however, a calibration expert must use good judgment when calibrating for feel.

The "feel" variable, an independent error, can create errors of considerable magnitude that may compensate for the instrument errors or be so significant that none of the other errors can even compare

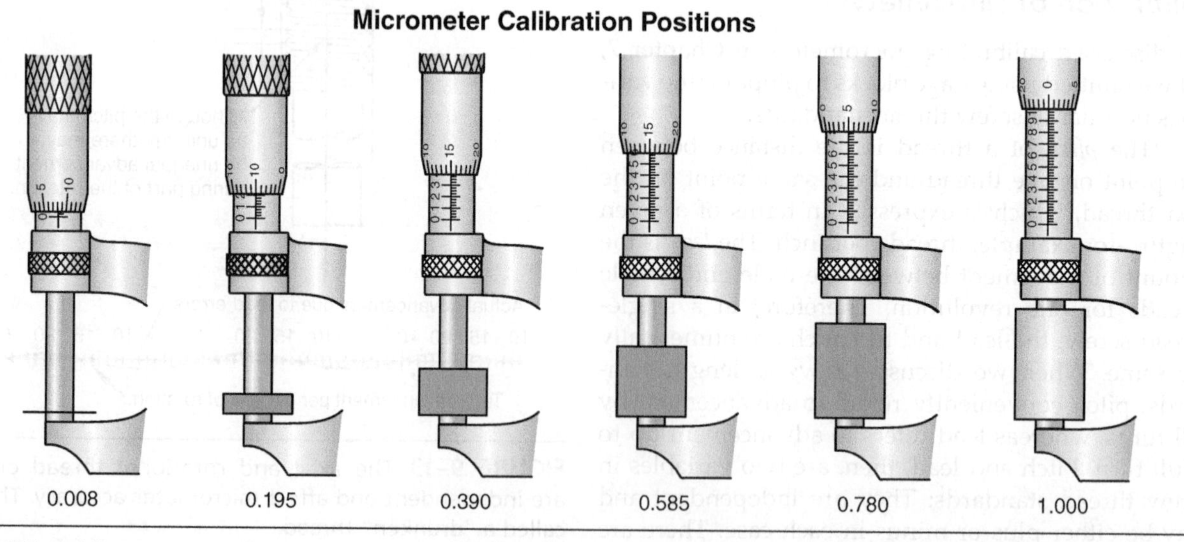

FIGURE 9–14 This is one of many sequences of calibration positions that check both the pitch and lead variables.

with them. Furthermore, who determines the feel that should be used? If we could quantify the ideal feel, how could you guarantee the instrument is set to this feel every time?

The precalibration checklist (see Figure 9–7) is supposed to eliminate errors so that we can apply the full value of the gage blocks to specific calibrations. Introducing unnecessary variables is not a sound metrological practice, and it is vital that we make all measurements with as nearly the same observational uncertainty, including feel, as possible. Fortunately, changes in feel have nearly the same effect as zero setting: Once the instrument has been calibrated, we can compensate for the feel of the individual user by zero setting the instrument for that individual. We cannot achieve reliability unless the calibrated instrument is adjusted to the observer or the observer is trained specifically on the use of the calibrated instrument although such thorough training is rarely done.

What can we do to minimize feel (or torque) differences in the total observational variable? Using a ratchet stop helps, but a better method is to apply the wringing technique. As you close the micrometer, slide and turn the flat micrometer contact surfaces to achieve the effect of wringing. You must be careful to keep the sharp edges of the contact surface from digging into and marring the precision surfaces of the setting standards or gage blocks, and you must be especially careful if these surfaces are carbide tipped. If we have to be so careful, why are the edges of the contact surface not rounded? A rounded edge has a tendency to ride up and over dust particles, and even though it requires more careful handling, a sharp edge pushes dust particles ahead of it as it slides along.

There are other observational variables, including parallax error and bias. Parallax can have an effect even on micrometer measurements, even though modern micrometer design minimizes it (see Figure 7–9).

Interpolation, perhaps the most serious observational error, can be eliminated with vernier-equipped micrometers. We made a very strong case against interpolation, but we need to interpolate when we try to determine readings to 0.0002 mm (0.0001 in.) increments with a plain micrometer. We must be able to read to this precision in order to calibrate the instrument, so a magnifying glass and an awareness of your bias are necessary to calibrate this micrometer to this level.

Internal Instrument Calibration

The calibration principles for internal measuring instruments are exactly the same as for external instruments; however, there are several additional error variables we must understand and control. First, gage blocks are male standards, so we can use them alone only for female measurements. If we add accessories, we can use gage blocks for male measurements.

For most situations, you can use standard gage block holders with caliper bar end standards. This setup provides a controlled wringing interval that helps you calibrate reliably. Unfortunately, this process takes time. You can minimize the time by making up the outside caliper assembly for maximum size, and then wringing blocks to the inside of one of the extension caliper bars for the smaller increments you need.

Of course, you cannot justify the interior method's time-saving performance if the process introduces substantial errors. In the example, we are calibrating an instrument with a discrimination of 0.02 mm (0.001 in.), and even with experience and interpolation between graduations, you cannot consider the instrument reliable for measurements finer than 0.05 mm. If each wringing interval is 1 μm (2 μin.) smaller, you probably would not need more than five intervals, which would be 1/20 of the finest reading—4 μm (0.0002 in.). The amount of error we introduced would not seriously affect the instrument's reliability for most measurements.

When you do a significant amount of internal calibration, you need a specialized instrument—a rigid frame with two internal reference surfaces that are spaced apart by gage blocks. This frame also helps spread wear over relatively large reference surfaces.

Even for the most skilled metrologists, very accurate measurement of inside diameters can be difficult. We must also be careful when high precision is required for the calibration of inside diameter gages or ring gages. One method we use is to allow the ring gage to rest on a laterally floating table so the contacts for the reference and measured points protrude through the opening. The antifriction support of the

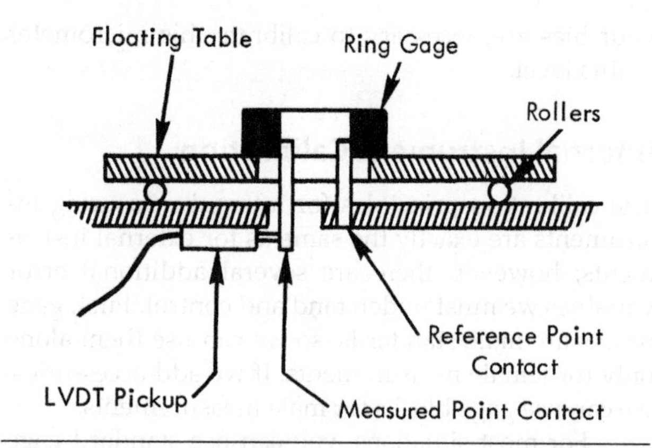

FIGURE 9–15 The floating table of the internal comparator minimizes friction that would impair aligning the gage with the axis of the instrument.

table lets you align the line of measurement of the ring gage precisely with the measurement axis of the instrument (see Figure 9–15).

Calibration of Working Standards

Working standards are gages used to set or calibrate other gages and, unlike fixed gages, working standards are used for extremely precise work. They are often used to set indicator gages with discrimination to 1 μm (0.0001 in.). Under perfect conditions, the rule of 10-to-1 would require that we hold the setting standards to within 0.1 μm (10 μin.). For practical purposes, working standards are calibrated with the master standards (gage blocks) as closely as the available instrumentation, personnel, and environment permit.

You must follow all of the precautions for the calibration of the high-amplification comparators not once, but twice. You should zero set the instrument to the gage blocks and measure the working standard to the same degree of precision.

Calibration of Master Standards and Traceability

You can achieve the degree of precision required of master standards only in the best-equipped metrology laboratories. Laboratory calibration technicians use interferometers to compare master gage blocks with grand masters, and, in some cases, they measure the blocks being calibrated directly with light waves.

Even though instructions on the calibration of master standards will not apply to most real-life situations you will encounter, we do want to emphasize how important calibrating master standards is. All accuracy derives from the International Standard (see Figure 9–1), and only a series of calibrations brings this "standard" into the measurement that you make. A manufacturer's master standards (usually a set of gage blocks) are the company's direct link to the international standard. Therefore, any time the master standards are not in current calibration, all measurement instruments are also out of calibration, which can lead to costly mistakes in production. Many production facilities, to be on the safe side, maintain two complete sets of master standards: one set that can be at the NIST or a metrology laboratory for calibration, and another set to be kept on-site for day-to-day calibrations.

Calibration is so important to the performance of manufacturing facilities that it has even become part of our legal system. Contractors who are working on government contracts must be able to prove traceability—that a series of calibrations associated with each instrument can be traced back to the standards at the NIST. Because so many subcontractors are involved with government projects, these traceability requirements affect much of the industry in the United States. In addition, small or close tolerances are called for on many commercial contracts, which also means that equipment must be kept in calibration at all times.

9–3 RECORD AND CORRECT CALIBRATION READINGS

Except for the lowest grades of gage blocks, you will have calibration data on each individual block, which states the amount that the block varied from its basic size at its most recent calibration. This amount, an independent systematic error, is stated as plus or minus.

If you find errors in only one or two blocks and are following the rule of 10, you can ignore these errors. If, as is often the case in calibration, you cannot follow the rule, the error may still accumulate. If a large number of blocks is involved, their combined errors may alter the direction of the rounding off that you will finally do, which can change a significant figure in the measurement.

We do not want to add rounding off errors or any other errors, so it is a good practice to systematically record and correct all readings. You should never correct the readings mentally and jot down answers, because you must always have a way to recheck your adjustment in case of a mistake.

Rounding off Calibration Results

In Chapter 2, we discussed a statistically sound method for rounding off unnecessary decimals. Two other methods are used in calibration that are not statistically sound at all and actually are exact opposites. These methods do not have generally accepted names, so we refer to them by their predominant characters: conservative and liberal.

If you use the rounding off of significant figures in calibration to affect the conclusion, the rounding off is liberal in nature. However, if you use the rounding off to ensure even greater separation from *observational error*, the rounding off is conservative. In both cases, you are manipulating the wear variable in the instrument.

Wear Considerations

Wear always has an effect on measurement and almost always causes trouble. Every time you use a gage or measurement instrument, the instrument wears; every time you measure a part, the part wears. The wear is minuscule but measurable, so we must take it into consideration. Unfortunately, in many cases, wear is misunderstood. For example, literally tens of thousands of very expensive class XX plug gages are purchased. In minutes of beginning to use these gages, the XX plug gages wear into class X tolerance. However, if moderately priced class X gages were purchased, they would have performed adequately and could have saved money for the manufacturer. Wear is so important that gage specifications require a standard, which is the *wear allowance.*

In a simple but typical case of wear (see Figure 9–16), we measured an inside diameter and noted its maximum and minimum metal limits. The difference between these limits is the tolerance zone. We use the gage to ensure that no bad parts pass, a process that is still widely used, even though statistical quality control makes it anachronistic. You measure

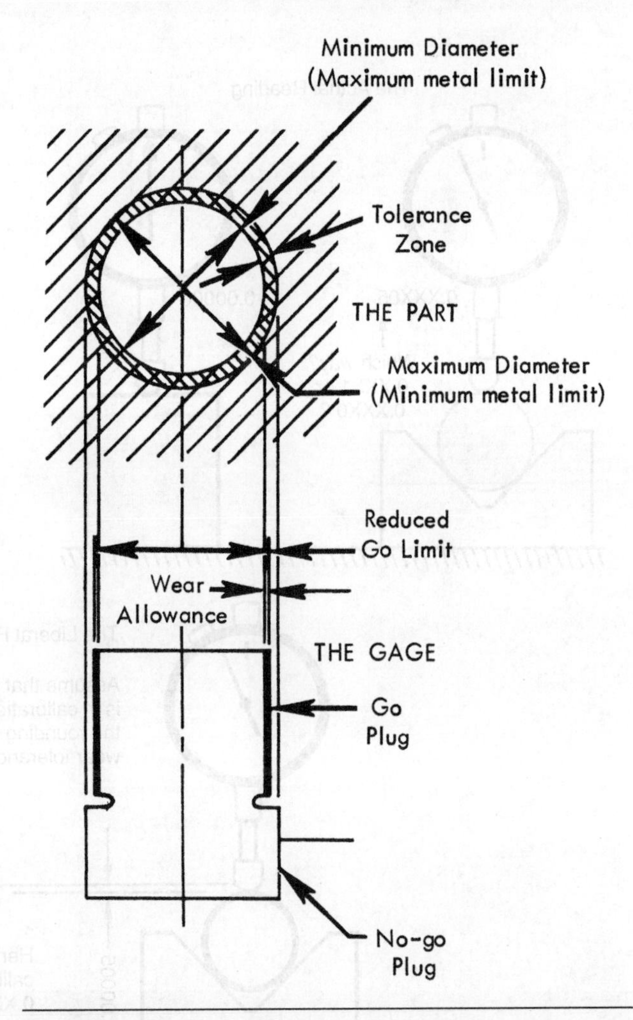

FIGURE 9–16 The go plug should enter the hole in the part. The no-go plug should not. The wear allowance on the go plug reduces the tolerance specified for the hole.

the diameter of the go plug at exactly the smallest size permitted by the tolerance, but each time you use the plug to check a part, it wears slightly. What starts out as an imperceptible wear will eventually allow you to pass undersized parts. If, however, the go plug has a wear allowance, you can use the gage for a satisfactory period of time while the allowance wears away gradually. When calibration eventually shows that the instrument is at its tolerance limit, you should replace the instrument.

In contrast, the no-go end should not go into very many parts; therefore, the wear is slight. Furthermore, no amount of wear on the no-go end will let you pass out-of-tolerance parts.

You calibrate a go/no-go gage periodically to check its condition. When the last decimal place reaches a five (5), you need to decide what to do with the instrument (see Figure 9–17). This figure does not show the consequences of your decision; however, it does give arguments to support the decision you make.

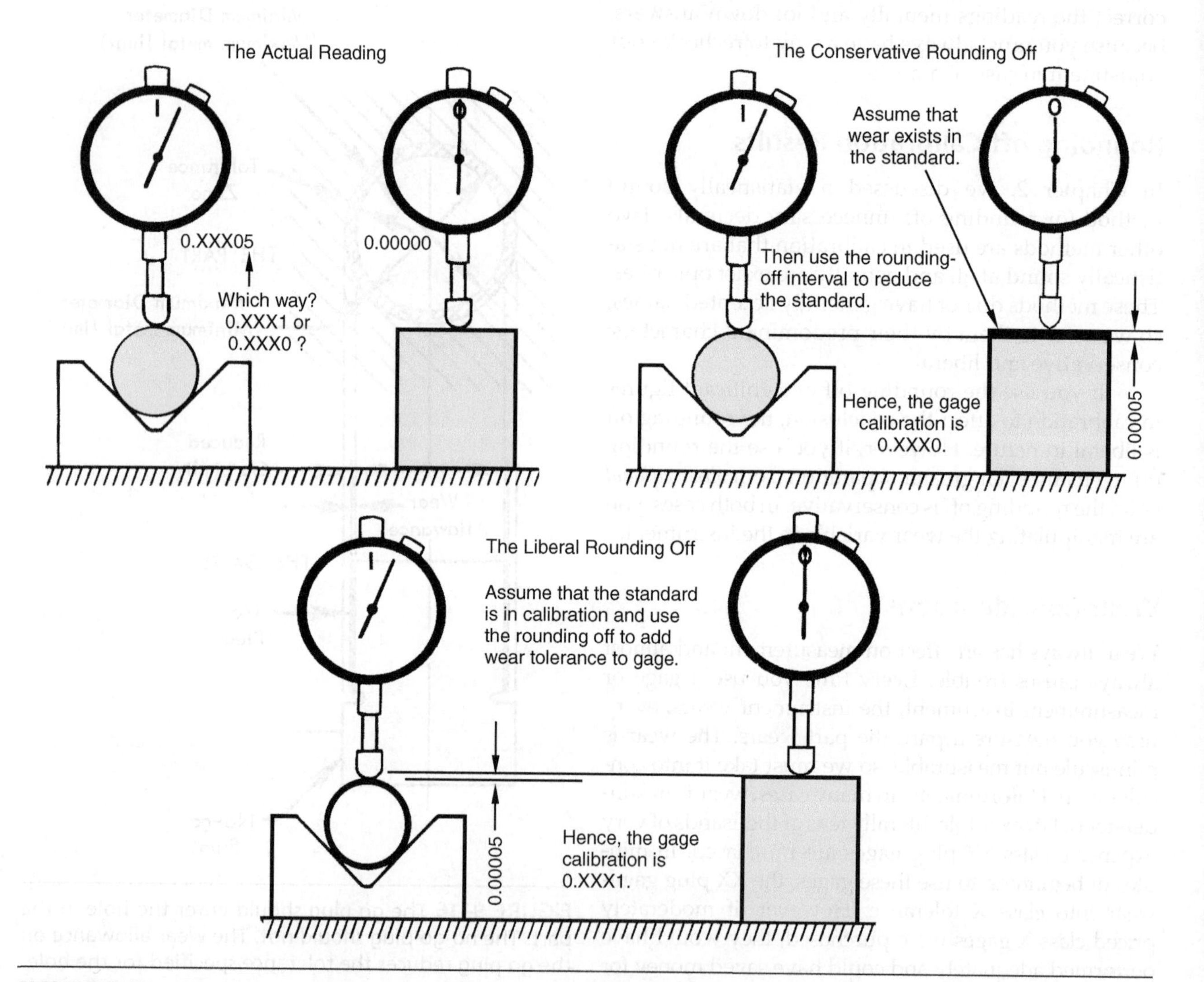

FIGURE 9–17 Take your pick. Both methods for rounding off have merit.

With the conservative approach, you assume that the standard may have worn since its last use. With this assumption, the zero setting represents a figure less than the presumed basic dimension; therefore, you add the rounding-off interval to bring the standard back to size. Adding the rounding-off standard is the same as subtracting from the gage you are calibrating. In our example, the rounding off drops the five from the reading, leaving the digit before the same.

In the liberal approach, you optimistically assume that the instrument is in calibration. We use the rounding-off interval exclusively to affect the gage itself, so it is logical to round up, adding the interval to the gage, which is the same as adding wear tolerance to the gage. Whether we use a conservative or liberal approach, however, we have not affected the metal of the instrument—all we have done is manipulate numbers.

SUMMARY

- Calibration is the measurement of measurement instruments and standards. Its purpose is to establish their relation to the international standard and their accuracy. It also establishes their other metrological features, such as precision and linearity. Because it relates to outside considerations, calibration requires the consideration of the line of authority in measurement, usually called traceability.
- Error is the enemy in measurement. The various errors can be distinguished in several ways. Random errors affect precision. Systematic errors affect accuracy. There are also errors that should not happen at all: the illegitimate errors. In calibration, we attempt to separate the various errors, determine their magnitude and occurrence, and then determine their interaction. These errors may be either dependent or independent.
- Gage blocks are the practical standard for most calibration. They are used in comparison with the instrument being calibrated. Efficient calibration is that which uses the fewest comparisons to determine the required metrological data. Efficiency is obtained by selecting calibration points that check more than one variable at the same time. Generally, this is expedited by bridging the area being investigated, and then by successive halving, zeroing in on the specific area.
- Because the metrological characteristics of instruments are very different, no general procedure can be given for calibration. It is necessary to study the instrument carefully before the beginning. Fortunately, the general procedure has been formalized for most instruments.
- Instrument calibration cannot be separated from instrument use. This insinuates human considerations into all precision calibration. The user of the instrument must be trained to the calibrated instrument, and vice versa. Furthermore, all of the elements of the measurement system enter into the calibration. This becomes particularly important for instruments such as comparators. Calibrating the final stage alone is not enough.
- Calibration accuracy depends on the subtler aspects of measurement. Cleanliness is absolutely essential, and temperature control is only slightly less important. When setting standards are calibrated, extreme attention to these details is tantamount to accuracy.
- Indicators, micrometers, and calipers can be calibrated in your own shop using gage blocks. Ideally, they would be calibrated before each use to ensure accurate measurements. More practically, gages are calibrated at least every few months, depending on use and care. Gage blocks and standards must be sent out to a laboratory that specializes in standards calibration.
- To comply with various ISO requirements, label each tool with a serial number and keep written records of when, where, and how often they are calibrated.

END-OF-CHAPTER QUESTIONS

1. Which of the following is the primary purpose of calibration?
   a. To detect wear and adjustment malfunctions in measurement instruments
   b. To determine whether or not an instrument that is inadequate for the requirement is being used

c. To detect errors due to parallax and other careless procedures
d. To determine the adherence of an instrument or process to standards

2. What is the basic principle of calibration?
a. Comparison with a higher standard
b. Examination of repetitive measurements
c. Measurement under constant temperature conditions
d. Use of a high-amplification comparator to check a lower-amplification instrument

3. What is the standard on which industrial calibration is based?
a. Wavelengths of light
b. Johannson's original length standards
c. The National Institute of Standards and Technology
d. A 1-meter length of iridium alloy at 68°F

4. Which of the following is the largest potential error in calibration?
a. Temperature differences
b. Parallax
c. Hysteresis
d. Human error

5. To which of the following does the scalar principle best relate?
a. The ratings of calibration instruments from least to most precise
b. The ratings of standards from least to most precise
c. The succession of steps in the calibration process
d. The succession of steps by which authority proceeds from the lowest to the highest authority

6. To which of the following do random errors relate most closely?
a. Precision
b. Parallax
c. Accuracy
d. Dirt
e. Loose elements in setup

7. To which of the choices in question 6 do systematic errors relate most closely?

8. If an error can be evaluated by systematic process, what type is it?
a. Random error
b. Indeterminate error
c. Systematic error
d. Determinate error

9. Which of the following is the factor that determines whether a measurement is determinate or indeterminate?
a. Use of a calibration instrument with 10 times the discrimination of the product's tolerance
b. The availability of standards
c. Whether it's defined by tolerance or not
d. The recurrence of random errors

10. An error that results from carelessness is called which of the following?
a. Random error
b. Parallax error
c. Cramping error
d. Illegitimate error

11. Which of the following is not a characteristic of systematic errors?
a. Recurrence
b. Regularity
c. Periodicity
d. Similarity

12. Which of the following does not describe the way errors interact?
a. Cancel each other out
b. Multiply each other
c. Independent of each other
d. Dependent on each other

13. If one gage block is out of calibration in an otherwise properly calibrated set, what type of errors does it create when the set is in use?
a. Dependent errors
b. Correlated errors
c. Calibration errors
d. Magnitude

14. If a steel tape with a kink in it is used for a series of measurements which then were totaled, which of the errors listed in question 13 could result?

15. Which of the following are the basic acts required for calibration?
   a. Determination of variables, determination of the error from each variable, ascertaining the interaction of the variables
   b. Elimination of all controllable variables such as dirt, loose joints, drafts, etc.; normalizing of the calibration setup; repetitive measurements
   c. Selection of instruments that obey the 10-to-1 rule; verification of the calibration traceability of those instruments; rechecking of all calculations
   d. Plan the procedure in advance; determine that the necessary instruments and standards are available; determine traceability of the selected standards back to the National Institute of Standards and Technology

16. What is the best way to check the functional characteristics of instruments?
   a. By the use of gage blocks
   b. By comparison with equivalent instruments
   c. By statistical analysis of repetitive measurements
   d. By visual inspection for worn references, loose members, and other signs of abuse

17. Which of the following would be the best characteristic for a standard to have when calibrating an instrument used for production inspection?
   a. Ten times the accuracy
   b. New or never used since last calibration
   c. Permanently assigned to this particular calibration
   d. Same size, shape, and material

18. Which of the following describes literally the meaning of a complete calibration?
   a. Performed only at the National Institute of Standards and Technology or a NIST accredited metrology laboratory
   b. Performed only by a professional metrologist using traceable standards
   c. Calibration at zero, the maximum size, and the midpoint
   d. Calibration at each point at which a reading might be expressed

19. How is the necessity for checking many points reduced?
   a. By watching for patterns
   b. By reducing the number of points and determining their root mean squares (RMS)
   c. By the use of laser measuring
   d. By successive doubling of size intervals

20. Which of the following is the most practical method that can be used to detect patterns?
   a. Using a trace oscilloscope with memory
   b. Using a PC
   c. Graphing on paper
   d. None of the above

21. What is an error that is restricted to a limited portion of the range called?
   a. Wear error
   b. Reference error
   c. Topical error
   d. Periodic error

22. If an error cannot be explained by normal deterioration, which of the following best describes it?
   a. Instrument error
   b. Illegitimate error
   c. Parallax error
   d. Stray error

23. Which of the following does the bridging technique refer to?
   a. Beginning calibration at both ends of the usable range and working toward the middle
   b. The use of C-shaped instruments such as the micrometer and caliper
   c. Support of long instruments such as vernier calipers so that they do not sag
   d. Successively narrowing the gap between measurements until the high or low point is found

24. What characteristic governs the selection of calibration points for most instruments?
    a. Even intervals for ease of graphing
    b. Whole numbers for ease of computation
    c. Even numbers to eliminate long decimals that result when calculating with odd numbers
    d. Selected numbers based on variables sought
25. Which of the following is an essential step in calibration yet supplies little information compared with the rest of the process?
    a. Zero setting
    b. Normalizing
    c. Selection of standards
    d. Checking for loose joints
26. The effect of errors of feel is similar to which of the following?
    a. Zero setting errors
    b. Loose member errors
    c. Parallax errors
    d. Temperature errors
27. Which of the choices in question 26 are considered to be observational errors?
28. Which of the following is interpolation?
    a. Averaging a series of measurements
    b. Checking the instrument reading against the specified dimension
    c. Approximation between two scale divisions
    d. The discrimination of an instrument
29. Which of the following involves the most variables?
    a. Length measurement
    b. The measurement of spheres
    c. Vertical measurements
    d. External measurements
30. What are measurements called that are identical in every respect except time?
    a. Repetitive
    b. Asynchronous
    c. Single sample
    d. Calibration
    e. Multiple sample
31. Which of the choices in question 30 is a measure of sensitivity?
32. Which of the following is the most nearly correct?
    a. Sometimes the measurement act distorts the object measured.
    b. Sometimes the measurement act distorts the measurement instrument.
    c. Absolute accuracy can be obtained only with high-amplification comparators.
    d. Absolute accuracy only exists until the first use of an instrument or standard.
    e. All measurement distorts the object and the instrument.
33. If a standard or part is simply placed on the reference surface and held there by gravity, which type of contact is involved?
    a. Point
    b. Assisted
    c. Manual
    d. Wrung
    e. Casual
34. What is the name for the amount of uncertainty that can be expected under standard conditions?
    a. Work tolerance
    b. Uncertainty
    c. Discrimination
    d. Reliability
    e. Single-sample variable
35. Which of the following can be reliably calibrated under shop conditions if adequate care is taken?
    a. High-amplification comparators but not highest scales
    b. Micrometer instruments
    c. Vernier instruments
    d. Gage blocks
36. Which of the following statements about wear allowance is correct?
    a. It enables a gage to have reasonable life before passing out-of-tolerance parts.
    b. It passes in-tolerance parts and rejects out-of-tolerance parts.

c. It is important for no-go gages but not for go gages.
d. It violates the principles of reliable measurement.

37. The purpose of a calibration schedule is to:
    a. meet customer requirements
    b. detect the deterioration of accuracy
    c. test quality personnel
    d. practice measurement procedures

38. From which of the following does the 10-to-1 rule derive its name?
    a. The value of the instrument is at least 10 times greater than the value of the particular measurement or gaging operation.
    b. The scale range must be at least 10 times the required length of the maximum measurement.
    c. It is a holdover from an earlier period in which it was considered that a gagemaker or toolmaker could produce work 10 times as precisely as a machinist.
    d. The gage should be 10 times as accurate as the part feature.

39. Which of the following best explains the wide acceptance of the rule of 10, which is another name for the 10-to-1 rule?
    a. Availability of relatively low-cost electronic instruments that provide multiple scales
    b. Recognition of errors that are caused by extraneous factors such as parallax, temperature differences, and vibration
    c. Facilitation of the selection of inspection instruments that minimize instrument errors in practical situations
    d. The increasingly fine tolerances in modern manufacturing

40. The 10-to-1 rule cannot be applied in all cases. Aside from a lack of suitable instrumentation and standards, how is this explained?
    a. It requires excessive time, particularly when time for temperature normalization is considered.
    b. When several successive stages are required, it may require finer discrimination than is attainable in an industrial situation.
    c. Semiskilled inspectors have difficulty understanding it.
    d. The cost of paperwork to substantiate the rule often costs more than the scrap savings achieved.

41. What should be done when the 10-to-1 rule cannot be followed?
    a. Document that fact so that subsequent users of the part will be forewarned.
    b. Change to an eight-to-one, or even lower, rule.
    c. Scrap the parts.
    d. Upgrade instruments to those of greater sensitivity and standards to those of higher accuracy.

42. Which of the following is the best method for scale selection with multiple-scale instruments?
    a. Use the next highest scale that was prescribed for the measurement.
    b. Select the scale that divides tolerance into 10 parts.
    c. Zero set at a high scale, and then return to the lowest scale on which the tolerance can be read.
    d. Select the most sensitive scale that will contain the tolerance.
    e. Always use the highest scale available.

CHAPTER TEN

MEASUREMENT BY COMPARISON

Man is a tool-using animal. Weak in himself and of small stature, he stands on a basis of some half-square foot, has to straddle out his legs lest the very winds supplant him. Nevertheless, he can use tools, can devise tools; with these the granite mountain melts into light dust before him; seas are his smooth highway, winds and fire his unwearying steeds. Nowhere do you find him without tools. Without tools he is nothing, with tools he is all.

Thomas Carlyle (1795–1881)

LEARNING OBJECTIVES

- Explain the role of amplification in measurement.
- Describe the difference between direct measurement and comparison measurement.
- Explain the relationship of precision, accuracy, and reliability.
- State practical applications for low-amplification comparator instruments.
- Compare analog (dial) and digital instruments.

OVERVIEW

All measurement requires that an unknown quantity be compared with a known quantity, called a *standard*. The three elements of measurement (see Figure 10–1) are all required in fundamental measurements: time, mass, and length.

The standards for rules, verniers, micrometers, and most optical instruments are not only built in, they are also calibrated. Another way you can measure is with a standard that is separate from the instrument (see Figure 10–2). Either way, measurement is made by comparing the unknown length with the standard, so we call it *comparison measurement*. Precision of comparison measurement depends on the discrimination of the standard and the means for comparing the standard and the unknown.

When the instrument is the standard, you are directly relating the standard and the unknown, so we call it *direct measurement*. In direct measurement,

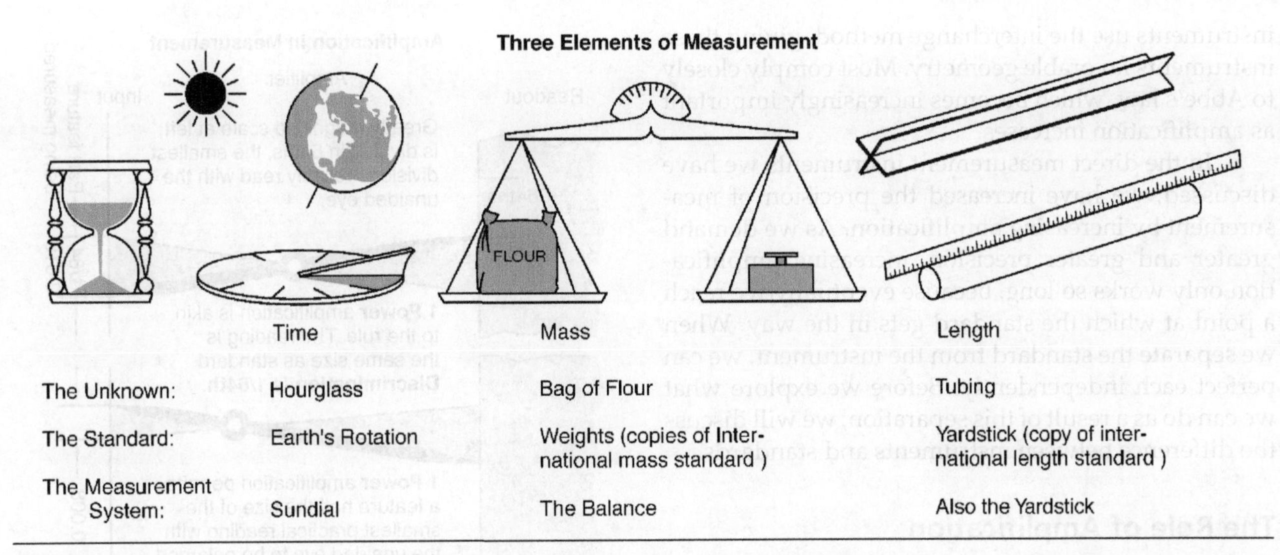

FIGURE 10–1 All measurement is made up of measurement of time, mass, and length. In each of these cases, three elements are involved: the known, the standard, and a system for comparing them. Only in crude length comparison are the standard and the system the same object.

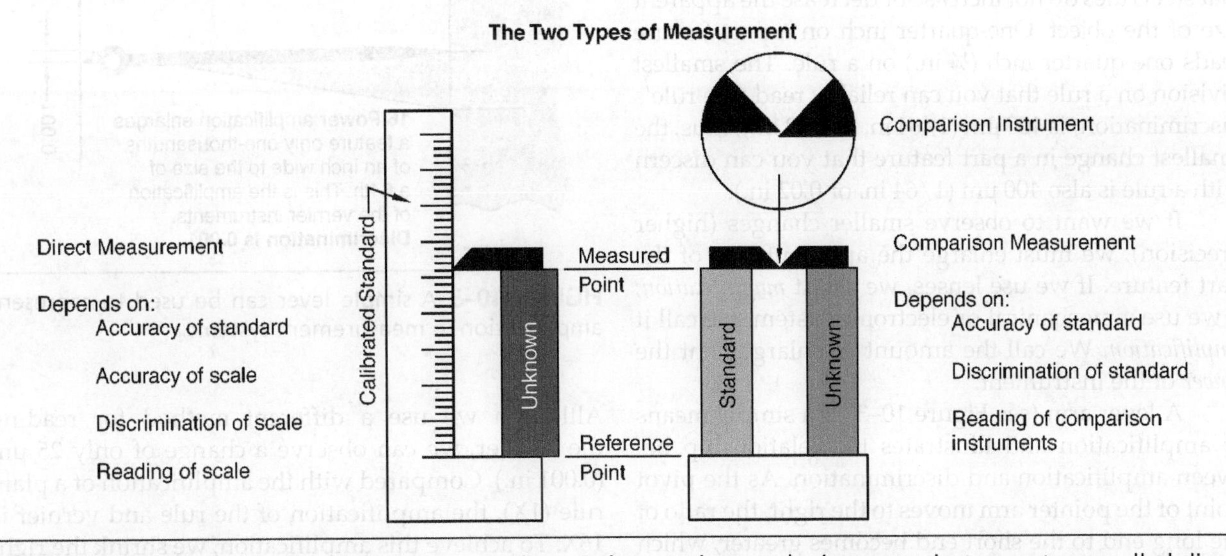

FIGURE 10–2 All measurement requires comparison. Rules, verniers, and micrometer instruments are called direct measurement because they do not require the intercession of another element.

precision depends on the discrimination of the scale and the means for reading it.

In contrast, accuracy depends on other factors, including important geometric considerations. In Figure 6–9, we demonstrated that all measurement is accomplished by either the interchange method or the displacement method. Verniers and micrometers measure by the displacement method; comparison instruments use the interchange method, giving these instruments favorable geometry. Most comply closely to Abbe's law, which becomes increasingly important as amplification increases.

In the direct measurement instruments we have discussed, we have increased the precision of measurement by increasing amplification. As we demand greater and greater precision, increasing amplification only works so long, because eventually we reach a point at which the standard gets in the way. When we separate the standard from the instrument, we can perfect each independently. Before we explore what we can do as a result of this separation, we will discuss the difference between instruments and standards.

The Role of Amplification

Steel rules have one power (written 1X)—an amplification factor of one, a mechanical advantage of one, or a leverage ratio of one to one. These expressions all mean that steel rules do not increase or decrease the apparent size of the object. One-quarter inch on a part feature reads one-quarter inch (¼ in.) on a rule. The smallest division on a rule that you can reliably read (the rule's discrimination) is 400 μm (1/64 in. or 0.02 in.); thus, the smallest change in a part feature that you can discern with a rule is also 400 μm (1/64 in. or 0.02 in.).

If we want to observe smaller changes (higher precision), we must enlarge the apparent size of the part feature. If we use lenses, we call it *magnification;* if we use a mechanical or electronic system, we call it *amplification.* We call the amount of enlargement the *power* of the instrument.

A lever arm (see Figure 10–3) is a simple means of amplification and illustrates the relationship between amplification and discrimination. As the pivot point of the pointer arm moves to the right, the ratio of the long end to the short end becomes greater, which increases the amplification or power.

The one-to-one relationship, like on a rule, is shown at the top. In the center drawing, the 2X amplification enlarges a distance of 200 μm (0.008 in.) to span 400 μm (1/64 in.) on the scale; therefore, we can read it.

The bottom drawing represents the amplification we gain when we add a vernier scale to a rule.

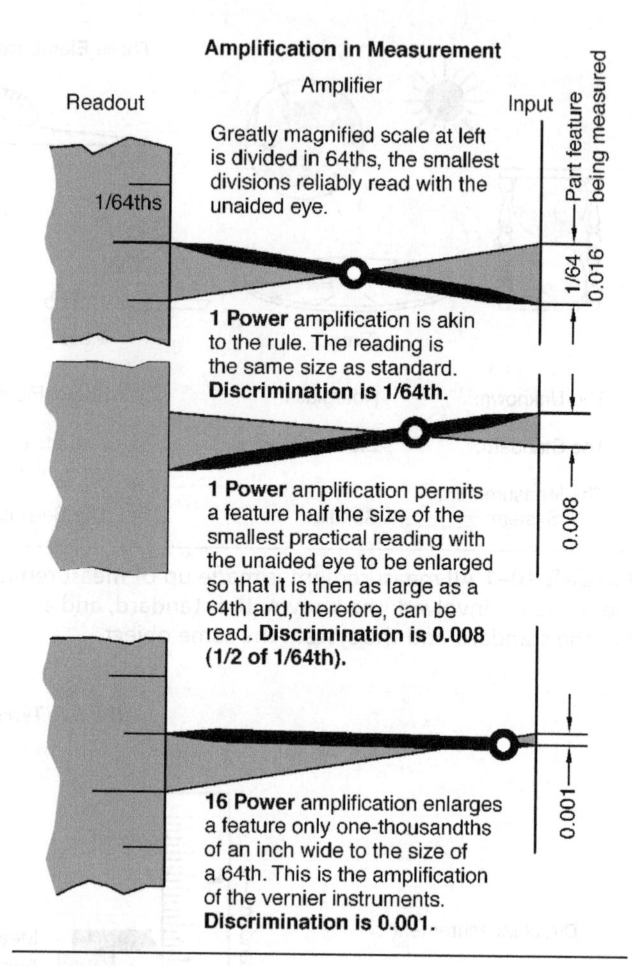

FIGURE 10–3 A simple lever can be used to represent amplification in measurement systems.

Although we use a different method for reading the vernier, we can observe a change of only 25 μm (0.001 in.). Compared with the amplification of a plain rule (1X), the amplification of the rule and vernier is 16X. To achieve this amplification, we shrink the right end of the pointer to a nubbin.

There is a limit to the increased amplification we can achieve, which is obvious when we study the micrometer. In Figure 10–4, we have redrawn the pointer. At the top, the graduations on the left represent one division of the thimble scale. At the far right end, the invisible movement of the pointer would actually be

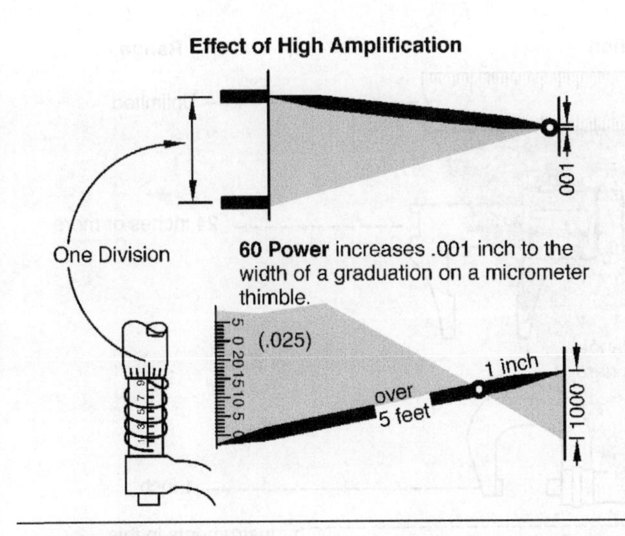

FIGURE 10–4 The micrometer provides the equivalent of an enormous lever arm.

25 μm (0.001 in.); therefore, the thimble divisions represent 25 μm (0.001 in.). The thimble graduations vary in separation because they are on a bevel. At the reading end, they measure about 1.5 mm (0.060 in.) for 60X power amplification.

If we blow up 1 mil (0.001 in.) to 60 mil (0.060 in.) so that we can read it reliably, the one thousand divisions it takes to make up an inch would stretch out 5 ft. We fit a 5 ft. scale into the small confines of a micrometer by wrapping the scale around using the principle of the screw.

If you trace the path of 0 on the thimble from full closed to full open, it makes a long spiral, crossing the index line every time the spindle has moved 0.025 in. a total of 40 times. Each time it passes the index line, the thimble has rotated through 25 spaces of approximately 0.060 in. each and has paced off a scale 1½ in. in length. After the 40th rotation, you have created a scale of 5 ft. (in reality, the scale is closer to 62½ in. for most micrometers), and you have stretched 1 in. into 5 ft.

Practical limits to this principle are caused by the built-in standard. Every increase in power causes a parallel decrease in another feature: scope, range, convenience, economy, and so forth.

A rule may have only 1X, but you can carry it in your shirt pocket. A vernier scale increases the rule amplification to 16X, but if it stays locked in its velvet-lined case, it is useless. The micrometer increases the amplification to over 60X. At the same time, the measuring range shrinks to 1 in., and, even for this 1 in., we need a wraparound scale 5 ft. long.

If we continue along these traditional routes to increase precision, we run into trouble. For example, we can create an instrument with 10 times the discrimination of present micrometers by simply adding a vernier scale to a micrometer. However, we have already demonstrated that, although a vernier improves the precision, it does not improve the inherent accuracy of the instrument.

To try another route, we could scale down a micrometer to one-tenth its size, creating a 400-pitch thread (impractical in large versions). Each revolution of the thimble would close the micrometer 0.0025 instead of 0.025 in. Even if we had to decrease the number of divisions on the drum in order to read them clearly, the new instrument would still be more precise, more accurate, and more reliable.

However, this instrument has a serious fault: When we trade range for power, the instrument is no longer practical. The new range is only 0.1 in.—the length of the screw standard. The short range of the "micro-micrometer" would limit the instrument to very few practical applications, because even when we put a large frame on a micrometer, the instrument's size is increased, not its range.

When an instrument incorporates its own standard:

- The measuring range is limited by the length of the standard.
- Discrimination is limited by the system we use to divide the standard into units and enlarge these units to a readable size.
- Accuracy is limited by the care the manufacturer takes in making it.

Separation of the Standard

If we could achieve high-amplification precision without sacrificing the length of the standard, we could measure at almost any power, which is exactly what

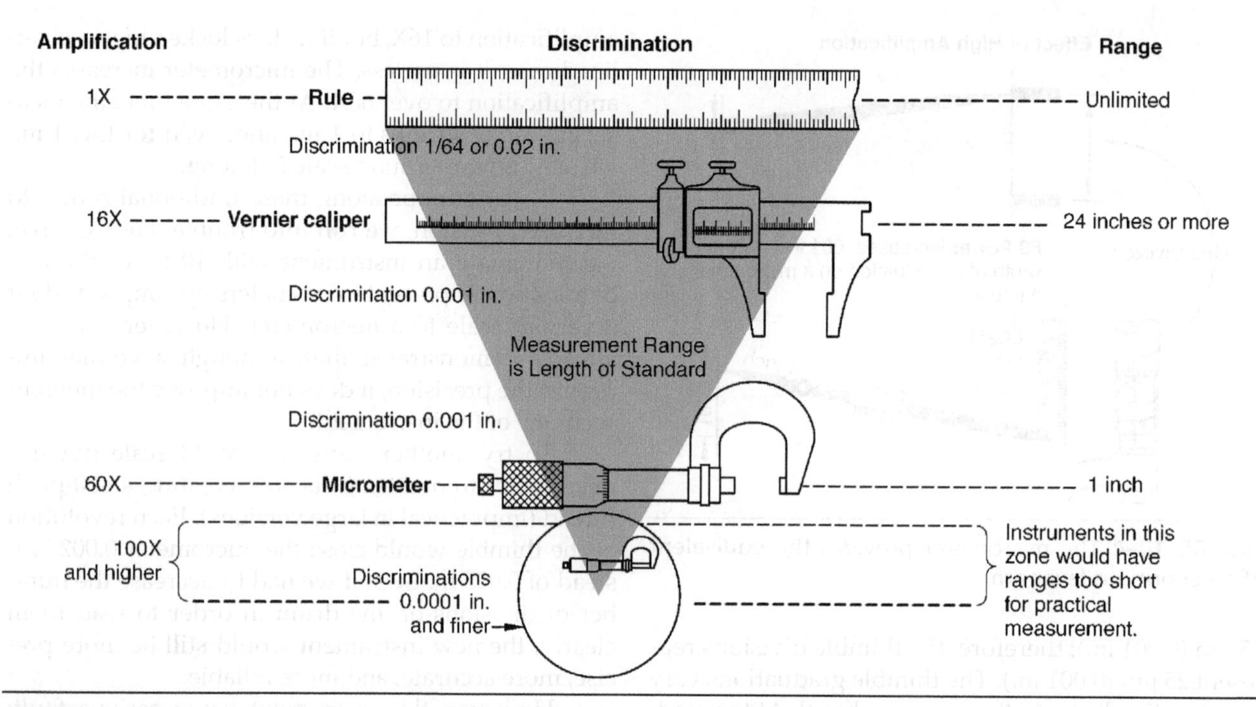

FIGURE 10–5 When the standard is part of the instrument, amplification limits range.

we can do when we separate the standard from the measuring instrument.

Without the standard, we are left with an instrument that simply compares the standard with the unknown length. Every linear measurement, even with a rule, requires a measurement system to compare the standard with the unknown distance. When the standard and the instrument—major components of any measurement system—are combined, changes in one component affect the other. When these components are separate, each can be refined independently of the other.

When we created the micrometer from the rule, we sacrificed range (see Figure 10–5); however, throughout the process, the micrometer showed us just how precise measurement could be. After we separate the instrument and the standard, new opportunities for measurement also become obvious. For example, first we find that the new comparison instrument not only can make linear measurements, it can also provide checks for concentricity, flatness, roundness, and position. We can use all ordinary means to achieve amplification: mechanical, electronic, optical, and pneumatic. This group of instruments is so large that we cannot describe them all, but if you understand the principles you will understand the instruments, even when the designs are different (see Figure 10–6).

10–1 THE DIAL INDICATOR

To understand measurement by comparison, you should know about the familiar comparison instruments called *dial indicators*. They were not the first comparison instruments, nor can they compare with the versatility, discrimination, and reliability of the electronic instruments. The dial indicators also cannot be used at all for some applications for which we use pneumatic comparators (air gages) and optical

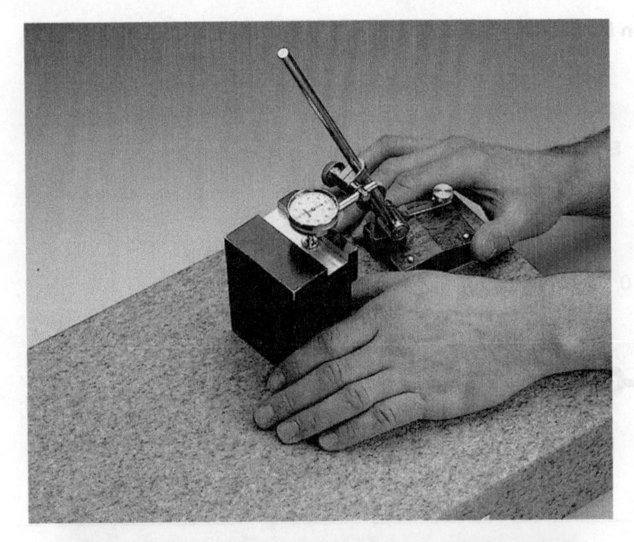

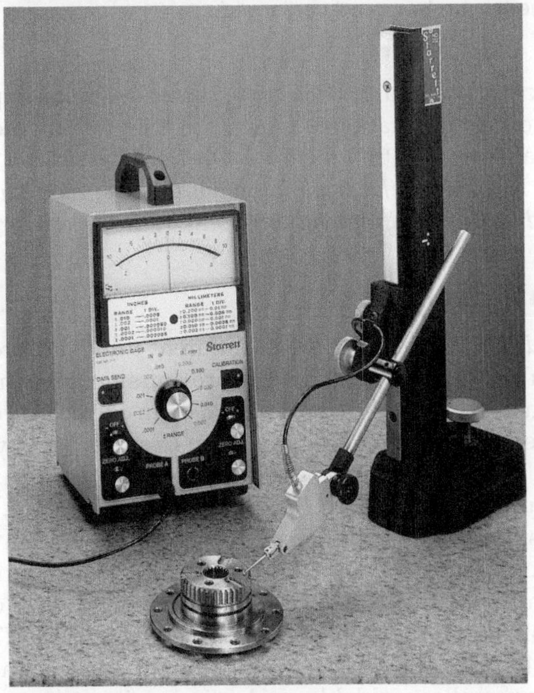

FIGURE 10–6 Indicator instruments range from low-amplification test indicators, used for setup and in process inspection, to highly precise electronics instruments with discrimination in millionths of an inch or meter. *(Courtesy of the L. S. Starrett Company)*

instruments. However, they are familiar and easy to use when you are learning the principles of comparison measurement.

The simplest indicator is a lever (see Figure 10–7), which is often used as a pointer. The typical indicator lever has a ball on the short end so we can more easily measure distance changes through contact with this ball. The ratio (see Figure 10–7A) is 1 to 10 (written 1:10) or an amplification of 10X.

If we need greater amplification, we can change the ratio of the ends, for example, to 1:16 (see Figure 10–7B). This new ratio is a 0.001 in. change to 1/64 in. Of course, this proportion of the lever might make the instrument impractical, which is why we usually use compound levers to gain high amplification. Figure 10–7C illustrates how we can achieve an amplification of 100.

We achieve greater precision by lengthening the scale through which the pointer swings with respect to the change of length being observed. If the swing of the pointer in A is the maximum practical amount, then the range of measurement in B would have to be smaller, and the range of C smaller yet. In fact, the range in C could only be 1/10 that of A, which would reduce greatly the practical applications of the measuring instrument.

Gearing provides the solution to practicality, because we can achieve the same mechanical advantage as levers but not limit the scale range. When two gears mesh, the angular movement of the driven gear is determined by the amount the driving gear turns, and by the ratio of the gear teeth (see Figure 10–8A). In our example, the driving gear has 10 times as many teeth as the driven gear. With every revolution of the driving gear, the driven gear goes through 10 revolutions. In Figure 10–8B, we have added two more gears, making the ratio 100:1.

To use this amplification for measurement, we need a way to make the driving gear rotate in proportion to changes in length and a means of reading the amount of rotation of the last driven gear. In Figure 10–8C, we added a spindle, which has a rack to engage the driving gear and a contact to engage the part we are observing. To complete the measurement,

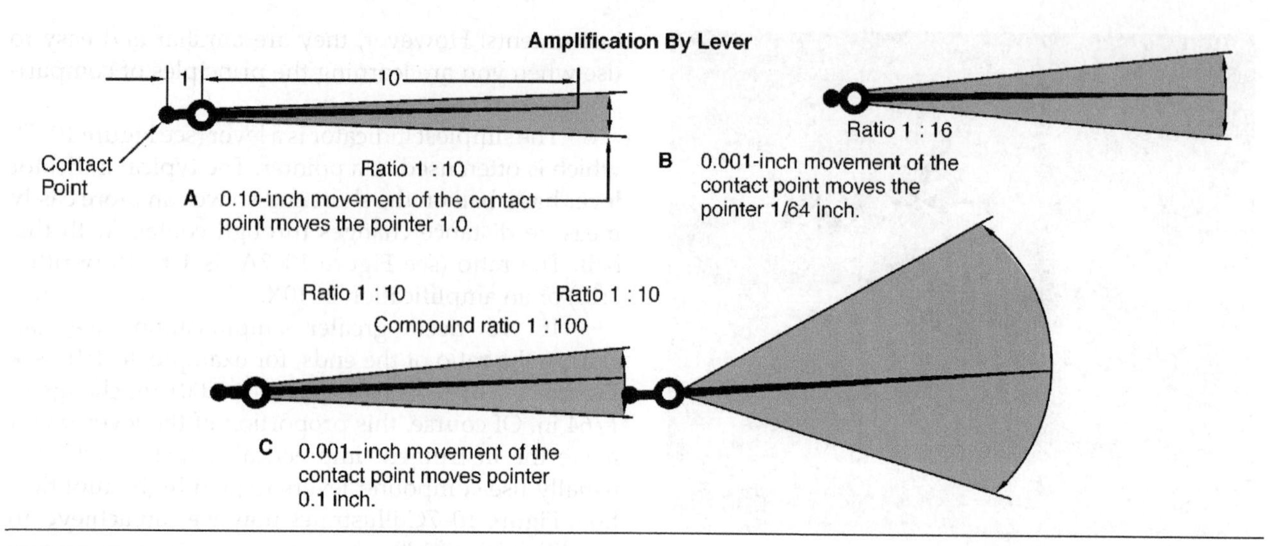

FIGURE 10–7 By means of compound levers, very small movements can be greatly enlarged.

we must have a standard; otherwise, all we have is gear motion, much like a football game without yard markers and end zones.

In Figure 10–8C, black dots show the "at-rest" position of the gears. In Figure 10–8D, we raised the contact slightly; therefore, the black dots show the rotation of the gears caused by the displacement of the spindle. Even though the first gear rotated only slightly, the final gear rotated 100 times as much as the first gear.

This gear movement does not mean that the pointer has moved 100 times as far as the spindle was displaced. The pointer itself provides amplification. One point on the pitch diameter of the last driven gear will have moved 100 times as far as the spindle moved, but it will have moved along an arc.

In Figure 10–8A (bottom), the gear is shown at rest, and B represents the position after the spindle has been displaced. The pointer has turned through the same angle but moved twice as far as the point on the pitch line of the gear. The pointer moves twice as far because it is twice the distance from the center of rotation, giving it double the amplification of the gear train.

We can design instruments with pointers and gears for as long as the use of these instruments is practical. In Figure 10–8C, the pointer's relationship to the center of rotation is four times the pitch diameter for an amplification factor of 4X. *The amplification of a dial indicator is the amplification of the gear train multiplied by the amplification contributed by the pointer.*

Scale Discrimination

Read and *discriminate* mean different things in precision measurement. If letters are missing in words, such as "dir-y d-g," you can read them correctly; your experience with reading fills them in automatically.

Reading does not work in precision measurement, because we need the exact values. Therefore, we must be able to distinguish, or discriminate, each division of measurement. How easily we can recognize each division is called *resolution,* and unless we can resolve each division, the instrument is useless. That is why we define the discrimination of an instrument as the smallest readable graduation on its scale.

Functional Features

As shown in Figure 10–8, the graduations on A and B are satisfactory for reliable discrimination; therefore, the same spacing could have been used for C. We

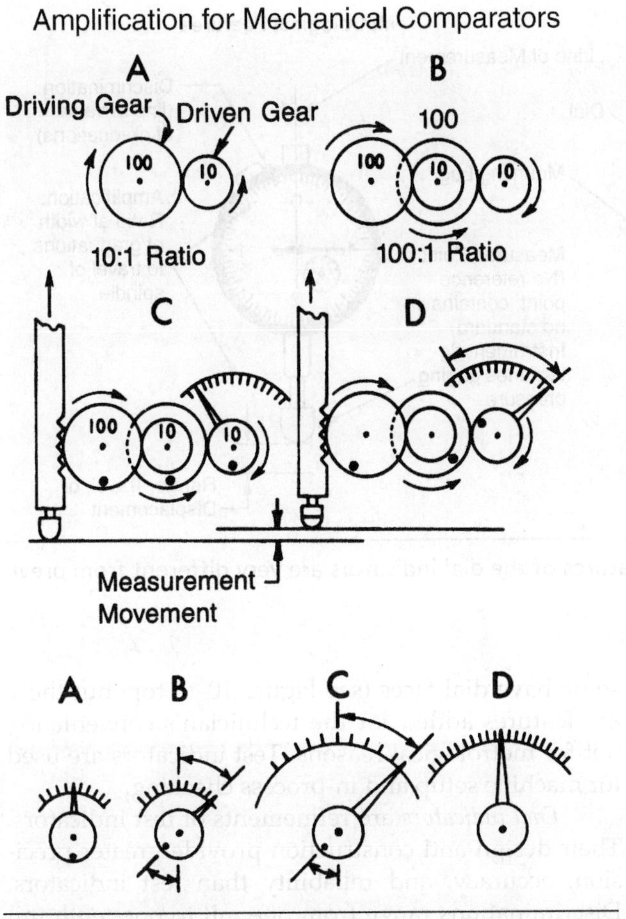

FIGURE 10–8 Gear trains (top) provide the mechanical advantage of levers but without the limitation of range. The length of the pointer (bottom) enters into the overall amplification. In a dial indicator, the pointer is often called the *hand*.

have shown this spacing at D on an instrument with twice the discrimination as C; therefore, increasing the amplification permits finer discrimination.

Dials, their sizes, and mounting dimensions are standardized typically as in Figure 10–9. Although most of the instruments we discussed have generally had the same features, the internal construction of dial indicators varies considerably among manufacturers. The dial is typically attached to the bezel, which has a knurled outer edge to allow you to turn the dial by hand. The bezel clamp locks the dial in position.

The spindle and rack of the dial indicator are usually one piece. The rack spring, which applies the gaging pressure instead of relying on the expertise of the technician, resists measurement movement and returns the mechanism to the "at-rest" position after each measurement. The hairspring (not the same as the rack spring) loads the gears in the train against the direction of the gaging movement and eliminates backlash that can be caused by gear wear. Various indicator backs are available, and the instrument is sometimes attached by the stem instead of the back, which may cause damage. The *telltale* is a revolution counter; when an indicator has a long range, the telltale helps you avoid errors caused by whole turns.

Metrological Features

In addition to their functional features, the metrological features of dial indicators make them different from all the instruments we have discussed. For example, dial indicators do not have a reference point. We could almost say that the dial indicator is a mechanized measured point looking for a place to be from.

Remember, the dial indicator has no standard that relates to the overall length of the part feature; however, there is a standard that relates to the observed change in length.

The first indicators were simple levers and only showed that a length was or was not the same size as another part (see Figure 10–10). These indicators did not tell us the size of the deviation, whereas modern indicators measure the amount of this deviation. As always, however, we cannot obtain a measurement without a standard. In this limited sense, indicators do incorporate a standard: The standard in the indicator measures change in length. An indicator cannot measure length unless the measurement is less than the total range of the indicator.

You need to understand the differences among three categories of indicators: test indicators, dial indicators, and *comparators*. *Test indicators*, the least precise, only indicate whether or not a point moves in one axis. Most test indicators have graduations, and

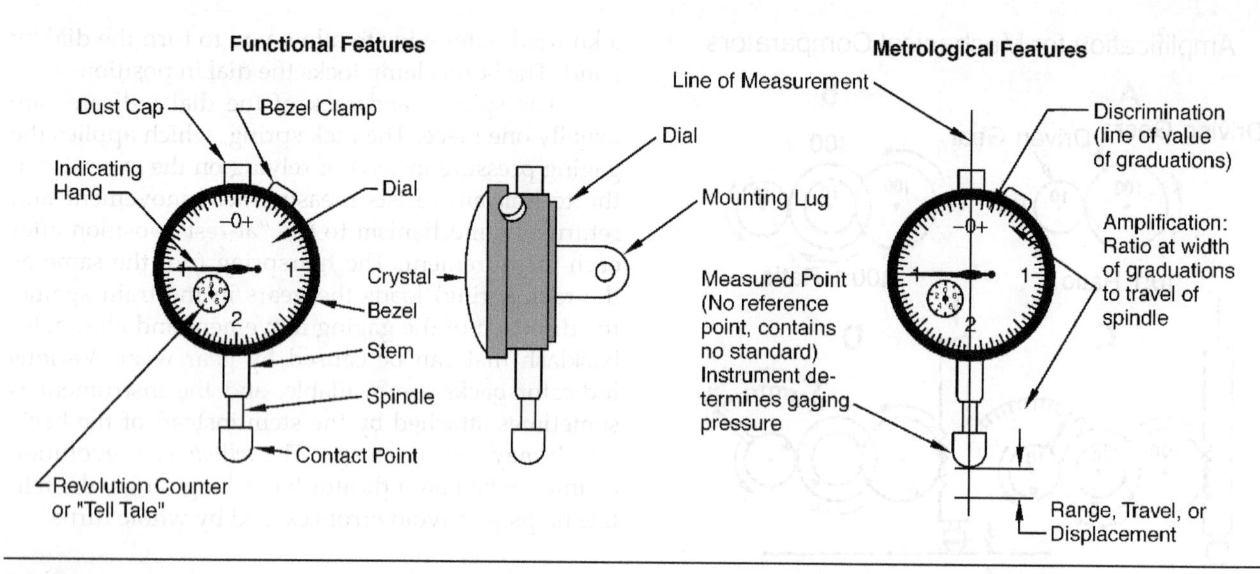

FIGURE 10–9 Both the functional and the metrological features of the dial indicators are very different from previous instruments.

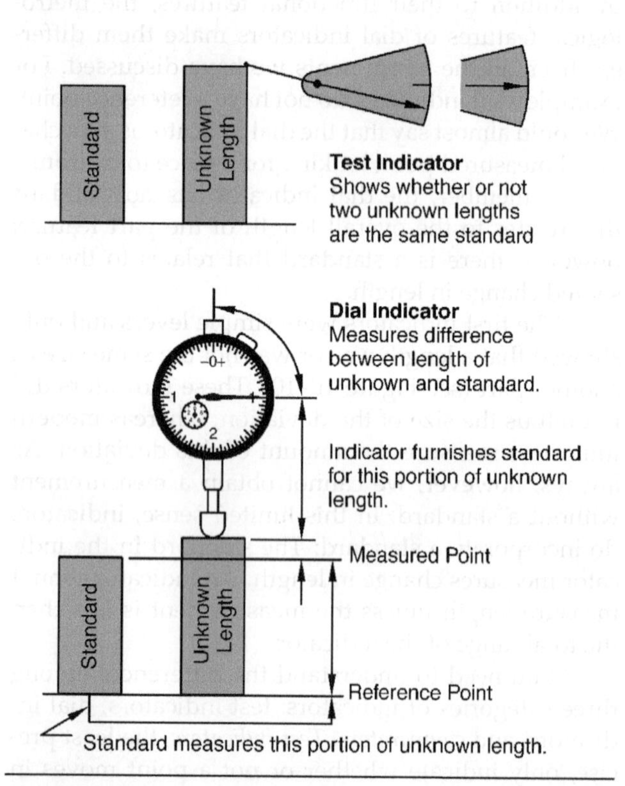

FIGURE 10–10 The dial indicator measures change in length, not length itself.

some have dial faces (see Figure 10–6) top, but these are features added for the technician's convenience, not for metrological reasons. Test indicators are used for machine setup and in-process checking.

Dial indicators are refinements of test indicators. Their design and construction provide greater precision, accuracy, and reliability than test indicators. Discriminations range from one mil to one-tenth mil (0.001 to 0.0001 in.).

We use other instruments and other principles for amplification, including electronic instruments, which provide higher amplifications than geared instruments. Electronic indicators are referred to as comparators, and they provide one additional benefit: a digital display of the measurement readings, which can be an advantage in many cases.

The dial comparator (Figure 10–11) displays spindle displacement, but electronic readouts make it possible to incorporate internal function calculations. With fixtures, measurements such as inside diameter and radius can be obtained without conversion tables (see Figure 10–12).

Even though digital reading instruments are the "newest" development, their presence does not mean that all other indicators are obsolete. We can use either

FIGURE 10–11 The dial indicator measures change in length, not length itself. *(Courtesy of the L. S. Starrett Company)*

dial indicators or electronic digital instruments to explore the metrological considerations of comparison measurement, although knowledge of solid-state electronics might come in handy when we discuss the electronic instruments. We will continue our discussion with mechanical instruments so you have a better understanding of the underlying mechanics that have gone into the development of our sophisticated, modern instruments.

Figure 10–13 lists various instruments and clarifies terminology; Figure 10–14 compares the principal metrological features, discrimination, and range; and Figure 10–15 makes it easier for you to understand the comparisons among these instruments.

In Figure 10–15, the bars to the right of the baseline compare the discrimination of the four instruments in Figure 10–14, and the bars to the left of the baseline compare their ranges. The bars are drawn to scale to show their relative proportions; therefore, the smallest reading (discrimination) of the highest amplification instrument—Instrument D, which is the electronic comparator—is hardly visible. The next most powerful one—Instrument C, a mechanical comparator—has one-fifth the discrimination but greater range.

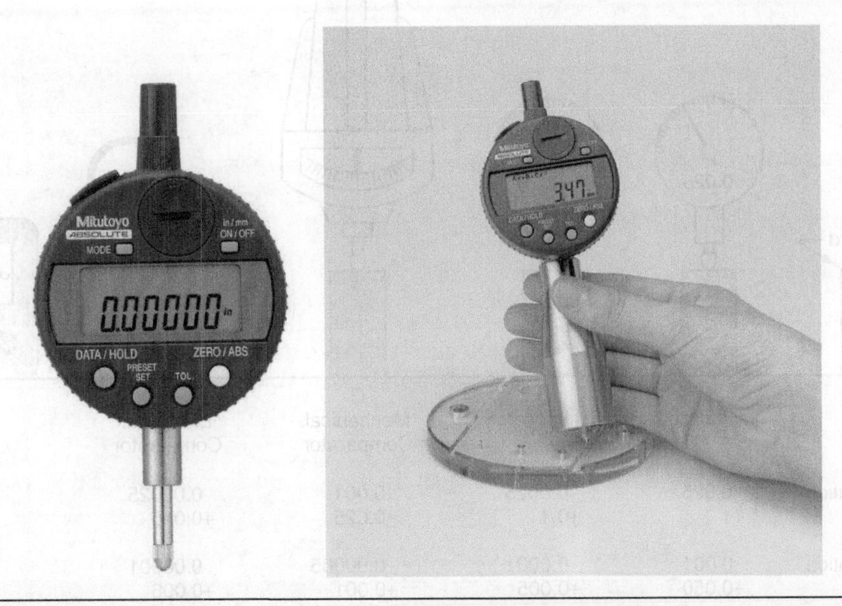

FIGURE 10–12 This indicator performs internal calculations using the formula Ax + B + Cx − 1 (assuming spindle displacement as x) while the specified coefficients A, B and, C can be set with respect to the purpose of measurement or dimensions of the fixtures. This allows you to read your measurements directly, without the need for conversions. *(Courtesy of Mitutoyo American Corporation)*

Comparator Instruments			
Name	**Alternate Name**	**Use**	**Discrimination**
Test Indicator	Dial test indicator	Setup in process checking	Not intended for measuring
Dial Indicator	Dial comparator, dial gage	Comparison gage measurement, alignment and positional measurement	0.01 to 0.002 mm 0.001 to 0.0001 in.
Comparator	Mechanical comparator, electronic comparator, air gages, and many are referred to by trade name	Comparison measurement of precise parts and for gage calibration	0.001 to 0.00002 mm 0.0001 to 0.00001 in.

FIGURE 10–13 These are the terms as generally used. They are ambiguous. For example, all are comparators, yet comparator alone usually refers to the higher-precision instruments. The test indicator and dial indicator are mechanical comparators, yet those terms are generally used for the more precise types only.

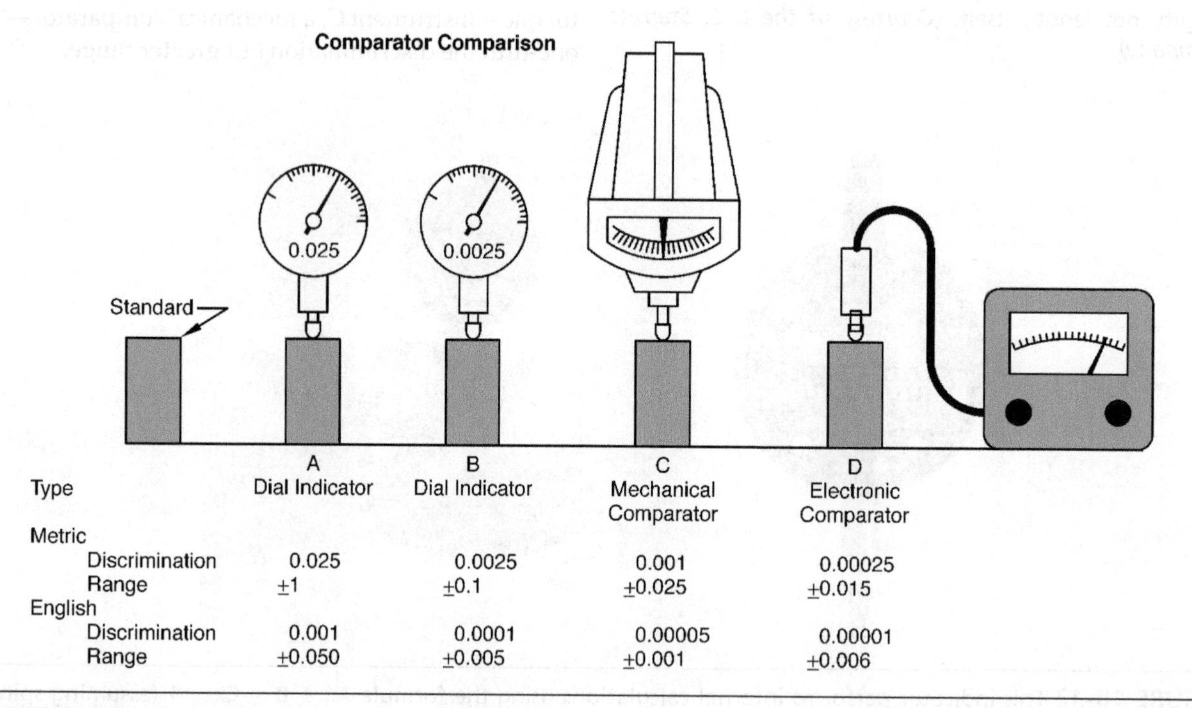

FIGURE 10–14 To grasp the significance of these comparisons, see Figure 10–15. (Note: Both C and D instruments are available in other ranges, D in multirange as well as digital models.)

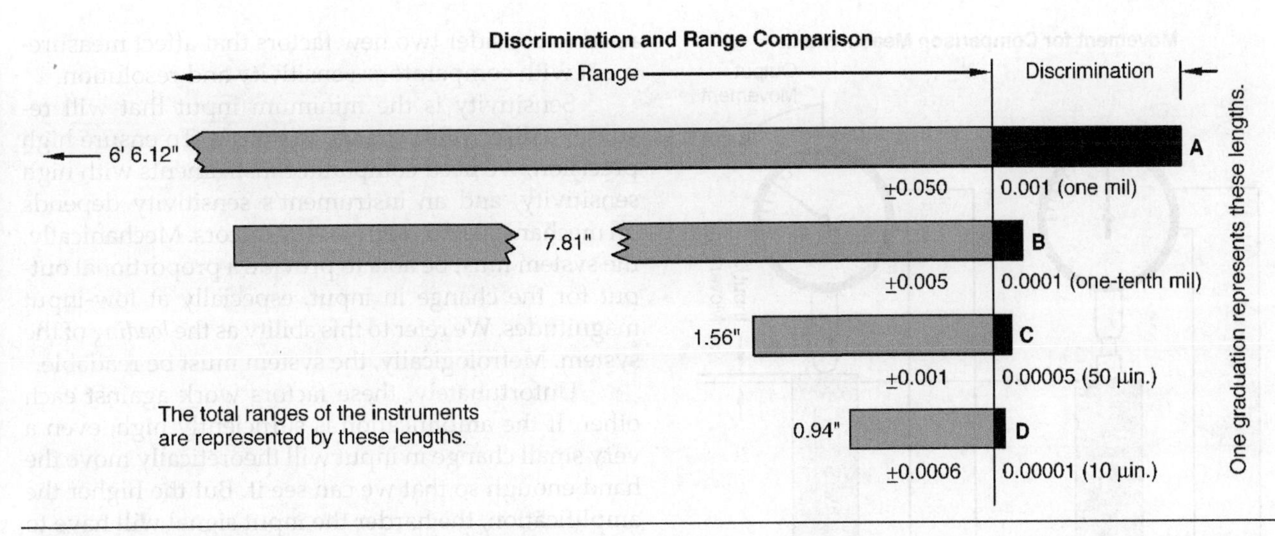

FIGURE 10–15 These bars, drawn to scale of 10 μin. to 1/64 in., show the tremendous differences between the comparators in Figure 10–14.

For the dial indicator in B, the discrimination is reduced another fifth, but the range is extended, drawn to the same scale; it is over 7 inches. The least precise instrument is the mil-reading dial indicator. When we lower the amplification of this instrument, we not only reduce the discrimination to 1 mil, but we also push the range out to over 6 feet in comparison with the range of the other instruments. Again, we have demonstrated that in order to gain precision, we must sacrifice range. (We will discuss how our illustration may be misleading in relation to the length of the ranges later.)

Movement in Measurement

All measurement requires movement. Of course, movement can include correctly aligning a standard and a part feature, which applies to both direct measurement and comparison movement for each of the two methods of measurement. In direct measurement, you align the reference and measured points of the part feature with a scale. In cases like measurement with a rule, you can see the movement; your eyes count the minor graduations between the reference point and the next major division (see Figure 10–16). In the case of the vernier and micrometer instruments, you physically change the instrument to duplicate the distance being measured. You then read the change in the instrument from the scale or scales. In all of these cases, you are not really reading the movement; you are reading the results of the movement.

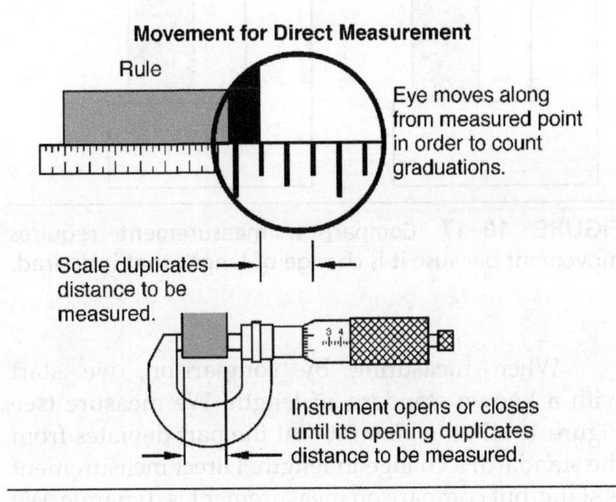

FIGURE 10–16 All measurement requires movement. Indirect measurement consists primarily of correlating the distance to be measured with the instrument's built-in standard, after which the scale is read.

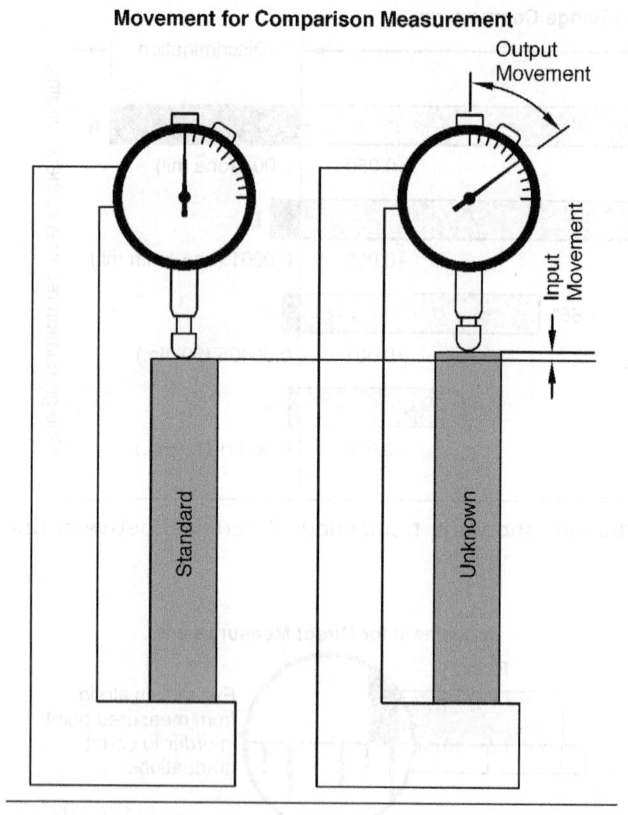

FIGURE 10–17 Comparison measurement requires movement because it is change of length that is desired.

When measuring by comparison, we start with a known standard of length. We measure (see Figure 10–17) the amount that the part deviates from the standard, a change in length. Direct measurement is static, but comparison measurement is dynamic. We refer to the ability of the instrument to detect and measure this change as the *sensitivity* of the instrument.

Sensitivity

Many comparator instruments have relatively low amplification and discrimination no greater than vernier or micrometer instruments. Comparators, on the other hand, are capable of very large amplification—up to 20,000X and more. At such amplification, we need to consider two new factors that affect measurement with comparators: sensitivity and resolution.

Sensitivity is the minimum input that will result in a discernible change in output. To ensure high precision, we need comparator instruments with high sensitivity, and an instrument's sensitivity depends on mechanical and metrological factors. Mechanically, the system must be able to provide a proportional output for the change in input, especially at low-input magnitudes. We refer to this ability as the *loading* of the system. Metrologically, the system must be readable.

Unfortunately, these factors work against each other. If the amplification is sufficiently high, even a very small change in input will theoretically move the hand enough so that we can see it. But the higher the amplification, the harder the input signal will have to work to push its way through all the gears, shafts, and springs to reach the hand.

Figure 10–18A shows an indicator with a very small input signal being applied to the spindle. It appears that nothing has happened, but actually the hand has moved so slightly we could not see it.

If we simply increase the amplification 10 times, we should be able to see the movement of the hand, but there actually is no improvement (see Figure 10–18B). To understand why we get this result, consider the hairspring, which loads the gear train against backlash. Part of the input is energy, which creates a torque to counteract the torque of the spring. Although the mechanics of the instrument provide a large swing of the hand, those mechanics also require an amplified force to counteract the torque of the spring. In Figure 10–18A, the force is multiplied 100 times by the gear train (assuming 4X amplification from the hand). When we add extra gearing (see Figure 10–18B), we must increase the force 1,000 times, and thus create additional "wind up" of the shafts and additional frictional loads on the bearings, rack, and gears.

The other two reasons why the hand did not move as we should have expected are inertia and compressibility. Therefore, we should consider other options to make the hand movement more noticeable. In Figure 10–18C, we did not increase the amplification over illustration A, but we can see the movement of the hand because of the instrument's improved resolution.

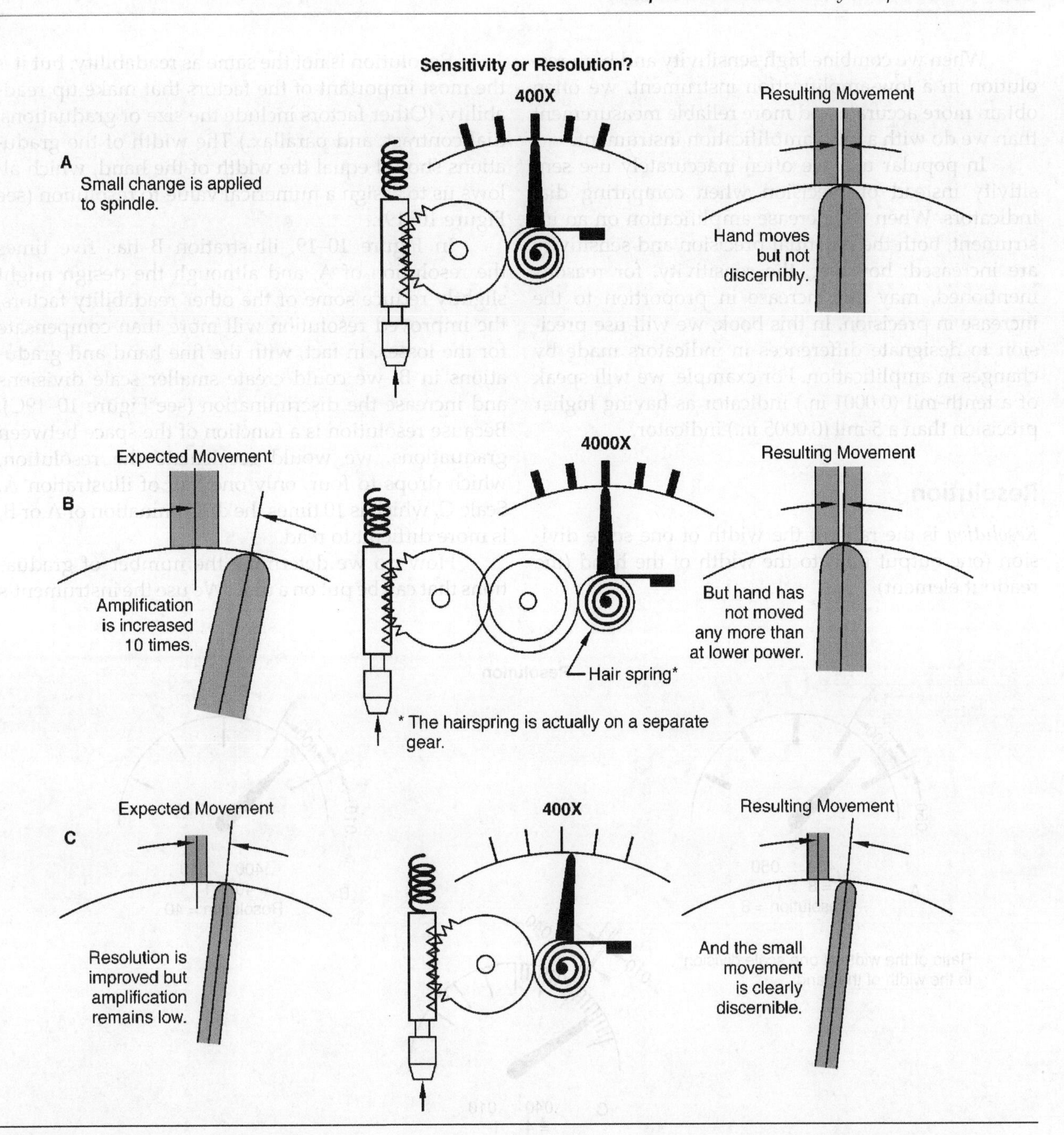

FIGURE 10–18 Increasing amplifications may not be as effective for improving sensitivity as better resolution.

When we combine high sensitivity and high resolution in a low-amplification instrument, we often obtain more accurate and more reliable measurement than we do with a high-amplification instrument.

In popular use, we often inaccurately use sensitivity instead of precision when comparing dial indicators. When we increase amplification on an instrument, both the potential precision and sensitivity are increased; however, the sensitivity, for reasons mentioned, may not increase in proportion to the increase in precision. In this book, we will use precision to designate differences in indicators made by changes in amplification. For example, we will speak of a tenth-mil (0.0001 in.) indicator as having higher precision than a 5-mil (0.0005 in.) indicator.

Resolution

Resolution is the ratio of the width of one scale division (one output unit) to the width of the hand (the readout element).

Resolution is not the same as readability, but it *is* the most important of the factors that make up readability. (Other factors include the size of graduations, dial contrast, and parallax.) The width of the graduations should equal the width of the hand, which allows us to assign a numerical value to resolution (see Figure 10–19).

In Figure 10–19, illustration B has five times the resolution of A, and although the design might slightly reduce some of the other readability factors, the improved resolution will more than compensate for the losses. In fact, with the fine hand and graduations in B, we could create smaller scale divisions and increase the discrimination (see Figure 10–19C). Because resolution is a function of the space between graduations, we would recalculate the resolution, which drops to four, only one half of illustration A. Scale C, which is 10 times the discrimination of A or B, is more difficult to read.

How do we determine the number of graduations that can be put on a dial? We use the instrument's

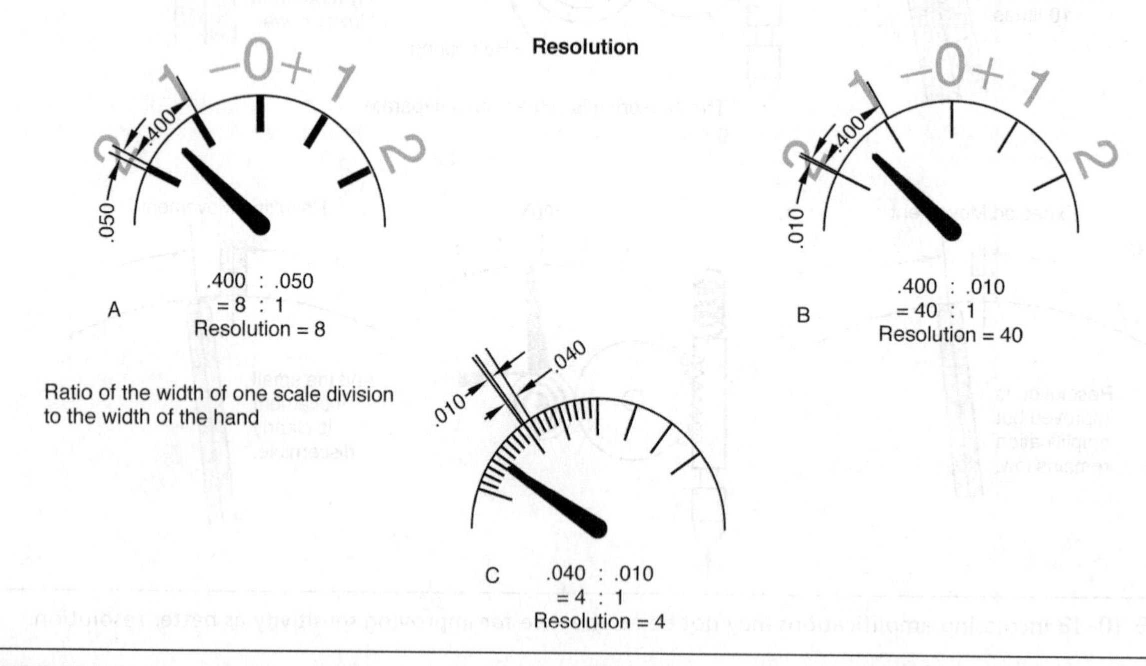

FIGURE 10–19 The larger the resolution factor (without impairment of other readability factors), the greater the reliability.

precision, which is the measure of the dispersement of repeated measurements. Remember, precision is essential for accuracy, but it is not a measure of accuracy. Precision and accuracy are both essential for reliability, but they are not the only factors in a measure of reliability.

Figure 10–20A is an enlargement of one division from Figure 10–19A, but we show five readings instead of one. Although the precision of the instrument leaves something to be desired, we would clearly associate all five readings with only one graduation. When an instrument is highly readable, the observed reading is also highly reliable.

In Figure 10–20B, the scale has 10 times the discrimination, but we need all five readings to reliably choose 0.0019 in. as the reading. How reliable would our measurement be if we only had one reading to go by?

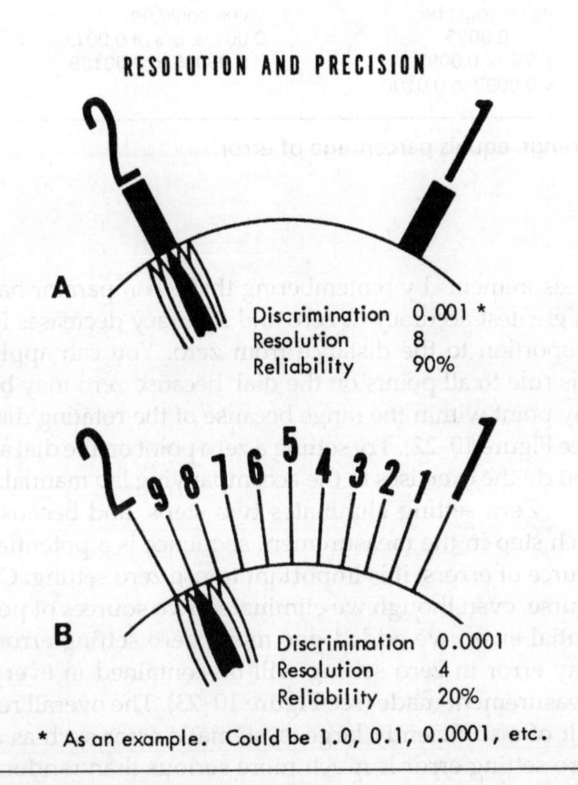

FIGURE 10–20 High resolution will not substitute for high precision.

Although discrimination and precision are poor, they may still be usable, but when we demand higher precision, other measurement factors must also increase. Generally, resolution increases with an increase in amplification and/or discrimination. If your instruments have low resolution, you should also suspect that they do not have the high precision, accuracy, sensitivity, or reliability that much of your measurement requires.

Comparator Accuracy

We use the terms *repeat accuracy* and *calibrated accuracy* with indicators. Although repeat accuracy is generally used, we should accurately use the term *repeatability.* Calibrated accuracy is redundant: We can only prove the accuracy of an instrument through calibration. We simply should use the term *accuracy.*

When speaking of comparators, accuracy is not stated in terms of parts of a unit of measurement, as two-tenth mil per inch (0.0002 in. per in.). It is stated in terms of dial graduations. Frequently, accuracy for dial indicators is stated as plus or minus one dial division with repeatability within one-quarter of one dial division.

These expressions of accuracy and precision are inaccurate because they fail to take into consideration that the accuracy decreases as the distance traveled increases. Therefore, we should state accuracy as a percentage of the full-scale range (see Figure 10–21).

In Figure 10–21, we used a comparator with a discrimination of 1 mil (0.001 in.) and a range of 20 mil (0.020 in.). In A, the comparator is zeroed on a 1.0000 in. gage block standard. In B, we replaced that standard with a 1.0200 in. block in order to move the hand through the full range. Instead of a correct 1.0200 in. reading, the comparator reads 1.0190 in.—1 mil (0.001 in.) short. If we take the error (0.001 in.), divide by the range (0.020 in.), and multiply the result by 100, we can determine the percentage of error: 5% error in this example.

This percentage is error, not accuracy. Comparator accuracy is not a function of the percentage of error, no matter how often a manufacturer states that an

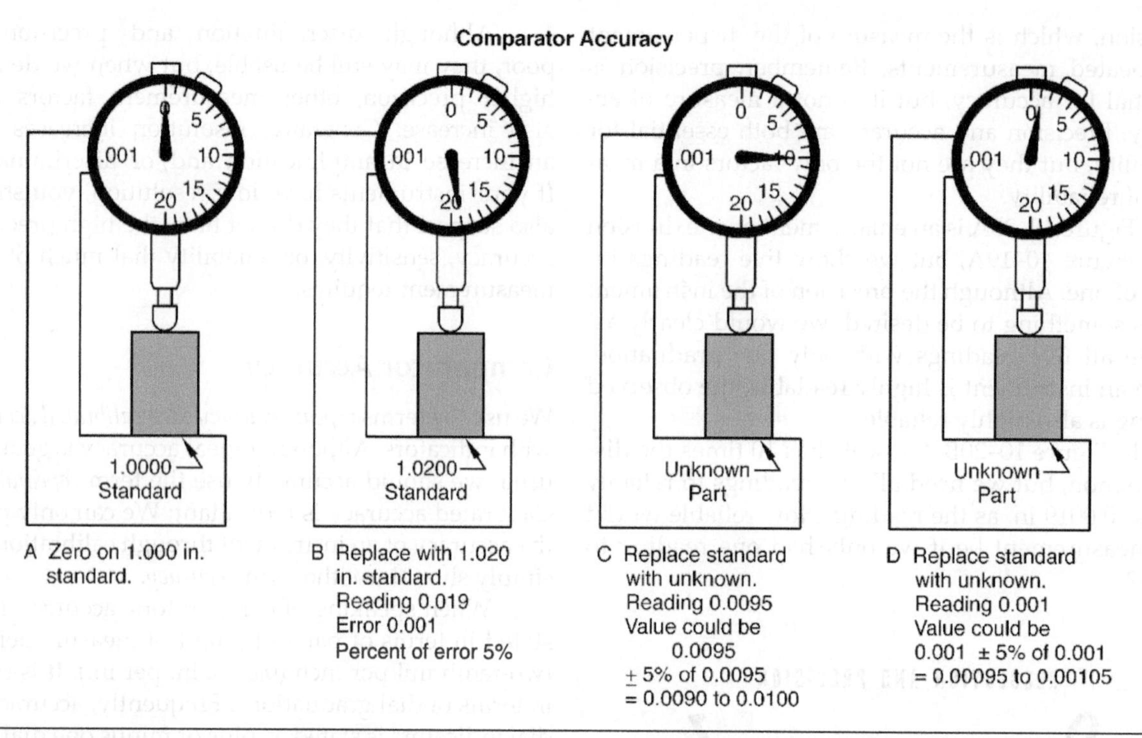

FIGURE 10–21 Measurement error divided by measurement range equals percentage of error.

instrument has "an accuracy of 62% of full scale." The American Standards Association states:

> Accuracy of an instrument is the limit, usually expressed as a percentage of full-scale value, which errors will not exceed when the instrument is used under referenced conditions.

In Figure 10–21A and B, the error is 5%; in C and D, we substituted unknown parts for the standard so you could see the advantage of basing accuracy on a percentage. We need to make a substantial adjustment in C, because the distance measured is so large. In D, where the distance measured is small, we only need to make a small correction.

By using the percentage method, we can see that the error in the measurement decreases faster than the distances; theoretically, at zero distance, we have zero error. We can improve the reliability of comparator measurements by remembering that a comparator has its greatest accuracy at zero and accuracy decreases in proportion to the distance from zero. You can apply this rule to all points on the dial, because zero may be any point within the range because of the rotating dial (see Figure 10–22). Try setting a zero point on the dial as you do the exercises in the accompanying lab manual.

Zero setting eliminates two steps, and because each step in the measurement sequence is a potential source of errors, it is important to use zero setting. Of course, even though we eliminated two sources of potential error, we added one more: zero setting error. Any error in zero setting will be contained in every measurement made (see Figure 10–23). The overall result of a sufficiently large, systematic error such as a zero setting error is much more serious than random error, because it applies to every measurement made until the error is detected and corrected.

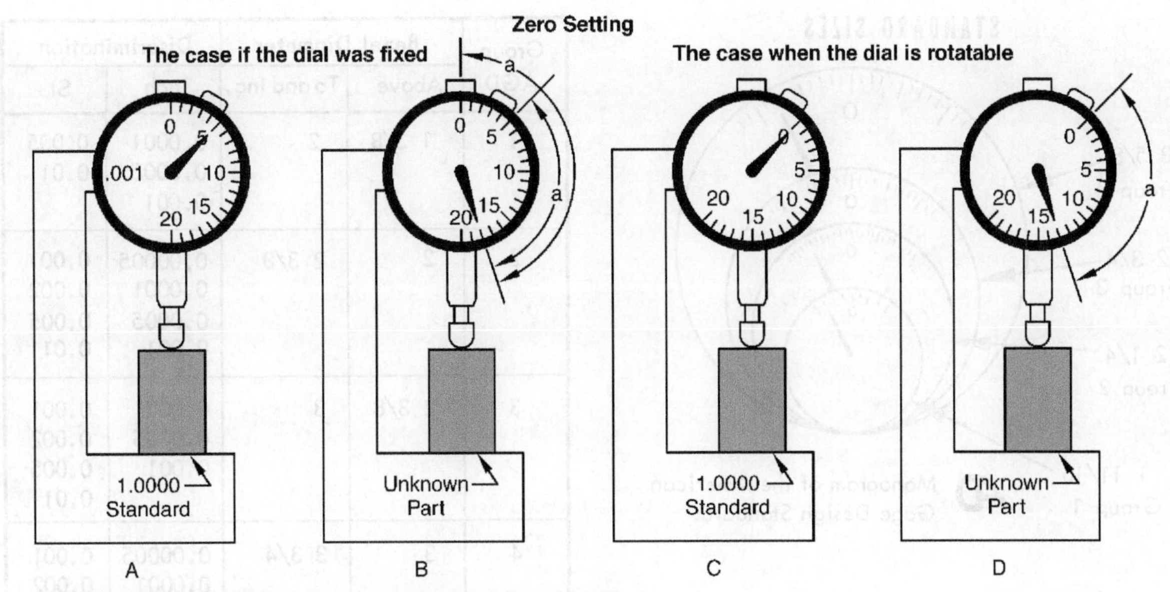

A. In this case, the 1.0000 standard provides a reading of 0.005.
B. The unknown (a) reading is 0.018. The difference is 0.013. Adding 0.013 to 1.0000 gives the length of the unknown as 1.013.
C. In this case, the dial is turned to provide a zero reading on the 1.0000 standard.
D. The unknown part provides a 0.013 reading directly. The part size is 1.013. There is no question that it is the travel of the hand to which the error correction applies.

FIGURE 10–22 The dial rotation permits zero setting. This eliminates computational errors.

Zero Setting Adjustment

Errors Eliminated:

1. Memory error
2. Computational error

These are random errors and **will not** be contained in all measurements.

Error Added:

Zero setting error

This is a systematic error and **will** be contained in all measurements.

FIGURE 10–23 Every act in measurement is a source of error. When the errors created are less than those reduced, the act is justified. These are the errors added and eliminated by zero setting.

Indicator Dials

In order to begin a general discussion of comparator instruments, we need to consider scales first. The scales for dial indicators, called *dials,* are quite different from those of other comparator instruments.

Dials, dimensions, and mountings have been standardized for dial indicators. Standard dials have four discriminations: *mil* (0.001 in.), *five-tenth mil* (0.0005 in.), *two hundred fifty microinch* (250 µin.), and *one hundred microinch* (100 µin.). With these specific discriminations, dial indicators must be made in a range of amplifications. Closely related to discrimination are dial size (see Figure 10–24) and range. If we need a certain spacing for readability, we also need a larger dial and longer range. Conversely, the higher the amplification, the higher the possible discrimination—if the dial is large enough that we can read it (see Figure 10–25).

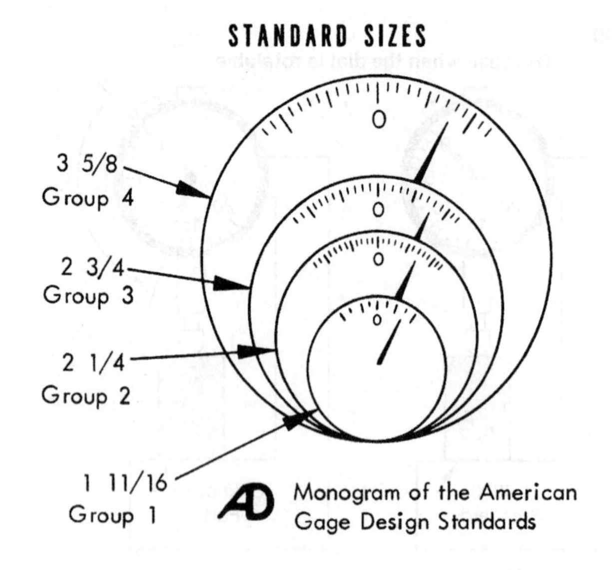

Group AGD	Bezel Diameter		Discrimination	
	Above	To and Inc.	Inch	SI
1	1 3/8	2	0.0001 0.0005 0.001	0.005 0.01
2	2	2 3/8	0.00005 0.0001 0.0005 0.001	0.001 0.002 0.005 0.01
3	2 3/8	3	0.0001 0.0005 0.001	0.001 0.002 0.005 0.01
4	3	3 3/4	0.00005 0.0001 0.0005 0.001	0.001 0.002 0.005 0.01

FIGURE 10–24 These four standardized dial sizes are available in a wide variety of graduations.

Design and Metrological Factors

Have the following permissible effect on these:

An increase in this factor:	Amplification	Dial Size	Discrimination	Readability	Range
Amplification		Increases	Increases	Increases	Decreases
Dial Size	Decreases		Increases	Increases	Increases
Discrimination	Increases	Increases		Decreases	Decreases
Readability	Increases	Increases	Decreases		Decreases
Range	Decreases	Increases	Increases	Increases	

FIGURE 10–25 Standard indicators compromise these factors for the greatest range of practical applications.

There are two general types of dials: *balanced* and *continuous.* Continuous dials have graduations starting at zero that extend to the end of the recommended range (see Figure 10–26), rotating both clockwise and counterclockwise. These dials correspond to the unilateral (one-way) tolerances used in dimensioning. Balanced dials have graduations that run in both directions from zero. These dials correspond to the use of bilateral (plus and minus) tolerances; generally, comparators with high amplification only have balanced dials.

TYPES OF DIALS

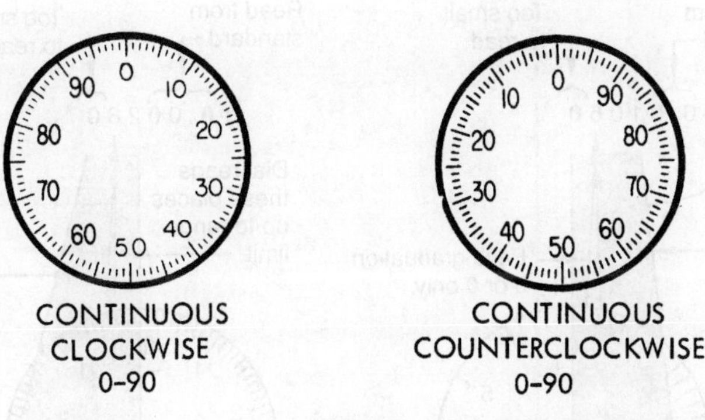

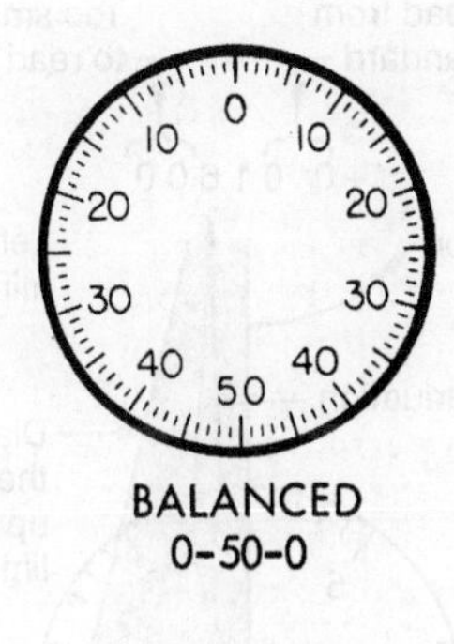

METHOD FOR DESIGNATING NUMBERS

FIGURE 10–26 These are the three types of dials used on dial indicators and the method for designating numbers.

Balanced Indicator Dials

A large variety of dial sizes and graduations are available for both balanced and continuous dials. The general principles of reading balanced dials are shown in Figure 10–27. The value of the smallest division—not the value of the numerals around the dial—is always shown on the dial to help you in counting. The value usually represents 10 divisions, but it also may represent 2, 5, or 20 divisions around the dial.

When you read a balanced dial indicator, you are reading, but the indicator is comparing. The indicator reading in itself means nothing. To have meaning, the reading must be added to or subtracted from the size of the standard to which the instrument has been calibrated.

Revolution counters probably do not have a place on balanced indicators, even though some indicators are manufactured with them. The conventional indicator has a total travel of 2½ turns. The hand can revolve completely through the plus range, backward through the minus range, and into the plus range again. The revolution counter, or "telltale," warns you when this overrotation has happened.

However, this warning is often misinterpreted. When you read the revolution counter correctly, you are actually deciding whether you are using the correct instrument for the measurement, with two exceptions: when the instrument has to travel a long distance to get to the point from which the measurement is referenced, and when you measure the starting tension of the instrument.

We still recommend the use of revolution counters on balanced indicators, but only when they are used as warning signals. You should use them to show that something in the setup has shifted, that something is radically out of control, or that you originally chose an overly sensitive indicator for the measurement (see Figure 10–28).

Continuous Indicator Dials

A continuous dial has double the range in one direction of an equivalent balanced dial. When you need even greater range, indicators are available with very long racks that pass completely through the cases (see Figure 10–29). You read these indicators much the same way you read balanced scale indicators, but there are no minus readings and larger numbers (see Figure 10–30).

In some circumstances, the great ranges of the continuous dial indicators are extremely valuable. Unfortunately, the same long ranges mean that these indicators are often used for the wrong purposes. You have to remember that these instruments are

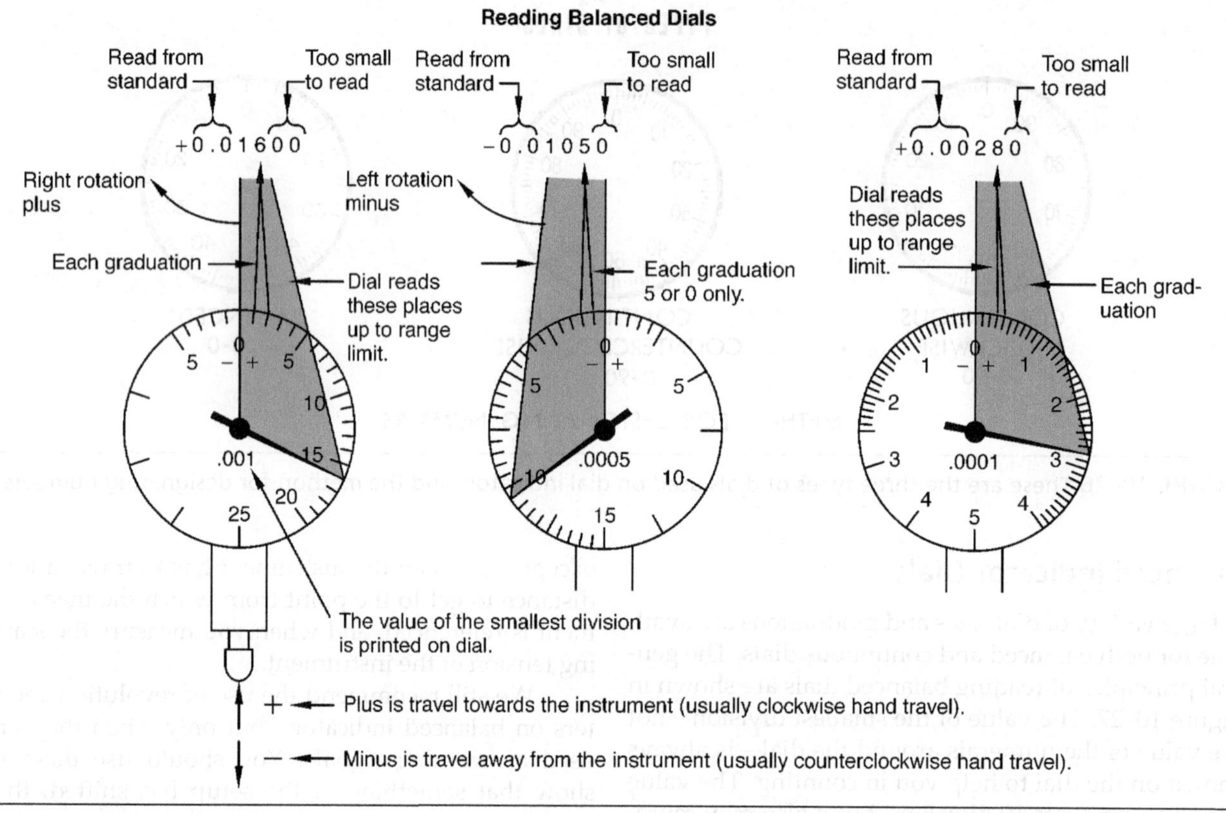

FIGURE 10–27 Although the dials may vary in size and graduations, these general principles apply.

Revolution Counters on Balanced Dial Indicators

Not Reliable for measurement.

Reliable for:

1. Showing that you are past the overtravel and within the measurement range.
2. Warning that an indicator with too high an amplification is being used.
3. Warning that something in the setup has shifted.
4. Warning that the part being measured is **out of control**.*

*Out of control means that the previous operation has been completely missed or that the machine is completely out of adjustment.

FIGURE 10–28 Balanced dial indicators are reliable for relative or comparative measurement.

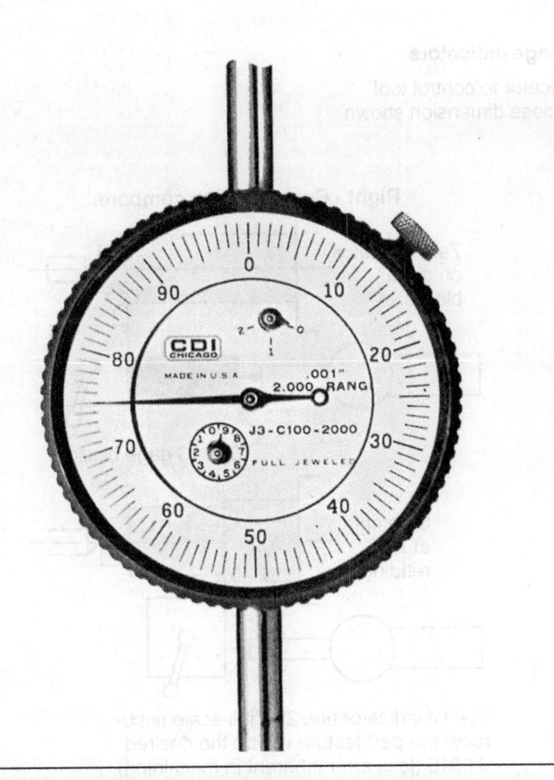

FIGURE 10–29 Long-range indicators have revolution counters, and very long-range ones have inch counters. *(Courtesy of Chicago Dial Indicator Co.)*

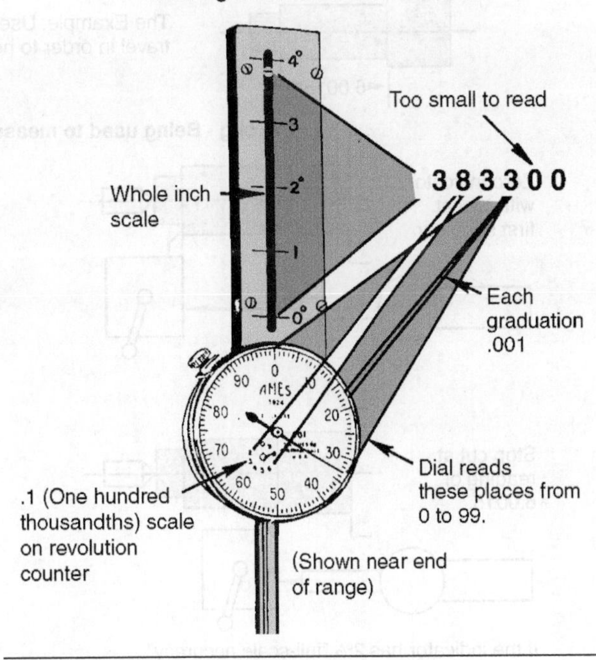

FIGURE 10–30 Continuous dial indicators have ranges up to 10 in. *(Courtesy of B. C. Ames, Inc.)*

comparators, not direct-measurement instruments, and that there is a relationship between the percent of full-scale accuracy and the length traveled (see Figure 10–31).

In Figure 10–31, we simplified the use of a long-range indicator: It is controlling the tool travel on a lathe to cut a shoulder of accurate length. The popular, but wrong, use is shown on the left; after roughing cuts, the lathe faces the shoulder to define one end of the part feature. The tool is backed off to the finished diameter of the next cut, and the long-range indicator, which is attached to the lathe bed, is zeroed somewhere against the carriage. When the long cut is made, the indicator warns the machinist when he or she is approaching the desired measurement. Unfortunately, when you multiply the small percentage of error in the indicator by the long travel, you can see how serious the error can become: a 2% error could result in the part being off from 5.8815 to 6.1215 in. We get such bad results because the comparator was used as a direct-measurement instrument instead of as a comparator.

To properly use the long-range comparator (see Figure 10–31 right), withdraw the tool to the diameter of the next cut after the shoulder on the small diameter at the right end of the part has been turned. Instead of zero setting the indicator against the carriage, we insert gage blocks for the dimension between the spindle and the carriage. This dimension represents the desired position of the carriage when the tool is at the end of the cut. We zero the indicator, which permits the machinist to carefully follow the progress of the cut and stop the machine accurately.

In the latter example, any errors in the finished part are primarily a result of machining (overtravel of the carriage, and so forth), because the distance

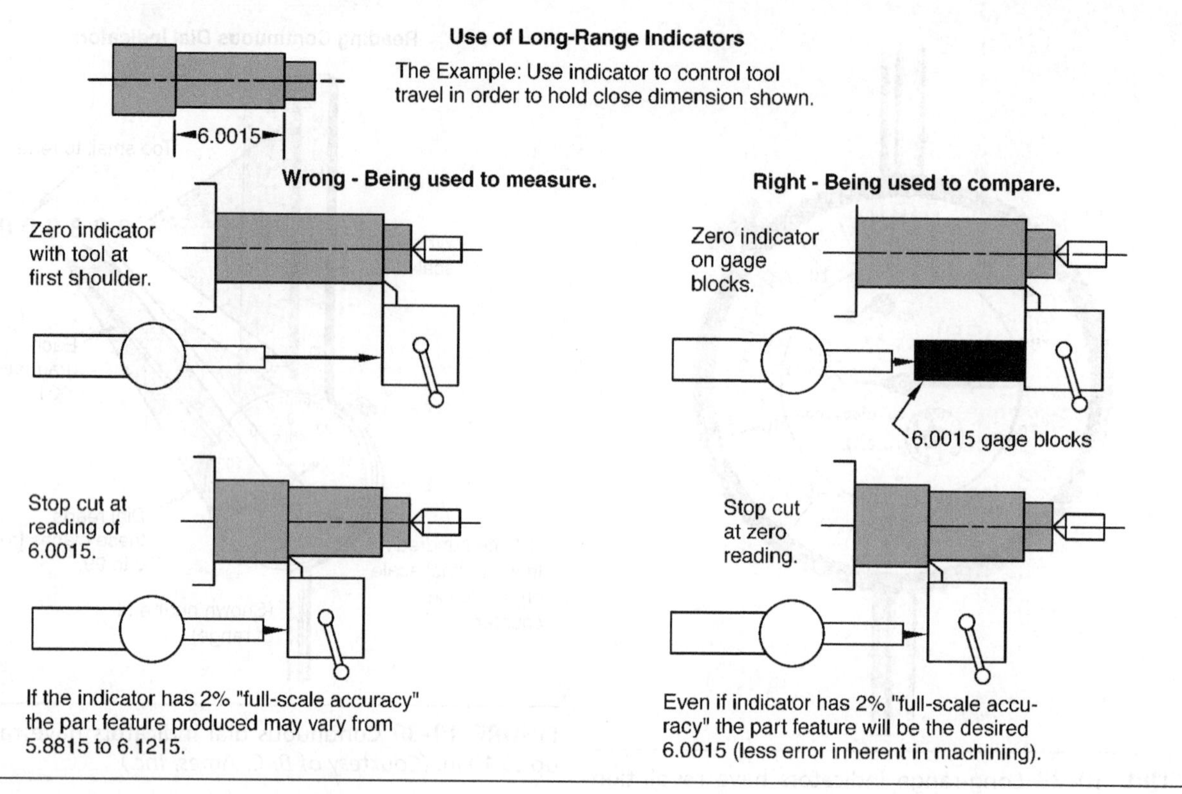

FIGURE 10–31 When used properly, the long-range indicator is not only convenient but reliable as well.

traveled by the lathe carriage was measured by the gage block standards. The indicator only showed when that distance was reached; therefore, we only used the indicator for its intended function—as a comparator.

Counterclockwise Dial Indicators

Our last example also shows how convenient continuous dials are because they read in either direction. In the example, a counterclockwise indicator is more convenient, and the so-called zeroing consists of setting the indicator to read the finished dimension when the gage blocks are in place. As the cut progresses, the indicator reads the distance traveled directly as shown.

Sometimes, a counterclockwise dial is a necessity (see Figure 10–32). The part feature we are measuring has a *unilateral tolerance* of 2.000 to 2.005 in. (2.000 + 0.005 in.). At A, a conventional indicator with a clockwise dial is zeroed on a 2.000 in. standard, and at B, the part that replaces the standard is at the high limit of the tolerance. As expected, the indicator reads +5 mil (0.005 in.), which we add to the original setting as established by the standard to provide the measurement of 2.005 in.

How do we work within the same tolerances (see Figure 10–32C, D, and E) if the shape of the part is very different? Theoretically, the phantom indicator at C could be exactly the same as in A and B, so the readings also would behave the same. Of course, you cannot punch holes in parts in order to measure them; therefore, we run the indicator in from the bottom (see Figure 10–32D). The bezel permits the dial to be turned, and, as far as the indicator is concerned, it does not matter how much the dial is turned. The dial position at D would be normal, and the zero setting is correct.

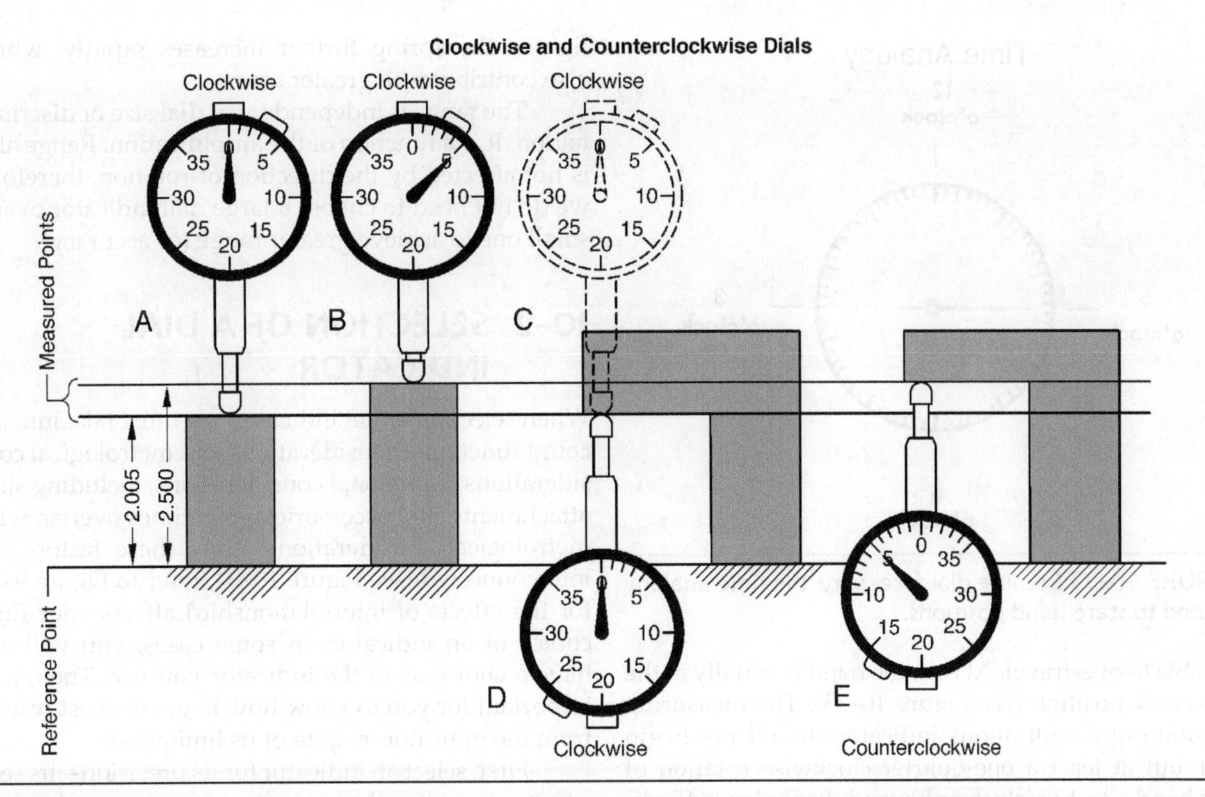

FIGURE 10–32 The dial rotation depends on the side of the measured point that the indicator is on.

One problem is that as the part feature becomes longer, the indicator travels backward so that the extreme reading would be 35 mil (0.035 in.) instead of 5 mil (0.005 in.). If you remember this "backward" rotation, you could simply calculate the proper reading, but the counterclockwise dial is a better solution. The hand in Figure 10–32E travels in the "normal" direction, but the scale is reversed. Now you can directly add the readings to the base dimension as at A and B.

You should use a *clockwise* indicator when the indicator is on the opposite side of the measured point from the reference point. You should use a *counterclockwise* indicator when the indicator is on the same side of the measured point as the reference point.

If you refer to Figure 10–31, you can see how these rules make measuring convenient. If you use the indicator the wrong way, the reference and measured points are interchangeable. When you do it the right way—with gage blocks—you clearly identify the reference point in the first step. Then it is easy to see that the indicator and the reference point both are on the same side of the measured point. Therefore, you should use a counterclockwise indicator.

Indicator Range

Before discussing indicator range, we need a method to describe the position of the hand on an indicator. Referring to the hand position by the reading is confusing because there are so many dial sizes and types. Therefore, we often state the position of the hand in relation to the position of the hands on a clock (see Figure 10–33).

Some indicators have long ranges, but most indicators have a working range of 2½ turns. The full range is even greater because the indicator needs to

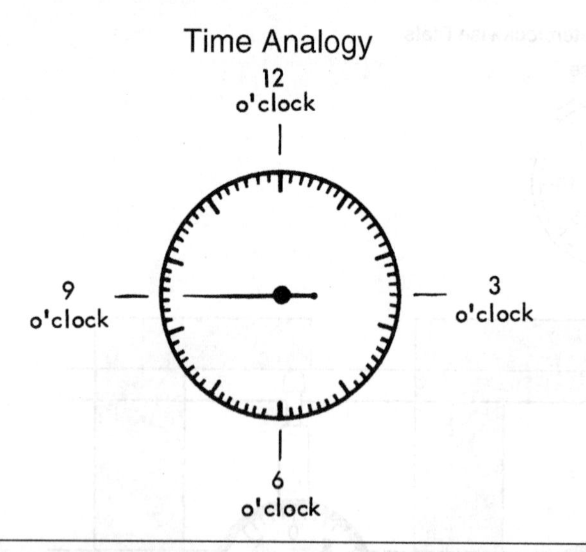

FIGURE 10–33 Because dial faces vary, the time analogy is used to state hand positions.

be able to overtravel. At rest, the hand is usually at the 9 o'clock position (see Figure 10–33). The measuring position of a continuous indicator should not begin without at least a one-quarter clockwise rotation of the hand—to 12 o'clock or beyond. For balanced indicators, the measuring should not begin until the hand has been rotated to the center of its working range—or 1¼ revolutions. Initial travel is independent of the zero position on the dial.

Usually, overtravel is a safety precaution for the occasional measurement that may run beyond the scale. For dial indicators, overtravel is important because it preloads the mechanical movement for starting tension. (In Figure 10–18, the hairspring preloads the gear train to take up backlash.)

If you examine Figure 10–10 closely, you can see that this spring is actually on a separate large gear and that it is in mesh with a small gear on the spindle. Only a small amount of spring tightening is required for the entire range, which provides nearly uniform torque for the mechanism.

To get within the range of uniform torque and correspondingly uniform preloading, we need overtravel, but the spring also is one factor that limits the indicator range. For any given proportion of spring, we reach a point after which the force required to tighten the spring further increases rapidly, which also contributes to greater errors.

The range is independent of dial size or discrimination. It is a function of the amplification. Range also is not affected by the direction of rotation; therefore, we do not need to choose a large dial indicator over a small one to achieve greater range (or accuracy).

10–2 SELECTION OF A DIAL INDICATOR

When selecting a dial indicator, we must take into account functional considerations and metrological considerations. Functional considerations, including size, attachments, and accessories, sometimes overlap with metrological considerations. How these factors are interconnected (see Figure 10–34; refer to Figure 9–25 for the effects of interrelationship) affects your final choice of an indicator. In some cases, you will not have a choice as to the indicator you use. Then, it is important for you to know how to get the best results from the indicator in spite of its limitations.

First, select an indicator for its precision—its sensitivity, magnification, and/or minimum graduation. In a dial indicator, precision is measured by discrimination and refers to four general classifications: one mil (0.001 in.), five-tenth mil (0.005 in.), two hundred fifty microinch (250 μin.), and one hundred microinch (100 μin.) (see Figure 10–35).

In general, you should select a comparator instrument with the highest precision for the range required, whether the comparator is mechanical or electronic. Select the shortest range that contains the tolerance: the shorter the range, the higher the permissible amplification, and the higher the precision.

For dial indicators, we use a special selection method (see Figure 10–36) because dial indicators lose accuracy with relatively long travel, and they lose sensitivity at high amplifications. From the available precision classes (minimum graduations), select the indicator that most nearly spreads the tolerance over 10 dial divisions—the discrimination should be 10% of the tolerance. In Figure 10–36, the tolerance is 6 mil (0.006 in. or 60.003 in.), so the discrimination should be five-tenth mil (0.0005 in.)

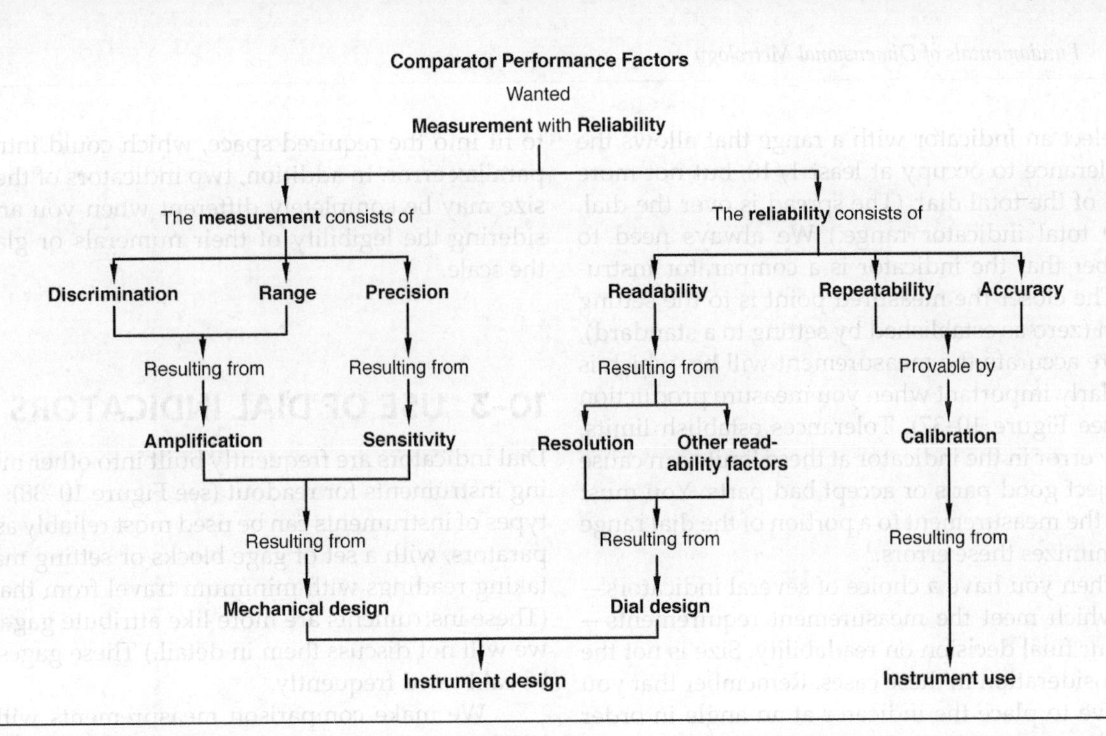

FIGURE 10–34 Although oversimplified, this chart shows the maze of interconnecting relationships that governs dial indicator design, selection, and use.

Dial Indicator Ranges

Numbering		Class of Precision Instrument				
B*	C**	0.001	0.0005	0.00025	0.0001	0.00005
0-50-0	0-100	0.250				
0-25-0	0-50	0.125	0.125			
0-20-0	0-40	0.100	0.100			
0-15-0	0-30	0.075	0.075	0.075		
0-10-0	0-20	0.050	0.050	0.050		
0-5-0	0-10			0.025	0.025	
0-4-0	0-8				0.020	
0-2-0	0-4				0.010	

*Balanced **Continuous

Notes:

1. Based on 2 1/2 turns range
2. Same for all dial sizes
3. Same for counterclockwise rotation
4. See manufacturers' catalogs for ranges of long travel types

FIGURE 10–35 The range, as well as the sensitivity, must be considered when selecting an indicator.

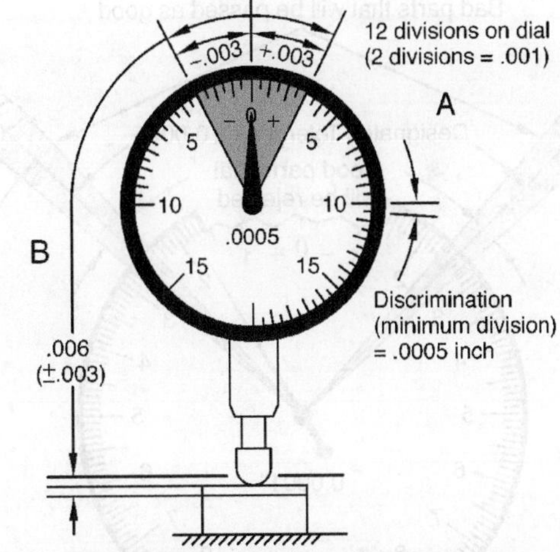

A. **Discrimination** – approximately 10% of tolerance (total tolerance should cover about 10 divisions).

B. **Range** – total tolerance more than 10% of dial and less than 25% of dial.

C. **Readability** – select largest dial that satisfies A and B and physically meet the requirements.

FIGURE 10–36 These steps ensure a practical selection for most measurement requirements.

Select an indicator with a range that allows the total tolerance to occupy at least 1/10, but not more than ¼, of the total dial. (The spread is over the dial, not the total indicator range.) We always need to remember that the indicator is a comparator instrument. The closer the measured point is to the setting position (zero as established by setting to a standard), the more accurate the measurement will be, which is particularly important when you measure production parts (see Figure 10–37). Tolerances establish limits, and any error in the indicator at these limits can cause us to reject good parts or accept bad parts. You must confine the measurement to a portion of the dial range that minimizes these errors.

When you have a choice of several indicators—all of which meet the measurement requirements—base your final decision on readability. Size is not the only consideration in these cases. Remember that you may have to place the indicator at an angle in order to fit into the required space, which could introduce parallax error. In addition, two indicators of the same size may be completely different when you are considering the legibility of their numerals or glare off the scale.

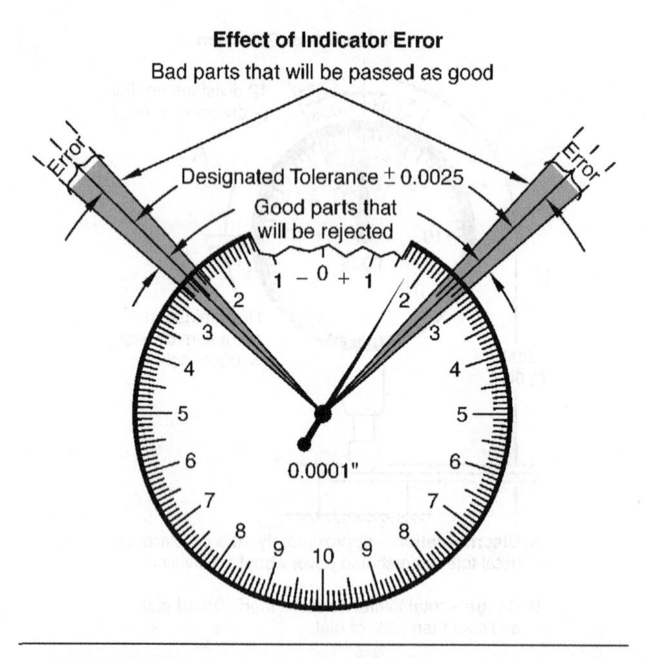

FIGURE 10–37 This exaggerated drawing shows that indicator error causes bad parts to be accepted and good parts to be rejected when used for production inspection.

10–3 USE OF DIAL INDICATORS

Dial indicators are frequently built into other measuring instruments for readout (see Figure 10–38). These types of instruments can be used most reliably as comparators, with a set of gage blocks or setting masters, taking readings with minimum travel from that size. (These instruments are more like attribute gages, and we will not discuss them in detail.) These gages must be calibrated frequently.

We make comparison measurements with dial indicators, most often surface plate measurements and inspections. (We will discuss positional relationships of comparator instruments later.)

Everything that we have said about cleanliness applies to dial indicators and is even more important. Whenever we increase the amplification of an instrument, the effects of dirt and vibration are correspondingly increased. A dial indicator may be a rugged instrument, but it is also a maze of moving parts. Keeping it clean and away from abrasive dust and cutting fluid extends its life and ensures its precision.

You must be careful when you are using a dial indicator because the slender spindle can be easily damaged. Avoid sharp blows and side pressure, and do not overtighten the contact points. Overtightening actually bulges the spindle sufficiently to interfere with its free travel in the spindle bearing. Clamping on the indicator stem may similarly cause binding. When you are stem mounting an indicator, it should be equipped with a standard threaded stem that screws into a hardened steel adapter.

Rigidity is also important to the proper use of a dial indicator, and you may have trouble realizing the "rubbery" action of steel until you measure to a tenth mil or closer. You can take precautions (see

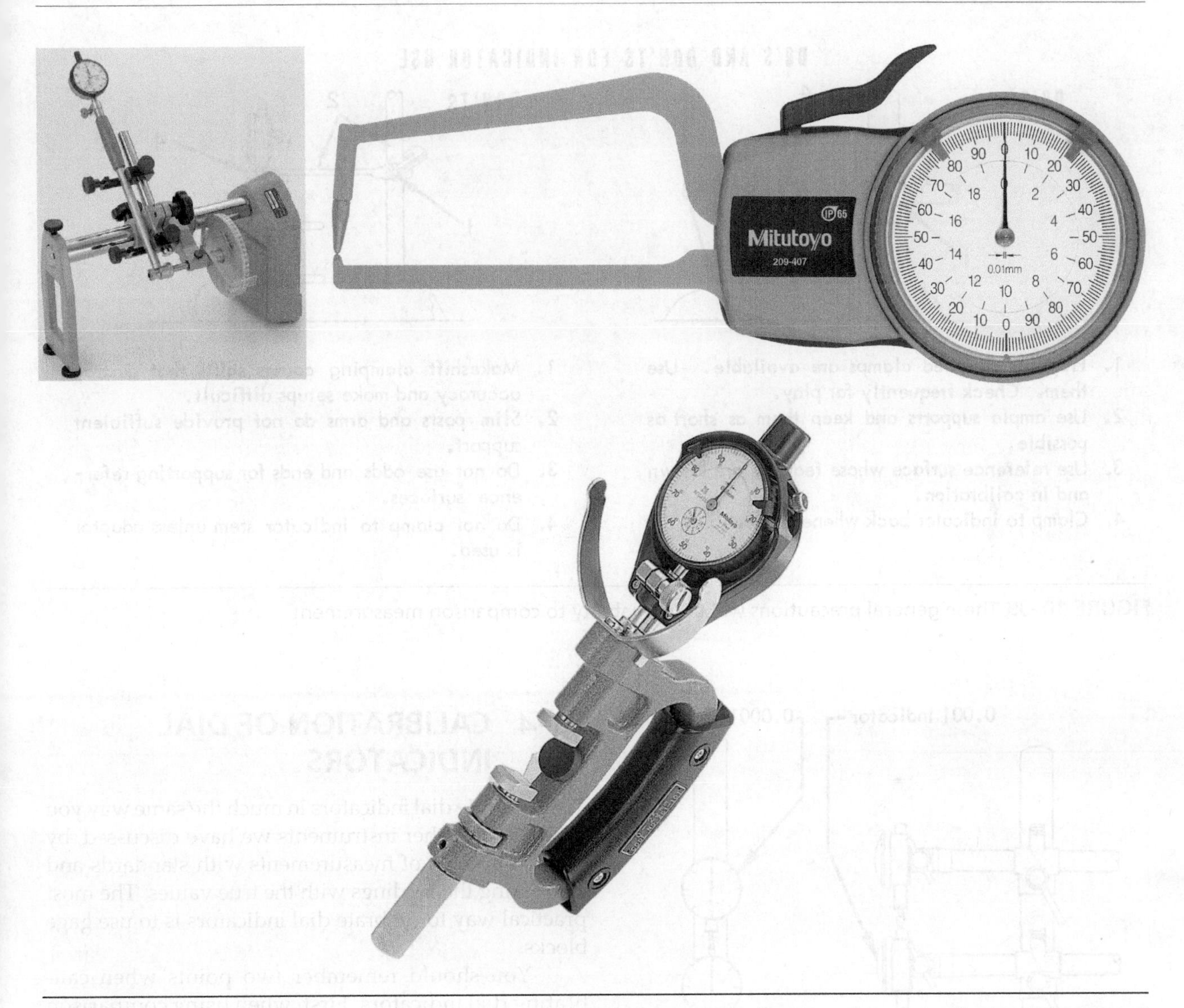

FIGURE 10–38 Dial indicators are often used as the readout portion of gages. They are reliable when treated as special forms of comparators rather than as direct measurement instruments. *(Courtesy of Mitutoyo American Corporation)*

Figure 10–39) when making comparison measurements. We can demonstrate the importance of rigidity using a standard indicator stand and accessories (see Figure 10–40).

We will not cover instructions for specific uses of dial indicators. Minor design differences can be important in use, so you would always consult the instrument manual.

DO'S AND DON'TS FOR INDICATOR USE

DO'S

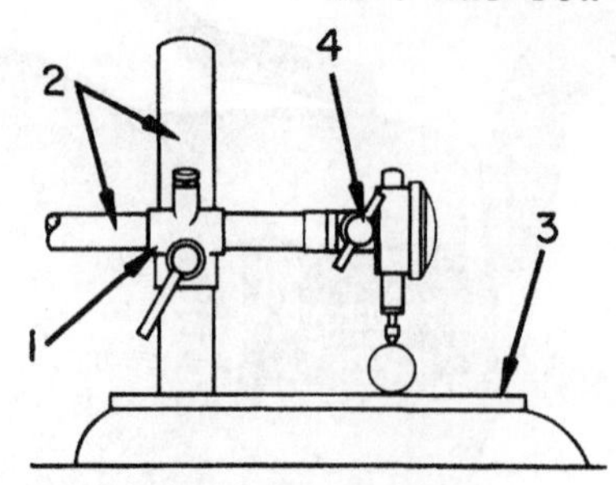

DON'TS

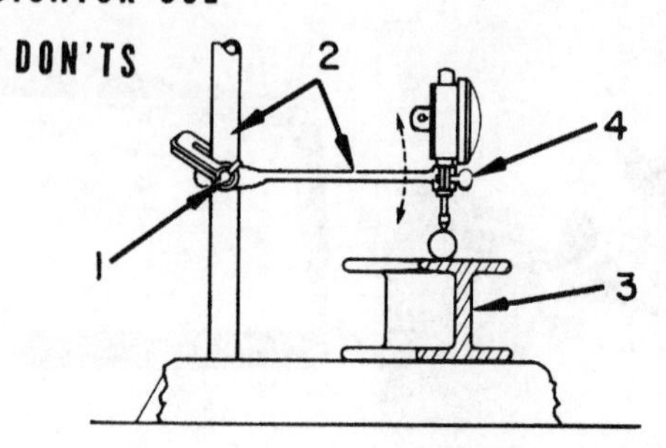

DO'S

1. Properly designed clamps are available. Use them. Check frequently for play.
2. Use ample supports and keep them as short as possible.
3. Use reference surface whose features are known and in calibration.
4. Clamp to indicator back whenever possible.

DON'TS

1. Makeshift clamping causes shifts that destroy accuracy and make setups difficult.
2. Slim posts and arms do not provide sufficient support.
3. Do not use odds and ends for supporting reference surfaces.
4. Do not clamp to indicator stem unless adaptor is used.

FIGURE 10–39 These general precautions will add reliability to comparison measurement.

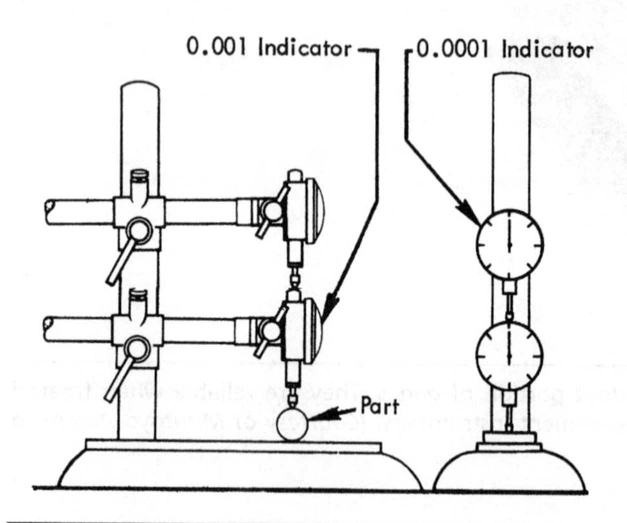

FIGURE 10–40 This simple demonstration will show vividly the deflection in gaging setups. First, zero the thousandths indicator on the part. Remove the part. Zero the tenths indicator on the thousandths indicator. Replace the part and notice the deflection on the tenths indicator. That is only part of the deflection because the tenths indicator is also being deflected.

10-4 CALIBRATION OF DIAL INDICATORS

You calibrate dial indicators in much the same way you calibrate the other instruments we have discussed, by making a series of measurements with standards and comparing the readings with the true values. The most practical way to calibrate dial indicators is to use gage blocks.

You should remember two points when calibrating dial indicators. First, when using comparison instruments, you must know that both the instrument and the standard by which they are set are in calibration. Both are required for reliable measurement. Second, you should try to duplicate the conditions under which the measuring is to take place as closely as possible for calibration. You can remove the indicator from the setup, return it to the gage department for calibration, and then replace it. It is a good start on calibration, but it is not a calibration of the complete setup, nor is it a calibration under the conditions of use.

10–5 ACCESSORIES AND ATTACHMENTS

There is a wide variety of modifications, attachments, and accessories for the instruments we discussed. Manufacturers' literature and operating instructions provide the best explanation of their uses and benefits.

Dial indicators are versatile instruments. We use their mountings to adapt them to many supports and can use them in many measurement situations because of their interchangeable contact points.

Contact points are available in a variety of hard, wear-resisting materials, including boron carbide, sapphire, and diamond. In general, we assume that a chromium-plated point lasts 10 times as long as a hardened steel point and that a carbide or harder point will last 100 to 1,000 times as long.

Points are available flat and rounded (see Figure 9–41). The pros and cons of flat points are shown in Figure 10–42. Spherical contacts, usually preferred to flat points, present point contact to the mating surface whether it is flat or cylindrical, as long as you take some precautions (see Figure 10–43). Excessive gaging pressure can indent the part. When cylindrical parts or standards are used, you must be careful to pass them through the spindle: the highest reading will be the diameter, and all other readings are chords.

Using contact points on spherical surfaces presents special problems (see Figure 10–44). You should only use a flat point unless the sphere is large and you take special care to find the highest reading. Even then, a flat point is unreliable, unless it is accurately square with the indicator spindle and parallel to the support surface. It can be difficult to attain this alignment, except when you use a comparator stand built for the purpose.

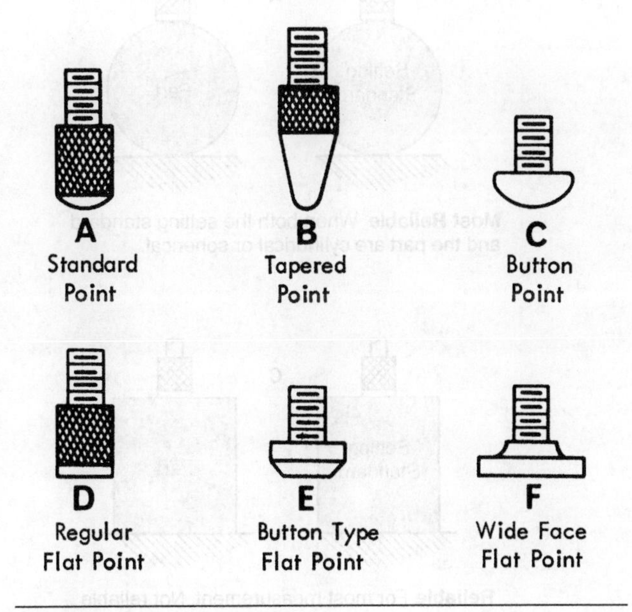

FIGURE 10–41 These standard contact points are interchangeable and available in a variety of wear-resisting materials.

Indicator Stands

The dial indicator, although important, is only one part of the measurement system, and really just a calibrated measured point. The accuracy of measurements depends as much on the positional relationships as upon the indicator. That is why indicator stands can be so useful.

There are basically two types of stands, which we loosely refer to as *loose joint* and *fixed travel*. When we use an indicator stand on a reference surface, we call it a *test stand* (see Figure 10–45). These stands are used on machine tools, on surface plates, on the parts being inspected, and even within the interior of mechanisms; their loose-joint construction gives them needed versatility. When we attach a reference surface to a dial, we call it a *comparator*.

The branch of mechanics that deals specifically with movement and position is called *kinematic mechanics* or simply *kinematics*. Although we do not use kinematic terms everyday in the shop, these terms do provide the simplest, clearest, and most concise way to discuss positional relationships in gaging.

In Figure 10–46, three axes have been constructed through a test stand. The left-to-right axis is conventionally called "x," the front-to-back axis "y," and the up-and-down axis "z." Every object potentially can rotate around these axes and move along the *axes*; however, they cannot rotate or move if they are constrained from doing so. The adjustments of the

Flat Contact Points

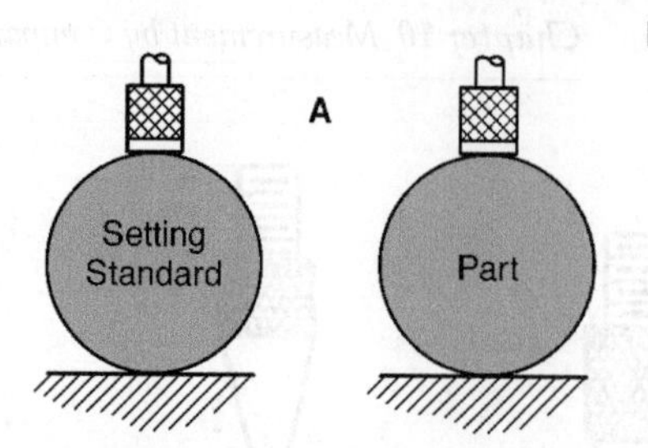

Most Reliable When both the setting standard and the part are cylindrical or spherical.

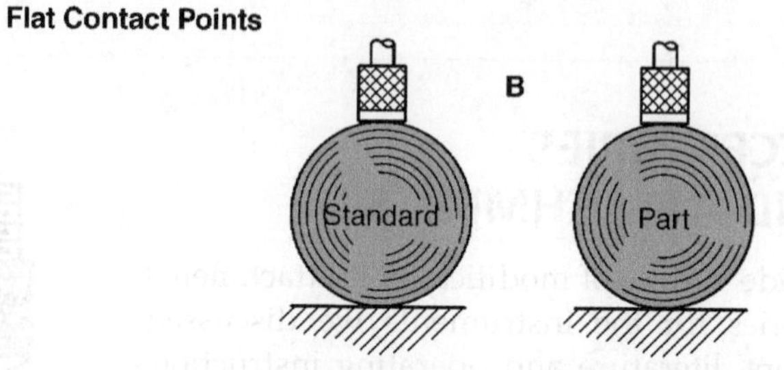

Reliable When either the standard and/or the part are spherical, providing that the contact areas are square to the spindles.

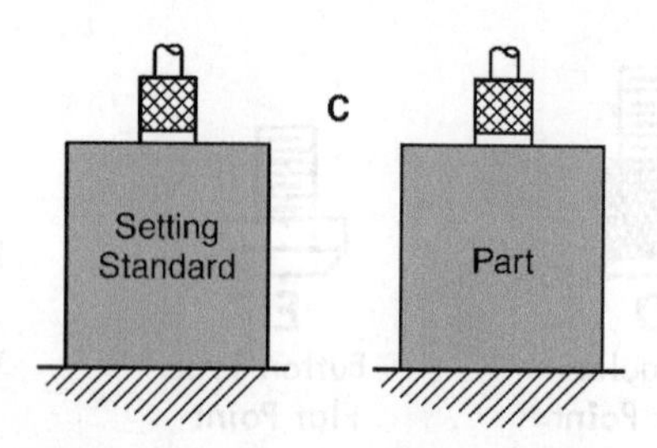

Reliable For most measurement. Not reliable closer than .0001 inch because of varying air and oil film.

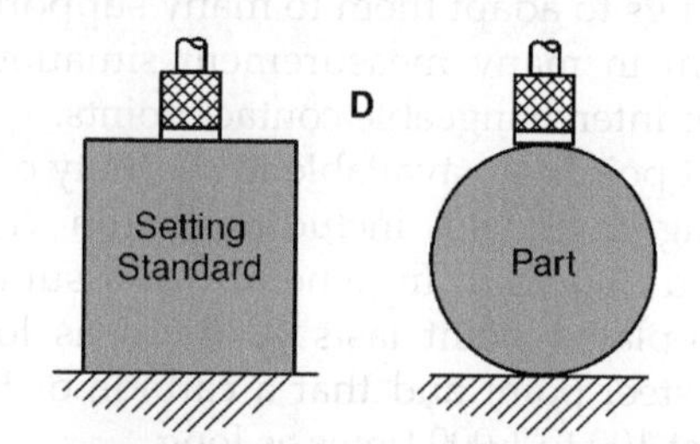

Least Reliable When zeroed on flat surface because of air film on flat and absence of film on cylinder.

FIGURE 10–42 Flat contact points have an important role but can be a serious cause of error when used improperly.

Spherical Contact Points

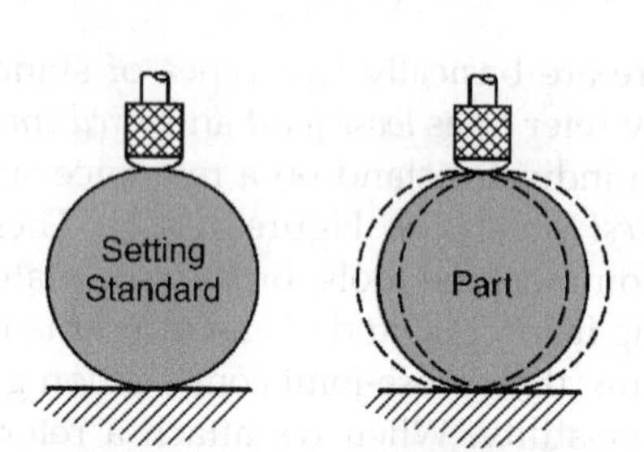

Reliable Providing both standard and part are passed completely under contact point and readings are taken at high points.

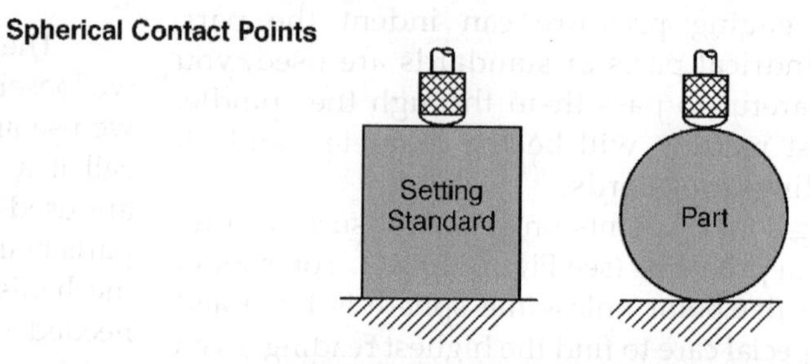

Reliable Point contact of spherical contact should rupture air and oil film on flat surface. Same with flat standard and flat part.

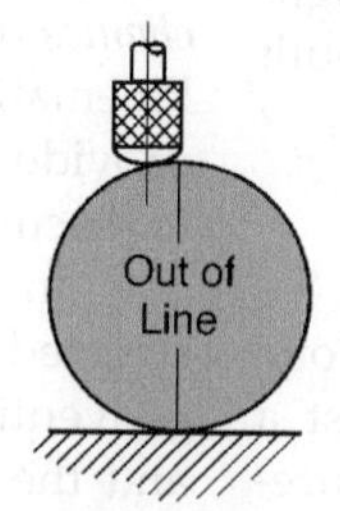

Not Reliable Sphere-to-sphere contact makes high point difficult to find. Out-of-line measurement from failure to pass cylinder under contact measures chord instead of diameter.

FIGURE 10–43 Excluding the measurement of balls, spherical contacts usually provide greater reliability than flat contacts. At very high amplifications, however, or with heavy gaging force, they present problems.

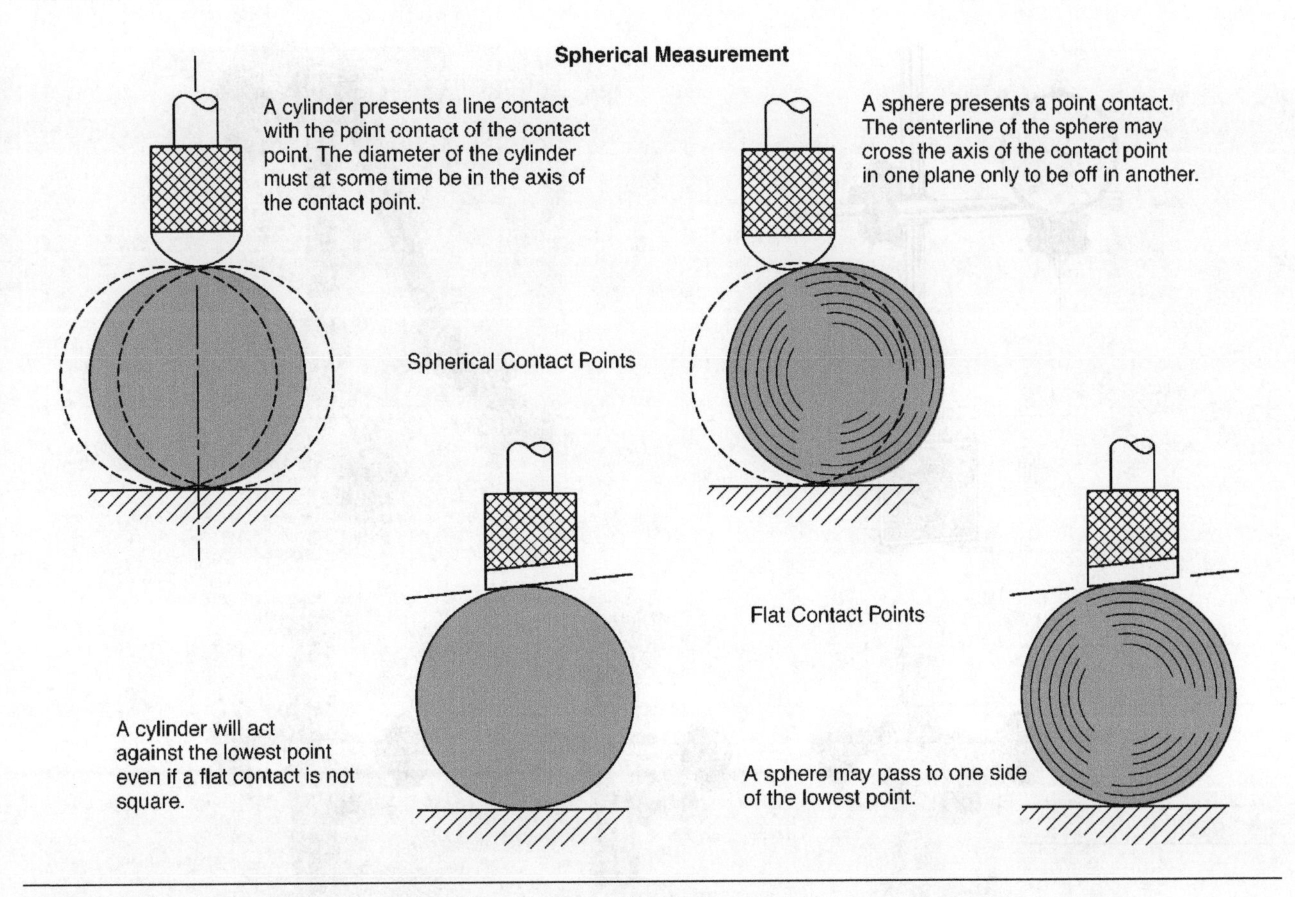

FIGURE 10–44 A sphere presents a problem even to a flat contact point.

test stand in Figure 10–46 allow the indicator to have all six of these degrees of *freedom*.

With these degrees of freedom in the indicator, the critical point (the measured point) also has six degrees of freedom. In order to make an accurate measurement, you must position and constrain the part and the indicator so there is only one degree of freedom remaining—that degree is the one we measure (see Figure 10–47A). If either the indicator or comparator is not sufficiently constrained or shifts during measurement, Figure 10–47B shows what happens.

So far, we have only considered two of the elements. If we add the standard and the reference surface, we have the complete measurement situation, and, of course, a range of errors (see Figure 10–48).

To eliminate possible misalignments, we can design them out, and then the test stand becomes the comparator. In the comparator, the reference surface is attached to the stand and thereby becomes part of it (see Figure 10–49). If the instrument has been satisfactorily manufactured, the positional relationship of the reference surface is no longer of concern.

The loose-joint support arrangement for the dial indicator permits the full range of movement. If the instrument is used only for comparison measurements, some of these movements are not needed. Any unneeded movement, even if provided with a lock, is a potential source of error (see Figure 10–49).

The kinematic considerations (see Figure 10–50) require that all movement within the instruments and stands be constrained except along the z axis. You

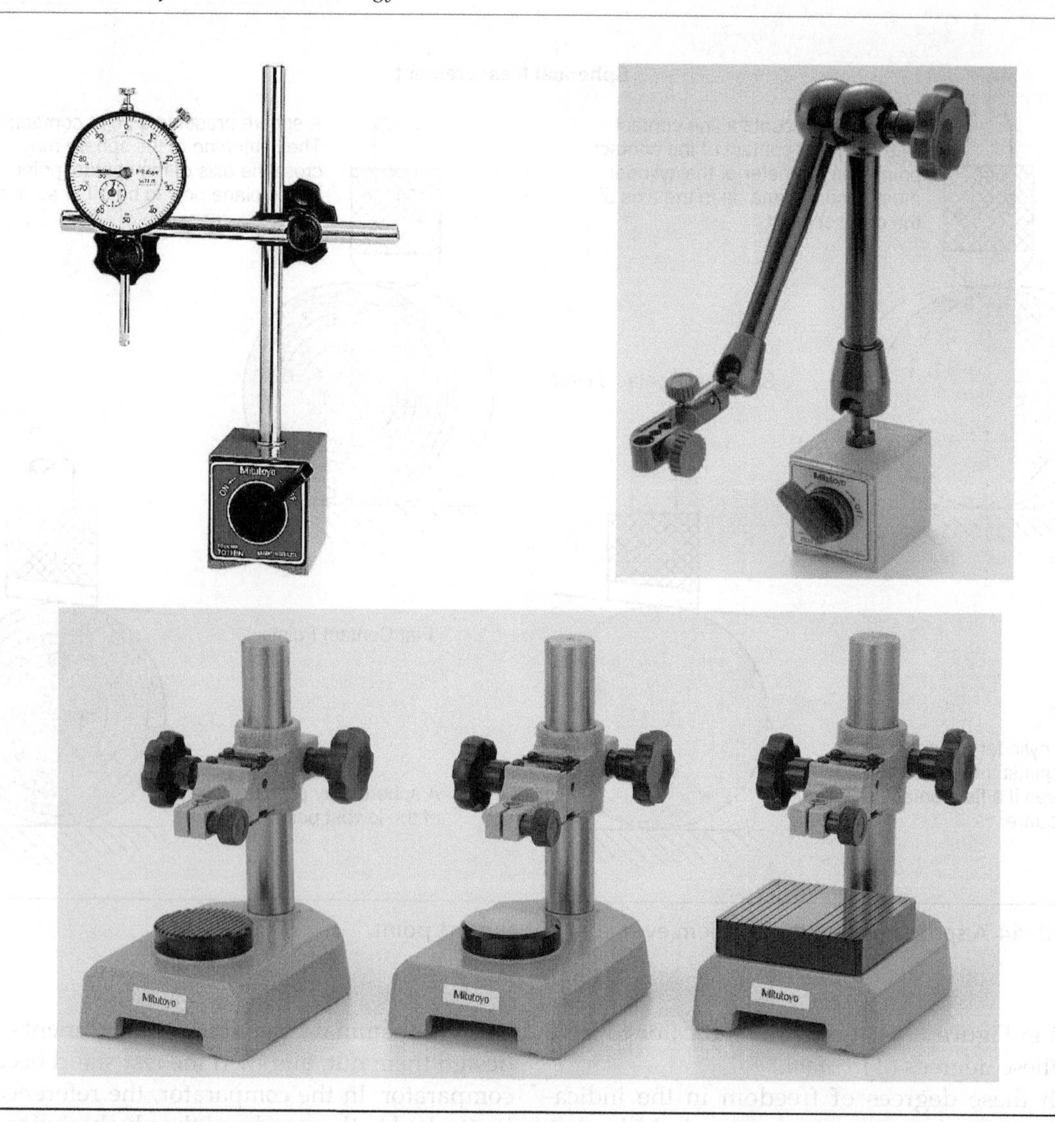

FIGURE 10–45 These indicator stands permit adjustment in all three axes. This gives them both versatility and unreliable positioning. *(Courtesy of Mitutoyo American Corporation)*

can raise or lower the head in that axis for measuring parts of various heights, and the actual measurement movement takes place in that axis.

The platen, anvil, or table must be precisely square to the z axis, the line of measurement, because serious errors can result from a lack of squareness (see Figure 10–51).

In our example, the axis of the indicator is one degree out of alignment with the line of measurement of the part. The indicator is zero set with a 2 in. standard, and then the part placed under it. We read a 50-mil (0.050 in.) difference on the indicator. When we multiply this result by the cosine of the angle, we get an actual change in length of 10 microinches

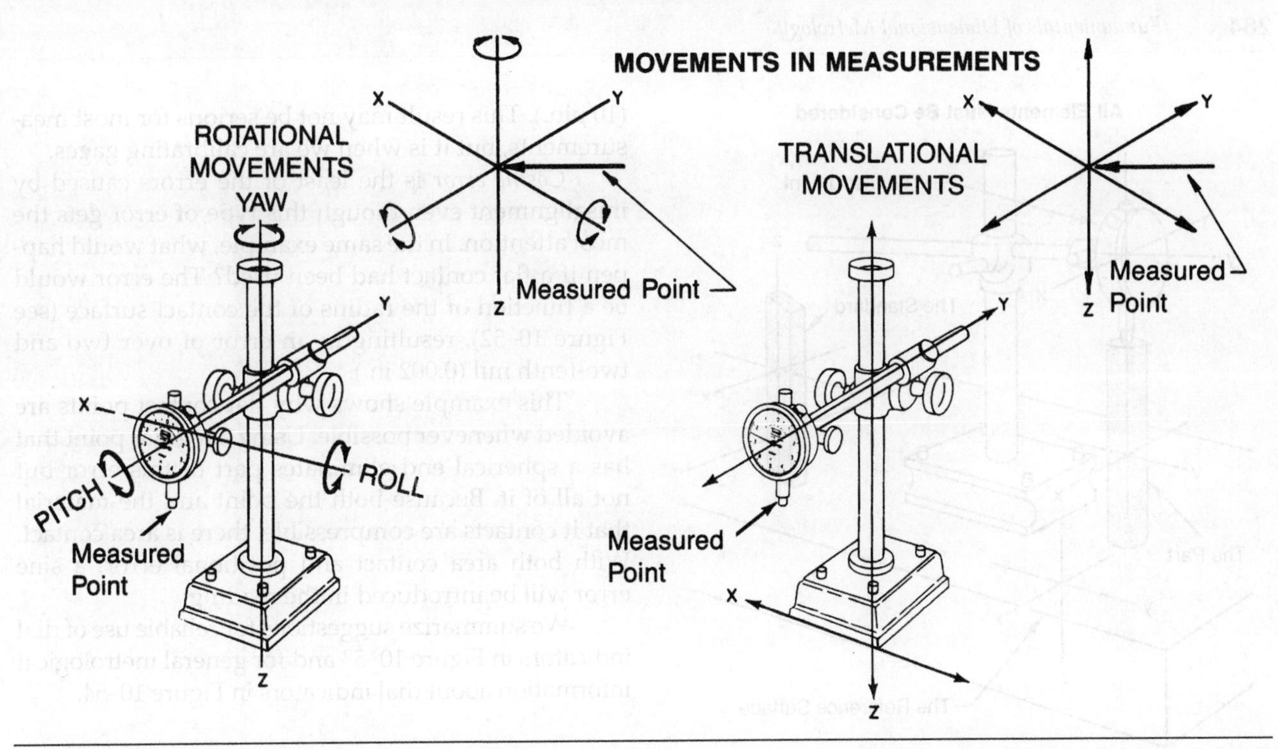

FIGURE 10–46 The loose-jointed test stand shows six separate movements that the measured point may have. Measurements around the rotational axis are known as polar movements. They are often referred to by nautical terms, yaw, roll, and pitch. Those in the translational planes are known as orthogonal measurements.

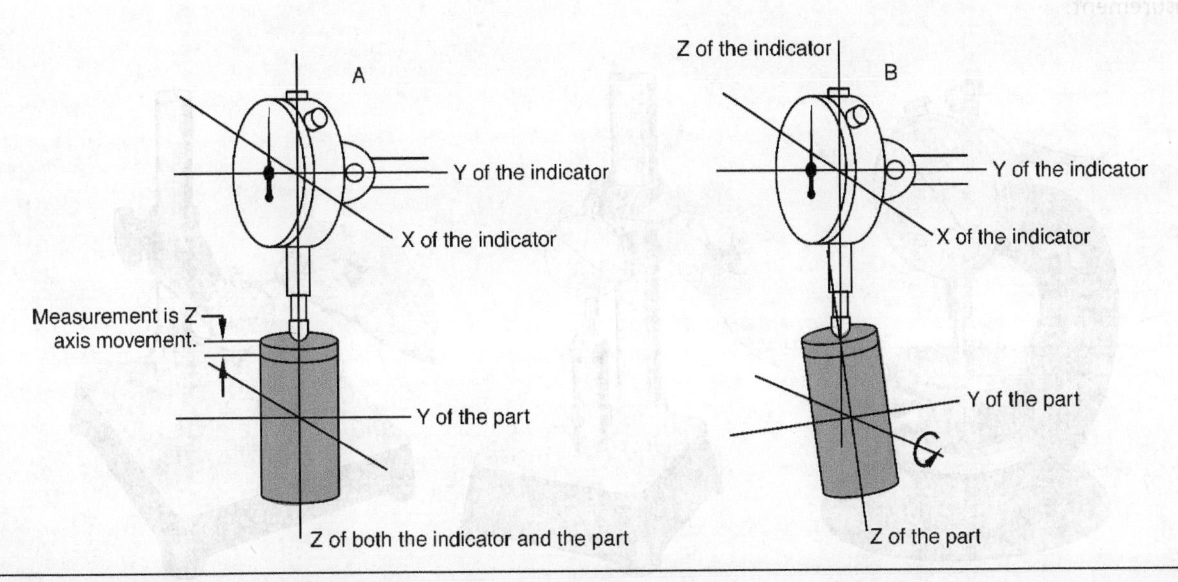

FIGURE 10–47 During setup, both the part and the indicator must be constrained from all x and y movement and their z axes brought in line. The indicator then measures movement along the z axis. If x of the part is not z of the indicator, the measurement will not be accurate as shown on the right.

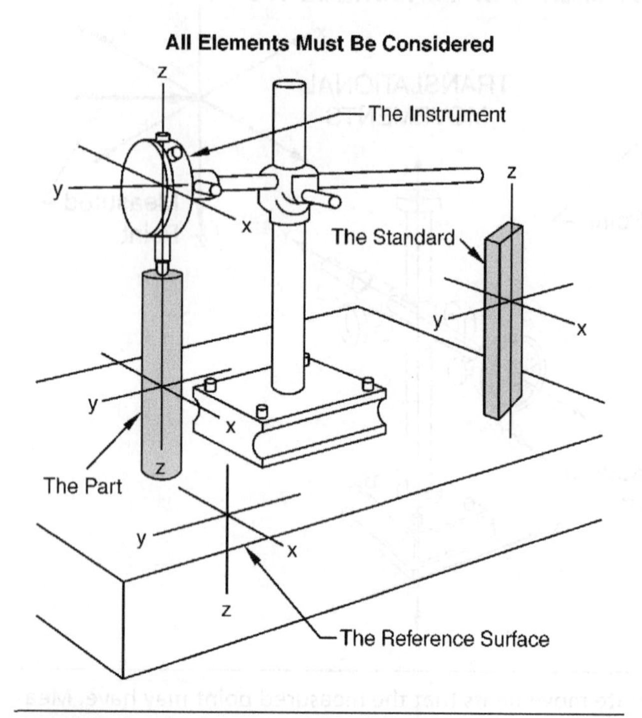

FIGURE 10–48 A rotation around any one of four separate x and y axes will destroy the accuracy of the measurement.

(10 μin.). This result may not be serious for most measurements, but it is when we are calibrating gages.

Cosine error is the least of the errors caused by misalignment even though this type of error gets the most attention. In the same example, what would happen if a flat contact had been used? The error would be a function of the radius of the contact surface (see Figure 10–52), resulting in an error of over two and two-tenth mil (0.002 in.).

This example shows why flat contact points are avoided whenever possible. Using a regular point that has a spherical end eliminates part of this error but not all of it. Because both the point and the material that it contacts are compressible, there is area contact. With both area contact and positional error, a sine error will be introduced in the reading.

We summarize suggestions for reliable use of dial indicators in Figure 10–53 and for general metrological information about dial indicators in Figure 10–54.

FIGURE 10–49 Dial comparators are indicators with stands that incorporate their own reference surfaces. Note that all the indicators shown have spindle-lifting levers. *(Courtesy of B. C. Ames, Inc.)*

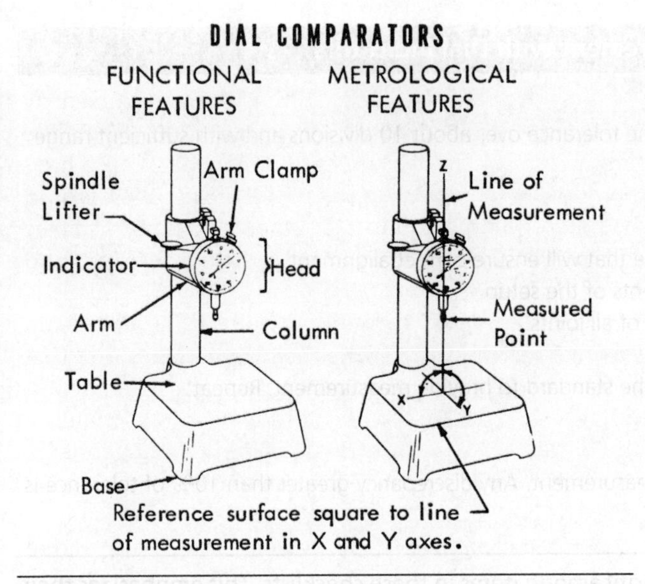

FIGURE 10–50 The dial comparator has the general C shape of all the outside diameter instruments.

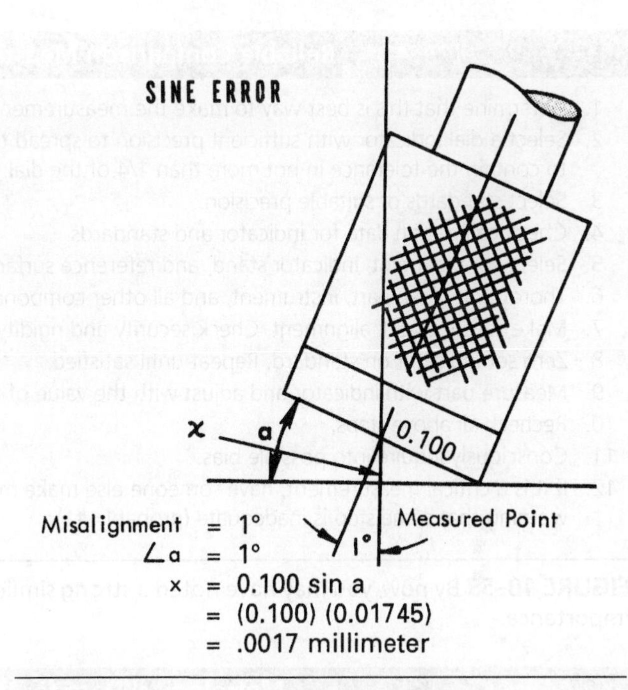

FIGURE 10–52 The sine of the angle caused by misalignment with a flat contact is a much more serious error than the cosine error. Note: A misaligned micrometer or snap gage would have this error at each contact surface.

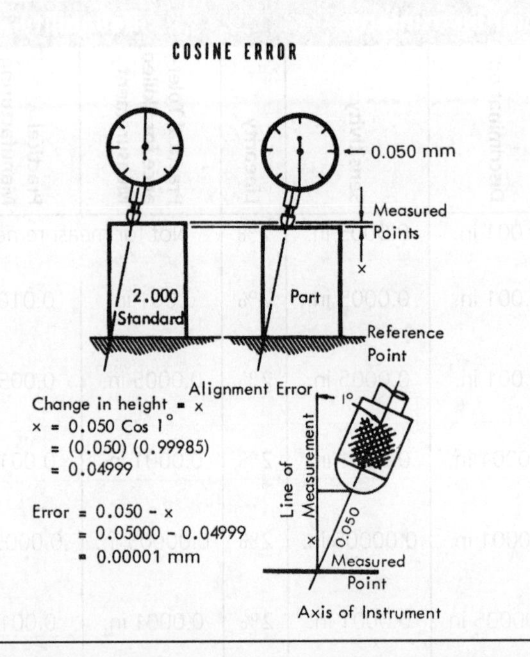

FIGURE 10–51 The cosine error caused by 1° applies only to the travel of the indicator and figures out to an error in length of only 10 μin. Note that the same misalignment with a vernier height gage creates an error based on the full length and would be 300 μin.

10–6 CONSTRUCTIVE USE OF ERROR

Error can be its own worst enemy. If you can isolate error, you can devise a way to compensate for it or even put the error to work for you. A case in point is when we use sine and cosine errors to square a comparator table with the axis of measurement. You can use the error caused by misalignment to show the direction, magnitude, and method to correct the misalignment.

Let us assume that either the indicator head or the base is adjustable, and we want to square them before using the setup. In Figure 10–55A and B, if a gage block is pushed uphill, the indicator registers an increase in height (sine error). If the block is pushed downhill (see Figure 10–55C and D), the reading does not change. These readings show both the direction of the slope (up or down) and its sign (plus or minus).

Reliable Comparison Measurement with Dial Indicators

1. Determine that this is best way to make the measurement.
2. Select a dial indicator with sufficient precision to spread the tolerance over about 10 divisions and with sufficient range to contain the tolerance in not more than 1/4 of the dial.
3. Select standards of suitable precision.
4. Check calibration data for indicator and standards.
5. Select contact point, indicator stand, and reference surface that will ensure correct alignment.
6. Thoroughly clean part, instrument, and all other components of the setup.
7. Make setup. Check alignment. Check security and rigidity of all joints.
8. Zero set indicator on standard. Repeat until satisfied.
9. Measure part with indicator and adjust with the value of the standard to provide measurement. Repeat.
10. Recheck all above steps.
11. Consciously inquire into possible bias.
12. If it is a critical measurement, have someone else make measurement. Any discrepancy greater than 10% of tolerance is warning that some step is inadequate (probably #1).

FIGURE 10–53 By now, you may have noted a strong similarity among items in these checklists. This emphasizes their importance.

Metrological Data for Vernier Instruments							Reliability	
Instrument	Type of Measurement	Normal Range	Designated Precision	Discrimination	Sensitivity	Linearity	Practical Tolerance for Skilled Measurement	Practical Manufacturing Tolerance
Test indicators	Comparison	0.030 in.	0.001 in.	0.001 in.	0.0005 in.	2%	Not for measurement	
0.001 indicators:								
on height gage stands	Comparison	0.250 in.	0.001 in.	0.001 in.	0.0005 in.	2%	0.001 in.	0.010 in.
on comparator stands	Comparison	0.250 in.	0.001 in.	0.001 in.	0.0005 in.	2%	0.0005 in.	0.005 in.
0.0001 indicators:								
on height gage stands	Comparison	0.050 in.	0.0001 in.	0.0001 in.	0.0001 in.	2%	0.0001 in.	0.001 in.
on comparator stands	Comparison	0.050 in.	0.0001 in.	0.0001 in.	0.00005 in.	2%	0.00005 in.	0.0005 in.
0.00005 indicators:								
on height gage stands	Comparison	0.010 in.	0.00005 in.	0.00005 in.	0.0001 in.	2%	0.0001 in.	0.001 in.
on comparator stants	Comparison	0.010 in.	0.00005 in.	0.00005 in.	0.00005 in.	2%	0.00003 in.	0.0003 in.

FIGURE 10–54 A practical comparison of dial indicators cannot be made without considering the means of supporting the indicator.

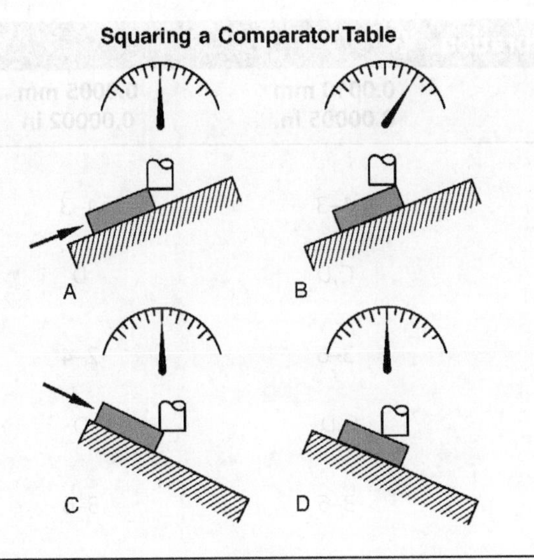

FIGURE 10–55 Passing a gage block between spindle and anvil along several paths can serve to detect and correct the misalignment.

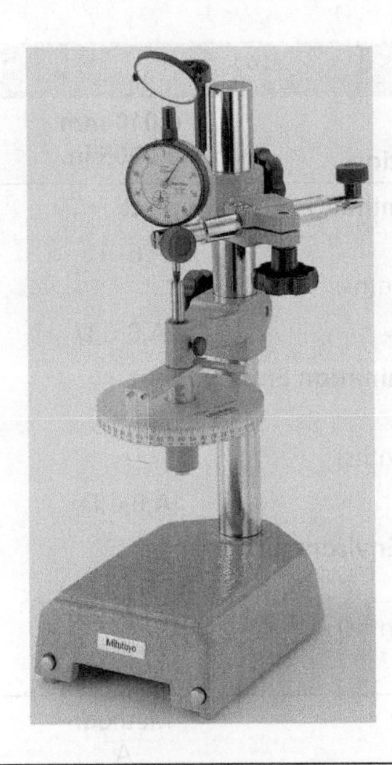

FIGURE 10–56 This instrument is comparing an indicator with a large drum micrometer. It is fast but limited to the accuracy of the micrometer. *(Courtesy of Mitutoyo American Corporation)*

If we repeat this process from various paths beneath the spindle, we can find the axis in which the deviation is greatest, and then the setup can be adjusted and retested until we achieve the desired alignment.

You need to remember that a stationary reading might mean:

- That squareness has been attained
- That the movement is from high to low
- That the axis of the movement is along the one square line that always exists (discussed further in Chapter 12)

Adjustments can completely destroy all hope we have of making precise measurements, or when properly checked and rechecked, they can ensure that every measurement we make is as precise as possible. You should always remember to repeat the test after the final clamping because all too often you will need to make additional adjustments.

Calibration of Dial Indicators

Throughout this discussion, we included all digital equivalents in the group of "dial indicators" that can be calibrated separately or as part of a measurement setup. Discrimination, range, and amplification of dial indicators are interrelated and determined by design; therefore, they are not subject to calibration. We calibrate instead the sensitivity and accuracy of a dial indicator.

In most cases, dial indicator calibration is haphazard, even when the instrument is calibrated by the manufacturer. This calibration usually consists of running the indicator through its range using a micrometer head (see Figure 10–56) and noting any differences, which are errors. This process generally violates the 10-to-1 rule, but even if it did not, it creates the potential for a large number of observational errors. On the other hand, this method is quick, so you can perform it frequently (see Figure 10–57).

Generally, the sensitivity of a dial indicator is determined by a variable made up of friction, inertia, and hysteresis. The instrument's accuracy is determined by a variable that combines gear rack error and the errors in the various gears of the gear train. The individual errors in each of these variables are independent,

Frequency of Calibration				
Dial graduation	**0.010 mm 0.0005 in.**	**0.002 mm 0.0001 in.**	**0.0010 mm 0.00005 in.**	**0.0005 mm 0.00002 in**
High Contamination Environment				
Frequency of checking (months)	6–9	3–6	1–3	1–3
Method	A,B,C,D	B,C,D	C,D	D
Low Contamination Environment				
Frequency of checking (months)	9–12	4–9	3–6	2–4
Method	A,B,C,D,	B,C,D	C,D,	D
Laboratory Environment				
Frequency of checking (months)	12	12	3–6	3–6
Method	A,B,C,D	B,C,D	C,D	C,D

Method:	Comparison to:
A	Large drum micrometer (see Figure 9–15)
B	Wrung gage block
C	Magnified large drum micrometer
D	Point contact gage block (see Figure 9–17)

FIGURE 10–57 Frequency of calibration depends on the discrimination and the frequency of use of the instrument as well as the available equipment.

but the general action of the variables on each other is correlated. For practical purposes, you can calibrate these variables as separate characteristics.

You check sensitivity by single-sample measurements that are identical in every respect except *time*. Because single measurements never provide a real test of accuracy, you must make *multiple-sample measurements*, measurements made in as many ways possible, by many observers, at separate times, and under different conditions. These tests can only approximate the accuracy of the instrument. Absolute accuracy can never be achieved because the act of measurement itself changes both the measurement system and the object being measured.

The results of single-sample measurements are usually called the *repeatability* of the instrument. It is difficult to make the same measurements of one instrument identical, and any differences between the measurements show up as error in the instrument being calibrated. Because these are independent errors, the final result may make the instrument appear better or worse than it actually is.

To calibrate the sensitivity of a dial indicator, you repeatedly measure one gage block at a point that is beyond the starting tension point of the instrument but less than the overtravel. To minimize observational error, you should zero the indicator, even though this step has nothing to do with the sensitivity of the instrument. Pass a gage block repeatedly between the indicator and the reference surface, which is actually where we create a problem, because the difference between the contact of the gage block and reference surface with each successive pass may be greater than the error being measured.

There are three general contact methods: casual, wrung, and point, which are all determined by the type of reference plane available. *Casual contact* is formed simply by laying the clean part on a clean reference surface, but this method is only suitable for measurement with large tolerances—0.02 mm (0.001 inch) or greater. You can have appropriate contact when the indicator stand is attached to the reference surface and the stand is not massive. Casual contact is not suitable for any sensitivity calibration but is suitable for coarse accuracy calibration.

More reliable than casual contact, *wrung contact* requires that the reference surface and the standard have a sufficiently good surface finish to permit wringing. In addition, the indicator support must not introduce sufficiently large independent variables to compensate for the sensitivity variable, which can be ensured by using the indicator in a fixed stand.

Wrung contact can be used to check the sensitivity of 0.005 mm reading dial indicators when you:

1. Carefully wring the gage block to the reference surface.
2. Raise the spindle.
3. Slide the gage block under the spindle.
4. Gently lower the spindle to the block.
5. Zero set the instrument.
6. Raise the spindle.
7. Slide the gage block somewhat.
8. Lower the spindle on the gage block again.

Note the reading and repeat the operation again. After four or five repetitions, you should be able to identify any measurements that contain illegitimate errors, such as the spindle landing on a particle of dust, and reject them. The rest of the readings show the sensitivity of the instrument, but if the readings vary by more than one-quarter dial division, you should return the instrument to the manufacturer for repair.

Because sensitivity and uncertainty are correlated, there may be circumstances under which their errors compensate. Therefore, as a precaution, you should repeat the sensitivity calibration at another position within the working range of the instrument.

The *three-point anvil* is the most common form of *point contact,* and most often the preferred method for checking the sensitivity of tenth-mil (0.0001 in.) reading indicators and all high-amplification comparators. This point contact instrument takes advantage of kinematic principle. You only need three points to establish the reference plane, so you remove all other redundant points, which greatly reduces the contact variable. Differences that can be caused by changing the wringing intervals are nearly eliminated, as are the effects of oil film and dust trapped between the surfaces.

With the three-point contact, you can make sensitivity calibrations reliably without the error that is introduced by the wringing interval variable. Even though the gaging force is low, the small areas of the contact points develop high unit pressures, which break through the air film that remains when surfaces are wrung together. The three-point contact is usually used only for small parts of low mass and when you are using an instrument with light gaging force.

A three-point anvil attachment for a high-amplification comparator (Figure 10–48) is called a *three-point saddle.* It is fastened to the standard anvil by set screws after both have been cleaned well and are dust free. The contact points are tipped with hard, wear-resistant material, such as boron carbide, and are precisely lapped to the same shoulder-to-crown heights. Several tapped holes in the top triangular plate provide a choice of spacings, so when you need a larger working surface, you can remove the triangular plate.

The three points should be spaced widely apart, but not so far apart that the gage blocks or part sags. The center line (CL) of the gage head contact point must fall within the triangle (see Figure 10–57), preferably close to the line connecting the two closest points (see Figure 10–58A), so that the gaging force causes minimum bending of the block or part. Keeping the points close together (see Figure 10–58B) also minimizes bend. If the part or block is long, closer points may not provide a stable reference plane. You can measure very thin blocks and parts with little bending by using only one reference point (see Figure 10–58C). Using only one reference point requires that the block be held precisely perpendicular to the travel of the gage head. You can determine if the block is perpendicular by spinning the block: If there is no wobble, the block is perpendicular.

You can compromise between a flat reference plane of the three-point contact when you use a serrated anvil. The serrated anvil reduces the wringing interval

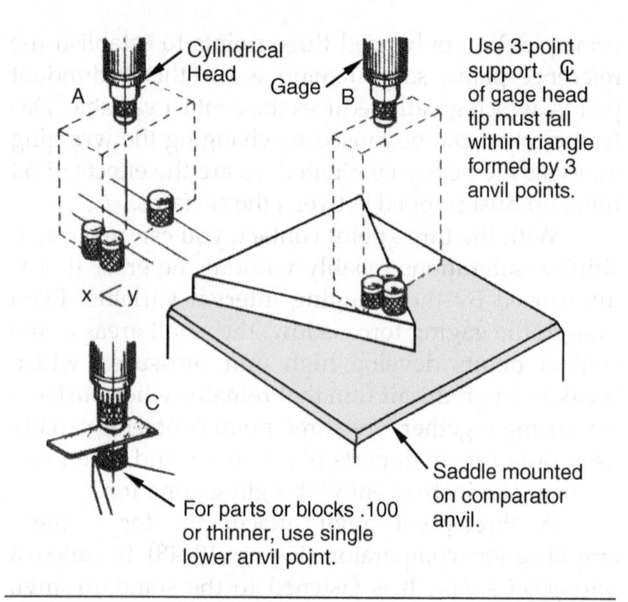

FIGURE 10–58 The contact of the head should always be within the triangle formed by the points. When it is near the line connecting the closest points, such as at A, bending is minimized. For least bending, only one reference point is used, as at C.

because of its higher contact pressure, and it is easier to remove dust from the contact surfaces. In addition, the serrated anvil provides more support to the block or part than does a three-point contact and can, in some cases, be more accurate than the three-point contact.

Accuracy Calibration

In addition to sensitivity, you can calibrate the accuracy of indicator-type instruments with gage blocks. But what is accuracy? Earlier, we said that *uncertainty* would be a better term to use. Accuracy is a measure of perfection, but uncertainty is a measure of error. Accuracy is the degree to which an observed value agrees with the true value, error is any deviation, and uncertainty is the maximum amount of error we can expect under the "standard" conditions—the normal circumstances under which a skilled person would make the measurement. These conditions will sometimes vary, so we also sometimes need to express the uncertainty in terms of the new conditions. These new conditions will then become part of the standard conditions for that particular measurement. In other cases, specific rules define the standard conditions for the measurement. These rules may be informal understandings within a plant, formal standards set up by trade or industry associations, or bylaws established by government bodies.

Accuracy Calibration of a Dial Indicator

You should calibrate sensitivity first because if you set the instrument for an accuracy that is finer than the sensitivity of the instrument, your readings will have little practical use. Accuracy is important for the design of the instrument—for example, if a steel rule has a discrimination of 0.5 mm, an accuracy error of 0.5 mm would be intolerable. You also cannot consider a dial indicator that will repeat no closer than one-half division accurate to any measurement that must be made to less than one-half division. We calibrate the instrument in order to refine the accuracy.

We calibrate that portion of the range from the end of the tensioning travel to the beginning of the overtravel. It is easier to calibrate an ordinary indicator with two turns of travel than it is to calibrate a long-range indicator.

Each gear in the indicator may be a source of error. These errors will be periodic, repeating every time the gear is in the same position. However, because these errors are also dependent, they will compensate. The rack error will also compensate for the other errors, but rack error is a random error.

When we calibrated the micrometer, we chose uneven intervals so that the periodic errors would not be encountered at the same magnitude of compensation. This process might seem logical for dial indicators, but it is not. To the calibration technician, there is no discernible relationship between the position of the internal elements and the pointer along the scale. Even intervals have as good a chance to randomly compensate for errors as uneven intervals do. To calibrate a dial indicator, you only need to check a point at the beginning of the range, one at the end, and one or two in between.

You should mount the indicator securely to a comparator stand that has an integrated reference surface (the anvil). You should only use a height gage stand that rests on a reference surface as a last resort because it adds errors to the calibration process.

Because it takes less time than wringing gage blocks, you should use a three-point anvil. If a three-point anvil is not available, wring gage blocks to the anvil. Even if the surface finish of the anvil is not good enough for wringing, you need to go through the act of wringing to ensure intimate contact between the blocks and anvil.

You should use the principle of bridging an error and then zeroing in on it. If any of the readings contain errors that are greater than the sensitivity of the instrument, you should make a record of that reading and the next reading. If that reading shows an error, you should then make a reading on the opposite end. Continue this process, reading between two points of error, until you locate the maximum error or until you find an error greater than the permissible maximum.

If you do find a reading that is greater than the permissible maximum, you should return the instrument for repair or replacement. If the maximum error in the calibration readings is not sufficient to remove the instrument from service, you should note this maximum on the calibration record. A continuing record of calibration helps speed up each successive calibration and helps the user understand problems that may arise in the use of the instrument.

If you are calibrating a long-range indicator (see Figure 10–31), you need to calibrate the portion of the range that produces usable readings for accuracy, but you calibrate for sensitivity at the beginning of the travel end and in the usable range. If you do not, a sensitivity error could be systematic and affect the entire range like a zero setting error affects a micrometer. If you use a long-range indicator throughout its range, you should check the indicator at several intervals—for example, 1 inch—and then zero in on any errors you uncover.

SUMMARY

- This chapter marked an important milestone—measurement by comparison. Hitherto, the standard of length was built into the instrument. Here it is separate. The instrument is nothing more than a means for comparison of a known with an unknown. By separating the standard from the means of comparison, we have broken the bonds on amplification. Measurement involves a change of length. With comparison instruments, this change can be magnified many times, thus making very small changes discernible.
- Dial indicators (calibrated measured points) are by far the most important comparison instruments. They achieve their amplification by a rack and gearing plus their pointer length. They are generally available with discriminations as fine as one-tenth mil (0.0001 in.). In themselves they provide no reference point. They must always be used in conjunction with a stand or holder. This has a C-frame configuration for external measurement and a straight frame for internal measurement. Both conform with Abbe's law and are inherently accurate.
- Because a dial indicator is loaded by the measurement movement, the energy to operate the instrument is limited. Amplification beyond a certain point does not increase sensitivity and is without value. To fully use an instrument's sensitivity, it must have good resolution, which is part of its readability. This can be tested by repeated measurements. The dispersion of readings is the precision, not the accuracy.
- The accuracy is the difference between the readings and the true values. All comparators have some error. The amount that gets into measurements is proportional to the travel. Thus, the dial indicator has its greatest accuracy at the place it is zero set. That can be anywhere in the working range. At that point, accuracy is reduced only by sensitivity. All comparison should be as close to zero as possible.
- It is always desirable to get a usable reading directly without computational steps. Balanced and continuous dials conform with bilateral and unilateral tolerances, respectively. Dials reading clockwise and counterclockwise are available.
- Long travel indicators have revolution counters. They are guaranteed troublemakers when used to read the distance traveled from a reference. They are highly reliable only when used to show that a reference has been reached. The useful measuring range for all indicators starts after the starting tension and ends before the overtravel. This

range is independent of size or discrimination. Both discrimination and range are involved when selecting an indicator. The discrimination should spread the tolerance over 10 dial divisions. Select the range so that the total tolerance occupies at least 1/10 but no more than ¼ of the total dial.

- Indicator versatility derives from the large number of contact points and mountings available. All of the latter are to be mistrusted. For every joint or adjustment in an indicator setup, double your distrust. Fortunately, the errors in stands can be analyzed systematically, which permits some corrections and compensations. The pattern of this book has been to move sequentially to instruments of higher amplification. That is why this chapter stopped at dial indicators. They are not the only comparator instruments. The next chapter takes up the high-amplification types.
- The discussion of comparison measurement has been based on the mechanical dial indicator. This is not a slight to the currently popular mechanical digital and electronic digital instruments. The physics involved for all types is the same, but it is easier to demonstrate using mechanics. The lack of how-to instructions is not important. The important message concerning practical applications is always to consult the manufacturer's manual for the instrument being used.

END-OF-CHAPTER QUESTIONS

1. Which of the following are the two general types of measurement?
   a. Line and end standards
   b. Sealed instruments and dial instruments
   c. Movement measurement and direct measurement
   d. Direct measurement and comparison measurement
2. Which of the following is the most important consideration in which comparison measurement differs from direct measurement?
   a. There is no practical limit for length with comparator instruments.
   b. More accurate standards are incorporated in comparison instruments than in direct measurement instruments.
   c. The standard is part of a direct measurement instrument, but a comparison instrument does not have a standard.
   d. Parallax error is less of a problem with comparison measurement.
3. On which of the following does the reliability of direct measurement depend?
   a. Accuracy of the standard and the scale
   b. Accuracy of the scale and its material
   c. Accuracy of the reading of the scale and its discrimination
   d. The scale discrimination and its lack of parallax error
4. Reliability of comparison measurement depends on which of the following?
   a. Accuracy and discrimination of the standard and accuracy of the reading
   b. Gage block material
   c. Discrimination of the standard and its accuracy
   d. Discrimination of the instrument and its amplification
5. All measurement, including time and mass as well as length, requires certain common factors. Which of the following are they?
   a. Gravity, time, and temperature
   b. The measurement system, the unknown measurement, and the calibration of the system
   c. Constant temperature and freedom from parallax error
   d. The standard, the unknown measurement, and the measurement system
6. Which of the following best describes "resolution"?
   a. The ability to separate precision from accuracy
   b. The smallest graduation on a scale or dial
   c. The visual separation between divisions
   d. The difference between the reading and the actual dimension
7. How is precision determined for a dial comparator?
   a. Comparison of the readings with gage blocks
   b. The dispersement of a series of readings

c. From the manufacturer's specifications
d. The distance between dial graduations

8. How is the accuracy of a dial comparator determined?
a. Comparison of the readings with gage blocks
b. The dispersement of a series of readings
c. From the manufacturer's specifications
d. The distance between dial graduations

9. How is precision increased in most measurement instruments discussed thus far?
a. By bringing the instrument and workpiece to 68°F
b. By using finer scale discrimination
c. Amplification
d. By getting the eye closer to the scale and avoiding parallax errors

10. Which of the following occur with an increase in amplification?
a. The range decreases but the scope increases.
b. The scope increases but the cost increases.
c. The importance of temperature increases.
d. The scope and range decrease as the cost increases.

11. When an instrument contains its own standard, which of the following limits its discrimination?
a. The system that divides the standard into units and enlarges the units to readable size
b. The system that moves the reading end of the standard from the reference end
c. The speed at which the system responds to a change in part length
d. The inevitability of parallax and temperature errors

12. What is the chief benefit of the separation of the standard from the instrument?
a. The temperature of the instrument does not affect the temperature of the standard and workpiece.
b. The standards can be carefully stored separately.
c. The standard does not get in the way of the workpiece during measurement.
d. Increased amplification does not reduce the permissible length of the standard.

13. Why do dial comparators use gear trains instead of levers to achieve amplification?
a. To house them in round cases
b. To achieve greater range
c. Because levers are more subject to mechanical errors
d. Because lever actions are difficult to keep clean

14. What is the relationship between indicators and comparators?
a. They are the same.
b. Indicators provide a reading of dimensional change, whereas comparators provide readings of deviations from standards.
c. Indicators are more precise than comparators.
d. Indicators are limited to 0.001 in. discrimination, whereas comparators may be much finer.

15. The use of a digital readout on an instrument shows which of the following?
a. That it is more precise than nondigital instruments
b. That it uses electronic instead of mechanical amplification
c. That it costs more than its nondigital equivalent
d. Taken by itself, it says nothing.

16. What is the difference between test indicators and dial indicators?
a. Dial indicators have round faces.
b. Both indicate change in one axis, but dial indicators are more precise.
c. Only dial indicators are available with a digital readout.
d. If used in a machine shop for setup, the instrument is known as a test indicator. If used for the inspection of parts, it is a dial indicator.

17. What is the relationship between indicators and comparators?
a. All indicators are comparators.
b. Some indicators are also comparators.
c. No indicators are comparators.
d. They are the same.

18. In popular usage, *comparator* refers to which of the following?
    a. Electronic digital linear measurement instruments
    b. All electronic linear measurement instruments
    c. All high-precision digital instruments
    d. Electronic and mechanical instruments for the precision comparison of lengths
19. Which of the following occurs when amplification is increased?
    a. Precision is increased but sensitivity is decreased.
    b. Parallax becomes a more serious consideration.
    c. Sensitivity is increased.
    d. Windup and friction in the system exceed the increased sensitivity.
20. This group of characteristics is essential for one important capability. The characteristics are resolution, size of the graduations, dial contrasts, and parallax. Which of the following, if any, is the capability they determine?
    a. Abbe's law error
    b. Readability
    c. Precision
    d. Accuracy
21. What determines the number of graduations that can be put on an indicator dial?
    a. The diameter of the dial
    b. The circumference of the dial
    c. The precision of the instrument
    d. The length of the maximum dimension
22. How are precision and accuracy related to dial indicators?
    a. There is no connection.
    b. Accuracy is essential for precision.
    c. Precision is essential for fulfillment of accuracy.
    d. They are stated differently but are essentially the same.
23. What are the usual discriminations in which standard dial indicators are available (in inches)?
    a. 0.001, 0.0025, 0.005, and 0.010
    b. 0.001, 0.0005, 0.0025, and 0.0001
    c. 0.001, 0.0025, 0.0001, and 0.00005
    d. 0.002, 0.003, 0.004, and 0.005
24. What error is eliminated or minimized by the zero setting adjustment?
    a. Parallax error
    b. Bias
    c. Alignment error
    d. Computational error
25. Which of the following best describes the difference or similarity between balanced and continuous indicator dials?
    a. Balanced dials have the same internal loading regardless of which direction they travel.
    b. Continuous dials have the same precision at all places in their ranges, whereas balanced dials do not.
    c. Continuous dials have parallax error only in one direction.
    d. Continuous dials and balanced dials are analogous to unilateral and bilateral tolerances.
26. Why should the shortest range of indicator that will contain the tolerance be selected?
    a. Convenience. There are more short-range indicators.
    b. Ease of setup
    c. Larger dials get in the way of the setups.
    d. The shorter the range, the higher the precision.
27. How are the many surface situations accommodated by dial indicators?
    a. By careful honing of the surfaces with a conditioning stone
    b. By adjustment of the gaging pressure
    c. By careful avoidance of parallax error
    d. By having interchangeable contact points
28. How many different movements are possible with a loose-jointed indicator stand?
    a. None
    b. All
    c. 2^3
    d. 6

CHAPTER ELEVEN

REFERENCE PLANES

LEARNING OBJECTIVES

- State the historical importance of flatness.
- State the importance of flatness to industry.
- Explain the degradation of workmanship principle.
- Describe modern reference surfaces.
- Evaluate flatness.

OVERVIEW

Throughout our discussion, we have emphasized that every linear measurement starts at a reference point and ends at a measured point, but so far most of the instruments we have discussed only have their own reference points. Their reference points were usually small planes: anvils, bases, or jaws. When we needed an additional reference, we called it a reference surface or plane.

11–1 BACKGROUND

Sir Joseph Whitworth (1803–1887)

Although considered an unpleasant, crotchety denizen of the machine shops, or "engineering works" as they were known in his day, Joseph Whitworth was a major factor in manufacturing and machining. When he was 14, his father placed him in the office of a cotton spinner to learn the business, but Whitworth was more interested in the textile machines. He concluded these machines were poorly designed and ran away to London to begin his career as a mechanic.

He found employment with the leading shop of the time, Maudslay & Field, the same shop that made the "Lord Chancellor" micrometer (see Figure 7–1). Placed next to the shop's best workman, John Hampson, he began to challenge the status quo of machining, proving that the surface plates of the shop were not sufficiently accurate. He did not apply new methods to machining, but he did use new techniques that embodied his understanding of the importance of accuracy in the machining process.

Although Whitworth did not invent the three-plate method for generating flatness, he is credited with its invention. What actually happened, though, is that Whitworth helped a process evolve to a higher level of useful precision. Working three plates against each other was already the established practice, performed by lapping with abrasive particles. Henry Maudslay used the three-plate method to hand file and hand lap the finish on surface plates, and he was still refining this process when Whitworth became a mechanic. In 1840, Whitworth substituted lapping with scraping—removing minute layers of metal by drawing a hardened cutting edge along the surface. Scraping allowed Whitworth to perfect small areas of the plates that were missed by overall lapping.

Whitworth knew that manufacturing and machining must change and that precision manufacturing would be impossible without precision measurement. He also knew that, in order to achieve precision measurement, metrologists would need true reference planes. Throughout his lifetime, he pursued this goal—at 30, when he opened his own shop and, later, through his partnership with Sir William Armstrong.

The superiority of Whitworth's machines was recognized at the 1851 London Exhibition, but these machines could not have been improved without Whitworth's near-perfect reference surfaces. In 1856, he delivered a paper to the Institute of Mechanical Engineers on the importance of flatness and a systematic approach to attain flatness. In this paper, he stated, "A true surface instead of being in common use, is almost unknown."

In addition to his method for achieving flatness, Whitworth made other important contributions. He recognized the advantages of end standards over line standards—that is, if you magnify the sense of feel, you can apply it more easily than eyesight for most practical work; and a tumbler held by friction between parallel planes provides sensitivity in millionths of an inch. More than 100 years ago, Whitworth built a measuring machine and commented: "We have in this mode of measurement all the accuracy we can desire; and we find in practice in the workshop that it is easier to work to the ten-thousandth of an inch from standards of end measurement, than to one hundredth of an inch from lines on a two-foot rule."

Whitworth also recognized that huge masses of metal did not ensure strength or stability. He revolutionized machinery construction using hollow frame members, rounded corners, smooth surfaces, and even by painting appropriate areas. The popular standardization of the screw threads used in Great Britain is known by his name, and Whitworth is credited with creating plug and ring gages. When he died, the leadership in the area of flatness moved to machining centers in New England. However, Whitworth's pioneering in England helped that country achieve a precision in flatness and measurement that the United States has yet to meet.

11-2 FLATNESS

Flatness, the evenness or levelness of a surface, is the degree to which the *reference plane* corresponds to the theoretically perfect plane. These surfaces, whether large or small, are often called "datum planes," and the most important one is the surface plate. Our discussion of the surface plate will lead to a discussion of right angles, through which we will discover standards for the most essential angles: 0°, 90°, and 180°. These standards give linear and angular measurements, points of known return, which are as essential for positional accuracy as the length standard is for linear accuracy (see Figure 11–1).

Flatness seems so basic that you might assume that manufacturing and metrology were always based on it. However, unlike the other geometric forms (circles, regular polygons), flatness does not appear in the natural world. The ability to manufacture a kind of flatness marks a breakthrough that prepared the world for the Industrial Revolution. Before we could create "flatness," the mechanical devices we used depended on swinging and turning movements. Without the ability to slide two surfaces together, simple reciprocating devices were a maze of linkages. In fact, James Watt probably devoted more time to his "parallel motion" patent than he did to his initial steam engine patent.

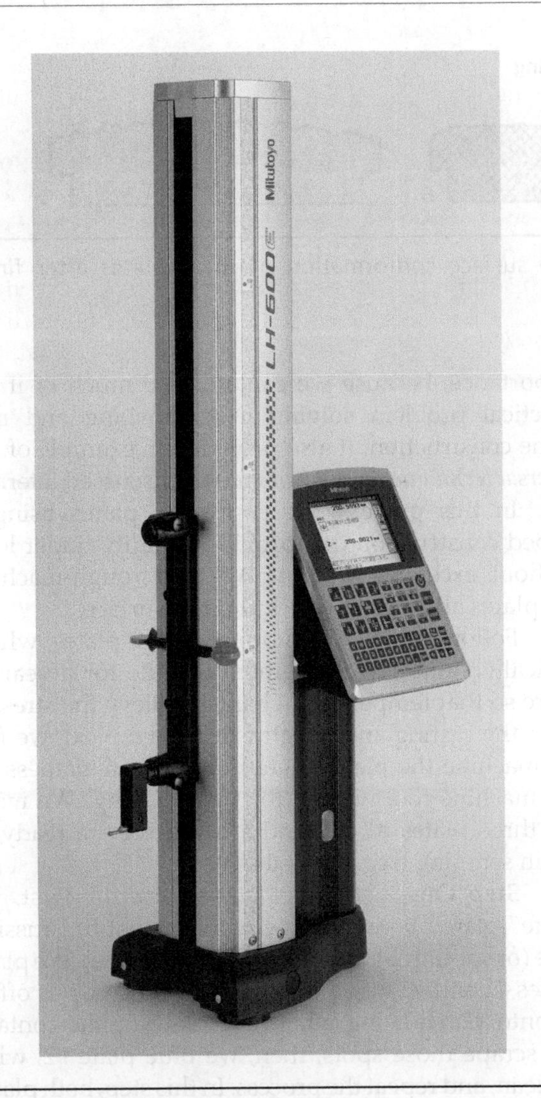

FIGURE 11–1 For positional accuracy, we must have a reliable reference plane. That is particularly true for height gage work, which magnifies the flatness errors. The great precision of a sophisticated height gage, such as the one shown, would be lost without an equally high-grade surface plate. This particular height gage is shown with an optional contact sensor. *(Courtesy of Mitutoyo American Corporation)*

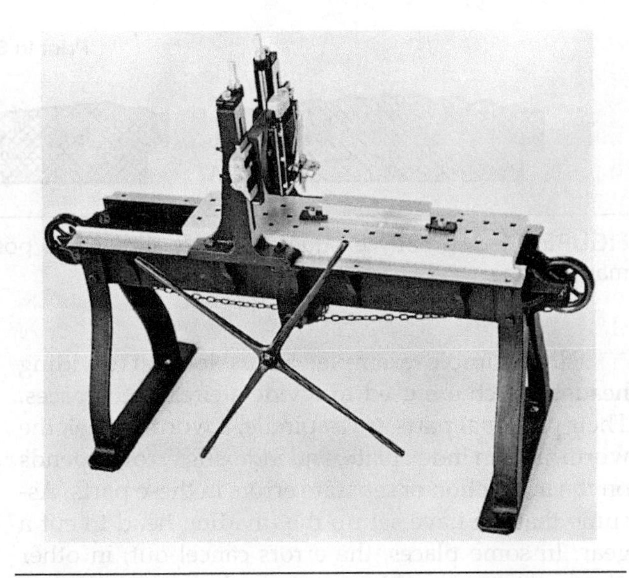

FIGURE 11–2 The planer invented by Richard Roberts in 1817 was the first machine tool that developed flat surfaces.

In 1817, Richard Roberts invented the planer (see Figure 11–2). Now, for the first time, parallel surfaces could be duplicated readily. This made sliding motions and flat surfaces possible, although at this point they were still inaccurate. On the other hand, we could not define accuracy, because there were not references to base flatness on. Then, in 1840, Sir Joseph Whitworth began to draw attention to the importance of flatness.

We do not use Whitworth's method for generating plane surfaces today. However, the method is still the best possible demonstration of one principle of flatness: the degradation of workmanship principle. This principle is one of the vital steps in metrology's progression to its current level of development.

Degradation of Workmanship

To understand this principle, you first need to understand what appears to be a contradiction: We can develop the plane surface to a high order of accuracy because we have no standard for it. This truism, which applies to all machine work, is called *"degradation of workmanship."* The products are not as precise as the machines; the machines are not as precise as the gages; the gages are not as precise as the standards, and so forth.

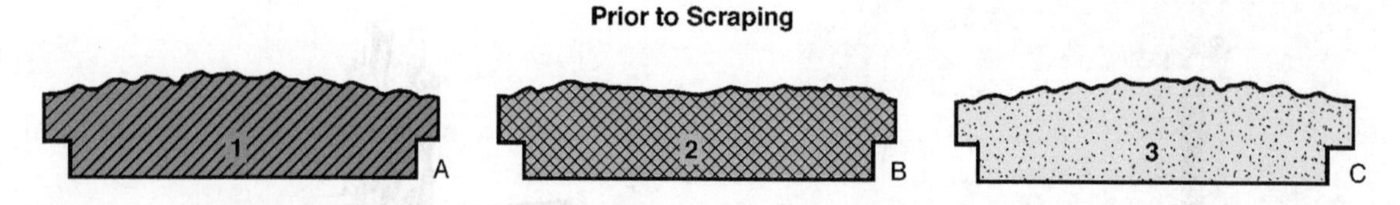

FIGURE 11–3 Greatly exaggerated, this shows the possible surface conformation of three plates after finish machining.

In a simple example, let us look at dividing heads), which are used to divide a circle into spaces. Their principal parts are a spindle, a worm wheel, the worm, and an index plate; and indexing error depends on the interaction of separate errors in these parts. Assume that we have set up the dividing head to cut a gear. In some places, the errors cancel out; in other places, the errors add or even reach their maximum. Next, we use the gear we just cut as a change gear on a lathe to cut a thread, and the errors from the dividing head now extend to the thread—and we also pick up from additional lathe errors. The thread we have cut will have greater errors than any of the previous threads because of the change gear we used.

Finally, to complete the circle, we use the thread as a worm for a dividing head, and the errors are passed on to the next generation of gears, threads, heads, and so forth. The more we repeat this cycle, the worse the errors get; therefore, if we reduce the number of repetitions, the fewer errors we pass on, and the better the final product is. In general, we can say that the less work we perform, the better. At the extreme, we get our most accurate work when we do not have to depend on any other previously manufactured items to do our work. Using this logic, a method of manufacturing plane surfaces that does not require any previously manufactured articles should be capable of perfection—theoretically.

The Three-Plate Method

In Whitworth's time, most surface plates were cast iron. Today, granite has replaced cast iron, and other methods, which we will discuss later, have replaced Whitworth's three-plate method for generating flatness. The technique still has a fundamental importance, because we can still use much of it for practical problem solving in toolmaking and machine construction. It also provides an example of the *reversal technique* or *reversal process*, discussed later.

In this process, we cast three plates using a ribbed construction that provides rigidity under load without excessive weight. We then rough-machine the plates along their edges and top surfaces.

Following this, we normalize the plates, which typically entails stacking them outside for a year or more so that temperature changes relieve the stresses from the casting and rough machining. Next, we finish machine the plates, determining their flatness by the machine tool we use for the finishing. We mark the three plates #1, #2, and #3, and we are ready to begin scraping (see Figure 11–3).

Step One, *Scraping Plates #1 and #2:* First, we "blue" plate #1—wipe it with a thin coat of Prussian blue (or special colors made for this purpose). We place plates #1 and #2 together, and the bluing comes off of #1 onto #2, showing where the plates make contact. We scrape those spots; then, we blue plate #2, wipe #1 clean, and repeat the process. In this step, both plates are scraped. We continue bluing and scraping plates until they are reasonably in agreement; however, we have no way of knowing as yet if we are scraping the plates more flat or less flat (see Figure 11–4). One could be convex and the other could be concave.

Step Two, *Scraping Plates #1 and #3:* For this step, we make an important change: we make plate #1 the control plate. We only apply bluing to plate #1, and we only scrape metal from plate #3 (see Figure 11–5). The plates will be more in agreement, but we still do not know if any of them are flat, and plates #2 and #3 now have the same amount of error as plate #1.

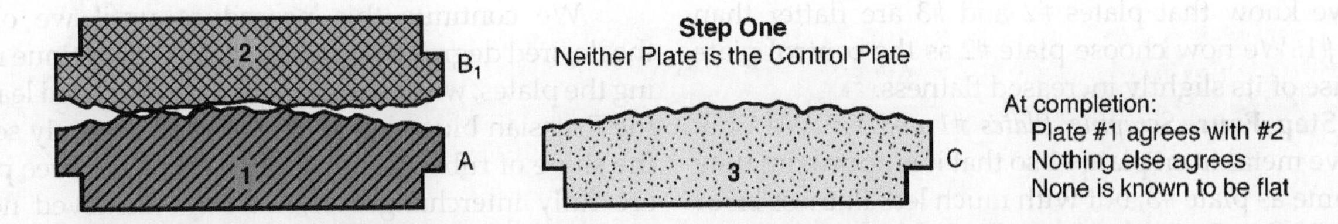

FIGURE 11–4 This step is carried only far enough to get general agreement between #1 and #2.

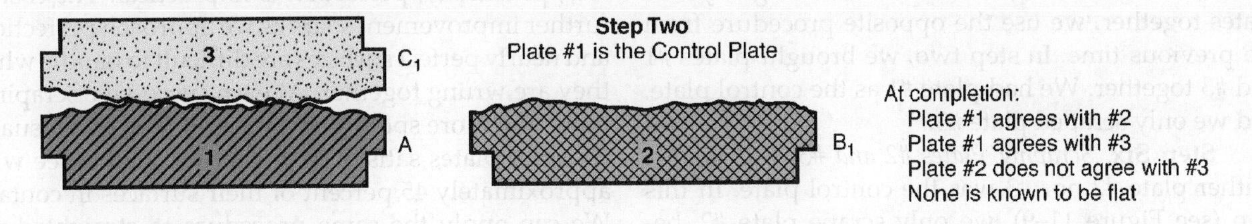

FIGURE 11–5 At the completion of this step, both #2 and #3 will have picked up #1's error.

To further our discussion of error, because the plates are laid on top of each other, the errors in plates #2 and #3 are in the same direction but are in the opposite direction of the errors in plate #1. In our next step, when we turn plate #3 over and place it on #2, the reversal process changes the relative direction of the errors in plate #3; therefore, the error is doubled. Plates #2 and #3 are out of agreement by twice the amount that the control plate #1 is concave or convex. In actual practice, remember that you do not know the direction of the error.

Step Three, *Scraping Plates #2 and #3:* We pair off plates #2 and #3 and work to reduce the doubled error by scraping both of the plates. It is impossible to know that we are removing metal uniformly from each; therefore, we scrape the plates until we have a general agreement between the plates (see Figure 11–6). We still cannot say that anything is flat,

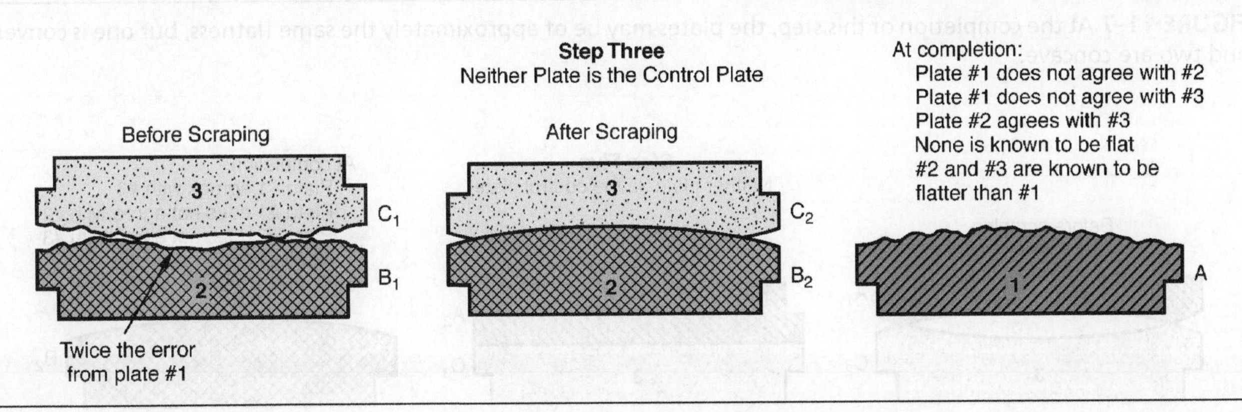

FIGURE 11–6 By scraping some of #1's error off of #2 and some off of #3, we get closer to flatness for these two plates.

but we know that plates #2 and #3 are flatter than plate #1. We now choose plate #2 as the control plate because of its slightly increased flatness.

Step Four, *Scraping Plates #1 and #2:* We only remove metal from plate #1 so that it is approximately the same as plate #3, but with much less flatness error than it had before (see Figure 11–7).

Step Five, *Scraping Plates #1 and #3:* We scrape both plates #1 and #3 in order to remove the doubled error (see Figure 11–8). Each time we bring any two plates together, we use the opposite procedure from the previous time. In step two, we brought plates #1 and #3 together. We had plate #1 as the control plate, and we only scraped plate #3.

Step Six, *Scraping Plates #2 and #3:* Previously, neither plate #2 nor #3 was the control plate. In this step (see Figure 11–9), we only scrape plate #2, because plate #3 already is in agreement with #1.

We continue this procedure until we obtain the desired degree of agreement. As we continue refining the plates, we eventually substitute dry red lead for the Prussian blue. At the final stages, we only scrape the shine of rubbed high spots. When the three plates are fully interchangeable, we have achieved nearly perfect flatness because two things that are equal to the same thing are equal to each other.

In reality, we can never achieve perfect flatness, and, practically, perfection is impractical. The cost of further improvement soars as we approach perfection, and nearly perfect surfaces are difficult to handle when they are wrung together. At each successive scraping, more and more spots "drop in"; therefore, we usually consider plates satisfactory when they all agree with approximately 45 percent of their surfaces in contact. We can apply the same procedure to straightedges. No matter how narrow it is, the straightedge must

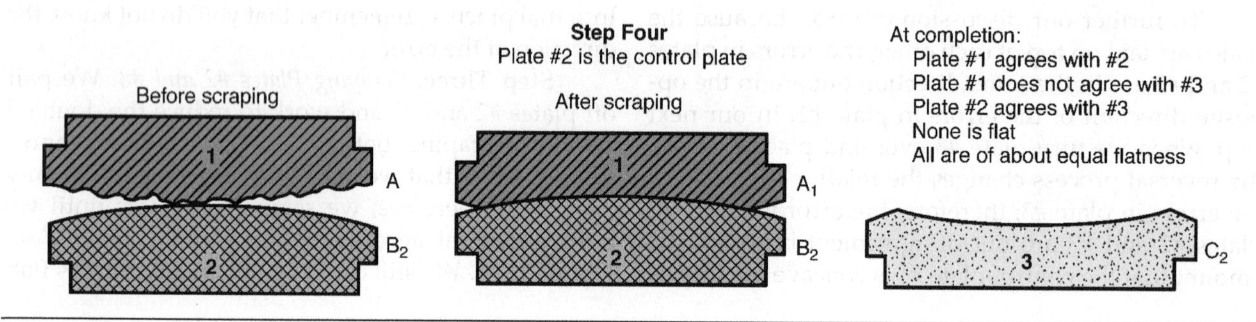

FIGURE 11–7 At the completion of this step, the plates may be of approximately the same flatness, but one is convex and two are concave.

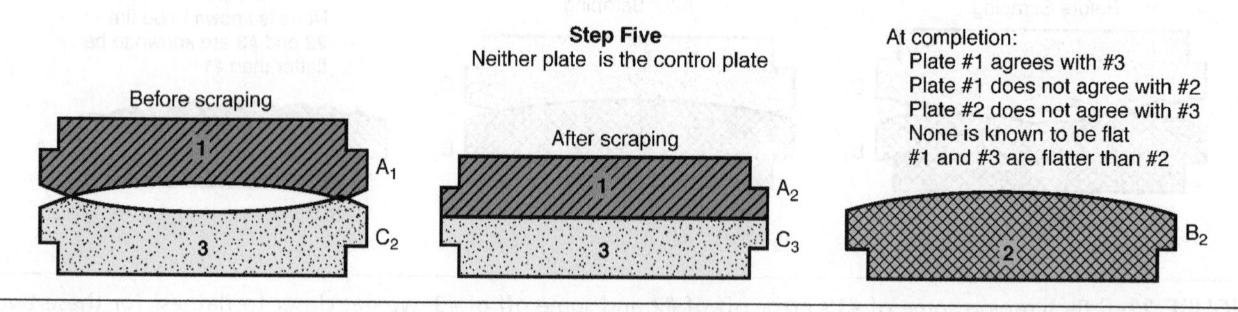

FIGURE 11–8 The first time plates #1 and #3 were brought together, only #3 was scraped. This time both were.

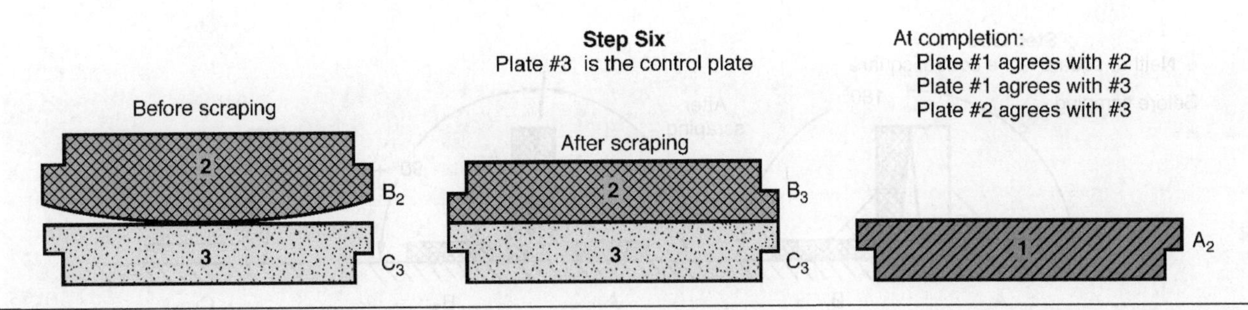

FIGURE 11–9 Although this step brings the three plates into approximate agreement, it does not mean they are flat. The entire procedure is repeated until the desired flatness is obtained.

have some width; therefore, straightedges are actually elongated surface plates. Surface plates and straightedges provide us with references and the standards for 0° and 180° angles.

11–3 PERPENDICULARITY

Next to flatness, the most important reference is perpendicularity, which, in the broadest sense, is simply a right angle. As with flatness, we can generate a right angle by interchanging three unknowns; however, the resulting angle will be slightly degenerate. While we can generate flatness without reference to any previously performed operations, we need a previous 180° angle, also called a *straight angle,* in order to generate a right angle. For this reason, the right angle will always contain some error from the straight angle.

We familiarly refer to right angles as squares; in other words, we use them to judge the *squareness* of other things. (We will discuss squares as standards in Chapter 12.) At this point, we will examine squares primarily as reference surfaces—reference surfaces that have greater areas on the contact surfaces.

Naturally, Whitworth's passion for reference surfaces also led to a study of perpendicularity, and although he did not invent the method of making squares, he formalized it. We make master squares beginning in about the same way as we make reference surfaces: casting, rough machining, normalizing, and finish machining. We then scrape the machined squares in pairs in the same order we use for plane surfaces (see Figure 11–10). As before, we continue to the desired degree of precision, but in contrast to previous reference surfaces, making squares depends on another reference—the plane surface. We can generate a "reference" on a convex surface (see Figure 11–11), but although everything matches, the resulting reference is not "square."

In these examples, we have been solving simultaneous equations: the three surfaces in each case are x, y, and z, and we solve the equation x = y = z. Of course, we only solve this equation when the coefficients—which are subtracted when we scrape away thicknesses of metal—of three surfaces are all the same. For the "math" involved, it does not matter if we are dealing with millimeters, inches, or cubits: the theory is the same (see Figure 11–12). In this example, we calibrated an inside square against a master square, which, because of the reversal process, we can develop at the same time that we use it for calibration.

In Figure 11–12A, we place the rough male square in the inside square, where its errors merge with the error of the inside square. Then, we remove metal until the rough square fits. In Figure 11–12B, we turn the rough square 90° and adjust it again until it fits. In Figure 11–12C, we repeat these same steps. Then, at the next 90° turn (see Figure 11–12D), we achieved a divergence (x) that is four times the combined errors: we now know both the magnitude and direction of the error, so we can correct it in the inside square. After this correction, we repeat the four

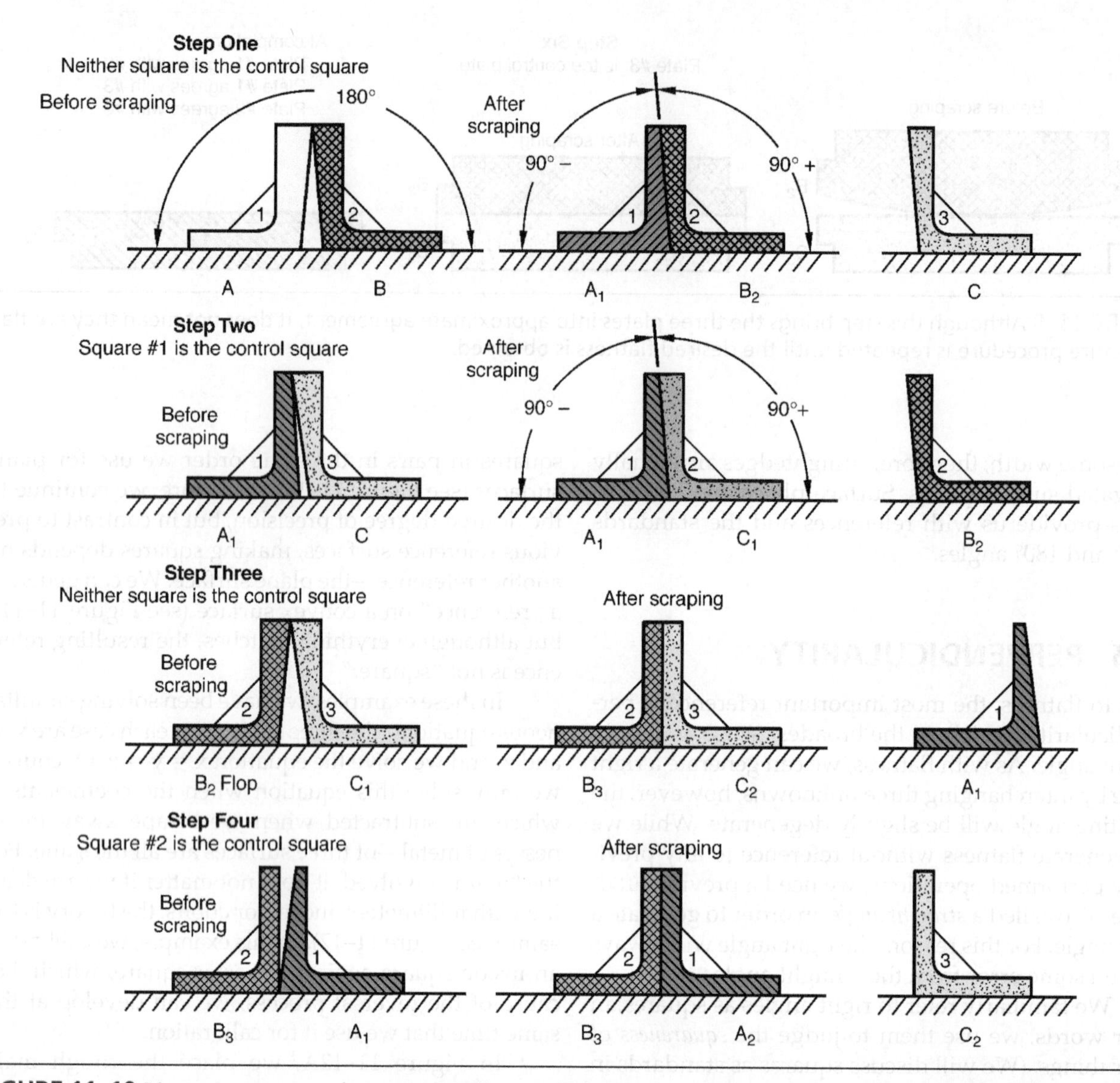

FIGURE 11–10 Master squares can be generated by much the same technique as was used for generating plane surfaces. However, the accuracy of the resulting squares is dependent on the accuracy of the reference surface on which the generation took place.

steps, and we again magnify the remaining error four times, locate and identify it, and correct it. We repeat this process, correcting the male and female squares at the same time, until we achieve the accuracy we want.

If we need a second inside square, all we have to do is use the finished male square to directly correct the new inside square. We would only scrape a square from scratch when the square is worn or if we doubt the integrity of the square we are using.

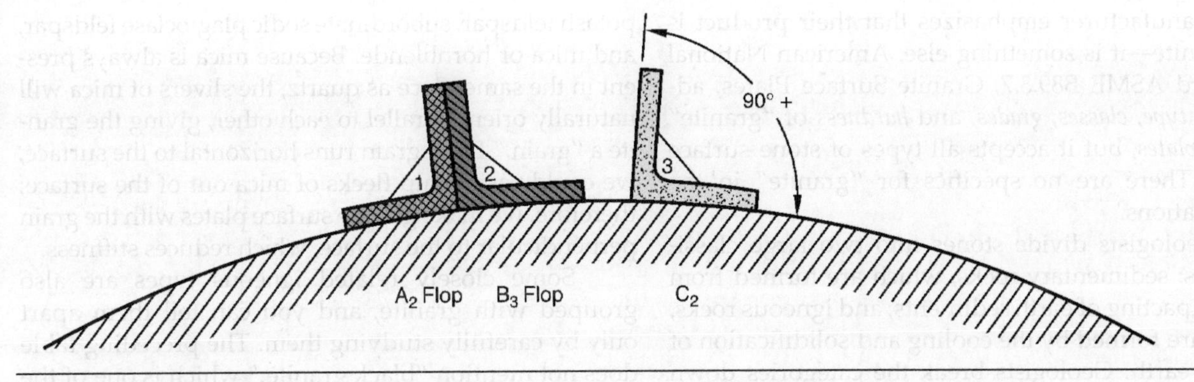

FIGURE 11–11 The generation of squares picks up the error of the reference surface as this example shows.

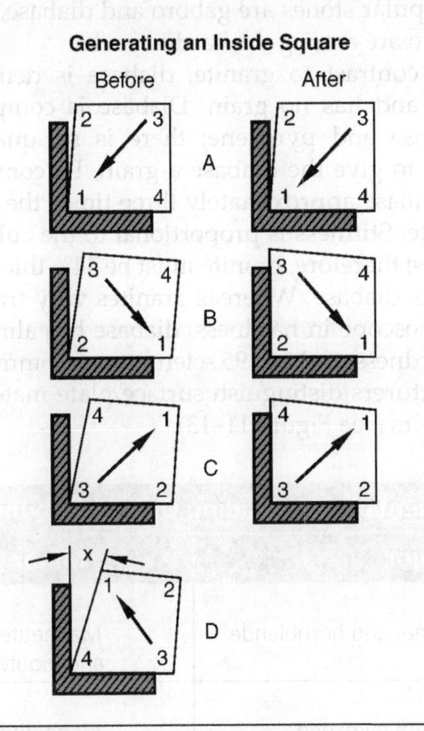

FIGURE 11–12 The same general technique can be used to simultaneously generate an inside square and a master square. X is the error magnified 4X.

11–4 MODERN REFERENCE PLANES

There probably will not be many circumstances when you will have to scrape your reference surfaces, and, today, commercial surface plates and right-angle plate accessories are generally granite instead of cast iron. Manufacturers use autocollimators to establish the geometry, and use lapping processes in place of hand scraping.

Granite Surface Plates

Granite surface plates provide superior reference surfaces, and manufacturers will attest to their superiority. Most large manufacturing facilities use them, and in plants that produce newer technologies, granite plates are used exclusively. Manufacturers rarely use metal for separate reference surfaces, except surfaces that are 15.24 cm (6 in.) in diameter or smaller, like toolmaker's flats. Built-in instrument or equipment references are usually metal, but granite is often used. To understand why, let us examine the properties of granite.

Americans first used granite for commercial surface plates around 1942, although it is impossible to say who used the first stone surface plate. Granite plates, however, are not just granite plates. There are:

- Several kinds of granite
- Several approaches to flatness
- Various functional features, such as ledges and inserts

Types of Granite

Although most metrological trade literature is a fairly dull collection of the same terms, definitions, and metrological data, no two manufacturers of granite surface plates consistently use the same terminology. The problem starts with the term *granite*, because each manufacturer emphasizes that their product is not granite—it is something else. American National Standard ASME B89.3.7, Granite Surface Plates, addresses *type, classes, grades,* and *hardness* of "granite" *surface plates,* but it accepts all types of stone surface plates. There are no specifics for "granite" in the specifications.

Geologists divide stones into two large classifications: sedimentary rocks, which are formed from the compacting of earth sediments; and igneous rocks, which are formed by the cooling and solidification of molten earth. Geologists break the categories down further:

Sedimentary	Igneous	
Limestone	Gabbro	Diabase
Marble		Hypersthene Gabbro
Slate	Granite	Biotite Granite #1
Shale		Muscovite-biotite granite gneiss
Sandstone		Biotite Granite #2
Miscellaneous		
Others		

Geologists define granite as an igneous rock in which you can individually distinguish the crystals with the naked eye, and which consists of quartz, potash feldspar, subordinate sodic plagioclase feldspar, and mica or hornblende. Because mica is always present in the same place as quartz, the slivers of mica will naturally orient parallel to each other, giving the granite a "grain." If the grain runs horizontal to the surface, we could easily chip flecks of mica out of the surface; therefore, we make granite surface plates with the grain perpendicular to the surface, which reduces stiffness.

Some closely related igneous types are also grouped with granite, and you can tell them apart only by carefully studying them. The preceding table does not mention "black granite," which is one of the most popular stones used for surface plates. The other most popular stones are gabbro and diabase, because these two are among the hardest rocks.

In contrast to granite, diabase is denser, less porous, and has no grain. Diabase is composed of plagioclase and pyroxene; there is no quartz and no mica to give the diabase a grain. Its composition gives diabase approximately three times the stiffness of granite. Stiffness is proportional to the cube of the thickness; therefore, granite must be 44% thicker to be as stiff as diabase. Whereas granites vary from 70 to 105 scleroscope in hardness, diabase has almost uniform hardness of about 95 scleroscope. Commercially, manufacturers distinguish surface plate materials by their colors (see Figure 11–13).

Class	Color	Texture	Mineral Constituents in Descending Order of Abundance	
			Major	Minor
1 Hypersthene gabbro	Dark gray	Fine-grained	Plagioclase, pyroxene, and hornblende	Magnetite and biotite
1 Diabase	Dark gray	Igneous	Plagioclase and pyroxene	Magnetite
1 Biotite granite	Pink	Medium-grained	Orthoclase with a small amount of microcline, plagioclase, quartz and biotite	Magnetite and garnet
1 Muscovite-biotite granite gneiss	Light gray	Medium-grained	Microline and orthoclase, oligoclase, quartz, rutile, muscovite, and biotite	Apatite
2 Biotite granite	Light gray	Medium-grained	Oligoclase, orthoclase and microline, quartz, and biotite	Apatite and zircon

FIGURE 11–13 Natural surface plate material.

Importance of Stiffness

Stiffness is important when we measure heavy workpieces to close tolerances. It measures the modulus of elasticity of the material (Hooke's law), making a ratio of stress to strain. The specification defines accuracy under load measured along a diagonal from opposite corners—for example, a normal load of 50 pounds for each square foot of surface area loaded in the center of the plate must not cause the plate to deflect more than one-half the flatness tolerance.

As we mentioned earlier, stiffness is a function of the cube of the thickness. Therefore, if we double the plate thickness, we increase the plate stiffness eight times, or it will deflect only 1/8 as much under a given load. If we increase the thickness 26%, we double the stiffness and cut the deflection in half. According to one manufacturer, they can double the stiffness of a surface for only a 10% increase in cost.

Surface Plate Selection

After evaluating all of the claims, we are left with a simple trade-off: strength versus wear life. The diabase and gabbro stones have no quartz and accompanying mica; therefore, they have no grain, which makes them stronger than true granites. In contrast, quartz is the most wear-resistant component in these stones. We are left to choose between thinner, less expensive diabase and gabbro stones that can carry the heavier loads and thicker granite stones that we resurface less often, thus saving money.

Surface plates from reputable manufacturers exceed the minimum requirements of GGG-P-463C(1); therefore, this standard is not much help when you are deciding on a stone. Fortunately, the standard does provide helpful guidelines for class, grade, and hardness.

Classes and Grades of Surface Plates

There are two *classes* of granite surface plates that are based solely on resistance to wear. Class 1 has the highest wear resistance, and Class 2 has medium wear resistance.

The specification divides these two classes into two *grades:* Grade A, specified for "high quality precision work," and Grade B, for "average quality precision work." Unfortunately, these work grades are not defined in the specification.

In practical use, we usually refer to Grade A plates as inspection grade plates. We refer to Grade B plates as toolroom grade or shop grade plates. Manufacturers also produce Grade AA surface plates, known as laboratory grade, which are not shown in the specification.

The Production of Surface Plates

Granite surface plates allow us to measure to closer tolerances—and save money. They also resist corrosion and rust, are not subject to contact interference, are nonmagnetic, hard, stable, long wearing, easy on the eyes, easy to clean, and have exceptional thermal stability. But what do these features do for us, the metrologists who use them?

Granite surface plates can be produced to flatness that we measure in millionths of an inch at comparatively low prices when compared with cast-iron and steel reference surfaces. Quarries cut boulders into slabs with wire saws. These saws are giant, ganged band saws that use twisted steel cable charged for cutting with abrasive slurry. We can only use a small portion of these slabs for surface plates, and plate manufacturers separate these portions with diamond-edged circular saws. They then send the slabs to honing beds where the slabs are roughly surfaced to within a few mm (25.4 mm or 0.001 in.). Then, the slabs are taken to temperature-controlled rooms where the rough plates are surfaced on rotary lapping machines, bringing them very close to the desired flatness. Manufacturers then use autocollimators to draw a profile of their surfaces and area lap, and recheck the surface until the high spots are eliminated and the plates are within tolerance. We do not use Whitworth's method in this process, but his work laid the foundation for the existence of any standards of flatness at all.

Because granite surface plates handle water so well, they do not rust or corrode, and you can keep your granite surface plates very, very clean. You can even scrub them, but we will go into maintenance later. Granites absorb so little moisture that you can leave even the finest metal surfaces on them for long periods of time and never have to worry about rust.

In contrast, cast-iron plates rust, and when we place iron or steel surfaces on them, cast-iron plates cause these surfaces to rust, too. For very fine metal surfaces, even the time from the end of your shift to the beginning of your next one—even overnight—can be too long to be in contact. You can control the harmful effects of cast-iron surface plates by using oil and by paying attention to every part of the plates at all times. Of course, oil collects dirt, and dirt adds measurable error and accelerates wear. It is a good idea to keep all precision surfaces covered, but you *must* keep cast-iron reference surfaces covered.

Granite plates have great thermal stability. In contrast, cast-iron plates respond quickly to drafts, bright lights, and the warmth of nearby personnel. Cast-iron plates normalize fairly quickly, whereas granite plates normalize more slowly, which, in some cases, outweighs the rust-free advantage of granite. If the humidity is high, moisture will condense on granite surfaces until the stone warms up, and this moisture will rust fine surfaces that are left on the granite plate. As you can see—and will need to remember to maintain the reliability of your measurement—there is no advantage for a metrologist that does not have a disadvantage.

The Hardness of Surface Plates

Hardness is an important consideration when you are selecting a surface plate. However, you cannot measure hardness with either the familiar Rockwell or Brinell hardness testers, because both of these testers are based on the deformation or flow of the material. Rock-type materials do not flow. They resist until the instrument reaches enough pressure to crush them. For rock-type materials, you need to use a scleroscope, which reads the calibrated height of the rebound of a hard test probe when you drop it from a known height.

We do not commonly use a scleroscope for hardness values in metrology. Therefore, to have a basis for comparison, we have provided the following table, which shows the range for commercial surface plates from the softest, gray granite, to the hardest, pink granite.

Scleroscope	Rockwell C
10	72
98	69
93	66
89	63
84	60
79	57

You can also test for hardness using the scratch test and Moh's Scale:

1. Talc
2. Gypsum
3. Calcite
4. Fluorite
5. Apatite
6. Feldspar
7. Quartz
8. Topaz
9. Corundum
10. Diamond

For example, if you can use the test specimen to scratch quartz but you can also scratch the test specimen with topaz, then the test specimen falls between the two on the Moh's Scale, so you would call it 7.5.

Hardness is not the only indication of wear resistance; density is often considered more important. Granite densities vary from 2611.0 kg/m^3 (163 lb./cu. ft.) for granites used for surface plates to above 3043.5 kg/m^3 (190 lb./cu. ft.).

Some people think that a disadvantage of granite is that you cannot wring to it, but you can, although the granite may not wring as strongly as metals and other materials. If you cannot wring to a granite plate, the plate is most likely poorly finished. However, even if the surface does wring, you cannot predict contact errors if you are making very precise measurements. We usually make these measurements with comparators with their own built-in references that have been lapped to a light band or less. To use most surface plates, you need to shift the work and the instruments around to make the measurement. In this case, unwanted wringing can be a nuisance, and wringing is rarely required because of contact error considerations. Granite plates provide casual contact (see Figure 9–19), which is precisely what we want for surface plate work.

Functional Considerations

Ledges are an accepted part of surface plate manufacturing. In fact, manufacturers actually denote the grade of a plate by its ledges: four ledges for laboratory grade, two ledges for inspection grade, and zero ledges for a toolroom grade. The number of ledges on a plate, however, does not ensure its reliability.

Tradition dictates many of the approaches we take to problem solving: an engine is in the front of an automobile because the horse used to be there. Ledges are still on surface plates for much the same reason. A ribbed cast-iron plate needed a ledge around the outer edges so the makers of the first granite plates—and the manufacturers of today—copied this design. Some people say that we need the ledges for clamping, but when we clamp only on the edge, we limit the wear to the edges of the plate and generate clamping error (see Figure 11–14).

Most importantly, edge clamping is unnecessary and, because ledges are cut with diamond saws, very expensive. It is more practical and less expensive to use threaded inserts imbedded in the surface and placed anywhere that best serves the intended use of the plate. You can attach quick-acting clamps to the inserts (see Figure 11–15). The T-slot can also eliminate ledges. The application for the plate will help you determine whether it should have a T-slot or inserts.

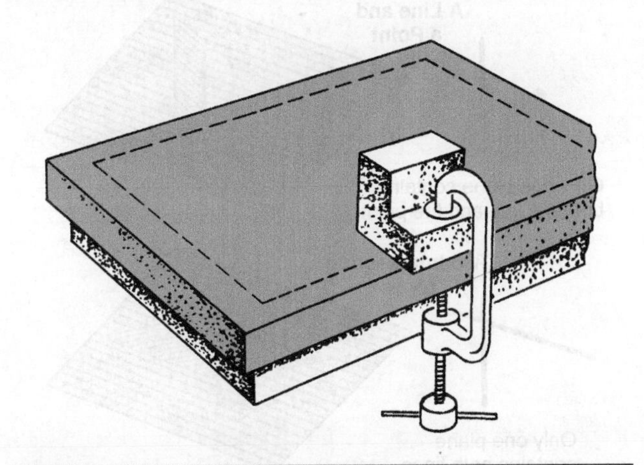

FIGURE 11–14 Edge clamping localizes wear although the reverse is desired.

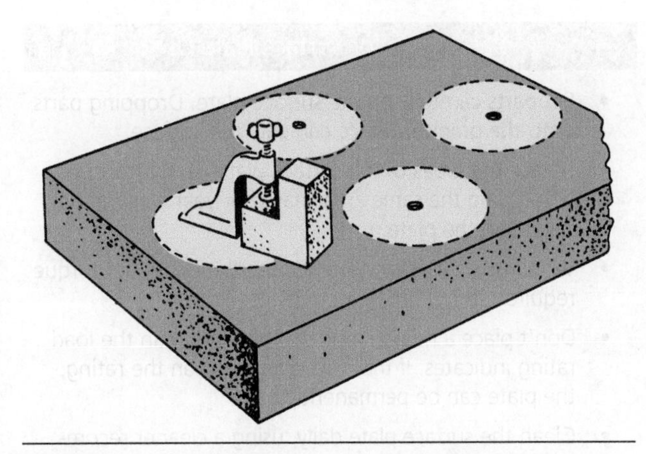

FIGURE 11–15 Threaded inserts in the plate permit quick-acting clamps to be used. 360° of the plate around the insert can be used, thereby spreading wear over a sizeable area.

Accessories provide both support and reference except for the straightedges, which are generally used for reference only.

Care and Maintenance

Surface plates do not require extensive care and maintenance. However, it is important to follow some basic rules. Keeping the surface clean and free from buildup of dust, dirt, grease, grime, and other foreign particles will maintain accurate tool readings and extend the life of the plate (see Figure 11–16). You will examine the surface and care of a granite surface plate in the lab manual exercises.

Do not use granite surface plates as workbenches or lunch tables. Dropping wrenches or hammers on plates can chip and nick the surface. Spilling food or drinks on a granite surface can leave microscopic pits since acids or grease dissolve the minerals in the plate.

Surface plates should be checked regularly for wear. Check the plate by setting an indicator gage on the plate and zero it at any point on the table. Move the gage over the plate and record the movements in the indicator. The plate should be checked from side to side and the indicator movements should be within the specified tolerance range. Movements greater than 0.000025" for AA plates indicate it is out

Basic Rules for Surface Plate Use

- Set parts carefully on the surface plate. Dropping parts onto the granite surface can chip the surface.
- Rotate the areas of the surface plate used for inspection. Using the same area year after year can wear a "hole" in the plate.
- When using threaded inserts, use the minimum torque required to hold the part in place.
- Don't place a heavier load on the table than the load rating indicates. If the load is higher than the rating, the plate can be permanently bent.
- Clean the surface plate daily, using a cleaner recommended by the vendor.
- If not in use, cover the plate with a surface plate cover to protect it from dirt and dust.

FIGURE 11–16 Following these rules will extend the life of your surface plate.

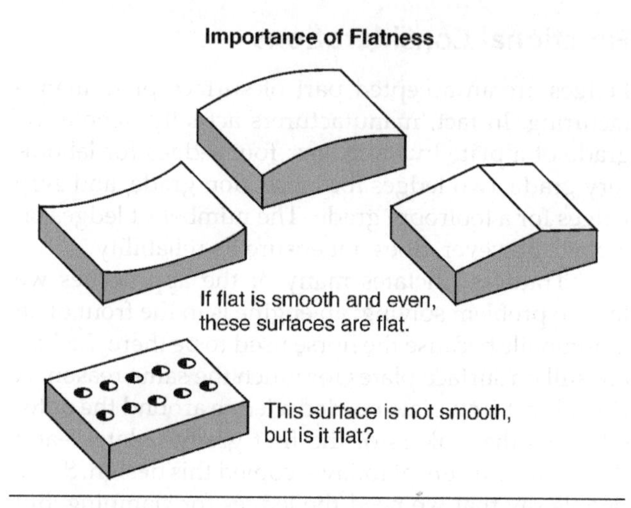

FIGURE 11–17 These exaggerated drawings show the confusion about flatness.

of tolerance. Out-of-tolerance plates can be relapped to restore the original accuracy.

Surface plates should be calibrated on an established timetable, depending on the frequency of use. Calibration of overall flatness traceable to NIST can be performed using an autocollimator.

11-5 HOW FLAT IS FLAT

The controversy still rages around the definition of "flat" (see Figure 11–17) and the dictionary cannot help metrologists with the definition. As a result, each manufacturer can claim that the configuration of their reference surface results from the method that generates true flatness—and has its own documentation to support its claims.

When we visualize a plane, we start with any three points in space. Only one plane can contain all three of those points, but because a line is an infinite series of points, the three points could define three lines. Each line could extend beyond the points that defined it. Therefore, to define a plane, you must have three points, one point and one line, or two lines that intersect (see Figure 11–18).

When we define planes, we use points and lines, which have no thickness. In fact, the plane itself has

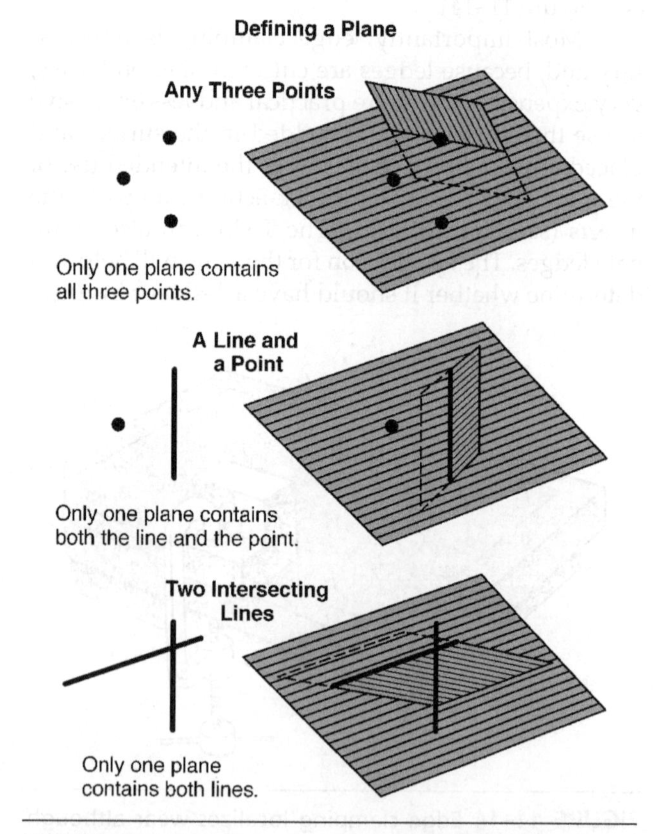

FIGURE 11–18 The small planes cutting the large planes show that only one in each case contains all of the elements.

no thickness, and no up or down, top or bottom, front or back. Because it is theoretical, this plane is the ideal reference plane.

In the real world, we must add substance to the theoretical plane so we can use it—and because we cannot make anything perfectly, we must build a tolerance into the surface. The tolerance of a reference surface is thickness—the amount that points on the surface lie outside of the theoretical plane. The range of the tolerance decreases as our skill in producing flat surfaces increases.

We can also choose to express the tolerance in different ways, which can create more problems because the reference plane itself is imaginary, so we cannot actually take measurements from it. We know about the plane only through our measurements of points, and we do not actually know that any of these points are exactly in the plane. Therefore, every time we talk about surface flatness, we are making a purely arbitrary selection of reference plane. Figure 11–19 shows that we can read one surface plate as zero deviation, x deviation, 2x, or even 4x.

The federal specification for flatness is clear: No point may deviate more than the specified tolerance from a mean plane (see Figure 11–20).

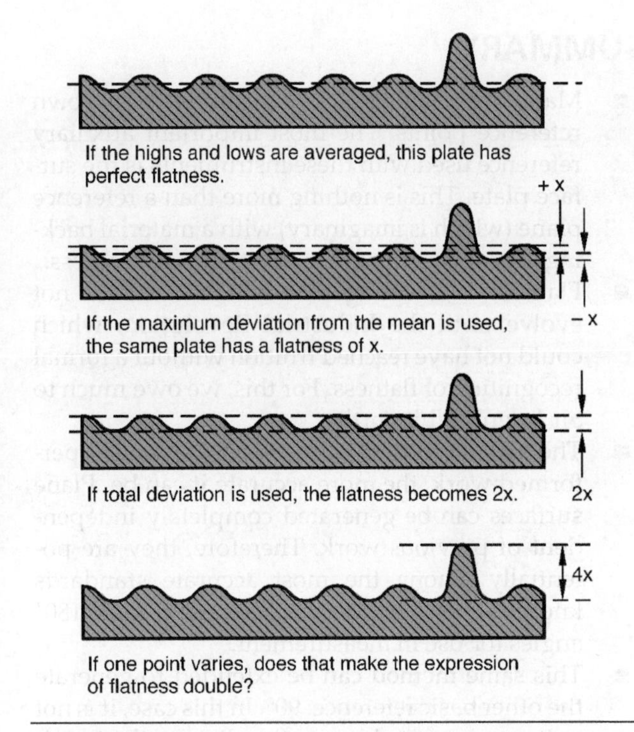

FIGURE 11–19 This illustration shows that the statement of accuracy must specify precisely how the accuracy is defined.

Size and Accuracy

Size				Work surface accuracy	
Work surface		Grade A thickness (minimum)	Grade B thickness (minimum)	Grade A tolerance plus or minus	Grade B tolerance plus or minus
Width	Length				
Inches	Inches	Inches	Inches	Inch	Inch
3 1/2	4	1	1	0.000025	0.0001
8	12	3	3	.000025	.0001
12	12	3	3	.000025	.0001
12	18	4	4	.000025	.0001
18	18	4	4	.000025	.0001
24	24	4	4	.000025	.0001
24	36	6	5	.000025	.0001
24	48	8	6	.000050	.00015
36	48	8	6	.000075	.00015
36	72	12	10	.00015	.0003
48	96	14	12	.0002	.0004
48	144	24	20	.0004	.001

FIGURE 11–20 This table is from Federal Specification GGG-P-463c(1). Note that the tolerance widens as the size increases.

SUMMARY

- Many instruments do not contain their own reference points. The most important auxiliary reference used with these instruments is the surface plate. This is nothing more than a reference plane (which is imaginary) with a material backing. The measure of its perfection is its flatness.
- Flatness is a stranger to nature and did not evolve until the Industrial Revolution, which could not have reached fruition without a formal recognition of flatness. For this, we owe much to Sir Joseph Whitworth.
- The less dependent work is on previously performed work, the more accurate it can be. Plane surfaces can be generated completely independent of previous work. Therefore, they are potentially among the most accurate standards known. They provide the essential 0° and 180° angles for use in measurement.
- This same method can be extended to generate the other basic reference: 90°. In this case, it is not quite so accurate because it relies on the previously created flat surface. Both inside and outside squares can be produced.
- Although government specifications only require two grades of surface plate, industrial standards require three: grades AA (laboratory), A (inspection), and B (shop).
- Granite surface plates are precision tools and should not be used as a workbench. Do not set your lunch, briefcase, or tools such as hammers and wrenches on the surface plate.
- Automotive and aircraft manufacturers use surface plates for body layout, to check for warped chassis, or to inspect small components for compliance to engineering standards. A flat reference plane is a vital tool for layout and inspection in every manufacturing industry.
- Not all granite is the same. Each type has its own particular physical characteristics, including hardness (wear resistance), stiffness (deflection under load), and density. The higher the quartz content, the better the plate.
- Surface plates are certified to Federal Specification GGG-P-463c(1) for both flatness and repeatability.

END-OF-CHAPTER QUESTIONS

1. In some cases, the measurement instrument does not contain the reference necessary from which to measure a part feature. In such instances, something else must be done. What?
   a. Revert back to the manufacturer for the specifications.
   b. Invert the instrument and subtract the reading from the total scale length.
   c. Use a hook rule to establish a reference.
   d. Use an auxiliary reference.

2. What is the most common reference plane in most shop work?
   a. The anvil of the micrometer
   b. The optical flat
   c. The surface plate
   d. The imaginary line that dissects the reference end of the part feature

3. Who was credited with recognizing the importance of flatness and screw thread standardization?
   a. Joseph R. Brown of Brown & Sharpe Manufacturing Co.
   b. Henry Maudslay
   c. Louis Polk of Sheffield
   d. Sir Joseph Whitworth
   e. Albert A. Michelson

4. Why was the old method of generation of flat surfaces discussed?
   a. It is historically significant.
   b. It is still the only method used to generate flat surfaces.
   c. It demonstrates the degradation of workmanship principle.
   d. It is fast and efficient and requires no expensive instrumentation.

5. What material is generally believed to have the best cost-effectiveness for surface plates?
   a. Chilled cast iron
   b. Normalized steel
   c. Industrial granite
   d. Liquid mercury

6. Which of the following materials are included in the classification "granite" surface plates?
   a. Diabase
   b. Hyersthene gabbro
   c. Biotite granite
   d. Muscovite-biotite granite gneiss
   e. Marble

7. Which of the materials has the greatest wear resistance?
   a. Diabase
   b. Hyersthene gabbro
   c. Biotite granite
   d. Muscovite-biotite granite gneiss
   e. Marble

8. Which of the following is the chief reason for the popularity of granite as a material for surface plates?
   a. Resistance to wear
   b. Flatness
   c. Low cost
   d. Imperviousness to water
   e. Stiffness

9. How is the flatness of the surface plate usually checked?
   a. Accumulated errors from repetitive measurements
   b. Use of the autocollimator
   c. Comparison with straightedges
   d. Three-plate matching

10. For surface plates, how is flatness defined?
    a. By the root mean square
    b. By the manufacturer's calibration
    c. By the separation between two planes that contain all high and low points
    d. By the vertical distance between the highest and lowest points for a unit of area

11. Which of the following features of a surface plate provides for the plate being a reliable horizontal reference surface?
    a. Smoothness
    b. Flatness
    c. Parallelism
    d. Length and width

12. Surface plates should be used for:
    a. A convenient lunch table
    b. A level surface to assemble small components
    c. For precision inspection and layout
    d. A clean surface to store tools

13. Which of the following statements is not true?
    a. All granites are the same.
    b. Surface plates require regular inspection.
    c. Surface plates should be calibrated according to a regular time schedule.
    d. There are three grades of surface plates.

14. Which of the following grades is a laboratory grade for surface plates?
    a. AA
    b. AAA
    c. A
    d. B

12

CHAPTER TWELVE

ANGLE MEASUREMENT

In mathematics I can report no deficiency, except it be that men do not sufficiently understand the excellent use of the Pure Mathematics.

Francis Bacon (1561–1626)

LEARNING OBJECTIVES

- Explain the importance of the circle in metrology.
- State the relationship of the circle to angle measurement.
- Explain the role of squares.
- Explain the role of levels.
- List the basic methods for precision angle measurement.
- Describe the inherent difference between linear and angle measurement.
- State the role of sine bars and plates.
- Apply the circle, trigonometric functions, and sine bars and plates to angular measurements.

OVERVIEW

All length and angle standards are arbitrary human inventions—even the light wave standard (2.99796×10^8 m/s or 186,284 mi/s)—because even though light is a natural phenomenon, man created a length standard out of it. One standard, however, is not an arbitrary creation of man: it actually exists in nature—the circle.

The circle can be the path of an electron around the nucleus of its atom or the circumference of a planet, but its geometry is always the same. The parts of the circle always have the same relationships to each other; therefore, the circle is a universal

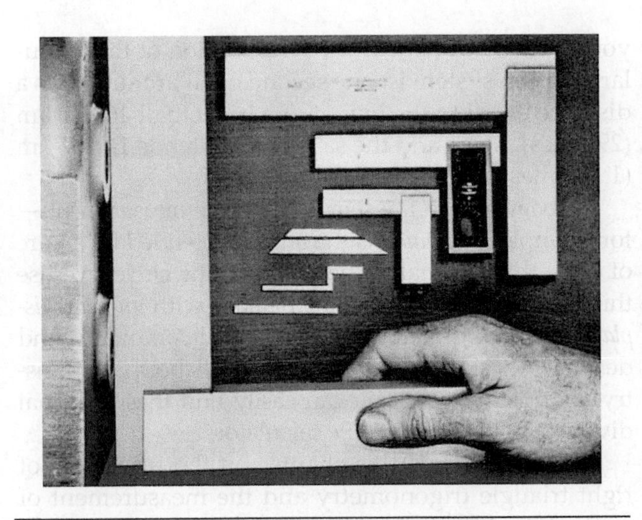

FIGURE 12–1 Squares, in all their diverse forms, are the most basic of the angle measurement instruments. *(Courtesy of the L. S. Starrett Company)*

standard that we can re-create anywhere at any time to measure angles. Angular measurement is inescapable in all technical endeavors, and is used in every phase of life, from botany and carpentry to billiards and marbles. Squares, in all their diverse forms, are the most basic of the angle-measurement instruments (see Figure 12–1).

We will discuss angular measurement instruments in much the same way you are likely to encounter them. There are several types of each basic instrument, and some have been so refined that it is hard to tell to which basic instrument family they belong. For this reason, we will concentrate on the principles. Exercises from the lab manual will show you practical use of a cylindrical square, bevel protractor, and a five-inch sine bar.

As we have emphasized before, you should always consult the manufacturer's manual when you are using any precision instrument. Remember, you can increase the precision of the instrument and your measurement when you use the instrument properly. The best way to learn to use an instrument properly is to read the manual, but for our purposes, we will explore the principles that form the basis for all instrument instructions.

Background

Historical evidence reveals that concepts of trigonometry date back to the ancient Greek and Egyptian engineers who used *trigonometry* to measure objects they could not measure directly. Today, trigonometry, which has evolved like any other branch of mathematics, is an invaluable tool in every phase of technology. When American physicist Albert A. Michelson (1852–1931) made more than 3,000 individual measurements from 1926 to 1929 and averaged the speed of light at 2.99796×10^8 meters/second, he was continuing the evolution of measurement.

The Circle

A *circle* is a curve consisting of points in a plane, all equally distant from a center point. It is different from all other curves because it is the same at all points. If we turn a circle around its center in the same plane, the circle appears exactly the same as it did before we turned it: all new positions are exactly like the original position, which is a characteristic of circles called *roundness*.

We form a circle by continuous motion of a fixed length around a point; therefore, the perfection of the circle is independent of the instrument we use to scribe it. In contrast, when we use a straightedge to create a line, we duplicate all the errors of the straightedge in the line.

Although some people imply that measurement has been a process of continual evolution from the Egyptian cubit to modern light wave standards, linear measurement did not evolve that way, unfortunately. The death of a particular culture also meant the death of its measurement system. The only conclusion we can accurately draw is that each culture creates its own system of measurements—except for angular measurement. The circle has been used as a standard probably as far back as biblical Babylon, and its use as a standard for angle measurement has been passed on to us in an unbroken line from the past.

The Babylonians used the sexagesimal system of notation for weights, time, and currency, which used a fraction with 60 as the denominator. Our "decimal system" would more accurately be called the "decimal-fraction" system, because you can write any number as a multiplier of ten (decimo) and/or a fraction. For convenience, we use a decimal point instead of a fraction with a denominator. There is a better way to refer to these numerical systems: the decimal system is the *base-ten system;* the binary system for computers is the *base-two system;* and the sexagesimal system is the *base-sixty system.*

Greek astronomers may have borrowed the base-sixty scale from the Babylonians, and then added their own twist, choosing 120 units to represent the diameter of a circle. They may have chosen this "base" because of the number of fractions they could create with the possible factors (2, 3, 4, 5, 6, 10, 12, 15, 20, 30, 40, and 60). At this time, π (pi) was defined as the whole number 3; therefore, the circumference of a circle was equal to three times the diameter and the circle was a set of 360 units.

People often assume that angular notations of minutes and seconds were derived from time or something to do with time, but the Latin root of "minute" is *minutus,* which means "small." Both time measurement and angular measurement derive a minor division from this root word. In the circle, each major division was divided by 60, becoming the first, second, third, and so forth "minute part" or "small part." With regular use, these divisions became simply minutes and seconds.

More recently, anthropologists have suggested that we have confused cause and effect: the Babylonians divided the circle into 360 parts in error, believing that each of the 360 sections was the distance the sun moved in one day. By the time the 5-day error was discovered—hundreds of years later—astrologers had based countless formulae in the easy-to-compute sexagesimal system, and it would have been too difficult to reexpress all those formulae based on 365. The circle remained based on 360 divisions; therefore, some anthropologists believe we use the sexagesimal system for the circle by chance, not by any systematic effort.

The 360 divisions of the circle were preserved through the Middle Ages by astronomers and have reached us intact. The notation is familiar: degrees (°), minutes ('), and seconds ("); the latter two, unfortunately, can be confused with feet and inches. To give you an idea how fine the discrimination of this angular notation system is: one second of an arc subtends a distance equal to the height of a basketball 46.671 km (29 miles) away, and the same arc subtends 1.8507 km (1.15 miles) on the moon.

Today, we do use some other angle measurements—for example, the *radian* is the angle subtended by an arc of circle that is equal to the radius of the circle. We use this measurement chiefly for dealing with *angular displacement,* like gear rotations. You will commonly find decimal divisions of the degree, used both in industry and science, and you can easily find these decimal divisions with any scientific calculator.

We will focus this chapter on the principles of right triangle trigonometry and the measurement of angles based on the circle.

12–1 ANGLES

We will mention a few basics about angles before we go on to angle measurement.

First, angles deal with directions—not spaces—because an angle is the relationship between two lines (see Figure 12–2). We can measure this relationship if we extend the lines until they intersect. The intersection is called the *vertex,* and the lines are *sides.* We can conveniently designate an angle using three letters, AOB in our example, and, frequently, we use the symbol ∠ to designate an angle, such as ∠ ABC. We do

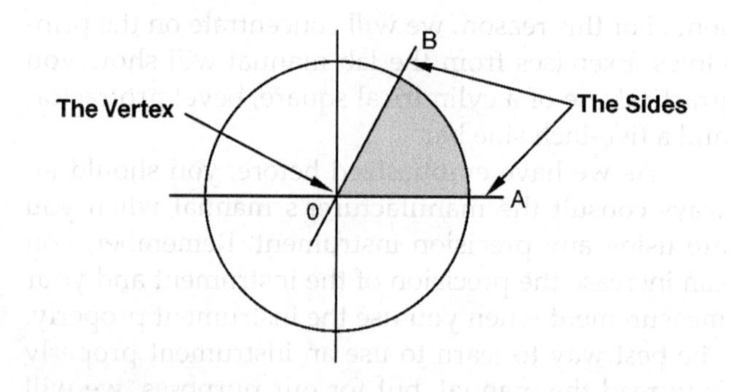

FIGURE 12–2 The angle is defined as AOB. It refers to the directions of the sides, not to the space between them.

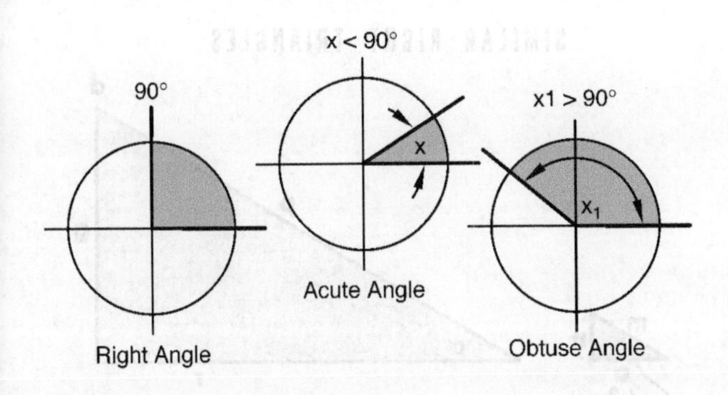

FIGURE 12–3 A right angle is 90°. If smaller, it is acute; if larger, it is obtuse.

need to be careful not to confuse the symbol notation for an angle with the mathematical notation for *larger than* (>) or *smaller than* (<).

A *right angle* is simply one-fourth of a circle or one *quadrant* (see Figure 12–3). If an angle measures less than 90°, it is an *acute angle;* and if the angle measures more than 90°, it is an *obtuse angle.* You can measure an angle from either direction (see Figure 12–4); for example, we could measure the angle on the left as 315° or 45° and the one on the right as 225° or 135°.

To minimize ambiguity, we use the terms *supplementary* and *complementary* angles. Two angles—one acute, which, unless specified, is always the "angle"; and one obtuse, which, unless specified, is always the supplement—that total 180° are *supplementary angles* (see Figure 12–5). Similarly, two angles that total 90° are complementary angles (see Figure 12–6).

You must observe two precautions with the measurement of angles. First, you must express the angles in terms of rotation in the same direction, either

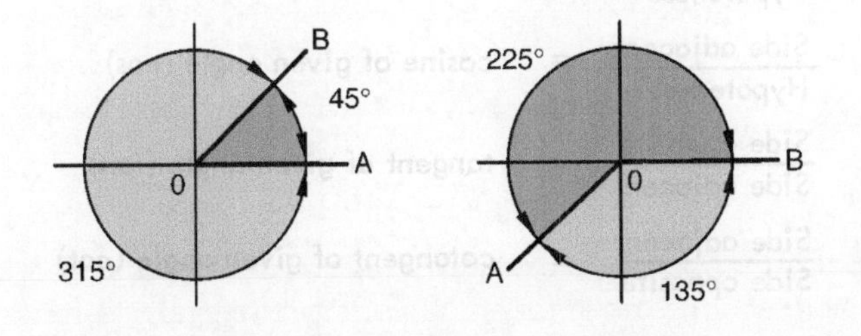

FIGURE 12–4 What is the measurement of ∠AOB in each case?

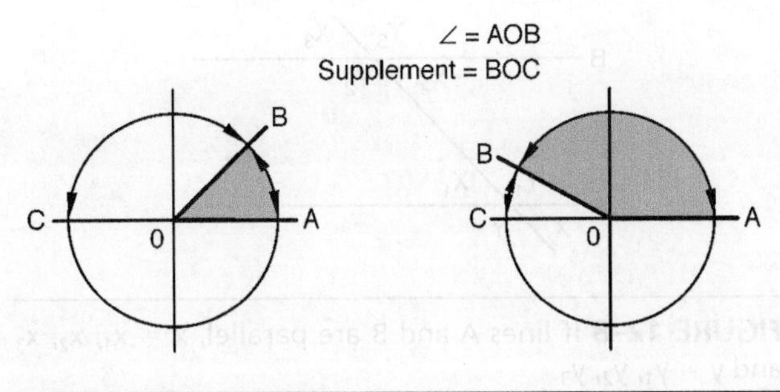

FIGURE 12–5 Supplementary angles total 180°.

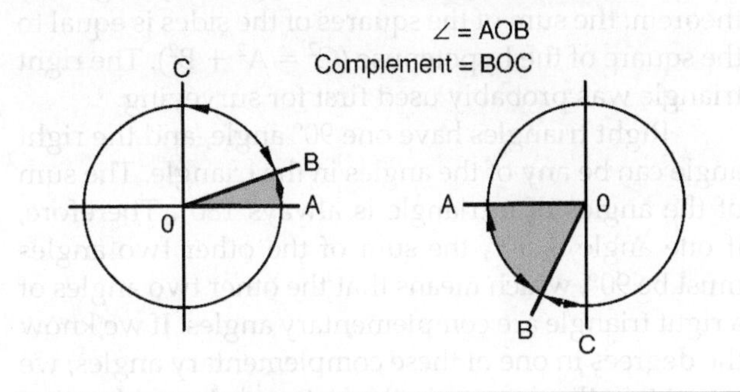

FIGURE 12–6 Complementary angles total 90°.

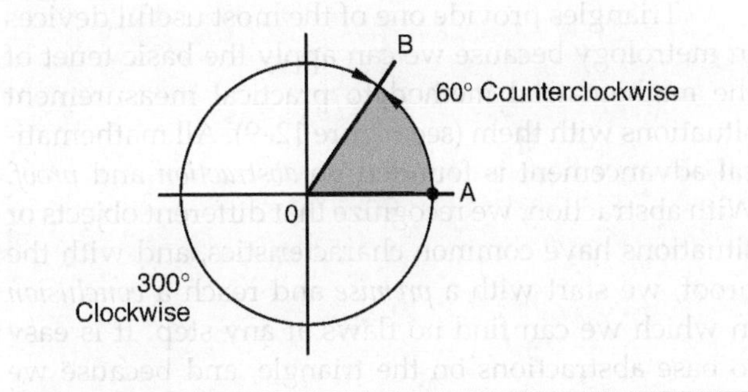

FIGURE 12–7 The clock rotation analogy is used to describe angles.

clockwise or counterclockwise (see Figure 12–7). The relationship of parallel lines and angles is shown in Figure 12–8.

Many measurement operations involve right triangles—so many, in fact, that the ancients attributed mystical properties to the right triangle. The Egyptians discovered that, if they knotted a string at three-, four-, and five-unit intervals, they could use the string to lay out a right triangle. Although the Egyptians

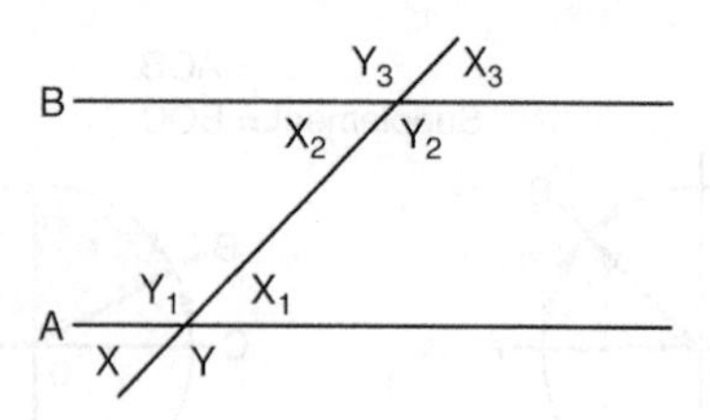

FIGURE 12–8 If lines A and B are parallel, $x = x_1, x_2, x_3$ and $y = y_1, y_2, y_3$.

did not know it, they were following the Pythagorean theorem: the sum of the squares of the sides is equal to the square of the hypotenuse ($C^2 = A^2 + B^2$). The right triangle was probably used first for surveying.

Right triangles have one 90° angle, and the right angle can be any of the angles in the triangle. The sum of the angles of a triangle is always 180°. Therefore, if one angle is 90°, the sum of the other two angles must be 90°, which means that the other two angles of a right triangle are complementary angles. If we know the degrees in one of these complementary angles, we can obtain the degrees in the last angle by subtracting the known quantity from 90.

Triangles provide one of the most useful devices in metrology because we can apply the basic tenet of the mathematical method to practical measurement situations with them (see Figure 12–9). All mathematical advancement is founded on *abstraction* and *proof.* With abstraction, we recognize that different objects or situations have common characteristics, and with the proof, we start with a *premise* and reach a *conclusion* in which we can find no flaws at any step. It is easy to base abstractions on the triangle, and because we define the angles in relation to a circle, triangles furnish almost irrefutable proof. If we use the example

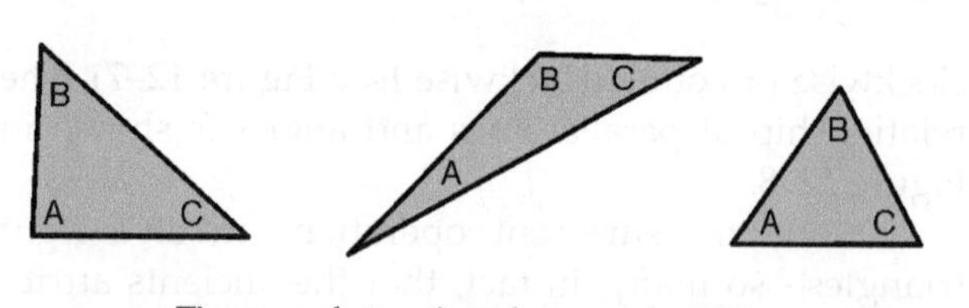

FIGURE 12–9 The sum of angles for all triangles is equal to 180°.

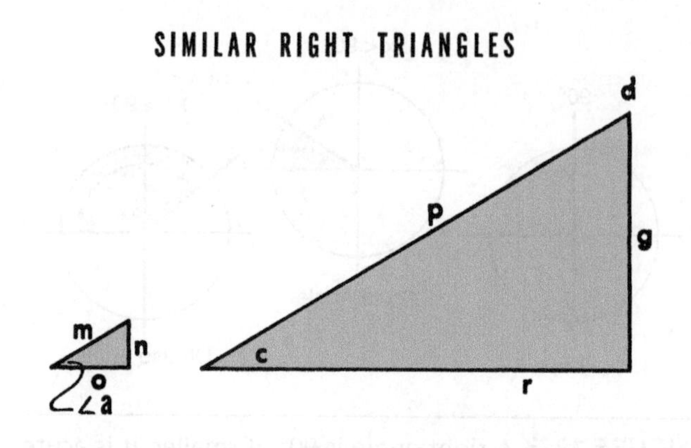

FIGURE 12–10 If ∠a of the smaller triangle equals ∠c of the large one, side m has the same relation to the side o as side p to side r.

of a 30°–60°–90° triangle, the proportions among the angles are exactly the same if the triangle is a minute printed circuit or a relationship among stars (see Figure 12–10). If the hypotenuse of the small triangle is 10 and the side opposite the given angle is 5, then in the larger triangle, if the hypotenuse measures 100, we know the side opposite the given angle must be 50.

Side opposite and *side adjacent* are the other important terms for triangles that we use throughout our calculation of trigonometric functions (see Figure 12–11). Trigonometric functions are important

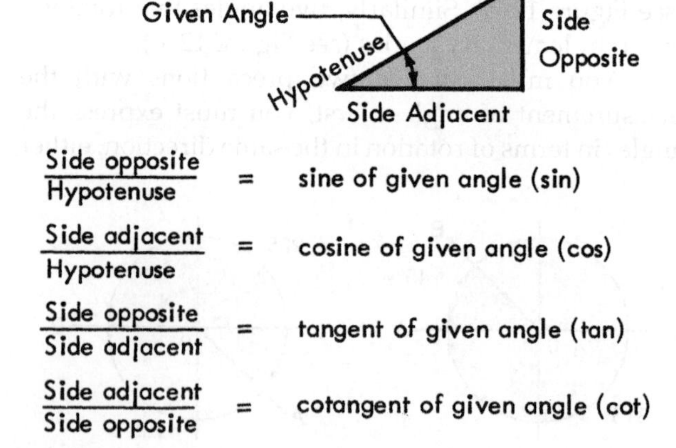

FIGURE 12–11 These are the most common of the trigonometric functions.

because they let us designate angles in relationship to the ratios of the sides of triangles. These ratios have been worked out to long decimals, so they are among the useful tools of measurement.

12–2 ANGLE MEASUREMENT

The instruments for angle measurement are equivalent to those for linear measurement and range from simple scaled instruments to highly sophisticated types using interferometry (see Figure 12–12).

You were probably introduced to the most familiar angle measurement instruments, the *simple protractor*, in grade school. With the addition of a vernier scale, we improve the discrimination of the protractor, creating the *vernier protractor*. Dividing heads and rotary table devices are protractors with additional mechanical support for greater range and/or reliability. The mechanisms can be very elaborate, but the basic principles are still the same.

There are even tools that measure angles in much the same way that we use the measuring microscope, called *optical tooling*. The most familiar instruments are the *collimator* and the *autocollimator*, which have the ability to measure over large distances. The most refined of all angle measurement instruments may be the *pointing interferometer*.

With angle measurement, we may not always be measuring angles—as strange as that may sound. Extremely sensitive instruments, like autocollimators, usually measure alignments, straightness, and flatness, and they generate angle readings, which they record as the measurements of error.

All of the preceding instruments are for direct measurement; however, like linear measurement instruments, there is also a group of instruments for the indirect measurement of angles. These indirect measurement instruments are based on the sine of angles, the geometry of regular polygons, and blocks similar to gage blocks. The most familiar instruments are based on the *sine* of an angle—the relationship of the height of a right triangle to its hypotenuse—and are appropriately called *sine bar, sine plate, sine table*, and so forth. We use all these instruments in conjunction with comparison measurement instruments, such as dial indicators or electronic comparators.

The Importance of Squares

When we bisect a circle, we automatically get two 180° angles; therefore, we may consider any point on a reference surface the vertex of a 180° angle. Much of our work in measurement involves distances from reference planes, so when the reference plane is *horizontal*—as it usually is in practical work—the distance is a height. Measurement reliability depends on the degree to which that height is perpendicular to the plane.

Perpendicularity is the measurement of an angle—a right or 90° angle. Remember, to have reliable measurement, you must have a standard, and the measurement must be provable. In Chapter 14, we demonstrated that we can generate a right angle

Equivalent Instruments

Linear Measurement	Type	Angular Measurement
Steel Rule	scaled	Plain Protractor
Combination Square	scaled	Protractor Head of combination set
Vernier Caliper	vernier	Vernier Protractor
Micrometer	mechanical	Index Heads
Gage Blocks	standards	Angle Blocks
Comparators	comparison	Sine Devices with comparators
Measuring Microscopes	optical	Autocollimators

FIGURE 12–12 The angular measurement instruments closely approximate their linear measurement equivalents.

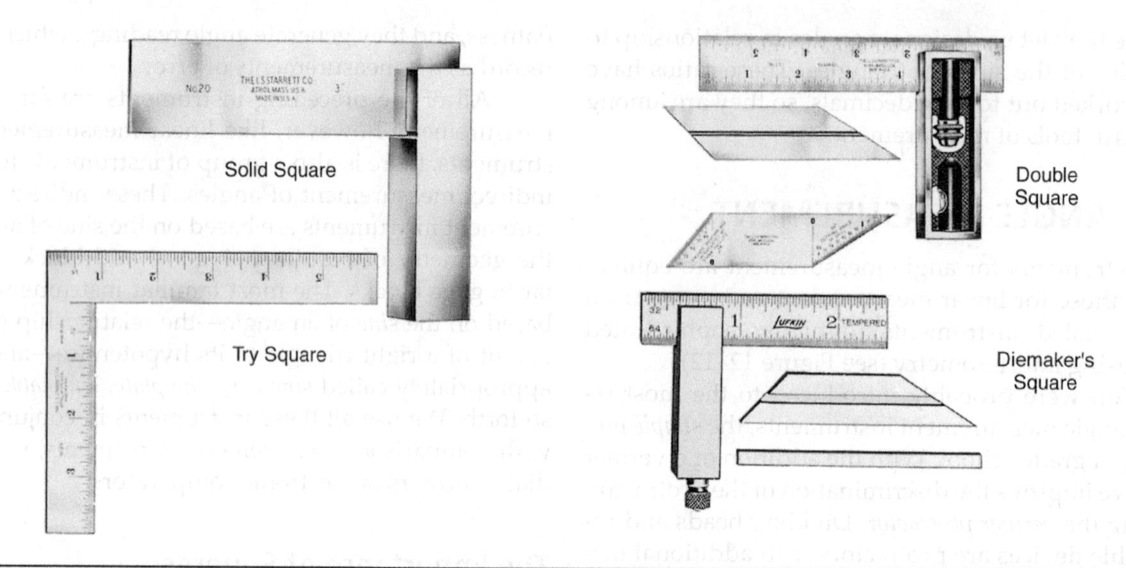

FIGURE 12–13 These are but a few of the many varieties of squares.

from nothing more than a plane surface, which is the most irrefutable proof possible. The right angle is one-fourth of our standard for angle measurement, the circle, and it is provable. Add four right angles together, and you are back to the circle; subtract two right angles, and you have no angle at all. The right angle is the logical place to begin a study of applied angle measurement, because it is self-checking.

Squares are hardened-steel right angles and are available in a wide range of sizes and shapes (see Figure 12–13), but we use the squares we will discuss in this chapter differently than other squares we have discussed. Previously discussed squares were large, rugged, and intended primarily to hold parts at right angles to a reference surface. The ones we will discuss now are measurement instruments used primarily to determine whether or not angles are right angles.

We discussed the most familiar right angle measurement instrument, the combination square, in Chapter 5 (terminology in Figure 12–14). We use either the inside or outside of these squares for comparison measurements. Most commonly, we simply

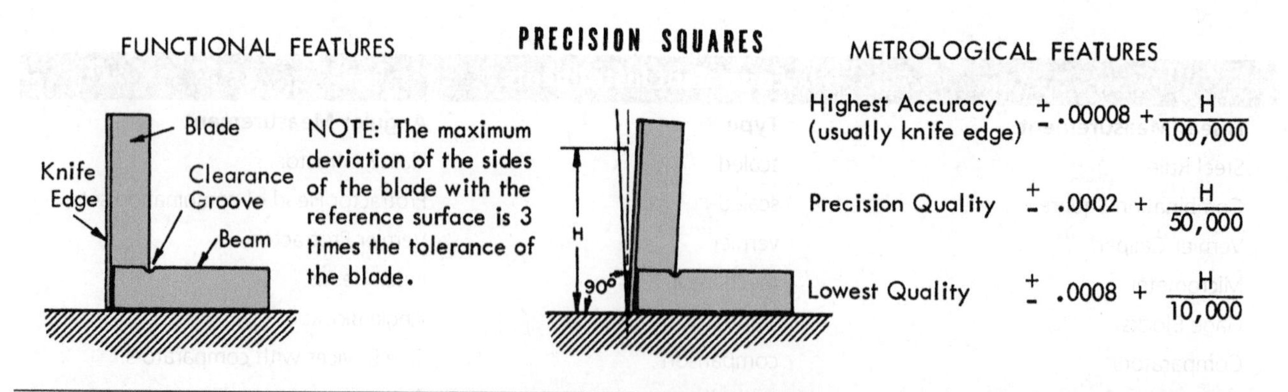

FIGURE 12–14 The precision square differs from many instruments in that its precision is built in and not determined by adjustments.

use observation, comparing the outside surfaces of the square with the part after placing both the square and the part on a reference surface. We usually do not pay much attention to the bases of the part and the square, but you should inspect both—and the reference surface—to make sure they are clean and free of burrs or dirt particles.

You should then slide the square into contact with the part feature you are checking. You should try to "wring" the square to the surface or at least go through the motions. You will usually be able to determine the "squareness" by simply looking, and if you need additional help, you can place a white sheet of paper behind the square and part. You will be able to see a space of 2.54 μm (one-tenth mil, 0.0001 in.) between the part and the square. If you need more precise measurement, you should use a magnifying glass.

For best results, you can place a light box or other good source of light behind the part and the square; again, you should be able to readily see a gap of 2.54 μm (0.0001 in.). If the space measures between 1.27 μm and 1.778 μm (50 μin. to 70 μin.), the light will look red to you. When the space is approximately 0.762 μm, the light will appear to be blue. The color of the light is the result of diffraction caused by the narrow slit, which is the same principle used in spectroscopy.

You can use another method to determine the space between the part and the square: insert narrow strips of paper between the part and the square at the top and bottom of the feature. When you push the square into the part, both strips of paper should be held in place; if they are, the part is square within the thickness of the paper. If you can pull the papers gently with the amount of tension, the part is square to a measurement less than the paper thickness. Cellophane is about 25.4 μm (0.001 in.) thick; therefore, some people consider this method accurate to 12.7 μm (0.0005 in.).

Some people will naturally turn the square at a slight angle to the surface they are checking in order to compare the part with a sharp edge (see Figure 12–15). However, only the narrow edges of the blade are accurately perpendicular to the beam, because the blade is unstable in its thin direction and can flex freely. If you turn the square, you add a portion of the inaccurate side to the accurate edge, but you will most likely see the error in squareness in the instrument and not the part. To eliminate this tendency, you can get a square with knife edges; however, you must remember that, as contact area decreases, the rate of instrument wear increases. You should reserve these fine-edged squares for highly precise work where you are using a crack of light to measure squareness. Another highly precise square is shown in Figure 12–16.

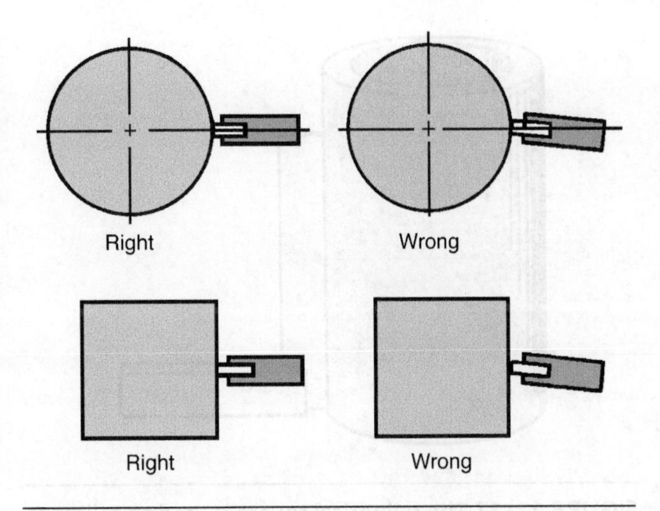

FIGURE 12–15 Squares are only accurate when used with the narrow edge of their blades along a diameter of round parts or perpendicular to the surface of flat parts.

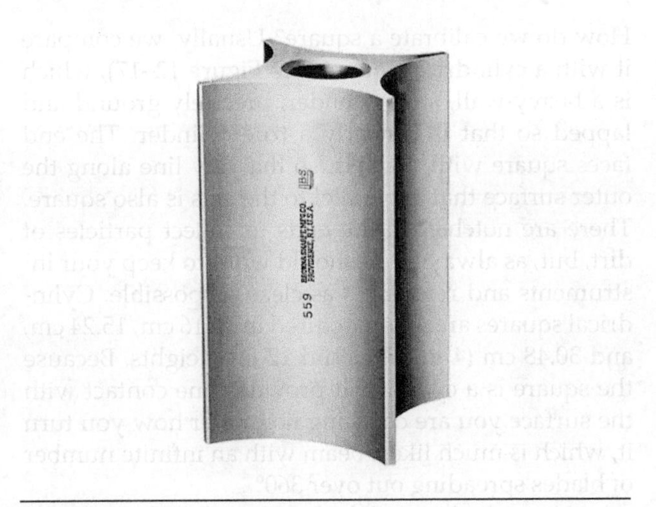

FIGURE 12–16 The master square provides a stable base and four reference edges.

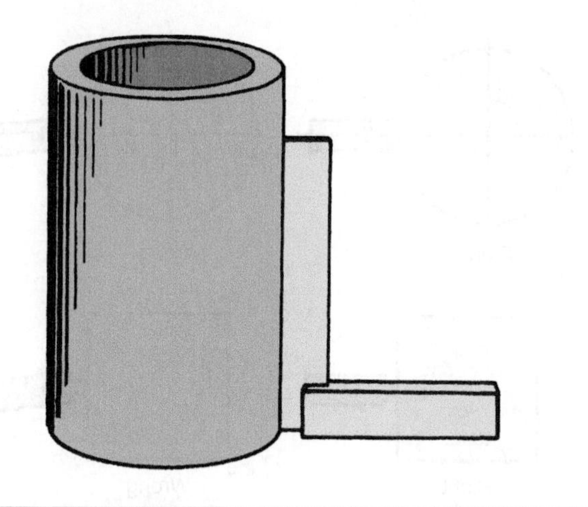

FIGURE 12–17 The cylindrical square is used to calibrate squares.

The Cylindrical Square

How do we calibrate a square? Usually, we compare it with a cylindrical square (see Figure 12–17), which is a heavy-wall, steel cylinder, precisely ground and lapped so that it is nearly a true cylinder. The end faces square with the axis, so that any line along the outer surface that is parallel to the axis is also square. There are notches on the ends to collect particles of dirt, but, as always, you should work to keep your instruments and references as clean as possible. Cylindrical squares are manufactured in 10.16 cm, 15.24 cm, and 30.48 cm (4 in., 6 in. and 12 in.) heights. Because the square is a cylinder, it provides line contact with the surface you are checking no matter how you turn it, which is much like a beam with an infinite number of blades spreading out over 360°.

You must be careful when you place the 30.48 cm (12 in.) cylindrical square on a surface plate, because it is very heavy and could damage the reference surface or injure you. This cylindrical square weighs more than 27.216 kg (60 pounds), and the very smooth surface of the reference plate makes this square difficult to handle.

If you lower a 30.48 cm (12 in.) cylindrical square vertically onto the plate (see Figure 12–18), it may slip. It is better to place the square in its storage box on the plate, because the square lies horizontally in the box and can be lifted out and placed horizontally on the plate. You can then raise the cylindrical square from one end, steadying the lower end to keep the square from sliding. If you need to move or reposition the square on a plate, you can most easily move it by rolling it up over the edge of the plate and then repositioning it by using the storage box.

The direct-reading cylindrical square is a variation of the standard cylindrical square (see Figure 12–19). It is a 15.24 cm (6 in.) chrome-plated square with a numerical readout of the amount that the part feature is out of square. In order to achieve this ability to read the amount out of square, we lap one end of the cylinder slightly out of square. We then calibrate the sides and lay them out in elliptical curves that are marked by small punch marks numbered at the top from 0 to 30.48 cm (0 in. to 12 in.) in steps of two. Each number represents 5.08 μm (0.0002 in.) (5.08 μm, 10.16 μm, etc., or 0.0002 in., 0.0004 in., etc.) and the zero deviation position is marked by a line of vertical punches.

You place the square in contact with the part feature and turn it until no light passes between the part and the square. You follow the highest curve to the top where the number shows the out-of-squareness in 5.08 μm (0.0002 in.) per 15.24 cm (6 in.) of height. You can make the reading at one of two places on the circumference, which makes the direct-reading cylindrical square self-checking. You can also use the direct-reading cylindrical square in the conventional manner, using the unnumbered end as the base.

The Degree of Squareness

For the erection of machinery and scientific work, squareness error is measured in seconds of arc. However, for machine shop work, we measure the linear amount of "out-of-squareness"—for example, the number of μm (mil) in a specified distance because it is more convenient. It is easy to determine squareness when you cannot see light passing between the square and the part: you can safely assume that the part is

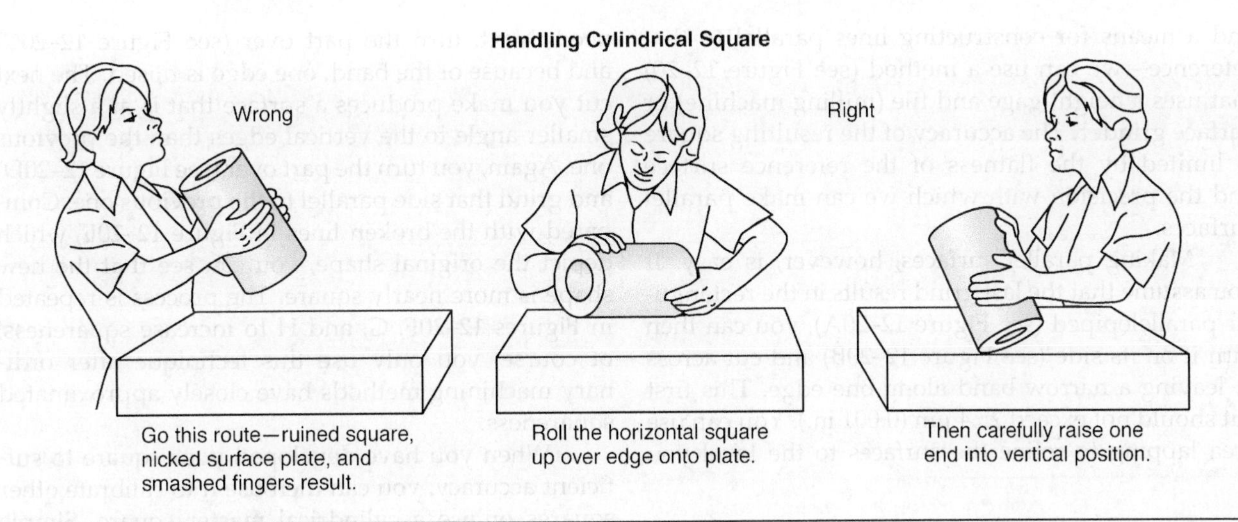

FIGURE 12–18 Handling a 12–inch cylindrical square requires care.

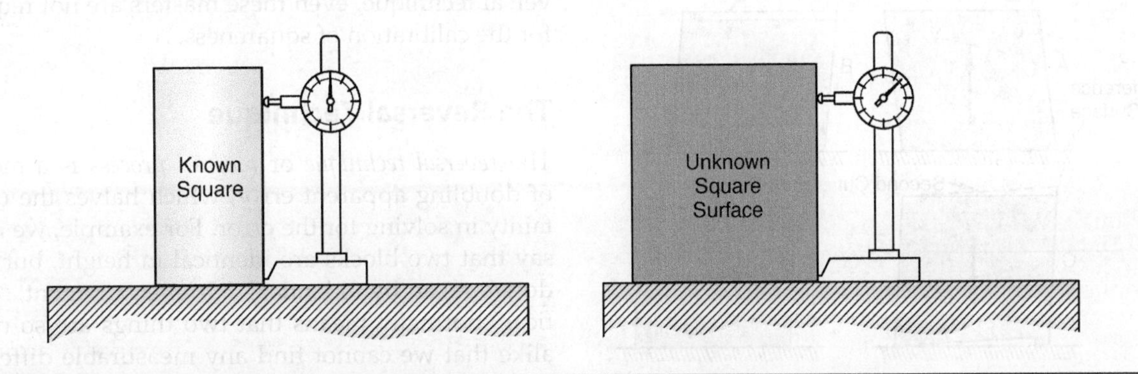

FIGURE 12–19 A crack of light is open to argument. So is an indicator reading—but not nearly as much.

square to within 10 times the squareness tolerance of the instrument. The part's squareness may be even more accurate, but it is difficult to be sure, because the reference surface affects the part and the square, which we will discuss later.

When the part is not square, we need a way to determine how much out-of-squareness a crack of light designates. For very narrow cracks, we can use color, but how do we distinguish between 25.4 μm (0.001 in.) and 50.8 μm (0.002 in.)? In the past, experts had no way of knowing what the difference was, so they often disagreed. Today, we can simply put together a setup with a dial indicator or electronic comparator that travels in the horizontal plane at some fixed distance above the reference plane. When we compare the horizontal readings of an unknown square with a known one, we obtain a numerical measure of squareness (see Figure 12–19).

From Parallelism to Squareness

Angular metrology starts with a standard for angle measurement that exists in nature. In order to create a square, all we need is a flat reference plane and a means for constructing lines parallel to that reference—we can use a method (see Figure 12–20) that uses a height gage and file (milling machine) or surface grinder. The accuracy of the resulting square is limited by the flatness of the reference surface and the precision with which we can make parallel surfaces.

Making parallel surfaces, however, is easy. If you assume that the last grind results in the rectangular parallelepiped (see Figure 12–20A), you can then turn it on its side (see Figure 12–20B) and cut across it, leaving a narrow band along one edge. This first cut should not exceed 25.4 μm (0.001 in.). You can use area lapping to refine the surfaces to the level you need. Next, turn the part over (see Figure 12–20C) and because of the band, one edge is raised. The next cut you make produces a surface that is at a slightly smaller angle to the vertical edges than the previous one. Again, you turn the part over (see Figure 12–20D) and grind that side parallel to the previous one. Compared with the broken lines in Figure 12–20E, which depict the original shape, you can see that the new shape is more nearly square. The process is repeated in Figures 12–20F, G, and H to increase squareness; of course, you only use this technique after ordinary machining methods have closely approximated squareness.

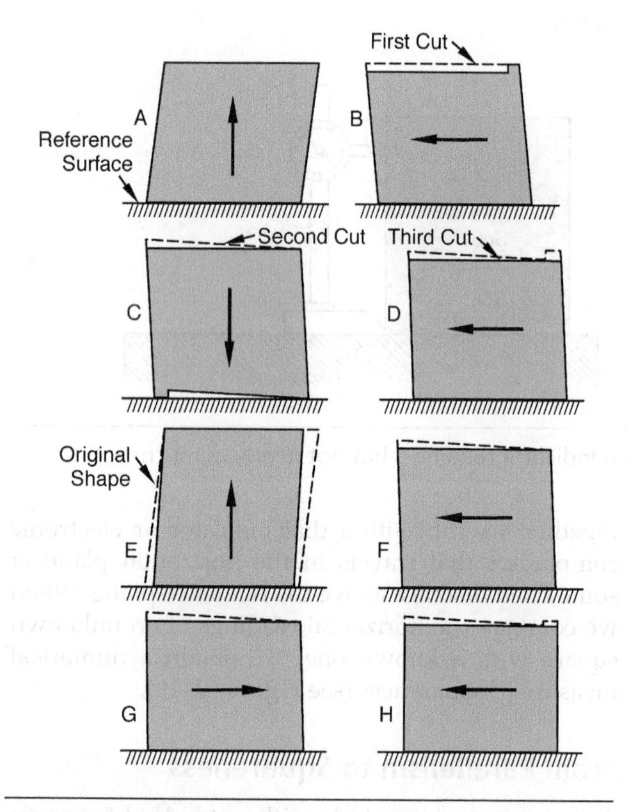

FIGURE 12–20 Method for attaining squareness, devised by the National Physical Laboratory in England, shows that only a flat reference surface is needed to create a right angle.

When you have developed your square to sufficient accuracy, you can then use it to calibrate other squares or use a cylindrical master square. Simple comparison provides the results, and thanks to the reversal technique, even these masters are not required for the calibration of squareness.

The Reversal Technique

The *reversal technique* or *reversal process* is a method of doubling apparent error, which halves the uncertainty in solving for the error. For example, we might say that two blocks are identical in height, but what do we mean by "identical"? In measurement, "identical" normally means that two things are so nearly alike that we cannot find any measurable difference with available instrumentation. If we measure the two blocks together (see Figure 12–21) and divide their combined length in half, each block will contain only half of the measured error, which is generally true even when we add the error of the extra step.

We can use the reversal process for square calibration, and all we need is a parallel—no master. The parallel must be relatively accurate; however, because any flatness error in the reference surface will cause calibration errors, it is vital that your reference surface be accurate.

First, you set up the parallel by clamping it to any stable support in a vertical position. Set the parallel approximately by sighting along the square that you are going to calibrate, and then place the square on the right of the parallel (see Figure 12–22). Hold

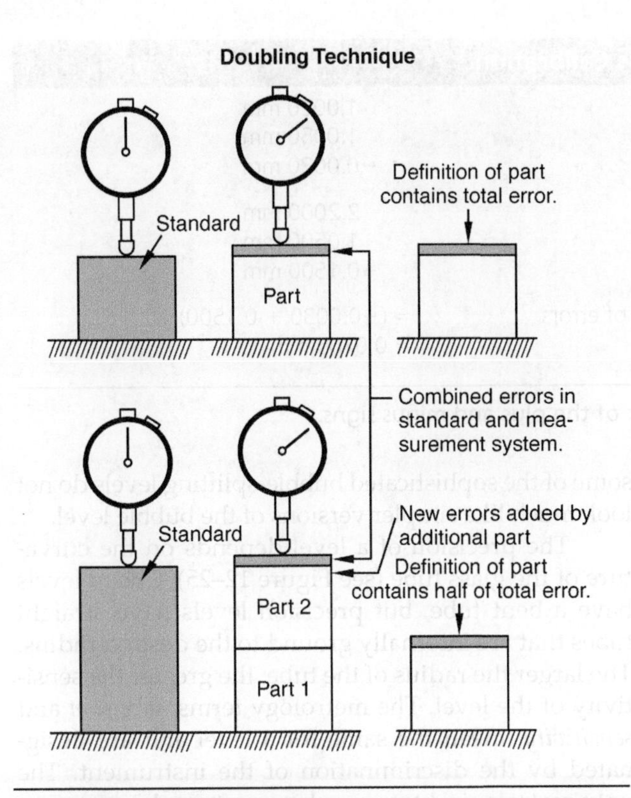

FIGURE 12–21 Measuring the additional part adds error. The resulting error in the measurement of the part is less than the error from the measurement of a single part.

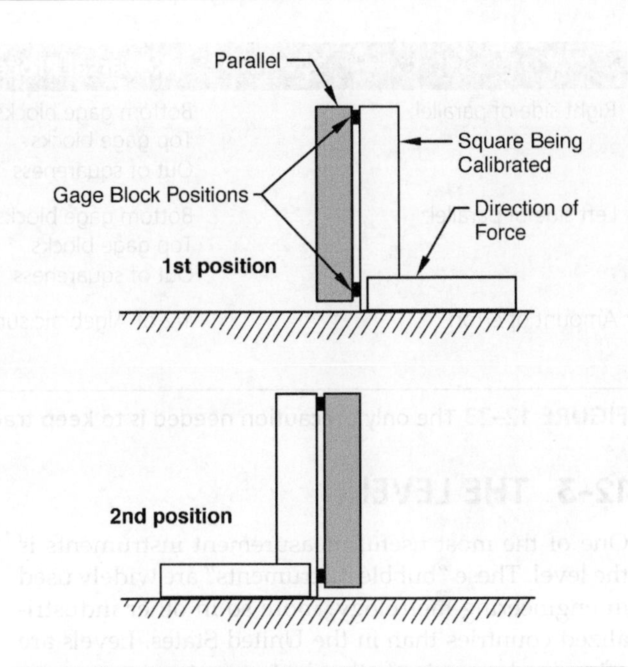

FIGURE 12–22 These two steps provide a measurement of double the squareness error.

the stock of the square firmly to the surface plate, and gently slide the square toward the parallel until it engages a 2.5527 mm (0.1005 in.) gage block. Apply only enough force to keep the block from falling under its own weight; you must not rest the block on the reference plate.

With the lower block in place, try to insert a 2.5502 mm (0.1004 in.) block in the top position, and if the fit is loose, try a 2.5552 mm (0.1006 in.) block. Continue this process until you find a block that is too large. You will know when the upper block is too large because it will cause the lower block to fall. Using the bridging technique, select a block halfway between the too-large block and the one tried before it. For example, if the 2.5502 mm (0.1004 in.) block is too small and the 2.5552 mm (0.1006 in.) block is too large, choose the 2.5527 mm (0.1005 in.) block.

If the selection of blocks you have to use is limited, try extending the range that your first block offers. You could use 5.588 mm (0.220 in.) for the lower position, made up from 2.7432 mm and 2.8448 mm (0.108 in. and 0.112 in.) blocks, which would leave the 2.54 μm (0.0001 in.) series blocks for the top position (for example, combined with 3.0226 mm, 3.048 mm, and 3.0734 mm [0.119 in., 0.120 in., and 0.121 in.] blocks). Generally, however, you do not need the wider range, and it is easier for you to handle a single gage block than a stack of wrung blocks. You must be careful not to apply more force than necessary to hold the square firmly on the plate and against the lower block.

Now, reverse the setup, being careful not to disturb the parallel, and repeat the entire process. You should not need to use so much trial and error to determine the blocks, because you now know the direction of the error. Typical calculations are shown in Figure 12–23.

Reversal Process Calibration		
Right side of parallel:	Bottom gage blocks	1.0020 mm
	Top gage blocks	1.0050 mm
	Out of squareness	−0.0030 mm
Left side of parallel:	Bottom gage blocks	2.2000 mm
	Top gage blocks	1.0500 mm
	Out of squareness	+0.1500 mm
Amount of error:	1/2 of Algebraic sum of errors	= (−0.0030 + 0.1500) 1/2
		= 0.0765 mm

FIGURE 12–23 The only precaution needed is to keep track of the plus and minus signs.

12-3 THE LEVEL

One of the most useful measurement instruments is the level. These "bubble instruments" are widely used in engineering metrology, more so in other industrialized countries than in the United States. Levels are often used as parts of other instruments.

In metrology, we use *precision levels, clinometers,* and *theodolites* for precision measurement. We use *bench levels* and *mechanic's levels* in the shop, but they are capable of high precision when you use them properly. All of these instruments use bubbles in fluid-filled tubes; the bubble is affected by gravity (see Figure 12–24). To keep the level from freezing, the tubes were once filled with "spirits of wine," which led to the term *spirit level.* Clinometers, theodolites, and some of the sophisticated bubble-splitting levels do not look much like simpler versions of the bubble level.

The precision of a level depends on the curvature of the glass tube (see Figure 12–25). Cheap levels have a bent tube, but precision levels have straight tubes that are internally ground to the desired radius. The larger the radius of the tube, the greater the sensitivity of the level. The metrology terms, *precision* and *sensitivity,* mean the same for a level and are designated by the discrimination of the instrument. The only metrological term we have not used is *accuracy,* which is not the same thing as precision in levels. Accuracy refers to the calibration of the level only; it is the amount that the reading of an angle varies from the true angle. You will most often use the term *sensitivity* in reference to levels.

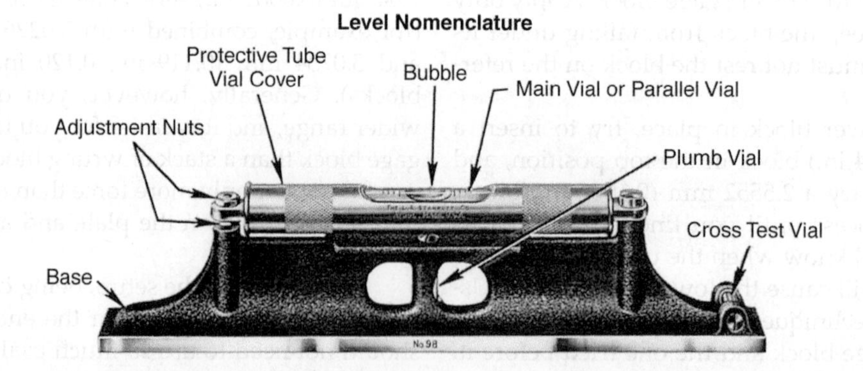

FIGURE 12–24 Most block levels are of this general type. The least precise ones do not have the test vials and have shorter bases. *(Courtesy of the L. S. Starrett Company)*

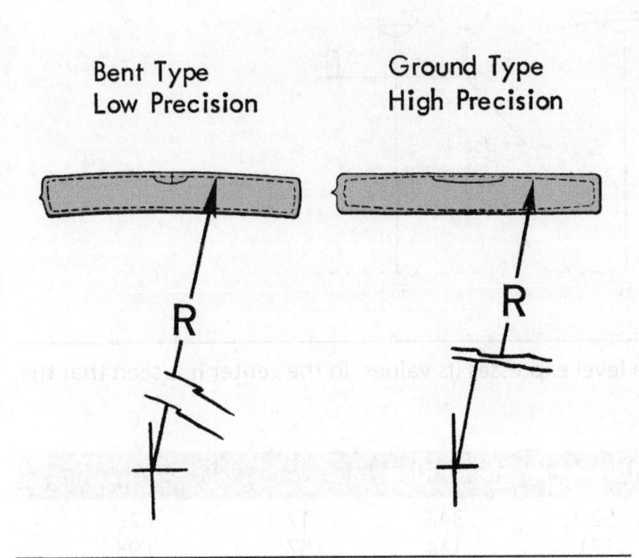

FIGURE 12–25 The longer the radius of curvature, the more precise the level will be.

The bubble and two references, the metrological features of a level (see Figure 12–26), create geometrical relationships between gravity and the center of curvature, and the scale and the bubble position ("fiducial plane") (see Figure 12–27A). When the reference planes are parallel, the bubble is centered. When the planes are not parallel, the bubble appears to move, even though Figure 12–27B shows that the bubble actually stays in place. With a few additions, like graduations on the vial to eliminate parallax (see Figure 12–27C), we have a practical instrument. We can connect the vial to a reference plane, which we also use as the base of the instrument. We can read any changes in the angle between the base plane and the gravity-related reference in the displacement of the bubble.

On most precision levels, each division is approximately 2.54 mm (1/10 in.) long, but sensitivity is designated as seconds per division. The most useful sensitivity for most precision measurement is 10 seconds per division, which requires a radius of 104.55 m (343 ft.) (see Figure 12–28).

FIGURE 12–26 This typical 10-second level has graduations that represent five-tenth mil (0.0005 in.) per foot. *(Courtesy of Mitutoyo American Corporation)*

You should always select an instrument or level that is the correct sensitivity for the measurement you are making. You must be very careful with an

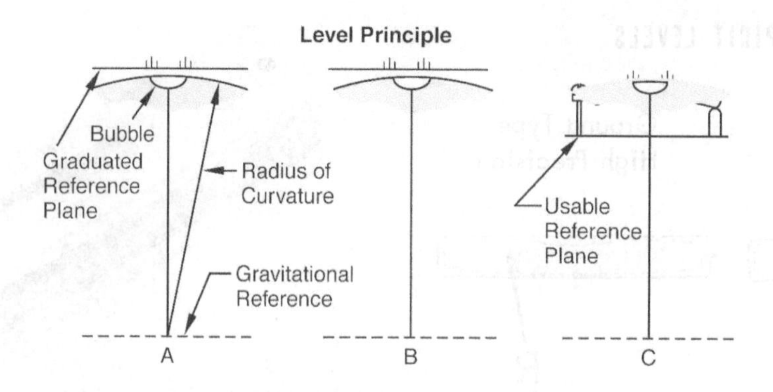

FIGURE 12–27 Nature provides the reference from which a level expresses its values. In the center it is seen that the graduated plane moves, and the bubble remains stationary.

Relationships of Curvature to Angle to Rise							
Radius in Meters	523.6	209.4	104.5	52.1	34.7	17.4	0.28
Radius in Feet	1718	687	343	171	114	57	0.93
Angle per division							
Seconds	2	5	10	20	30	60	3600
Minutes						1	60
Degrees							1
Rise mm/m	0.0083	0.0208	0.0417	0.0833	0.125	0.250	15.0
Rise mm/cm	0.0012	0.0030	0.0060	0.012	0.018	0.036	2.16
Rise in./ft.	0.00010	0.00025	0.00050	0.001	0.0015	0.003	0.180

FIGURE 12–28 The precision of a level is usually expressed as seconds per division. The actual rise per division will vary among instruments of various design, depending on width of graduations.

instrument with 5-second sensitivity, for example, because it is very sensitive to temperature change and takes a long time to settle down.

The square level (see Figure 12–26) is particularly useful for machine-tool inspection. Its body is a 20.32 cm (8 in.) square whose sides are square and parallel within one-half bubble division. Levels with 10-second sensitivity have 5.08 μm (0.0002 in.) squareness tolerance.

Thermal Error

As with other instruments, temperature affects levels—shrinking the bubble inside the tube. Like other high-amplification instruments, a high-sensitivity level is calibrated at the standard temperature (20°C or 68°F), so when the level is warmed, the liquid expands, reducing the length of the bubble. If the level warms up while you are making a series of measurements, the readings toward the end of the series will be smaller with respect to the true values than the readings at the beginning of the series.

You can compensate for this potentially serious source of error by reading the level properly. If you only read one end of the bubble, you should include the full temperature change error in the reading; however, if you read both ends and average the

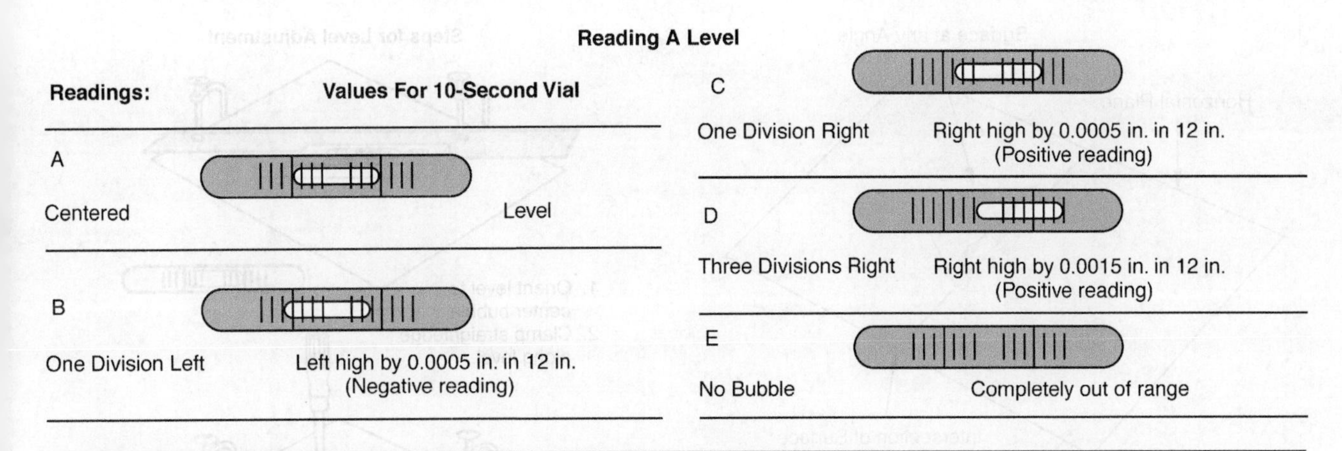

FIGURE 12–29 The readings are the number of divisions that the bubble moves. Having two ends, the reading is visually corroborated as read.

measurement, the error will cancel out. Of course, the bubble will fall short of the calibration markings, and there will be some residual observational error.

Sensitivity also depends on the viscosity of the fluid in the vial. Levels filled with low viscosity liquids also have high vapor pressure, which can result in some highly erratic readings when the vial is heated unevenly. You can see the effects of uneven heating when you place a finger on one end of the vial for a few minutes and watch the results. A precision level, fully normalized in a dark room, will actually respond to the heat of a flashlight directed at one end of the vial.

Reading Levels

You read a level as shown in Figure 12–29. The centered position is marked by long graduations—or sometimes a dot at each end—that bridge the bubble so this position can be established with just your unaided eye. The graduations at the end of the bubble duplicate each other and automatically corroborate the readings.

When we discussed the rule of 10-to-1, we did mention that there were some exceptions, and two of them have to do with reading levels. First, your eye is exceptionally able to resolve displacements of one line in relation to another. Second, you can increase the precision of measurement with a repeatability of 5 seconds, which is many times greater than predicted by the rule of 10-to-1. Figure 12–29E shows that if the surface is far out of the range of the level, you will not be able to see the bubble. You must be sure that you know which end of the vial conceals the bubble before you begin to adjust the level. You do not want to waste time adjusting the wrong way, just because you cannot see the bubble.

There is a convention for reading a level (see Figure 12–29): The left bubble in relation to the reader is negative, and the right bubble in relation to the reader is positive. This convention is the same even if you reverse the level.

Level Adjustment

All levels with a sensitivity of 20 seconds or greater have an adjustment screw (see Figure 12–26), which aligns the vial to the base. Adjustment is very simple for even the most high-precision measurement. All you need is a stable, flat surface, which does not need to be level. Regardless of the position of a flat surface, one direction across the surface is horizontal (see Figure 12–30).

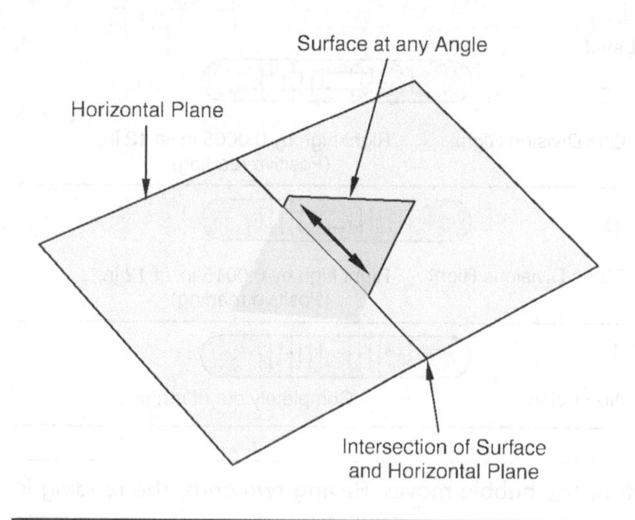

FIGURE 12–30 No matter what angle a surface may be, one direction across it is horizontal.

First, orient the level on the surface until the bubble is centered, and then clamp a straightedge to the surface along one side of the level (see Figure 12–31) (reading shown at A). Reverse the level and read the error as in Figure 12–31B. Repeat this process, orienting to a new location on the surface (see Figure 12–31D), until you have achieved the desired accuracy. Remember, you are adjusting the calibration accuracy, not the precision or sensitivity.

Again, we have used the reversal process (see Figure 12–32). For ordinary accuracy requirements, you should carry this process through to one-fifth of a division.

If the level is so far out of calibration that the bubble is off the scale, it may take long and tedious adjustments to make it useable again, but there is a trick that will help (see Figure 12–33). Select two lines across the surface that provide the same reading and bisect the angle they form. This midpoint line approximates the level line needed to "bring in" the instrument.

You can calibrate sensitivity by using a tilting table. Although we will not cover the mechanics of using this device, you should be aware that the sensitivity tolerance should not be greater than ±10%, which is two-tenths of a division.

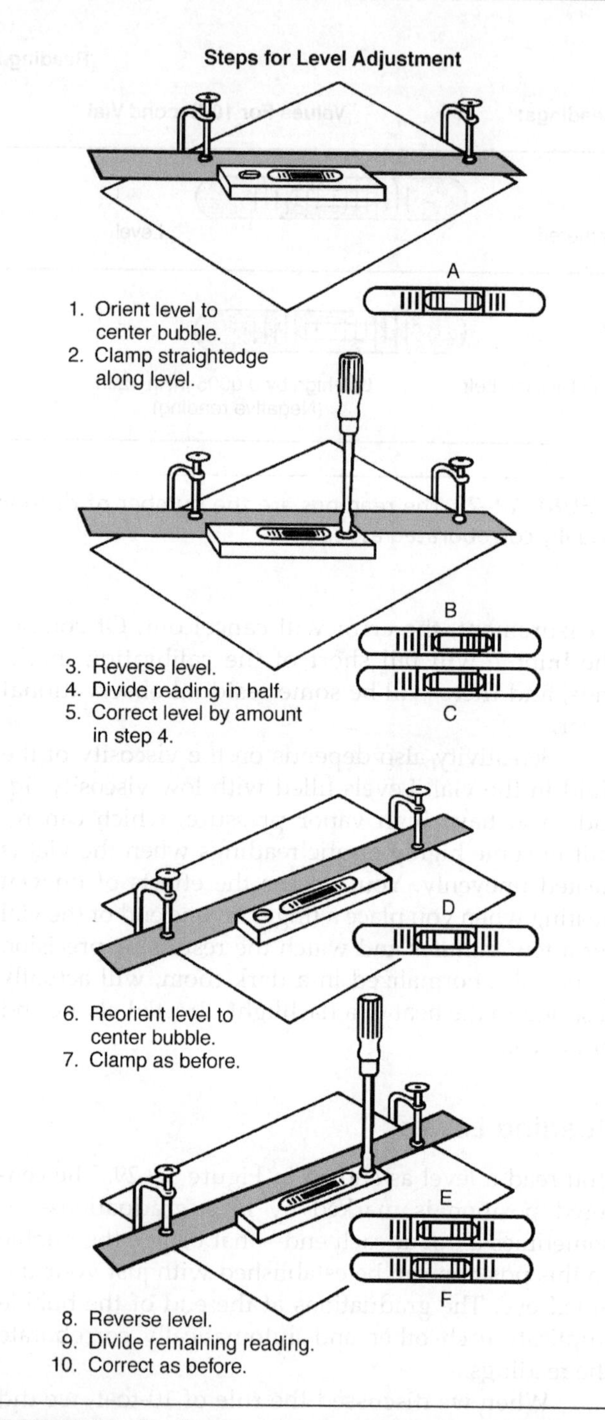

FIGURE 12–31 These steps are repeated until the desired accuracy setting is achieved.

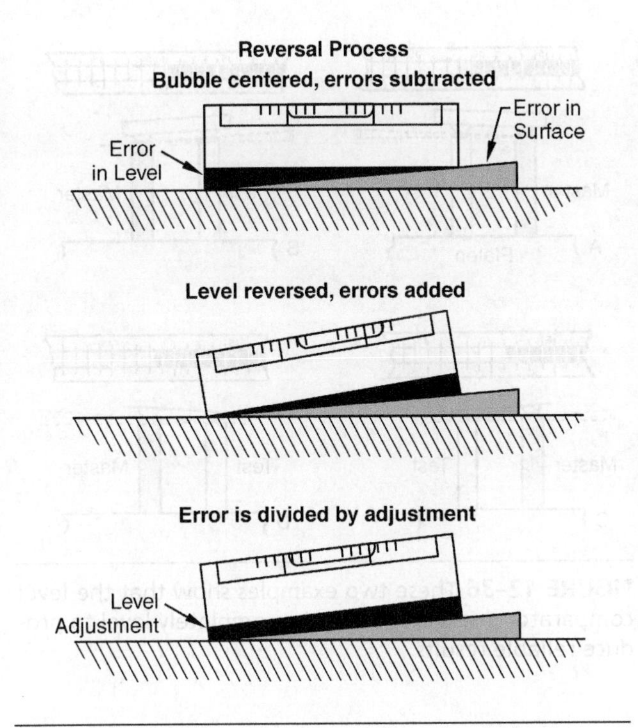

FIGURE 12–32 The reason the surface did not have to be level to adjust the level is figuratively shown here.

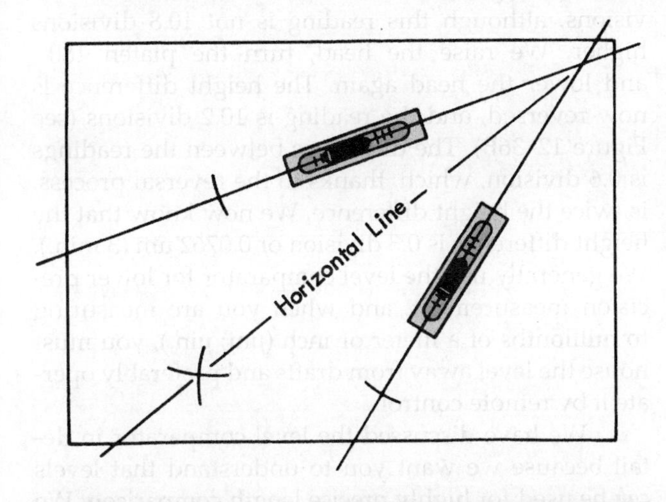

FIGURE 12–33 If the angle formed by two lines of the same level reading is bisected, a line approximately horizontal will be found.

The Reversal Technique for Levels

The reversal process not only helps adjust levels, it also provides exceptional reliability when we use them. Whenever you make precise measurements with a level, you should take readings at both ends of the bubble, then reverse the level and repeat the readings (see Figure 12–34). By averaging the four readings, you obtain a possible accuracy of one-tenth of a division. Of course, this accuracy is only possible if you are extremely careful when handling the levels and references.

In most practical work, it is more convenient to deal with very small angles in terms of linear displacement rather than in seconds. You must remember, however, that a level measures angles, even though we usually translate the angular measurement into linear measurements.

We express the linear displacement in terms of a measurement base (see Figure 12–28), which is an expression of the linear rise distance per division, generally 30.48 cm (12 in.) in the United States.

Linear sensitivity is inversely proportional to the length of the measurement base (see Figure 12–35). In the top example, a 12.7 μm (0.0005 in.) rise acting on a 25.4 cm (10 in.) measurement base moves the bubble one division. In the center example, when we reduce the measurement base to 12.7 cm (5 in.) and do not change the rise, the measurement registers as two divisions. In the bottom example, a 6.35 cm (2.5 in.) measurement base requires only a 10.16 μm

Reliability with Levels

For Precise Measurement:

1. Take readings from both ends of the vial.
2. Reverse level.
3. Repeat readings from both ends.
4. Average the four readings.
5. Repeat all steps for critical cases.

FIGURE 12–34 These steps enable level measurements to be relied on to one-tenth division.

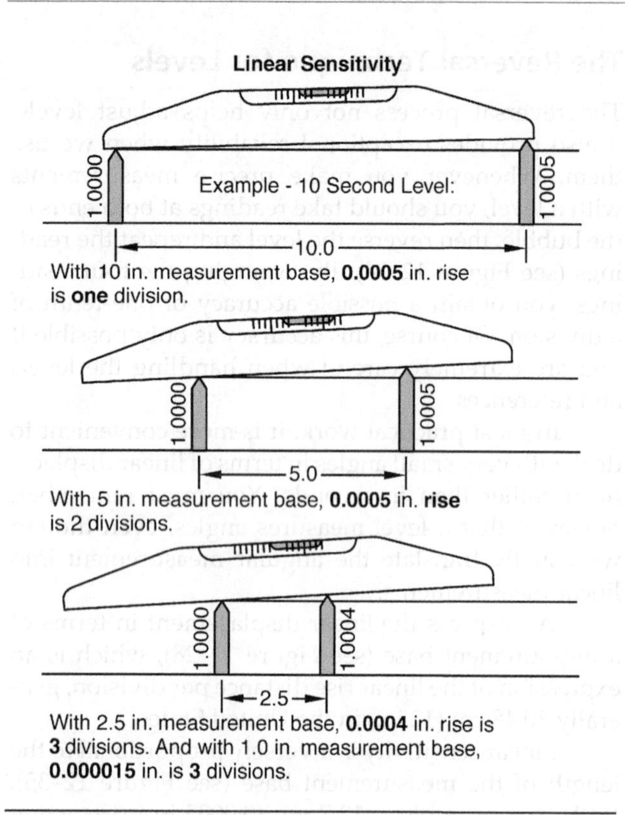

FIGURE 12–35 The shorter the measurement base, the greater the linear sensitivity.

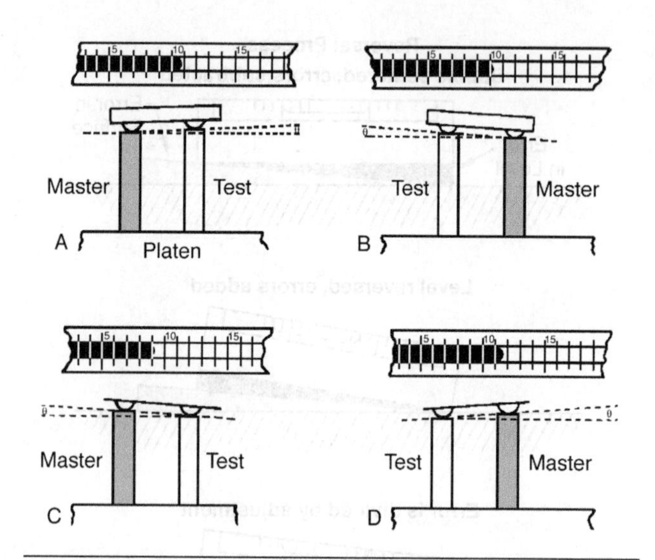

FIGURE 12–36 These two examples show that the level comparator does not have to be completely level to produce reliable results.

(0.0004 in.) rise for three divisions (full scale) displacement. If we reduce the measurement base further to 2.54 cm (1.0 in.), a rise of only 3.81 μm (0.00015 in.) also uses the full scale. In other words, a 2.54 cm (1 in.) measurement base increases the linear sensitivity to 1.27 μm (0.00005 in.) from the 12.7 μm (0.0005 in.) linear sensitivity of the same instrument with a 25.4 cm (10 in.) measurement base.

Use of Levels

Manufacturers have not overlooked the tremendous linear sensitivity of the precision level and the reliability of the reversal process. For proof, see the level comparator.

In our example of measurement with a level comparator (see Figure 12–36), we wrung the master and the unknown to the platen and lowered the head onto them. We read the scale and see that the unknown is higher than the master and reads 10.8 divisions, although this reading is not 10.8 divisions higher. We raise the head, turn the platen 180°, and lower the head again. The height difference is now reversed, and the reading is 10.2 divisions (see Figure 12–36B). The difference between the readings is 0.6 division, which, thanks to the reversal process, is twice the height difference. We now know that the height difference is 0.3 division or 0.0762 μm (3 min.). We generally use the level comparator for lower precision measurement, and when you are measuring to millionths of a meter or inch (μm; μin.), you must house the level away from drafts and preferably operate it by remote control.

We have discussed the level comparator in detail because we want you to understand that levels *can* be used for highly precise length comparison. We can also use level comparators on a surface that is not exactly level itself, so we can use them to calibrate surface plates. A precision level is used for many of the highest precision measurements, except in instances when we must use an autocollimator. Using a level for machine alignment and setup is so straightforward that we hardly need to discuss them. It is important to remember that levels can be used for more than work that is level.

There are four principal disadvantages of levels:

- General ignorance about their proper use and application
- The length of time required to settle down
- Their single sensitivity characteristic
- They do not produce an output that can be used as loading for a measurement system.

Level and telescope combinations have found a permanent place in the theodolites of metrology since World War II, and as the requirements for tighter tolerances increase, precision levels will certainly be used more often in dimensional metrology. Figure 12–37 lists the metrology data for levels and squares.

Metrological Data for Squares							Reliability	
Instrument	Type of Measurement	Normal Range	Designated Precision	Discrimination	Sensitivity	Linearity	Practical Tolerance for Skilled Measurement	Practical Manufacturing Tolerance
Combination square	comparison	none	none	not applicable	beyond accuracy	not applicable	30′	1°
Precision square	comparison	none	none	not applicable	beyond accuracy	not applicable	30″	1′
Surface plate square	comparison	none	none	not applicable	beyond accuracy	not applicable	10″	30″
Cylindrical square	comparison	none	none	not applicable	beyond accuracy	not applicable	5″	30″
Graduated cylindrical square	comparison	0. to 0.0012″	0.0001″ in 6″	0.0002″ in 6″	beyond accuracy	50 mike within 6″	0.0002″ in 6″	0.0004″ in 6″
Square and transfer stand	All factors limited by metrological data of transfer instrument							
Mechanic's level	direct	6°	1°	1°	30′	30′	1°	2°
Precision level	direct	1′20″	10″	10″	5″	5″	10″	30″

FIGURE 12–37 It is possible to use a slit of light to compare squareness with such a high degree of sensitivity that this factor is of relatively little importance when evaluating squares.

12-4 THE PROTRACTOR

For measuring angles, the simple protractor is equivalent to the rule for measuring lengths. Its discrimination and sensitivity are directly related to its finest division, which is usually 1°, except in the largest protractors.

Like a steel rule, the simple protractor has limited use. But mechanical additions to the rule resulted in the versatile combination square and adding a vernier resulted in the precise vernier caliper or height gage. If we make similar modifications on the simple protractor, we get the *universal bevel protractor* (see Figure 12–38).

The heart of this instrument is a dial, graduated in degrees, which are grouped into four 90° quadrants. The degrees are numbered to read either way: from zero to 90, then back to a zero, which is opposite the zero you started at.

The vernier scale is divided into 24 spaces, 12 on each side of zero, numbered 60-0-60. The 24 spaces equal 46 spaces on the dial; therefore, one vernier division equals 1/12 of 23 degrees or 1 11/12 degrees. The difference between one vernier division and two

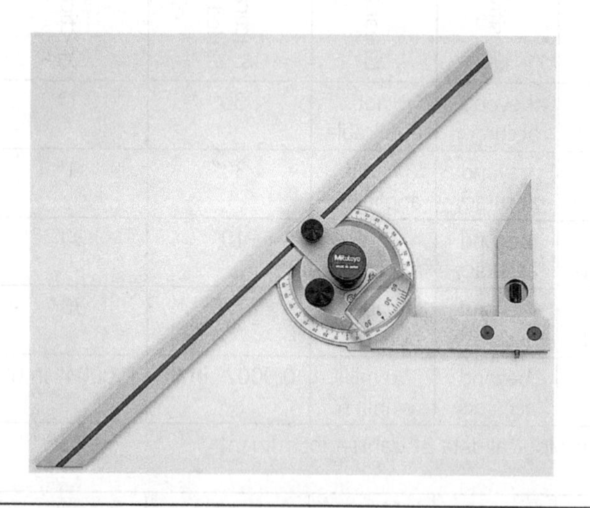

FIGURE 12–38 The universal bevel protractor is to angle measurement what the combination square is to linear measurement. On some instruments, the blade clamp and the fine adjustments are all located at the center. *(Courtesy of Mitutoyo American Corporation)*

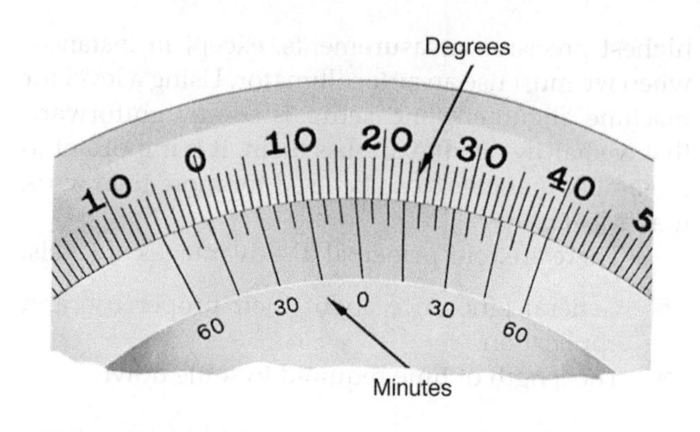

FIGURE 12–39 Degrees are read directly, but minutes.

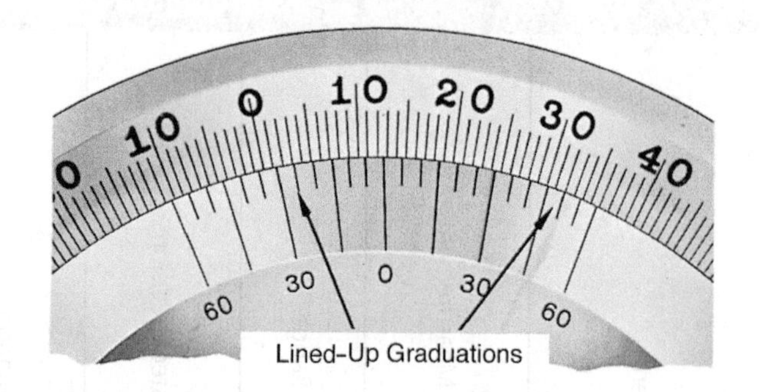

FIGURE 12–40 Care must be used to read the minutes from the correct pair of lined-up graduations.

dial divisions is 1/12 degree or 5 minutes. When the angle is an exact degree, the zero graduations of the vernier and the two 60-minute graduations line up with dial divisions (see Figure 12–39).

Sometimes people get confused about which direction they should read the vernier. For example, in Figure 12–40, is the reading 12° plus 50′ or plus 25′, or minus 25′? You can eliminate this confusion if you remember one rule: Always read the vernier in the same direction from zero that you read the dial and add the minutes to the dial degrees (see Figure 12–41). With this rule in mind, the reading in Figure 12–40 is 12° 50′.

Applications for Vernier Protractors

We can determine the angle or degrees in any arc with the universal bevel protractor—but we do need to

RULE FOR READING VERNIER PROTRACTORS

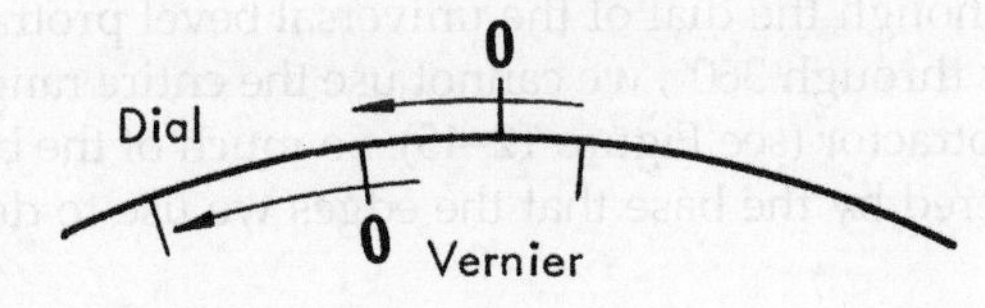

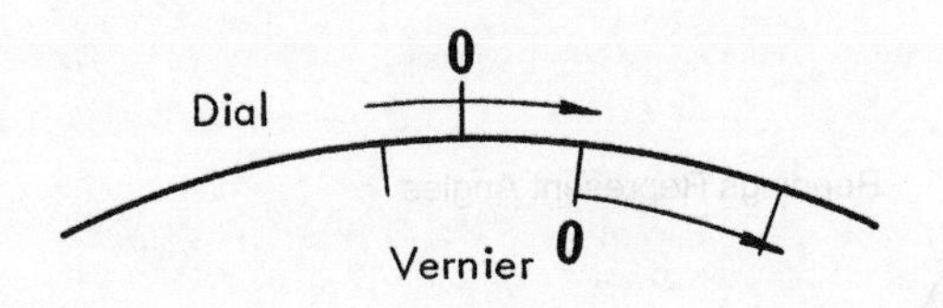

FIGURE 12–41 Always read the vernier in the same direction from zero that the dial is read and add the vernier minutes to the scale degrees.

be careful. When the blade is placed in one position, you are actually forming four angles with the blade: two that we can read on the dial and vernier, and two supplementary angles. At times, the blade may not be extended through the base; then, one angle and one supplement are not usable.

When the protractor is set at 90° (see Figure 12–42B), all four angles are as read. If you turn

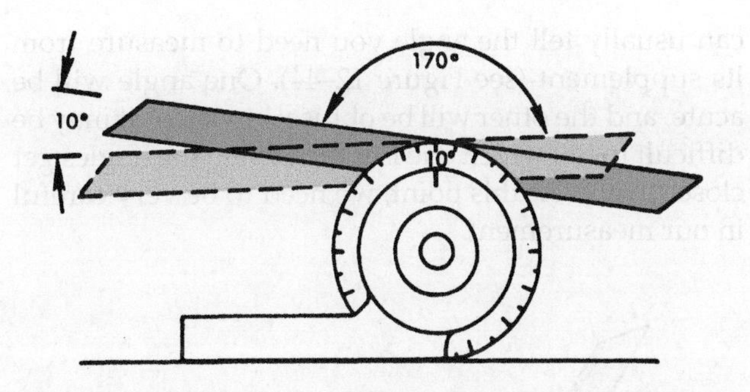

FIGURE 12–43 When reading from 0°, there is little danger of confusing the angle and its supplement.

the blade counterclockwise (see Figure 12–42A), the angle you read will represent only the angles formed from the blade to the base counterclockwise, which happens in two positions as shown. If you turn the blade clockwise (see Figure 12–42C), the angle read will be formed only in two places, which are always from the blade to the base rotating clockwise.

Unfortunately, we cannot use this relationship as a general rule because the reverse applies when you are reading from a straight angle (180° or 0°) (see Figure 12–43), but it is not a problem. You would not be likely to confuse a 10° angle with 170° supplement, so we do not really need a rule, because you

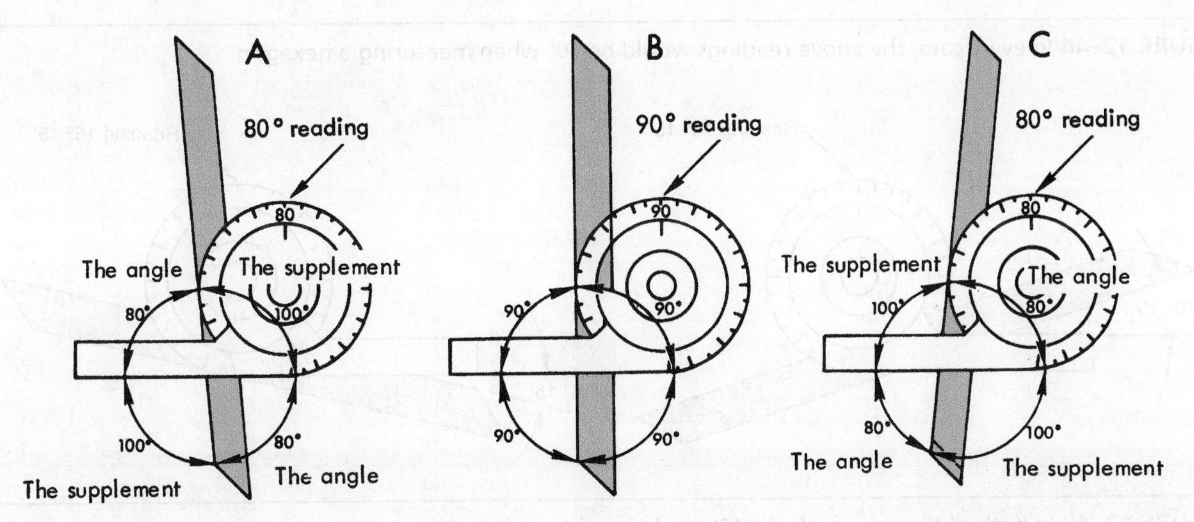

FIGURE 12–42 When reading from 90°, these are the positions where the angle and its supplement are found.

can usually tell the angle you need to measure from its supplement (see Figure 12–44). One angle will be acute, and the other will be obtuse; however, it may be difficult to tell which one is which when the angles get closer to 45°. At this point, we need to be very careful in our measurement.

Measuring Acute Angles

Even though the dial of the universal bevel protractor rotates through 360°, we cannot use the entire range of the protractor (see Figure 12–45). So much of the blade is covered by the base that the edges we use to define

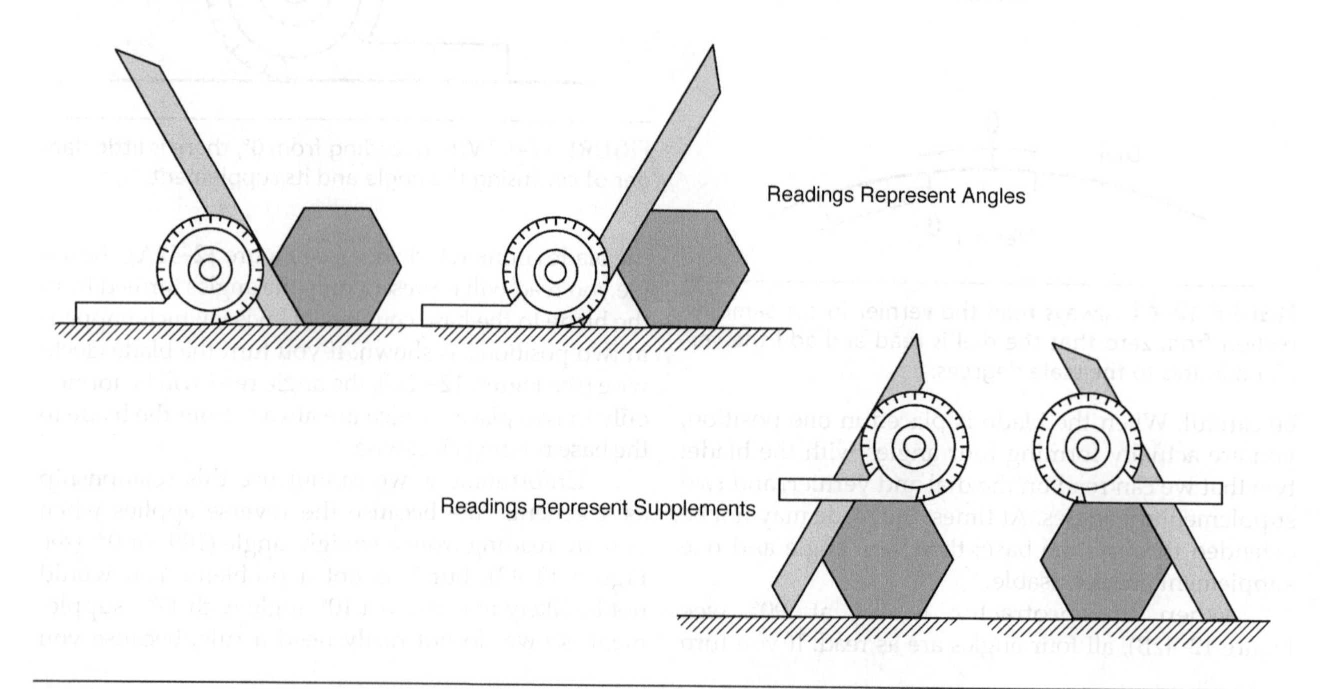

FIGURE 12–44 In every case, the above readings would be 60° when measuring a hexagon.

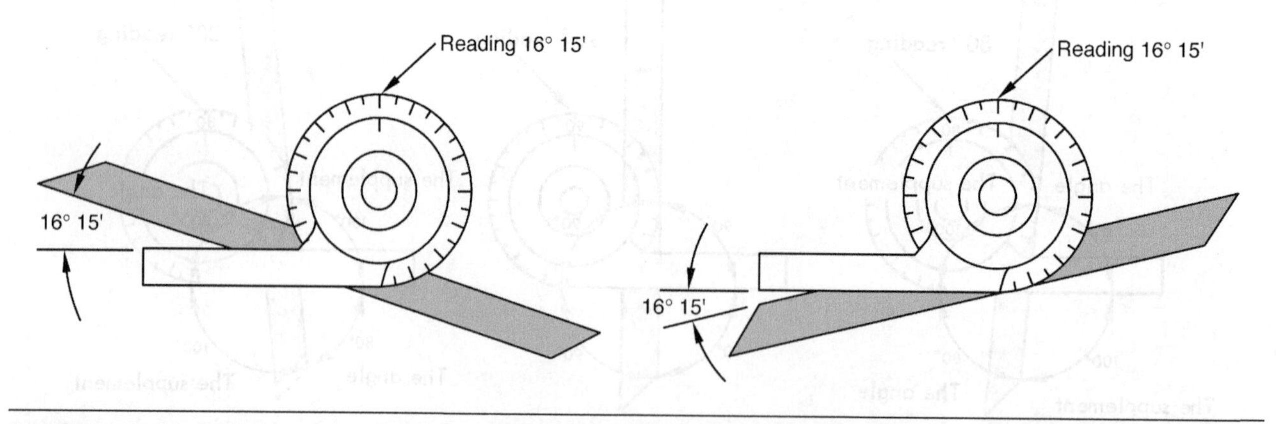

FIGURE 12–45 In both of these examples, a 15° angle has been set. Unfortunately, so little length of the blade and base enclose the angle that it is not useable for most measurement.

the angle are too short for most measurements. If we add an attachment to form a reference edge at 90° to the base, we have solved this problem (see Figure 12–46). Because the long edge is at a right angle to the base, any angle measured in this area is the complement of the angle we actually want to measure. We obtain the angle we are measuring by subtracting the attachment reading from 90°.

We can also create an acute angle measurement device by adding a parallel extension of the base to a protractor (see Figure 12–47). With this instrument, you can use the blade at the top of the dial for direct readings of angles. Any time an error—operational, manipulative, visual, computational, or any other kind—can be eliminated, you enhance the reliability of your measurement. This attachment, however, is often too bulky to be placed in the area of the angle.

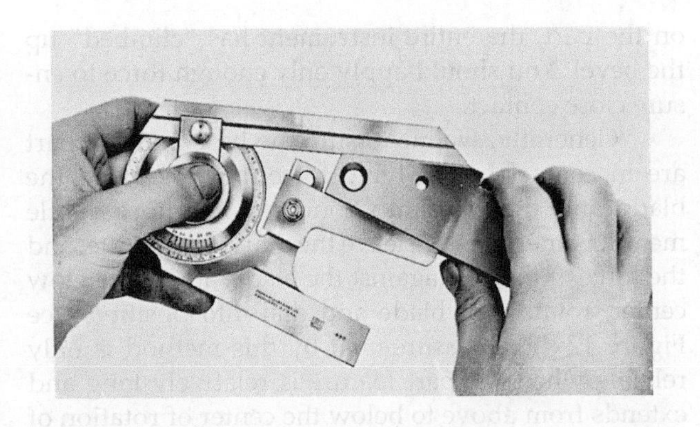

FIGURE 12–47 In this form of acute angle attachment, the readings are direct and need not be converted. *(Hexagon Metrology, Inc. TESA.)*

Precautions for Using the Universal Bevel Protractor

Remember, a universal bevel protractor does not measure the angle on the part—it measures the angle between its own parts. This instrument provides information about the feature of the part we are observing or inspecting only if the physical parts of the protractor are in intimate contact with features of the part—and the closer you can establish contact between the protractor and the part feature, the more accurate your readings will be. The protractor must be in the same plane—or a parallel plane—as the angle you are measuring.

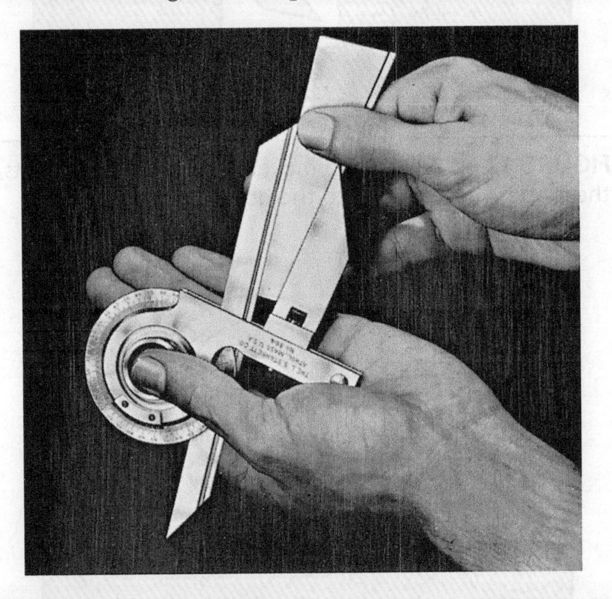

FIGURE 12–46 The acute angle attachment is being used here to measure a 10° angle. *(Courtesy of the L. S. Starrett Company)*

If you improperly place the protractor in contact with the part feature (see Figure 12–48B and C), you generate blade contact or base contact error. If you are careful and repeat your measurements, you can eliminate many of these errors; however, often surface obstructions and overconstraint also generate contact errors. Pay attention to how you are making your measurement and the general conditions under which you are making the measurement, as always.

Also, as always, you must ensure the cleanliness and proper preparation of all surfaces. When we shorten the measurement base, as with levels, we decrease the reliability of our measurements. The shorter the part features you are trying to measure, the more serious the effect that a burr or grain of dust have on the reliability of your measurements.

In an effort to achieve reliable results, you might constrain the instrument excessively and create error (see Figure 12–48C). You can see that in order to ensure that the blade is firmly in contact with the bevel on the part, the entire instrument has "climbed" up the bevel. You should apply only enough force to ensure close contact.

Generally, we make sure the base and the part are in contact by applying force, but we bring the blade and the part into contact with more subtle means. Sometimes, we leave the dial free to rotate, and the force of the part against the blade, above or below center, rotates the blade and dial into position (see Figure 12–49). Measurement by this method is only reliable when the part feature is relatively long and extends from above to below the center of rotation of the dial. If the part feature is short and/or entirely on one side of the center of the dial (see Figure 12–50), the measurement will contain significant error.

You can determine if the blade is in contact with the part by placing a light behind the part and protractor and adjusting the protractor accordingly. If the surface finish of the part is good, you will not be able to see light between the part and instrument, and if the finish is poor, you can adjust the instrument so that the light that "leaks" through is uniform. Using this method, you lock the dial in an approximate position and then fine adjust the setting. When you cannot place a light behind the blade, you can use the paper strip method we discussed with squares.

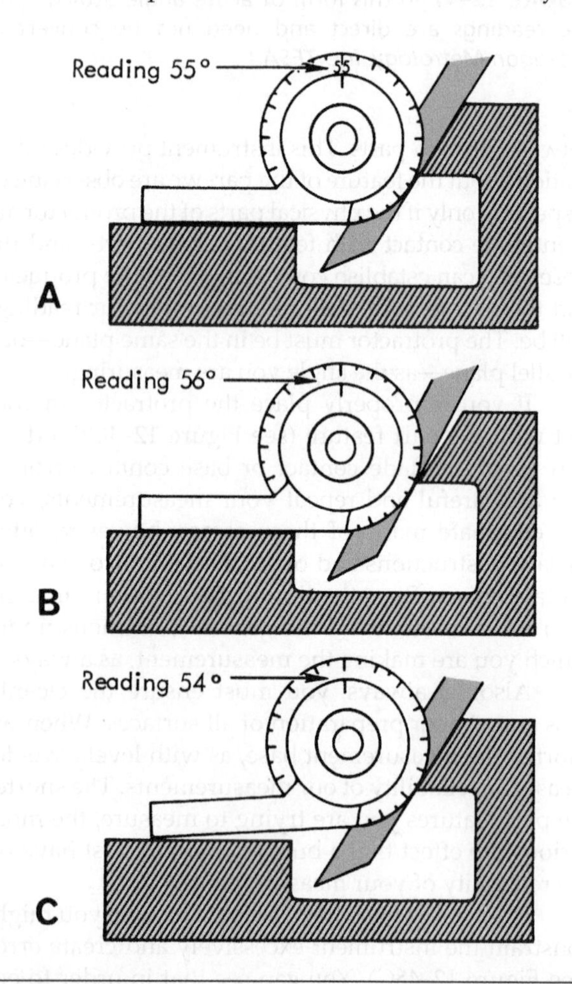

FIGURE 12–48 The instrument must duplicate the angle before it can measure the angle.

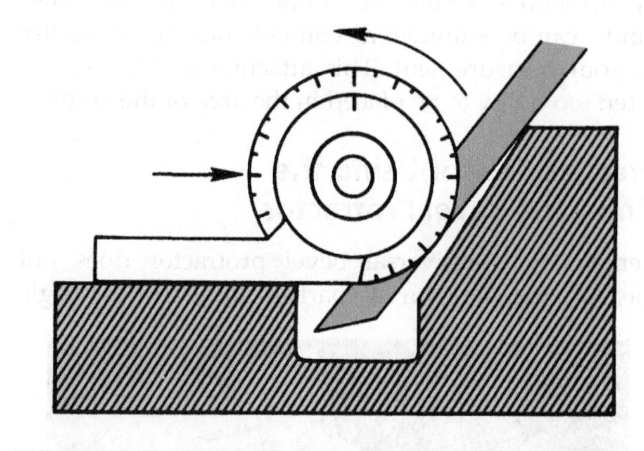

FIGURE 12–49 When the movement to the right halts, the blade will be mated to the bevel surface.

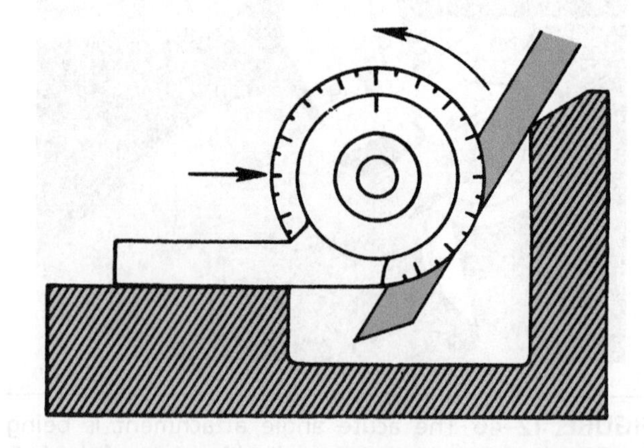

FIGURE 12–50 The blade has already passed the mating position, and further movement causes still greater error.

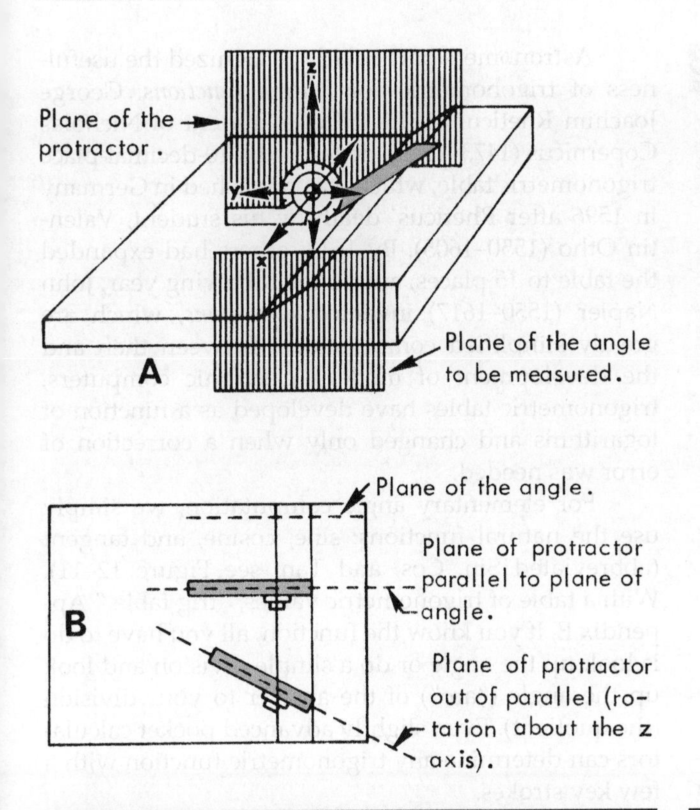

FIGURE 12–51 The true angle is measured only when the protractor is in a plane parallel to that of the angle.

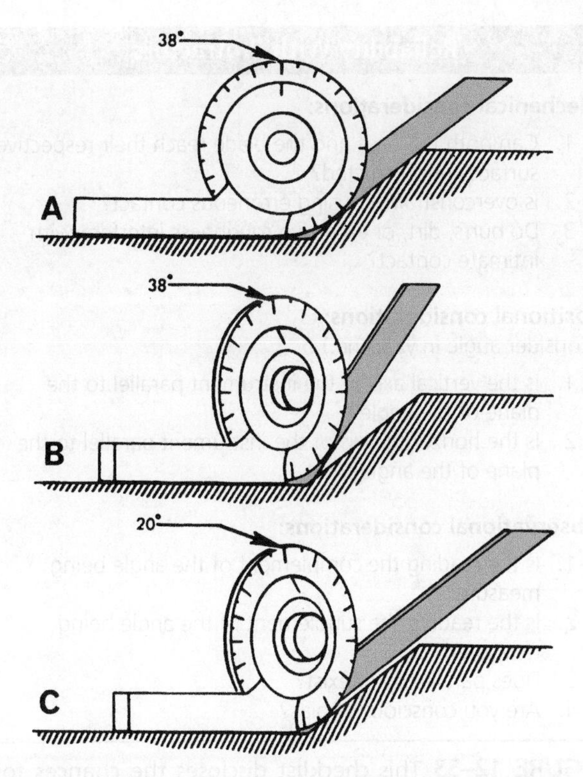

FIGURE 12–52 The importance of positional relationship in angle measurement is shown here.

You can only measure the true angle when the instrument is in a plane parallel to the plane of the angle (see Figure 12–51). When the protractor is resting on the part (see Figure 12–51) it is constrained from translation in the z axis (up and down), but it is free to move in the y axis away from the part feature and to slide back and forth along the part in the x axis. Because the rotation is constrained only in one axis, x, the blade's rotational position in that axis is the measure of the angle (see Figure 12–51B).

The effect of rotation around the z axis is demonstrated in Figure 12–52. We obtain a 38° reading when the instrument is parallel to the plane of the angle. However, if we rotate the instrument, the upper end of the blade moves away from the part feature even though the bottom remains in contact. If we rotate the dial so that the blade can be brought into contact again, we read a reduced measurement.

If we slant the instrument, we introduce exactly the same errors, but the rotation is around the y axis. These are independent errors, and when they exist, you never know whether they are adding or canceling. You do know that you have lost reliability in your measurement.

Reliability with the Universal Bevel Protractor

You can use a general checklist (see Figure 12–53) and the same general precautions that we recommend for any vernier class instrument when you use the universal bevel protractor. You must not expect greater precision than the instrument discrimination (5 min.); the 10-to-1 rule applies, tempered by the same practical considerations you use for linear measurement. If the only instrument you have available is the universal

Reliability With Protractors

Mechanical considerations:

1. Can both the base and the blade reach their respective surfaces unobstructed?
2. Is overconstraint causing erroneous contact?
3. Do burrs, dirt, or excessive roughness interfere with intimate contact?

Positional considerations:
(Consider angle in yz plane.)

1. Is the vertical axis of the instrument parallel to the plane of the angle?
2. Is the horizontal axis of the instrument parallel to the plane of the angle?

Observational considerations:

1. Is the reading the complement of the angle being measured?
2. Is the reading the supplement of the angle being measured?
3. Does parallax error exist?
4. Are you conscious of bias?

FIGURE 12–53 This checklist discloses the chances for error in angle measurement.

bevel protractor—or vernier protractor as it is usually called—and you must measure to 5 min., exercise every precaution possible, and then have someone else check your results.

These protractors are far more delicate by nature than any other instrument, so you must be extremely careful when you use them and maintain them (see Figure 12–54).

12–5 TRIGONOMETRIC FUNCTIONS

The trigonometric functions, formed by the sides of triangles, are among the most useful tools in practical angle measurement. Because the sides of a triangle are always straight lines, we can measure them with linear measuring instruments. Once we know two sides, we can easily express the fraction formed by them in decimal form.

Astronomers particularly recognized the usefulness of trigonometric—or *natural functions*: George Joachim Rheticus (1514–1576), a student of Nicholas Copernicus (1473–1543), developed a 10-decimal-place trigonometric table, which was published in Germany in 1596 after Rheticus' death by his student, Valentin Otho (1550–1605). By 1613, others had expanded the table to 15 places, and in the following year, John Napier (1550–1617) invented *logarithms*, which are greatly simplified computations. Between then and the development of modern electronic computers, trigonometric tables have developed as a function of logarithms and changed only when a correction of error was needed.

For elementary angle computation, we simply use the natural functions: sine, cosine, and tangent (abbreviated Sin, Cos, and Tan; see Figure 12–11). With a table of trigonometric values, "trig table," Appendix E, if you know the function, all you have to do is look up the angle or do a simple division and look up the angle ("arc") of the answer to your division (the quotient). Even slightly advanced pocket calculators can determine any trigonometric function with a few key strokes.

$$\text{Sin} = \frac{\text{Opp}}{\text{Hyp}} \qquad \text{Cos} = \frac{\text{Adj}}{\text{Hyp}} \qquad \text{Tan} = \frac{\text{Opp}}{\text{Adj}}$$

You can use this general procedure directly, and it is used in an entire family of instruments: the sine bars, sine plates, sine tables, and so forth. We will discuss the difficult direct method first.

Use of Tangent

One of the trigonometric functions used in practical angle measurement is *tangent*—abbreviated and pronounced *tan*. Typically, you use tangent when you are measuring the angle that the side of an external taper makes with its axis (see Figure 12–55). Set a tapered part on a reference surface and support two cylinders on either side. First, you set the cylinders on low stacks of gage blocks (see Figure 12–55A), or on the reference surface, and then on high stacks (see Figure 12–55B).

Care of the Universal Bevel Protractor

Before use:

1. Wipe off dust and oil.
2. Examine for visual signs of damage or abuse.
3. Run fingers along base and blade to detect burrs.
4. Check mechanical movement for freedom.
5. Check clamps for security.
6. Allow instrument to normalize.
7. Determine that the instrument has been recently calibrated.

During use:

1. Keep case nearby so that instrument may be placed in case rather than on hard surface when not being used.
2. Avoid excessive handling to minimize heat transfer.
3. Do not slide along abrasive surfaces.
4. Do not overtighten clamps.
5. Do not spring or bend by overconstraint.
6. Take precautions to avoid dropping instrument and to avoid dropping objects on it.
7. Avoid work near heat sources.

After use:

1. Clean thoroughly. Do not use compressed air, which could drive particles into instrument. Dip in solvent and shake dry if exposed to cutting fluids.
2. Lubricate moving parts.
3. Apply thin rust-preventative lubricant.
4. Replace in case.

FIGURE 12–54 These precautions require only minutes, but they ensure minutes of arc accuracy.

$$\text{Tan } a = \frac{\text{side opposite}}{\text{side adjacent}} = \frac{\frac{d_2 - d_1}{2}}{h_2 - h_1}$$

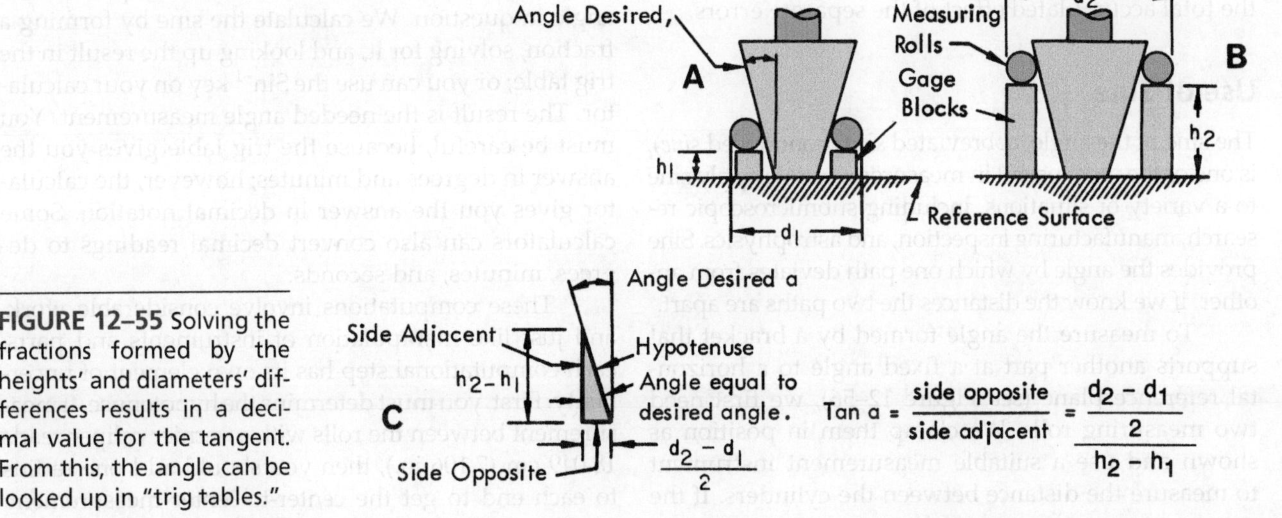

FIGURE 12–55 Solving the fractions formed by the heights' and diameters' differences results in a decimal value for the tangent. From this, the angle can be looked up in "trig tables."

You use a micrometer or other instrument to measure the separation of the cylinders at both positions. In practical machine shop work, we call these cylinders *measuring rolls* or *measuring rods.*

We must find the difference in height between the gage block stacks and the difference in the measurements over the cylinders to determine the triangle we will use to obtain the necessary value (see Figure 12–55C). In that triangle, we know the side opposite and the side adjacent but not the hypotenuse. In this triangle, the surface of the tapered part is the hypotenuse, the height dimension $h_2 - h_1$ is the side adjacent, and the side opposite is the dimension $(d_2 - d_1)/2$. For the fraction, the side opposite is the numerator, and the side adjacent is the denominator. The result is the tangent of the angle, a ratio of the dimensions. Note that we divided the horizontal measurement in half because we are concerned with the angle on one side of center.

We look up the result of our calculation in a trigonometry table and find the angle to as fine a measurement as minutes, with provisions for interpolating to seconds. You can also find the actual angle by using the $\tan^{-1}$ key on most calculators (usually a second function), but the calculator result does not mean you have measured the angle to seconds or even to minutes. The measurement can be no more precise than the least precise step you use to make the measurement. The uncertainty (lack of accuracy) will be the total accumulated effect of the separate errors.

Use of Sine

The sine of the angle, abbreviated *sin* (pronounced *sine*), is one of the most useful in measurement. We apply sine to a variety of situations, including submicroscopic research, manufacturing inspection, and astrophysics. Sine provides the angle by which one path deviates from another, if we know the distances the two paths are apart.

To measure the angle formed by a bracket that supports another part at a fixed angle to a horizontal reference plane (see Figure 12–56), we first need two measuring rolls. We clamp them in position as shown and use a suitable measurement instrument to measure the distance between the cylinders. If the diameters are the same, we add one diameter to the measurement to obtain the center distance, which is also the hypotenuse of the triangle.

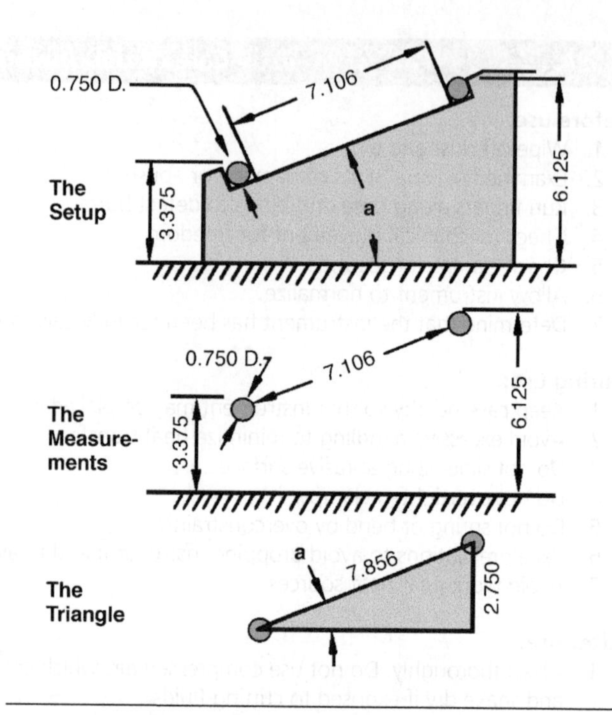

FIGURE 12–56 The setup must be viewed as a problem in trigonometry to find the angle.

If we subtract the height of each cylinder from the reference plane, we obtain the side opposite the angle in question. We calculate the sine by forming a fraction, solving for it, and looking up the result in the trig table; or you can use the Sin^{-1} key on your calculator. The result is the needed angle measurement. You must be careful, because the trig table gives you the answer in degrees and minutes; however, the calculator gives you the answer in decimal notation. Some calculators can also convert decimal readings to degrees, minutes, and seconds.

These computations involve considerable work, and just like manipulation of instruments and parts, each computational step has its own element of uncertainty. First, you must determine the hypotenuse. If measurement between the rolls with a vernier caliper reads 18.049 cm (7.106 in.), then you should add one radius to each end to get the center-to-center measurement:

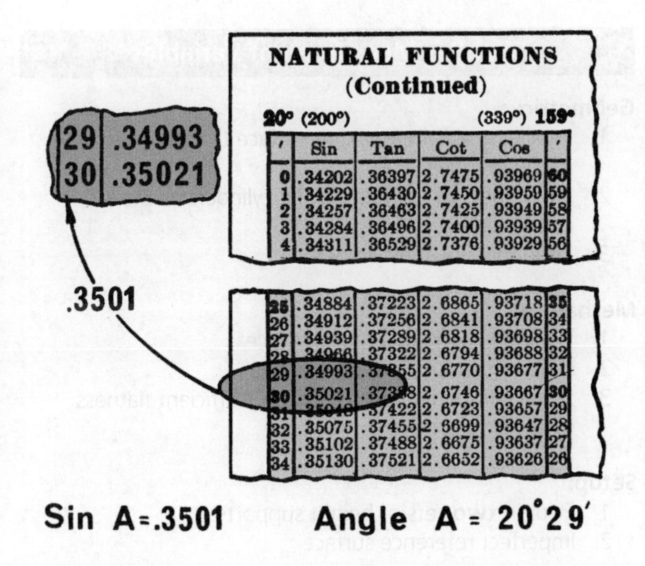

FIGURE 12–57 The sine from Figure 12–56 falls between the sine 20° 29′ and 20° 30′. Why was 29′ chosen instead of interpolating to seconds?

18.049 cm plus 2 × 9.525 mm = 19.954 cm (7.106 in. plus 2 × 0.375 in. = 7.856 in.). You find the side opposite by measuring the heights of the rolls above the reference surface and subtracting: 15.558 cm minus 8.5725 cm = 6.9855 cm (6.125 in. minus 3.375 in. = 2.750 in.).

Now we have the data we need to determine the sine of angle A: 6.9850 cm divided by 19.954 cm (2.750 in. divided by 7.856 in.). The resulting decimal is 0.35005 or 0.3501, completely independent of whether you are using English or metric measurement. The decimal is the ratio of one side of the triangle to another and would be the same if you made the original measurement in centimeters, rods, nautical miles, or Soviet archines.

Finally, you consult the trig tables and locate the angle (see Figure 12–57). To be consistent with your measurement instrument, you should round the angle off to the same decimal place that represents the precision of the measurement instrument. In this example, we used vernier calipers, so we round to minutes instead of interpolating to seconds or further.

12–6 SINE BARS AND PLATES

A sine bar (see Figure 12–58), or a similar sine instrument, helps us learn something about angles, using the same method we have been discussing, but part of the problem is built into the instrument and also resolved by the instrument.

The sine bar, a steel bar that has a cylinder near each end, forms a hypotenuse (see Figure 12–59), and the instrument is designed with a distance between

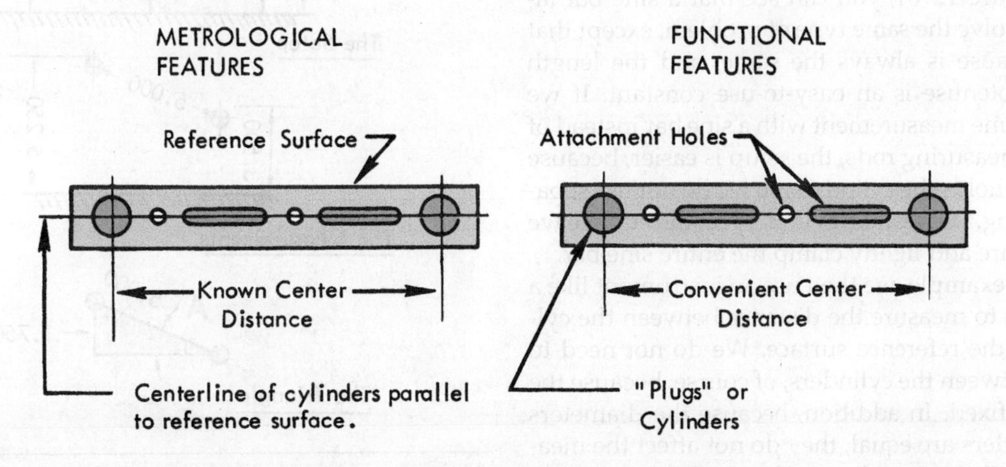

FIGURE 12–58 The sine bar is a hypotenuse of a triangle frozen in steel with a length selected to minimize computations.

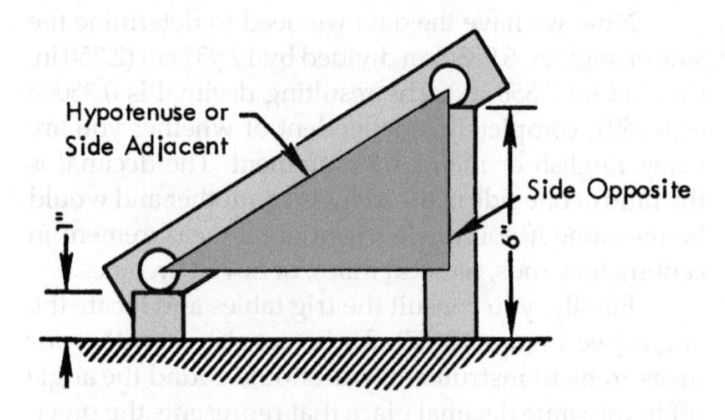

FIGURE 12–59 The sine bar forms a triangle with a hypotenuse of 5 in. The side opposite is the difference in the height of the cylinder supports.

the cylinders—usually 12.7 or 25.4 cm (5 or 10 in.)—that makes computation easy. The working surface is as parallel to the centerline of the cylinders as can be practically manufactured. Because sine bars, by their nature, introduce several elements of uncertainty (see Figure 12–60) they must be very carefully constructed.

When one of the cylinders is resting on a surface, you can set the bar at any desired angle by simply raising the second cylinder. You obtain the desired angle when the height difference between the cylinders is equal to the sine of the angle multiplied by the distance between the centers of the cylinders.

In Figure 12–61, you can see that a sine bar allows us to solve the same type of problem, except that the hypotenuse is always the same; and the length of the hypotenuse is an easy-to-use constant. If we make the same measurement with a sine bar instead of two loose measuring rods, the setup is easier, because the sine bar holds the cylinders, so we do not use separate clamping. We rest the cylinders on the surface we must measure and lightly clamp the entire sine bar.

In the example, we then use an instrument like a height gage to measure the distance between the cylinders and the reference surface. We do not need to measure between the cylinders, of course, because the distance is fixed. In addition, because the diameters of the cylinders are equal, they do not affect the measurement. When we subtract the heights, we obtain the triangle we need.

Sine Bar Measurement Variables

Geometric:

1. Parallelism of the working surface to the centerline of the cylinders
2. Squareness of the axes of the cylinders to the instrument
3. Roundness of the cylinders

Mechanical:

1. Error in center-to-center distance
2. Differences in cylinder diameters
3. Surface imperfections, such as insufficient flatness of working surface

Setup:

1. Error in **two** sets of height supports
2. Imperfect reference surface

FIGURE 12–60 These three sources of error should be recognized when using the sine bar.

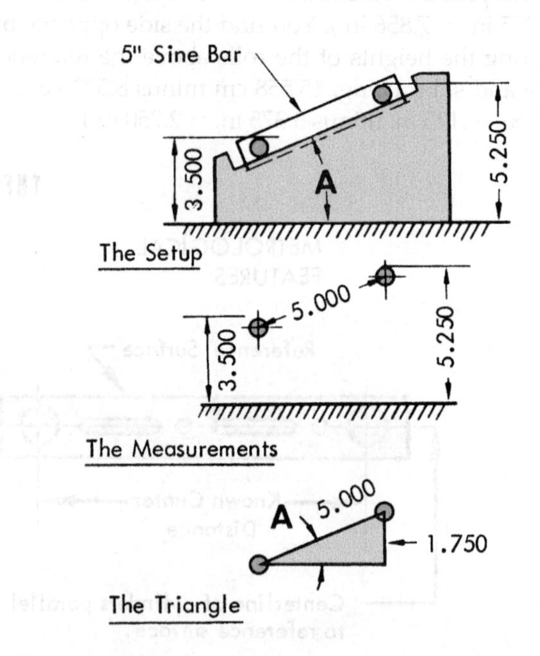

FIGURE 12–61 The sine bar greatly simplifies the angle measurement in Figure 12–56.

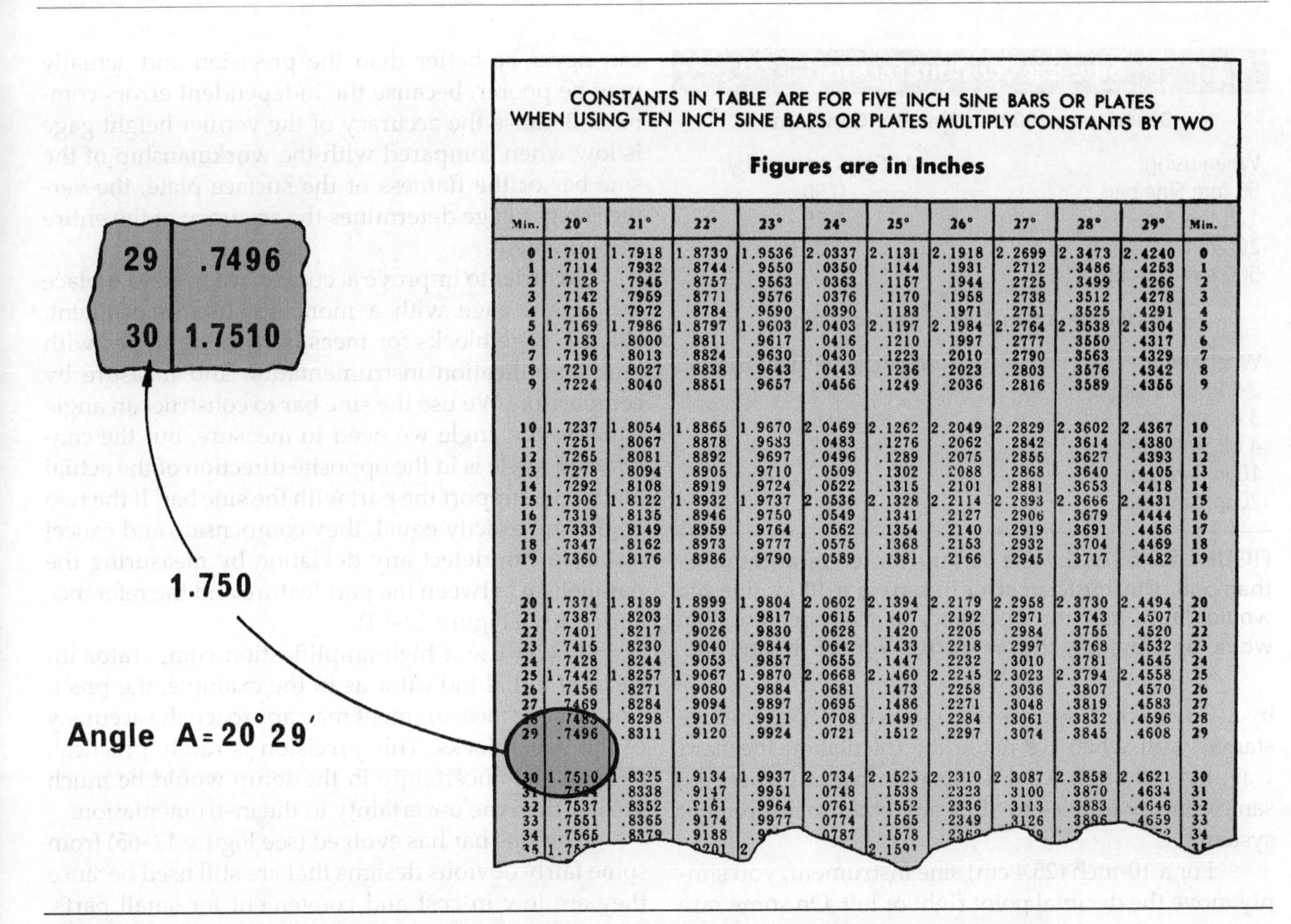

CONSTANTS IN TABLE ARE FOR FIVE INCH SINE BARS OR PLATES
WHEN USING TEN INCH SINE BARS OR PLATES MULTIPLY CONSTANTS BY TWO

Figures are in Inches

Min.	20°	21°	22°	23°	24°	25°	26°	27°	28°	29°	Min.
0	1.7101	1.7918	1.8730	1.9536	2.0337	2.1131	2.1918	2.2699	2.3473	2.4240	0
1	.7114	.7932	.8744	.9550	.0350	.1144	.1931	.2712	.3486	.4253	1
2	.7128	.7945	.8757	.9563	.0363	.1157	.1944	.2725	.3499	.4266	2
3	.7142	.7959	.8771	.9576	.0376	.1170	.1958	.2738	.3512	.4278	3
4	.7155	.7972	.8784	.9590	.0390	.1183	.1971	.2751	.3525	.4291	4
5	1.7169	1.7986	1.8797	1.9603	2.0403	2.1197	2.1984	2.2764	2.3538	2.4304	5
6	.7183	.8000	.8811	.9617	.0416	.1210	.1997	.2777	.3550	.4317	6
7	.7196	.8013	.8824	.9630	.0430	.1223	.2010	.2790	.3563	.4329	7
8	.7210	.8027	.8838	.9643	.0443	.1236	.2023	.2803	.3576	.4342	8
9	.7224	.8040	.8851	.9657	.0456	.1249	.2036	.2816	.3589	.4355	9
10	1.7237	1.8054	1.8865	1.9670	2.0469	2.1262	2.2049	2.2829	2.3602	2.4367	10
11	.7251	.8067	.8878	.9683	.0483	.1276	.2062	.2842	.3614	.4380	11
12	.7265	.8081	.8892	.9697	.0496	.1289	.2075	.2855	.3627	.4393	12
13	.7278	.8094	.8905	.9710	.0509	.1302	.2088	.2868	.3640	.4405	13
14	.7292	.8108	.8919	.9724	.0522	.1315	.2101	.2881	.3653	.4418	14
15	1.7306	1.8122	1.8932	1.9737	2.0536	2.1328	2.2114	2.2893	2.3666	2.4431	15
16	.7319	.8135	.8946	.9750	.0549	.1341	.2127	.2906	.3679	.4444	16
17	.7333	.8149	.8959	.9764	.0562	.1354	.2140	.2919	.3691	.4456	17
18	.7347	.8162	.8973	.9777	.0575	.1368	.2153	.2932	.3704	.4469	18
19	.7360	.8176	.8986	.9790	.0589	.1381	.2166	.2945	.3717	.4482	19
20	1.7374	1.8189	1.8999	1.9804	2.0602	2.1394	2.2179	2.2958	2.3730	2.4494	20
21	.7387	.8203	.9013	.9817	.0615	.1407	.2192	.2971	.3743	.4507	21
22	.7401	.8217	.9026	.9830	.0628	.1420	.2205	.2984	.3755	.4520	22
23	.7415	.8230	.9040	.9844	.0642	.1433	.2218	.2997	.3768	.4532	23
24	.7428	.8244	.9053	.9857	.0655	.1447	.2232	.3010	.3781	.4545	24
25	1.7442	1.8257	1.9067	1.9870	2.0668	2.1460	2.2245	2.3023	2.3794	2.4558	25
26	.7456	.8271	.9080	.9884	.0681	.1473	.2258	.3036	.3807	.4570	26
27	.7469	.8284	.9094	.9897	.0695	.1486	.2271	.3048	.3819	.4583	27
28	.7483	.8298	.9107	.9911	.0708	.1499	.2284	.3061	.3832	.4596	28
29	.7496	.8311	.9120	.9924	.0721	.1512	.2297	.3074	.3845	.4608	29
30	1.7510	1.8325	1.9134	1.9937	2.0734	2.1525	2.2310	2.3087	2.3858	2.4621	30
31	[illegible]	.8338	.9147	.9951	.0748	.1538	.2323	.3100	.3870	.4634	31
32	.7537	.8352	.9161	.9964	.0761	.1552	.2336	.3113	.3883	.4646	32
33	.7551	.8365	.9174	.9977	.0774	.1565	.2349	.3126	.3896	.4659	33
34	.7565	.8379	.9188	[illegible]	.0787	.1578	.2362	[illegible]	[illegible]	[illegible]	34
35	[illegible]	[illegible]	.9201	[illegible]	[illegible]	2.1591	[illegible]	[illegible]	[illegible]	[illegible]	35

FIGURE 12–62 Little computation is required when using the table of constants for the sine bar.

We do not need to divide out the sine; we simply refer to a table of constants for the 12.7 cm (5 in.) sine bar (see Figure 12–62). Reading across the degrees at the top, 1.750 is greater than 20° (1.7101) and less than 21° (1.7918); therefore, we go down the 20° column and find that 1.750 is closer to 1.7496 than to 1.7510. Reading the minutes in the left column, we determine that the angle is 20° 29′.

Calculations for 5-Inch Sine Bars

Five inches (12.7 cm) was chosen as the standard center distance for most sine instruments, because multiplying by 5 is equal to multiplying by 10 and dividing by 2. Computation with 5 is very easy, even when you do not have a table of constants available and must use a table of natural functions.

To set a 5-inch (12.7 cm) sine bar to an angle, look up the sine in a table of natural functions, move the decimal point one place to the right and divide by 2. The result is the height that you must elevate one cylinder above the other. To find the angle when you know the height difference, move the decimal point one place to the left, multiply by 2, and look up the result in the table.

To prove to yourself that these steps work, use Figure 12–61, where the height difference is 1.750. Move the decimal point one place to the left, 0.1750: multiply

Sine Bar Constant Factors

Based on a 100 mm Sine Bar Constant

When using:	Multiply constant by:
50 mm Sine bar	0.50
125 mm Sine bar	1.25
200 mm Sine bar	2.00
500 mm Sine bar	5.00

Based on a 5-inch Sine Bar Constant

When using:	Multiply constant by:
2.5 in. Sine bar	0.5
3 in. Sine bar	0.6
4 in. Sine bar	0.8
10 in. Sine bar	2
20 in. Sine bar	4

FIGURE 12–63 Many sine bars use center distances other than 5 in. Obviously an error in setting a 10 in. sine bar would cause half the inaccuracy in measurement that would be caused by the same error with a 5 in. sine bar.

by 2, 0.350, which is close enough to the 0.3501 that we started with when we made the calculation the hard way. Even though we did this example in inches, the same concept applies to SI or any other measurement system.

For a 10-inch (25.4 cm) sine instrument, you simply move the decimal point right or left. On some rare occasions when you use a 20 in. (50.8 cm) instrument, move the decimal point the same way, but multiply by 2 to locate the sine, and divide by 2 to determine the angle.

When you have a sine bar without a standardized length, you can still use a table of constants for a 5-inch sine bar to determine the sine of angles. You use the figures in the table and convert them by multiplying by constants (see Figure 12–63).

Comparison Measurement with Sine Bars

In the method we just discussed, the precision of your measurement is limited by the sensitivity of the instrument you use. Because we used a vernier height gage in our example, we automatically limited the precision to 25.4 μm (0.001 in.). In addition, the accuracy can never be better than the precision and actually may be poorer, because the independent errors combine. Because the accuracy of the vernier height gage is low when compared with the workmanship of the sine bar or the flatness of the surface plate, the vernier height gage determines the accuracy of the entire measurement.

In order to improve accuracy, we need to replace the height gage with a more sensitive instrument. We use gage blocks for measurement of angles with high-amplification instrumentation and measure by comparison. We use the sine bar to construct an angle equal to the angle we need to measure, but the constructed angle is in the opposite direction of the actual angle. We support the part with the sine bar. If the two angles are exactly equal, they compensate and cancel out. You can detect any deviation by measuring the parallelism between the part feature and the reference surface (see Figure 12–64).

If you use a high-amplification comparator instead of a dial indicator as in the example, the precision of your measurement may approach the accuracy of the gage blocks. This precision is rarely practical, because the uncertainty in the setup would be much greater than the uncertainty in the instrumentation.

The sine bar has evolved (see Figure 12–65) from some fairly obvious designs that are still used because they are low in cost and convenient for small parts. Older versions are not self-supporting and must be

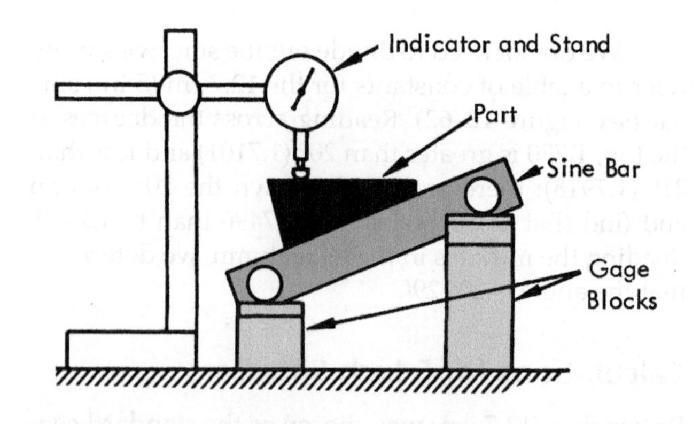

FIGURE 12–64 For comparison measurement, the sine bar is used to cancel out the angle being measured.

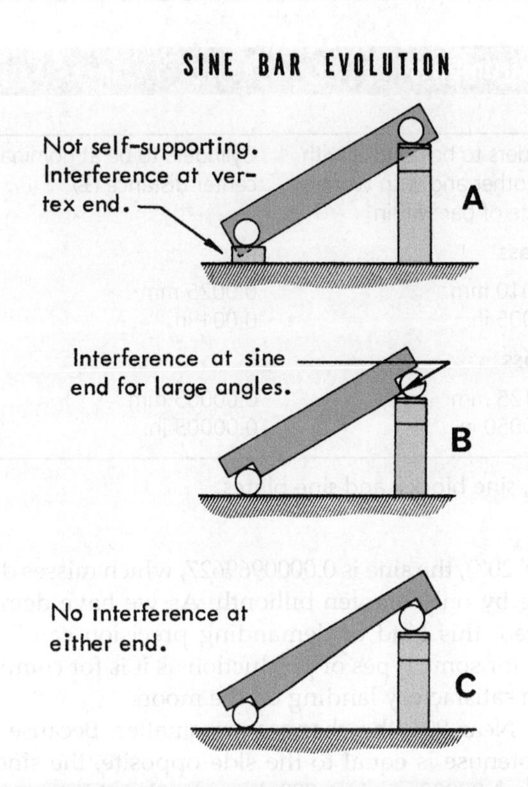

FIGURE 12–65 All three types of sine bars are in extensive use, although the more refined types clearly evolved after the simple type at A.

FIGURE 12–66 A sine plate is a sine block with an attached base. *(Courtesy of Suburban Tool, Inc.)*

clamped for vertical use, but the chief disadvantage is that we cannot rest the lower plug directly on the reference surface. In Figure 12–65B, the instrument is self-supporting, but it is more difficult to manufacture and has interference in part of its range. The instrument in Figure 12–65C eliminates the interference. In addition, while we must scrap any type B instrument that is lapped undersized, we can always correct an undersize condition in type C instruments by lapping the opposite plug-locator shoulder.

Sine Blocks, Sine Plates, and Sine Tables

Sine blocks are wide sine bars, *sine plates* are wider sine blocks, and *sine tables* are still wider. Manufacturers do not necessarily make distinctions among these instruments, and MIL-STD-120 only covers sine bars and sine plates, defining all sine instruments wider than 2.54 cm (1 in.) as sine plates. For practical purposes, a sine instrument wide enough to stand unsupported is a sine block. If the sine instrument rests on an integral base, it is a sine plate (see Figure 12–66). If the sine instrument is an integral part of another device, such as a machine tool, it is a sine table.

These instruments all differ from sine bars because the part rests on the instrument. You will only very rarely be able to set up one of these sine instruments so it rests on the part. The only practical way you can use these sine instruments to measure angles is by comparison; however, you can also use them to hold parts so that angles can be produced on them. In fact, some of these instruments are manufactured specifically for this purpose, as in Figure 12–66. Tolerances for these instruments vary. The tolerances for one high-grade manufacturer are shown in Figure 12–67.

Using the Complement of the Angle

As the angle you are measuring increases, the precision of sine measurement decreases (see Figure 12–68). The sine is the result of dividing the side opposite by the hypotenuse, where the hypotenuse is always the same for a particular sine instrument. At 1°, the side opposite is extremely small compared with the hypotenuse; therefore, the sine is very small (0.01745 for 1° angle). When we increase the angle by one degree, to 2°, the sine nearly doubles to 0.03490, which is a

Tolerance of Sine Instruments				
	Bar	**Cylinder**		
Size	Working surface to be flat, square with sides and parallel within	Cylinders to be alike, round and straight within	Cylinders to be parallel with each other and with working surface of bar within	Cylinders to be at nominal center distance (±)
		Commercial Class		
125 mm 5 in.	0.0005 mm 0.0002 in.	0.0025 mm 0.0001 in.	0.00010 mm 0.00005 in.	0.0025 mm 0.001 in.
		Laboratory Class		
125 mm 5 in.	0.00125 mm 0.00005 in.	0.00025 mm 0.00010 in.	0.00125 mm 0.000050 in.	0.00005 mm 0.00003 in.

FIGURE 12–67 These are typical tolerances for quality sine bars, sine blocks, and sine plates.

comparatively large increase. At 3°, the sine increases significantly again, although not nearly as much as the increase to 2°. As the size of the angle continues to increase, the increase in the sine decreases.

As the angles get even smaller, the effects of "doubling" the sine are even more significant. For example, the sine of 0° 0′ 10″ (a 10-second angle) is 0.0000484814.* If we make the angle 10 seconds larger

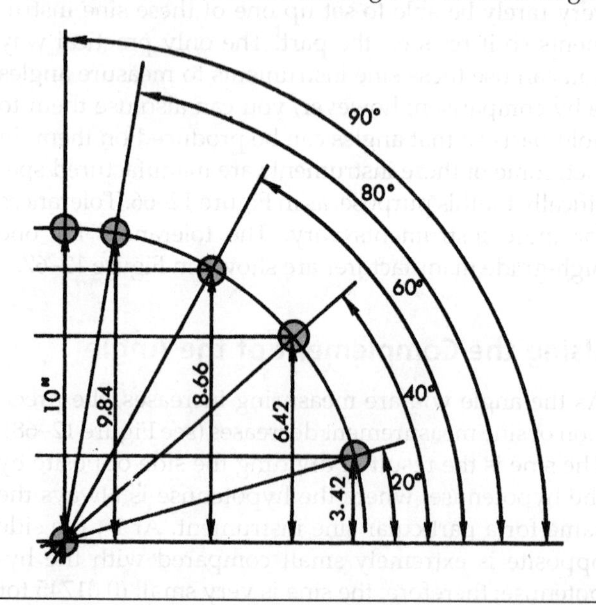

FIGURE 12–68 The sine changes rapidly near 0° and very slowly as it approaches 90°.

(0° 0′ 20″), the sine is 0.0000969627, which misses doubling by only one ten billionth. As we have demonstrated, this kind of demanding precision can be as vital for some types of production as it is for completing a satisfactory landing on the moon.

Near 90°, the changes are smaller. Because the hypotenuse is equal to the side opposite, the sine of 90° is 1.00000; and at 89°, because the side opposite has decreased only slightly, the sine is nearly 1.00000 (0.99985 for 89°). Between 89° and 88°, the sine changes by only 0.00046. At this end of the 0° to 90° quadrant, a large change in angle results in a very small change in the sine—exactly opposite from the other end of the quadrant.

To show how this affects measurement (see Figure 12–69), we are trying to determine if a part feature angle is 65°. In the example, because the angle is not exactly 65°, the indicator reads 127.3 µm (0.005 in.) lack of parallelism in 5.08 cm (2 in.). At B, we make a setup to measure the complement of the angle or 25°, and again the lack of parallelism is 127.3 µm (0.005 in.) in 5.08 cm (2 in.). In each case, the 127.3 µm (0.005 in.) in 5.08 cm (2 in.) indicator reading does not show that the angles have been read with precision, because the parallelism change at the steep angle (see Figure 12–69A) does not represent nearly the same angle change as it does in the complement (see Figure 12–69B).

As the angle becomes larger, this relationship becomes more important: at 80°, a 25.4 cm (10 in.) sine

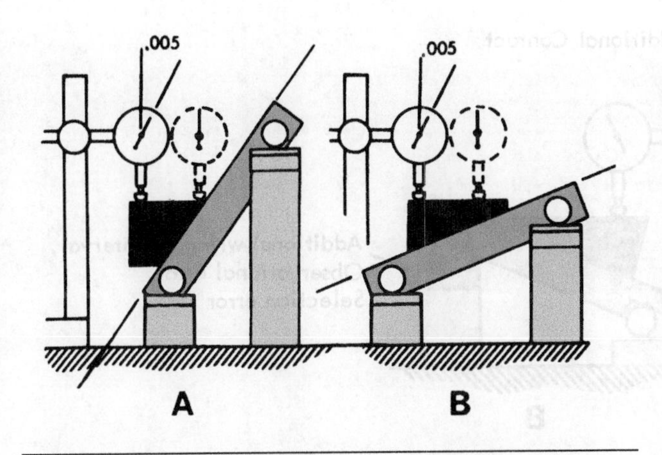

FIGURE 12–69 In A, the indicator shows that the out-of-parallel condition is 0.005 in 2 inches. In B, it reads the same. Does this mean that both angles are being read to the same precision?

bar requires only a 12.7 μm (0.0005 in.) height change to change the angle one minute. Comparatively, at 10°, a one-minute angle change is caused by a 71.12 mm (0.0028 in.) (71.12 μm) height change—more than five times the amount required at the complement. Anything that disturbs the measurement at 10° will have one-fifth the effect that it would have at 80°. When you are measuring an angle greater than 60° with sine instruments, you should make the measurement using the complement of the angle. Some authorities do not recommend using sine instruments at angles over 45°.

Use of Gage Blocks

You will encounter some problems when you try to use a sine instrument to measure an angle whose sine is smaller than the thinnest gage block. Because most sets start at 0.5 mm (0.100 in.), you will have thinner blocks infrequently, even though some sets are made to 0.2 mm (0.010 in.). As you might expect, the sine of the angle you are measuring is rarely exactly equal to one gage block; therefore, you must usually combine two or more blocks, raising the height of the stack even more.

When you use certain sine bars (see Figure 12–65 and Figure 12–69), you eliminate this problem. Because the lower cylinder must rest on a gage block or stack of gage blocks, the sine will always be the difference in heights of the stacks, which is the same principle that applies when you are using sine blocks.

While photographs of sine tables often show the gage blocks placed with their long dimension across the cylinders, metrologically, they should be placed in line with the cylinders. In the discussion of comparators, we mentioned the high unit pressures caused by a spherical surface. Similarly, the line contact of the cylinders of a sine instrument also creates high pressures, which slightly deform both the surfaces on which they rest and the cylinders themselves. If the line that the pressure is acting on is longer, the deformation will be less.

As a compromise, use square gage blocks whenever they are available. When you are making heavy setups, you should use double stacks of gage blocks, spaced for approximately equal loading.

We recommend you use gage blocks between the lower cylinder and the reference surface, even when they are not required mathematically to make up the sine of the angle. In addition to reducing the deformation of the reference surface, the additional blocks help when the line contact of the cylinder falls into low portions of the reference surface. It is better for the cylinder to rest on the very hard, dense surface of gage blocks.

Because large numbers of errors are common to different methods of measurement (see Figure 12–70)—for example, the compression of the cylinders when under weight load—we must be concerned only with the errors that are inherent in each specific case. In Figure 12–70A, the two important errors are the surface imperfections and compressibility of the reference surface. But in Figure 12–70B, we have minimized these sources of error by inserting a gage block between the lower cylinder and the reference surface. Unfortunately, when we minimize the errors in this manner, it is like we are adding another stack of gage blocks and all the errors that can be associated with the new block stack: imperfect contact between the added gage block and the reference surface; failure to correctly combine the higher stack; observational error reading the stack; and wringing error.

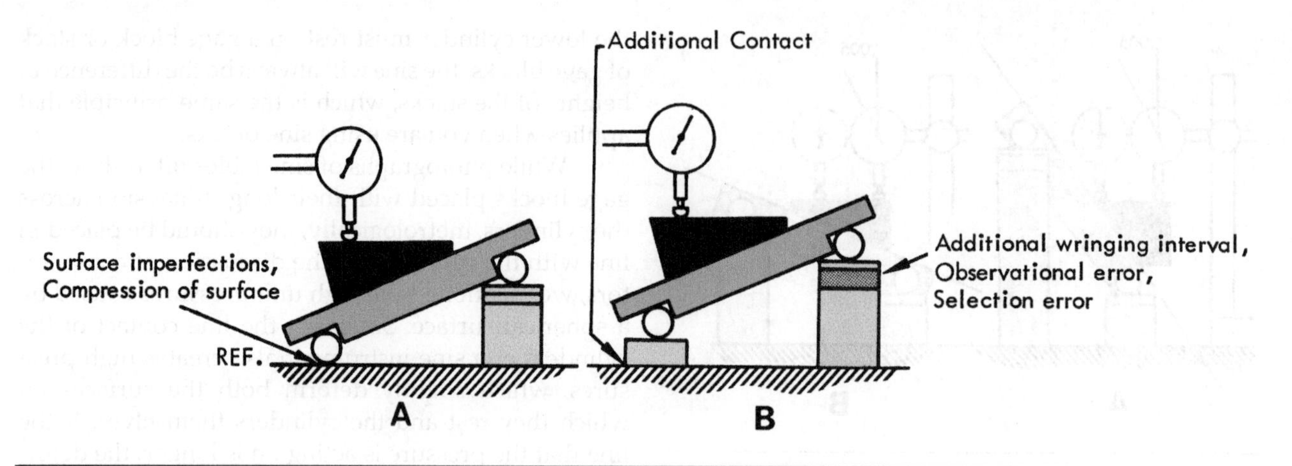

FIGURE 12–70 The comparator does not know the difference between A and B, except for the errors caused by the two setups.

If the reference surface is a toolmaker's flat, we do not gain any benefits from adding the block under the lower cylinder. If, on the other hand, the reference surface is a rough-machined portion of a large part, then all of the added sources of potential error are minor compared with the error we are working to minimize.

Compound Angles

While an in-depth study of *compound angles* is properly handled in a mathematics course, it is impossible to do much practical measurement in the shop or the laboratory without some basic knowledge of them.

While simple angles lie in one plane, compound angles are formed by the edges of triangles that lie in different planes. If these triangles lie in different planes, they define the boundaries of solids, real or imaginary, and their edges are the intersections of the planes.

In a solid formed by the intersections of planes, like a pyramid (see Figure 12–71), the angles on the surface planes are called *face angles.* There are four sets, one for each point, and around each point, there are three face angles. Around A, for example, there are ABC, ABD, and CAD.

A *dihedral angle* is the opening between two intersecting planes, and its *plane angle* is an angle whose sides are in two intersecting planes perpendicular to the line of intersection at the same point (see Figure 12–72). Where two planes intersect at line AB, the dihedral angle is the opening between the planes. In order to express the size of this angle, we construct the plane angle CDE, which lies in a plane perpendicular to the line AB.

Every pair of faces in any solid figure form a dihedral angle, and in the case of cubes and rectangular parallelepipeds, the face angles equal the plane angles. Whenever one or more sides are at any angle that is a multiple of 90°, the plane angles and the face angles will be the same (see Figure 12–73). Of the example's

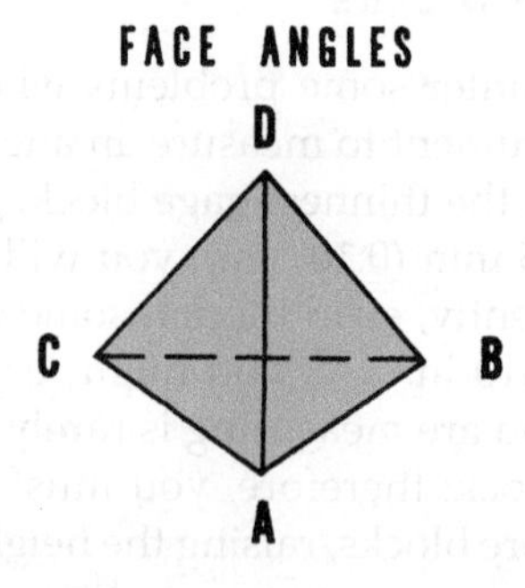

FIGURE 12–71 In this pyramid, there are four sets of face angles, three around each point A, B, C, and D.

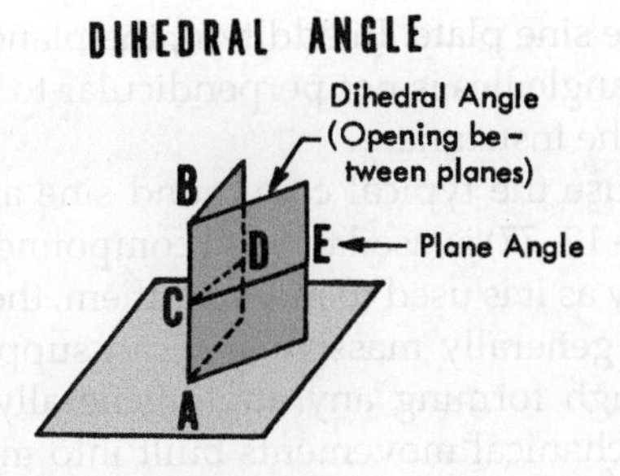

FIGURE 12–72 The plane angle lies in a plane perpendicular to the intersection of the two planes.

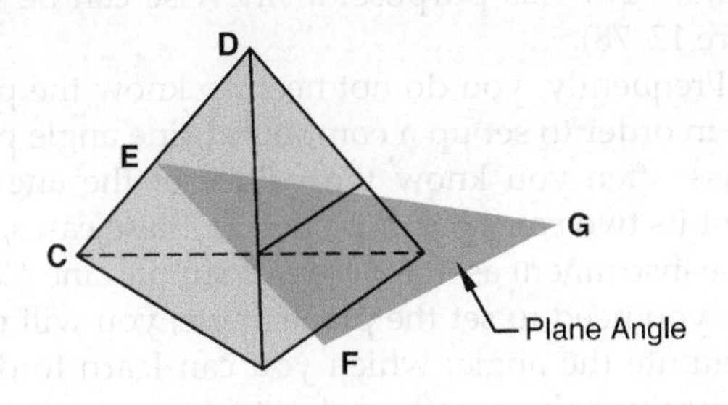

FIGURE 12–73 Each pair of sides of the pyramid has a plane angle. These plane angles differ from the face angles and determine the shape of the pyramid.

12 face angles, we can alter 10 without changing the plane angle FEG.

There are as many plane angles as there are edges; therefore, in Figure 12–73, there are six plane angles—half as many as there are face angles. Clearly, plane angles are a more efficient means for defining the shape of a solid than the face angles; therefore, we must deal with the plane angles when we are machining solid figures or measuring them.

We can always divide any triangle into right triangles. Similarly, we can subdivide any compound angle. Some combination of two of five types of angles (see Figure 12–74) will form every compound angle.

Each angle consists of two planes that are perpendicular to a third plane; without these planes, which provide the references, we could not measure the angles. The fourth plane, which closes the figure, is the one that creates the compound angles. Because

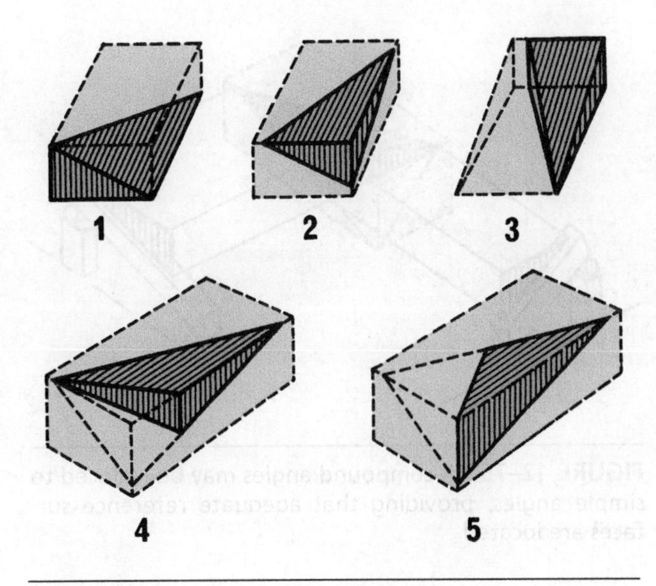

FIGURE 12–74 In order to find the compound angle, most solid figures reduce to one of these five types.

each pair of planes has a dihedral angle between them, the closing plane creates three compound angles, and in each case, one side of the dihedral angle is one of the reference planes.

Not all compound angle measurement problems resemble these five types. In addition, all compound angle measurement problems can be divided into two or more parts, each of which will fit one of the five general types. We can visualize all of these angle types—except type 3—circumscribed by a rectangle: type 3 angles are circumscribed by a wedge. These angle types also show why we can use sine plates to measure compound angles. Sine instruments do not measure angles directly. We set them up to cancel out the angle so that a comparator can check the parallelism of the angle to the reference surface.

To set sine instruments (see Figure 12–75) to the face angles of the figure, we project them to the sides of the imaginary rectangular parallelepiped that circumscribes the angles (type 4 in our example). To project the instruments, we place the pivoting axes of the sine instruments at right angles to each other; then we place them on top of each other. This process is the principle embodied in the *compound sine angle plate.*

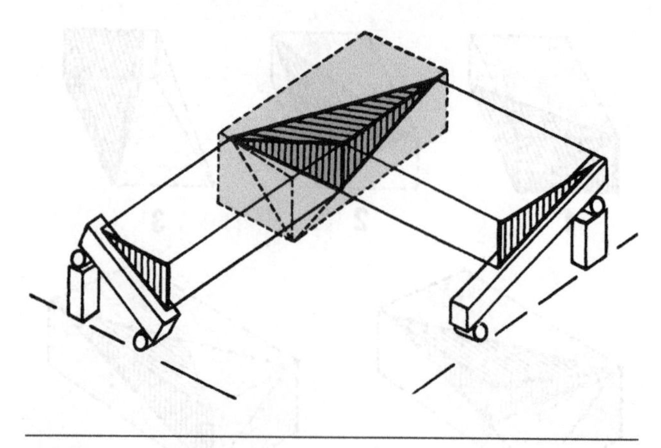

FIGURE 12–75 All compound angles may be reduced to simple angles, providing that adequate reference surfaces are located.

The Compound Sine Angle Plate

When two sine plates are superimposed (see Figure 12–76) the base creates one plane, and the top creates the second plane. Any two planes that are not parallel eventually intersect even if the intersection is just an imaginary line in space. The line along which these planes intersect passes through the intersection of the hinge lines. The plane angle, the compound angle we need to measure, lies in a plane perpendicular to the line of intersection of the two planes and is different from either of the angles set in the upper or lower parts of the sine plate. In addition, the plane in which the plane angle lies is not perpendicular to any of the planes of the instrument.

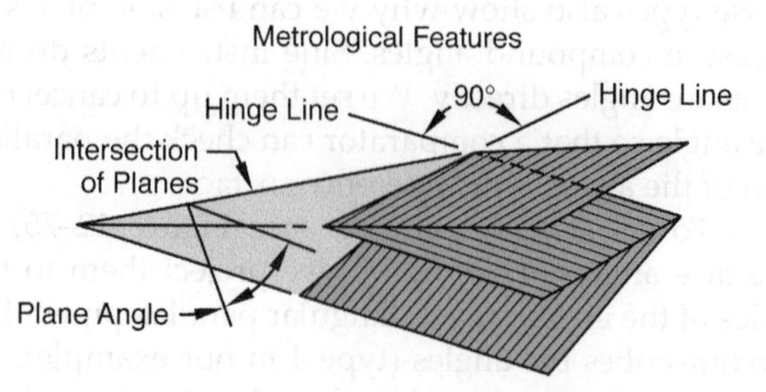

FIGURE 12–76 The plane angle is formed by the interaction of the two sine plates. It is neither of the angles set in the separate sections.

Because the typical compound sine angle plate (see Figure 12–77) is used to form compound angles as extensively as it is used to measure them, these instruments are generally massive and can support heavy loads. Rough forming any angle generally depends on the mechanical movements built into milling machines, boring mills, and other machine tools; therefore, sine plates are not required until the finishing operation, which is usually performed on grinding machines. For this purpose, a sine vise can be used (Figure 12–78).

Frequently, you do not need to know the plane angle in order to set up a compound sine angle plate, because often you know the values of the angle in each of its two component planes. In these cases, you use the instrument as if it were two simple sine plates. When you need to set the plane angle, you will need to compute the angle, which you can learn to do in most machine shop mathematics texts.

One useful shortcut we should mention is modeling clay. You can use it to help you visualize the angles. Simply model the part, turn it around as needed and look at it until you can determine the angle you

FIGURE 12–77 The compound sine angle plate is two simple sine plates with hinge lines at right angles to each other. (*Courtesy of Suburban Tool, Inc.*)

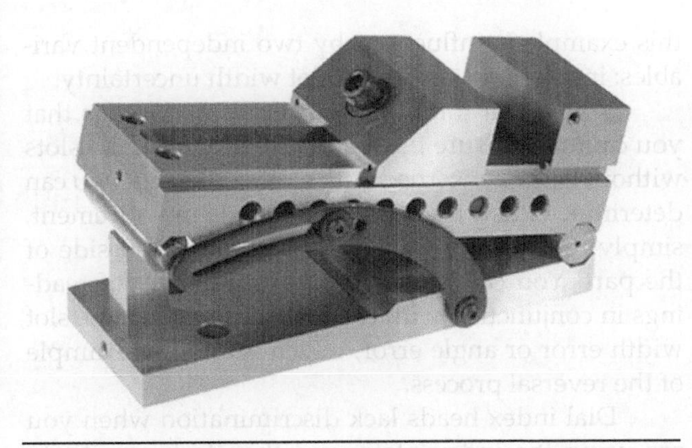

FIGURE 12–78 A sine vise. (*Courtesy of Suburban Tool, Inc.*)

need to measure. As a quick check, you can slice through the model perpendicular to any edge and find the plane angle. Although your tolerances may have to be ±5° or more, you will know whether or not you are on the right track.

12–7 MECHANICAL ANGLE MEASUREMENT

When we discussed compound sine angle plates, we mentioned that they are used as much in machining angles as in measuring angles. Another method of angle measurement, called *mechanical indexing*, and its related tools—the *dividing heads, indexing heads* or *index heads*—were developed specifically for machining rather than measurement. The devices are actually more machines than instruments; however, they are very valuable for angle measurement. You can find more information on these tools in textbooks on machine tool operation.

As the name implies, these devices were originally developed to divide circles into equal divisions; in practical application, they are used for milling teeth in cutters and flutes in reamers and for spacing bolt circles. Because the circle is the standard for angles, any division of a circle is angle measurement, whether we measure it in numbers of divisions or degrees and minutes.

There are three principal classes of index heads: dial, plain, and universal. Each of these can be divided into subclasses, particularly the universal heads, which can be extended to very sophisticated devices. Like other instruments we have discussed, the mechanical angle measuring devices range from low to relatively high amplification.

The Dial Index Head

The lowest amplification, the dial index head, is one power (1X); therefore, it has limited practical application today. The dial index head consists of a horizontal spindle mounted on a base that rests on a reference surface (see Figure 12–79). On one end of the spindle is a plate for clamping parts, and on the other end is a knob or wheel for turning the spindle and usually a ring calibrated in degrees. The plate, which is also an index plate, contains holes with a plunger that engages the stationary housing. The typical plate has 24 holes, which provides 360° rotation in 15° increments.

The two types of sensitivity and accuracy for this instrument provide an interesting contrast (see Figure 12–80), and reemphasize one point we made earlier: You cannot judge the accuracy of an instrument by its graduations. Using the graduations, you can set the head to 1°; however, if reliability is important, repeatability is limited to ½°—which leads to 30-minute sensitivity. In this case, the uncertainty in the sensitivity of the instrument is significantly greater than the uncertainty in the instrument itself, and the accuracy is limited to 30 minutes.

FIGURE 12–79 A dial index.

Dial Index Head Capability			
Method of Operation	**Discrimination**	**Sensitivity**	**Accuracy**
Graduations	1°	30′	30′
Index Plate	15°	30″	1′

FIGURE 12–80 The widely divergent precision of the two methods of operation shows the precision that can be built into mechanical devices. These values are typical but vary among manufacturers of dial index heads.

When you use the index plate, the instrument settings have a discrimination of 15°, but the only limit to sensitivity and accuracy is the care used in the design and manufacture of the head. Typically, design and the amount of wear in these instruments (see Figure 12–80) vary considerably among manufacturers.

Parts using the index head for machining or measurement have an axis of rotation and usually an open center, which means we can mount them to the plate rather simply. At the center of the plate is a hardened pin, and bushings are slipped into the pin to adapt it to the inside diameter of the part, which centers the part to the limit of accuracy of the pin, bushings, and part's inside diameter. We hold the pin to the plate with clamps, and then use the pin, in part, as a reference for measurement (see Figure 12–81). The measurement in this example is influenced by two independent variables: index uncertainty and slot width uncertainty.

You might think that our example implies that you cannot measure the angular positions of the slots without being concerned with slot widths, but you can determine either. If you need the angle measurement, simply repeat steps 2 through 4 from the left side of the part. You can then use the resulting set of readings in conjunction with the first set to determine slot width error or angle error, which is another example of the reversal process.

Dial index heads lack discrimination when you are using the index plate, and they lack sensitivity when you are using the instrument graduations. In order to improve either discrimination or sensitivity, we need to increase the amplification of the instrument. Manufacturers have used all the familiar additions: vernier scales, micrometer screws, and the microscope.

The Plain Index Head

With amplification mechanically applied, the dial index head evolved into the plain index head. In addition, when you are using the index plate of the dial index head, you are working with relatively high sensitivity and a low error variable. The discrimination, unfortunately, is limited to 15° increments, and

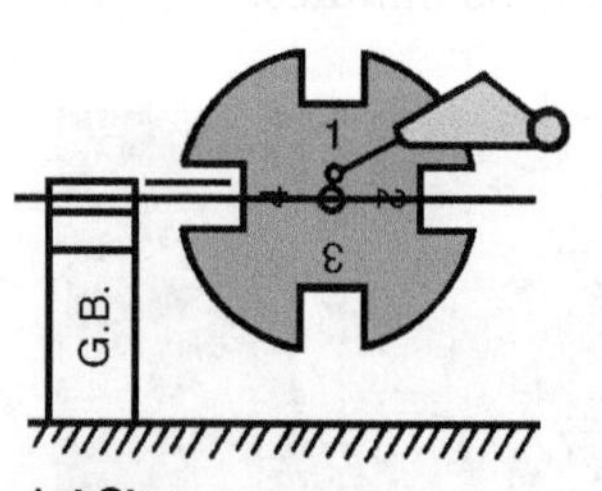

1st Step
Set head on zero. Loosely clamp the part. Zero indicator on pin. Using gage blocks or height gage, determine height of pin.

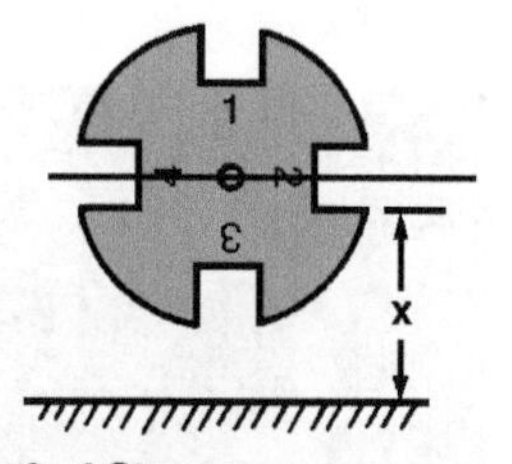

2nd Step
Subtract half of pin diameter from pin height. Subtract half of slot width in order to find height x.

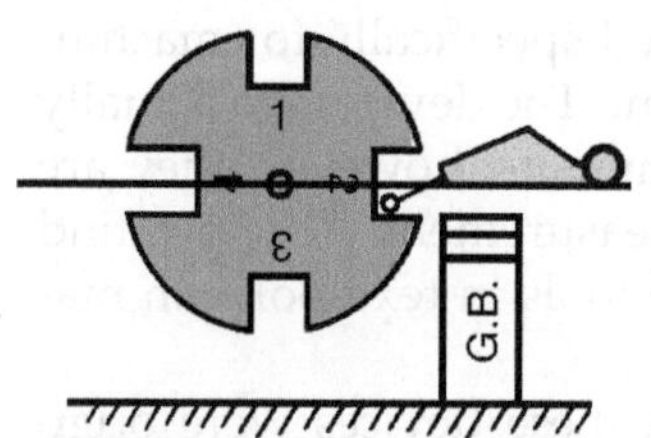

3rd Step
Zero indicator on dimension x. Indicate slot. Now note deviation from zero.

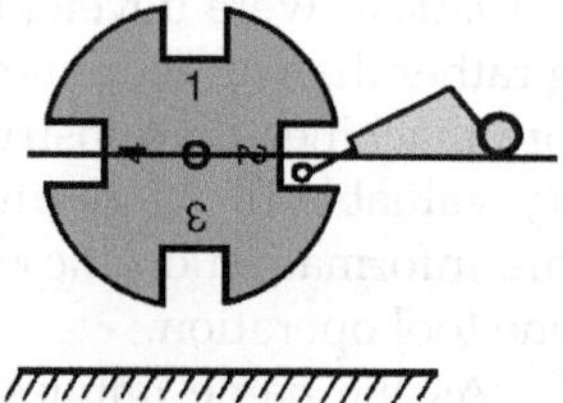

4th Step
Index to next slot. Indicate slot. Note deviation from zero. Continue for remaining slots.

FIGURE 12–81 For clarity, the dial index head is not shown in this sequence of operation steps. The deviation read by the indicator will contain both the index error and slot width error.

adding more indexes does not solve the problem. We would need a plate with 6.35 mm (0.25 in.) diameter holes that was at least 7.3152 m (24 ft.) in diameter in order to obtain a discrimination of one minute.

The true solution is gearing: We connect the index plate to the spindle and face plate so that the index plate makes two revolutions when the spindle makes one revolution. An index plate with 24 holes at 15° intervals would revolve twice to the face plate's one turn. The index plate would also make 48 stops through the revolution, which divides the face plate rotation into 7½° increments, doubling the discrimination.

The plain index head, which starts at a 2X amplification, can provide a discrimination of 40X when we add a worm and a gear (see Figure 12–82). One turn of the index plate turns the spindle 9°; therefore, if we are using the same 24-position index plate that we used with the dial index head, the discrimination increases to 0° 2′ 15″.

To make full use of the increased amplification, we fix the index plates, which are interchangeable, and turn the crank. The crank carries a pin, which we

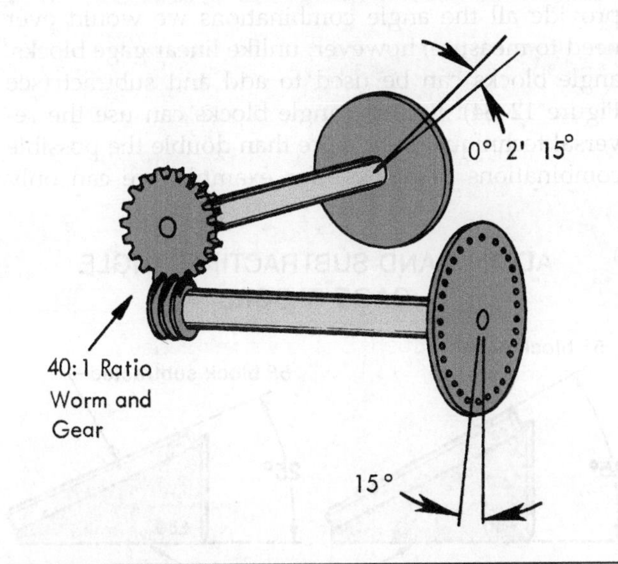

FIGURE 12–82 If 40-to-1 gearing is placed between the dial index head, index plate, and the face plate, the rotation of the face plate per index is reduced from 15° to 0° 2′ 15″.

can locate in the index plate holes. As with other instruments, a variety of constructions is available from manufacturers. Figure 12–83 is typical.

Universal index heads are even more sophisticated, and are best explained in textbooks on precision machining.

Metrology of Mechanical Indexing

Mechanical index heads were developed for the needs of the shop. These heads are rugged—made to withstand the forces of machining, not to bear the weight of heavy parts—and have very unique functional features. A high-quality head will have a spindle base error less than 6.35 μm (0.00025 in.), and you can align the spindle with the reference plane of the base so that there will be no more than 25.4 μm (0.001 in.) runout at the end of a 45.72 cm (18 in.) test bar.

In this textbook, we are primarily concerned with the metrological characteristics of index heads: accuracy, sensitivity, and precision. We usually state the accuracy of index heads as the total accumulative error. *Indexing* is like stepping off of a circle with dividers (refer to Figure 5–34). For a high-quality dividing head, we state the accuracy: "the accumulated error in indexing from one hole to the next through a complete circle must be within 38.1 μm (0.0015 in.) on a 30.48 cm (12 in. dia.)." The angular error is proportional to the size of the angle but not proportional to the number of divisions. In practical work, angles do

FIGURE 12–83 A rotary index table.

not necessarily stop and begin all over again at 360°, but often you will find it easier to consider angles of multiples of 360°. The accuracy as stated is equal to 457.2 μm (0.018 in.) in 3.6576 m (12 feet).

In certain cases, precision and discrimination are the same; however, they are different when the instrument reads in response to something done to it, like placing a gage block under the spindle. Precision and discrimination are the same when the instrument is the doer—for example, when we create an angle with a sine bar. Dividing heads create angles to which we can compare an unknown angle.

We define the precision of an index head as the fineness to which we can set it to angles, or the extent to which the head can divide the circle, which depends on the gear ratio (see Figure 12–82), and the number of holes in the index plate. Standard index plates have 11 rows of holes on each side. With all these combinations of full turns and hole spaces, many angles will be duplicated. Even after we have eliminated the duplicate angles, a large number of separate angles are still available, and we can form many others.

The amount of usable indexing discrimination is limited by the error. The maximum, 38.1 μm (0.0015 in.) in a 30.48 cm (12 in.) diameter, is one part in 25,133. Generally, it is unreliable to attempt a setting closer than 10 times the uncertainty, which reduces the usable discrimination to one part in 2,513 or approximately 8 minutes.

12-8 MEASUREMENTS TO SECONDS OF ARC

Seconds in angle measurement are similar to *mikes* in linear measurement. Some people have trouble knowing whether they are working in seconds or not; however, only the precision level of instruments can be used regularly for measurement in seconds.

We need to measure in seconds because of astronomy. In astronomy, a second sweeps through expanses of space—for example, one angular second on Earth subtends more than 1 mile on the moon. While you are probably not surprised that we need this kind of precision in the development, production, launch, and recovery of missiles and the shuttles, we also need this precision for the erection of very large, high-speed machinery like paper mills and continuous glass-producing machines.

For small work pieces, angle gage blocks allow us to measure angles to incredible precision fairly easily. For large measurement situations, optical methods provide measurement to seconds that are equally useful. For situations between these two, we can use various rotary table devices.

Angle Measurement with Gage Blocks

C. E. Johansson is credited with the introduction of angle gage blocks, even though angle blocks of lower precision than Johansson's were in use earlier. Other authorities credit Dr. G. A. Tomlinson of the National Physical Laboratory (UK) with recognizing the value of this technique and determining the sizes to provide the greatest number of combinations around 1939. A 10-block set provides all angle combinations between 0° and 180° in 5′ increments.

It might seem impossible that 10 blocks could provide all the angle combinations we would ever need to measure; however, unlike linear gage blocks, angle blocks can be used to add and subtract (see Figure 12–84). Because angle blocks can use the reversal technique, they more than double the possible combinations of blocks—for example, we can only

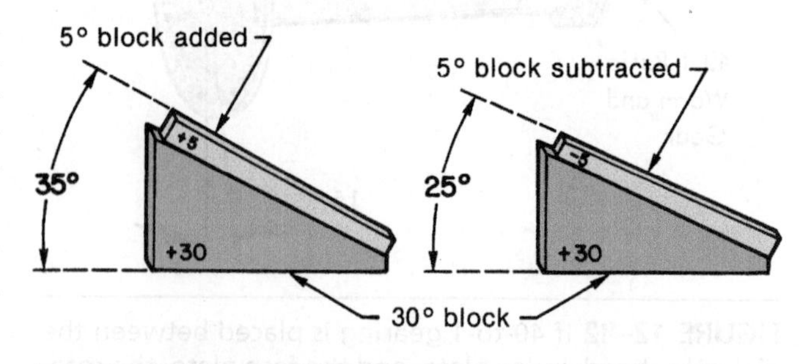

FIGURE 12–84 Many combinations can be made because each block combines in two ways.

form three different lengths with two length blocks, but we can form eight angles with two angle blocks. That is why we only need a small number of blocks to form a large number of angles.

Johansson applied the principles of gage blocks to angle blocks; however, his angle blocks did not win wide acceptance, perhaps because they were too short. In 1941, the National Physical Laboratory of the United Kingdom devised a set with 3 × 5/8 inch surfaces, which was the basis for the modern, 16-block sets (see Figure 12–85). These sets claim accuracy to 1/5,000,000th of a circle.

We can combine most angles in several ways, but as with linear gage blocks, it is better to use as few blocks as possible. Five blocks in the series (1, 3, 5, 20, and 30) can make up more combinations than 1, 3, 5, 15, and 30; however, the latter series is usually used for most practical work, because more working angles are based on 15° than 20°. Sets are also available that use decimal minutes, with the same 1, 3, 9, 27, and 41 series for degrees, minutes, and decimal minutes. We do not make a distinction for SI, because metric angle measurement proved impossible to enforce and was dropped.

In addition to the 16-block set (see Figure 12–85), blocks are available in sets of 11, for one-minute-of-arc discrimination, and 6, for one-degree discrimination.

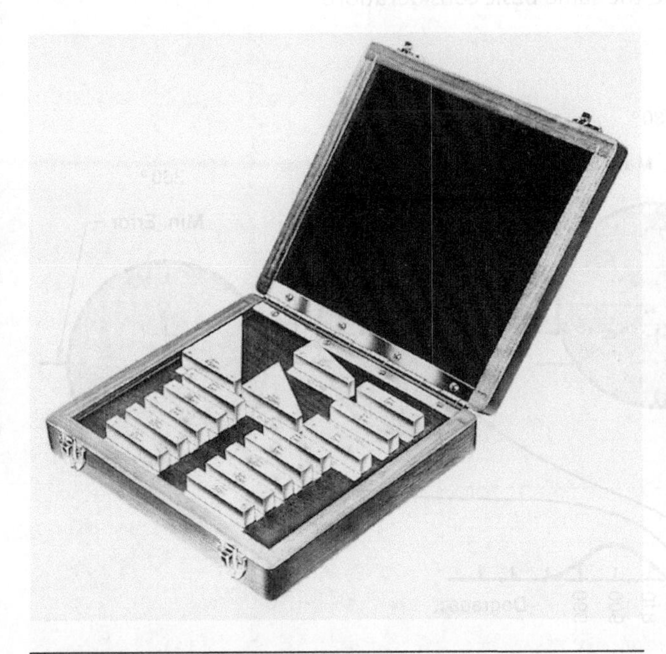

FIGURE 12–85 This set of 16 angle gage blocks forms all angles between 0° and 99° in one-second steps—a total of 356,400 combinations. These are laboratory master grade to an accuracy of 1/5,000,000th of a circle (¼ second). Also available are inspection grade to ½-second accuracy and toolroom grade to 1-second accuracy. *(Courtesy of the L.S. Starrett Company)*

12–9 ACCURACY AND PRECISION IN ANGLE MEASUREMENT

We have discussed the three general methods of angle measurement:

- Mechanical indexing with the dividing head, for example
- The divided scale with the vernier protractor
- Comparison with angle gage blocks measure

Measurement by any of these methods is based on our absolute standard for angle measurement—the circle.

Rotary tables are the most familiar way to set angles (see Figure 12–86); however, no matter how accurate their scales are, that accuracy may be degraded by the table's eccentricity. We get eccentricity when the center of revolution of the table is not exactly in the center of the scale (see Figure 12–87) and it doubles the error. For example, 2.54 μm (0.0001 in.) eccentricity in a 10.16 cm (4.0 in.) diameter circle results in an error greater than 20 seconds.

No matter how well constructed a rotary table is, there will always be some error. Fortunately, we can use the reversal process to cancel out the error (see Figure 12–88). By placing only two indices 180° apart, we can determine the readings and average them so that they cancel the error.

Instruments that establish measurement precision have the same sensitivity and discrimination; however, instruments that measure do not have the same sensitivity and discrimination. The circular division devices we have discussed establish angles,

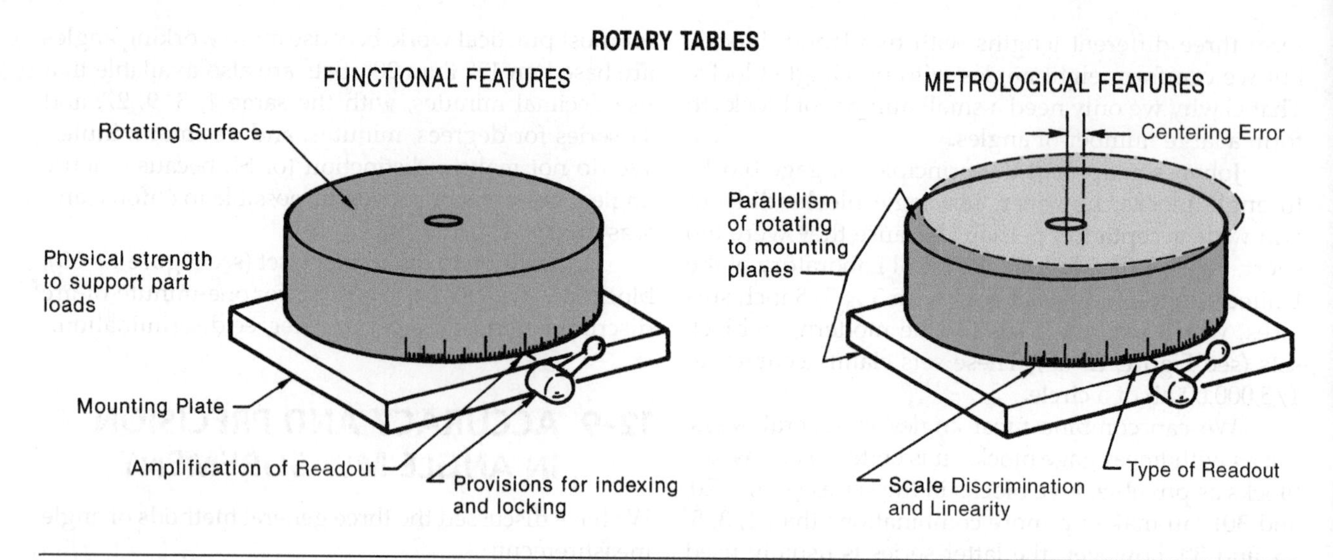

FIGURE 12–86 Simple or complex, all rotary tables involve the same basic considerations.

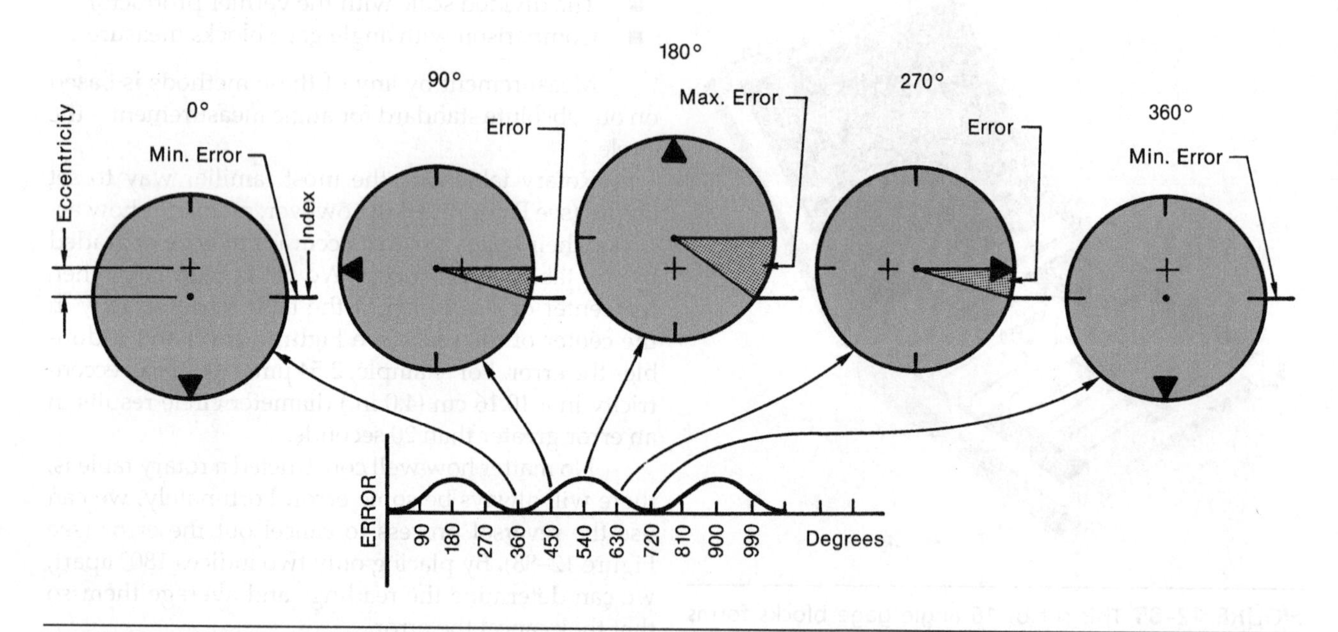

FIGURE 12–87 Eccentricity error rises and falls with rotation, repeating itself every 360°.

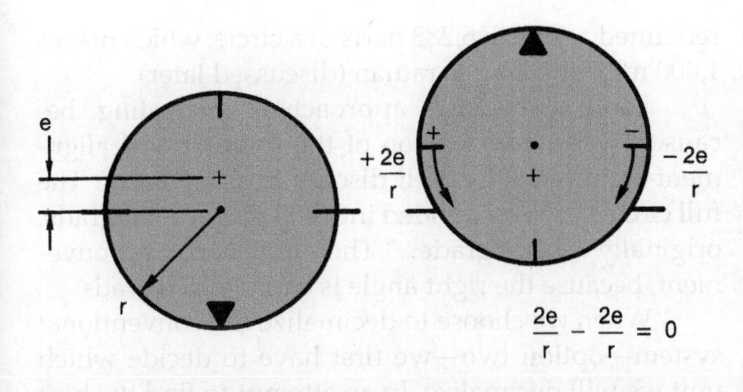

FIGURE 12–88 Eccentricity error may be cancelled out by the reversal process. Two observations 180° apart have equal error but opposing signs.

which is why we are only concerned with their accuracy and discrimination. When we add a readout system, however, the instrument measures; therefore, the addition of a readout system adds the characteristics of precision and sensitivity.

The simple direct reading rotary table measures only to degrees, but when we add a calibrated lead screw drive (see Figure 12–89), we can take readings to one second of arc with a claimed indexing accuracy of ±10 seconds. The table with the screw drive is more precise; however, the simpler table has fixed stops for each degree, so that we can locate an angle ±¼ second of arc (according to the manufacturer's claims). The low precision table, not the high precision one, is more accurate.

FIGURE 12–89 This rotary table is rotated by means of a worm drive.

The Caliper Principle

In Figure 12–87, we took two readings 180° apart and then eliminated the *eccentricity error* for a perfectly divided scale. Reversing this process, we can obtain a perfectly divided scale by correcting a series of readings because we start with the circle as our standard.

We make divided circles on machines called *dividing engines,* which have been developed to great accuracy and precision. You probably will not ever have to make a divided circle, but you do need to understand the principle behind the process, the caliper principle. With the caliper principle, we compare divisions with each other until we have established the number and uniformity we need.

For example, if we mount a plate to revolve around a precise center, we can scribe lines on it, lines 1 and 2, approximately 180° apart. We set two index positions, A and B, opposite these lines (see Figure 12–90 left), then rotate the disc so that line 1 coincides with index B (right). If we are lucky and lines 1 and 2 are exactly 180° apart, 2 and A will coincide also. The amount 2 and A differ equals twice the error from the true 180° position, and we can repeat this process to check our results.

Next, we move the index positions, A and B, approximately 90° apart (see Figure 12–91 left), and add a trial division line, 3. We bring line 1 into coincidence with B, and if line 3 is exactly 90° from 1 and 2, it will coincide with A. Again, the difference between 3 and 4 is twice the error (right), and the midpoint between

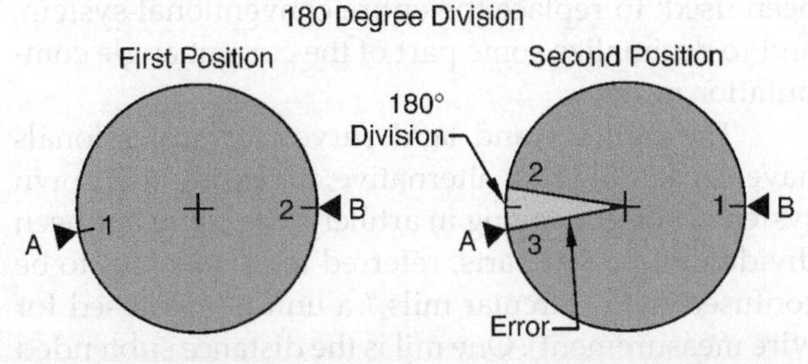

FIGURE 12–90 Uniform divisions are found by comparing divisions against each other.

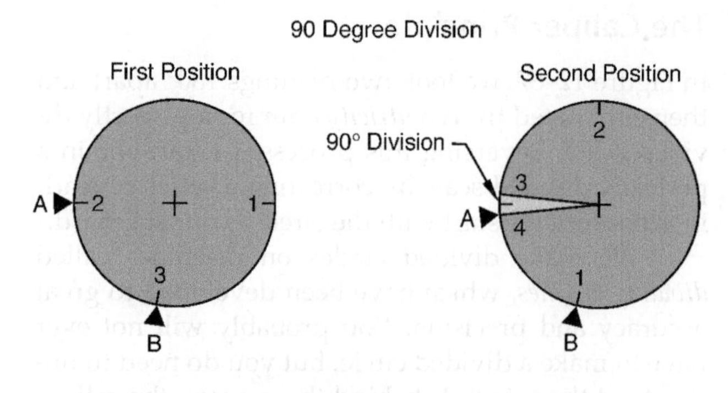

FIGURE 12–91 This process can be carried to any number of uniform divisions. Nonuniform divisions can be made up of combinations of uniform divisions.

the two lines is the 90° we want. You can follow these steps until you have the number of divisions you want, which also equals the degree of precision of the plate. If you add a microscope and machine-guided scriber, you can greatly increase the accuracy of your divisions.

Today, industry and science have a faster and even more accurate method to measure angles and to form them: the optical polygon with an autocollimator (see Chapter 16).

12–10 THE DEGREE, GRAD, AND GON

There have been many attempts to simplify computation with angles. Generally, two approaches have been used: to replace the entire conventional system, and to decimalize some part of the current angle computation system.

The military and land surveying professionals have chosen the first alternative, inventing their own systems. For gun laying in artillery, the circle has been divided into 6,400 parts, referred to as *mils* (not to be confused with "circular mils," a unit of area used for wire measurement). One mil is the distance subtended at a range of 1000 units—for example, 1 mil equals 1 yard displacement at a range of 1,000 yards. Through measurement compromised, the mil has been slightly redefined as one in 6,283 parts of a circle, which makes 1,000 mil equal to one radian (discussed later).

Land surveying's approach is interesting, because it uses an extension of the same optical alignment techniques we will discuss in Chapter 17. The full circle has been divided into 400 parts called grads, originally called "grades." These divisions are convenient, because the right angle is equal to 100 grads.

When we choose to decimalize the conventional system—option two—we first have to decide which unit we will decimalize. In an attempt to find the best choice, professionals have decimalized all the units available. The focus of decimalization has always been for ultraprecise measurement; therefore, the people making the calculations have concentrated on the second of arc, resulting in a system that divides the full circle into 1,296,000 seconds. On first hearing of this many divisions, most people believe that the numbers involved will be inconveniently large; however, when we take a closer look, the numbers are far more manageable than in the conventional system. For example, an angle that measures 163° 53′ 47″ has seven numerical digits plus three unit identifiers, but when we express it in seconds, the angle equals 590,027 seconds—six digits with no identifiers. If we need to measure to smaller divisions than the second, we use decimals of the second, which also reduces notational errors.

As SI's (metric system) creators addressed everything measured by man, they also addressed angle measurement, creating the common terms *radian, steradian,* and *grad.* All of the trigonometric functions are applicable in either the English (standard) or SI system of measurements, and the radian is most applicable to measurements dealing with rotation.

For angles, we distinguish between *plane angles* and *solid angles.* A plane angle lies in one plane and is defined as the length subtended along a circumference at a known distance from the vertex. A solid angle is three-dimensional; therefore, we define it as the area subtended on the surface of a sphere at a known distance from its vertex at the center of the sphere. In SI, the unit for the plane angle is the radian (rad), and the unit for the solid angle is the steradian (sr). You should not confuse the symbol *rad* with the symbol for a unit of radiation.

The radian is an angle whose arc length equals its radius, simply defined as 180° divided by pi, or, in grads, 200 divided by pi. Thus:

1 radian = 57.29578 degrees
= 3437.747 minutes
= 206,264.8 seconds
= 63.662 grads
1 degree = 0.017453 radians

Because the division of the circle into 400 grads is available as a system, we can decimalize it for very small angles. Some confusion may have resulted in Germany in the 1970s, because *Winkel-Grad* means *degree* in German. Grad was changed to *gon,* and gon is now used extensively in European trade literature and SI tables of equivalents, although there has never been an explanation of gon. The gon notation is:

1 gon = (pi divided by 200) radians
= 0.9 degrees
= 54.0 minutes
= 3240.0 seconds
1 mgon = 3.240 seconds

Although calculators have eliminated most of the need for calculations, tables for conversion of angles to linear measurement are still available and provide a reference for relative values (see Figure 12–92).

Angle to Linear Measurement

Angle	Per Inch	Per 10 Inches
1 sec.	4.85 μin.	48 μin.
1 min.	291 μin.	0.002 909 in.
1 deg.	0.017 46 in.	0.174 54 in.

Linear Measurement to Angle

Linear Measurement	Per Inch	Per 10 Inches
1 μin.	0.206 sec.	0.021 sec.
100 μin.	20.6 sec.	2.06 sec.
0.001 in.	206 sec.	20.6 sec
0.01 in.	2060 sec.	206 sec.

FIGURE 12–92 Converting Angle-Linear Measurement.

SUMMARY

- In this chapter, we discussed the unique status of the circle—a standard in nature. From division of the circle, all of our angular measurement is derived and provable. The ready convertibility back and forth from angular measurement and linear measurement permits us to measure angles linearly, and vice versa. To systematize this, there is geometry and trigonometry.
- Like linear measurement, the lowest amplification angular measurement is by a scaled instrument, the protractor. Following the same pattern, the vernier protractor adds amplification. A host of other devices extends the amplification up to and beyond those found in most linear-measurement instruments. Two special angles deserve attention. These are the right angle (90°) and the flat angle (180°). In measurement at least, squares (90° angle) are indispensable. They provide the axis for height measurement. They are self-checking. For most work, they can be visually compared to an unknown part feature. This can be done with great precision, less than 2.54 μm (one-tenth mil, 0.0001 in.). This means that they must be calibrated to even greater precision. Several methods can be devised that require no master. A faster method is comparison with a cylindrical square.
- Calibration without a master uses the reversal process, one of the most helpful measurement techniques. It is used to cancel out unwanted variables, and to use the geometry of the part as a reference in the measurement of that part.
- The horizontal (180° angle) is every bit as important as the vertical. The level is to it what the square is to the vertical. It is to angular measurement what the optical flat is to linear measurement. It provides high precision with almost unbelievable simplicity. Unfortunately, it is slow.
- The important lesson about levels is that they are not restricted to leveling. They can be comparators. Any two surfaces with the same bubble reading are parallel along that axis. Bubble readings are angular measurements, usually in seconds of arc. They are readily converted into

linear measurements. The shorter the measurement base, the higher the precision of these linear measurements. Like squares, precision levels can be calibrated without a standard. The standard is always available, being gravity; and every flat surface has one line across it, which is horizontal.

- Because of its versatility, many variations of levels are available. Square levels, for example, provide both horizontal and vertical references. In large-scale metrology, water levels are used to compare surfaces separated by large distances and with obstructions in between. The universal bevel protractor is to angle measurement what the combination square is to linear measurement. It is adaptable to many diverse measurements and has a discrimination, or least count, of 5 minutes. Although simple in construction and use, it requires careful attention to the complements and supplements of angles. The same considerations emphasized for reliable linear measurement apply to angular measurement: cleanliness, bias, temperature change, "cramping," and positional relationships.
- The next most frequently encountered angle measurement devices are the various sine instruments, of which the sine bar is the most common. These devices are based on a triangle of which the hypotenuse is constant. Changes in the side opposite the angle are converted into angles by means of tables. They can be used to set up an angle or to cancel out an angle. Thus, they only provide standards and depend on comparators for actual measurement. They are capable of discrimination in seconds but are affected by too many factors to be considered reliable finer than about 30 seconds. The huskier sine devices are known as sine plates and sine tables. Double sine plates with their hinges at right angles to each other form compound angles. The measure of a compound angle is its plane angle. All compound angles may be reduced to one or more of five types.
- An entirely different family of angle devices evolved from mechanically rotating spindles. One branch of this family achieves precision and accuracy by high-amplification readout devices, usually microscopes reading calibrated scales. The other branch of the family places little confidence in the ability of the observer but high reliance in the craftsmanship of the builder. It uses index plates with holes for positive stops with index pins. The lowest precision of these has no amplification: the index plate turning with the spindle. The higher precision ones have the index plate connected to the spindle through a worm gear so that many plate revolutions are equivalent to one spindle revolution.

END-OF-CHAPTER QUESTIONS

1. Which of the following best describes the term *angle?*
   a. Any intersecting lines or planes not at right angles
   b. The rate at which intersecting lines diverge
   c. The apex of a triangle
   d. The space between two intersecting lines
2. How is an angle defined?
   a. By the average of at least three protractor readings
   b. By its complement or supplement
   c. By the separation of its extreme ends
   d. By the portion of a circle it sweeps
3. In what field was angular measurement first developed?
   a. Navigation
   b. Railroad engineering
   c. Surveying
   d. Astronomy
4. What is the most important of the following precautions when making successive angle measurements?
   a. Avoidance of parallax error
   b. Selection of the correct vernier scale
   c. They must be made in the same direction.
   d. Subtract all readings over 180° from 180°.
5. Most angle measurement instruments have linear counterparts. Which of the following are valid comparisons?
   a. Simple protractor to vernier caliper
   b. Clinometer to micrometer
   c. Solid square to straightedge
   d. Steel rule to simple protractor

6. What do straightness, alignment, and flatness have in common with angle measurement?
   a. They are also precision measurements.
   b. They require angle instruments.
   c. They are not traceable to NIST like linear measurements.
   d. None is reliable if the standards are not accurate to ±0.001 inch.

7. The use of sine instruments to measure angles is best described as follows:
   a. Comparison measurement
   b. Angle gage block measurement
   c. Angle measurement of heavy workpieces
   d. Establishment of angles in contrast with measurement of angles

8. Which of the following is the most common method to check perpendicularity with a square?
   a. Examination for light gap
   b. Direct reading digital comparators
   c. Gage blocks with a sine bar or table
   d. Feel

9. What is the fastest method to calibrate a square?
   a. By electronic comparator
   b. NIST
   c. Against itself
   d. Against a master square

10. How does a direct reading cylindrical square differ from a standard cylindrical square?
    a. It has measurement readings on its surface.
    b. The digital readout is built in.
    c. The base is at a slight angle to its vertical axis.
    d. All readings are recorded in a log.

11. Which of the following methods allows errors to be compared with themselves?
    a. The reversal process
    b. Direct comparison with digital comparators
    c. The precision level
    d. Use of opposing sides of the universal square set

12. What is a level?
    a. An instrument that bridges areas on a surface plate
    b. An instrument that uses gravity to establish a reference plane
    c. An attachment for a universal square set
    d. A rotary table

13. In what units do levels discriminate?
    a. 0.0001 inch per inch of length
    b. 0.001 inch per unit
    c. Seconds
    d. Decimals of degrees

14. What is the greatest sensitivity usually found in precision work?
    a. 0.0001 inch per inch of length
    b. 10 seconds
    c. Double the flatness tolerance
    d. 1 second

15. Which end of a level is read?
    a. End away from observer
    b. Either end
    c. Average of both ends
    d. Lowest end

16. How does the 10-to-1 rule apply to levels?
    a. Not applicable
    b. Desirable, but less critical than for other instruments
    c. Essential
    d. Necessary, but must be converted to seconds

17. Which of the following features of flat surfaces relates to levels?
    a. Flatness tolerance is between two parallel planes.
    b. Every flat surface has a level line.
    c. If vertical instead of horizontal, a level cannot be used.
    d. Flatness has no theoretical limits for length and width.

18. Under what circumstance is a level on a flat surface rotated 360 degrees?
    a. To find the position with least parallax
    b. To find the low spot
    c. To ensure that there is good physical contact
    d. To locate the level line

19. What is the role of the micrometer water level?
    a. Cross between micrometer and level
    b. Permanent installation of levels

c. Measurement over extended distances
d. Determination of level of bodies of water

20. What phenomenon is utilized in reading the micrometer water level?
a. Refraction
b. Capillary action
c. Surface tension
d. Magnetic field

21. What is a universal bevel protractor?
a. Any instrument that measures all bevels
b. A protractor with a vernier scale and adjustable blades
c. A combined protractor and linear measurement instrument
d. An instrument that can use either gage blocks or a vernier scale to measure angles

22. What is the discrimination of a vernier protractor?
a. Depends on the size of the workpiece
b. Angles do not have discrimination
c. From 30 seconds to 10 minutes
d. 5 minutes

23. For a given setting of the vernier protractor, which of the following precautions is the most serious?
a. Determining which of the four readings should be used
b. Observing the back side of the instrument
c. Overheating from handling
d. Noting errors from overtightening the clamp

24. The vernier scale is always read in the same direction that the dial is read. Why?
a. To avoid parallax error
b. Because the vernier extends in both directions from zero
c. Numbers would be reversed if it were read in the other direction
d. To avoid the accumulation of error

25. At which of the following angles does the greatest problem in reading result?
a. 0° and 90°
b. Very small angles (under 10°)
c. 45°
d. Angles over 90°

26. Which of the following is the most important single consideration for angle measurement with the vernier protractor?
a. Absence of parallax error
b. Cleanliness of instrument
c. Adequately long scales
d. Intimate contact of the reference surfaces with part features

27. When placing a vernier protractor in contact with the workpiece, which of the following is of greatest concern?
a. Avoiding overconstraint and underconstraint
b. Having the instrument and workpiece at the same temperature
c. Using nondetergent lubricant
d. The adjustment of scale lengths

28. What must be done for measuring small angles?
a. Calculate from their complements.
b. Use shorter scales.
c. Use the acute angle attachment.
d. Prepare attribute gages.

29. How is the sine defined?
a. It is the function used for sine bars and sine plates.
b. It is the side opposite the angle divided by the hypotenuse of the triangle.
c. The square root of the sum of the squares of the two sides.
d. It is the first column in a table of trigonometric functions.

30. What instruments measure angles by means of their sines?
a. Universal vernier protractor
b. Spirit level
c. Acute angle attachments for the vernier protractor
d. Sine bars and sine plates

31. Which of the following is the best description of a sine bar?
a. It is a steel bar that holds cylinders at an accurate spacing.
b. It is a hypotenuse in physical form that can be set up with various sine lengths to form angles.
c. It is the combination of a sine instrument, gage blocks, and reference surface.
d. It is a bar that always holds cylinders at 5-inch centers.

32. The size of sine bars is stated in inches. What does this mean?
    a. It is the overall length ±¼ inch.
    b. It is the overall length of the bar.
    c. It is the overall length of the bar less 2 inches.
    d. It is the distance between the centers of the cylinders (rolls) of the sine bar.

33. The most popular size for sine bars is 5 inches. Why?
    a. Sine bars are not used for SI measurement.
    b. The ease of calculation from tables of trigonometric functions
    c. Sine bars tend to warp if over 5 inches.
    d. That is adequate for most workpieces.

34. What is the generally accepted reliable limit for sine instrument measurement?
    a. No limit
    b. 10 seconds
    c. Determined only by the gage blocks
    d. Usually 30 seconds

35. What is the basic difference between sine bars, sine plates, and sine tables?
    a. Widths
    b. Tolerance range
    c. Sine bars are for inspection, and the others are for machining
    d. Simple and compound angle applications

36. When should the setup be made for the complement of the angle?
    a. For all angles over 90°
    b. When a table of natural trigonometric functions is used
    c. When gage blocks of sufficient height are not available
    d. When the angle is greater than 60°

37. The sine bar is the hypotenuse of a triangle. The sine is the function of the angle formed between the base and the hypotenuse. How is it created physically?
    a. By the use of scaled rules with or without verniers
    b. By the use of screw adjustments
    c. By inserting the workpiece
    d. By inserting gage blocks to lift one end of the sine bar

38. How is the amplification achieved in mechanical indexing?
    a. Gearing
    b. Magnification
    c. Reversal technique
    d. Mechanical leverage

39. For measuring to seconds of arc, which of the following are commonly used?
    a. A universal vernier protractor with measuring glass
    b. Sine bars
    c. Sine plates
    d. Angle gage blocks

40. 4 degrees equals how many radians?
    a. 0.069812
    b. 0.69812
    c. 69
    d. 1

41. If the sine of Angle A = 0.35021, what is Angle A?
    a. 20° 29′
    b. 20° 30′
    c. 29° 20′
    d. 25° 30′

42. What combination of gage blocks is needed to set a 5″ sine plate at a 24° angle?
    a. 1.9771
    b. 2.0337
    c. 0.19771
    d. 1.5

43. What combination of gage blocks is needed to set a 5″ sine plate at a 20° 10′ angle?
    a. 1.9771
    b. 1.7237
    c. 2.0337
    d. 0.40674

13

CHAPTER THIRTEEN

SURFACE MEASUREMENT

LEARNING OBJECTIVES

- State the significance of surface metrology.
- Explain the essential differences between surface metrology and other forms of dimensional metrology.
- Describe the basis for the assessment of surface texture.
- Explain the role of standards for surface texture.

OVERVIEW

Surface metrology, which includes the commonly recognized elements finish and roundness, is an important consideration in manufacturing. As the speed of machines and an industry-wide need for better reliability and cost control increase, the need to control surface features increases logarithmically.

Surface metrology requires a new fundamental principle. The other fundamentals we have studied concern the relationship between a feature of a part or assembly and some other feature or the conceptual design. In contrast, surface metrology examines the deviation between one point (or points) on a surface and another point (other points) on the same surface.

Surface metrology is a concern of many branches of science and is widely involved in the world of commerce and manufacturing. Although many of the working applications of surface metrology are more for the look and feel of a product—and manufacturers set their own nondimensional, empirical standards—there are important standards set for items that contain parts that are subject to loads, that move in relation to one another, and that fit closely together, even if they do not move.

To meet performance requirements, the surfaces must be prepared for the job: for "rough" mechanisms with surfaces that have distinctive textures, the surfaces can be cast, forged, or rolled. For surfaces that contact other surfaces, they must be finished by machining or another process (see Figure 13–1). Machining, from rough planing and milling to fine diamond turning, creates textures that resemble one another—and look nothing at all like unmachined textures—but whose measurements vary greatly. We create the finest machined surfaces with abrading

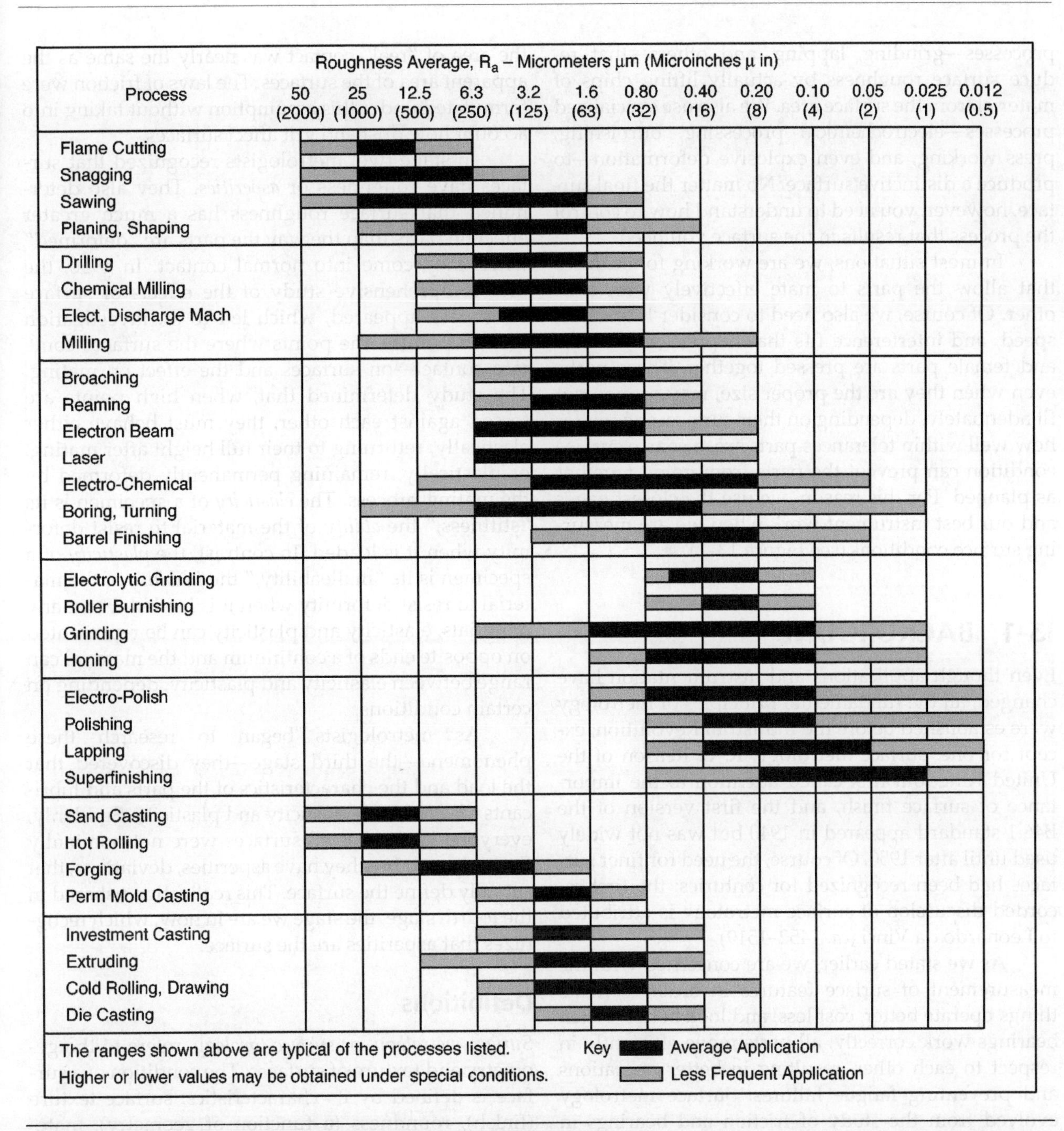

FIGURE 13–1 This chart is from ANSI B46.1-2002, Surface Texture. This is an official standard of the American National Standards Institute and is the basis for much information in this chapter. Note that the chart is in both microinches and micrometers. The chart shows that surface metrology is concerned with measurement that ranges from relatively large (0.002 in) to extremely minute (0.5 μin.)

processes—grinding, lapping, and others—that reduce surface roughness by actually lifting chips of material from the surface area. We also use specialized processes—electrochemical processing, burnishing, press working, and even explosive deformation—to produce a distinctive surface. No matter the final surface, however, you need to understand how to control the process that results in the surface you need.

In most situations, we are working for surfaces that allow the parts to mate effectively with each other. Of course, we also need to consider lubricants, speed, and interference fits that happen when male and female parts are pressed together. These parts, even when they are the proper size, may or may not fit adequately, depending on their surfaces: no matter how well within tolerances parts are, a coarse surface condition can prevent the parts from going together as planned. For this reason, we use the closest gages and our best instrument work when we are measuring surface conditions (see Figure 13–1).

13–1 BACKGROUND

Even though applications and instrumentation have changed, all the fundamental principles of metrology were established before the Industrial Revolution, except for one: surface metrology. R. E. Reason of the United Kingdom first called attention to the importance of surface finish, and the first version of the B46.1 standard appeared in 1940 but was not widely used until after 1950. Of course, the need for finer surfaces had been recognized for centuries: the first recorded discussion of surface metrology is attributed to Leonardo da Vinci (*ca.* 1452–1519).

As we stated earlier, we are concerned with the measurement of surface features in order to make things operate better, cost less, and look better. When bearings work correctly, all parts move efficiently in respect to each other, resulting in better operations and preventing fatigue failures. Surface metrology evolved from the study of friction and bearings in four distinct stages.

Stage one reaches back to Leonardo da Vinci, with renewed emphasis in the eighteenth century. Everyone assumed that contacting surfaces were flat; therefore, the area of "real" contact was nearly the same as the apparent area of the surfaces. The laws of friction were formulated under this assumption without taking into account how dust and grit affect surfaces.

In stage two, metrologists recognized that surfaces have roughness or *asperities*. They also determined that surface roughness has a much greater effect on parts than the way the parts are "deformed" when they come into normal contact. In 1950, the first comprehensive study of the effects of surface roughness appeared, which led to an investigation of high points—the points where the surfaces come into contact—on surfaces and the effect on mating. The study determined that, when high points are forced against each other, they must behave either elastically, returning to their full height after mating, or plastically, remaining permanently deformed by the mating process. The *elasticity* of a specimen is its "stiffness," the *ability* of the material to resist deformity when it is loaded. In contrast, the *plasticity* of a specimen is its "malleability," the *inability* of the material to resist deformity when it is loaded. For many materials, elasticity and plasticity can be represented on opposite ends of a continuum and the material can range between elasticity and plasticity, depending on certain conditions.

As metrologists began to research these phenomena—the third stage—they discovered that the load and the characteristics of the parts and lubricants affected both elasticity and plasticity. Suddenly, everyone knew that all surfaces were not nominally flat and smooth—they have asperities, deviations, that actually define the surface. This realization ushered in the fourth stage, the stage we are in now, which recognizes that asperities are the surface.

Definitions

Surface metrology, or *surface topology*, refers to the geometry and texture of surfaces. The condition of a surface is defined by its characteristics: surface texture (finish), roundness (a function of geometry), material, hardness, and surface metallurgy. In the United States, ANSI B46.1 (discussed later) creates a practical standard for the geometry and texture only. It does not apply to other surface characteristics.

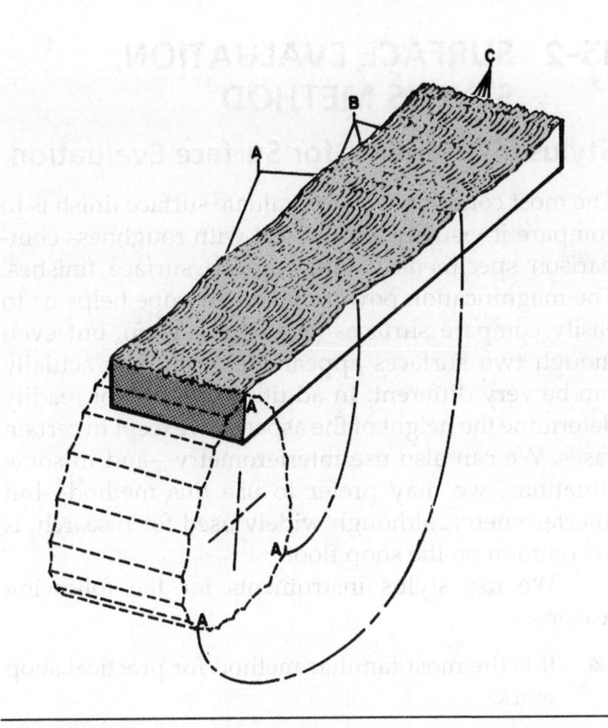

FIGURE 13–2 If the asperities in a machined surface were greatly magnified, they might appear like this. The chip lifting action causes the smallest ones as in C. Tool chatter marks will be perpendicular to those, as at B. Flexure of the machine, irregularities in the ways or other imperfections in the process appear as much more widely spaced asperities, A. If the part were cylindrical rather than flat, the same surface would be found wrapped around the center of rotation.

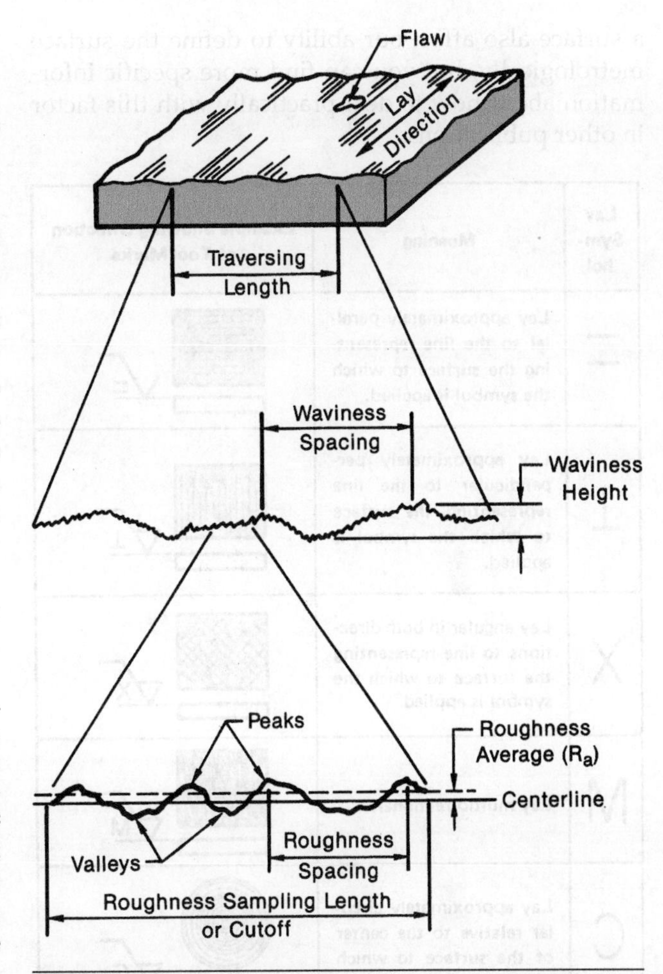

FIGURE 13–3 Certain terms for surface assessment are generally accepted. Roughness is the finest of the asperities, whereas waviness concerns the more widely spaced ones. Both of these are repetitive for most machine surfaces, but flaws occur randomly.

In Figure 13–2, we have shown three forms of asperity from a theoretically normal surface: *roughness, waviness,* and *error of form,* which all vary according to the length of spacing, or *wavelength,* as we refer to it in surface metrology. The fourth asperity is not distinguished by wavelength. It is a *flaw* (not included in measurement based on ANSI B46.1) in the surface (see Figure 13–3). Do not be misled by this illustration; the flaw is not the only problem with the surface. There is also an error of form, and roughness and waviness do not always have the same lay. Lay is the direction of the asperities (see Figure 13–4), which, in most cases, means that roughness and waviness are perpendicular to each other (see Figure 13–1).

You should remember that there are no necessarily good and bad surfaces—there is only the surface that is appropriate for the application. For example, a somewhat rougher surface may retain a lubricant better, and if we need better reflective qualities, we would probably be less concerned with the overall geometry than we would if we were producing a precision bearing. Resistance to corrosion and stress are other factors that determine the appropriate finish. Unfortunately, the metallurgical properties of a surface also affect our ability to define the surface metrologically, but you can find more specific information about how to deal practically with this factor in other publications.

Lay Symbol	Meaning	Example Showing Direction of Tool Marks
=	Lay approximately parallel to the line representing the surface to which the symbol is applied.	=
⊥	Lay approximately perpendicular to the line representing the surface to which the symbol is applied.	⊥
X	Lay angular in both directions to line representing the surface to which the symbol is applied.	X
M	Lay multidirectional.	M
C	Lay approximately circular relative to the center of the surface to which the symbol is applied.	C
R	Lay approximately radial relative to the center of the surface to which the symbol is applied.	R
P[3]	Lay particulate, non-directional, or protuberant.	P

FIGURE 13–4 Lay is the direction of the asperities. Most asperities recur regularly. This chart from ANSI Y14.36-1996 (Surface Texture Symbols) categorizes the various lay configurations and shows the standardized symbols used on drawings. ANSY Y12.36-1996 is bound with ANSI B46.1-2002.

13–2 SURFACE EVALUATION, STYLUS METHOD

Stylus Instruments for Surface Evaluation

The most common way to evaluate surface finish is to compare it visually and by feel with roughness comparison specimens having various surface finishes. The magnification power of a microscope helps us to easily compare surfaces (see Figure 13–5), but even though two surfaces appear identical, they actually can be very different. In addition, we cannot readily determine the height of the asperities, except in certain cases. We can also use interferometry—and in some situations, we may prefer to use this method—but interferometry, although widely used for research, is uncommon on the shop floor.

We use stylus instruments for the following reasons:

- It is the most familiar method for practical shop work.

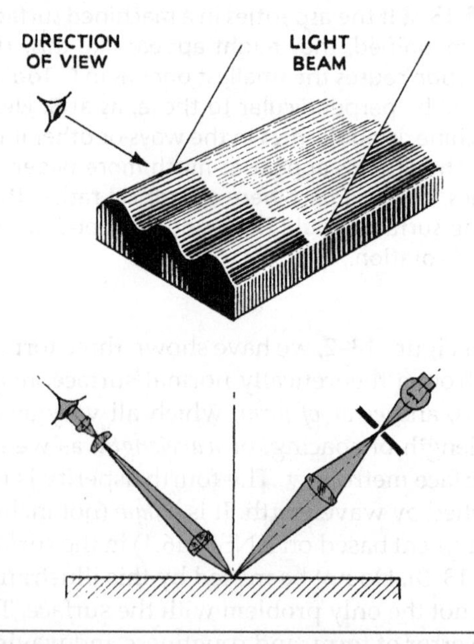

FIGURE 13–5 Examination of surfaces by microscope can be informative, but it does not usually allow the heights of the asperities to be determined without destroying the test part by cutting a taper through the surface. The method shown uses a line of light instead of a cut into the material. This is known as the Schmaltz Profile Microscope.

- It best demonstrates the fundamental principles of surface metrology.
- Even though B46.1 discusses several other methods, the standard is based on the *stylus method.*

Stylus instruments operate like a phonograph pickup: the stylus is drawn across the surface and generates electrical signals that are proportional to the changes in the surface (see Figure 13–6). With this method, we can achieve great increases in amplification (refer to Chapter 14) without the need to compensate for increases in pressure of the stylus against the workpiece because the power is supplied externally.

You can read the changes in height directly with a meter or on a printed chart, but you probably will not get much useful information. If the instrument can provide a graph of the stylus path along the surface, you gain more information, but you cannot consider this graph a complete assessment of the surface. In addition, any graph created by a stylus instrument does not represent the real surface. A graph will greatly exaggerate heights of the asperities in relation to the spacing, which is known as the *distortion ratio.* When you compare graphs (see Figure 13–7), you must always remember the distortion ratio: two graphs of the same surface made with instruments with different distortion ratios do not look the same at all.

FIGURE 13–6 Surface roughness measurement instrument. *(Courtesy of AMETEK Taylor Hobson)*

In all other measurement, a 0.0254 μm (millionth of an inch) is a small amount, but it is usually suspect, even for gage blocks. If we use a scale with 0.1524 μm to 25.4 mm (6 microinches to the inch), 25.4 mm (1 in.) on the scale corresponds to more than 4.0234 km (2½ miles).

During the rest of our discussion of surface metrology, we will provide SI and English values for all measurements except for asperity values. Ultimately, we use surface metrology to obtain surfaces that function properly for the task, which, in many ways, is not related to dimensions. We use the surface values simply for comparison; therefore, the principles are the same, regardless of the units we choose.

In our discussion, we will use the SI unit micrometer, abbreviated μ (lowercase mu): one micrometer (1 μm) equals one thousandth of a millimeter, which equals 39.37 microinches. One microinch equals 254 angstroms, and one angstrom (Å) is 1×10^{-10} m.

As you study the relationship between the units, you will see that we use very small units to measure asperity heights; therefore, we need to create a graph that enlarges these measurements so we can see them. Asperities are usually widely spaced; however, the flanks of asperities rarely exceed 5°. If we use the same height and length magnification, we would create a graph that extended for many feet in order to see the asperity heights. In fact, the graph would be so long that it would be hard for you to see all of it at once.

Clearly, the solution is to use much greater magnification for the height than for the length. The electronic stylus makes creating such distinctions easy—an instrument's height magnification is shown in relationship to its length magnification. For example, for a 3000/150 instrument, the height is multiplied 3000 times, and the length is multiplied only 150 times.

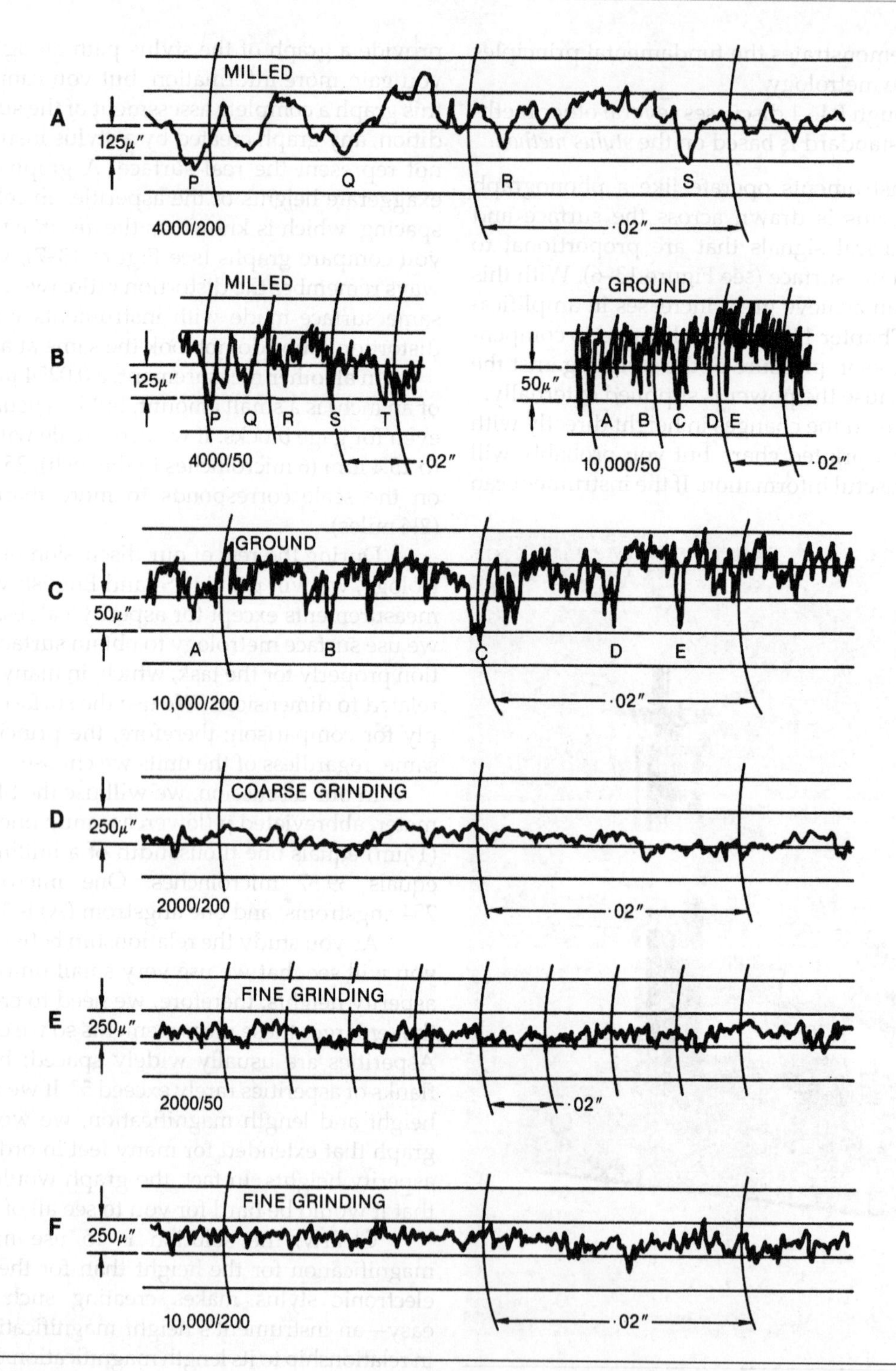

FIGURE 13–7 Magnification ratios can cause deceiving appearances. The milled surface in A appears quite different from the ground surface in C. But when the distortion ratio of the milled surface is reduced as in B, it is almost indistinguishable from the ground surface in C.

The Stylus and the Datum

Because the asperities we measure are so small, does not the instrument we use need to be even smaller so it can accurately measure the heights? For the primary texture or roughness, the tip radii are usually between 1.27 μm and 12.3 μm (50 and 500 μin.). With such a fine point, you might expect scratches in the surface you are measuring. Only very soft surfaces, like aluminum, are visibly affected by the stylus point, and even then, the surface just looks marred. The scratches are actually so small that we can only measure them in a laboratory. In most cases, for a stylus with a 0.254 μm (10 μin.) tip radius, you can use working pressures up to 120 milligrams scratch free. The standard stylus radius in ANSI B46.1 is 400 micro-inch (0.0004 in. or 10.16 μm).

We can see with an electron microscope that no surface is perfectly smooth. Even the asperities of a surface will have asperities, which will have asperities, and so on, and so on. Therefore, we need to use the appropriate instrument to measure surface asperities only as precisely as the final application requires. Electronics makes this selectivity possible, which we will discuss later.

There are two types of stylus instruments: *true-datum,* also called *skidless* instruments, and *surface-datum* or *skid-type* instruments. With a true-datum instrument, we draw across the surface in a very precise, mechanically controlled movement, and we make an assessment of the surface based on the path of the datum forms (see Figure 13–8) (early model used to demonstrate this principle). The advantage of this measurement method is that the resulting graph is nearly a true representation of the surface along that one line, showing roughness, waviness, errors of form, and flaws. The disadvantage is that it is very difficult to set the instrument up; you must precisely align the surface being assessed with the path (datum) of the instrument.

In contrast, you can easily set up surface-datum instruments because they use the surface being assessed as the datum. A supporting slide—either a skid, a rounded member fixed to the head, or a shoe, a flat pad in a swivel mount in the head (see Figure 13–9) rests on the surface and slides the stylus pickup along. Skids may be located in front of, behind, or on the opposite side of the stylus, and typical

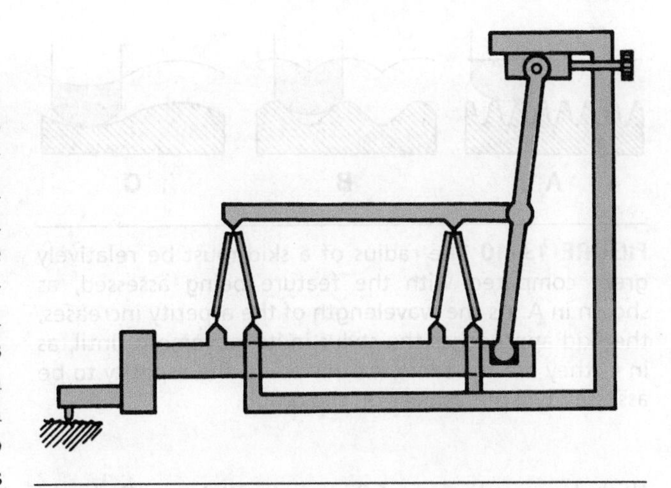

FIGURE 13–8 True-datum stylus instruments move the stylus across the part along a reference datum established by the instrument. An early true-datum instrument is shown in this schematic.

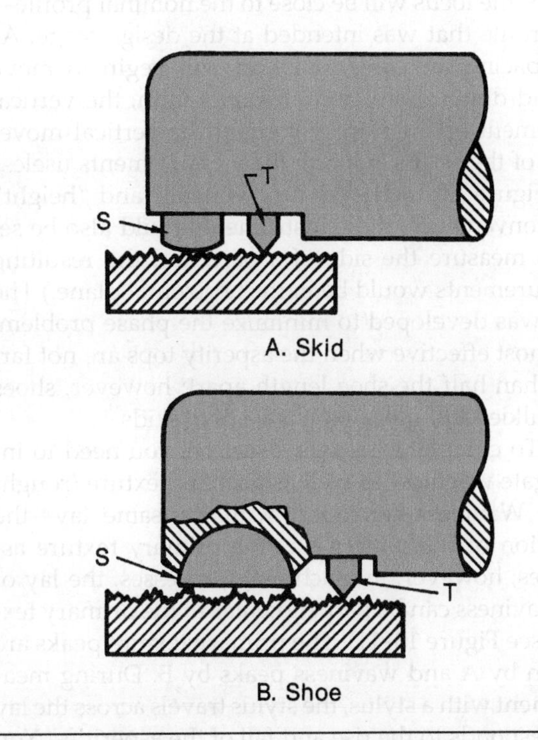

FIGURE 13–9 Surface-datum instruments create their own reference datum. This is done by supporting the stylus, T, by a member S, that is sufficiently wide to slide along the surface. Two types are in use, skids as in A and shoes as in B.

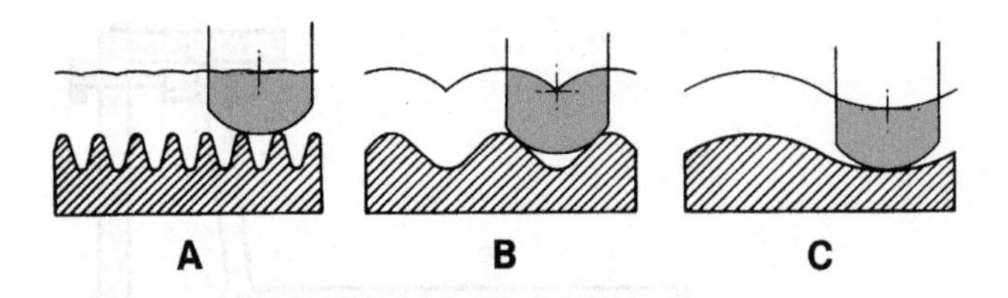

FIGURE 13–10 The radius of a skid must be relatively great compared with the feature being assessed, as shown in A. As the wavelength of the asperity increases, the skid approaches the stylus in its movement until, as in C, they are the same. At this point, the asperity to be assessed will not appear on the graph.

skids have radii from 6.35 mm to 50.8 mm (¼ in. to 2 in.) (see Figure 13–10).

As a skid slides along the surface, it creates datum, which represent the locus of the skid's center of curvature. When the tops of the asperities are close together, the locus will be close to the nominal profile—the profile that was intended at the design stage. As the spacing increases, the skid will begin to move up and down more, and at some width, the vertical movement of the skid will equal the vertical movement of the stylus, making the measurements useless (see Figure 13–10C). (We use "vertical" and "height" as a convenience; these instruments could also be set up to measure the side of a part, and the resulting measurements would be in the horizontal plane.) The shoe was developed to minimize the phase problem. It is most effective when the asperity tops are not farther than half the shoe length apart; however, shoes are bulkier and more expensive than skids.

To completely assess a surface, you need to investigate waviness as well as primary texture (roughness). Waviness can occur with the same lay—the direction of the pattern—as the primary texture asperities; however, in machining processes, the lay of the waviness can be perpendicular to the primary texture (see Figure 13–11). The primary texture peaks are shown by A and waviness peaks by B. During measurement with a stylus, the stylus travels across the lay and responds to the rise and fall of the asperities. You need a pointed stylus for the primary texture, which travels in direction a; and to measure the waviness

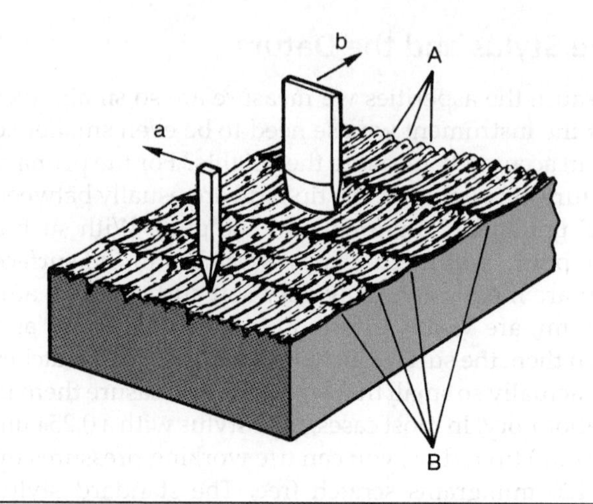

FIGURE 13–11 If it is desired to assess primary texture (roughness), shown as A, a pointed stylus is needed. It must move in direction a. If A was not a concern, but the chatter marks, B, were to be investigated, a wide stylus traveling in direction b would not reveal the roughness, but would produce a graph of the chatter marks.

while moving in direction b, you need a blunt stylus to separate the two surface characteristics.

You must be careful when assessing waviness that you choose both the best stylus form and the proper direction of travel. Even on very finely machined surfaces, you can easily recognize the primary surface lay, and chatter marks (see Figure 13–11) are also easy to recognize, even when they are very slight. Most waviness, however, is difficult to see. There is one test you can make for the direction of waviness: rub the surface lightly with a flat oilstone, and it will usually reveal the crests of waviness.

Wavelength, Frequency, and Cutoff

Skids simplify surface assessment with stylus instruments; however, they also create distortion through the phase relationship between the stylus and the skid. The readings of the instrument and the resulting graphs show the change in height of the stylus with respect to the skid—not in relationship to the nominal profile. In addition, the skid, like the stylus, rises and falls according to the surface asperities; therefore, the skid distorts the stylus height measurements.

When the stylus is connected to an electronic sensor in the head of the instrument, we are using the same principle we use for electronic comparators. Therefore, when the head of the stylus instrument crosses the surface at a uniform speed, it generates an electrical frequency, which can be described by:

$$f = \frac{S}{\lambda}$$

where f is the frequency in hertz, S is the speed of traverse, and λ (lowercase lambda) is the wavelength.

The amplitudes of the waves correspond to the heights of the asperities; but for each surface and instrument head, the waves are distorted from the actual asperity heights by the phase relationship between the stylus and the skid. In Figure 13–12, we have depicted a surface with both roughness and waviness. In Figure 13–12A, the stylus and skid are in phase; therefore, there is a minimum distortion of the heights of the roughness asperities. In Figure 13–12B, the stylus and skid are extremely out of phase, resulting in maximum distortion of the roughness asperity heights. Most often, the distortion will fall between the extremes, as in Figure 13–12C, but a very small change in wavelength can go from an in-phase condition to an out-of-phase condition. As we said before, you need more than a graph of roughness from a surface-datum instrument to make a practical assessment of the surface (see Figure 13–13).

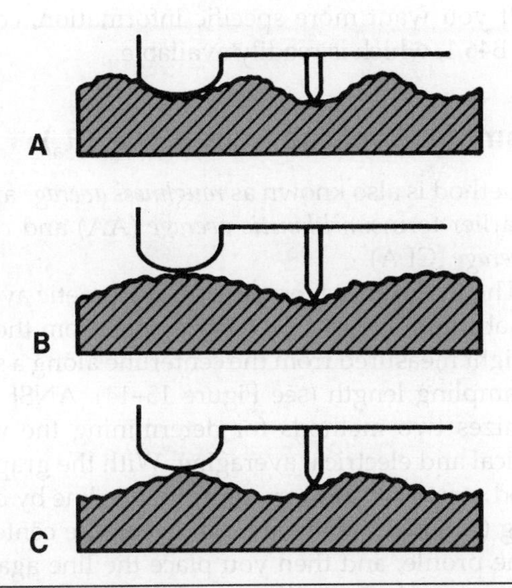

FIGURE 13–12 In A, the stylus and skid are in phase. Therefore, the primary texture (roughness) will be relatively undistorted. In B, they are out of phase. In this situation, the waviness appears in the roughness readings and is very misleading. Conditions such as C are more common and difficult to detect.

Measuring surface conditions by changing the datum into electrical frequency is a very popular method of surface assessment, especially because it allows us to select the surface characteristics we will assess. Distinguishing among the characteristics—roughness, waviness, and errors of form—can be difficult. For example, asperities of roughness may also carry their own surface asperities, and if we use enough magnification, we will be able to see both areas of asperities. Similarly, it may be difficult to distinguish very long wavelengths of waviness from errors of form.

If we are assessing roughness only, we look at the frequencies generated by the wavelengths of the roughness asperities—we would not want to consider the asperities of the asperities. Even if we use a stylus that is small enough to reveal the asperities of asperities, they would be easy to identify; they would have a higher frequency because their λ (in $f = S/\lambda$) would be smaller than that of the asperities you want

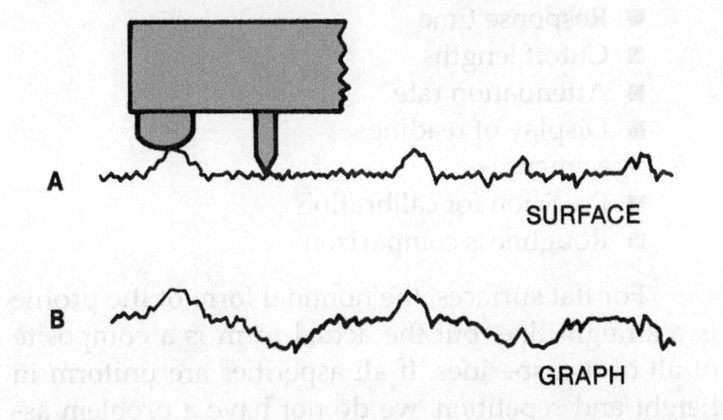

FIGURE 13–13 The effect of the out-of-phase condition on the graph is shown in B. If the in-phase condition had occurred, the graph would appear the same as the surface in A.

to measure. We can easily eliminate these higher frequencies electronically, in much the same way you adjust the tone control of an audio system. For electronic surface metrology instruments, this adjustment is called the *cutoff*.

13-3 NUMERICAL VALUES FOR ASSESSMENT

A surface texture standard, like all standards, ensures that various observers, using various instruments under a wide range of conditions all produce nearly the same values. They can never be exactly the same, but they must be close enough to result in a device that can be affordably manufactured and maintained. As always, the "close" required compromise, because a part that works fine for consumer electronics may be unsuitable for a missile. A standard attempts to identify all the variables and give them limits in relationship to the uses they will have. These variables, included in ANSI B46.1 include:

1. Identification of surface characteristics
2. Establishment of nonambiguous terminology
3. Tracer head characteristics
   - Stylus form, force, support
   - Skid form, support
4. Traversing length
5. Instrument characteristics
   - Response time
   - Cutoff lengths
   - Attenuation rate
   - Display of readings
6. Specimens
   - Precision for calibration
   - Roughness comparison

For flat surfaces, the nominal form of the profile is a straight line, but the actual form is a composite of all of its asperities. If all asperities are uniform in height and repetition, we do not have a problem assigning a datum for the assessment of surfaces; however, one exceptionally high crest or low valley in the sampled length of surface could distort the entire assessment. In order to minimize the problem, a number of surface assessment methods have been investigated, and 10 have been standardized by the major industrial countries.

Only one, however, has been standardized for the United States by the American National Standards Institute (ANSI) and the American Society of Mechanical Engineers (ASME) in ANSI B46.1. It is known as Roughness Average and is updated as necessary (the current version is ANSI B46.1-2002, although we always assume that we are using the latest version of the standard and do not list the year).

ANSI B46.1 reviews the other nine assessment methods standardized for other countries at last publication. In addition, because some surfaces cannot be adequately expressed by any recognized standard, ANSI B46.1 reviews other methods that manufacturers regularly use. We will discuss these methods in the same order as B46.1. We will introduce each method by its most familiar name, which is used in ANSI B46.1, but most also have other names. The symbols used to identify the methods are nearly uniform, so they are a convenient way to refer to the methods.

If you want more specific information, consult ANSI B46.1, which is readily available.

Arithmetic Average Roughness (R_a)

This method is also known as *roughness average* and by two earlier terms: *arithmetic average* (AA) and *centerline average* (CLA).

The roughness average is the arithmetic average of the absolute values of the deviations from the profile height measured from the centerline along a specified sampling length (see Figure 13–14). ANSI B46.1 recognizes two methods for determining the value: graphical and electrical averaging. With the graphical method, you establish a graphical centerline by determining (solving for) the areas between the centerline and the profile, and then you place the line again so that the areas above the line equal the areas below the line. As you can imagine, making these calculations is tedious even with a computer, so we generally use the electrical averaging method. With this method, you use instruments that automatically average the readings. These two methods generate different values,

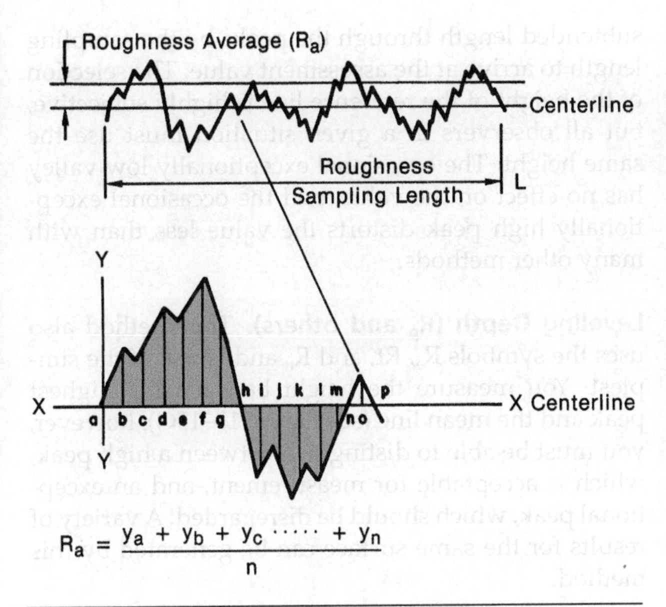

FIGURE 13–14 The top profile is a typical roughness sampling length. The lower profile is an enlargement of part of that profile and has been divided into equally spaced segments. The y distances from the centerline are averaged without regard to sign. This is an approximation of the R_a. The more closely spaced the segments, the closer the approximation will approach the true value.

because the graphical centerline is a straight line and the electrical averaging centerline is wavy. This error is known as methods divergence, and although it can be as much as 10%, it is typically within 1% or 2% of the R_a value.

The calculations required for the graphical method may be tedious, but it eliminates cutoff requirements in the electrical averaging method, which, in turn, reduces the sampling length and traversing length. In order to meet the standard for the graphical method, you can analyze as few as five consecutive roughness sampling lengths, which reduces calculation time.

For the electrical averaging method, you must select a cutoff that contains the assessment information you want—typically roughness or waviness—which determines the *sampling length.* Atypical portions of the profile can cause bias in the resulting R_a value, so you need to lengthen the *traversing length* so you have enough information to eliminate bias. Traversing lengths range from 20 to over 60 times the cutoff: five cutoff wavelengths are standardized. Manufacturers prefer 0.8 mm (0.0314 in.) for most work, and metrologists assume you are using that cutoff unless another is specified. You must also be concerned with the response of the system and the attenuation rates for both short and long wavelengths.

You will find the derivation of these values and tables for their application in ANSI B46.1, along with a discussion of calibration and specimens. Only four industrial countries have only one surface texture standard: Canada, Netherlands, Switzerland, and the United States, which all use the R_a assessment method. All other industrial countries have adopted R_a but also use others—for example, France has seven additional standards.

The R_a method for surface texture assessment is applied easily and, for the most part, is easily understood. One exception is that we sometimes fail to recognize that the R_a value is an index for surface comparison, not a dimension. The R_a value is always much less than the peak-to-valley height, and in most cases, the value is approximately 25% of the extreme differences.

Other Standardized Assessment Methods

Root-Mean-Square Roughness (R_a or RMS). The RMS assessment is closely related to the roughness average (R_a). Instead of averaging the distances from the centerline to the profile, you square the distances, average them, and determine the square root of the result. The resulting value is the index for surface texture comparison and is usually 11 percent higher than the R_a value.

Because determining this result requires extra efforts and provides no additional value, RMS was dropped in the United States in 1955, but it is still in use in a few industrial countries.

Maximum Peak-Valley Roughness (R_{max} or R_t). The R_{max} assessment determines the distance between the lines that contact the extreme outer and inner points on the profile. Unless you eliminate any exceptional peaks and valleys, they can distort the result (see Figure 13–15A); therefore, the judgment of the observers results in different values for the same surface. This method is the second most popular in industrial use.

Ten-Point Height (R_z). The R_z assessment averages the distance between the five highest peaks and five deepest valleys within the sampling length (see Figure 13–15B). In contrast to the R_{max} method, it reduces the effect of exceptional peaks and valleys.

Average Peak-to-Valley Roughness (R or H or H_pl). With the R assessment method, you average the individual peak-to-valley heights (see Figure 13–15C), which sounds similar to the R_a method. However, you use the heights between adjacent peaks and valleys, not measured from a centerline to peaks and valleys.

Average Spacing of Roughness Peaks (A_r or A_R). The A_r assessment method, also known as the average wavelength of roughness peaks, averages the distances between the peaks without regard to their heights (see Figure 13–15D). This average depends on what we determine a peak is; therefore, if it is to be useful, all the observers in a particular manufacturing situation must agree on the value of the peak.

Swedish Height of Irregularities (R or H). Also known as the *Profiljup Method* (see Figure 13–15E), this method is only a standard in Sweden (H) and Denmark (R), even though the method appears to be particularly useful for wearing surfaces such as bearings. It assumes that, in a wear situation, the peaks are affected by wear, but the valleys are not. You determine the distance between two parallel lines that cut the profile: the upper line through the peaks so that 5% of the material is above the line; and the lower line through the valleys so that 90% of the material is above the line.

Bearing Length Ratio (T_p and others). Based on the same assumption as the previous method, you create a reference line through some of the peaks. This line is at a predetermined height from the mean line (see Figure 13–15F), and you then divide the subtended length through the peaks by the sampling length to arrive at the assessment value. The selection of the height of the reference line is highly subjective, but all observers in a given situation must use the same height. The occasional exceptionally low valley has no effect on the value, and the occasional exceptionally high peak distorts the value less than with many other methods.

Leveling Depth (R_p and others). This method also uses the symbols R_e, Rt, and R_u and is one of the simplest. You measure the height between the highest peak and the mean line (see Figure 13–15G); however, you must be able to distinguish between a high peak, which is acceptable for measurement, and an exceptional peak, which should be disregarded. A variety of results for the same surface can be generated by this method.

Waviness Height (W). You use this method to assess waviness without regard to roughness by determining the peak-to-valley distance of the total profile within the sampling length (see Figure 13–15H). It is less likely that you will find an exceptional peak or valley that could distort the waviness value than you would when assessing roughness.

Other Assessment Methods

Particular industries have developed a number of methods to deal with their unique situations, and although not recognized as a national standard, they embody some useful principles and have resulted in recognized methods for surface texture assessment. We will introduce you to them briefly in this text to give you an idea of where developments in surface metrology may occur in the future.

Amplitude Density Function (ADF). This assessment method summarizes the amplitude properties of a profile—amplitude probability and amplitude density. The numerical value is the probability that, at a given ordinate value (height), the profile amplitude will be within a slice height Δy (delta y) of the sampling length (L) (see Figure 13–16A). You can calculate this value for any height by moving the Δy slice. The significance of this method, shown by the graph to the

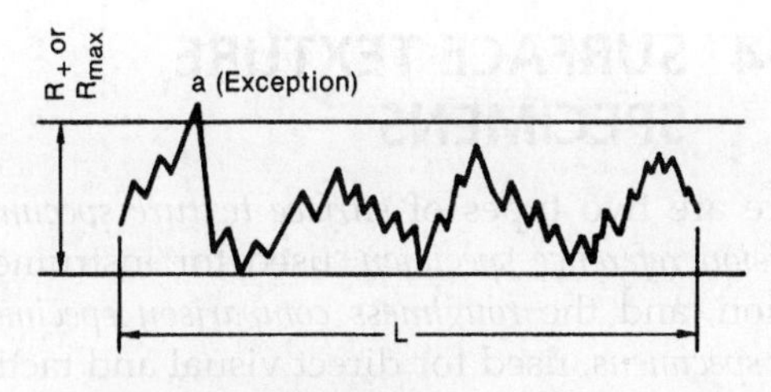

A. MAXIMUM PEAK-TO-VALLEY ROUGHNESS HEIGHT (R_t OR R_{max})

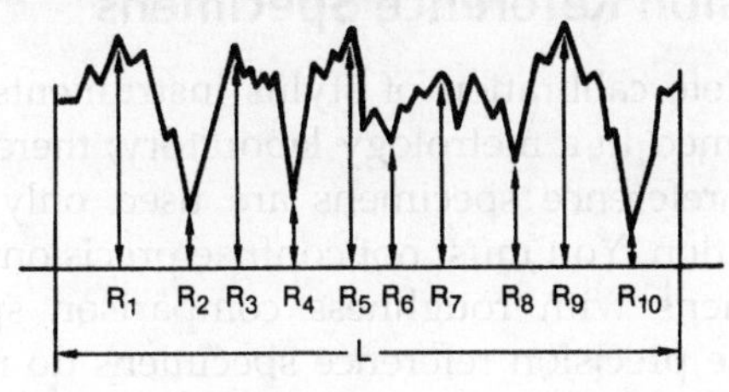

B. TEN-POINT HEIGHT (R_Z)

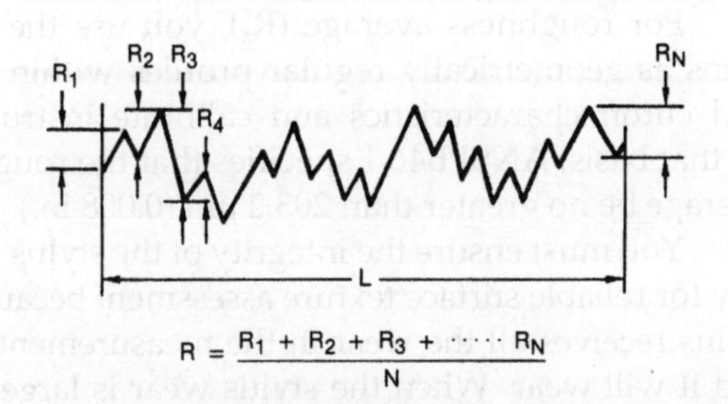

C. AVERAGE PEAK-TO-VALLEY ROUGHNESS (R AND OTHER SYMBOLS)

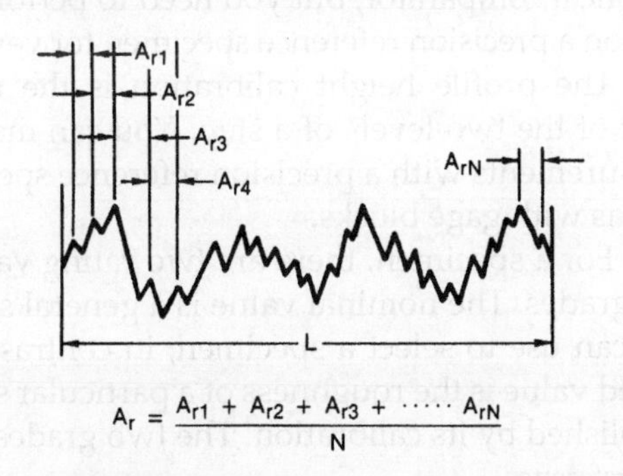

D. AVERAGE SPACING OF ROUGHNESS PEAKS (A_r OR A_R)

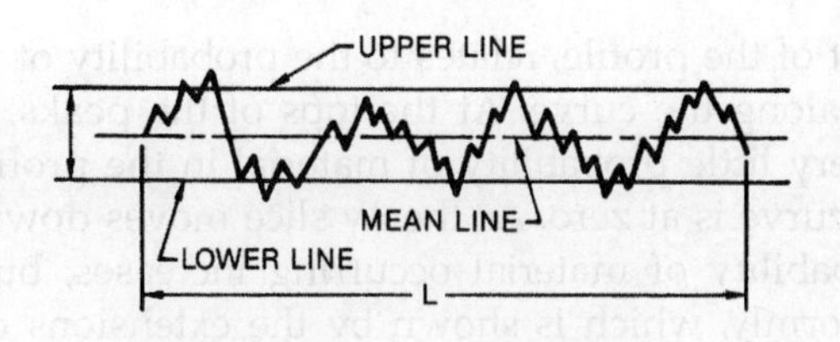

E. SWEDISH HEIGHT OF IRREGULARITIES, PROFILJUP (R OR H)

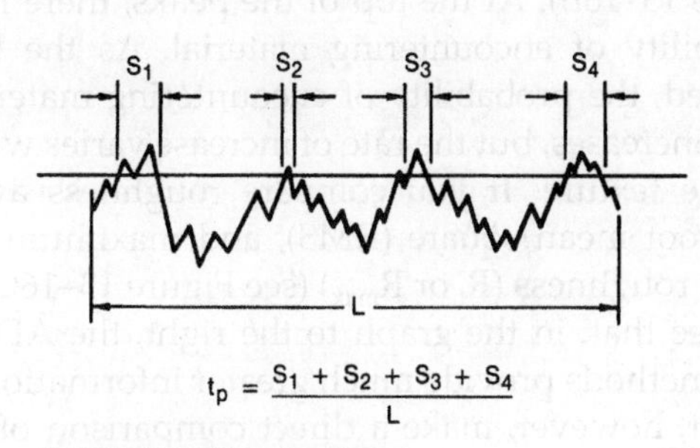

F. BEARING LENGTH RATIO (t_p AND OTHER SYMBOLS)

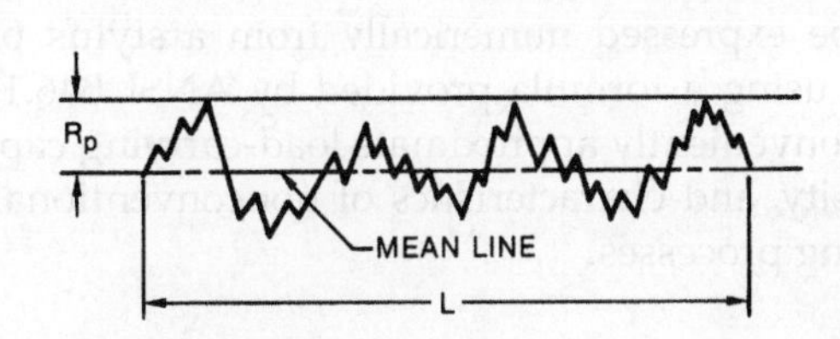

G. LEVELING DEPTH (R_p AND OTHER SYMBOLS)

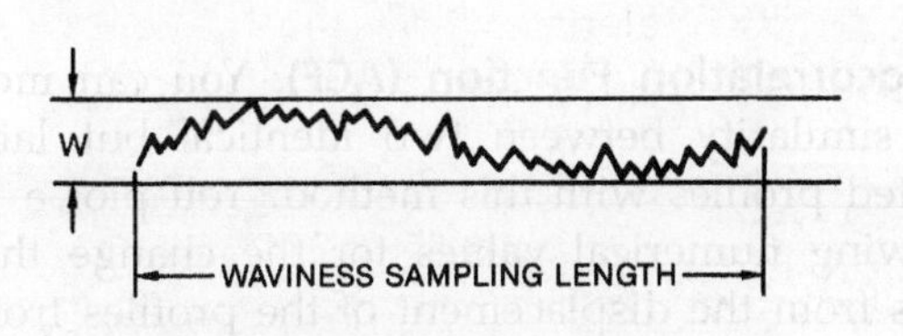

H. WAVINESS HEIGHT (W)

FIGURE 13–15 Measurement of surface irregularities.

right of the profile, relates to the probability of material along the curve. At the tops of the peaks, there is very little probability of material in the profile, so the curve is at zero. As the Δy slice moves down, the probability of material occurring increases, but not uniformly, which is shown by the extensions of the curve to the right.

Bearing Length and Bearing Curve (BAC). In a related probability assessment, you use a line to cut the profile instead of cutting a slice of the profile (see Figure 13–16B). At the top of the peaks, there is little possibility of encountering material. As the line is lowered, the probability of encountering material always increases, but the rate of increase varies with the surface texture. If you compare roughness average (R_a), root mean square (RMS), and maximum peak-valley roughness (R_t or R_{max}) (see Figure 13–16C), you will see that, in the graph to the right, the ADF and BAC methods provide much greater information. You cannot, however, make a direct comparison of these graphs, because the R_a, RMS, and R_t are numerical indices, whereas the ADF and BAC are probabilities.

Skewness. This measure of symmetry of the profile revolves about a mean line (see Figure 13–16D), and can be expressed numerically from a stylus profile trace using a formula provided by ANSI B46.1. You can conveniently approximate load-carrying capacity, porosity, and characteristics of nonconventional machining processes.

Kurtosis. This term is borrowed from statistics where it means "the quality of peakedness." Applied to the ADF, it measures sharpness. Using the formula in ANSI B46.1 (see Figure 13–16A), a perfectly random surface has a kurtosis of 3, but an actual surface may be greater or less.

Autocorrelation Function (ACF). You can measure the similarity between two identical but laterally shifted profiles with this method. You plot a graph showing numerical values for the change that results from the displacement of the profiles from the zero-shift position. These graphs make it easier to distinguish the characteristics of surfaces formed by different methods.

13-4 SURFACE TEXTURE SPECIMENS

There are two types of *surface texture specimens*: the *precision reference specimen*, used for instrument calibration, and the *roughness comparison specimens* and *pilot specimens*, used for direct visual and tactile comparison of surfaces.

Precision Reference Specimens

Complete calibration of stylus instruments must be performed in a metrology laboratory; therefore, precision reference specimens are used only for field calibration. You must not confuse precision reference specimens with roughness comparison specimens, because precision reference specimens do not duplicate the appearance or the feel (tactility) of actual surfaces. We calibrate three primary characteristics with precision reference specimens: roughness height (R_a), stylus check, and profile height.

For roughness average (R_a), you use the specimens as geometrically regular profiles within specified cutoff characteristics and calibrate instruments on that basis. ANSI B46.1 specifies that the roughness average be no greater than 203.2 μm (0.008 in.).

You must ensure the integrity of the stylus geometry for reliable surface texture assessment because the stylus receives all the wear in the measurement act—and it will wear. When the stylus wear is larger than the tolerance for its radius, it will ride over the finest asperities. You can inspect the stylus with a microscope or optical comparator, but you need to perform actual tests on a precision reference specimen for verification.

The profile height calibration is the measurement of the two levels of a step. You can make these measurements with a precision reference specimen as well as with gage blocks.

For a specimen, there are two rating values and two grades. The nominal value is a general statement you can use to select a specimen; in contrast, the assigned value is the roughness of a particular specimen established by its calibration. The two grades are *shop* and *reference*.

According to ANSI B46.1, the assigned value of shop grade specimens shall not vary from the nominal value by more than 0.050 μm (1.9685 in.) or 6%,

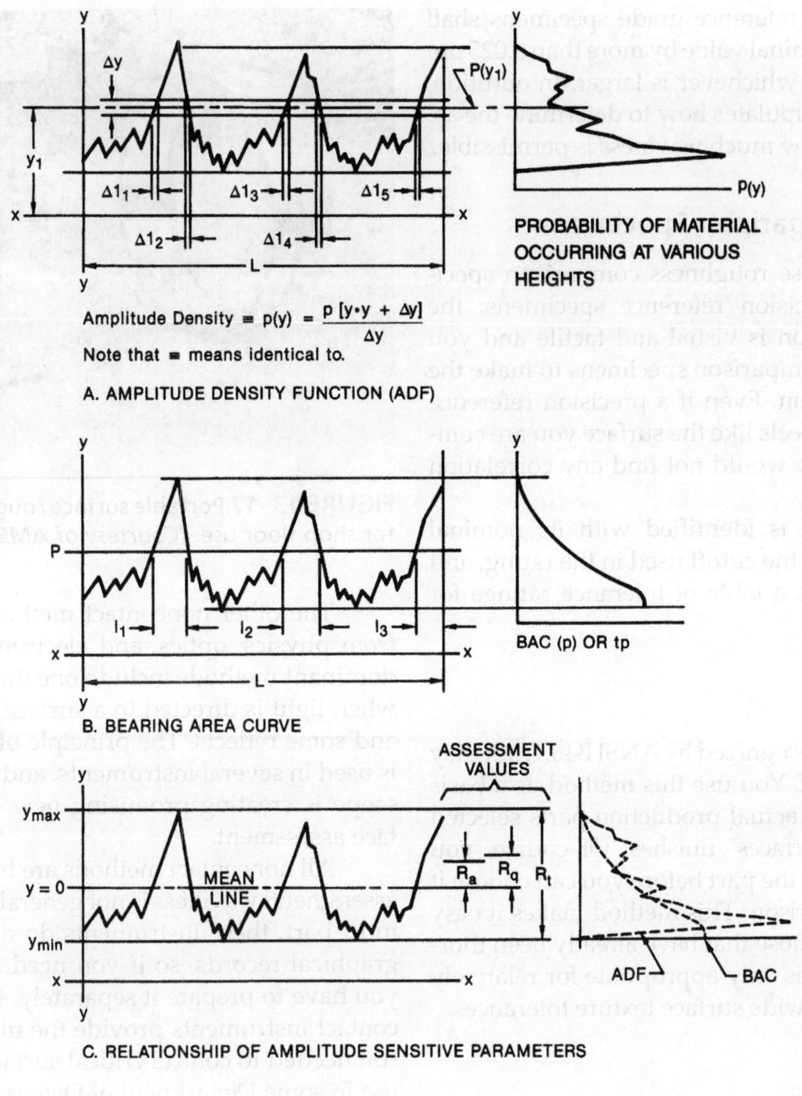

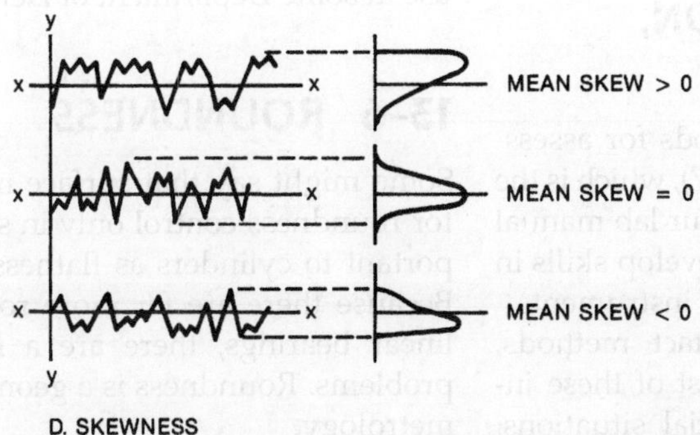

FIGURE 13–16 A surface texture analysis.

whichever is larger: reference grade specimens shall not vary from the nominal value by more than 0.025 μm (0.98425 μin.) or 2%, whichever is larger. In addition, the ANSI standard stipulates how to determine the assigned values and how much waviness is permissible.

Roughness Comparison Specimens

You must not confuse roughness comparison specimens with the precision reference specimens: the roughness comparison is visual and tactile and you use the roughness comparison specimens to make the roughness comparison. Even if a precision reference specimen looks and feels like the surface you are comparing, you probably would not find any correlation of assessment values.

Each specimen is identified with its nominal roughness value and the cutoff used in the rating, and ANSI B46.1 provides a table of tolerance ratings for roughness.

Pilot Specimens

Pilot specimens are recognized by ANSI B46.1, but they are not covered by it. You use this method as a basis of comparison using actual production parts selected with the desired surfaces' finishes. Of course, you must carefully assess the part before you can choose it as a piece for comparison. This method makes it easy to compare parts to those that have already been thoroughly tested, but it is only appropriate for relatively coarse surfaces with wide surface texture tolerances.

13–5 SURFACE EVALUATION, OTHER METHODS

We have been discussing stylus methods for assessment of surface texture (see Figure 13–17), which is the only contact method in general use. Your lab manual contains exercises that will help you develop skills in the application of surface measurement instruments.

There are a variety of noncontact methods, all using optical-type instruments. Most of these instruments are recommended for special situations. The one that is in general use is known as the Schmaltz method, or light section microscopy (see Chapter 17).

FIGURE 13–17 Portable surface roughness measurement for shop floor use. *(Courtesy of AMETEK Taylor Hobson)*

The other noncontact methods use phenomena from physics, optics, and electronics. The three predominant methods include one that uses the fact that, when light is directed to a surface, some of it scatters and some reflects. The principle of the interferometer is used in several instruments, and the electron microscope is creating promising new approaches to surface assessment.

All noncontact methods are limited to roughness assessment: waviness is not generally included. For the most part, these instruments do not directly provide graphical records, so if you need a graphical record, you have to prepare it separately. In some cases, noncontact instruments provide the most direct comparison needed to control critical surfaces, and they are in use in some Department of Defense specifications.

13–6 ROUNDNESS

Some might say that surface metrology is important for roundness control only in spheres, but it is as important to cylinders as flatness is for plane surfaces. Because there are far more rotational bearings than linear bearings, there are a number of roundness problems. Roundness is a geometric aspect of surface metrology.

At the highest grade of shop work, such as making gages, you must be aware of roundness considerations.

Unfortunately, roundness in bearings is often overlooked in routine work. If you follow a careful program of statistical quality control, you will be able to detect a gage that is sufficiently out of round to affect the control of production parts. In some cases, even an in-round bearing can have a short life because of lubrication problems or vibrations.

Lobes and the V-Block Method

Figure 13–2 shows the relationship between nominally flat and nominally round surfaces. "Waviness" on flat parts is referred to as "lobes" on cylindrical parts. We assess the higher frequency asperities on round parts in much the same way we assess them on flat parts, and they behave in much the same manner. You can use surface-datum instruments; however, they will not reveal the lobes, which we need to be especially concerned about in roundness. To detect the lobes, you need to use true-datum instruments.

Except at the extremes—very fine precision or unusually coarse surfaces—roughness does not affect the dimensions of a part; in contrast, lobes regularly affect perceived dimensions (see Figure 13–18). In both examples, a measurement taken across AA will be greater than one taken across BB, but you cannot say that BB is the "right" dimension and the AA dimension is an aberration you should ignore. If, for example, the samples in the figure are plug gages, BB would be the dimension that would determine the acceptance or rejection of parts.

Early in the machine age, metrologists adopted V-blocks to improve measurement. Today, they are usually used with comparators as an easy method of measurement, but they more often add to the errors generated by comparators (see Figure 13–19). For the same part (10-lobe) in two comparator measurement setups, the lobe heights at A are in phase and are compounded; however, at B, we used a different angle on the V-block. The lobes at B are out of phase; therefore, between the lobes, the lobe height and the valley depth cancel. The measurements would not reveal the *lobing* condition.

You can always find an angle for the V-block that provides the desired measurement, usually the in-phase condition. The problem is that, unless you can use better instrumentation, it may be extremely difficult to determine the number of lobes. You can replace the comparator with a true-datum stylus instrument and analyze a graph that is produced as the part is rotated.

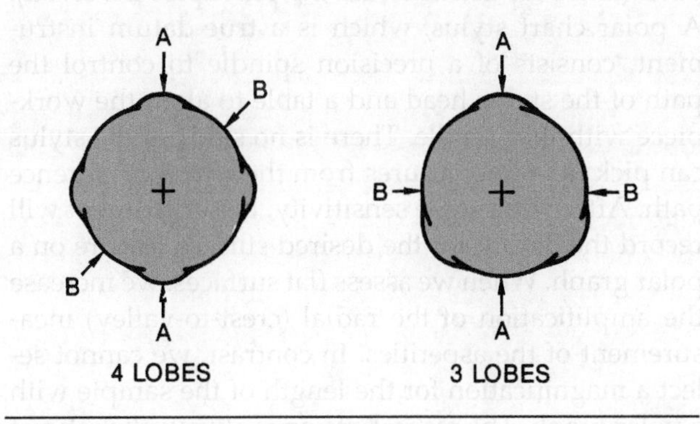

FIGURE 13–18 Caliper-type measurements such as AA and BB can provide enough data to calculate the height of the lobes—if their number is known. Unfortunately, it is nearly impossible to determine the number using only diametral measurements.

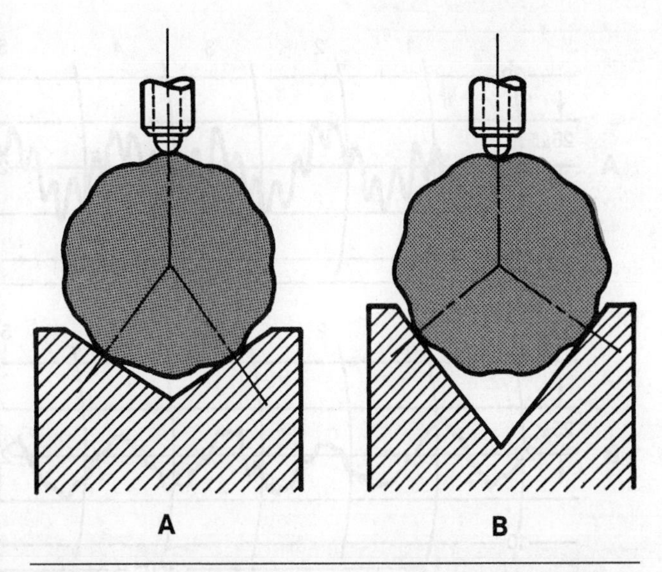

FIGURE 13–19 In V-block setups, the accuracy depends on the use of the proper included angle for the number of lobes. In A, the heights are in-phase. The same part in a V-block of smaller included angle results in the out-of-phase condition shown in B.

The Polar Chart Method

Manufacturers have developed a number of optical inspection methods that are practical for a variety of applications; however, stylus instruments that produce polar charts (see Figure 13–20) are most commonly used (and best demonstrate the principles involved). A polar chart stylus, which is a true-datum instrument, consists of a precision spindle to control the path of the stylus head and a table to align the workpiece with the spindle. There is no skid, so the stylus can pick up all departures from the circular reference path. After you set the sensitivity, the instrument will record the datum for the desired surface feature on a polar graph. When we assess flat surfaces, we increase the amplification of the radial (crest-to-valley) measurement of the asperities. In contrast, we cannot select a magnification for the length of the sample with a polar graph. The magnification is always the ratio of the part size to the closed circular path of the graph.

The advantage of the polar graph is shown in Figure 13–21. When we use the proper V-block for a five-lobe part, we get the results in Graph A—five lobes with 26 shorter asperities superimposed. If we use a V-block

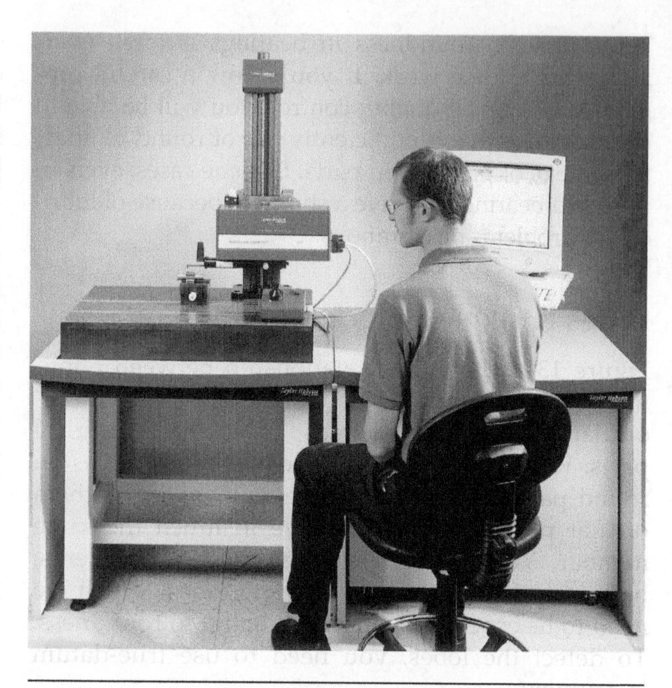

FIGURE 13–20 Precision surface and form measurement is combined with computer software to analyze diffraction patterns. *(Courtesy of AMETEK Taylor Hobson)*

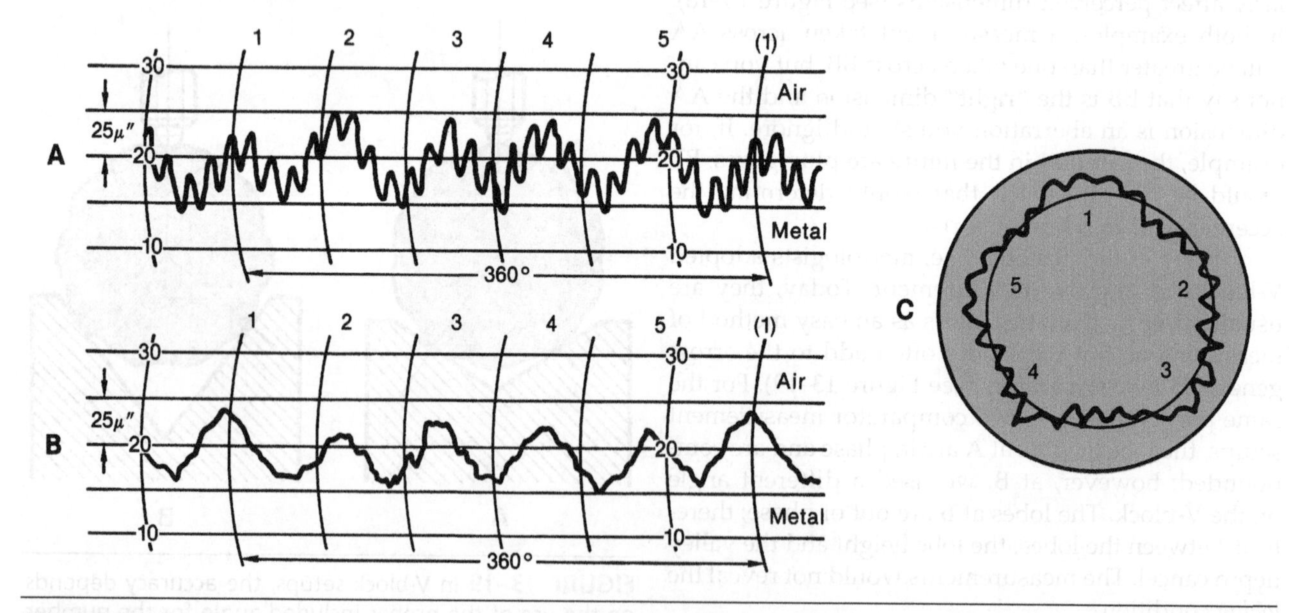

FIGURE 13–21 The polar graph in C is easier to understand for round parts than the strip chart type, A and B. Graph A was made with proper V-block for a five-lobe part. B shows the graph resulting from an out-of-phase V-block. The polar graph, C, shows the actual profile.

that causes out-of-phase cancellations, we generate Graph B, which still reveals the lobes, but at a reduced height, and the shorter asperities almost have been lost. A polar graph of the same part, Graph C, is much easier to comprehend than the linear graphs, and you can clearly distinguish both the lobes and finer asperities.

You must remember that, like other graphing instruments we have mentioned, the polar graph is not the shape of the part: the graph is a representation of the departures from the true shape. Because the radial magnification is so great, the polar graph often has concave profiles (see Figure 13–22). In A, the departure is shown as x, and if we double the radial magnification, the departure is represented in B. By the time we quadruple the magnification, the sides appear as in a hexagon C, and any further radial magnification results in a concave profile, D, which is at eight times magnification. These varying magnifications demonstrate how important it is to remember that graph is a representation.

The standard for roundness is found in ANSI Y14.5—Dimensioning and Tolerancing.

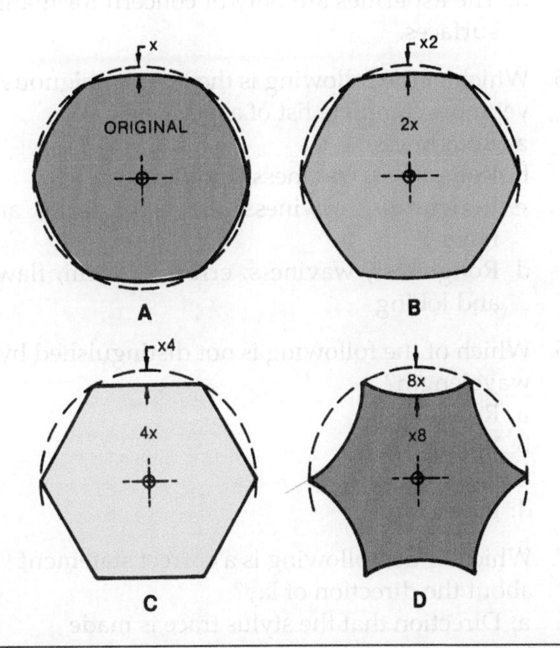

FIGURE 13–22 One of the problems with polar charts is that the distortion resulting from radial magnification may create wrong impressions. These charts show that concavity may appear on the graph, even though it does not exist on the part.

13-7 NOTATION FOR SURFACE ASSESSMENT

The best way to become proficient in surface metrology is to practice with a variety of instruments, carefully following the instructions in the manual. We hope, however, that we have given you an understanding of the principles you will need as you work with surface metrology instruments, even if you are not involved in the assessment process. Notations we use to designate surface features have been standardized, because so many people can be involved in surface assessment. The standardized notations are found in ANSI Y14.36-1996, and we have shown some examples in Figure 13–4 and a handheld surface texture instrument in Figure 13–23.

FIGURE 13–23 Surface finish and form measurement in a portable handheld unit. *(Courtesy of AMETEK Taylor Hobson)*

SUMMARY

- Measurement can be no more precise than the precision of our knowledge about the surfaces involved. These include the measuring surfaces as well as those of the workpieces. Furthermore, the characteristics of surfaces can be as important to their satisfactory use as their measurements. This is particularly true for mating parts, such as bearings of all kinds, that move in relation to each other.
- Unlike linear measurement, which cannot have more than one value for any part feature, surface measurement may have several values for the same feature. These are measurements of asperities—deviations from theoretical perfection. Those that are usually involved are roughness, waviness, and error of form. In most situations all three are not required. However, it is essential that all that are needed are included in the assessment. For example, the roughness of a shaft might be less than required, but it could still gall in its bearing if it is out of round.
- The most common instruments for surface metrology use a stylus drawn over the surface to create an electrical analogy of the surface irregularities. By electronic processing, the particular characteristics desired may be separated and examined. Nearly any stylus instrument will have adequate precision. It is important, however, that an instrument is selected that distinguishes the characteristics required.

END-OF-CHAPTER QUESTIONS

1. Which of the following explains the increased importance of surface metrology?
   a. The higher prices of consumer and industrial goods
   b. Greater operating speeds
   c. The need for higher reliability
   d. The better appearance of products
2. Most concern about surface metrology has occurred in which of the following time frames?
   a. The Industrial Revolution
   b. The Renaissance
   c. Since work began on missiles and rockets
   d. Since about 1950
3. Which of the following efforts promoted the early study of surface metrology?
   a. Development of sliding ways for machine tools
   b. Development of rotational bearings
   c. Perfection of gage block surfaces
   d. Improvements in grinding machines
4. Which of the following is most nearly correct?
   a. The asperities of a surface are deviations from the majority of the surface area.
   b. The asperities are sharp, pointed peaks that protrude from a surface.
   c. The asperities are the surface of the part of the workpiece being examined.
   d. The asperities are only of concern for mating surfaces.
5. Which of the following is the least ambiguous yet most complete list of asperities?
   a. Roughness
   b. Roughness, waviness, and flaws
   c. Roughness, waviness, errors of form, and flaws
   d. Roughness, waviness, errors of form, flaws, and lobing
6. Which of the following is not distinguished by wavelength?
   a. Roughness
   b. Waviness
   c. Errors of form
   d. Flaws
7. Which of the following is a correct statement about the direction of lay?
   a. Direction that the stylus trace is made
   b. Direction of the asperities
   c. Perpendicular to the asperities
   d. Any selected straight line taken as a reference

8. Which of the following is the most common method of surface evaluation?
   a. Statistical analysis of failure rate
   b. Inspection by stylus instruments
   c. Comparison to specimens by appearance and feel
   d. Microscopic examination
9. Which of the following types of instruments is most extensively used for surface examination?
   a. Interferometers
   b. Glossmeters
   c. Microscopes
   d. Contact instruments
10. Which of the following is the best analogy for the trace of a stylus instrument?
    a. Topographical map
    b. Phonograph
    c. Altimeter of aircraft
    d. Rolling ball
11. What is meant by "distortion ratio"?
    a. The difference between the true value and the measured value
    b. The difference between a surface texture value obtained with the exceptional peaks and valleys included, compared with the same surface with them eliminated
    c. The difference in the sampling length taken along the lay as compared with it perpendicular to the lay
    d. The difference between the magnification of the length as compared with that of the height
12. Which of the following is usually correct?
    a. Asperity heights are minute compared to their lengths.
    b. Asperity lengths are minute compared to their heights.
    c. Asperity lengths usually equal their heights.
    d. Asperity lengths are usually from 5 to 20 times greater than their heights.
13. Which of the following statements about the stylus are valid for most assessments?
    a. It must be small enough to pick up all asperities.
    b. It must not be sharp enough to scratch the surface.
    c. It must be softer than the surface being assessed.
    d. It must be sized to pick up the smallest asperity being investigated but ignore smaller ones.
14. From which of the following does a surface-datum instrument arrive at the reference line?
    a. The peaks and valleys
    b. The speed of the travel of the stylus
    c. The surface on which the skid or shoe rests
    d. A length 50 times the sampling length
15. Which of the following best describes "cutoff"?
    a. End of travel of the stylus across the surface being examined
    b. Highest or lowest frequency allowed to pass into the profile reading
    c. Highest or lowest asperity peaks allowed to pass into the profile reading
    d. Portion of profile on paper tape cut off for record of assessment
16. Which of the following is the correct statement?
    a. There are many methods of surface assessment, but only one has been standardized.
    b. Each country has its own standard.
    c. It is only permissible to use a standardized method.
    d. The United States has one official standard but recognizes the use of others.
17. Which of the following is standardized for use in the United States?
    a. Root-mean-square roughness
    b. Average peak-to-valley roughness
    c. Arithmetic average roughness
    d. Roughness average
18. Which of the following comes closest to describing roughness average?
    a. Arithmetic average of the height deviations from a centerline
    b. Algebraic average of the height deviations from a centerline

c. Arithmetic average of the height deviations divided by the number of asperities
d. Arithmetic average of the height deviations divided by the average length between asperities

19. Which of the following best describes the relationship between sampling length and traversing length?
a. Traversing length may be any length longer than the sampling length.
b. Traversing length must be 20 times the sampling length for roughness under 50 μin. and increases for finer textures.
c. Traversing lengths may be between 20 and 60 times the cutoff.
d. There is no rule for traversing length in comparison to sampling length.

20. Which of the following best describes the numerical values for the roughness average?
a. The value is a dimension for the asperities.
b. The value can be converted into a dimension by the application of certain formulae.
c. The chief purpose of the value is to show confirmation of a standard.
d. The value is an index to facilitate comparison of surfaces.

21. What is the status in the United States of the assessment methods that have been standardized in other countries?
a. Forbidden
b. Used but not legal
c. Permissible when better results occur than with roughness average
d. Obligatory when better results occur than with roughness average

22. Which of the following are essential for the calibration of stylus instruments?
a. A metrology laboratory
b. An optical comparator
c. Precision reference specimens
d. A high-amplification comparator

23. What is the role for roughness comparison specimens?
a. Calibration of stylus instruments
b. Setup of surface assessment evaluation instruments
c. Use with noncontact instruments
d. For comparison by appearance or feel

24. Which of the following best describes roundness?
a. Similarity to the revolution of the earth about the sun
b. Ability of mating male and female parts to rotate without interference
c. The opposite of flatness
d. A surface, all of whose points are equidistant from the same center

25. Which of the following best describes the V-block method?
a. The flanks of the V-block are compared with the lobes of the cylindrical workpiece.
b. The included angle of the V-block magnifies the effect of lobing.
c. The axis of rotation varies according to the lobing.
d. Only 450 and 600 V-blocks may be used.

26. Which of the following describes the effect of lobing when using the V-block method?
a. For 450 blocks, add 25%; for 600 blocks, add 33.3%.
b. For 450 blocks, subtract 25%; for 600 blocks, subtract 33.3%.
c. Effect is dependent on the number of lobes and the included angle of the V-blocks.
d. Errors cancel out with V-blocks so that lobing is not included in readings.

27. Which of the following best describes the difference between the V-block and polar chart methods?
a. Except for long wavelengths, the results are the same.
b. The *polar chart method* provides true-datum readings.
c. Lobes cancel out with the polar chart method.
d. Lobes are only read on polar chart instruments when texture is cut off.

28. Which of the following best describes a polar chart?
a. A polygon, all points of which are equidistant from a center
b. Deviations from a nominal circle

c. Temperature recording
d. The surface profile of a cylindrical part

29. How many degrees are there in a polar chart?
a. 0 to 180
b. 90 to 360
c. 360 maximum
d. 360

30. Which of the following best describes the relationship between the radial measurement (asperities) to the angular rotation?
a. As set by the cutoff of the instrument
b. Radial measurements are slightly magnified to be more visible
c. Between one and five radial measurements per degree of rotation
d. Radial measurements may be of 1,000 times greater magnification than asperities.

31. Which of the following terms correspond with respect to polar charts?

a. Waviness	Asperities
b. Lobing	Waviness
c. Roughness	Asperities
d. Lobing	Asperities
e. Asperities	Rotation

32. Complete the following sentence: A polar chart is ______.
a. the magnified profile of the surface of a cylinder.
b. the shape of a cylindrical workpiece.
c. the departures from the desired shape of a cylindrical workpiece.
d. the depiction of the roughness of a cylindrical workpiece.
e. the depiction of the waviness of a cylindrical workpiece.

CHAPTER FOURTEEN

14

HIGH-AMPLIFICATION COMPARATORS

LEARNING OBJECTIVES

- Explain the distinction between measuring instruments and gages.
- State the reason why power amplification makes a radical departure from mechanical instruments.
- Describe the practical advantages of multiple scale selection.
- Explain the basis of the 10-to-1 rule.

OVERVIEW

So far, we have discussed dial indicators for comparison measurement. Dial indicators are inexpensive, widely used, and reliable within their capability and inherent shortcomings. When we use them as comparators, they are very reliable, especially when we are measuring as close to the zero setting as possible. The inherent shortcoming of dial indicators comes from their chief method of amplification: gear trains.

Theoretically, there is no limit to the amplification possible with gear trains. Practically, gear errors, friction, inertia, and hysteresis limit the potential of gear trains. Although some dial indicators have discrimination of 1.2 μm (50 μin.), these instruments are the exception: the practical limit of discrimination is 2.5 μm (0.0001 in.). We lose more in sensitivity than what is gained in discrimination when we exceed this practical limit.

Because of these mechanical limitations, manufacturers have developed better means of amplification for comparison measurement:

- Other mechanical actions
- Optical
- Pneumatic
- Electronic (see Figure 14–1)

Electronic instruments are most widely used in industry today, so why are we studying all the other high-amplification mechanical instruments? The basic principles of electronic and mechanical instruments are the same, and these principles are easier to understand with mechanical instruments because you do not need extensive electronics training. We

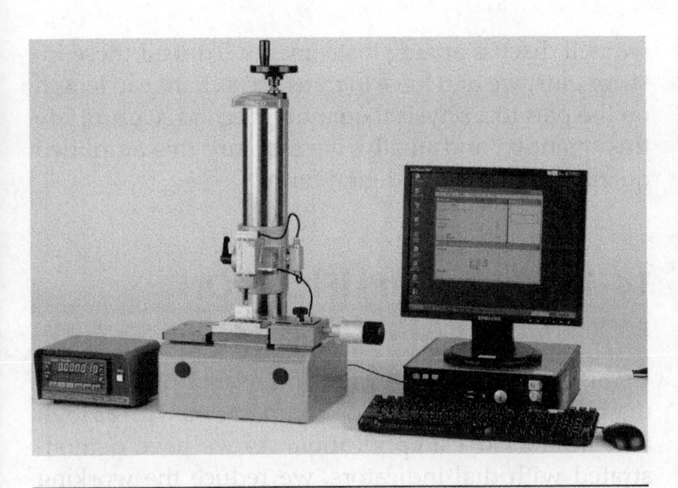

FIGURE 14–1 Gage block comparator. *(Courtesy of Mitutoyo American Corporation)*

are working on the fundamentals, so you need to refer to manufacturers' manuals for specific, practical information about the use and maintenance of your instruments.

Gages and Measurement Instruments

Even though the terms *gages* and *measurement instruments* are popularly used interchangeably, there are important distinctions we need to understand. For example, air gages are high-amplification instruments, but they are omitted from this chapter. Obviously, we need to distinguish between gages and other measurement instruments.

Various purchasing departments of the federal government have done much to advance standardization (see Figure 14–2). Note that, by definition, gages are used primarily to check or inspect. Although someone did considerable measurement to design and build the gage, the user of the gage is concerned only with the reaction of the gage to the part feature being checked—a right or wrong, go or no-go decision. Even when the readout has numbers, they are arbitrary units, useful to the quality control department, but with little direct meaning to the person checking.

Measurement instruments, or measuring equipment (see Figure 14–2), provide numerical information that we can use immediately. Often measurement instruments, which are highly versatile, are used for gaging, but gages, which are highly specialized, are rarely used for measurement.

The *air gages*—properly called pneumatic comparators—are gages: They compare parts with masters. Air gages are extremely valuable and important for modern quality control, particularly when you are trying to inspect tiny hole diameters. Even though air

Gages and Measuring Equipment

1.1.3.1	**Length Standards.** Standards of length and angle from which all measurements of gages are derived.
1.1.3.2	**Master Gages.** Master gages used for checking and setting inspection of manufacturers' gages.
1.1.3.3	**Inspection Gages.** Inspection gages used to inspect products for acceptance.
1.1.3.4	**Manufacturers' Gages.** Manufacturers' gages used for inspection of parts during production.
1.1.3.5	**Nonprecision Measuring Equipment.** Simple tools used to measure by means of line graduations.
1.1.3.6	**Precision Measuring Equipment.** Tools used to measure in thousandths of an inch or finer.
1.1.3.7	**Comparators.** Precision measuring equipment used for comparative measurements between the work and a contact standard such as a gage or gage blocks.
1.1.3.8	**Optical Comparators and Gages.** Optical comparators and gages are those which apply optical methods of magnification exclusively.

FIGURE 14–2 These definitions are condensed from Military Standard, Gage Inspection, MIL-STD-120. They show the distinctions among the classifications of equipment involved in metrology. Note that the high-amplification comparators discussed in this chapter are not independent instruments but require the use of length standards, another one of the classifications.

gages are becoming more versatile as measurement instruments, they are not commonly used as measurement instruments, so we will discuss them with pneumatic measurement instruments (see Chapter 15). We will also refine our definitions of certain terms later.

Traditionally, the words *gage* and *gauge* have confused students of metrology. A very simple explanation is that *gauge* is the British spelling; however, both spellings are commonly used in the United States. The terms were clarified somewhat in the 1983 interim standard for coordinate measuring machines (Chapter 18):

Gage—A mechanical artifact of high precision used either for checking a part or for checking the accuracy of a machine.

Gauge—A measuring device with a proportional range and some form of indicator, either *analog or digital*.

Analog and Digital Instruments

Modern consumer electronics have propelled us from a completely analog world into the world of digital readouts. These two systems of readout are valuable for describing more than the time; they are perfect for a variety of measurement activities, particularly for some of the more complex instruments. Until now, we have studied analog instruments exclusively, but today many gages are digital.

Analog instruments use almost any physical quantity to correspond to numbers. If a block of wood stands for 25.4 mm (or 1.6 km or one dozen roses), then two blocks represent 50.8 mm (3.2 km or two dozen roses). Numbers are represented by measurable quantities. For dial indicators, the quantity is measured by the sweep of the pointer across the dial. For electronic instruments, the quantity is measured by the electrical signal, and for pneumatic comparators, the quantity is measured by an air stream.

In contrast, a digital instrument deals with numbers directly. For example, when a block of wood is placed in a certain slot, it might represent 10, regardless of its length. In short: *An analog instrument measures; a digital instrument counts.* The gas gage on an automobile is an analog instrument, whereas the mileage gage is a digital odometer. In this chapter, we will discuss analog instruments. To read these instruments, we first have to translate a change in length on the part to a physical quantity. Second, we amplify this quantity, and finally, we measure this amplified quantity in analog or digital form.

14-1 HIGH-AMPLIFICATION COMPARATORS

Using mechanical, optical, and electronic devices, manufacturers have created comparators with as high as 1 million to 1 amplification. As we have demonstrated with dial indicators, we reduce the working range as we increase the amplification. To have a usable range, most practical instruments have much lower amplification: 1,000 to 1 to 20,000 to 1. This range is still very large compared with the range of dial indicators.

In spite of some individual features, dial indicators are all rather similar; however, high-amplification instruments are not. Electronic instruments have a unique feature: power amplification. Unlike all previously discussed instruments, many electronic instruments can provide two or more scale ranges. Versatility plus relatively low cost means that these electronic instruments are almost universally used for everything from line inspection of production parts to master gage calibration.

We will discuss mechanical instruments first, because their development shows how these instruments could naturally evolve to the more sophisticated instruments. They also show us how good it is to have options in measurement. We will also discuss reliable measurement to μm (tenth-mil) increments (0.0001 in.) and finer. Before we discuss measuring reliably to such small increments, review the terms that apply to these measurements, in Figure 14–3 and Appendix D.

High-Amplification Mechanical Comparators

As we tried to achieve higher amplification on dial indicators, these instruments have also been refined (see Figure 14–4).

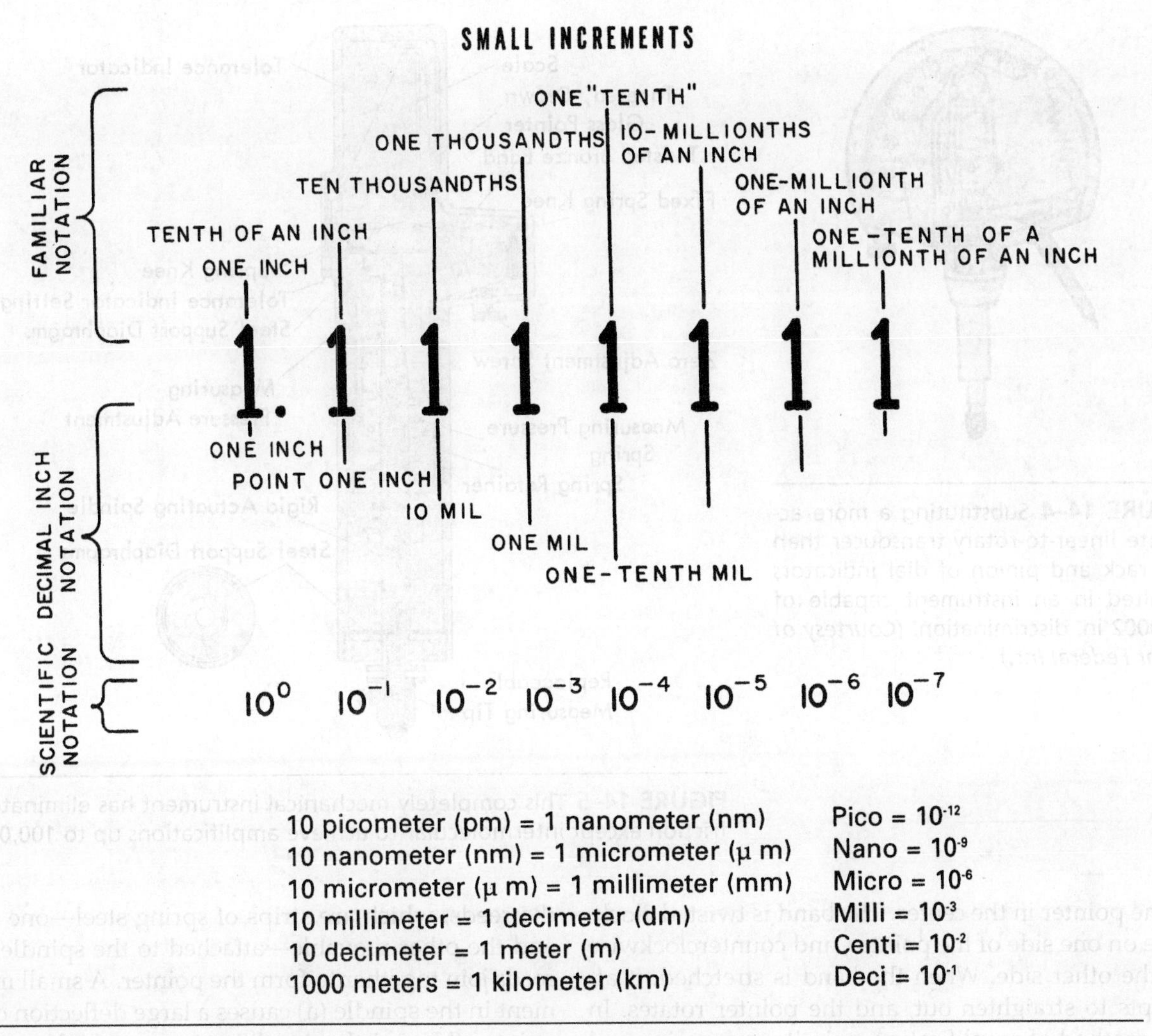

FIGURE 14–3 At high amplification, it is easy to misplace a decimal point.

A dial indicator's mechanism is a cascade of stages, each one at a higher amplification. An error in the gearing at the final stage has a relatively small effect on accuracy, whereas an error in a stage close to the input has a tremendous effect on accuracy. The weakest stage in a dial indicator is the rack and pinion gear. The gear set in Figure 14–4 has been replaced with a precision steel ball rolling along a finely lapped sapphire surface. Extremely minute changes in position of the spindle are directly translated to the gear train. In this design, the mechanical loading of the parts is independent of the contact pressure, so that windup and other types of distortion are minimized. In the first two gear stages, the movements are so slight that the mechanism uses only segments of gears.

Typically, the discrimination of these instruments is available to 0.5 µm (20 µin.) increments, but manufacturers have come up with some ingenious systems to further increase amplification mechanically. In Figure 14–5, all friction is eliminated, except intermolecular friction (molecular particles sliding against each other when the twisted band is stretched).

The twisted band (see Figure 14–5) is fixed at one end, connected to the spindle at the other, and attached

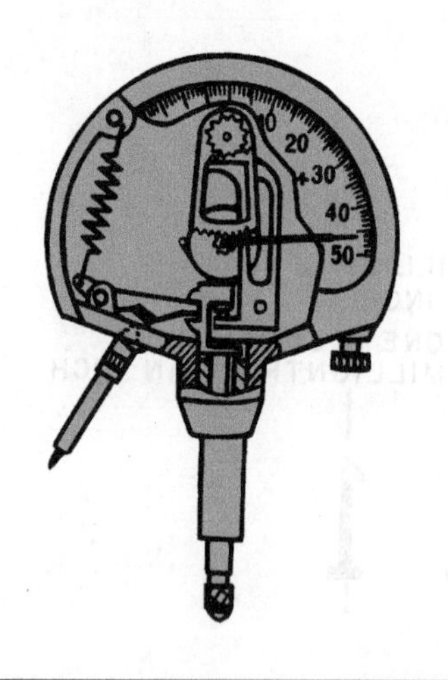

FIGURE 14–4 Substituting a more accurate linear-to-rotary transducer than the rack and pinion of dial indicators resulted in an instrument capable of 0.00002 in. discrimination. *(Courtesy of Mahr Federal Inc.)*

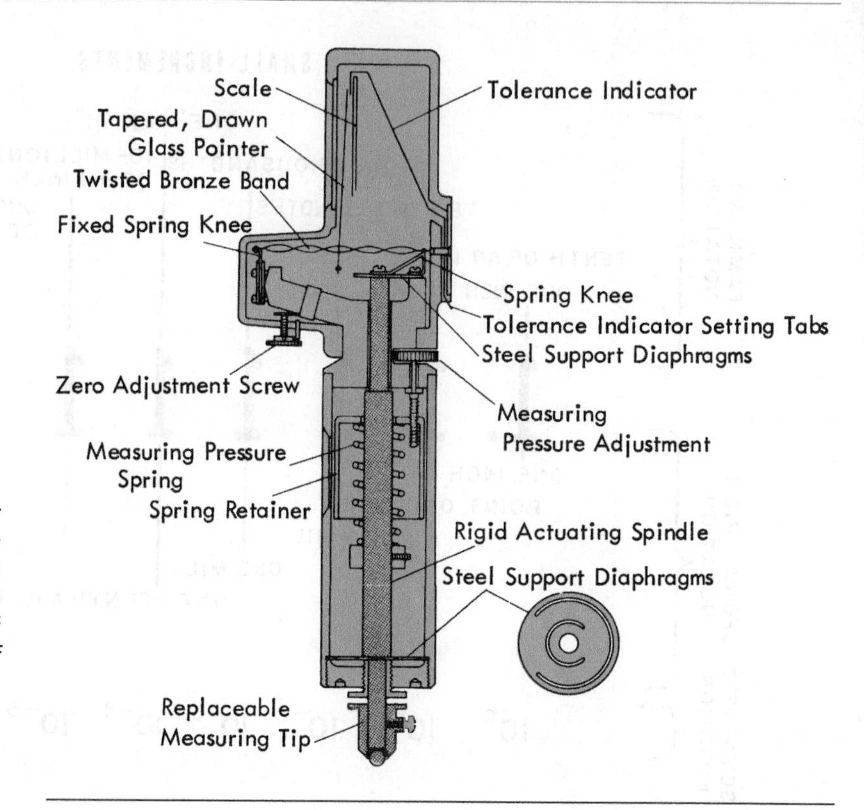

FIGURE 14–5 This completely mechanical instrument has eliminated all friction except intermolecular to achieve amplifications up to 100,000X.

to the pointer in the center. The band is twisted clockwise on one side of the pointer, and counterclockwise on the other side. When the band is stretched, it attempts to straighten out, and the pointer rotates. In this method of amplification, spindle suspension and twisted band drive are also affected by spring suspension. There is no rubbing friction, which not only results in high sensitivity and repeatability, but also ensures low maintenance.

Reed-Type Comparator

The reed-type comparator (see Figure 14–6) is similar to the previous instrument because it requires only intermolecular friction to achieve its amplification and to support all of its moving parts. What makes it interesting to us is the method it uses to achieve additional amplification: the *optical lever.*

At the right in Figure 14–7, we show the principle of the reed mechanism. Two blocks are attached to the reeds, which are strips of spring steel—one fixed and the other movable—attached to the spindle. The reeds join together to form the pointer. A small movement in the spindle (a) causes a large deflection of the pointer (b), and the amplification equals the ratio of (b) to (a). The action is directional; if the direction of movement of (a) reverses, so does (b).

To further amplify this action, we use an optical lever (see Figure 14–7 left). The pointer moves an aperture, called the *target,* across a focused beam of light. This light beam has the same effect as a lever arm, but the beam is weightless and frictionless.

In Chapter 10, we demonstrated that the amplification of the hand or pointer multiplies the amplification of the mechanism. Although dial indicators have short hands, the effective hand of the reed comparator can be much longer. The reed comparator's hand is the length of the reed plus the optical lever. If the reed mechanism provided 50X, the hand leverage would only have to be 40:1 to increase the total amplification to 2000X.

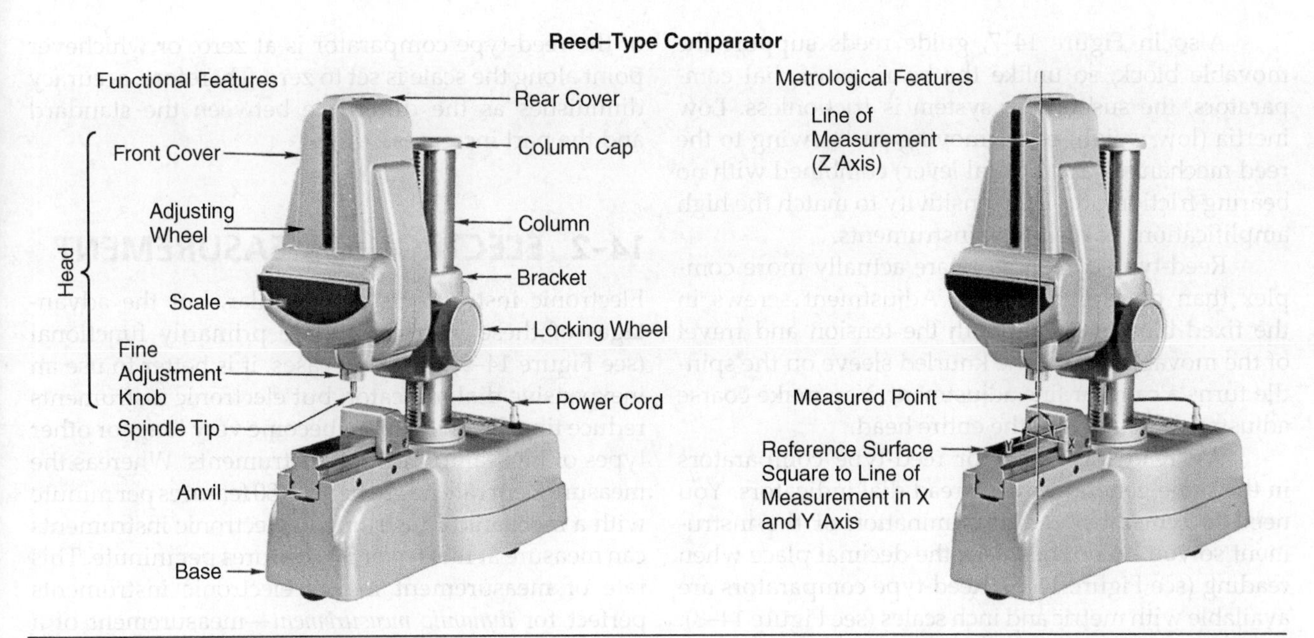

FIGURE 14–6 The metrological features of this reed-type comparator are the same as the other high-amplification mechanical instruments. They are relatively inflexible in configuration compared with the great flexibility of the new electronic instruments. However, there is never any doubt about their geometry. Their adherence to Abbe's law is clear, although it may not be with their electronic counterparts. *(Courtesy of CE Johansson)*

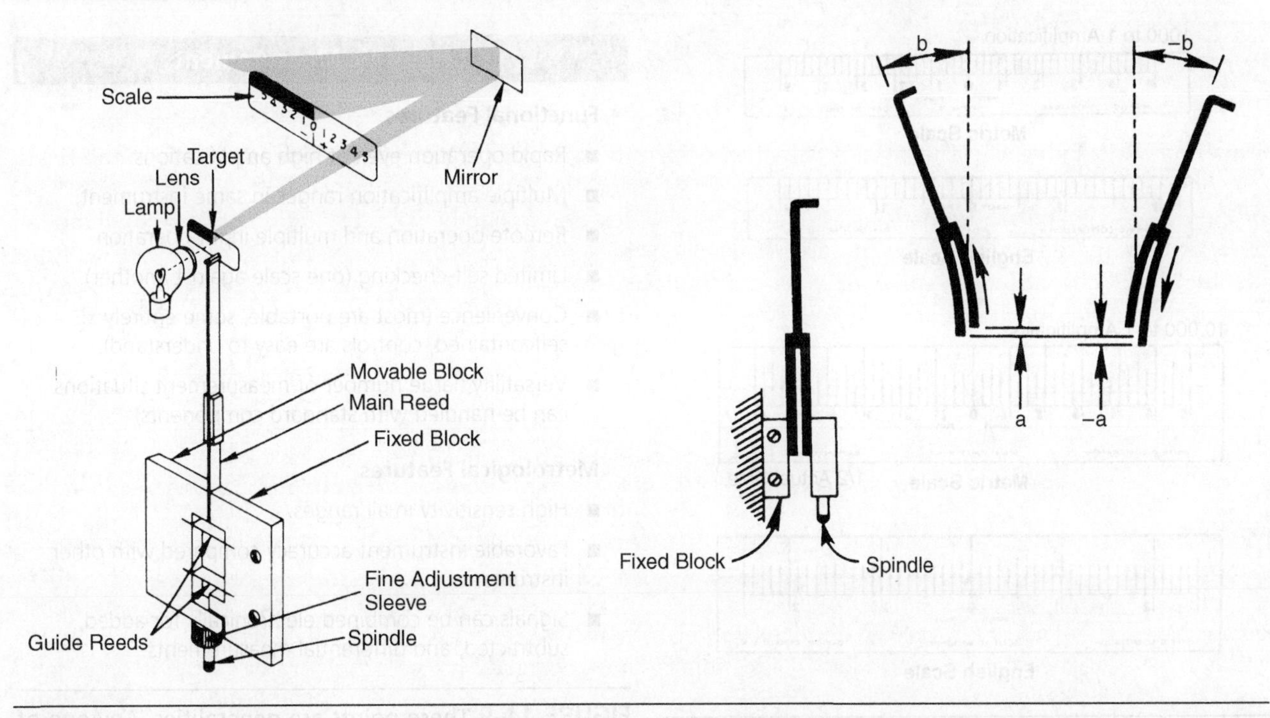

FIGURE 14–7 An optical lever is used to amplify the reaction. In the actual instrument, the optical path is longer than is shown here and a prism is used instead of a mirror.

Also in Figure 14–7, guide reeds support the movable block, so unlike the bearings of dial comparators, the suspension system is frictionless. Low inertia (low weight of the moving parts owing to the reed mechanism and optical lever) combined with no bearing friction provides sensitivity to match the high amplifications of reed-type instruments.

Reed-type comparators are actually more complex than these illustrations. Adjustment screws in the fixed block regulate both the tension and travel of the movable block. The knurled sleeve on the spindle turns a cam for fine adjustment. You make coarse adjustment by moving the entire head.

We read the scales for reed-type comparators in the same general way we read dial indicators. You need to remember the discrimination of the instrument so you do not misplace the decimal place when reading (see Figure 14–3). Reed-type comparators are available with metric and inch scales (see Figure 14–8).

Reed-type comparators are more sensitive than dial indicators, and you can use the entire scale of reed-type comparators, which do not require pretensioning. Like a dial indicator, the greatest accuracy of the reed-type comparator is at zero, or whichever point along the scale is set to zero. Therefore, accuracy diminishes as the difference between the standard and the part increases.

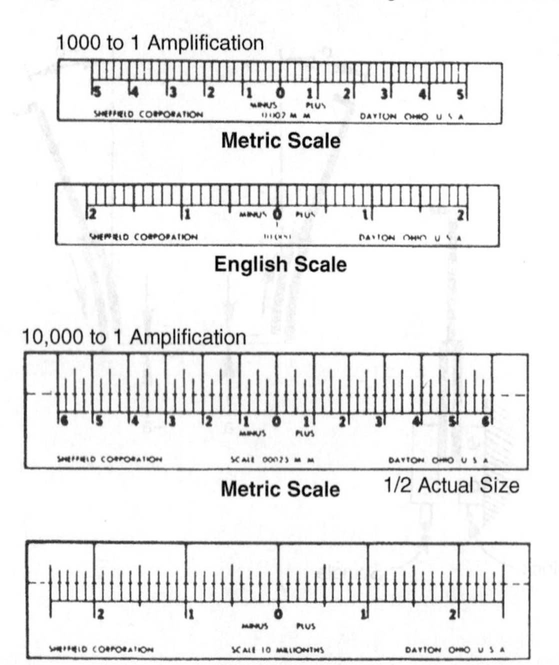

FIGURE 14–8 Reed-type instruments are available in a range of amplifications with either metric or English scales.

14-2 ELECTRONIC MEASUREMENT

Electronic instruments are popular, but the advantages of these instruments are primarily functional (see Figure 14–9). In many cases, it is better to use an inexpensive dial indicator, but electronic instruments reduce time lag, which can become very long for other types of high-amplification instruments. Whereas the measurement rate might be 50 to 80 features per minute with a mechanical instrument, electronic instruments can measure at rates over 500 features per minute. This rate of measurement makes electronic instruments perfect for *dynamic measurement*—measurement of a moving feature or a changing value. Measuring the thickness of a strip passing through a rolling mill is a moving feature, and monitoring deflection under a varying load is a changing value.

Electronic Measurement

Functional Features

- Rapid operation even at high amplifications.
- Multiple amplification ranges in same instrument.
- Remote operation and multiple input operation.
- Limited self-checking (one scale against another).
- Convenience (most are portable, some entirely selfcontained, controls are easy to understand).
- Versatility (large number of measurement situations can be handled with standard components).

Metrological Features

- High sensitivity in all ranges.
- Favorable instrument accuracy compared with other instruments.
- Signals can be combined electronically for added, subtracted, and differential measurements.

FIGURE 14–9 These points are generalities. Any one of them might be sufficient reason to accept or reject an electronic instrument for a particular measurement.

Power Amplification for Measurement

As we discussed, inertia and friction can severely limit mechanical instruments, but electrical systems do not have these problems. More importantly, like no other instrument, their *power* amplification—the ability to convert a mechanical motion into an electrical signal—can be electronically amplified.

As we studied instruments from low to high amplification, the only movement that has been amplified in every case is the measurement movement. All of the energy required to operate the instrument had to be generated by the action of the part on the instrument, called *loading*. Loading decreased steadily between the instrument's input stage and its readout stage because of energy losses in the instrument, most caused by friction. We reach a point in the design of dial indicators when so much of the loading is used to drive the mechanism function to wind up shafts, overcome bearing friction, and cause gear teeth to slide over each other that not enough energy remains to move the readout element. High mechanical amplification often leads to poor sensitivity.

Creative designers have generated a number of ways to conserve loading, and the reed-type comparator is just one example. In the long run, however, we always reach a mechanical loading limit as we increase amplification. Electronic instruments, in contrast, use power from an outside source, which not only makes it possible for us to generate records of the measurements, but also to use the readout signal to control machines or processes. Without electronic instruments, automation or numerical control of machines would be impossible.

Generalized Measurement System

All measurement instruments are described by the generalized system: length, mass, and time (see Figure 14–10). For the instruments we have discussed

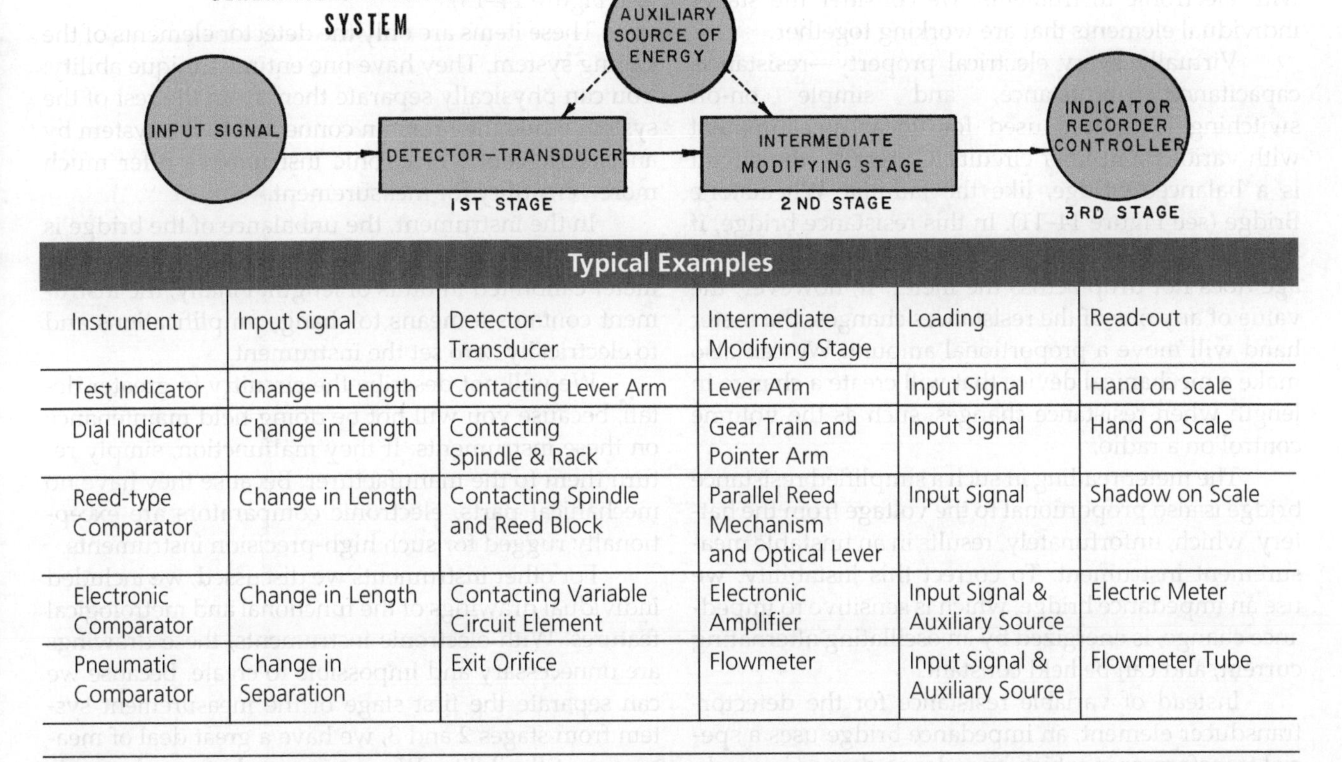

Typical Examples

Instrument	Input Signal	Detector-Transducer	Intermediate Modifying Stage	Loading	Read-out
Test Indicator	Change in Length	Contacting Lever Arm	Lever Arm	Input Signal	Hand on Scale
Dial Indicator	Change in Length	Contacting Spindle & Rack	Gear Train and Pointer Arm	Input Signal	Hand on Scale
Reed-type Comparator	Change in Length	Contacting Spindle and Reed Block	Parallel Reed Mechanism and Optical Lever	Input Signal	Shadow on Scale
Electronic Comparator	Change in Length	Contacting Variable Circuit Element	Electronic Amplifier	Input Signal & Auxiliary Source	Electric Meter
Pneumatic Comparator	Change in Separation	Exit Orifice	Flowmeter	Input Signal & Auxiliary Source	Flowmeter Tube

FIGURE 14–10 All measurement instruments, including those that measure mass and time, as well as length measurement, fall into the generalized system.

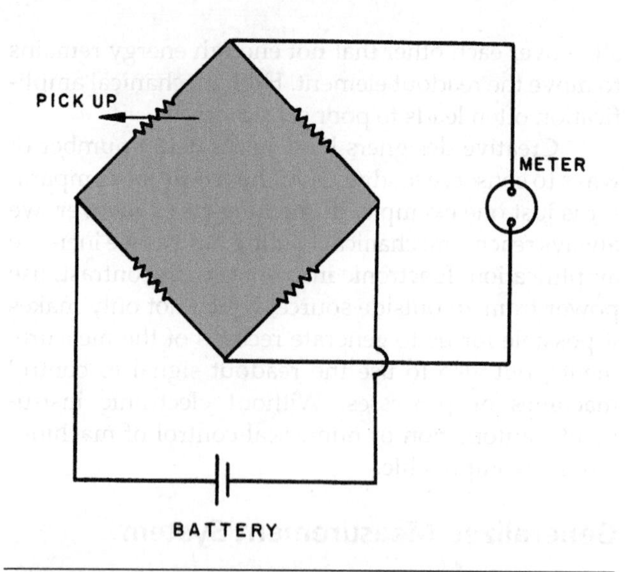

FIGURE 14–11 This simplified bridge circuit is similar to those used in electronic comparators.

so far, it is not easy to separate the stages; however, with electronic instruments, we consider the stages individual elements that are working together.

Virtually every electrical property—resistance, capacitance, inductance, and simple on-off switching—has been used for linear measurement with various amplifier circuits. Generally, the circuit is a balanced bridge, like the familiar Wheatstone Bridge (see Figure 14–11). In this resistance bridge, if the value of each of the four resistances is equal, voltage does not drop across the meter. If, however, the value of any one of the resistances changes, the meter hand will move a proportional amount. We can also make a mechanical device that will create a change in length when resistance changes, such as the volume control on a radio.

The meter reading in such a simplified resistance bridge is also proportional to the voltage from the battery, which, unfortunately, results in an unstable measurement instrument. To correct this instability, we use an impedance bridge, which is sensitive to impedance change, is energized by an oscillating alternating current, and can be held constant.

Instead of variable resistance for the detector-transducer element, an impedance bridge uses a special transformer in which impedance change is nearly

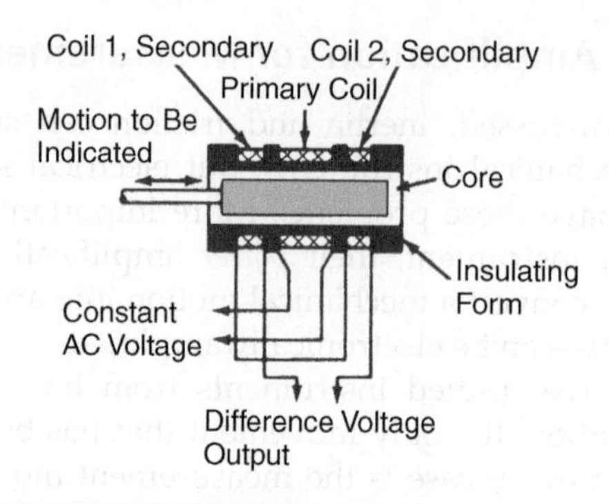

FIGURE 14–12 This is an LVDT. The LVDT gives an output voltage that is proportionally linear to linear changes in core position.

proportional to a change in its armature position. This special transformer is called a *linear variable differential transformer* (LVDT). An LVDT's armature is called the *core* (see Figure 14–12). When placed in a housing and provided with a suitable contact point (a variety of models is available), the LVDT is called the *gage head* (see Figure 14–13).

These items are only the detector elements of the gaging system. They have one entirely unique ability: You can physically separate them from the rest of the system while they remain connected to the system by an electric cable. Electronic instruments offer much more versatility for measurement.

In the instrument, the unbalance of the bridge is amplified electronically and connected to a sensitive meter calibrated in units of length. Finally, the instrument contains a means to change amplifications and to electrically zero set the instrument.

We will not describe the circuitry in greater detail, because you will not be doing field maintenance on these instruments. If they malfunction, simply return them to the manufacturer. Because they have no mechanical parts, electronic comparators are exceptionally rugged for such high-precision instruments.

For other instruments we discussed, we included individual drawings of the functional and metrological features. With electronic instruments, these drawings are unnecessary and impossible to create. Because we can separate the first stage of the measurement system from stages 2 and 3, we have a great deal of measurement flexibility. We can incorporate an electronic

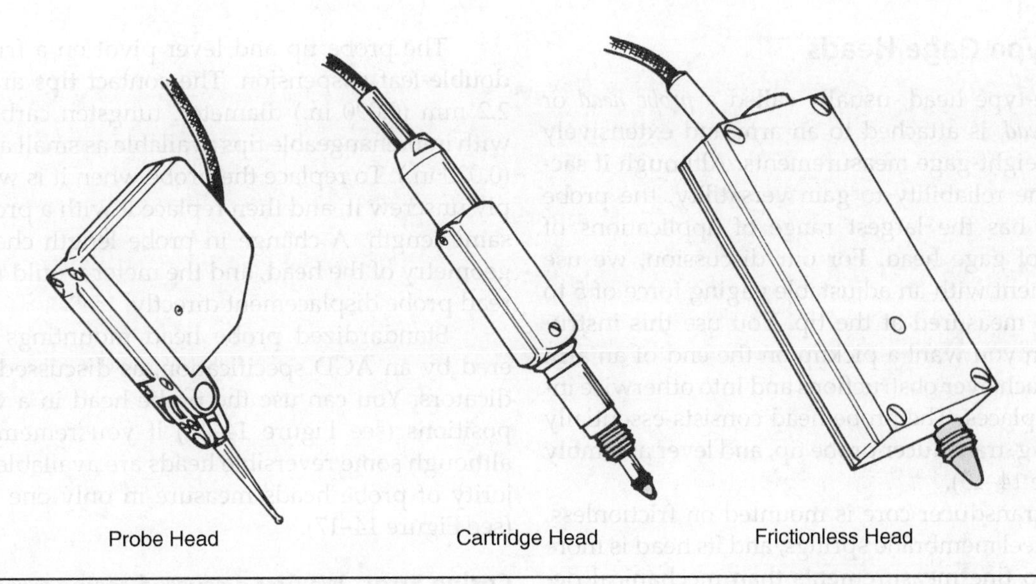

FIGURE 14–13 These are the three most popular configurations for gage heads using LVDT transducers.

Comparison of Gage Heads

	Probe Type	Cartridge Type	Frictionless Type
Repeatability	0.000001″	0.0000008″	0.0000004″
Gaging Pressure	10–20 grams	20–30 grams	Continuously adjustable between 15 and 80 grams
Total Spindle Travel	.024″ ± .004″	.043″ ± .004″	.043″ ± .004″
Spindle Travel to Zero Position	.012″	.012″	.012″

FIGURE 14–14 The relative sensitivity of these gage heads is shown by their repeatability. Note that the cartridge type is somewhat more sensitive than the probe type, but the frictionless type is twice as sensitive as the probe type. These particular values apply to the gage heads shown in Figure 10–13. Those of the other manufacturers may be different.

measurement head into caliper-type instruments, height gage setups, and comparators equally well. In this chapter, we will discuss comparators primarily, because they use electronic components most completely.

The Gage Head

There are three main designs for gage heads (see Figure 14–13) of which the frictionless type is the least common but the most sensitive. This gage head is not totally without friction—there is still intermolecular friction in the reeds. Gage heads have some typical characteristics (see Figure 14–14) and the three major gage heads all use LVDTs as transducer elements.

The frictionless gage head has limited application only. The versatile probe head is useful for routine inspection of individual parts and small production lots, but it also requires relatively more operator skill than the other two. We prefer the cartridge head for gaging setups; the reliability of this head is based more on its design and manufacture than on the user's skill.

Probe-Type Gage Heads

The probe-type head, usually called a *probe head* or *indicator head,* is attached to an arm and extensively used for height-gage measurements. Although it sacrifices some reliability to gain versatility, the probe head also has the largest range of applications of any type of gage head. For our discussion, we use an instrument with an adjustable gaging force of 5 to 100 grams measured at the tip. You use this instrument when you want a pickup on the end of an arm that can reach over obstructions and into otherwise inaccessible places. The probe head consists essentially of a housing, transducer probe tip, and lever assembly (see Figure 14–15).

The transducer core is mounted on frictionless, stainless steel membrane springs, and its head is more sensitive to fine measurements than mechanical devices. The probe's friction-clutch mounting allows a wide choice of gage positions, and the probe tip's measuring range is 1.5 mm (0.060 in.) with 50.8 mm (0.002 in.) overtravel.

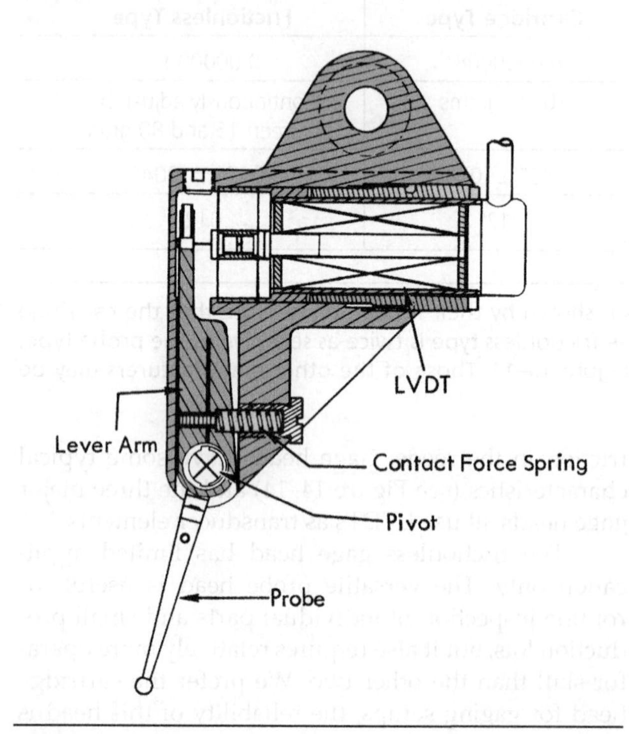

FIGURE 14–15 Probe heads differ considerably but all have similar features. The adjustable probe acts on an LVDT cartridge in a housing for convenient mounting.

The probe tip and lever pivot on a frictionless, double-leaf suspension. The contact tips are usually 2.2 mm (0.090 in.) diameter, tungsten carbide balls, with interchangeable tips available as small as 500 mm (0.020 in.). To replace the probe when it is worn, simply unscrew it, and then replace it with a probe of the same length. A change in probe length changes the geometry of the head, and the meter would no longer read probe displacement directly.

Standardized probe head mountings are covered by an AGD specification, as discussed with indicators. You can use the probe head in a variety of positions (see Figure 14–16) if you remember that, although some reversible heads are available, the majority of probe heads measure in only one direction (see Figure 14–17).

Cosine Error. We first discussed cosine error with indicator stands (see Figure 10-51 , but it affects probe head measurements, too. As the angle increases, the amount of vertical movement required to encompass the same arc increases (see Figure 14–18). If we assume that the signal generated by the head is proportional to the angle, we also must take into consideration the effect of the angle on the probe to determine the amount of cosine error (see Figure 14–19), your lab manual has measurement exercises that will help you understand and evaluate cosine error.

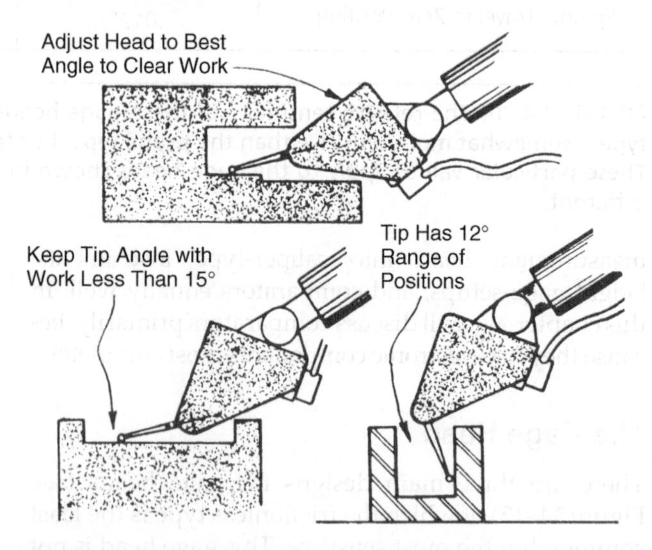

FIGURE 14–16 Many positions are possible with the probe head.

In Figure 14–20, a plus error in the instrument results in a plus error in the reading, but a minus error in the actual part size. The reverse is also true.

As we study cosine error, we should also reemphasize one point: A comparator's greatest accuracy is at zero. Whenever possible, use a probe as a comparator, not as a direct-measurement instrument. In some cases, the angle in which you have to place the probe will produce a large cosine error. When you zero the comparator close to the actual part size, you minimize the cosine error in the reading, which you can reduce even further by using the *10-to-1 rule* (see Figure 14–21).

The steps that we recommend for minimizing cosine error can be used to reduce other types of error, too. As a general rule, you should keep the probe angle as close to zero as possible and never greater than 15° in either direction.

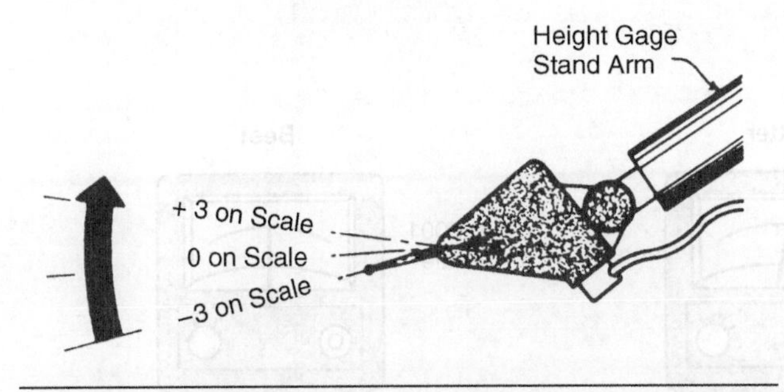

FIGURE 14–17 Although some probe heads have reversible action, most do not. The direction of movement must be remembered. To operate in the opposite direction, simply turn the head 180°.

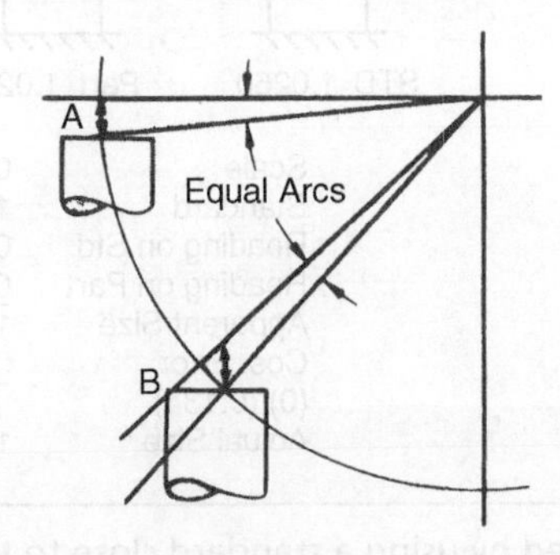

FIGURE 14–18 It takes considerably greater travel of the contact in B to produce the same arc that was produced by the travel of the contact in A.

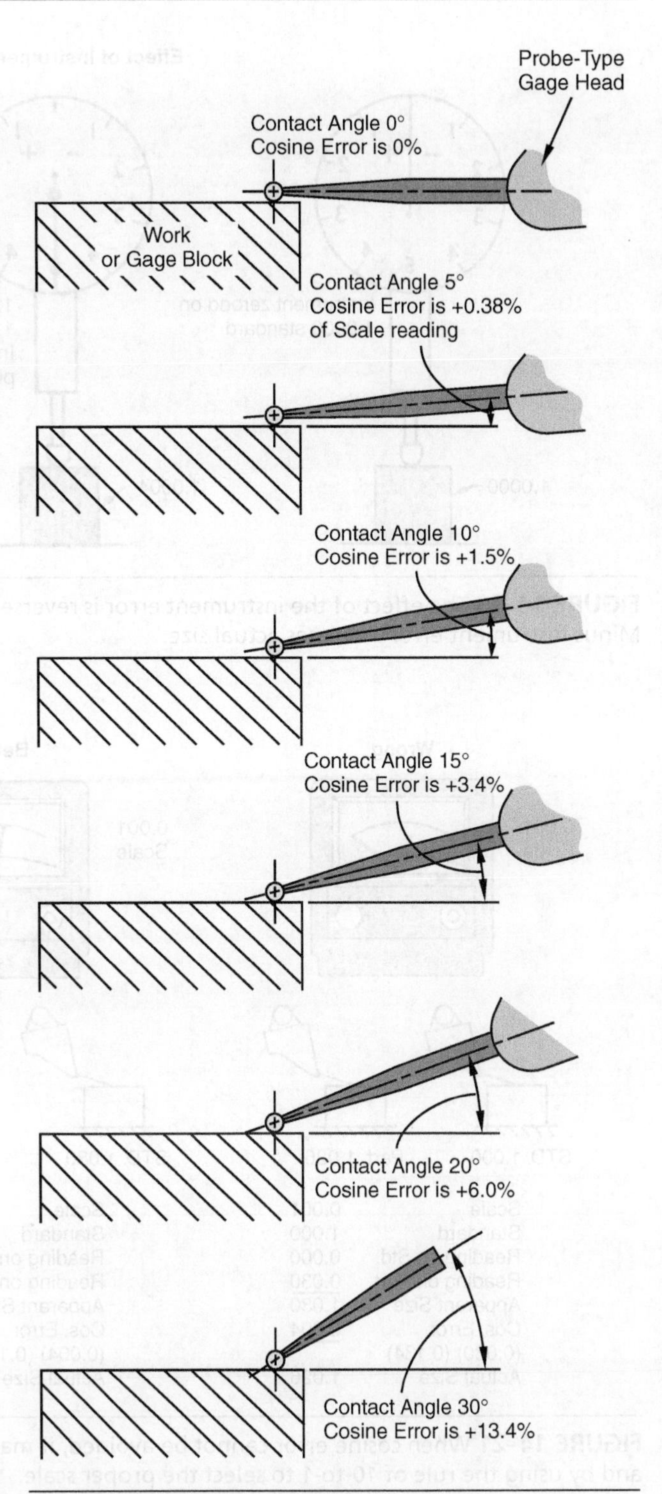

FIGURE 14–19 The cosine error is dependent on the probe position.

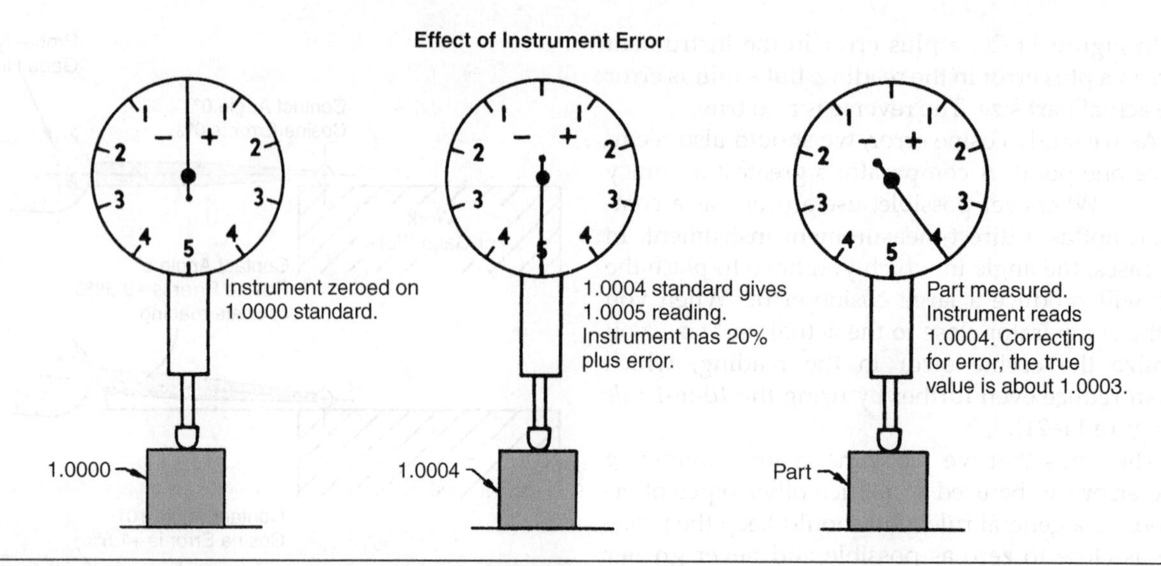

FIGURE 14–20 The effect of the instrument error is reversed on the part. Instrument error also decreases actual size. Minus instrument error increases actual size.

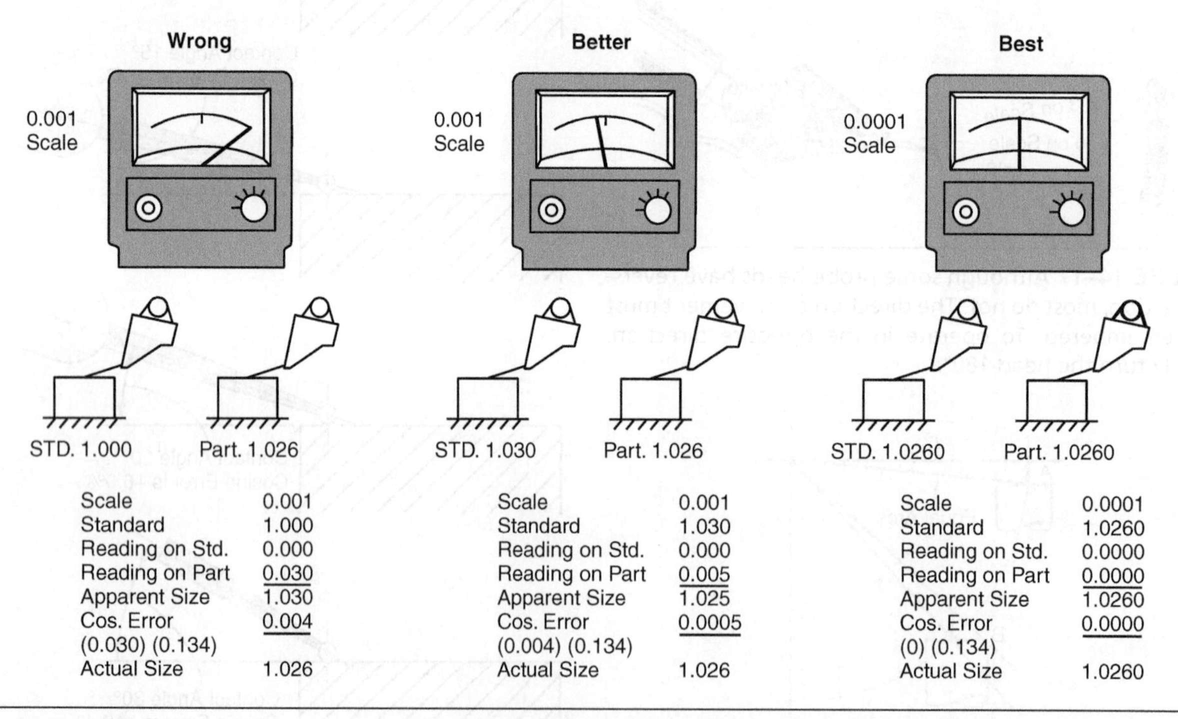

FIGURE 14–21 When cosine error cannot be avoided, it may be minimized by using a standard close to the part size and by using the rule of 10-to-1 to select the proper scale.

Gaging Force. You need to remember the distinction we made between *force* and *pressure* when we discussed micrometer ratchet stops (see Figure 7–26). Sometimes, when people say "gaging pressure," they mean it. More often, however, they mean "gaging force."

We need to reemphasize that you must keep gaging force low. You must use enough force to ensure positive contact between the instrument contact and the part, but when you make measurements to 2.5 μm (0.0001 in.) or finer, adding even a few grams of force will cause measurable distortion, bending, and/or compression. Instruments are designed so that you can adjust the force that is applied to the tip of the probe head. This force is proportional to the displacement: When you decrease the force, you also decrease the displacement. You can set the instrument for minimum force or maximum displacement (see Figure 14–22).

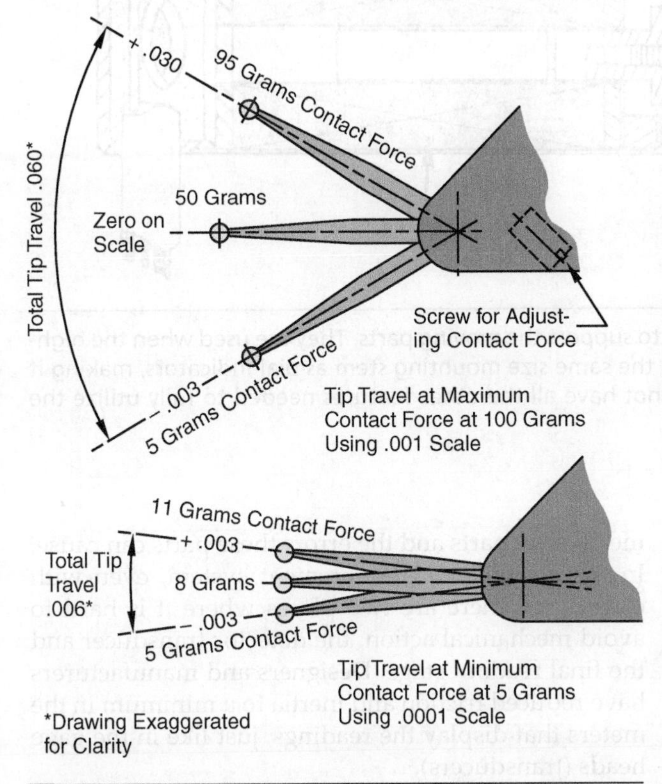

FIGURE 14–22 The probe head may be set either for maximum travel or minimum gaging force.

Normally, the head is factory adjusted for maximum displacement. If you are adjusting the instrument for minimum force, set the comparator on the desired scale. Then, with the probe not deflected, back off the adjusting screw until the hand is just off the scale on the minus side.

Frictionless Gage Head

The frictionless gage head is the most sensitive gage head that uses LVDT transducers. We use it most often for calibration because this head is not as convenient as probe and cartridge heads. All the moving parts are in one continuous axis (see Figure 14–23) supported by two parallel flexure reeds.

The construction of the frictionless gage eliminates the need for linear bearings at the same time that it ensures that the spindle axis is always normal to the displacing force. Movement away from the at-rest position causes the spindle to move sideways so that when the spindle travel is short, the side movement is slight. Because the total gap in the magnetic field remains the same, the electrical change is negligible.

Cylindrical Gage Head

The cylindrical head is almost as versatile as the probe head (see Figure 14–24). Because they have the same diameter as the AGD standardized stem of dial indicators, you can substitute a cylindrical head for a dial indicator in gages that were originally designed for dial indicators.

Originally, cylindrical heads used a sliding action. The contact attached to the spindle moved the core in the LVDT, and a spring returned the moving parts. Cylindrical heads were intended only as an improved replacement for dial indicators. When we need to make comparator measurements to the highest precision, we use a frictionless head (see Figure 14–23). These heads completely conform to Abbe's law. In fact, their reliability comes from this conformity.

Although both the sliding-type cylindrical head and the reed-suspension frictionless head are still widely used, manufacturers have combined their advantages in the Bellville-type of suspension

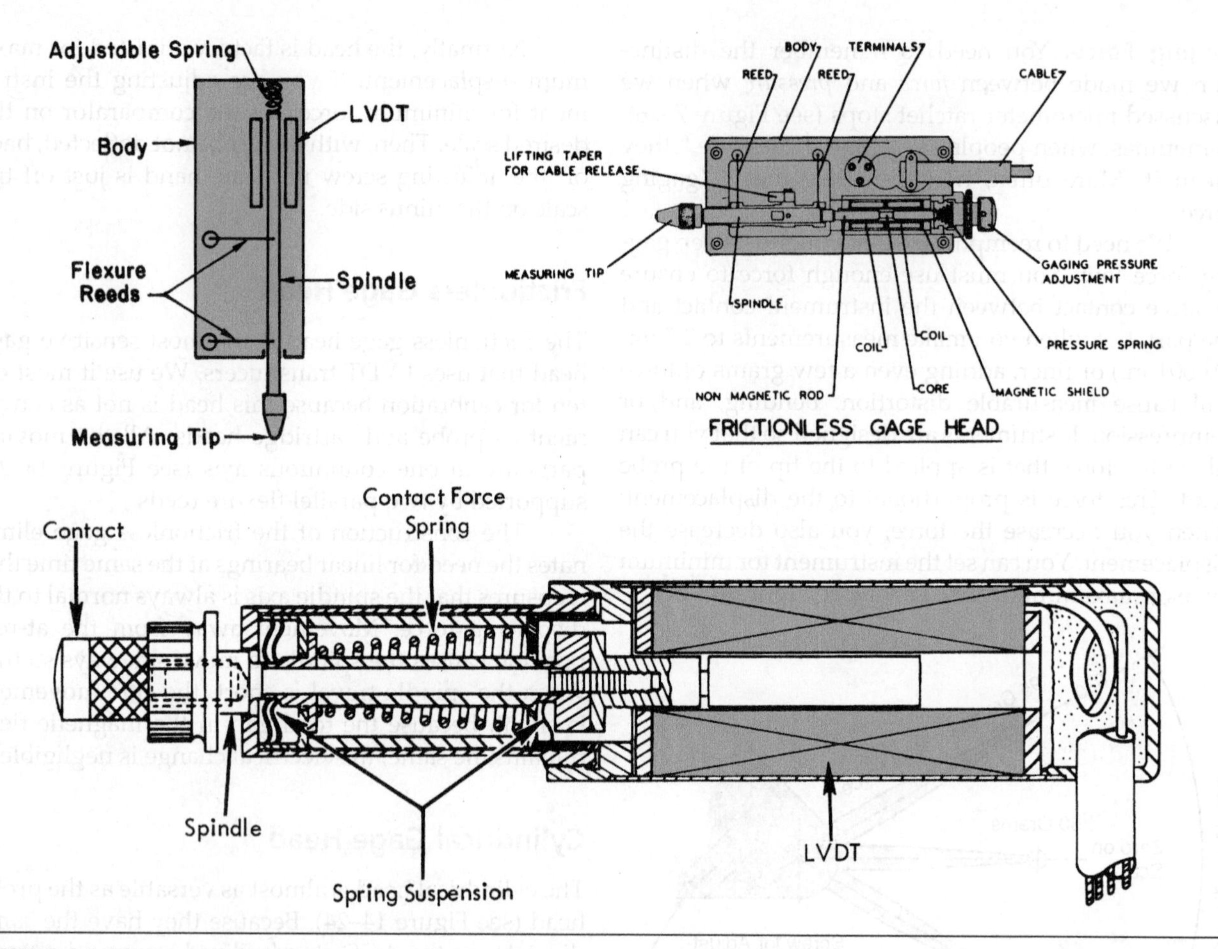

FIGURE 14–23 Frictionless gage heads use flexure springs to support the moving parts. They are used when the highest precision is required. The cylindrical type (bottom) has the same size mounting stem as dial indicators, making it interchangeable in indicator setups. However, these may not have all the other features needed to fully utilize the capability of the frictionless heads.

springs. For this spring, the gaging force is 6 ounces or 170 grams, which is greater than the force of the probe head but is nearly uniform throughout the travel of 1.65 mm (0.065 in.). The transducer components in the probe and cylindrical heads are the same; therefore, the heads are interchangeable.

Electronic Amplifier and Meter

The principal advantage of an electronic measurement system is the minimization or even absence of mechanical parts and the errors these parts can cause. In the generalized measurement system, even with electronics, there are two places where it is hard to avoid mechanical action: the detector transducer and the final readout stage. Designers and manufacturers have reduced friction and inertia to a minimum in the meters that display the readings, just like in the gage heads (transducers).

In our illustrations, so far, we have shown a traditional readout, which is a sensitive meter. Its moving element is a tiny armature connected to the long,

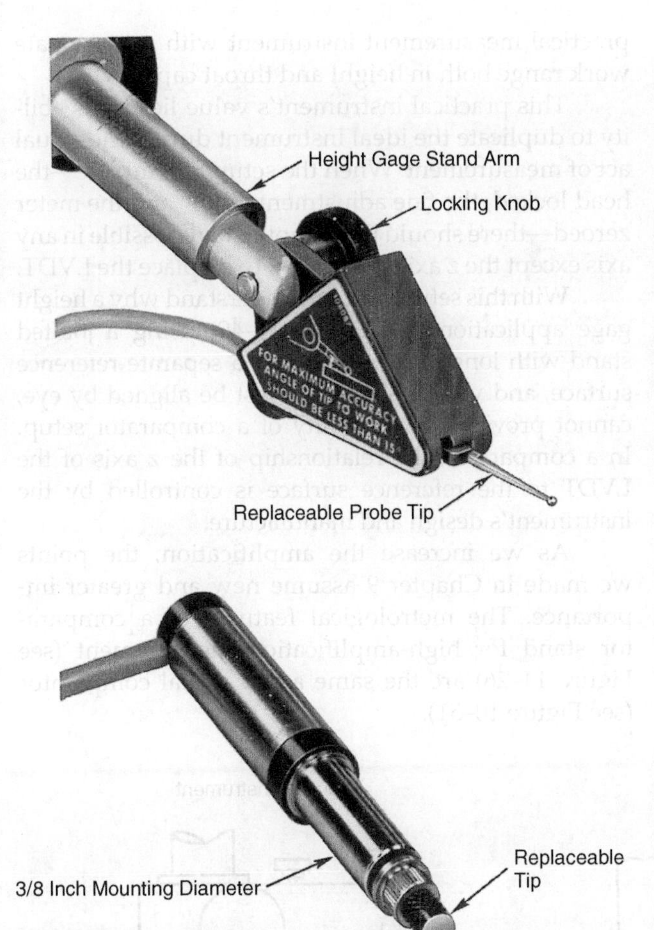

FIGURE 14–24 The probe-type head (top) and cylindrical head (bottom) fulfill nearly all requirements.

slender hand. Traditionally, such meters have used jeweled bearings, like the ones used in wristwatches.

Manufacturers sometimes use a torque band—a taut, steel band held between two end blocks—to reduce friction even more, and this band is attached to the armature and hand assembly. The amount of twist to the band is proportional to the magnitude of the electrical signal acting on the armature.

The torque band serves several functions:

- It eliminates the friction of a jeweled movement.
- It reduces the inertia of moving parts by substituting a thin band for the shaft.
- It eliminates the hairspring and its attachment arrangements.
- It provides a resilient support that is nearly insensitive to shock.

Even though you should always be careful when using precision instruments, in practical industrial use, these instruments suffer more abuse than they ever receive in the testing laboratory.

Electronic displays also help eliminate friction, which we will discuss later. You need to understand the principles of the mechanical instruments, because the same principles apply to fully electronic instruments.

Stands for Electronic Comparators

For most high-amplification measurement instruments, all three elements of the generalized measurement system are combined into one unit, restricting them to comparator-type measurements only. In the electronic systems, the detector-transducer is independent of the remainder of the system, which allows you to use them in fixed production gages and for both comparator and height gage applications (discussed with surface plate measurements).

You can make the most complete use of the high amplification of electronic instruments when you use them as comparators, but the term *comparator* needs to be clarified before we continue our discussion. Although all of the dial indicators we have discussed thus far are used for comparison measurement (and thus could be called comparators), *comparator* usually means a rigid-frame instrument that furnishes both reference and measured points. When the head of an electronic comparator is installed in a loosely jointed stand that rides on a reference surface, it is known as a height-gage setup, or height-gage instrument.

The higher the amplification of the instrument, the more critical the positional relationships become. In an ideal situation (see Figure 14–25A), the standard would be very nearly the same length as the part, would also have the same shape, and would be constructed of the same material. The instrument would hold the standard and the part in the same position and would have no movable members, making the only possible movement the displacement of the LVDT along the line of measurement, or z axis.

This setup would have limited use and would also be difficult to arrange. To create a more usable setup, we could add a fine adjustment of the head (see Figure 14–25B), which is an extension of the z axis movement. Immediately, the addition creates a chance for positional and other errors and lowers the reliability of our measurements. Next, we could arrange the head so the instrument would have a greater range of measurement (see Figure 14–25C). Although this change greatly extends the z axis, it also greatly reduces the reliability—for example, the stop on the frame that locates the parts and standard in A and B could no longer be used when the parts differ widely in height.

Finally, we could make an arrangement that recognizes that the standard and the parts will rarely be the same shape and that no one arrangement for the part support will give adequate versatility. Figure 14–25D has moved away from our original ideal, but it is a practical measurement instrument with an adequate work range both in height and throat capacity.

This practical instrument's value lies in its ability to duplicate the ideal instrument during the actual act of measurement. When the setup is complete—the head locked, the fine adjustment made, and the meter zeroed—there should be no movement possible in any axis except the z axis movement to displace the LVDT.

With this setup, you can understand why a height gage application (see Figure 10–49) using a jointed stand with long arms, relying on a separate reference surface, and whose elements must be aligned by eye, cannot provide the reliability of a comparator setup. In a comparator, the relationship of the z axis of the LVDT to the reference surface is controlled by the instrument's design and manufacture.

As we increase the amplification, the points we made in Chapter 9 assume new and greater importance. The metrological features of a comparator stand for high-amplification measurement (see Figure 14–26) are the same as for a dial comparator (see Figure 10–51).

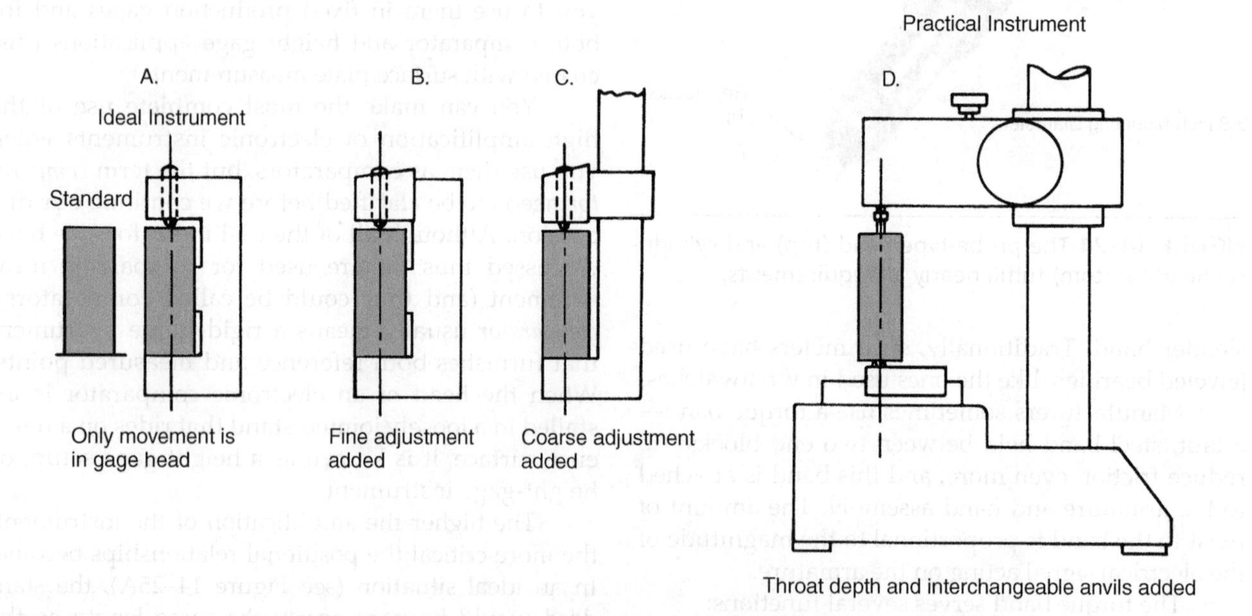

FIGURE 14–25 The theoretically ideal instrument gives up reliability to gain range and versatility.

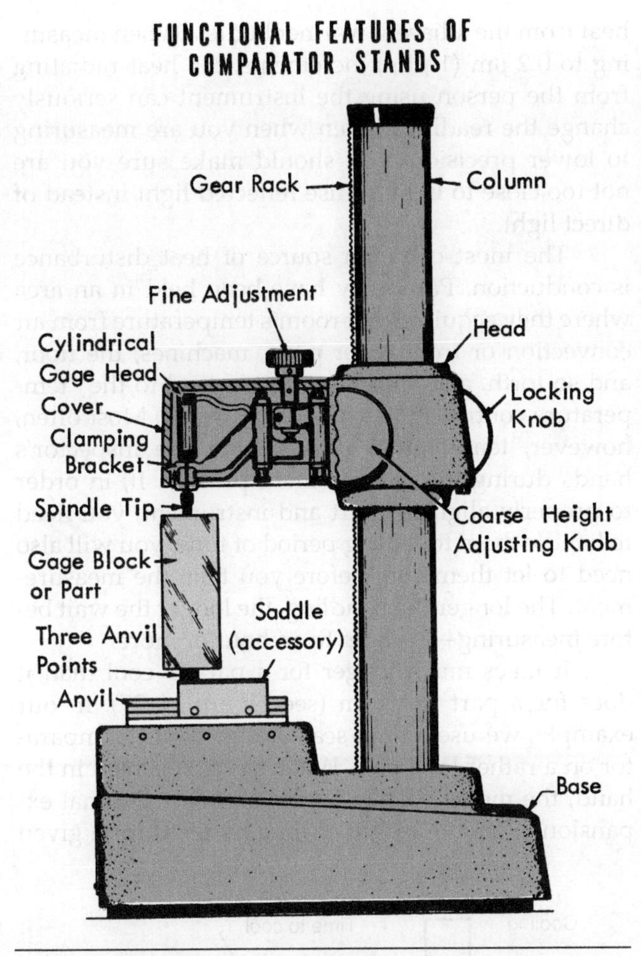

FIGURE 14–26 High-amplification comparators require stands that combine rigidity with ease of operation.

Although construction details vary, they must combine three qualities:

- The stand must provide positional relationships sufficiently accurate for the measurement to be performed.
- It must be sufficiently rigid that the ideal instrument is approximated during measurement.
- It must be readily adjustable so that its part size may be changed quickly and without danger of damage.

It is not easy to meet these requirements. For example, at high amplification on a 0.25 μm (10 μin.) scale, the instrument will commonly be swung to the out-of-scale range when the head clamp is tightened. Instruments have been manufactured to overcome this problem and vary in design.

Reference Surfaces. Although dial comparators often use a simple, flat platen as the reference surface (see Figure 10–50), these platens are not suitable for high-amplification comparators, which usually use separate anvils that are attached to the stands.

A typical anvil (see Figure 14–26) is a hardened, ground, and precisely lapped steel block, 32 mm wide × 38 mm high × 108 mm long (1¼ in. wide × ½ in. high × 4¼ in. long). It is held in place by set screws that bear against the V-slots along the anvil's sides, firmly forcing the bottom of the anvil into contact with the stand.

The top and bottom surfaces—one flat, the other serrated—are parallel within 60.25 μm (610 μin.) per 25 mm (1 in.) of the length and width. You can use either surface as the reference surface, but you should use the serrated side for flat parts because of air film (discussed further in Chapter 9). Use the flat side for cylindrical, spherical, or other parts that present line or point contact with the reference surface.

Effect of Temperature. We need to reemphasize two axioms for reliable measurement: You can only make reliable measurements if measurement conditions are as clean as possible; and you must maintain the correct positional relationship between the instrument and the part. If you cannot follow these rules, you cannot measure reliably. A third rule you should follow is to use an instrument of much higher precision than the tolerance of the part, preferably 10 times higher.

At this point, we need to add a fourth axiom: "keep cool." Nearly all engineering materials expand when their temperature rises. To complicate matters, they expand at very different rates—different coefficients of expansion. The coefficient of expansion is usually expressed as μin. per degree (1 μin. per degree) of temperature increase for 1 inch of material length. Similarly, the coefficient of linear expansion is defined as the change in length per unit of original length per degree change in temperature.

International standards are established for one temperature, 68 degrees Fahrenheit (68°F) or 20 degrees Centigrade (20°C). Metrologists perform the highest precision measurements in laboratories maintained at those temperatures.

In most industrial applications that require high-amplification measurement, manufacturers use steels, all of which have nearly the same coefficients of expansion. You must take one precaution: Make sure that all elements of the measurement are at the same temperature. At a precision of 2.5 μm (0.0001 in.), temperature can begin to generate errors, and when required precision reaches 1.2 μm (50 μin.) and finer, temperature can become a major problem. We must take steps to reduce the errors generated by inconsistencies in temperature, but first we must identify the ways heat is transmitted: by *convection, radiation,* and *conduction.*

Convection is the transfer of heat by air currents, so drafts can have an effect on precision measurement. You can shield your instruments and setups from drafts and reduce convection errors.

Radiation is the transference of heat by rays such as light, and every cool body receives radiant heat from the adjacent warmer bodies. When measuring to 0.2 μm (1 μin.) increments, the heat radiating from the person using the instrument can seriously change the readings. Even when you are measuring to lower precision, you should make sure you are not too close to light, or use reflected light instead of direct light.

The most common source of heat disturbance is conduction. Parts may have been held in an area where they acquired that room's temperature from air convection or from other parts, machines, the floor, and so forth, and were then transferred to the "temperature controlled" measurement room. Most often, however, temperature passes from the inspector's hands during the measurement process. If, in order to properly align the part and instrument, you need to handle them for a long period of time, you will also need to let them cool before you take the measurement. The longer the handling, the longer the wait before measuring—even up to an hour or more.

It takes much longer for a part to cool than it does for a part to warm (see Figure 14–27). In our example, we use a high scale and zero the comparator on a rather long part. When the part is held in the hand, the meter quickly begins showing thermal expansion. Note the length of time required for a given

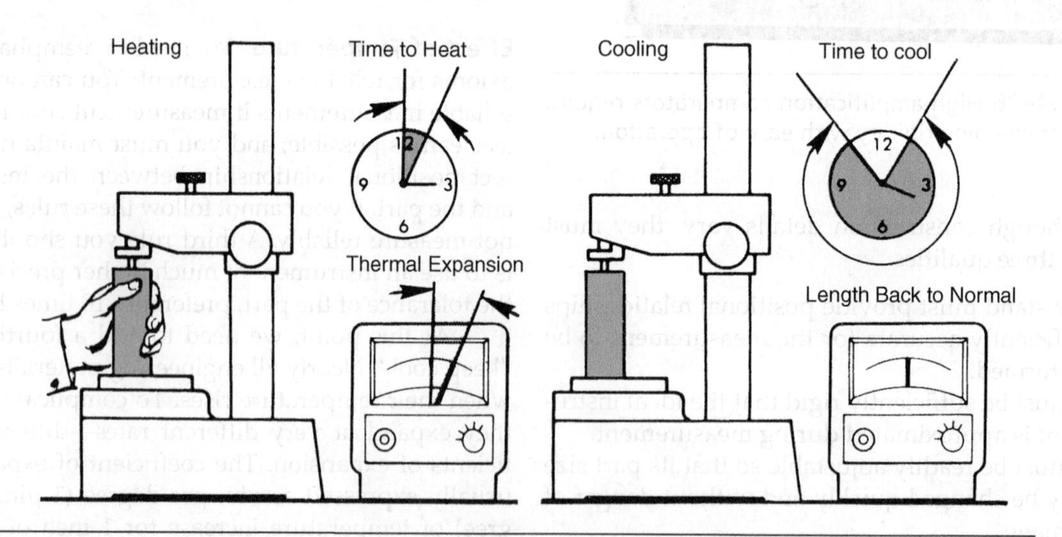

FIGURE 14–27 It takes much longer for a part to cool than to heat.

expansion, say 0.25 μm (10 μin.). When we release the part, it begins to normalize—returns to the *ambient temperature* or the same temperature as the instrument and the standard—but notice the amount of time it takes.

When you work with materials with different coefficients of expansion, all of these heat problems are increased. For example, a steel part 25.4 mm (1 in.) long will change 1.5 μm (60 μin.) with a 212.2°C (10°F) temperature change; a 25.4 mm (1 in.) brass part will change 2.3 μm (90 μin.); and an aluminum part will change 3.3 mm (130 μin.).

One way to speed the normalization of parts is to use a heat "sink." A heat sink is a relatively massive metal surface with a high rate of heat transfer and a good surface finish. It is located alongside the comparator, and we can place parts and standards on it before we measure them.

When problems arise, you should examine all the factors that can result in errors (see Figure 14–28), including heat considerations.

14-3 APPLICATIONS UNIQUE TO ELECTRONIC MEASUREMENT

Differential Measurement

Because electronics are so flexible, manufacturers can adapt the electronic comparators for uses that would be impossible for mechanical instruments. One example is the *differential comparator,* which provides measurement information from the algebraic sum of the signal from two pickup heads. Typically, differential comparators measure concentricity, roundness, parallelism, and similar part features with relatively high precision and without expensive special fixtures.

The amplifier for differential measurement is similar to ones we have discussed; however, this amplifier has places for two input channels and two additional switches to select the mode of operation. Typically, one switch is marked A, and the other is A-B. In the A position, only one channel is used; in the A-B position, both the A and B channels are used.

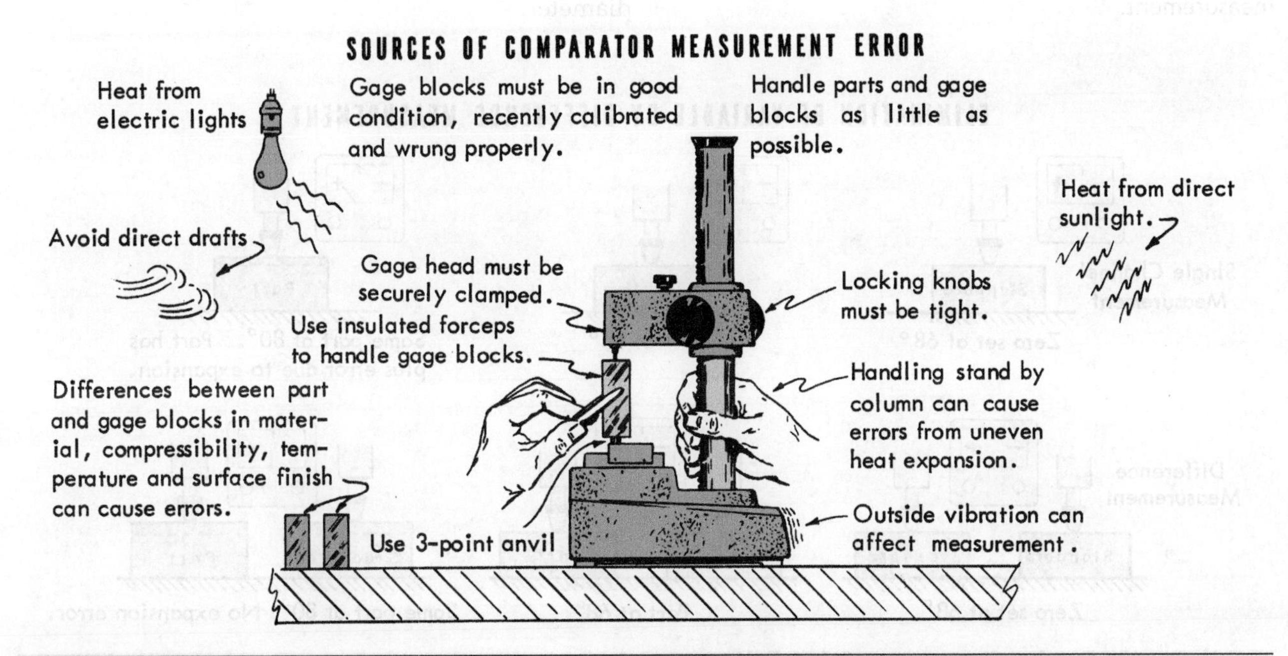

FIGURE 14–28 All sources of potential errors should be consciously investigated whenever a high-precision measurement is made.

The second switch controls the way the two channels are combined. When set at plus (+), the channels are combined as +A-B; and when at minus (−), they combine as −(A + B) or −A-B. Plus and minus indicate the direction of the pointer movement for a given movement of the contact in the gage head.

Difference Measurements

Generally in *difference measurements* (see Figure 14–29), the two gage heads (A and B) are aligned along parallel lines of measurements on the same side of the part. With the reversal switch at plus and the selector switch in the A position, you place a standard between head A and the reference surface, and the instrument is electronically zeroed. Then, the selector switch is turned to A-B, and the head B set to a standard. In this case, the instrument is zeroed mechanically so you do not disturb the adjustment of head A. You should recheck the settings before measurement begins, then replace either of the standards with the part to be measured. The reading is the difference between the two heads, or the amount that the part differs from the standard.

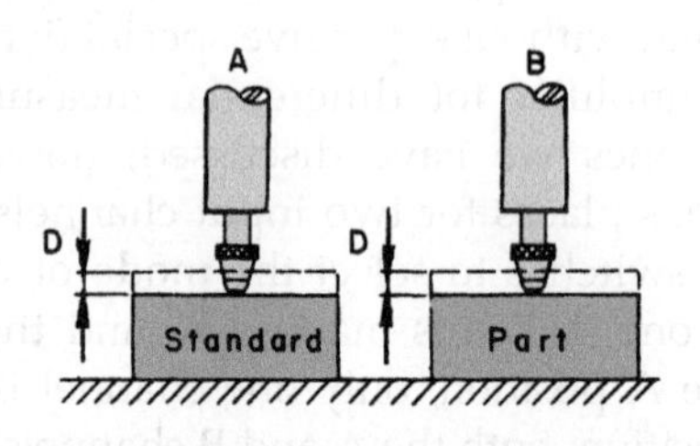

FIGURE 14–29 This is the general setup for difference measurement.

Of course, we could take the same measurement with an ordinary comparator setup, but this alternative measurement method minimizes temperature expansion and contraction (see Figure 14–30). We can also eliminate other errors, as in Figure 14–31. This setup measures concentricity between two surfaces, but we can make the measurement without worrying about all the variables that could affect the measurement if we only used one channel. Similarly, in Figure 14–32, we checked a taper without regard to diameter.

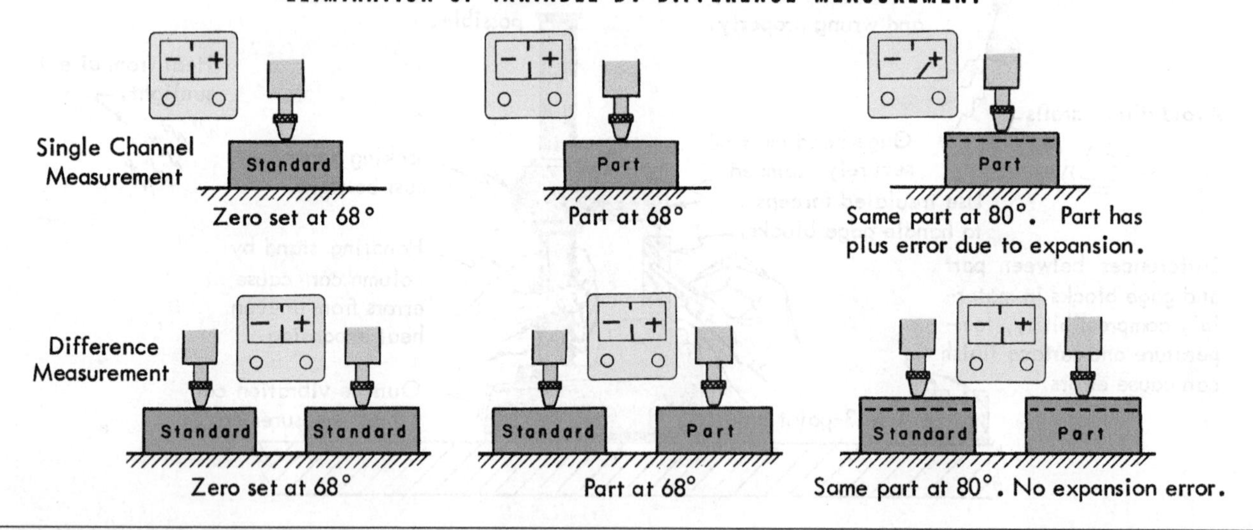

FIGURE 14–30 Thermal expansion provides an example of the elimination of a variable by means of difference measurement.

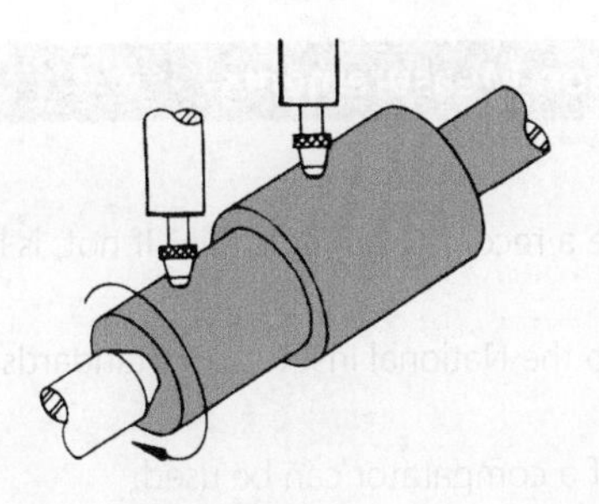

FIGURE 14–31 With the setup shown, only the difference in concentricity is shown on the indicator. If both parts are out of round in the same amount, the reading is not affected.

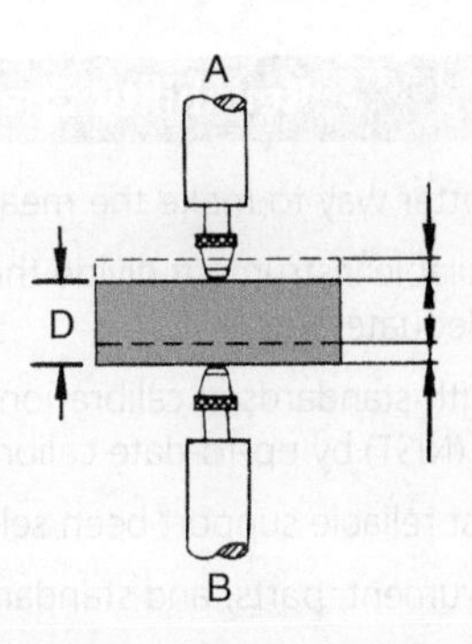

FIGURE 14–33 This is an example of sum measurement using opposing bands.

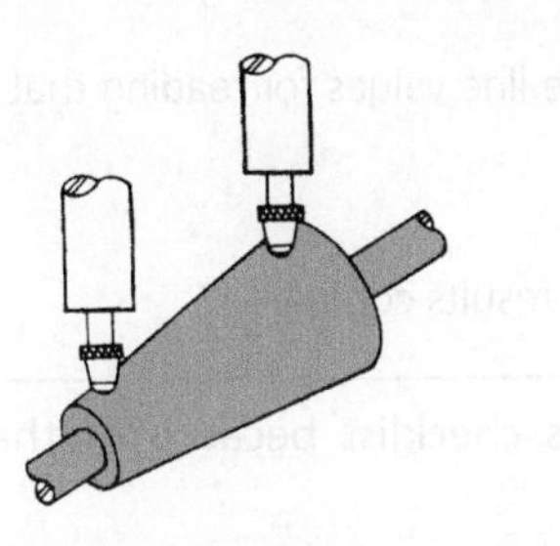

FIGURE 14–32 By placing two gage heads parallel on a tapered part, it is possible to check the degree of taper without regard to its diameter. The taper must be set to a master, of course.

Sum Measurements

In *sum measurements* (see Figure 14–33), you usually align the two gage heads (A and B) along the same line of measurement but on opposite sides of the part. You place the selector switch in the A position, as in the previous case, but you set the reversal switch at minus. You then zero head A electronically on the standard, turn the selector switch to A-B, and with the standard in place, mechanically zero the instrument again. After rechecking these settings, substitute the part for the standard. The pointer reads the sum of gage heads A and B.

You can move the part in this setup perpendicular to the line of measurement without changing the measurement (see Figure 14–33). The part feature (D) reads zero on the instrument, and if D moves up a distance x, then a 1x is added to head A and a 2x to head B. The net result is no change of reading. Figure 14–33 shows a sum measurement setup for comparing gage pins. In all of these examples, differential measurement is helpful, but it does not solve all measurement error problems. In the last example, centering error could exceed the scale range, and if the diameter is very small, the centering error could displace the line of measurement from the true diameter enough to cause a minus error. In addition, differential measurement requires extra steps, each of which can generate an extra error, so you need to follow all the general precautions for reliable comparison measurement (see Figure 14–34) and be aware of all the potential sources of error.

14–4 METROLOGICAL ADVANTAGES OF MULTIPLE SCALES

An instrument with several scales is versatile and can take the place of a number of separate instruments (see Figure 14–35) for comparator applications.

The multiple-scale feature saves in two ways that are more important than equipment costs: It simplifies the process of getting into scale range when you are using a high-amplification instrument; and it lets you choose the best scale for every measurement.

Reliability Checklist for Reliable Comparison Measurements

1. Is there a better way to make the measurement?
2. Can the available instrument divide the tolerance into 10 parts? Is there a record to support this? If not, is lower precision adequate?
3. Are the length standards in calibration i.e., is their accuracy traceable to the National Institute of Standards and Technology (NIST) by up-to-date calibration records?
4. Has the most reliable support been selected? Don't use a height gage if a comparator can be used.
5. Are the instrument, parts, and standards scrupulously clean?
6. Are all parts of the setup locked and secured to eliminate all movement except displacement in the gage head?
7. Has the environment been checked for drafts, direct light, vibration, and other error-causing disturbances?
8. Have the instrument, parts standards, work holders, and reference surfaces (whichever apply) been fully normalized?
9. Has the best scale been selected (electronic instruments only) and are the line values for reading that scale understood?
10. Has the measurement been repeated as a check?
11. If critical, has the measurement been repeated by someone else and the results compared?

FIGURE 14–34 Temperature considerations have been added to this checklist because of their importance in high-amplification measurement.

As mentioned in Chapter 10, dial indicators are available with a discrimination of 2 mm (50 millionths), but they are not considered to be general application instruments, in part because it is difficult to zero set them. Even with a fine adjustment, you can pass the entire scale range so quickly that it is impossible to tell whether the hand has turned part of a turn, one turn, two turns, or more (see Figure 14–36). With a multiscale instrument (see Figure 14–37), you can easily use the lower scales to find the higher scales you need.

You *always* need to use the most appropriate scale for the measurement. The reasons why this statement is true will also help explain our insistence on a 10-to-1 ratio, as mentioned earlier.

In Figure 14–38, we have enlarged a portion of the scale of a comparator with a discrimination of 2.5 µin. (100 µin.) and used it to measure a part with a tolerance of 5.0 min. (200 min.). However, this comparator can only divide the tolerance of the part into two parts.

When we zero set the hand, the gray area represents the *uncertainty* of the instrument. Uncertainty means *repeat error* or dispersion of repeated readings, based on *sigma 3.* In statistical quality control, sigma 3 means that over 99% of all readings will fall into the zone. In an actual instrument, the zone of uncertainty should not be more than one fourth of a division, which you can see more clearly in Figure 14–38. When the zone of uncertainty is within the tolerance of the part, it does not create problems. At B, for example, the part could be on the low tolerance or on the basic dimension, but it is within tolerance.

When we use the tolerance as read on the scale to evaluate parts (see Figure 14–39), the hand is on the low limit (A) of the tolerance. The arrow shows that actual part size is well below that limit: the part would be bad, but it would be passed as good. In example B, the hand is in the same position, but the part is well within tolerance, and example C shows an extremely out-of-tolerance part that would pass as well within tolerance. Although these cases have been at the low

Typical Scale Selections		
Scale	**Comparator Applications**	**Height Gage Applications**
0.02 mm 0.001 inch	*Production:* appliance parts, agricultural machinery parts, builders' hardware parts, plastic parts, small castings, and small forgings. *Tool and Gage Inspection:* not recommended.	*Plate Inspection:* heavy equipment parts, marine equipment parts, heavy engine parts, forging, and castings. *Tool and Gage Inspection:* assembly jigs and fixtures, patterns and templates. *Setup Measurements:* large planer and boring millwork.
0.01 mm 0.0005 inch	*Production Inspection:* truck and automotive parts, motor shafts, machinery shafts, bushings, bearings (not antifriction), small gears, and precision hardware. *Tool and Gage Inspection:* wide tolerance production gages and small cutting tools.	*Plate Inspection:* machinery and machine tool housings and parts, engine parts, motor, and generator parts. *Tool and Gage Inspection:* machining and inspection jigs and fixtures, precision assembly jigs and fixtures, templates, cams, and large cutting tools. *Setup Measurements:* milling machine and boring millwork, table positioning, and rough surface grinding.
0.002 mm 0.0001 inch	*Production Inspection:* high-production gages and high-speed engine parts, pump parts, small precision gears, and firearm parts. *Tool and Gage Inspection:* production gages and precision cutting tools.	*Plate Inspection:* plastic and injection molds, dies, precision machining and inspection jigs and fixtures, aircraft parts, and large instruments. *Tool and Gage Inspection:* precision machining and grinding. *Setup Measurements:* finish surface grinding, chemical process setup.
0.001 mm 0.00005 inch	*Production Inspection:* instrument and control parts, electronic components, and antifriction bearings. *Tool and Gage Inspection:* close tolerance production gages and master gages.	*Plate Inspection:* precision aircraft and missile parts. *Tool and Gage Inspection:* precision inspection fixtures and instruments. *Setup Measurements:* jig borer, measuring machine, and precision boring machine work.
0.0002 mm 0.00001 inch	*Production Inspection:* high-precision hydraulic and electric parts. *Tool and Gage Inspection:* height accuracy master gages.	*Plate Inspection:* not generally applicable. *Tool and Gage Inspection:* not generally applicable. *Setup Measurements:* not generally applicable.

FIGURE 14–35 There are electrical comparators with a wide range of scales. Most will be between those shown in this table. It is essential to use the best scale in the range of the instrument for each particular situation.

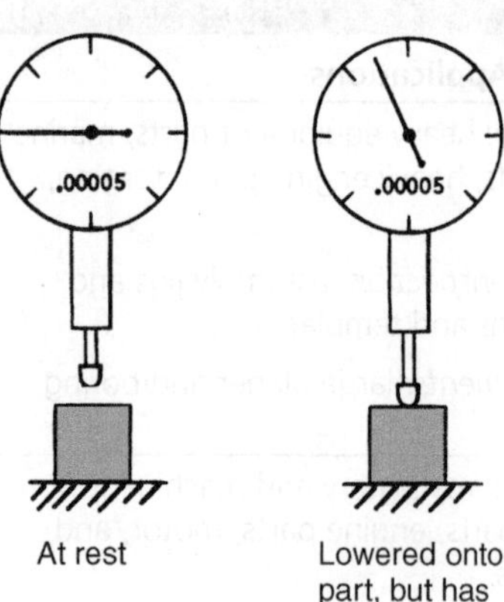

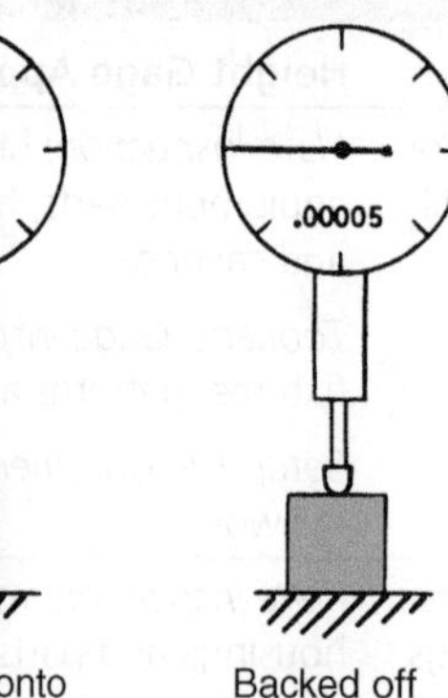

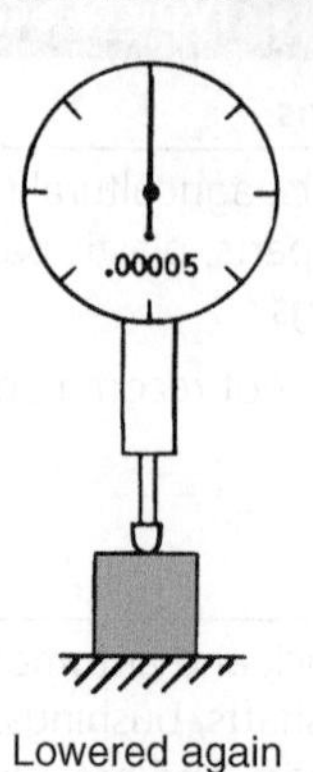

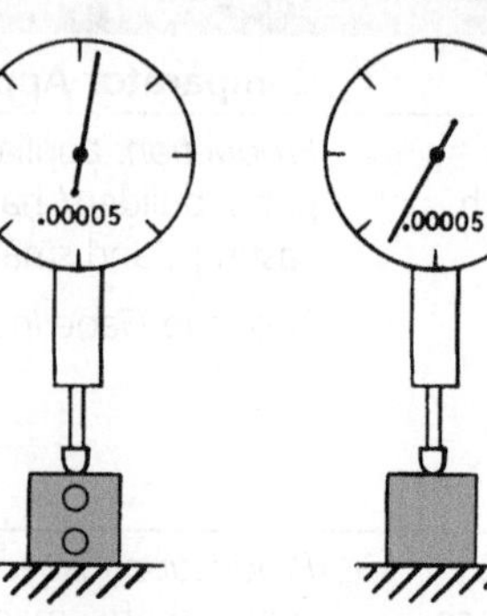

FIGURE 14–36 The problems of zero setting a high-amplification comparator that has only one scale are often even more difficult than this example.

FIGURE 14–37 Multiple-scale selection provides the best amplification for each role in which the instrument is used.

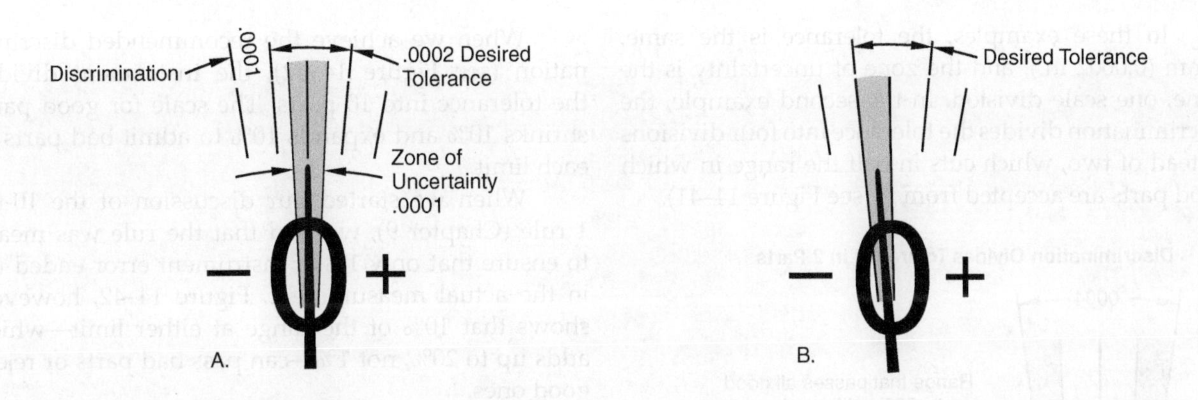

FIGURE 14–38 The dispersion of readings does not cause a problem as long as it is entirely within the tolerance.

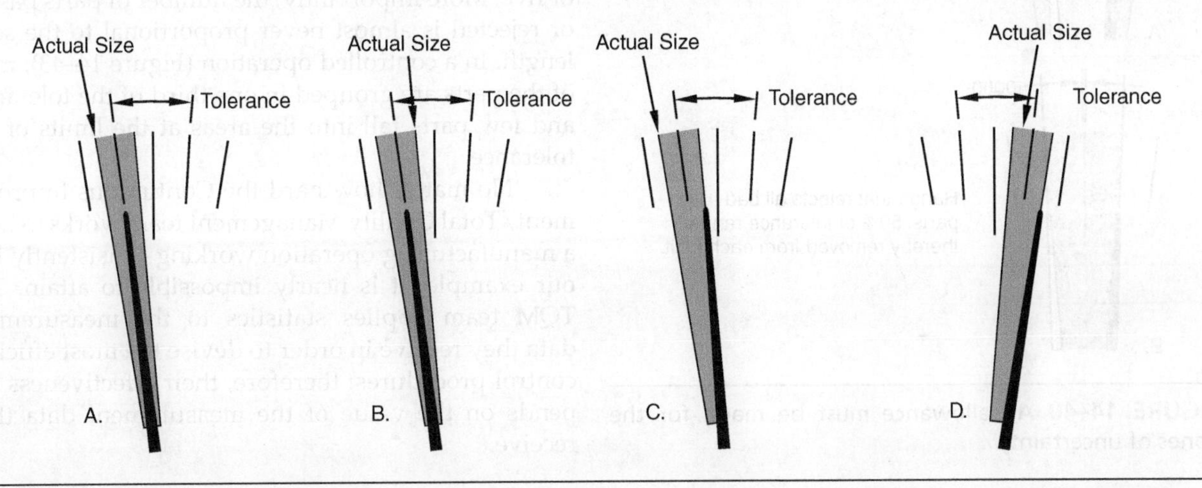

FIGURE 14–39 These examples show that the apparent reading may be very different from the actual size as a result of the instrument uncertainty.

limit, they could all be repeated at the high limit. For instance, example D shows a good part being rejected as being past the high-tolerance limit.

In order to achieve reliable measurement, we must compensate for the zone of uncertainty. The objectives may be not to reject any good parts or not to accept any bad parts, and the results are shown in Figure 14–40. The reading must be double the tolerance (A) to push the zone of uncertainty out of the tolerance range, if we are able to accept all good parts. If we are not taking any chances on the tolerance, we should only accept parts that fall between the zones of uncertainty. In this case (B), no parts pass, so we need a finer discrimination.

In these examples, the tolerance is the same, 5 mm (0.0002 in.), and the zone of uncertainty is the same, one scale division. In the second example, the discrimination divides the tolerance into four divisions instead of two, which cuts in half the range in which good parts are accepted from A (see Figure 14–41).

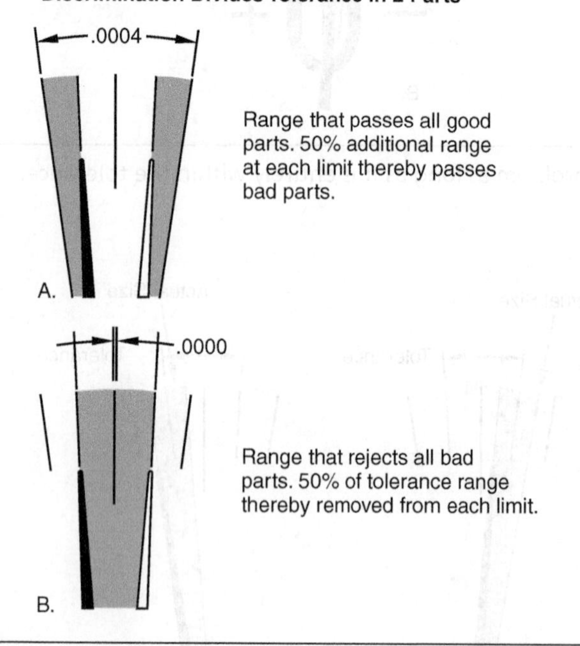

FIGURE 14–40 An allowance must be made for the zones of uncertainty.

When we achieve the recommended discrimination (see Figure 14–42), the instrument divides the tolerance into 10 parts. The scale for good parts shrinks 10% and expands 10% to admit bad parts at each limit.

When we started our discussion of the 10-to-1 rule (Chapter 9), we said that the rule was meant to ensure that only 1% of instrument error ended up in the actual measurement. Figure 14–42, however, shows that 10% of the range at either limit—which adds up to 20%, not 1%—can pass bad parts or reject good ones.

To clarify the definition, we first drew the zone of uncertainty four to five times larger than we would normally expect, which drops the percentage to four or five. More importantly, the number of parts passed or rejected is almost never proportional to the scale length. In a controlled operation (Figure 14–43), most of the parts are grouped in one third of the tolerance, and few parts fall into the areas at the limits of the tolerance.

No matter how hard the Continuous Improvement/Total Quality Management team works to keep a manufacturing operation working consistently like our example, it is nearly impossible to attain. The TQM team applies statistics to the measurement data they receive in order to devise the most efficient control procedures; therefore, their effectiveness depends on the value of the measurement data they receive.

Discrimination Divides Tolerance in Four Parts

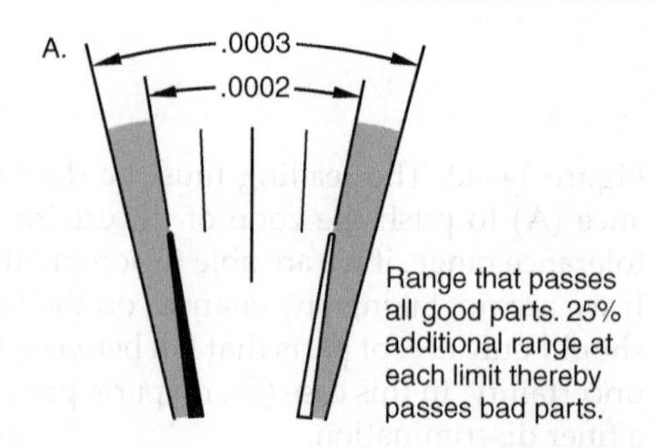

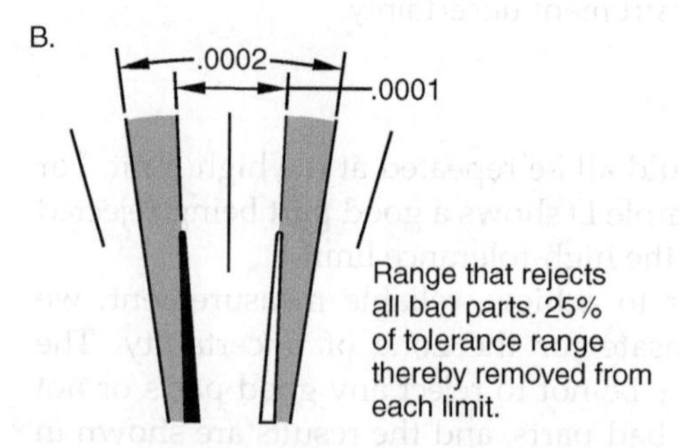

FIGURE 14–41 The higher discrimination reduces the lost range for good parts and the additional range for bad parts.

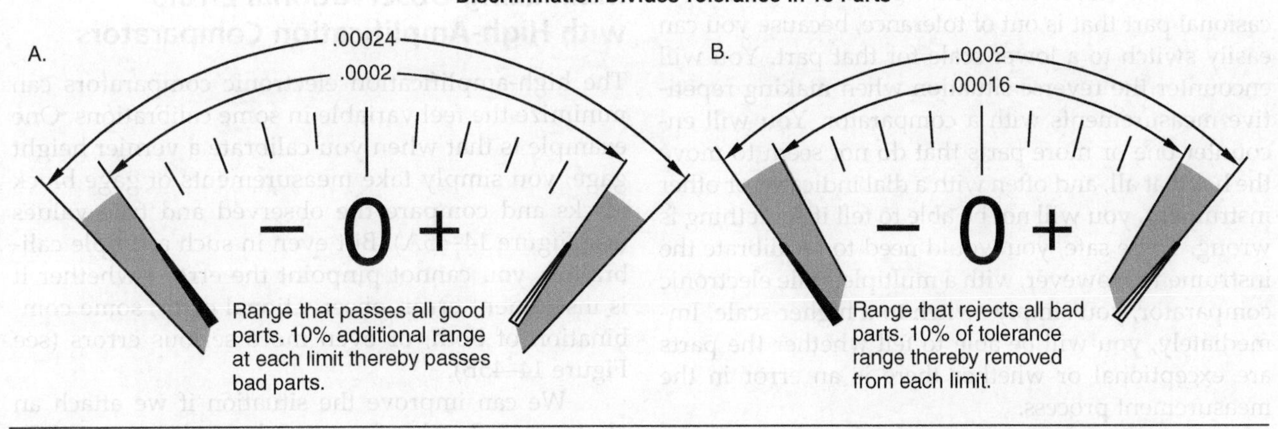

FIGURE 14–42 When the instrument has sufficient discrimination to divide the tolerance into 10 parts, only 10% of range gives trouble at each limit.

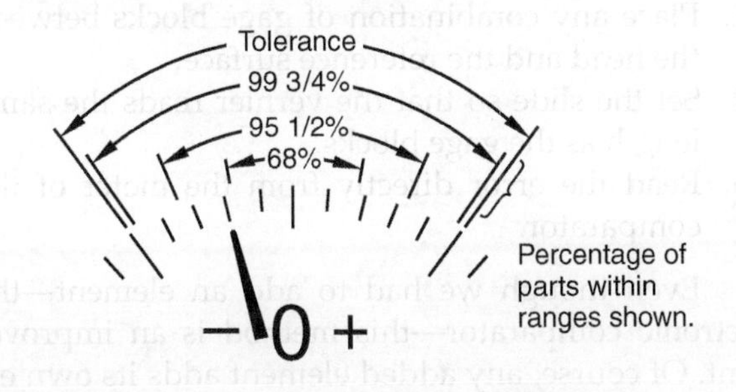

FIGURE 14–43 In a typical manufacturing operation, most of the parts will be grouped close to an average size. The purpose of quality control is to ensure that the average is close to the basic desired size.

Best Scale for the Measurement

Theoretically, you should always select the most sensitive scale that completely contains the tolerance of the dimension to be measured.

Although we cannot follow this rule with dial indicators, as demonstrated in Chapter 10, we can follow it with electronic comparators. For our example instruments, the three scales provide ranges from 76 μm to 7.6 μm (0.003 in. to 0.0003 in.), which cover most tolerances likely to occur in precision measurement. Similar instruments have scales with discrimination up to 0.025 μm (1 μin.), but because such scales must be used under laboratory conditions, these instruments are used primarily for gage calibration.

In one situation, you may not want the highest precision that will provide the necessary range: If an unskilled worker is performing routine production inspection, the great sensitivity of a high-amplification scale may create the impression that the instrument "has gone crazy" and cause the worker to "cramp" the instrument, try to adjust the instrument, or worse. Proper training is the best way to correct this problem, but as a temporary measure, the worker can sometimes use a lower scale.

Digital Instruments

High-amplification comparators are more vulnerable to use by unskilled workers when they are converted to digital readout, as are other self-contained systems from sensors to standards to displays. The manufacturer can display only the highest useful digit, and with an electronic system, the signal conditioning and display unit are independent of the sensor, standards, and setup. Except in the most precise applications, some displayed digits are meaningless, because they are beyond the reliability of the overall situation.

You do not need to allow overtravel for the occasional part that is out of tolerance, because you can easily switch to a lower scale for that part. You will encounter the reverse situation when making repetitive measurements with a comparator. You will encounter one or more parts that do not seem to move the hand at all, and often with a dial indicator or other instrument, you will not be able to tell if something is wrong. To be safe, you would need to recalibrate the instrument. However, with a multiple-scale electronic comparator, you simply switch to a higher scale. Immediately, you will be able to tell whether the parts are exceptional or whether there is an error in the measurement process.

As we discussed with the 10-to-1 rule (Chapter 9), the greater the precision of the instrument with respect to the tolerance, the smaller the zones of uncertainty at each of the limits. With a multiple-scale instrument, you can take full advantage of these smaller zones of uncertainty (see Figure 14–44).

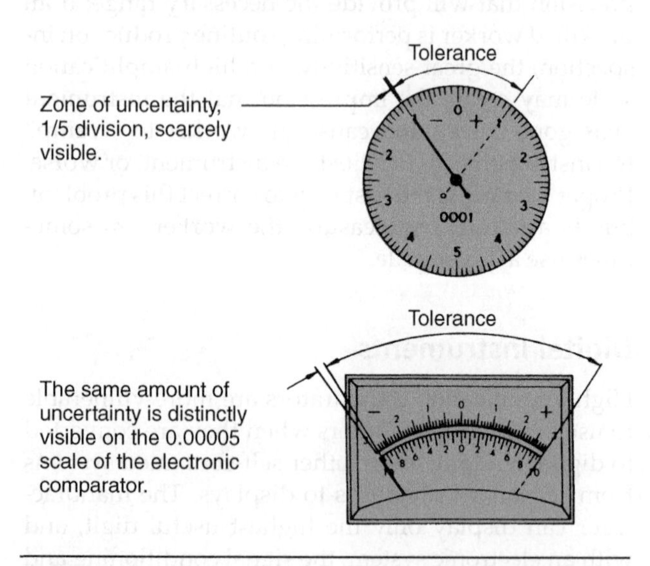

FIGURE 14–44 Both of these instruments fulfill the requirements of the 10-to-1 rule. The electronic gage provides readings in the uncertainty zone of the dial indicator.

Minimizing Observational Errors with High-Amplification Comparators

The high-amplification electronic comparators can minimize the feel variable in some calibrations. One example is that when you calibrate a vernier height gage, you simply take measurements of gage block stacks and compare the observed and true values (see Figure 14–45A). But even in such a simple calibration, you cannot pinpoint the error—whether it is instrument error, observational error, some combination of both, or even more serious errors (see Figure 14–45B).

We can improve the situation if we attach an electronic pickup head to the slide (see Figure 14–45C). We can easily zero the pickup against the reference surface. Calibration then follows these steps:

1. Place any combination of gage blocks between the head and the reference surface.
2. Set the slide so that the vernier reads the same length as the gage blocks.
3. Read the error directly from the meter of the comparator.

Even though we had to add an element—the electronic comparator—this method is an improvement. Of course, any added element adds its own errors, but if these additional errors are much smaller than the errors that the additional element eliminated, you should choose the new method. However, if the added errors are nearly equal to the errors eliminated, you should not use this method. We can recommend the use of high-amplification comparators, because they do eliminate significant errors.

We use the preceding method of calibration more often with dial indicators, and you can get favorable results if you use an indicator of sufficient discrimination and sensitivity. You will remember from the sensitivity section in Chapter 10 that this is not an easy combination to find. However, we do not lose sensitivity with increased discrimination in an electronic or pneumatic comparator. When you have a choice, you should therefore use an electronic or pneumatic comparator.

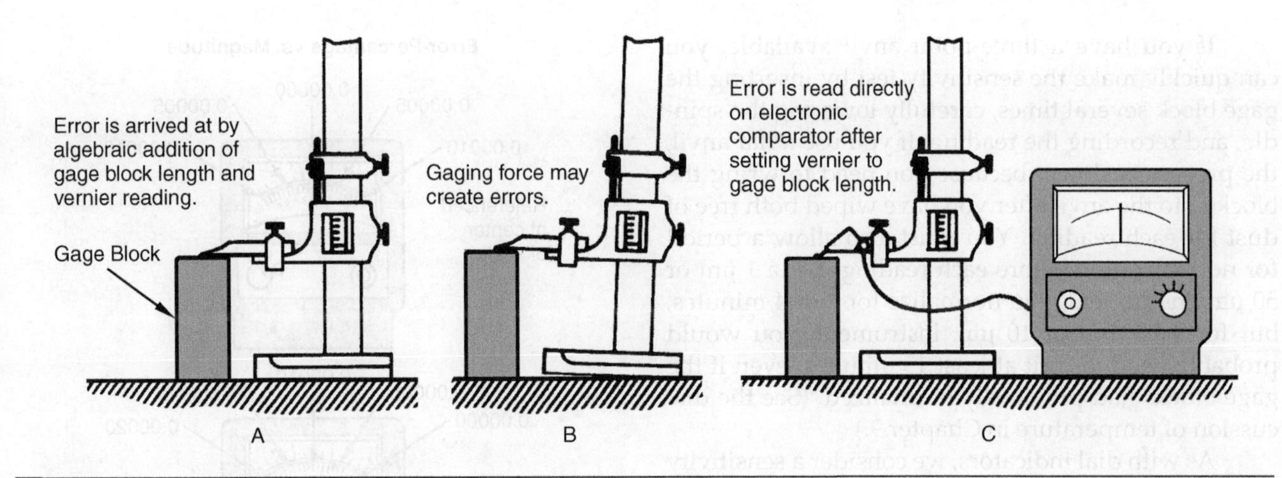

FIGURE 14–45 High-amplification comparators can be used in some calibration procedures to minimize errors caused by feel and eliminate computational errors.

Calibration of High-Amplification Comparators

You calibrate digital and analog high-amplification comparators the same way you calibrate dial indicators, but you must be more careful. You must be sure that the gage blocks are accurate enough for the instrument and have not been damaged since their last calibration. In these calibration situations, you must maintain cleanliness and try to control the effects of temperature as carefully as you control all the other steps in the calibration procedure. Finally, if you are calibrating the instrument, you need the following information about the positional relationships:

- They need to be known.
- They must be within adequate tolerances.
- They must be secure.

If you are calibrating for size differences from zero to 0.1 μm (0.000000 to 5 μin.), you *cannot* calibrate the instrument under shop conditions. Even though the gage blocks are accurate enough and the instrument has a discrimination of 0.1 μm (1 μin.), you cannot say that the instrument readings are accurate to thousandths of a millimeter or millionths of an inch. The instrument's readings are in ten thousandths, not millionths, and you most likely will not be able to trace the source of those ten thousandths. They might come from any of the complex assortment of intermingled variables.

If the calibration involves readings in the 0.0005 to 0.0010 range, you may be able to make them under ordinary shop conditions but not reliably, and it will be a slow, tedious process. For example, when you calibrate within this range in the gage room, you would use a heat sink to normalize the gage blocks only for the last hour. However, in the shop, you would need to normalize a single gage block and the instrument at least 20 minutes before making a reading, and at least one hour for two or more wrung blocks. Obviously, the time you would need for a careful calibration—and all the necessary readings—would be prohibitively time consuming.

If you are calibrating an electrically powered instrument, make sure it warms up at least as long as is recommended in the manufacturer's handbook before you make the test. Then, leave the instrument powered up to ensure its stability throughout the rest of the calibration process.

If you have a three-point anvil available, you can quickly make the sensitivity test by inserting the gage block several times, carefully lowering the spindle, and recording the reading. If you use a flat anvil, the process is slower because you need to wring the block(s) to the anvil after you have wiped both free of dust for each reading. You must also allow a period for normalization before each reading. For a 1 μm or 50 μin. instrument, you normalize for 3 or 4 minutes, but for a 0.2 μm or 10 μin. instrument, you would probably need to wait at least 15 minutes, even if the gage blocks had previously normalized. (See the discussion of temperature in Chapter 9.)

As with dial indicators, we consider a sensitivity of ¼ scale division satisfactory, but if you encounter a wider deviation between readings, do not pull the instrument from service immediately. You should check the variables that might add compensating errors, one by one, and determine which errors are causing some readings to be larger than the instrument sensitivity and some to be smaller.

Because the total range of most high-amplification instruments is less than one scale division of a dial indicator, their accuracy calibration is limited to successive 1 μm increments (10 μin.). The 1.001 to 1.009 mm gage blocks are helpful for high-amplification calibration.

We generally state the accuracy of high-amplification comparators as a percentage of "full-scale deflection," or the range. The percentage must be evaluated (see Figure 14–46) in terms of magnitude of error. Figure 14–47 shows how the calibration points vary for multiple-scale instruments.

You must calibrate the scales of multiple-scale electronic comparators separately, even though, at a glance, it might appear that if the highest amplification scale is calibrated, all the lower scales should also be within calibration. If you refer back to the generalized measurement system (see Figure 14–10), you see that the electronic comparator consists of several distinctly separate elements that work together; therefore, each of the elements is subject to its own errors—dependent, independent, and correlated. The total error for each element may also be dependent, independent, or correlated with respect to the other elements.

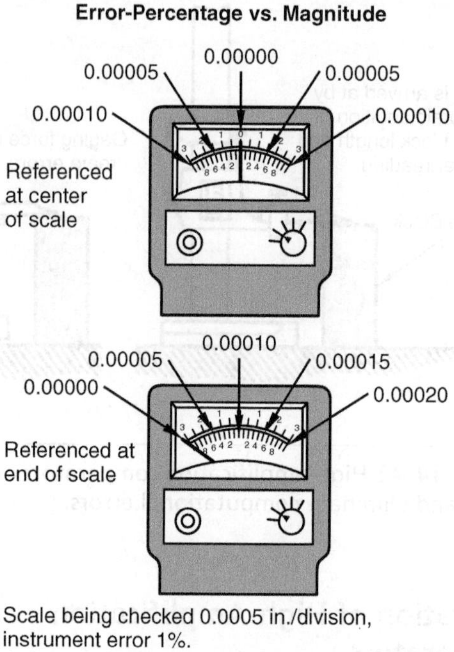

FIGURE 14–46 The percentage of error remains the same, but the magnitude varies with total distance traveled. In calibration, it is desirable to stretch out travel to find the errors (bottom), whereas in the use of the instrument the reverse is true (top).

To illustrate, assume that a gage (the head detector-transducer stage) has a massive error when its probe is between 5° and 5½° with the horizontal. If the amplifier (the intermediate-modifying stage) and the meter (the indicator-recorder-controller stage) are perfectly linear, the gage head error would react differently according to the amplification of the scale you select. Using the lowest scale, the error would appear as uncertainty in one portion of the scale only (see Figure 14–48A). This error would be random from the standpoint of the instrument, because it occurs only in one area with no pattern or predictability. From the standpoint of measurement with the instrument, it would be a systematic error, because you could always predict whether a measurement would or would not contain this error. If either the amplifier or the meter also contained errors, the gage head error would be independent and compensate.

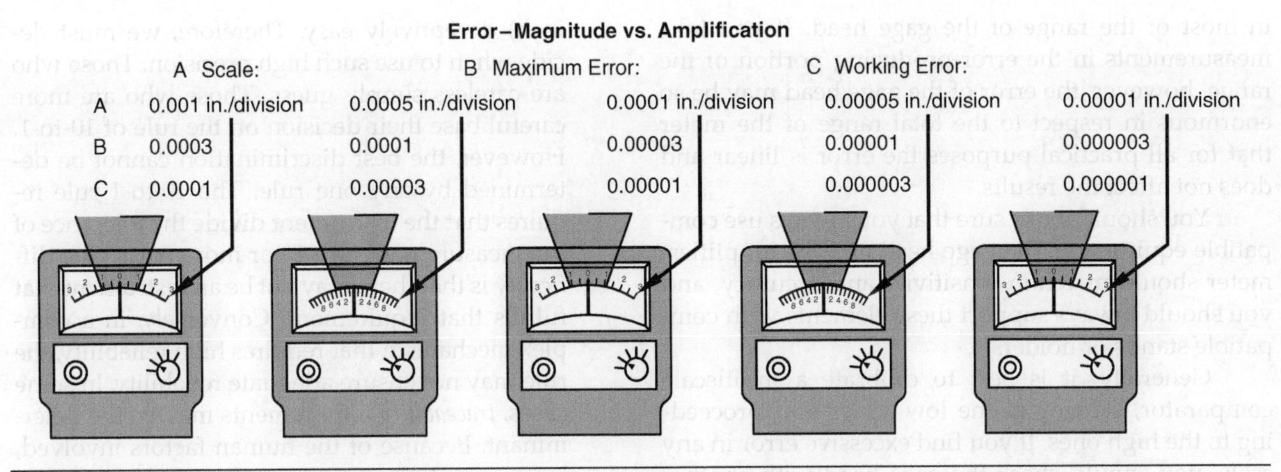

FIGURE 14–47 In a multiple-scale instrument, the percentage of error will be the same for all scales, but the magnitudes will vary in proportion to the amplification. Note that if measurement is restricted to one third of the total scale range (shaded portion), instrument capability exceeds general measuring capability in all scales except the lowest amplification.

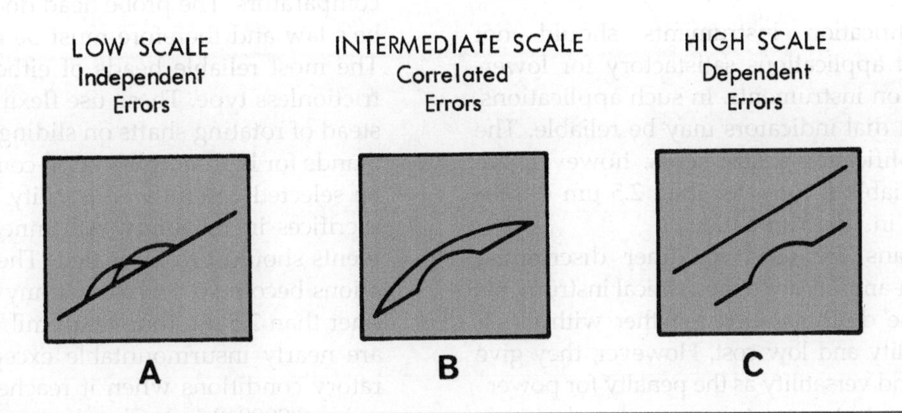

FIGURE 14–48 This example assumes an error in one portion of the gage head range, a linear amplifier and a meter with an error in one portion of its range. The relationship of these errors changes with the amplification of the system as shown.

On a higher-amplification scale, the gage head error can encompass the entire scale range (see Figure 14–48B). In this case, the error of the head is independent of the amplifier and meter errors, but each occurs on top of the others. These errors are independent because they compensate. If you take the measurement on either side of the gage head error, the error does not affect the amplifier or the meter error, and the errors are correlated.

Figure 14–48C shows the result of the gage head error at the high-amplification scales, where a dependent error condition now exists. You can avoid error in most of the range of the gage head. If you take measurements in the error-producing portion of the range, however, the error of the gage head may be so enormous in respect to the total range of the meter that for all practical purposes the error is linear and does not affect the results.

You should make sure that you always use compatible equipment. The gage head and the amplifier-meter should match in sensitivity and accuracy, and you should always support these elements with compatible stands or holders.

Generally, it is best to calibrate a multiscale comparator, starting at the low scales and proceeding to the high ones. If you find excessive error in any scale, you should check the gage head with another amplifier-meter and check the amplifier-meter with another gage head. You should check reversible gage heads in both directions.

SUMMARY

- High-amplification instruments should not be used in applications satisfactory for lower-amplification instruments. In such applications, low-priced dial indicators may be reliable. The lower-amplification instruments, however, are seldom reliable for greater than 2.5 μm (tenth-mil, 0.0001 in.) discrimination.
- Other means are used for finer discriminations. High-amplification mechanical instruments achieve fine discrimination together with excellent reliability and low cost. However, they give up range and versatility as the penalty for power.
- Of next greatest importance are the electronic instruments. Study of these requires recognition of the generalized system that exists for all measurement. For electronic instruments, the intermediate-modifying stage is different than for the previous instruments. They have power amplification. They are not limited by loading during the act of measurement. This gives them both their versatility and their high sensitivity.
- Electronic instruments make measurement in 0.25 μm (mike) increments (millionths of an inch) deceptively easy. Therefore, we must decide when to use such high precision. Those who are careless simply guess. Those who are more careful base their decision on the rule of 10-to-1. However, the best discrimination cannot be determined by any one rule. The 10-to-1 rule requires that the instrument divide the tolerance of the measurement into 10 or more parts. The difficulty is that there may not be an instrument that fulfills that requirement. Conversely, in a complex mechanism that requires high reliability, the rule may not ensure adequate reliability. In some cases, *traceability* requirements may be the determinant. Because of the human factors involved, lower-sensitivity instruments may sometimes produce better results than higher ones.
- There are two general styles of gage heads used with electronic gaging: probe and cylindrical. The probe is used in height gage setups. The cylindrical is used in C-frame instruments of all types, from handheld snap gages to ultra-precise comparators. The probe head does not obey Abbe's law and therefore must be used with care. The most reliable heads of either style are the frictionless type. These use flexing members instead of rotating shafts on sliding surfaces.
- Stands for high-amplification comparators must be selected carefully. Versatility is achieved by sacrifices in reliability. All unnecessary adjustments should be eliminated. Thermal considerations become a problem at any discrimination finer than 2.5 μm (one-tenth mil, 0.0001 in.) and are nearly insurmountable except under laboratory conditions when it reaches 0.52 μm (ten mike, 0.000010 in.). Cleanliness and contact between parts similarly takes on new importance at high amplifications.
- A distinct advantage of electronic measurement is the ease with which an electrical signal can be manipulated. This makes possible differential measurements. These require two heads whose outputs either add or subtract. Both are used to eliminate unwanted variables from measurement. They find good applications in checking roundness, concentricity, and parallelism.

Metrological Data for High-Amplification Comparators								Reliability	
Instrument	Type of Measurement	Normal Range	Designated Precision	Discrimination	Sensitivity	Linearity	Practical Tolerance for Skilled Measurement	Practical Manufacturing Tolerance
Mechanical (5000 X) Comparison	Metric English	0.025 mm 0.001 in.	0.0006 mm 0.000025 in.	0.0006 mm 0.000025 in.	0.00025 mm 0.00001 in.	2% 2%	± 0.0005 mm ± 0.00002 in.	±0.006 mm ±0.00025 in.
Electronic 0.0001 Scale Comparison	Metric English	± 0.06 mm ± 0.0024 in.	0.0025 mm 0.0001 in.	0.0025 mm 0.0001 in.	0.0013 mm 0.00005 in.	2% 2%	± 0.0013 mm ± 0.00005 in.	± 0.0025 ±0.0001 in.
0.00005 Scale Comparison	Metric English	±0.004 mm ± 0.0016 in.	0.0013 mm 0.00005 in.	0.0013 mm 0.00005 in.	0.0005 mm 0.00002 in.	2% 2%	±0.0010 mm ± 0.00004 in.	± 0.013 mm ±0.0005 in.
0.00001 Scale Comparison	Metric English	±0.0061 mm ± 0.00024 in.	0.00025 mm 0.00001 in.	0.00025 mm 0.00001 in.	0.0013 mm 0.00005 in.	2% 2%	± 0.00020 mm ± 0.000008 in.	± 0.00025 mm ± 0.00001 in.

FIGURE 14–49 At very high amplifications, the many variables to measurement accuracy are even more important than the inherent capability of the instrument itself; hence, these conservative data.

- Reliable high-amplification measurement requires careful attention to details. Every part of the measurement system, including the observer, must be checked for suitability, precision, and accuracy. Both knowledge and skill are needed. Intelligent application, patience, and complete objectivity are prerequisites. The data in Figure 14–49, although general, will help. Another large class of high-amplification instruments is the pneumatic type. They are discussed in Chapter 15.

END-OF-CHAPTER QUESTIONS

1. Which of the following best explains the use of electronics, optics, and pneumatics for comparators?
   a. Need for discrimination finer than 0.001 in.
   b. Desire for greater portability
   c. Advantages of remote reading
   d. Discrimination finer than 0.0001 in.

2. What is the chief advantage of reed-type instruments over dial indicators?
   a. Low cost
   b. Infinite range
   c. Automatic zero setting
   d. Absence of gear trains

3. In the progression from low- to higher-amplification instruments, which of the following is always a fact?
   a. The size of the measurement system always increases.
   b. The size of the transducer always decreases.
   c. The importance of parallax error increases.
   d. The measurement movement is amplified.

4. Operation of all instruments requires energy. What is the source of the energy for dial indicators, reed comparators, and other mechanical instruments?
   a. Heating that results from the compression of the workpiece and instrument parts
   b. Static electricity
   c. The windup of shafts
   d. It is manually applied through the actuation of the mechanical parts.

5. The basic circuit for electronic instruments is called the balanced bridge. Which of the following best explains this?
   a. A bridge is a structure across an opening. On a smaller scale, this is exactly the same structure as most gages with the reference at one end of the opening and the transducer at the other end.
   b. It resembles a balance in which the unknown is in one pan and the reference balance is in the other pan. This is another example of the parallel between linear measurement and measurement of mass.
   c. In an electrical bridge, any change in resistance in any one branch of the bridge will move the meter hand a proportional amount. That amount can be expressed as linear displacement.
   d. If a meter is substituted for any one of the three branches of a bridge, it can be calibrated to read in linear displacement.

6. In most electronic measurement instruments, the electrical characteristic used in the bridge is not resistance. Which of the following is the best explanation for this?
   a. Capacitance is used because electronic condensers can store energy.
   b. Impedance is used so that interference that might get picked up by the cables of resistance devices is minimized.
   c. Impedance allows the use of alternating current in the transducers, which can be operated at higher voltages than direct current transducers.
   d. Impedance allows the transducer to be a transformer, which is responsive to smaller displacements than is practical with variable resistance devices.

7. Which of the following are functional benefits derived from the multiple scales of electronic instruments?
   a. The use of one instrument for a variety of measurement and gaging applications
   b. Ease of locating the desired range when working at high amplification
   c. Improved reliability by zero setting at a scale higher than the one required for the measurements
   d. Lighter and more consistent gaging pressure

8. Which of the following are metrological benefits derived from the multiple scales of electronic instruments?
   a. The use of one instrument for a variety of measurement and gaging applications
   b. Ease of locating the desired range when working at high amplification
   c. Improved reliability by zero setting at a scale higher than the one required for the measurements
   d. Lighter and more consistent gaging pressure

9. It is recommended that zero setting be conducted at a scale different from the one that will be used for the measurements. Which of the following best explains this?
   a. It ensures that all circuits are functional.
   b. A lower scale is easier to locate.
   c. It is good practice for the user.
   d. It prevents catastrophic setup errors, such as being off the scale completely.

10. When distinguishing between dial indicators and electronic instruments for a given application, which of the following are the most important considerations?
   a. Practicality—the greatest cost-effectiveness for the application
   b. Ease of use and time required for use
   c. Availability of higher scales
   d. Inherent accuracy of the instrument

11. What is the basic rule for setting up gaging operations?
   a. The locations of the units and the observer must minimize parallax error.
   b. A reserve instrument must be on hand.
   c. The gage should be 10 times as accurate as the part.
   d. The gage should be calibrated immediately before using.

12. What is usually meant by "zone of uncertainty" in a measuring instrument?
   a. The amount of dispersion of repeated readings
   b. The zone of parallax errors from shifts in the observer's position
   c. The excuse used to explain operator error
   d. One tenth of the tolerance of the part feature as displayed on the readout of the instrument

15

CHAPTER FIFTEEN

PNEUMATIC MEASUREMENT

LEARNING OBJECTIVES

- Explain why pneumatic metrology has such an important role in industry.
- Describe how amplification may be increased without as great a reduction in range as is required for mechanical and electronic instruments.
- List the advantages of eliminating metal-to-metal contact.
- Explain the value of measuring several features simultaneously.

OVERVIEW

In pneumatic measurement, we use pressurized air forced through a port whose size we control, and then changes in the measured part feature create corresponding changes in the calibrated airflow. The methods we use to measure using air pressure are called *pneumatic gaging, air gaging*, or, more appropriately, *pneumatic metrology.*

The equipment we use in pneumatic metrology varies in operation and in metrological characteristics. We will discuss only the most commonly used pneumatic metrology equipment in this chapter. You should base your selection of equipment on a thorough study of manufacturers' data, and as always, you should consult the manual for how to use a specific instrument.

15-1 BACKGROUND

Pneumatic metrology is successful because it frequently eliminates *metal-to-metal contact,* and it provides *power amplification* along with linear amplification. In addition, the instruments are low in cost when compared with other methods with similar features.

Pneumatic comparators have power amplification, just like electronic comparators. The change-in-length signal is amplified, and the power to operate the measurement system comes from an external source. These comparators do not depend on the energy of the pickup element through contact with the part. Therefore, high amplification can be obtained with high sensitivity.

When there is no metal-to-metal contact, pneumatic comparators usually provide advantages we cannot obtain with other instruments. In addition, pneumatic comparators are not limited to applications with air sensing; we can add contact-type pickup elements for other measurement needs.

15–2 FROM HISTORY TO SEMANTICS

Although the principles of pneumatic metrology can be traced back to 1917, most of the advances in technology have occurred since World War II. Throughout the world, many methods for using fluid flow to solve particular measurement problems were developed, ranging from regulating paper manufacture to gaging rifle barrels. The first publicized use of pneumatic measurement was in France, controlling the size of automotive carburetor jets, a use which was expanded and refined during World War II.

Originally, the application of pneumatic measurement was only for production inspection or "gaging" of the finished product. Because it is well suited to gaging several features at once, pneumatic measurement is still an important means of production inspection today, but we also use pneumatic measurement instruments for a variety of measurement applications.

In pneumatic metrology, we need more specific definitions of three terms we have been using loosely: *measurement*, *gaging*, and *inspection*.

As we discussed, language is probably the *least* precise thing about measurement. We can measure to 0.2 µm (10 µin.), but if we say we are "gaging" the part, the measurement we "gaged" means something completely different to the quality control personnel, the machine operator, the plant manager, and the research scientist (Figure 14–2 summarizes part of MIL-STD-120). In pneumatic metrology, popular use has made "gaging" and "measurement" almost synonymous; but ambiguity can lead to errors. We will use the terms more precisely throughout this book.

15–3 PRINCIPLES OF PNEUMATIC INSTRUMENTS

Back-Pressure Instruments

Pneumatic comparators are analog devices that use a very abundant resource—air—to create change in a

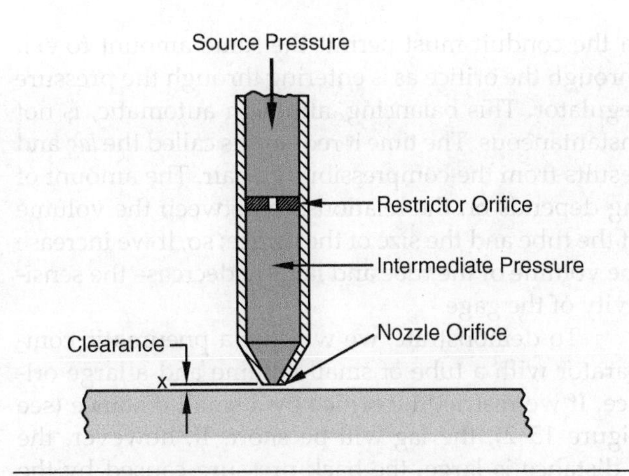

FIGURE 15–1 This is the double orifice system upon which back-pressure pneumatic comparators are based.

gage as a part feature is measured. These comparators measure two possible changes: pressure, which was developed first, and flow rate.

The double orifice pneumatic comparator (see Figure 15–1) is the basis for all other pneumatic comparators. In it, the intermediate pressure depends on the source pressure and the pressure drops across the restrictor orifice and the nozzle orifice. A loss in pressure between any two points in a fluid system, usually caused by eddies in the flow, causes a pressure drop. The nozzle orifice restricts the flow in direct relation to the clearance distance, x: when the clearance distance changes, the intermediate pressure also changes. This change is measured with an analog of the change in clearance. If we move the plate in the illustration closer to the orifice, we restrict the flow of air, and as a result, less air escapes into the atmosphere, which increases the back pressure. We read the measurement on the pressure gage, which we call the *back-pressure gage*.

With these instruments, you might expect that the constant pressure from the pressure regulator would result in a constant flow through the nozzle orifice. It does not, which creates a limitation of the back-pressure air gage: the back pressure changes the rate of flow.

The air that enters the system must equal the amount that escapes through the orifice. The pressure in the conduit must permit the same amount to exit through the orifice as is entering through the pressure regulator. This balancing, although automatic, is not instantaneous. The time it requires is called the *lag* and results from the compressibility of air. The amount of lag depends on the relationship between the volume of the tube and the size of the orifice; so, if we increase the volume of the tube and lag, we decrease the sensitivity of the gage.

To demonstrate, we will use a pneumatic comparator with a tube of small volume and a large orifice. If we restrict the orifice by a small distance (see Figure 15–2), the lag will be short. If, however, the x distance is large, the back pressure caused by the tube diameter affects the pressure gage more than the changes in the orifice restriction (x distance). This air gage has high sensitivity but a short range.

In the reverse case, we use a large tube and a small orifice; then, the changes of restriction of the orifice (changes in x distance) predominantly affect the pressure gage. We can extend the range, but we will also extend the lag; therefore, back-pressure air gages are limited in both sensitivity and range.

The back-pressure pneumatic comparator in Figure 15–2 is built with two readouts. One readout is the bourdon-tube pressure gage which most often has a "best" amplification range of 1,000X to 5,000X. Although it is a rugged and practical instrument, use of the bourdon-tube pressure gage wears the moving parts in the final stage. We can use a water column to help eliminate wear, but instruments using this solution have not caught on for popular use.

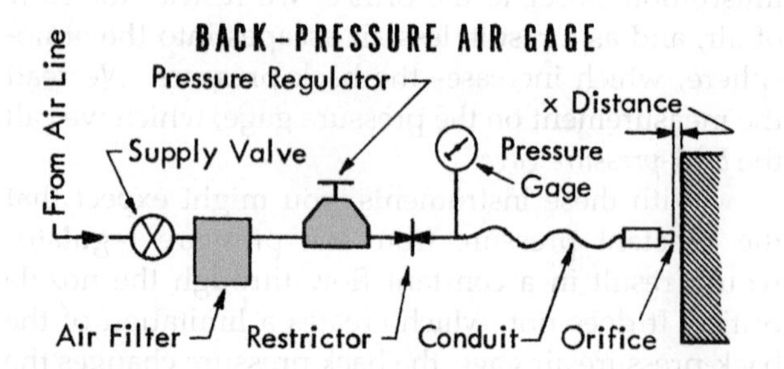

FIGURE 15–2 In the basic back-pressure air gage, a change in x alters the conduit pressure and is read on the pressure gage.

Manufacturers have refined the design of back-pressure pneumatic comparators to extend their range and sensitivity, but lag limits the design of all of these comparators to short tube lengths. We use the *balanced-type* system and the *rate-of-flow* system to reduce these limitations.

Balanced Systems

The balanced system (see Figure 15–3) starts like the basic back-pressure system shown in Figure 15–2. We divide the flow of air into two channels with two restrictors, one for reference channel and one for the zero setting valve. The measuring channel continues through the air hose to the gaging spindle, and between the two channels is a differential pressure gage. We can draw similarities between the balanced system and the bridge circuit in electronic comparators.

The back-pressure system shown in Figure 15–2 shows changes in the gaging pressure with respect to the atmospheric pressure, but there is no way for us to control atmospheric pressure. In the schematic (see Figure 11–3) with the balanced-system instrument, the measuring channel pressure changes with respect to the reference channel pressure. The reference channel is controlled by the equalizing jet restrictor, and the measuring channel is controlled by the master jet restrictor; therefore, we have complete control of the system.

With these additions, we can eliminate some of the problems with back-pressure systems and gain helpful features. Unlike other systems, the balanced-system, back-pressure instrument requires only one master and one zero setting adjustment. Manufacturers also claim that wear of the spindle does not change the amplification of the system. We show the relationship among the amplification, discrimination, and measurement range for one line of these instruments (see Figure 15–4) and a unique application of this instrument (see Figure 15–5). Your lab manual exercises will have you set up and balance a system based on the one-half inch master and jet plug.

Rate-of-Flow Systems

To compensate for the limitations of back-pressure instruments, the *constant-pressure, rate-of-flow,* or *velocity*

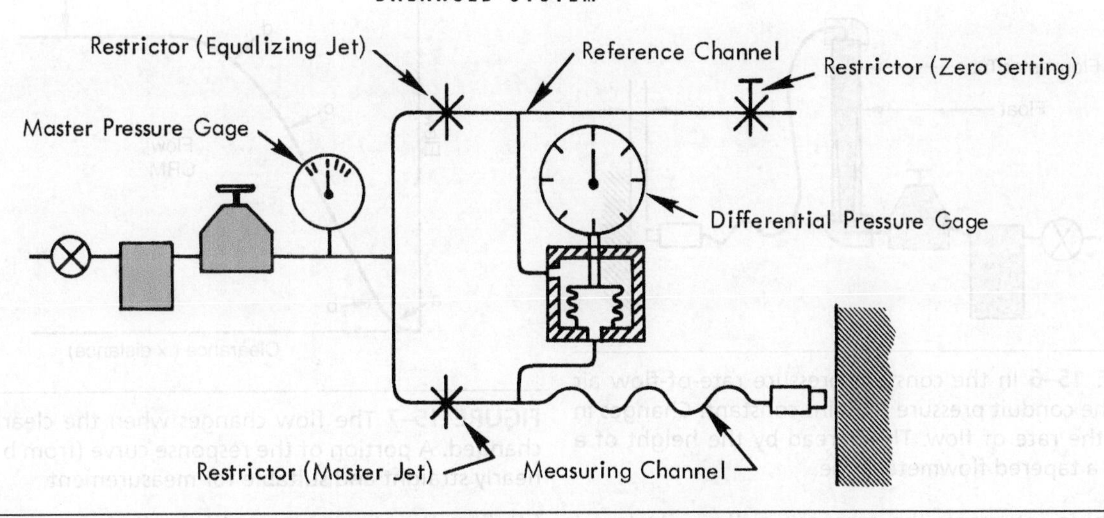

FIGURE 15–3 The balanced system has fixed amplification. The only adjustment is zero setting.

Balanced System Data		
Amplification	**Discrimination**	**Measuring Range**
1250:1	0.0025 mm 0.0001 in.	0.150 mm 0.006 in.
2500:1	0.00125 mm 0.00005 in.	0.075 mm 0.003 in.
5000:1	0.00050 mm 0.00002 in.	0.035 mm 0.0015 in.
10000:1	0.000125 mm 0.000005 in.	0.015 mm 0.0006 in.
20000:1	0.000125 mm 0.000005 in.	0.0075 mm 0.0003 in.

FIGURE 15–4 The balanced systems are available in a wide selection of amplifications and with relatively wide measuring ranges.

FIGURE 15–5 This flatness gage uses a balanced system. The part is placed on the surface plate and passed over the air jet. Deviation from flatness is read directly from the instrument dial.

pneumatic comparator was developed. In this instrument, the pressure is constant, but a change in flow rate shows a change in the x distance. In the basic circuitry (see Figure 15–6), the flow through the tube is not deliberately restricted, as in the back-pressure instrument. The air passes as rapidly through the tube as the restriction at the orifice (x distance) allows. The flow rate is shown by the level of the float in the internally tapered flowmeter tube.

With this system, we can use a much higher airflow rate; consequently, we can extend the working range if the instrument is much longer than back-pressure instruments. Because the system itself causes only slight restrictions, the instrument's lag is also slight.

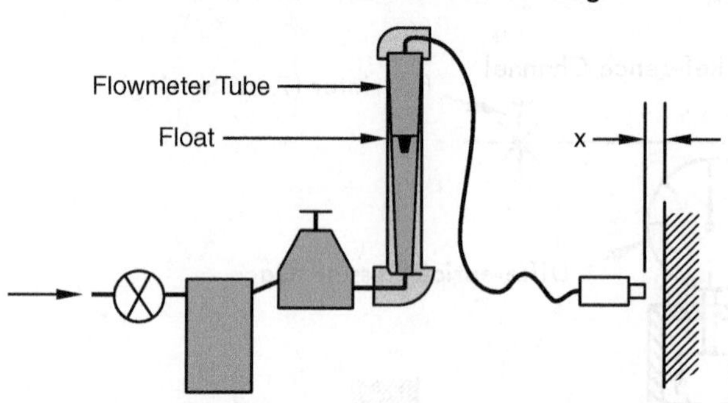

FIGURE 15–6 In the constant-pressure rate-of-flow air gage, the conduit pressure remains constant. Changes in x alter the rate of flow. This is read by the height of a float in a tapered flowmeter tube.

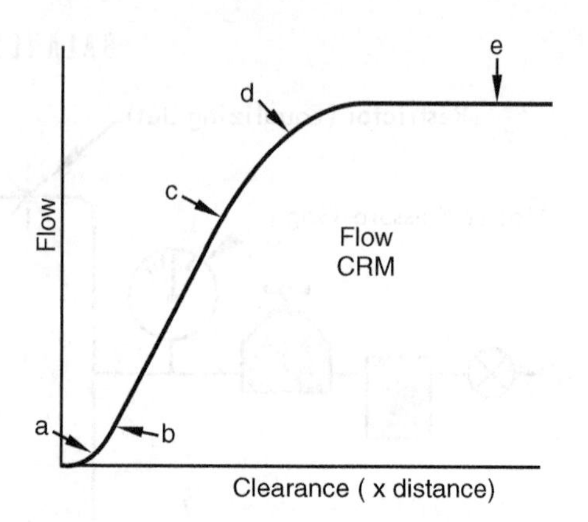

FIGURE 15–7 The flow changes when the clearance is changed. A portion of the response curve (from b to c) is nearly straight and suitable for measurement.

If anything interrupts the escaping airflow, it almost instantaneously slows flow throughout the system.

In the pneumatic comparator (Figure 15–7), the y axis is airflow, and the x axis, starting at zero, is the clearance or distance between the gage and the part. The first small increase in the clearance shows no flow (a), but, as the clearance increases, the flow change per unit-of-clearance change becomes greater until it reaches the maximum slope (b). The flow-to-clearance ratio remains constant for a linear portion—the part that is usable for measurement—of the curve (c). When the clearance is sufficiently large, changes in clearance no longer cause equal changes in the flow (d). The flow increases until the instrument reaches a point (e), after which increases in flow no longer change the readout.

The column-type flow circuit pneumatic comparator (see Figure 15–8) directly provides amplifications of more than 50,000 times. With some additional equipment, we can increase the amplification to 100,000 times. This instrument is:

- Completely free of the wear and *hysteresis* of mechanical systems
- Immediately responsive
- Simple to calibrate

Features of Pneumatic Metrology

Functional Features:

1. No wearing parts
2. Rapid response
3. Remote positioning of gage heads
4. Self-cleansing of heads and parts
5. No hysteresis
6. Adaptability to diverse part features
7. Small size of gage head

Metrological Features:

1. No direct contact
2. Minimum gaging force
3. Inspection of attribute (go and no-go gaging) or variable (size) measurement
4. Range of amplification
5. Adaptability to several modes of measurement (length, position, surface topography)

FIGURE 15–8 These features recommend air gaging for small holes, highly polished parts, fragile or easily determined parts, remote measurements, and the inspection of multiple part features. This is based primarily on the rate-of-flow type of air gage with flowmeter columns, but many of the features apply to other air gages as well.

Like other pneumatic comparators, the column-type flow circuit pneumatic comparator lets us perform many difficult measurements easily and reliably. The unique gage heads we can use with pneumatic gaging make pneumatic instruments very versatile.

15–4 APPLICATION OF PNEUMATIC METROLOGY

The Gaging Element

In pneumatic metrology, the first step of the generalized measurement system (see Figure 15–9) is called a *gaging element,* which corresponds to the *gage head* for high-amplification comparators. In the three general gaging elements:

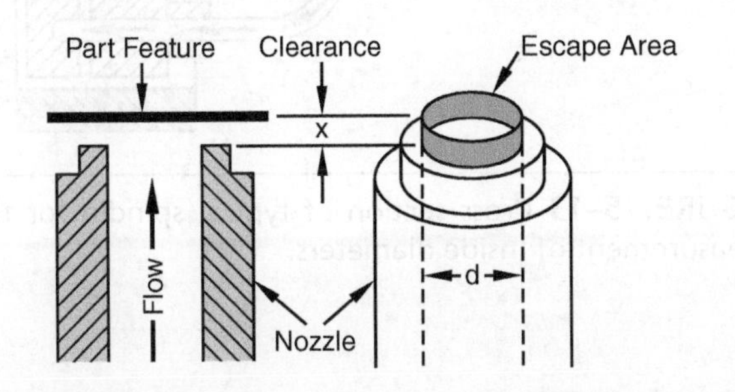

FIGURE 15–9 The airflow through the gaging element is proportional to the exit nozzle cross section and the clearance or x distance.

- Type 1—the hole being measured is the exit nozzle of the gaging element.
- Type 2—an air jet that does not contact the part is the gaging element.
- Type 3—the air jet is mechanically actuated by contact with the part.

The first type of element is only suitable for inside measurement, so it has limited applications. We use it when we need to control the cross-sectional area rather than the shape of the part, most often for carburetor jets and nozzles for the production of synthetic fibers.

The second type of gaging element is the simplest. The rate of flow depends on the cross-sectional area of the nozzle and the clearance between the nozzle and the part feature (see Figure 15–9). A single air jet, placed close to the part, forms the basic gaging element (see Figure 15–10).

When we discussed electronic gaging, we also discussed the advantages of differential measurement, which consists of electronically combining the output of two gage heads. Differential measurement makes it possible for us to make measurements more reliably than we could with a single-head instrument. Pneumatic gaging makes it easy to apply the principle of differential measurement. In fact, unlike with electronic gaging, we can make a differential measurement entirely in the gage head without altering the rest of the system and the instrument is not limited to two inputs. With a thorough understanding of the

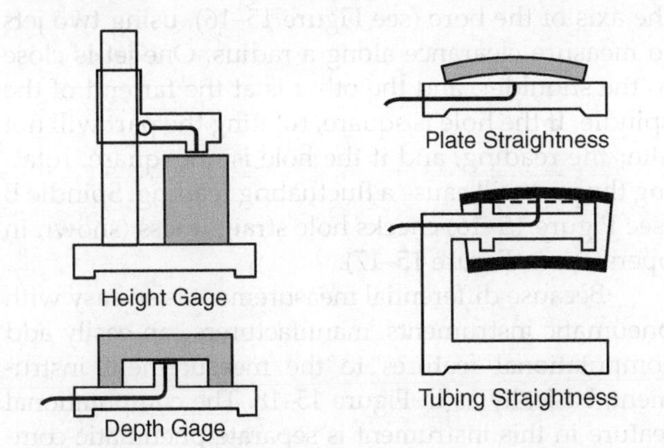

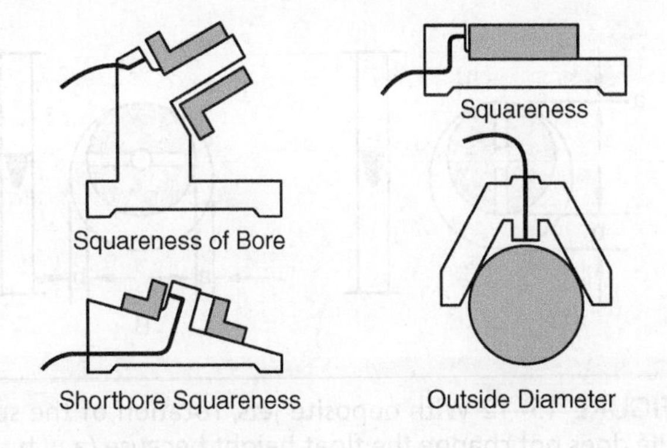

FIGURE 15–10 Single jet nozzles form the basic gaging element as in these examples.

principle and creativity, metrologists can create a variety of combinations for reliable differential pneumatic measurement.

Pneumatic gaging is especially important for the inspection of holes. The gaging elements, also called *spindles,* can be adapted to measure nearly any feature of the hole, including diameter, roundness, and straightness. If we use a spindle with one jet on one side (see Figure 15–11), the float height in the flowmeter tube will rise and fall as the spindle rotates and alternately opens and closes the one jet. If we use two opposed jets, we have an instrument for differential measurement (see Figure 15–12) where clearance A plus clearance B equals the same equivalent x distance.

We show a cross section of a spindle (see Figure 15–13) that would be used for determining roundness and diameter. By rotating the spindle, we can measure any changes in the diameter. If we want to measure the average diameter, we would use three or four jets, and we can also use them in different patterns (see Figure 15–14).

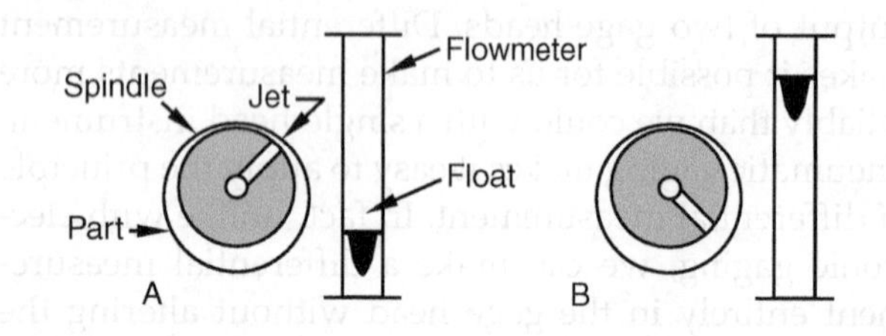

FIGURE 15–11 Rotation of the single jet spindle changes the height of the float.

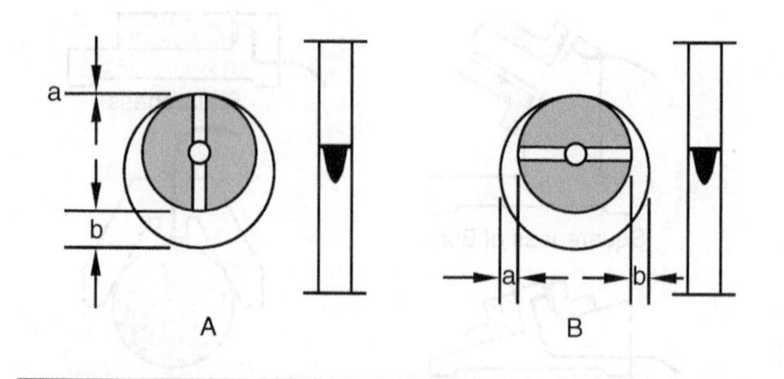

FIGURE 15–12 With opposite jets, rotation of the spindle does not change the float height because (a + b = x) for any position.

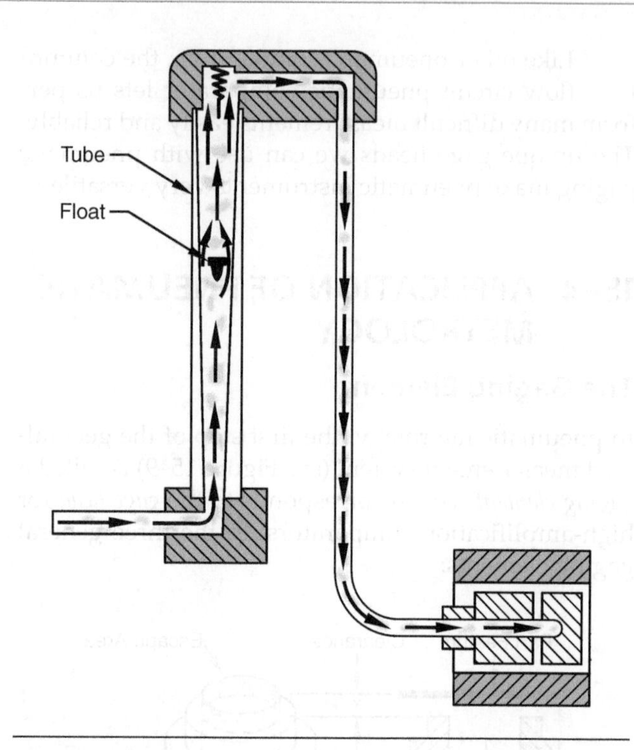

FIGURE 15–13 Cross section of typical spindle for the measurement of inside diameters.

If we modify the basic spindle, we can use it to measure other characteristics of a hole. In Figure 15–15, a mirror shows the front and back of two spindles. Item A measures the squareness between a face and the axis of the bore (see Figure 15–16), using two jets to measure clearance along a radius. One jet is close to the shoulder, and the other is at the far end of the spindle. If the hole is square, rotating the part will not alter the reading, and if the hole is not square, rotating the part will cause a fluctuating reading. Spindle B (see Figure 15–16) checks hole straightness (shown in operation in Figure 15–17).

Because differential measurement is so easy with pneumatic instruments, manufacturers can easily add computational features to the measurement instrument, basically as in Figure 15–18. The computational feature in this instrument is separate pneumatic comparators for the large and small diameters, plus an

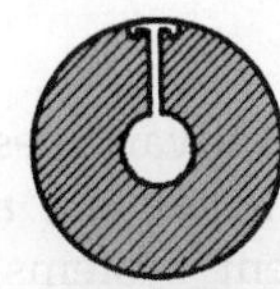

SINGLE-JET PLUGS—check concentricity, location, squareness, flatness, straightness, length, depth.

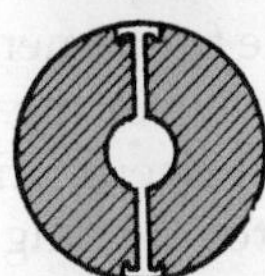

TWO-JET PLUGS—check inside diameters, out-of-round, bell-mouth, taper.

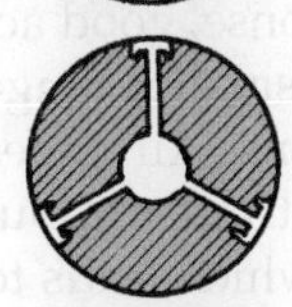

THREE-JET PLUGS—for checking triangular out-of-round.

FOUR JETS—are used to furnish average diameter readings.

SIX JETS—will show average determinations for both two-jet and three-jet conditions.

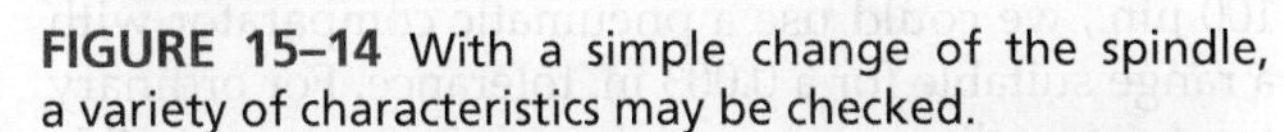

FIGURE 15–14 With a simple change of the spindle, a variety of characteristics may be checked.

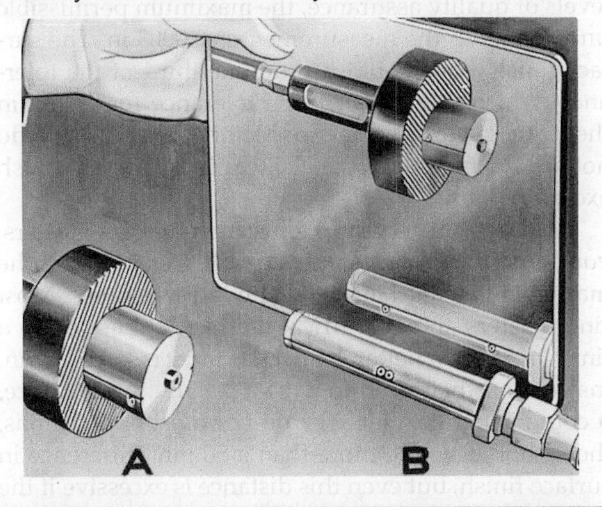

FIGURE 15–15 The mirror shows the back side of the spindle for checking squareness (A) and another for checking straightness (B).

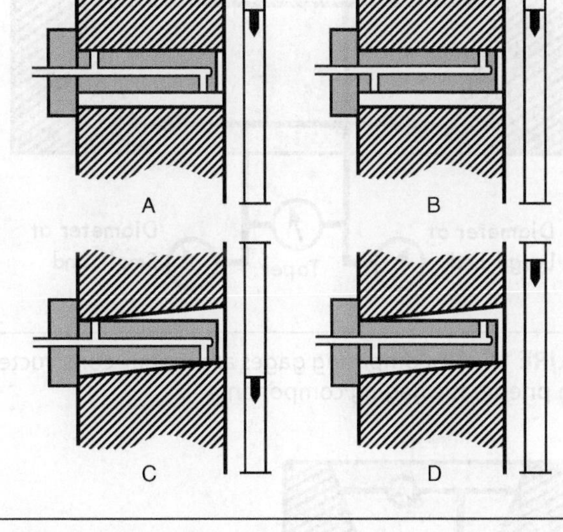

FIGURE 15–16 If the axis of the hole is square to the shoulder, rotating the spindle does not alter the height of the float as in A and B. Lack of squareness is shown by fluctuating float height as in C and D.

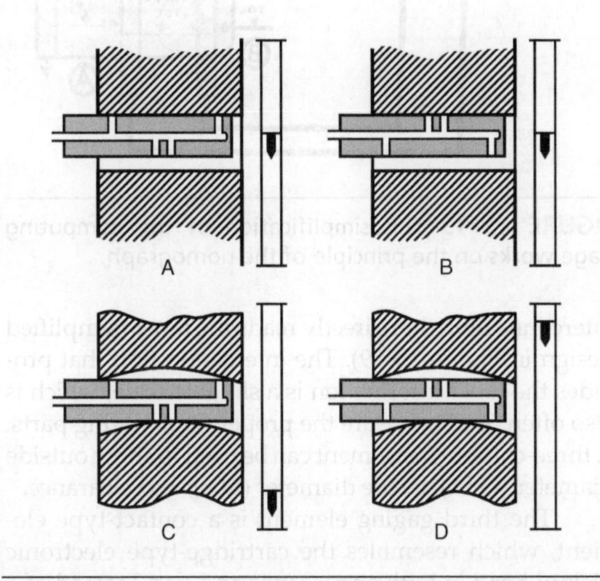

FIGURE 15–17 This spindle checks straightness. Rotating the spindle from A to B does not change the combined clearances if the hole is straight. If it is not, a maximum and minimum reading results.

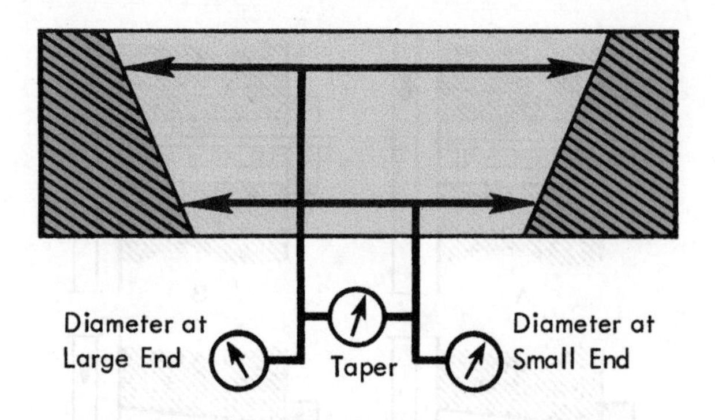

FIGURE 15–18 Computing gages are readily constructed with pneumatic gaging components.

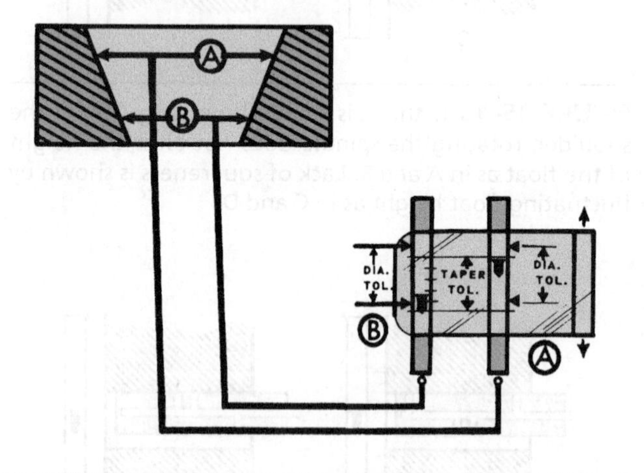

FIGURE 15–19 This simplification of the computing gage works on the principle of the nomograph.

interconnection that directly reads the taper (simplified design in Figure 15–19). The interconnection that provides the taper information is a slidable scale, which is also often used to ensure the proper fit of mating parts. A three-column instrument can be used to show outside diameter (O.D.), inside diameter (I.D.), and clearance.

The third gaging element is a contact-type element, which resembles the cartridge-type electronic pickup head in both appearance and use. Instead of a linear variable differential transformer (LVDT) transducer, an air valve is used that changes the air flow in proportion to the linear change.

Surface Finish Considerations

If pneumatic comparators provide the advantages of no metal-to-metal contact, why should we use a contact-type gage? Like all measurement systems, however, pneumatic metrology obeys one fundamental law of measurement: to gain something in measurement, you must give up something. A pneumatic comparator with the type 2 element provides high amplification, high sensitivity, fast response, good accuracy, and other benefits, but it has a smaller range and is less sensitive to surface conditions. Air gages have a smaller range than electronic instruments, but air gages provide superior resolution, which leads to exceptional readability.

For surface finish, the measured point for the type 2 gaging element is the *pitch line* of surface roughness. We can define the pitch line as the mean value between the highest and lowest points of the surface finish (see Figure 15–20), but this point is not the same one we would use if we were measuring with a contact-type instrument. The difference between the two measuring techniques is the potential source of error.

For example, if we gage a surface finish of 100 μin., we could use a pneumatic comparator with a range suitable for a 0.003 in. tolerance. For ordinary levels of quality assurance, the maximum permissible uncertainty in the measurement is 0.0003 in. The surface finish alone could use up one-third of the tolerance, which would not leave tolerance for errors in the instrument or by the inspector. Generally, we do not use the type 2 gaging element if the surface finish exceeds 60 μin.

When you set a type 2 instrument to its masters, you must remember surface finish, except when the master and the part have the same surface finish. Most ring masters and cylindrical masters are lapped to a finish of 5 μin. or better. If an instrument is set to a 5 μin. master and then used to gage a 200 μin. part feature, it could generate a large error. For most applications, there should not be more than a 50 μin. difference in surface finish, but even this distance is excessive if the tolerance is 0.001 in. or closer. Therefore, it is another good reason, summarized in Figure 15–21, for using type 3 gaging elements.

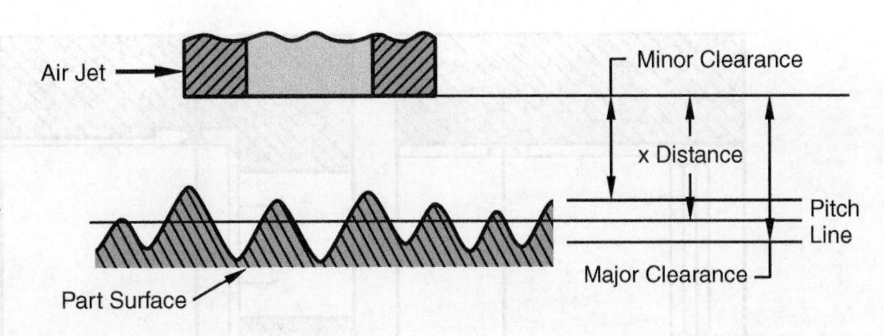

FIGURE 15–20 With the type 2 gaging element, the reference line is the pitch line between the crests and valleys of the surface.

Selection of Gaging Elements

Element	Operation	Measurement
Type 1	Part forms own exit nozzle	Cross-sectional area
Type 2	Clearance between part and nozzle	Tolerances not greater than 0.075 mm (0.003 in.) on parts with surface finishes of 65 μin. or finer
Type 3	Mechanical contact with regulates flow	Tolerances greater than 0.075 mm (0.003 in.), or parts with surface finishes rougher than 65 μin., or parts whose surface finish differs from the master by 50 μin. or more

FIGURE 15–21 One of these three gaging elements will reliably accommodate nearly every measurement requirement.

Type 3 gaging elements are preferred for a variety of applications (see the examples in Figure 15–22). When the surface width is less than 0.100 in., we can rarely use type 2 instrumentation reliably, and serrations or grooves in a surface can affect the reliability of these elements. In these cases, and in multiple inspection setups, we can use type 3 elements to check length, width, thickness, and taper simultaneously (see Figure 15–23).

"Mastering" the Pneumatic Comparator

Although type 2 and type 3 instruments are very different, we set them up for measurement similarly using three steps:

1. Set the instrument to a master at the high or low limit.
2. Adjust the amplification.
3. Set the instrument to a master at the second limit.

You will notice that two of these steps are different than setting up comparators (see Chapter 14). The amplification is subject to both selection and adjustment, and because of this amplification subjectivity, both limits are set to masters. Opponents of pneumatic metrology say that setting the instruments to two masters is a disadvantage but, actually, using two masters is an advantage of pneumatic metrology because it materially reduces the uncertainty that is inherent in setting the pneumatic comparator. Although other comparators are often used for direct measurement, it is nearly impossible to use a pneumatic comparator for direct measurement.

For inspecting small holes, gaging elements are set to ring gages. We must set the gaging element to gage block calibration fixtures when we inspect large inside measurements or outside measurements.

When we measure to 0.0001 in. or finer, we must be careful that the instrument is *mastered* as closely

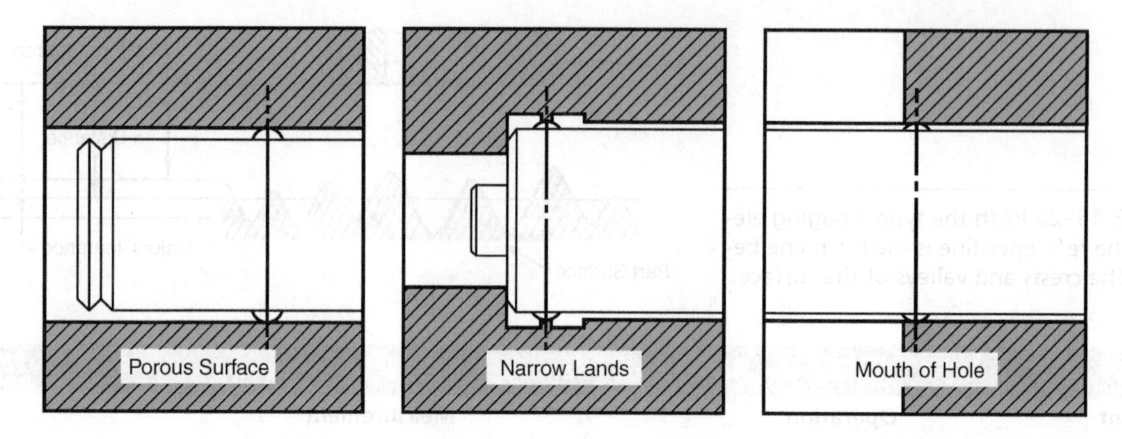

FIGURE 15–22 For these situations, the type 3 gaging element, providing metal-to-metal contact, is preferred to the air sensing of type 2.

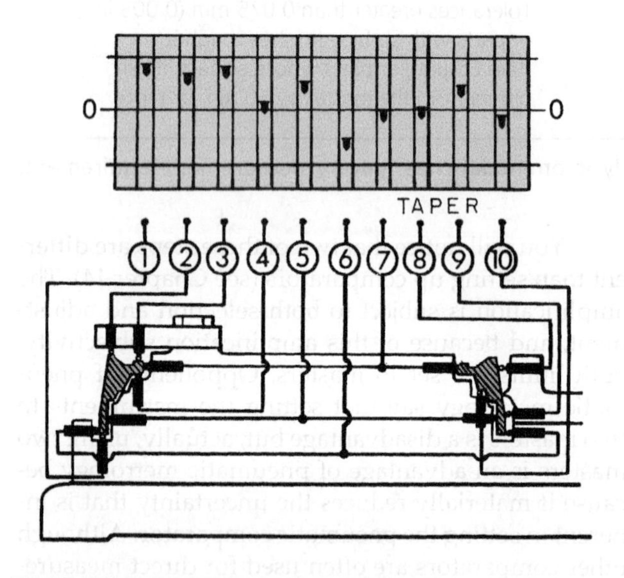

FIGURE 15–23 The vertical flowtubes of pneumatic comparators particularly suit them for multiple inspection setups. Either type 2 or 3 gaging elements may be used. This example shows length, width, thickness, and taper dimensioning being checked simultaneously with 10 type 3 gaging elements.

to the way we use it for measurement. For example, if the jets of the gaging element are in a vertical line when you are checking parts, you should make sure they are vertical when you set the instrument to its master. As we mentioned, you should be sure that the masters and the part have as near to the same surface finish as possible.

15-5 METROLOGICAL ADVANTAGES OF PNEUMATIC COMPARATORS

There are other questions we can ask, especially about the metrological features of pneumatic comparators. What amplification and precision relationships govern pneumatic comparators? In the beginning of the chapter, we talked about the large selection of pneumatic metrology equipment. As we continue our discussion, we will base our data on the instrument shown in Figure 15–24.

In Figure 15–25, we show the discrimination and range for several amplifications. The scale lengths are

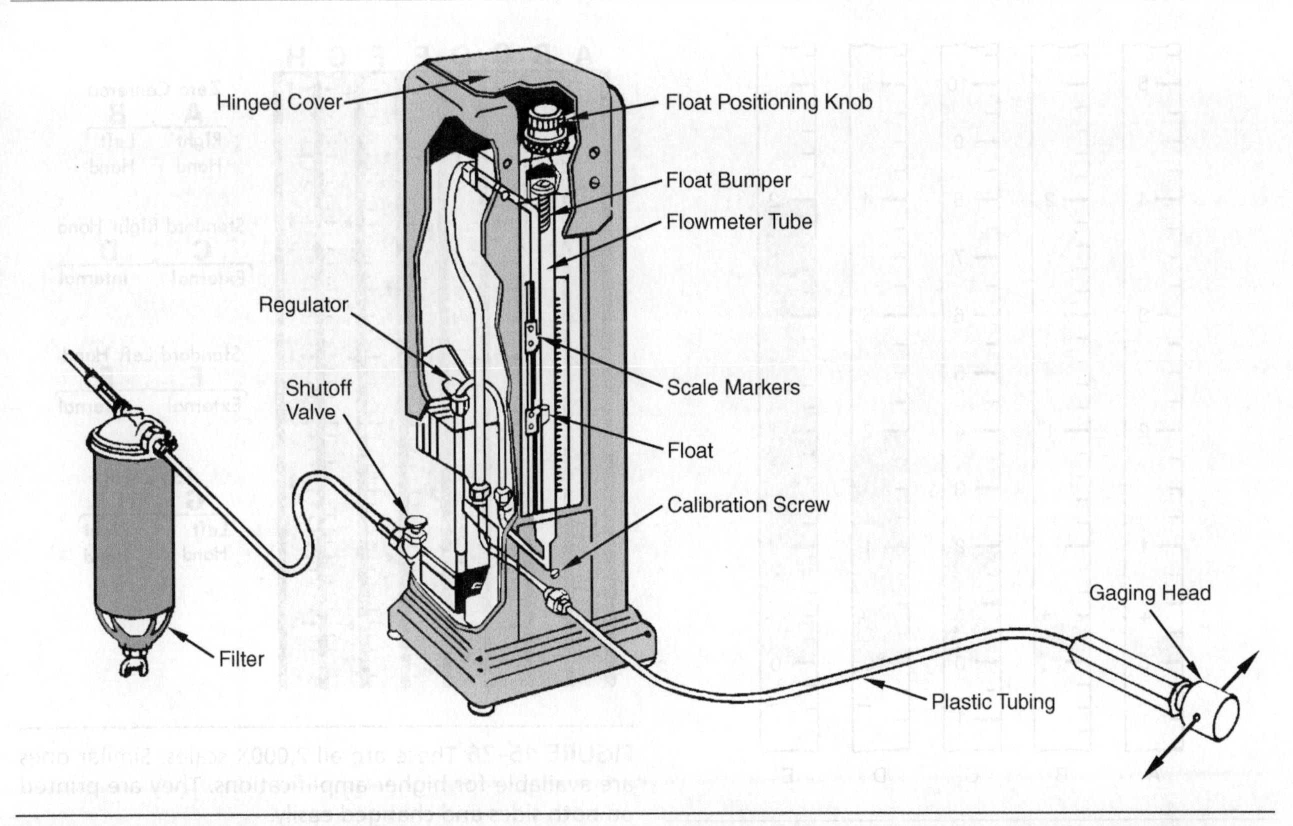

FIGURE 15–24 This flow diagram shows the simple circuit and adjustments required for a typical rate-of-flow-type pneumatic comparator.

longer than those for other forms of comparators with similar amplifications, which is one important reason that the pneumatic comparator provides exceptional reliability.

Before we continue, let us review the metrological points made in previous chapters. The accuracy of an instrument is its adherence to some standard, and the precision is the fineness with which it subdivides the standard. Whereas its fidelity or truthfulness is measured by its accuracy, its practicality is measured by its precision, which is limited by the sensitivity—the smallest input signal that produces a discernible output signal. Precision is increased by amplification, which is any means for stretching out the scale so that the output signal will be greater in magnitude than the input signal; however, precision that cannot be read is not useful. Resolution, the ratio of the width of a scale division to the width of the readout element, determines readability.

In Chapter 10, we showed that increasing the amplification of mechanical comparators does not produce an equivalent increase in sensitivity; therefore, we can only stretch the scale out so far. In pneumatic comparators, this relationship is true, too, but not to the same extent. We can make the scale very long, and a long scale means we can have useful, wide graduations for theoretically high sensitivity. Wide graduations are also easier to discern with your eye—they are at a higher resolution, increasing the readability of the instrument.

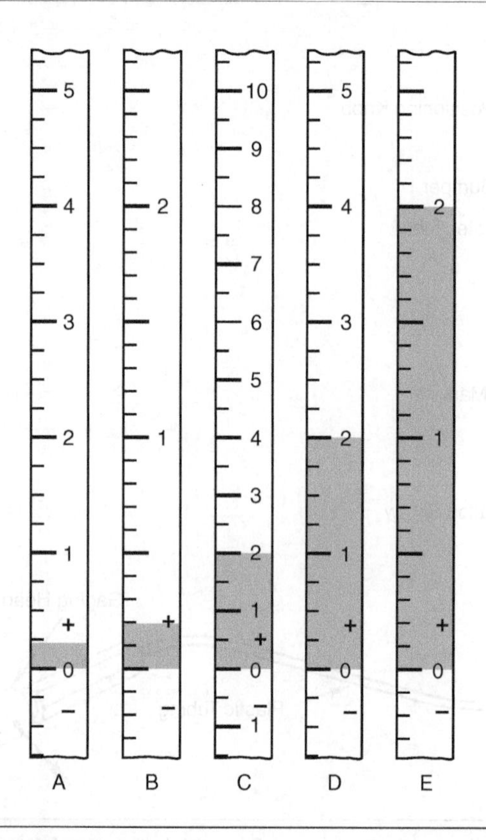

FIGURE 15–25 With pneumatic comparators, tolerances can be spread over great scale lengths, thus minimizing observational error.

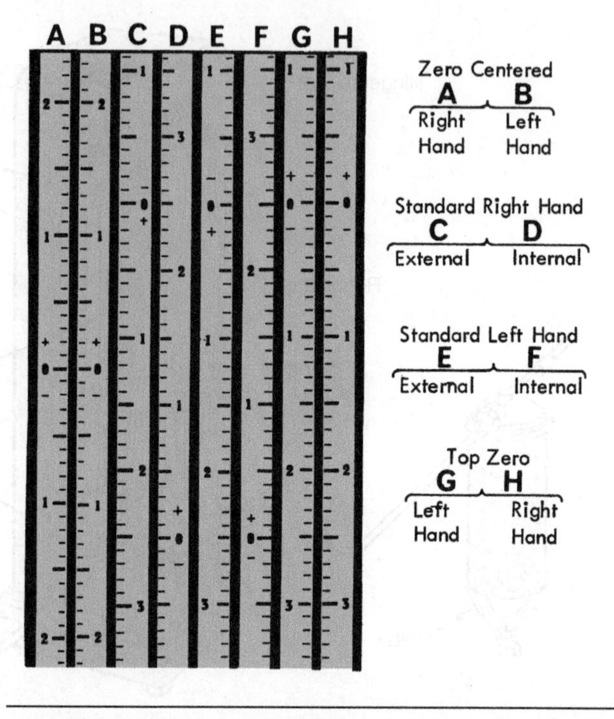

FIGURE 15–26 These are all 2,000X scales. Similar ones are available for higher amplifications. They are printed on both sides and changed easily.

15-6 READING THE PNEUMATIC COMPARATOR

The basic pneumatic comparator can accommodate a variety of scales, which vary in both style and amplification. The scales in Figure 15–26 have the same amplification, 2,000X, and the same discrimination, 0.0001 in. Right hand and left hand refer to the side of the flowmeter tube along which the scale is placed.

For pneumatic measurement, we need some new terminology (see Figure 15–27), but most of the pneumatic measurement terms basically mean what they sound like they mean. For example, the float "comes up" as the element approaches the part tolerance, and it does "go beyond" when the part exceeds tolerance.

The sign of the value is reversed for inside and outside measurements (see Figure 15–28). At A, the gaging element is in a large hole, and the air that escapes through the jets in the gaging element is only slightly restricted. The air races through the system carrying the float high on the scale. In B, a smaller hole restricts the airflow, and the float stops low on the scale. In A, the larger hole produces a positive reading, and in B, we would expect the small hole to produce a negative reading.

In C and D, the part feature is a diameter, and the jets are arranged in a ring around the part. In C, the airflow is rapid when the feature is small, which is the reverse scenario of the A/B case. In D, the larger part restricts the flow, so that, for external measurement, minus represents a large flow and plus represents a small flow.

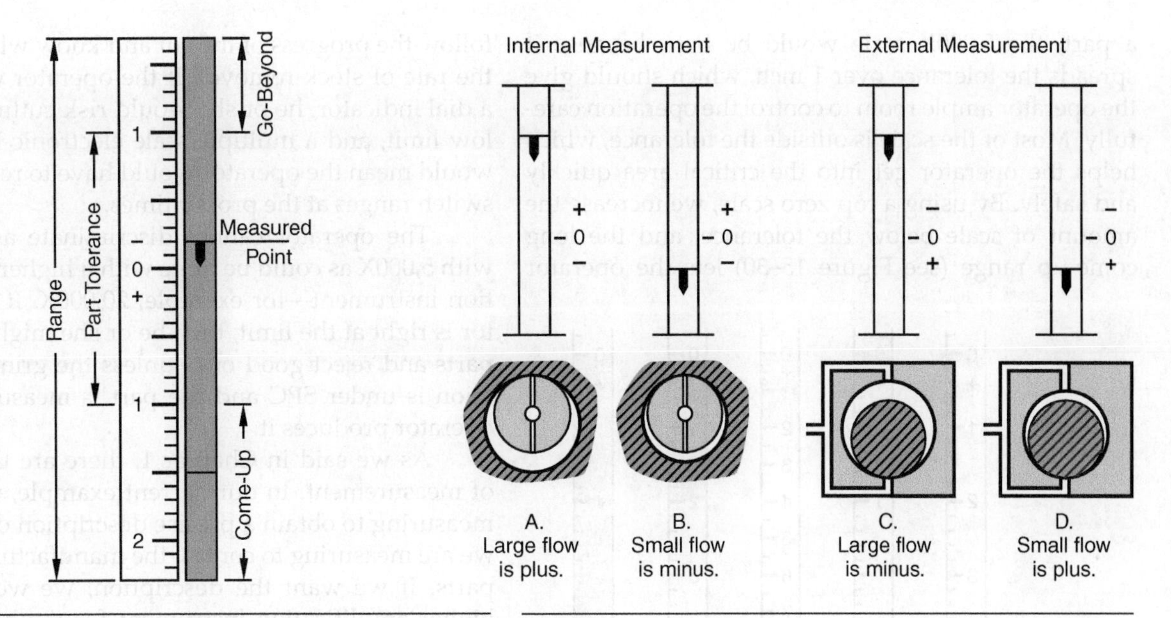

FIGURE 15–27 "Come-up" and "Go-beyond" are terms particular to pneumatic comparators.

FIGURE 15–28 Separate scales are needed for internal and external measurement because the relationship of flow rate and the plus-minus sign is reversed.

In all measurement, there is a difference between inside and outside measurements, but only pneumatic comparators allow us to change the scale to adapt to inside or outside. We make this change of the sign in setup by changing the scale, not by making computations for each measurement, which eliminates much uncertainty in the resulting measurement. The elimination of computation makes the selection of the right scale for a measurement with a pneumatic comparator very important.

Reading the scales of pneumatic comparators is very simple, because all are marked with amplification, discrimination, and range. The float's flat top provides a clear reference, and, because you can easily select higher amplifications, you do not need to interpolate the measurement.

Scale Selection

You make two sets of decisions when selecting the proper scale: a set to determine the magnification, and a set to determine the style of scale.

The general rule for scale selection is: "Always select the most sensitive scale that will contain the tolerance of the dimension to be measured entirely." In Chapter 10, we saw that we could not always follow this rule with dial indicators because of their inherent limitations, but in Chapter 14 we showed that multiple-range electronic comparators do follow this rule. In some cases, however, following this rule is unnecessary. These are cases in which we can use the long scales of pneumatic comparators.

When space on a scale is limited, scale divisions must be crowded, resulting in poor readability and decreased reliability. Spreading out the scale improves this situation, but there must be a limit to how much improvement is possible. The pneumatic comparator can spread out a tolerance well beyond the amount required for maximum readability (see Figure 15–27). The amount of spread required depends on the reason for the measurement.

In a typical example, a grinding operation requires a 0.0002 in. tolerance in different spreads (see Figure 15–29). If the operator is cylindrically grinding a part, the 5,000X scale would be best, because it spreads the tolerance over 1 inch, which should give the operator ample room to control the operation carefully. Most of the scale is outside the tolerance, which helps the operator get into the critical area quickly and safely. By using a top zero scale, we increase the amount of scale below the tolerance, and the long come-up range (see Figure 15–30) lets the operator

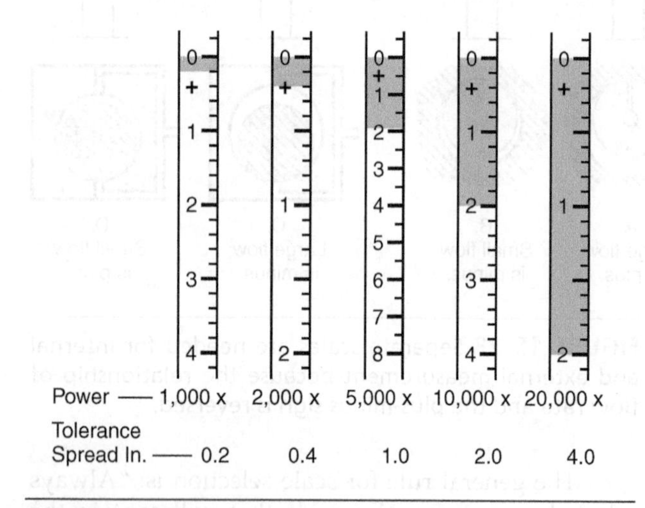

FIGURE 15–29 The effect of amplification on tolerance spread is shown by this comparison of a 0.0002 in. tolerance on five scales.

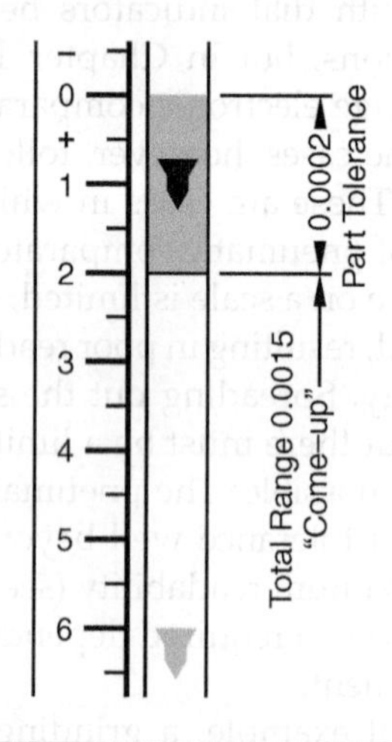

FIGURE 15–30 The long come-up range speeds the grinding operation.

follow the progress of the cut and know when to slow the rate of stock removal. If the operator were using a dial indicator, he or she would risk cutting past the low limit, and a multiple-scale electronic instrument would mean the operator would have to remember to switch ranges at the proper times.

The operator cannot discriminate as precisely with 5,000X as could be done with a higher amplification instrument—for example, 20,000X. If the operator is right at the limit, then he or she might pass bad parts and reject good ones unless the grinding operation is under SPC and the part is measured as the operator produces it.

As we said in Chapter 1, there are three states of measurement. In our current example, we are not measuring to obtain a precise description of one part; we are measuring to control the manufacture of many parts. If we want the description, we would use a higher amplification instrument for the best results. For manufacturing control, measurement keeps the ground parts within their tolerance and away from the limits, which the added come-up range helps achieve better than additional discrimination.

If we change conditions, does that change the scale selection? If we substitute for the manually operated one, the come-up range is unimportant. We may be using the scale mainly to contain the tolerance, so we could use a larger amplification such as 10,000X. The additional precision makes the inspection of borderline cases more difficult, but it also permits a higher degree of control so we can use the automatic grinder to its fullest ability.

"Top Zero" Spindle Technique

These two examples show zero-centered and top zero scales, but, as with most benefits, we sacrifice something for them. When the float is raised when *mastering* (setting to a standard) the comparator, the flow rate is increased, but because the instrument must operate in the linear portion of its performance curve (see Figure 15–7) any increase in flow cannot exceed linearity. We can also increase the flow by increasing the clearance between the gaging elements and the part; therefore, to take full advantage of top

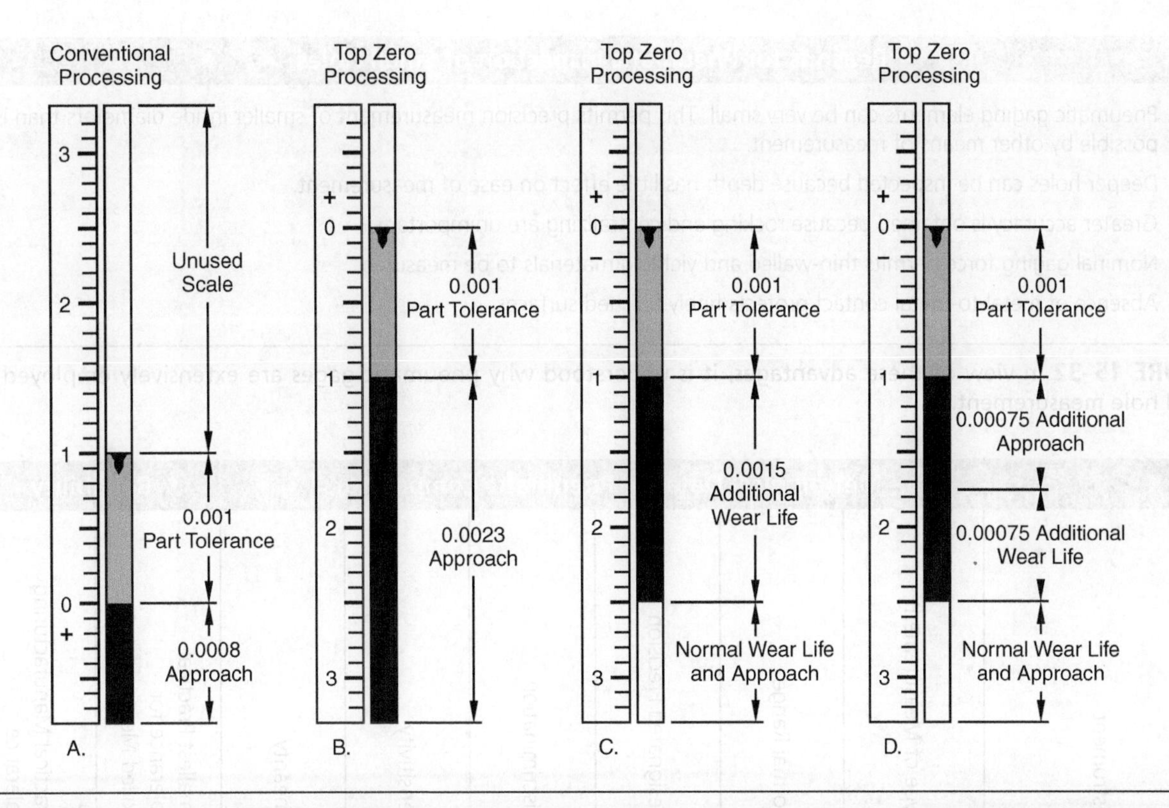

FIGURE 15–31 Special gaging elements and scales are available to modify operation (A) in order to provide maximum come-up range (B), to provide additional wear-life (C), or provide an optimum combination of these (D).

zero scales, we need spindles with greater than usual clearance.

To understand this need in pneumatic metrology, remember that the gaging element has a seriously limited range and the gaging element wears. Although the metal-to-metal contact is eliminated in the mechanism, a pneumatic instrument suffers wear from being inserted and removed from the workpiece.

This wear will increase the clearance, which permits greater flow and results in a higher position of the float along its scale. Frequent mastering helps control this problem. At each mastering, the float is moved slightly by reducing the flow, but after this process has been repeated many times, the clearance will be beyond the measuring range. Special gaging elements and scales (see Figure 15–31) automatically compensate for this change.

SUMMARY

- Most production gaging equipment developed from measuring instruments. Pneumatic comparators are the reverse. They are measuring instruments that developed from production gaging equipment, the field in which they still enjoy their greatest popularity.
- Pneumatic comparators are clearly the preferred method for the measurement of small holes (0.500 in. and smaller) when precision in 0.001 increments is required (see Figure 15–32). Of course, they have many advantages for larger holes as well. They provide a remarkable combination of desirable features: high amplification without loss of sensitivity, easily selected scale ranges, differential measurement, excellent readability, no

Reasons for Pneumatic Inspection of Small Holes
• Pneumatic gaging elements can be very small. This permits precision measurement of smaller inside diameters than is possible by other means of measurement. • Deeper holes can be inspected because depth has little effect on ease of measurement. • Greater accuracy is obtained because rocking and centralizing are unimportant. • Nominal gaging force permits thin-walled and yielding materials to be measured. • Absence of metal-to-metal contact protects finely finished surfaces.

FIGURE 15–32 In view of these advantages, it is understood why pneumatic gages are extensively employed for small hole measurement.

Metrological Data for Pneumatic Comparators							Reliability	
Instrument	Type of Measurement	Normal Range	Designated Precision	Discrimination	Sensitivity	Linearity	Smallest Practical Tolerance for Skilled Measurement	Practical Manufacturing Tolerance
I Back pressure, dial-type air gages. Similar to Figure 11–4	Comparison							
62 1/2X type 3 gaging elements to 7,500X type 2 or 3		0.060 in. 0.005	0.002 in. 0.0005	0.002 in. 0.00002	0.0004 in. 0.00002	0.001 in. 0.00003	0.005 in. 0.0001	0.010 in. 0.0003
Flowrate with type 2 gaging elements 1,000X	Comparison	0.0075	0.0002	0.0002	0.00004	0.0001	0.001	0.003
40,000X		0.00022	0.000010	0.000005	0.000002	0.000005	0.00005	0.00015
100,000X		0.000090	0.000005	0.000002	0.000002	0.000002	0.00002	0.00005
Flowrate type with type 3 gaging elements 62 1/2X	Comparison	0.080	0.003	0.003	0.0005	0.0015	0.008	0.016
5,000X		0.0018	0.0001	0.0001	0.00005	0.00005	0.0002	0.0006

FIGURE 15–33 These data are based on representative instruments. Some instruments will vary from them considerably. The actual instrument specifications should be studied closely before selecting a pneumatic comparator.

metal-to-metal contact of sensing element, freedom from critical gaging geometry, no mechanical parts, few opportunities for operator error, and low cost for the basic instrument. They traditionally suffer from three serious limitations: short measurement range, sensitivity to surface finish, and the need for expensive gaging elements and masters that offset the low instrument cost.

- Gaging elements with air jets may be considered suitable for amplifications up to 100,000X, although 40,000X is generally the highest. This permits discrimination of 0.000002 in. under ideal conditions. They may be used at as low an amplification as 1,000X, which provides 0.005 in. In the average case, a 0.0001 in. tolerance can be measured adequately with a pneumatic gage having 5,000 to 1 amplification. This provides 0.0002 in. discrimination and an operating range of 0.0015 in. A surface finish of 65 microinches is required even for the low-amplification ranges. For the high ranges, it must be proportionately better and must be nearly the same in the part and the masters.
- Many limitations are eliminated by mechanical contact air gaging elements. These are used when the surface finish is poor, when the area to be measured is small, and when greater range is required. They are available with ranges from 0.001 in. to 0.040 in. The highest amplification generally used is 5,000X.
- This is summarized in the metrological data table (see Figure 15–33). When consulting the table, remember that there is a diversity of equipment available. The values selected are representative, but they are not all inclusive. Nothing in this chapter can substitute for a thorough study of the manual furnished with each pneumatic comparator.

END-OF-CHAPTER QUESTIONS

1. Which of the following best describes pneumatic metrology?
   a. Measurement by means of compressed air
   b. Measurement comparators applied to the gaging of production parts, particularly holes
   c. The use of change of airflow to show change in length
   d. The use of calibrated tubes to extend comparator scales
2. What other terms are used for pneumatic metrology?
   a. Air gaging
   b. Hole gaging
   c. Coordinate gaging
   d. Mastering
   e. Production gaging
3. What is the chief feature for most applications of pneumatic metrology that is unique to this method?
   a. No electric outlet or batteries are needed.
   b. The air blows away the dust that could cause errors.
   c. There is a long range of the scales even at high amplification.
   d. There is no metal-to-metal contact.
4. Which of the following characteristics do electronic and pneumatic comparators have in common?
   a. Insensitivity to temperature of the workpiece
   b. Absence of parallax error
   c. Low initial cost coupled with relatively high tooling costs
   d. Power amplification and linear amplification
5. Which of the following describes "gaging" as a special form of measurement?
   a. Measurement to determine whether a part feature is between its high- and low-tolerance limits
   b. The use of an instrument for which the sensing head may be located away from the rest of the instrument
   c. Any measurement for the production of parts
   d. Measurement by comparator-type instrument
6. Why is a pneumatic comparator considered to be an analog measuring instrument?
   a. Because it is capable of precision to 0.0001 in. and higher
   b. Because the amplification and scale may be changed

c. Because the change in flow is proportional to the change in feature size
d. Because a vertical scale is usually used instead of a dial or digital scale

7. What two properties of air are subject to calibrated change for linear measurement?
a. Density and barometric pressure
b. Rate of flow and pressure
c. Humidity and temperature
d. Nitrogen–oxygen ratio and CO_2 percentage

8. What were the first types of pneumatic instruments called?
a. Air pressure comparators
b. Back-pressure instruments
c. Pneumatic pressure instruments
d. Pressure measurement instruments

9. What is "pressure drop"?
a. The difference between the high and low readings
b. The equivalent of parallax with a dial indicator
c. The difference in pressure between two points in the system
d. Error from sudden loss of line pressure

10. Which of the following best describes the difference in the time delay between the length change and the change in the reading?
a. Hysteresis
b. Impedance match
c. Lag
d. Volumetric discrepancy

11. Why was the balanced system developed?
a. For higher amplification
b. To operate on less supply air
c. Increased repeatability
d. To eliminate lag

12. What is the chief difference between back-pressure and balanced systems?
a. The pressure must be higher with back-pressure systems in order to include both high and low limits.
b. Straight tubes are used in balanced systems, whereas tapered tubes are necessary for back-pressure systems.
c. Bourdon-tube measurement displays may only be used with the back-pressure type.
d. Balanced systems have a reference channel against which the measuring channel is read.

13. How does the operation of a rate-of-flow system differ from both the back-pressure and balanced systems?
a. Larger gaging elements can be used.
b. Smaller gaging elements can be used.
c. The system is independent of differences in workpiece temperatures.
d. No restrictions are required.

14. What are the differences between the way rate-of-flow, balanced, and back-pressure instruments are read?
a. There are no differences.
b. Balanced and back-pressure instruments are read by pressure gages calibrated in linear measurement, whereas rate-of-flow instruments use flowmeters.
c. Pressure gages calibrated in linear measurement units may be used for all instruments.
d. Rate-of-flow instruments are digital, whereas the others are analog.

15. What in pneumatic measurements corresponds to the detector-transducer of the generalized measurement system?
a. The gaging element
b. The upper- and lower-tolerance limit masters
c. The combination of the element in the workpiece feature at the proper gaging pressure
d. The masters

16. What is meant by "differential measurement"?
a. The scale expansion between the low limit and the high limit of the tolerance
b. The combined results of measurement with two or more elements
c. The proportional increase of the high measurement as the low measurement is increased
d. The simultaneous adjustment of the zero setting for both high and low limits

17. An advantage of pneumatic measurement is to measure directly such features as roundness, diameter, straightness, and squareness. Which of the following best describes that capability?
   a. Ease of inserting the gaging element
   b. Differential measurement
   c. Wide gap for airflow around the gaging element
   d. Multiple measurements after averaging

18. Which of the following is the most important limitation of pneumatic metrology?
   a. Availability of compressed air
   b. Limited scale range
   c. Surface finish of workpieces
   d. Need for both upper- and lower-limit masters

19. Pneumatic gaging utilizes special terminology. Which of the following expressions refers to the calibration of a gaging setup?
   a. High- and low-limit setting
   b. Mastering
   c. Top zero setting
   d. Come-up calibration

20. "Come-up" refers to which of the following?
   a. Bringing the workpiece into the gaging setup
   b. Determining the next highest scale needed for the measurement
   c. Regulating the line pressure to get the reading onto the scale
   d. The travel of the float before it reaches the tolerance

21. Which of the following best describes the precision for practical measurement with pneumatic comparators?
   a. Up to 0.0001 in. available, but an average working range to 0.001 in.
   b. Up to 200,000 power, but 40,000 is the usual maximum
   c. Up to 200,000 power, with a practical tolerance control to 0.000005 in.
   d. Tenth mils (0.0001 in.) is considered average.

16

CHAPTER SIXTEEN

OPTICAL FLATS AND OPTICAL ALIGNMENT

LEARNING OBJECTIVES

- Explain the importance of light waves as length standards.
- Explain the phenomenon of interference and the formation of fringe bands.
- Describe how fringe bands are manipulated for linear measurement.
- Explain the significance of the reversal technique.
- State the role of those instruments that rest on the work being examined.
- State the role of those instruments that rest on stands independent of the work being examined.
- Measure flatness with an optical flat.
- Describe applications for alignment telescopes, optical squares, and theodolites.

OVERVIEW

Our discussion started with rules, which have one power amplification, and has progressed in order of amplification. The instruments we will now discuss have the highest amplification. Using light waves for measurement, we can achieve amplification as high as one million power and still have an instrument that is practical for daily use.

With other instruments, every time we tried to achieve higher amplification, we made the instrument more expensive, less familiar, and more delicate. In contrast, the equipment for measuring with light waves is inexpensive, familiar, and very rugged (see Figure 16–1).

The primary instrument that uses light, the *optical flat*, is a transparent plate with one face finished to near-perfect *flatness*. When we place this face on

FIGURE 16–1 A remarkable phenomenon of light provides fringe patterns—the contour maps for micromeasurements. *(Courtesy of Carl Zeiss, Inc.)*

another nearly flat surface, we can see light bands, or *fringes*—also known as interference—which correspond to the distance separating the two surfaces. Because we know the wavelength of *light*, we can determine the separation by counting the fringes.

For metrology, there are four primary applications of optics:

- *Magnification* (see Figure 16–2), the visual enlargement of an object so we can examine and measure it more precisely
- *Alignment*, using *light rays* to establish references, such as lines and planes
- *Interferometry*, using the phenomenon of light to measure
- Using the known wavelength of light as the absolute standard of length

Optical alignment is not a "new" technique for metrologists and manufacturers. Ancient structures, including Stonehenge and the pyramids of Egypt, depended on sighting methods. A crude instrument used by the Egyptians to lay out right angles was refined by the Romans and called the "groma." By the fifteenth century, Arab navigators were using the astrolabe, an early sextant, and Europeans began using the plane table not long thereafter.

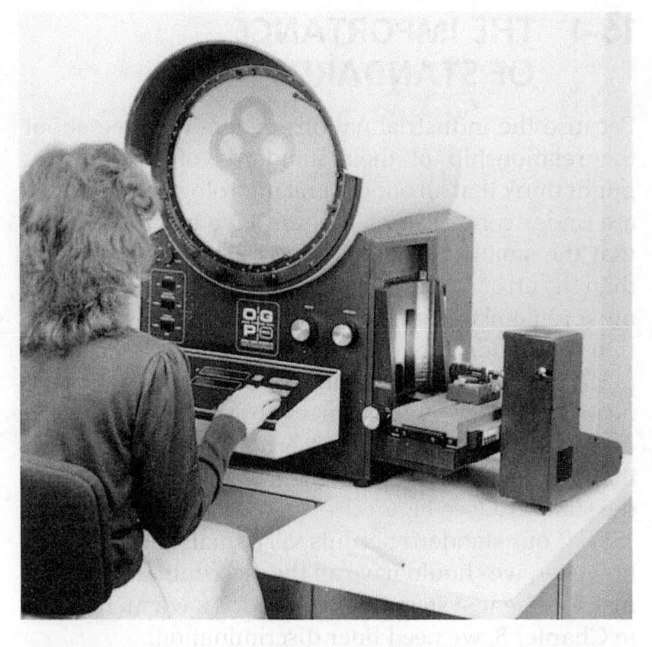

FIGURE 16–2 Optical magnification is now recognized as one of the most useful assists to metrology. One of the first widely adopted optical instruments was the optical comparator. An instrument such as the one shown here can inspect many features in the time that would formerly have been required for only one. *(Photo Courtesy of Certified Comparator Products)*

Surveying's fundamental geometry and mathematics were fully developed by the eighteenth century. We have since added some methods of surveying, such as aerial surveying, but the fundamentals and the primary instrument, the theodolite, are the same. During the Industrial Revolution and until World War II, mechanical instruments were widely used for many "surveying" purposes—for example, a taut piano wire was used to align large machinery.

For most other *macrometrology* (very large measurement), some major components already existed. It might be the casting of a large machine bed or weldments of a giant crane, or not uncommonly, the hull of a ship into which the propeller shaft bearings had to be aligned. It was natural in such cases to use the workpiece itself to establish the references.

16-1 THE IMPORTANCE OF STANDARDS

Because the industrial nations generally agree about the relationship of their standards of length, you might think that all our general metrological problems are under control. They are, until we reach the point that the smallest division of any standard is greater than the error to which the standards are defined. You must remember that, no matter how precisely we define the standard, we cannot ever reach the absolute perfection of measurement. Therefore, whenever we need to make a measurement to a subdivision of a unit that is smaller than the error known to exist in the standard, that measurement does not conform to the standard (see Figure 16–3).

If our standard permits very small subdivisions, however, we should have all the discrimination in our measurement system that we need. As we mentioned in Chapter 8, we need finer discrimination:

- As the speed of the mechanism increases
- As the number of component parts increases
- As the importance or cost of the end result increases
- For international trade
- Because of the complexity of systems

Manufacturing cannot continue to economically supply the demands of society without constantly improving the means of measurement, and it was this pressure that forced the adoption of light wave standards.

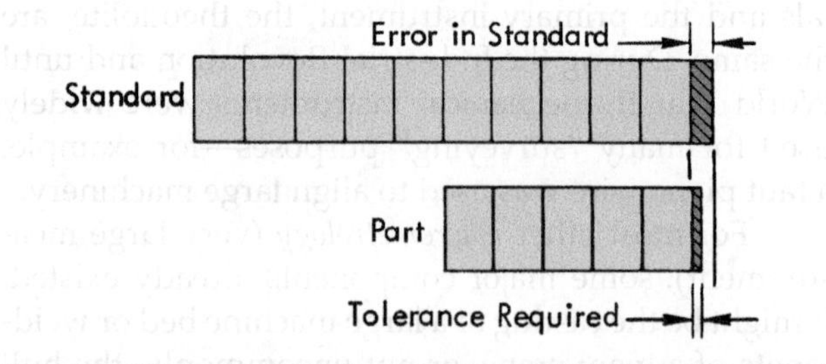

FIGURE 16–3 Whenever a measurement is made requiring a subdivision of the standard that is smaller than the known error in the standard, a measurement out of conformation to the standard results.

16-2 LIGHT WAVES AS STANDARDS

The value of light waves as a length standard was recognized long before light waves were adopted as a practical standard. Light was recognized as a useful standard when Albert A. Michelson (1852–1931) and Edward W. Morley (1838–1923) were actually testing the theory of ether drift, which is that outer space behaves more like a transparent liquid than complete emptiness. To test this theory, they used an interferometer to measure the path difference between part of a light beam passing through a tremendous distance through space compared to another portion traveling only a short distance.

To improve the precision of the experiment, Michelson and Morley had to measure the wavelength of a particular color of light, choosing the red cadmium spectral line. They used an interferometer and, in the experiment, measured the wavelength in terms of the meter, which meant that they could also express the meter in terms of the wavelength of that particular color or light.

Until Michelson's experiments, the unit of length and the standard of length had always been separate. The meter bar was never really equal to the meter; it was simply as close a representation of the meter as we could construct. As long as the standard and unit were separate, there would always be a discrepancy between them. With Michelson's work, we have for the first time a standard of length that was also a unit of length.

Metrologists immediately recognized the advantages:

- The wavelength of light was more stable than any "material" that had been used in the construction of a standard.
- The wavelength was so short that, at the time at least, no one thought it would have to be divided into smaller units to increase the discrimination.
- The light was relatively easy to reproduce anywhere.

Using the red cadmium light, we can measure length displacement to 6.43849×10^{-10} m (0.000,000,025,348,388,280 in.). During the 50 years following Michelson's original determination of the meter, metrologists have made eight other determinations using red cadmium light, and the average of the nine observations was 6438.4696 angstroms.

$\text{Å} = 1 \times 10^{-10}$ meter
$\text{Å} = 1 \times 10^{-4}$ µm
$\text{Å} = 0.1$ nm
1 angstrom (Å) $= 3.9374 \times 10^{-9}$ inch

Red cadmium light was limited in the sharpness of the fringes it provided, and metrologists began to search for a better light source. The definition of the International Meter Bar with red cadmium light is still an accepted standard.

In the search for a better light source, researchers returned to the green line of mercury with which Michelson had started. By bombarding gold in a nuclear reactor, we obtain an isotope of mercury without atomic spin, which results in a measurement to one part in 100 million. In 1960, researchers tried isotope 86 of krypton gas, which proved to be a valuable choice and became the accepted standard. With this krypton gas isotope, we define an inch as 41,929,399 wavelengths.

In spite of the capability to measure with light, the differences among international inches were not reconciled until 1959, and the differences among SI systems were not reconciled until 1960. The U.S., Canadian, and British inch are now all equal to 25.4 mm.

At the General Conference on Weights and Measures in Paris in October 1983, delegates agreed on a new definition of the meter, which was based largely on research by the National Institute of Standards and Technology (NIST) in the United States. Under the new method, the meter is defined as the distance light travels in a vacuum during 1/299,792,458 of a second, which is 10 times more accurate than the krypton standard. This new standard for the meter also correlates length with time, creating the most accurate measurement capability we have.

Application of this standard is a highly sophisticated undertaking and, except at the highest level of international standardization, has little relevance. For example, to apply the method using isotope 86 of krypton gas, you must take steps like immersing the lamp in a bath of liquid nitrogen. The steps may be difficult, but these methods and others, like using helium light for shop work, do have practical applications at all but the most extreme levels of precision.

16-3 MEASUREMENT WITH OPTICAL FLATS

To make practical measurements with optical flats and complete the exercises in your lab manual, you will need:

- The part you are going to measure
- An optical flat
- A monochromatic—or one wavelength—light
- A suitable work surface

White light is made up of a combination of colored light, but not all white light uses colored light in the same combination. Each color of light has its own separate and distinct wavelength. The visible spectrum—the one that we can see with our eyes—are those wavelengths between approximately 370 nm and 750 nm. Visible light is designated by color:

Violet	(370–455 nm)
Blue	(456–492 nm)
Green	(493–577 nm)
Yellow	(578–597 nm)
Orange	(598–622 nm)
Red	(623–750 nm)

For the international standard, metrologists and other scientists very carefully chose the best particular color of light at a definite wavelength—a wavelength that could be reproduced exactly anywhere in the world. When krypton 86 gas is excited electrically, it emits a light, and we can reproduce the steps that create this light fairly easily. Therefore, electrically excited *krypton 86* gas light was chosen as the standard.

For practical use, we must also consider other factors: the definition of the fringe bands, how easily we can see them, cost, and convenience. The light emitted by helium gas has proved most practical when balanced among all the considerations. Helium light fits somewhere in the middle of the spectrum (see Figure 16–4).

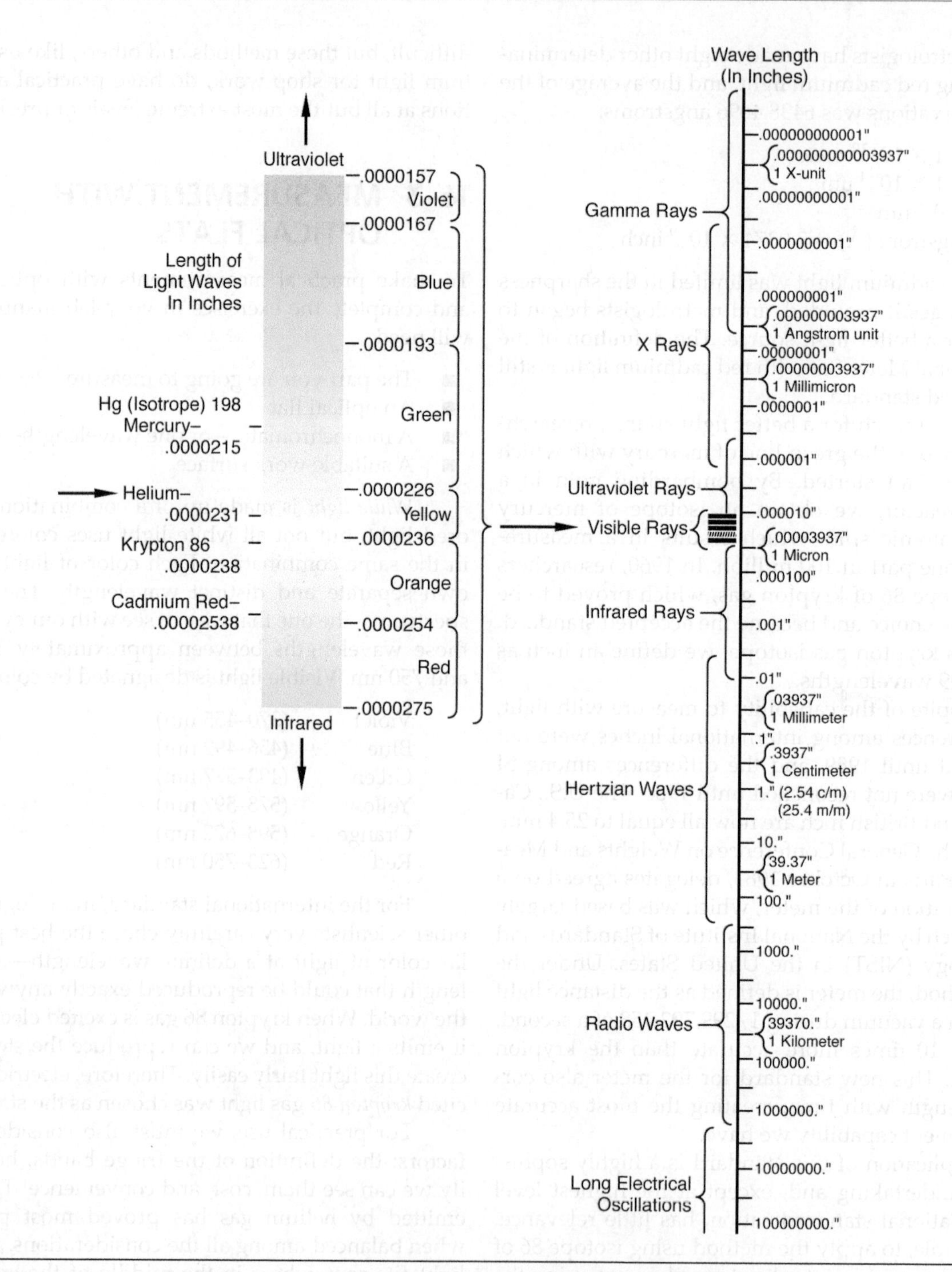

FIGURE 16–4 Light waves represent only a very small part of the electromagnetic spectrum.

Optical flats are made in a range of shapes, materials, and sizes up to 10 inches in diameter. You are most likely to encounter 25 to 75 mm (1 to 3 in.) diameter flats in materials ranging from inexpensive glass to very expensive sapphire. The majority are made from high-quality optical quartz.

The quality of the measuring surface is determined by its degree of flatness. *Working flats* are 0.10 μm (4 μin.), a unilateral tolerance that means no point deviates in height from any other point by more than that amount. *Master flats* are 0.05 μm (2 μin.), and *reference flats* are 0.03 μm (1 μin.). Some manufacturers also provide a *commercial grade,* 0.2 μm (8 μin.).

For an extra charge, you can purchase an optical flat with a *coated surface*—a thin film, usually titanium oxide—to reduce the amount of light lost by reflection. The less light that is lost by unwanted reflection, the clearer the *fringe bands* (see Figure 16–5). You can see that the coating is so thin that it does not affect the position of the fringe bands. Unfortunately, a coated flat requires even greater care. In some relatively rare cases, uncoated flats provide better fringe bands—for example, for carbon seals.

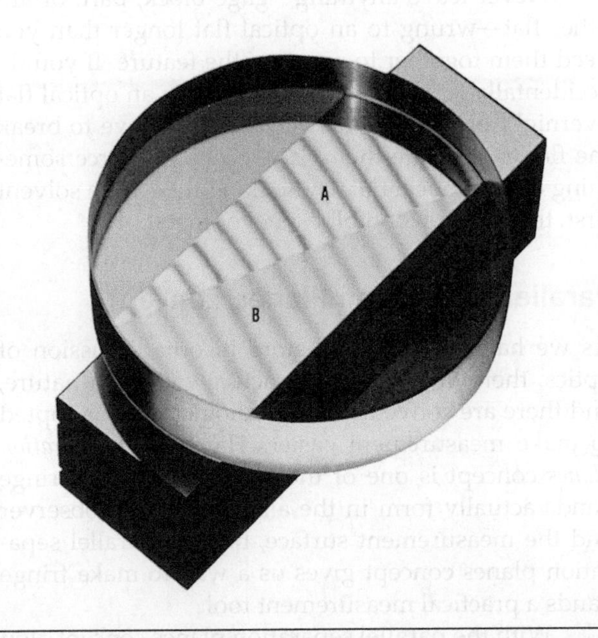

FIGURE 16–5 Coated flats provide sharper fringe bands in most cases. In this photograph, the flat has been half coated. A is coated and shows the fringes much better than B.

You can make measurements with an optical flat on just about any clean, rigid platform in most cases when you need to measure the changes on one surface. If, however, you need to compare two surfaces, the supporting surface limits the precision of measurement: the precision of the measurements can be no better than the precision of the supporting surface.

Because of the need for comparison, optical flats are often used as support for the part, but steel flats, known as *toolmaker's flats* or *gagemaker's flats,* are also available. These steel flats let you wring parts and gage blocks to their surface for measurement; of course, you can use other precision-finished surfaces, such as comparator anvils. You must remember, however, that the errors contributed by the supporting surface are independent errors. They may combine with measurement errors, increasing the uncertainty of the overall system.

Setup for Measurement

Throughout this text, we have emphasized that precision measurement setup starts with cleanliness. A stray particle of dust that lands on the part before the flat is placed over it can completely destroy any chance for reliable measurement. For some borderline measurement cases with optical flats, one particle of dust can actually prevent the formation of fringe bands. You should have a camel hair brush available so that, at the last moment before placing the flat on the part and/or gage block, you can whisk it off. Use a camel hair brush because it does not shed.

You will notice the effects of temperature changes more than you will with any other instrument. Fortunately, most optical flat measurement involves relatively small parts that normalize quickly, and most flats have a lower coefficient of conduction than the metal parts we measure with them. Optical flats are not heated as rapidly by handling, but they also take longer to regain the ambient (surrounding) temperature after they are heated or cooled.

You can see the effects of temperature by wringing a gage block to an optical flat until three or four bands appear and then place the gage block and flat on a surface under a *monochromatic light*. Next, drip four or five drops of solvent on the flat near one edge, and blow on the solvent so it evaporates rapidly, cooling one part of the flat. You will not see an immediate change in the fringe pattern. However, when you return in approximately 5 minutes, you will find a changed fringe pattern. Because of the contraction in the cooled part of the flat, the flat bends. If you let the flat stand for an hour, the fringe pattern will change again as the flat normalizes, but it will not return to the same configuration as the beginning of the experiment.

To view an object properly through an optical flat (see Figure 16–6), you should be as nearly perpendicular to the surface as possible for the more accurate measurement.

To achieve maximum clarity, you should have the measurement surface as close to the light source as convenient.

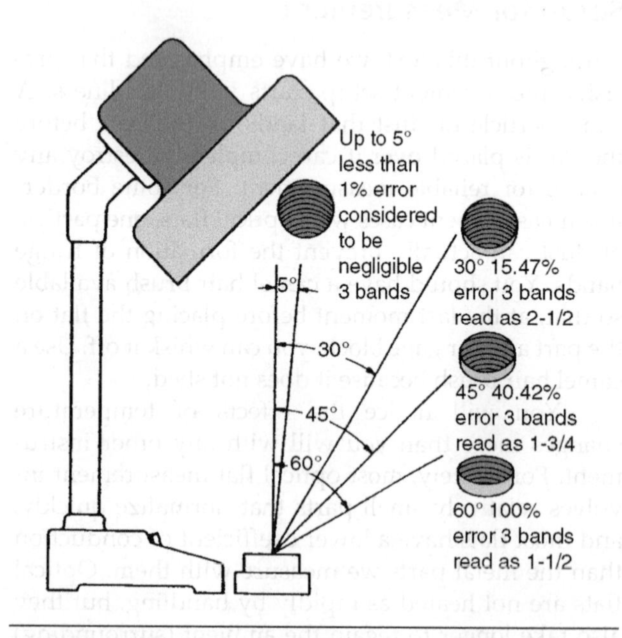

FIGURE 16–6 The smaller the angle, the more accurate the measurement will be.

You wring a gage block or part to a flat much the same way as you wring it to another block. It will begin to wring almost immediately unless:

- The surface is insufficiently flat.
- The surface finish is insufficiently fine.
- The surfaces are improperly cleaned.

Excessive wringing wears the flat unnecessarily.

Metrologists disagree about the degree of surface finish necessary for wringing because measuring very fine surface finishes is difficult. Some require a 0.1 μm (4 μin.) number 1 (AA) surface; others require surfaces as fine as 0.03 μm (1.2 μin.) in publications. To compare, the best gage blocks have less than 0.0254 μm (1.0 μin.) number 1 (AA) surface finish.

As soon as the block begins to wring, you will see fringe bands, and as you continue to wring, you can cause the bands to run in any direction across the part. (With practice, you will be able to maneuver the bands as you want, as will be discussed later.)

Never leave anything—gage block, part, or another flat—wrung to an optical flat longer than you need them together to measure the feature. If you do accidentally leave something wrung to an optical flat overnight or longer, you may actually have to break the flat to separate them. If you need to force something off an optical flat, try soaking the two in solvent first, then use a wood block, never metal.

Parallel Separation Planes Concept

As we have mentioned before in our discussion of optics, there are things that actually exist in nature, and there are conventions metrologists have adopted to make measurement easier. The *parallel separation planes* concept is one of the conventions. The fringe bands actually form in the air between the observer and the measurement surface, but the parallel separation planes concept gives us a way to make fringe bands a practical measurement tool.

With the parallel separation planes concept, you imagine a set of planes that are all parallel to the working surface of the flat and one-half wavelength apart (see Figure 16–7). The intersections of the "planes" and the part create dark fringe lines; therefore, the

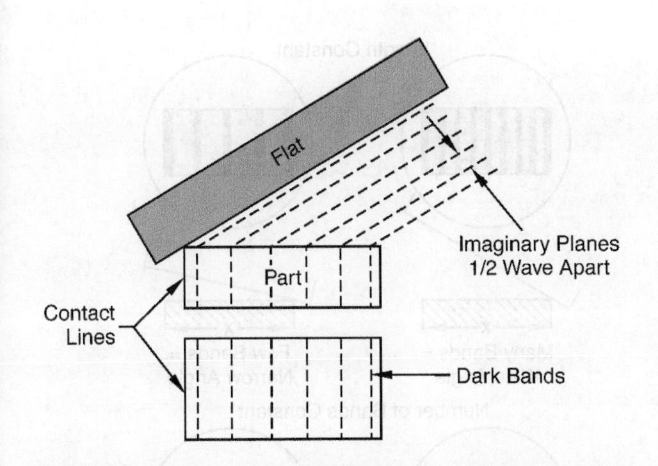

FIGURE 16–7 The parallel separation planes concept envisions planes parallel to the working surface of the flat and one-half wavelength apart. Dark fringes occur at their intersection with the part.

number of fringes represents the separation between the surfaces in units of one-half wavelengths.

Cosine error does exist, but in a real case, not like our exaggerated drawing (see Figure 16–7), the two surfaces are so nearly parallel (1 to 10 seconds of arc) that the error is in micrometers (billionths of an inch).

The Basic Air-Wedge Configuration

You can conveniently practice surface inspection with optical flats using gage blocks, but you can also use any other sufficiently flat part such as refrigeration compressor seals and automatic transmission seals. Because they will be readily available to you, we will continue our discussion using gage blocks, but as always, the principles apply to any suitable part.

After you thoroughly clean the gage block and optical flat, wring them together and work with the wringing until the fringe pattern crosses the block (see Figure 16–8). The five dark bands (fringes) indicate that there is an air wedge of five one-half wavelengths height separating the flat from the gage block. We do not know which way the wedge is facing (either the right or left end may be open), but we can find out which is open by applying force to the ends. If you do

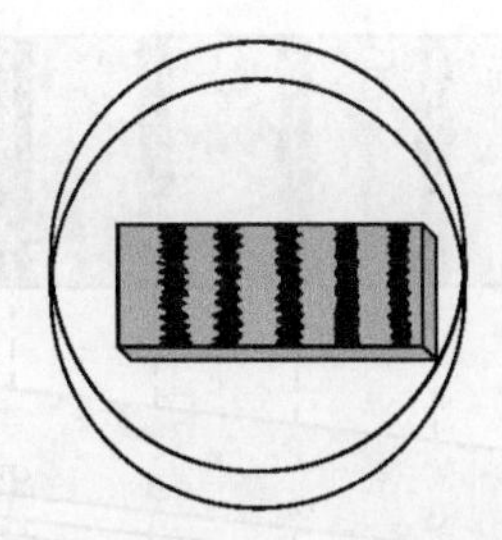

FIGURE 16–8 The five fringes represent an air wedge with a height of five one-half wavelengths. They do not show which end is the open one.

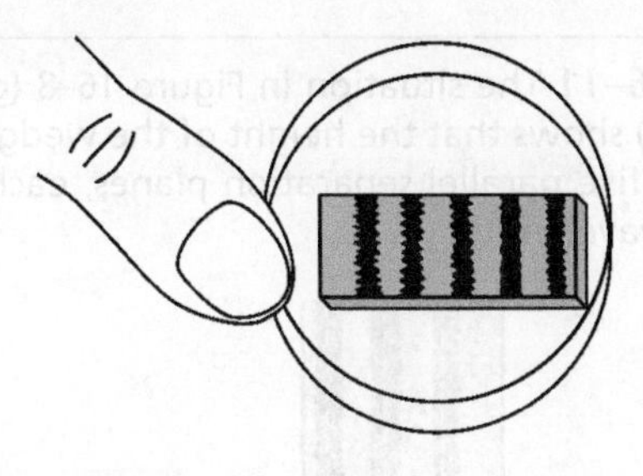

FIGURE 16–9 If force at one end does not change the fringe pattern or causes very little change, that end is in contact with the gage block or part.

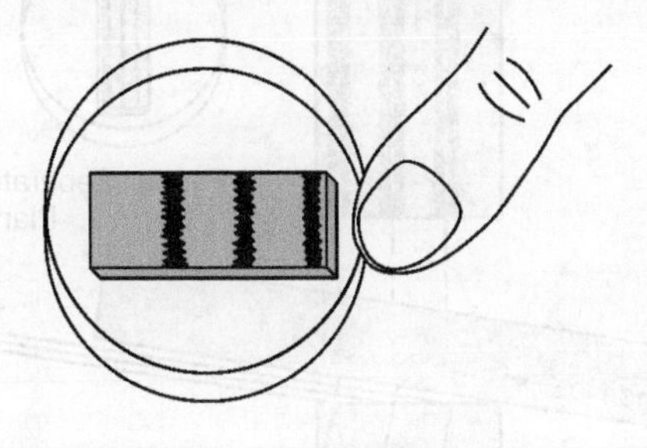

FIGURE 16–10 When force does widen the fringe bands, you know you are squeezing the air wedge closed.

not see much change, you are pushing along the line of contact (see Figure 16–9). When the fringes spread out, you are forcing the wedge closed (see Figure 16–10), and pushing against the open end. For illustration, we have exaggerated the situation in Figure 16–11.

The air wedge forms the basic configuration for fringe patterns, and we can form just as useful fringe

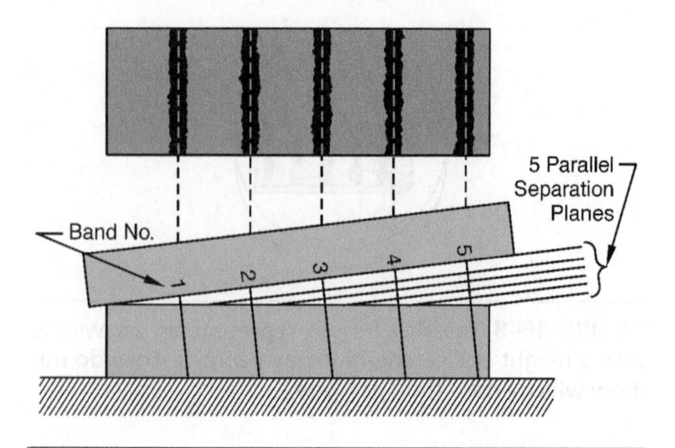

FIGURE 16–11 The situation in Figure 16–8 (greatly exaggerated) shows that the height of the wedge is represented by five parallel separation planes, each equal to one-half wavelength.

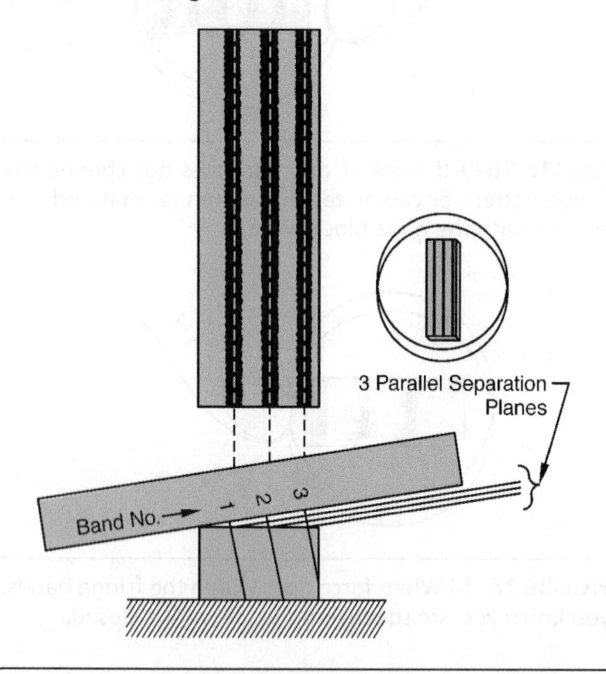

FIGURE 16–12 If the fringe bands had run lengthwise, the air wedge would still exist and the contact could be found by applied force.

patterns if we orient the bands lengthwise along the block (see Figure 16–12). We can always find the contact by applying force and then determine the height at the widest point by multiplying the number of

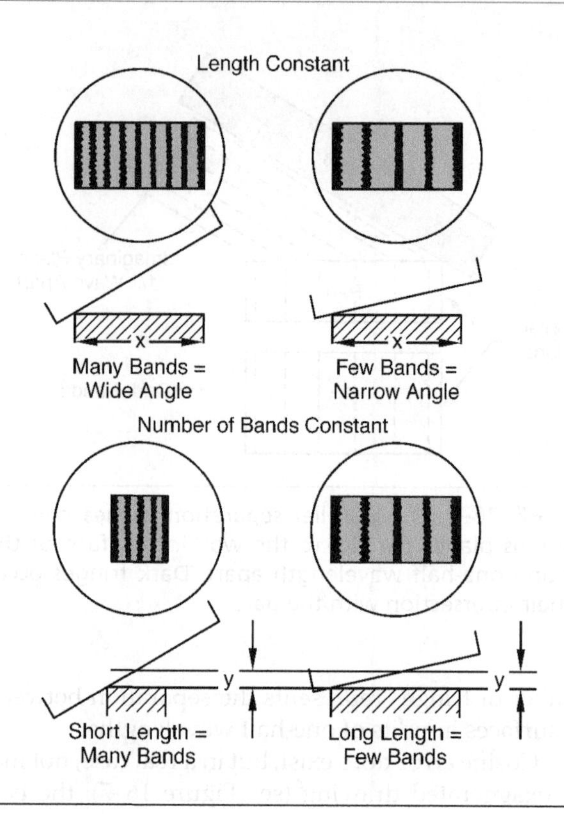

FIGURE 16–13 These relationships apply to air wedge measurements.

fringes by one-half the wavelength of the light we use—295 nm (11.6 µin.) for helium light. In general, two relationships (see Figure 16–13) always apply: the fewer the bands, the narrower the angle; and the more numerous the bands, the greater the angle.

You must understand that the number of bands is a measure of height difference—not of absolute height. As in all measurement, you must have a reference from which you can express the length.

Surface Inspection

In surface inspection, the reference is easy: We arbitrarily pick some part of the surface as the reference from which we express the other parts. Referring to Figure 16–12, we can tell from the fringe pattern that the surface is flat because the fringe bands are straight and uniformly spaced.

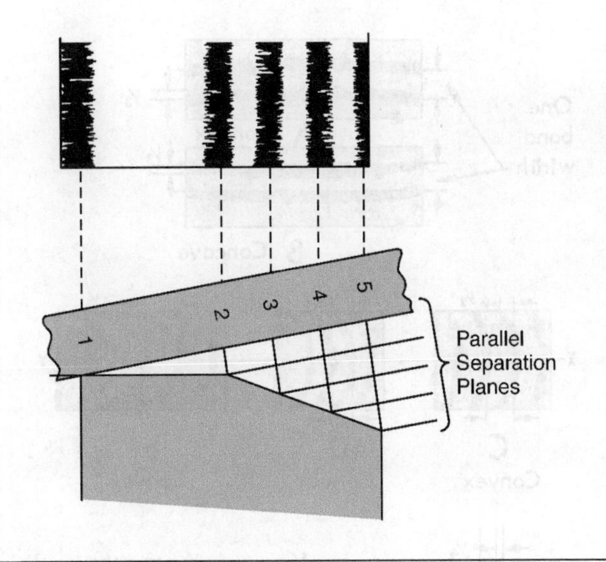

FIGURE 16–14 The sharp drop-off is clearly shown by the close bands on the right.

If there had been a sharp drop-off, the pattern would have changed (see Figure 16–14) and we would know from the closely spaced bands that the angle is larger along one edge than along the other. Again, we cannot make a measurement unless we know a reference, which, in this instance, is the left edge of the block.

To extend the principle (see Figure 16–15A), we use the lower edge of the block as a reference line because we know that the block is nearly flat along its length. In Figure 16–15B, the bands curve toward the line of contact, showing that the surface is convex and high in the center. We have shown a concave and low in the center surface in Figure 16–15C. Figure 16–15D shows that the surface drops off at the outer edges. In this case, you should notice how important the reference line is: if the reference had been at the top of the surface, the same fringe pattern would have told us that the surface rises at the outer edges.

We show a surface that is flat at one end but becomes increasingly convex in Figure 16–15E, and F shows a surface that is progressively lower toward the bottom left-hand corner. You should notice that the bands turn toward the line of contact and get progressively farther apart. From lower right, in Figure 16–15G, to upper left, this surface is flat, but because the bands

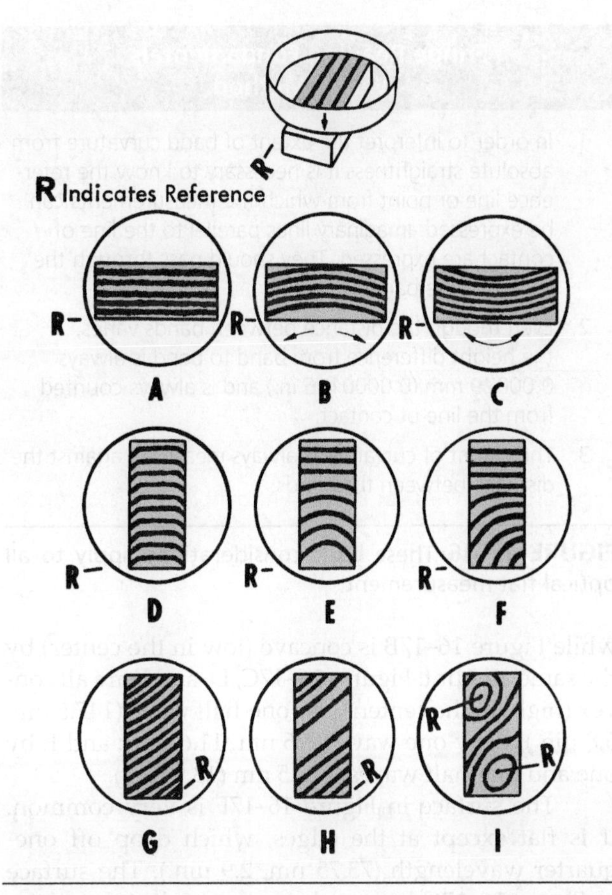

FIGURE 16–15 Fringe patterns reveal surface conditions like contour lines on a map.

curve slightly away from the line of contact, we know that the surface is slightly concave. In Figure 16–15H, the surface is flat in the direction that the bands run; however, the bands are widely spaced diagonally across the center, so we know the surface is lower at the ends. You will run into the contact in Figure 16–15I often. It shows two high points surrounded by lower areas.

Once you can recognize surface configurations from their fringe patterns and understand the fundamentals of measuring with fringes (see Figure 16–16), it is an easy step to measurement of the configurations.

In Figure 16–17A, the surface is convex (high in the center) by one-third of a band, which is equal to one-third of 295 nm (11.6 μin.) or 98.3 nm (3.866 μin.),

Fringe Reading Fundamentals Wedge Method

1. In order to interpret the extent of band curvature from absolute straightness it is necessary to know the reference line or point from which the measurements can be expressed. Imaginary lines parallel to the line of contact are expressed. They should pass through the center of the bands.
2. Even though the distance between bands varies, the height difference from band to band is always 0.00029 mm (0.0000116 in.) and is always counted from the line of contact.
3. The extent of curvature is always measured against the distance between the bands.

FIGURE 16–16 These basic considerations apply to all optical flat measurement.

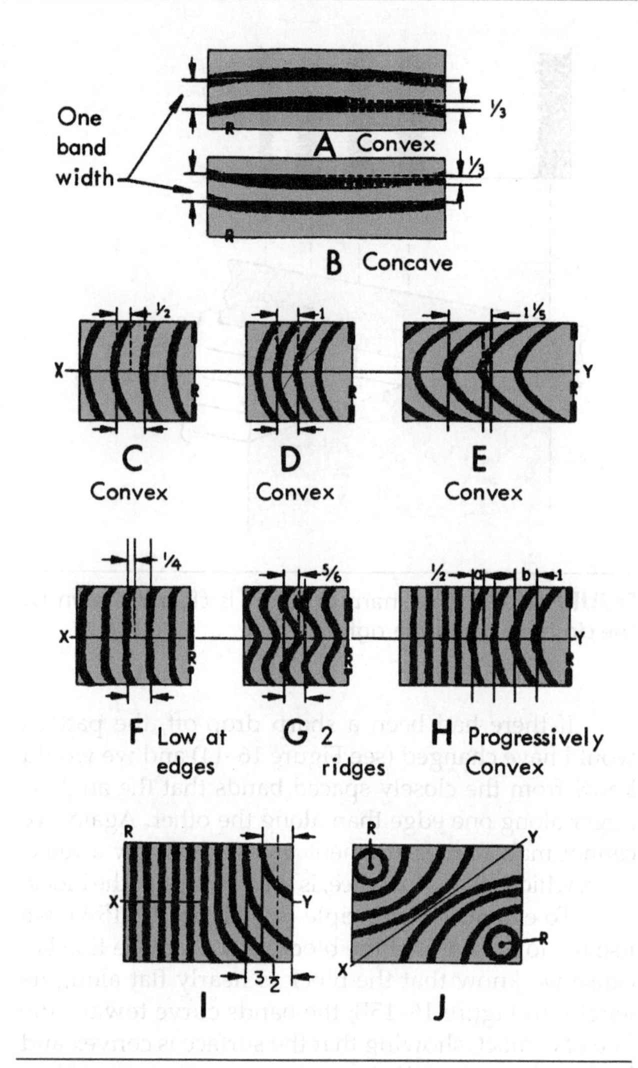

FIGURE 16–17 Measurements are the deviation of bands compared with the distance between bands.

while Figure 16–17B is concave (low in the center) by the same amount. Figures 16–17C, D, and E are all convex (high) in the center: C by one-half wave (147.5 nm, 5.8 µin.); D by one wave (295 nm, 11.6 µin); and E by one and one-half waves, 442.5 nm (13.9 µin.).

The surface in Figure 16–17F is very common. It is flat except at the edges, which drop off one-quarter wavelength (73.75 nm, 2.9 µin.). The surface in Figure 16–17G has two low troughs, but the center and edges are the same height: the ridges are five-sixths of 295 or 245.8 nm (11.6 or 9.7 µin.) high.

In the real world, few surfaces are as uniform as our examples. Most change from end to end, typically as in Figure 16–17H, which starts with a flat surface and becomes progressively more convex toward the right. At (a), the surface is 0.15 µm (one-fourth wavelength, $\lambda/4$, 5.8 µin.) convex; then, at b, it is 295 nm (one-half wavelength, $\lambda/2$, 11.6 µin.) convex. In Figure 16–17I, the surface is flat near the reference line but rises at the right edge, so that the top right edge is approximately one wavelength high while the lower right edge is about three and one-half wavelengths high.

In Figure 16–17J, the surface has two high points and between them is a low trough that we approximated with line XY. In addition, you should notice the four convex bands on each side of the high points, which indicate that the trough is four and one-half wavelengths or 1.33 nm (52.3 µin.) low.

Which Way Should the Bands Run?

In each of the examples we have shown, we could have wrung the surfaces and the optical flat so that the fringe pattern was perpendicular to the pattern depicted. These fringe patterns would have created just as useful

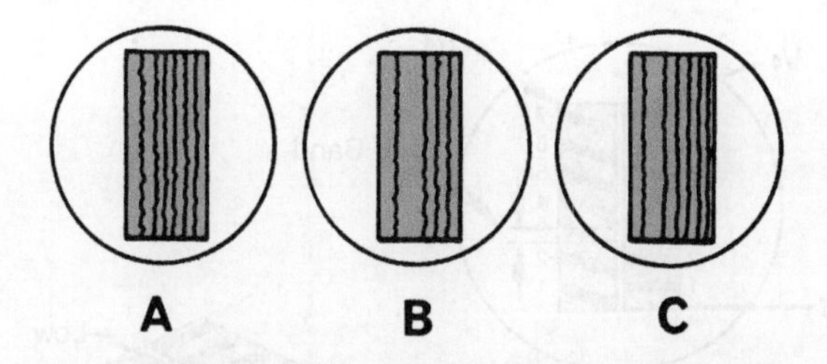

FIGURE 16–18 Spacing of the bands depends on the angle of the air wedge. A shows the same pattern as in Figure 13–14. B is the same surface but with a smaller angle. C is a larger angle.

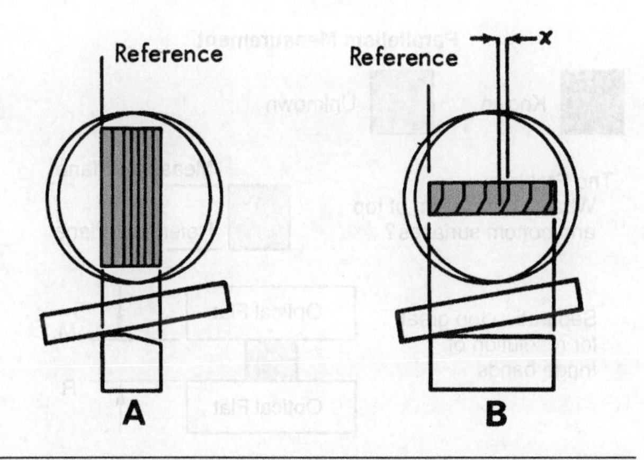

FIGURE 16–19 Both patterns show the surface conformation, but B permits direct measurement of the drop-off.

contour maps, showing the surface configuration, but measurement would have been more difficult because in cases like Figure 16–18A, we would have had many more bands to count to make our measurement.

Remember, the number of bands is a measure of height difference, not of absolute height (see Figure 16–18). All three fringe patterns show that the surface is straight along its length but drops off along one edge. The band spacing varies because the angle of the air wedge in B is smaller than in A. In Figure 16–18C, the air wedge is greater than at A. From these patterns, we could calculate the amount of drop-off; however, it would be easier to calculate the drop-off if we reoriented the pattern (see Figure 16–19). You can see the relationship between the two fringe patterns in this exaggerated drawing, where Figure 16–19A corresponds to Figure 16–19B. In B, we determine the amount of drop-off by calculating what fraction of one band is represented by (x).

In summary, if we know the contact point for any surface, the fringe pattern shows the surface conformation. If the elevation changes, it is easy to measure the change by using the bands that cross that area of the surface. We can measure the change by the amount that the bands deviate from straightness.

Measurements of Parallelism

We can measure the parallelism of planes to great accuracy with optical flats; however, there is a severe limitation to the measurement of parallelism. Fringes are only visible with the unaided eye when the separation distance is very short—less than 10-fringe bands (2.54 μm, 0.0001 in.). The Fringe Conversion Table (see Figure 16–29) can be used throughout this section.

We can use interferometers to make it easier to see greater separation distances, and, again, gage blocks can provide invaluable help. Because we can make up more than 200,000 dimensions to 1 μm (0.0001 in.) with a standard set of gage blocks, you never need to have separations beyond the resolution (sensitivity) of optical flats.

For example, if we want to check the parallelism of a questionable gage block, we can wring it to a flat, along with a known block. We know that the wrung surface of each gage block is parallel to the other to the limit of flatness of the optical flat. If the known block is in calibration, we have constructed a plane of known parallelism to the reference plane (see Figure 16–20). In addition, the top surface of the known block is almost parallel to the top surface of the unknown block.

Lack of parallelism may be along two of the three coordinates (see Figure 16–21). Therefore, to measure parallelism, we must determine the rotation about both of these axes. With optical flats, we can determine the rotations simultaneously. Thanks to gage blocks, we can construct the substitute reference plane very near the top plane of the unknown part;

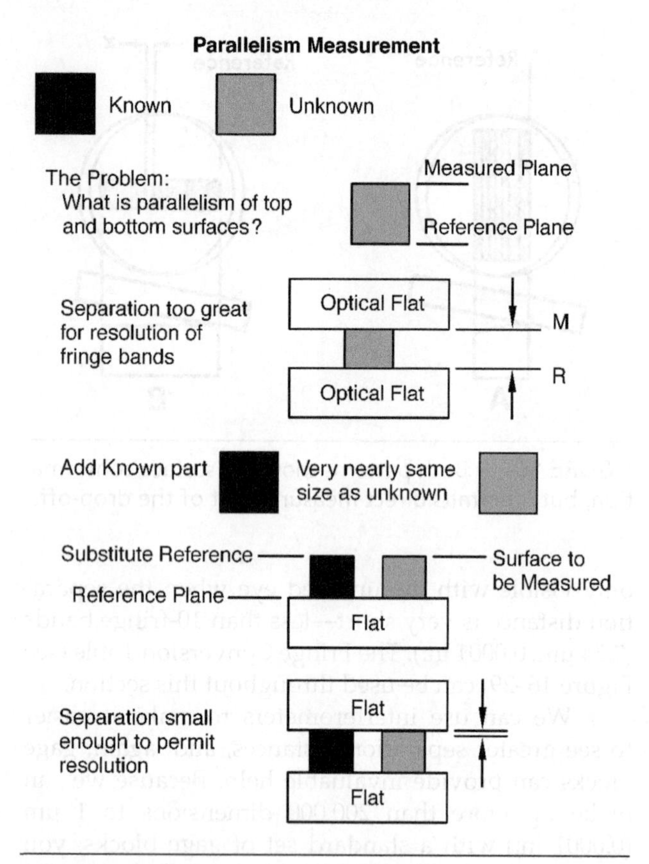

FIGURE 16–20 With gage blocks, a surface can always be constructed within the resolution range of optical flats.

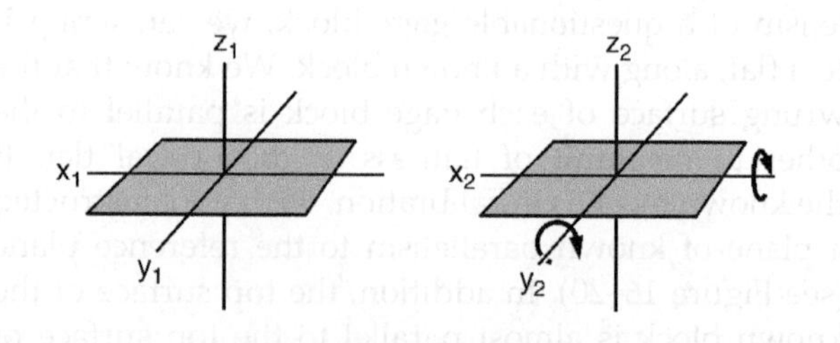

FIGURE 16–21 Parallelism of two planes is a measure of their relative rotation about the x and y axes. Rotation about the z axis and translation (movement) along any axis does not disturb parallelism.

therefore, we can see the fringe patterns. When we compare the fringe patterns of the known block to the unknown block's fringe patterns, we have measured the parallelism.

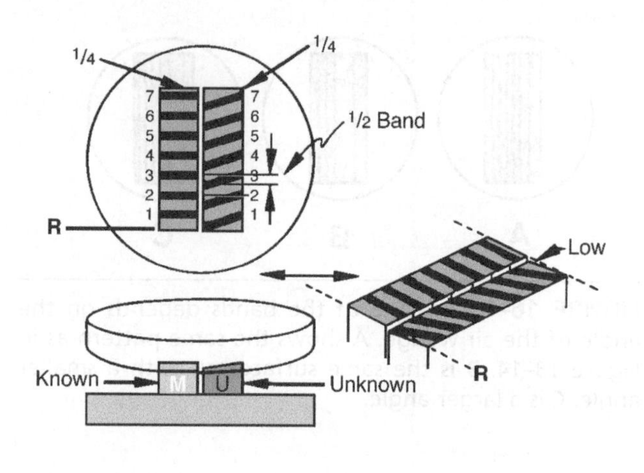

FIGURE 16–22 In this example, both surfaces have the same number of bands but the bands on the unknown surface appear at an angle.

Checking for parallelism (see Figure 16–22) reveals:

- That the unknown surface is parallel longitudinally to the known surface because it produces the same number of bands.
- That the surfaces are parallel across the width because the bands on the unknown are at an angle to those on the known surface.
- That the unknown is out of parallel to the known by one-half band in one width.

It also answers questions (see Figure 16–23) about:

- The relationship of the planes in the x axis
- The relationship of the planes in the y axis
- How much the planes are out of parallel

In this case, the fringe bands are parallel on both surfaces, which shows that the surfaces are parallel across their widths (x axes). Because there are more bands on the unknown surface than on the known one, we know that the air wedge between the unknown and the flat must be greater than the air wedge between the known and the flat. There are two more bands on the known block than on the unknown, which indicates that the unknown is out

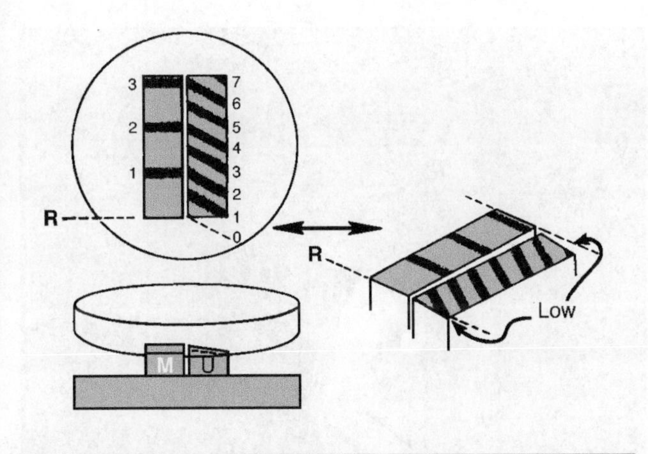

FIGURE 16–23 More bands in this example are seen on the unknown than on the known surface.

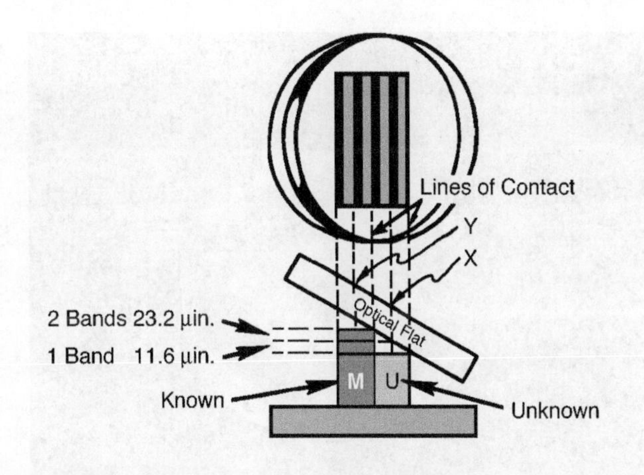

FIGURE 16–25 Counting the bands on the unknown part provides the measurement of the height difference. X and Y are points to apply force to find the lower part.

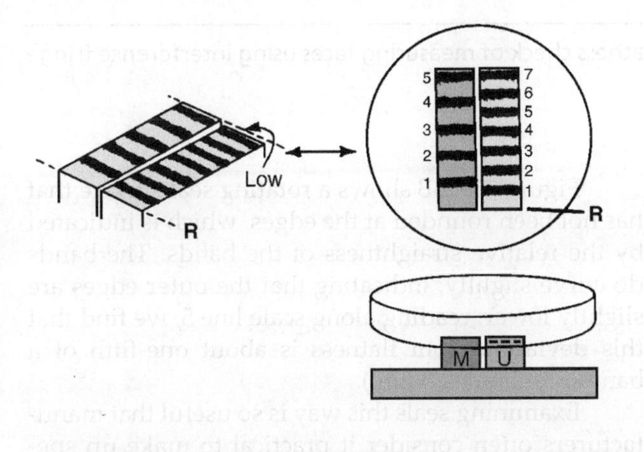

FIGURE 16–24 Surfaces can be out of parallelism in two axes simultaneously. This example shows the resulting fringe pattern for one case.

of parallel by two bands in the length of the block. We could also say that the unknown is 586.74 nm (23.1 μin.) lower at its far end. (See Figure 16–29 for the values.)

Parallelism can exist in the x axis (see Figure 16–23), the y axis (see Figure 16–24), and in both axes at the same time (see Figure 16–25). In this example, the unknown surface is out of parallel by one band in the x axis and three bands in the y axis.

16-4 APPLICATIONS OF OPTICAL FLAT MEASUREMENT

Laboratories and manufacturers frequently prefer optical flats for certain applications—for example, they are unsurpassed for close examination of surface configurations, like ring seal inspection (see Figure 16–26). Figure 16–26A shows the band pattern produced by a rotating seal that is not flat and has its greatest variation from flatness approximately across the center.

The spot at the bottom is the point of contact, and from this point, we have established the seven scale lines as shown. Across the center, we construct line a-b, which terminates at the ends of band number 5. Then, we construct line x-y at right angles to the scale lines. We can now use this scale to interpret the flatness of the rotating seal at various points and measure variations in flatness. You should number all bands, and for easy reference, you can draw white dashes on the photo to direct the eye to the continuation of each band crossing the bore.

First, we will determine the degree of flatness across the center. Looking at the figure, you will notice that the outer ends of band 5 join reference line a-b and that across the center, line a-b is about halfway between scale lines 3 and 4 (or is at scale 3.5). You will

FIGURE 16–26 (A) Optical flat and square gage block. (B) Flatness check of measuring faces using interference fringe pattern. *(Courtesy of Mitutoyo American Corporation)*

also see that, at scale line x-y, band 5 coincides with scale line 5. The difference between the outer ends of the band and the center, therefore, is 1½ scale lines (442.5 nm, 17.4 µin.).

If we follow line x-y from the x end and compare the distance between bands with the distance between scale lines, we see that the bands are closer together toward the y end, which means that the surface is falling away from the reference plane established by the scale lines. Note that band 8 and scale line 7 coincide; therefore, the surface at this point is low by one band (295 nm, 11.6 µin.).

Bands 2, 3, 4, and 5 curve relatively more sharply as they approach the outer edges and the edge of the bore, which indicates that the edges have been lapped down more rapidly than the main part of the surface—most likely because loose abrasives piled up on the edges on the laps. We can determine the amount of wear by constructing profile lines as previously described. Note that band 2 travels around the bore and curves toward the point of contact, which indicates that the edge of the bore is rounded off—low about 178 nm (7 µin.).

Figure 16–26B shows a rotating seal surface that has not been rounded at the edges, which is indicated by the relative straightness of the bands. The bands do curve slightly, indicating that the outer edges are slightly lower: reading along scale line 5, we find that this deviation from flatness is about one-fifth of a band (58.42 nm, 2.3 µin.).

Examining seals this way is so useful that manufacturers often consider it practical to make up special optical flats where necessary (see Figure 16–27). In our example, Figure 16–27A has an extra thick flat, which is used to inspect a recessed seal. We need the additional thickness for mechanical reasons, not optical reasons, and we use the protruding portion to manipulate the flat.

When the recess is too deep for a thick flat, you can use a flat with gripping surfaces (see Figure 16–27B). In our example, we handle the flat with tweezers. In many instances, the seal is on a shaft; therefore, some part of the shaft must pass through the flat (see Figure 16–27C).

You can even check a large surface, such as a high-precision bench plate (see Figure 16–28), for

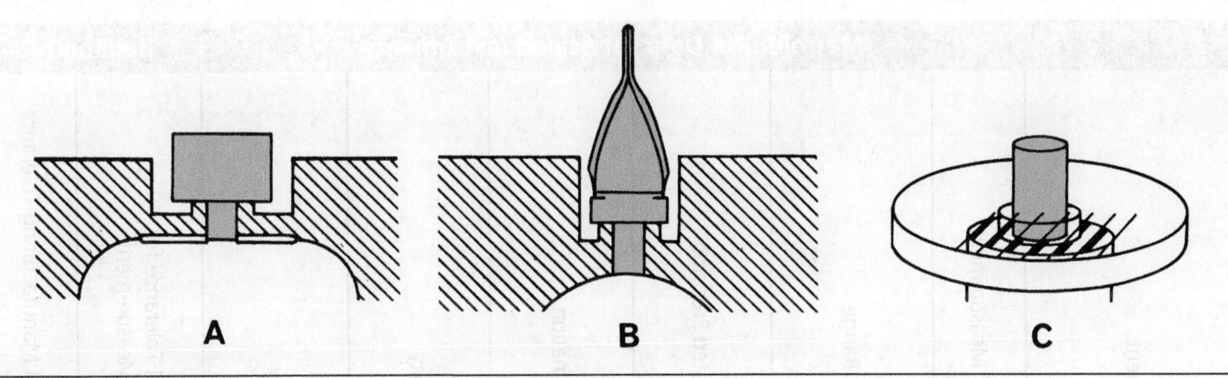

FIGURE 16–27 Special optical flats are used to examine surfaces inaccessible with standard ones.

Number of Bands	Metric Nanometer(nm)	Microinches (μin.)
0.1	29	1.2
0.2	59	2.3
0.3	88	3.5
0.4	118	4.6
0.5	147	5.8
0.6	176	6.9
0.7	206	8.1
0.8	235	9.3
0.9	264	10.4
1.0	294	11.6
2.0	588	23.1
3.0	881	34.7
4.0	1175	46.3
5.0	1469	57.8
6.0	1763	69.4
7.0	2056	81.0
8.0	2350	92.5
9.0	2644	104.1
10.0	2938	115.7
15.0	4407	173.5
20.0	5876	231.3

FIGURE 16–28 Fringe Conversion Table

flatness with a relatively small optical flat. If we use a 50 mm (2 in.) flat, and the band pattern reveals an error in flatness of 100 nm (4 μin.) across its 50 mm (2 in.) width, we can assume that this rate of error continues in the direction in which the bands run. For this example, the error will be 610 nm (24 μin.) across 30.48 cm (12 in.) of the surface. However, you should recheck before making any assumptions by taking successive readings as you move the flat 50 mm (2 in.) at a time in the direction in which the bands appear. Using the same technique, you can check the entire surface area and determine its degree of flatness. Although this method is practical and is a very precise method for small areas, it is generally too time consuming for surface plates. The Fringe Conversion Table (see Figure 16–29) is referenced throughout this section.

Interferometers

An extension of optical flats is the use of interferometers, which have long been used for dimensional measurement in the physical sciences and are rapidly coming into industrial use. Their principle is the same as optical flat measurement, but the mechanical design minimizes time-consuming manipulation. Because they have greater stability, you can use magnifying optical systems that permit resolution over longer distances and discrimination to fractions of a band. Additionally, the use of lasers has greatly extended the potential range and discrimination of the interferometers.

Metrological Data for Scaled Instruments							Reliability	
Instrument	Type of Measurement	Normal Range	Designated Precision	Discrimination	Sensitivity	Linearity	Practical Tolerance for Skilled Measurement	Practical Manufacturing Tolerance
Optical Flats Flatness Measurement	direct	to 0.0001 in.	1 to 10 µin.	1 to 10 µin.	1 to 10 µin.	1 to 10 min.	1 to 3 µin.	3 µin.
Surface Finish Measurement	direct	to 0.0001 in.	1 to 10 µin.	1 to 10 µin.	1 to 10 µin.	1 to 10 min.	1 to 3 µin.	3 µin.
Length Comparison Measurement	direct	to 0.0001 in.	1 to 10 µin.	1 to 10 µin.	1 to 10 µin.	1 to 10 µin.	3 to 5 µin.	20 µin.

FIGURE 16–29 These values are highly arbitrary because of the difficulty in making a science of the skill of measuring with optical flats. For example, a 2-band pattern usually permits one band to be divided into 10 parts, but a 10-band pattern on the same part permits virtually no interpolation.

16–5 PRINCIPLES OF OPTICAL METROLOGY

The measurement of manufactured components is an essential part of production quality control. Many optical measurement instruments, including optical flats, require a reference surface of known form with which to compare the surface being measured.

Aircraft production, even though it seems different from other manufacturing, has the same basic requirements for the linear dimensions and geometry of the features, which are straightness, flatness, plumbness, and squareness (see Figure 16–29). We gain a great advantage when we use optical methods to examine these requirements because of:

- The magnification possible with the telescope
- The reversal technique
- The ability to use gravity as a reference

In Chapter 17, we will discover that there is really little difference between a microscope and a telescope. Both use the same phenomenon: That for measurement purposes, you can treat a real *image* formed by a lens exactly as if it were the real object; therefore, the image can be magnified by a lens system. In addition, both can have a reticle in the plane of the image that you can view along with the image. In a telescope, the point in the object where the center of the crosslines appears is on the *line of sight* or the *line of collimation,* which is absolutely straight in undisturbed air. We can aim the crosslines at a target, we can place a target on the line of sight, or we can make measurements based on the line of sight. You can expect to be able to make measurements to an accuracy of 1:200,000, and the reference itself is accurate to 50–70 µm at 30 m (100 feet). For optical metrology, you must be able to check the alignment within each instrument by the reversal technique, doubling the error so that we can eliminate it.

Gravity provides the basic reference for optical metrology. As discussed in Chapter 12, the spirit level can produce a reference of great accuracy as well as precision. In optical metrology, "vertical" means in the direction of gravity, and "horizontal" or "*azimuth*" means perpendicular to gravity.

Geometry of Optical Alignment

Optical metrology provides advantages over contact measurement. For the results of optical alignment to be useful, they must be unambiguous and repeatable; therefore, they must follow some basic principles, with which you should be familiar. They include:

- A straight line is the shortest distance between two points, and any two points establish a line. There can only be one line established by any two points, but there is no limit to its length.
- Three points on the same line are *collinear*. If any point is not on the line created by two other points, you can describe the relationship among the points in three ways (see Figure 16–30). You must be able to recognize which line (and points along that line) is the desired reference and which are errors (see Figure 16–31).
- Any three points establish three straight lines. In addition, they establish a triangle with three inside and nine outside angles (see Figure 16–32). If you know the distances between the points, you can establish the angles by trigonometry. Even more useful in optical alignment, their gradients are proportional.
- A plane is defined by three points that are not collinear (see Figure 16–33) and an infinite number of planes may revolve around one line (AB), but only one specific plane can contain one line (AB) and any other point.
- Two straight lines may be parallel in one plane but not in another plane (see Figure 16–34).
- Light is reflected from a mirror at an angle that is twice the angle at which the mirror is set to the light source (see Figure 16–35).

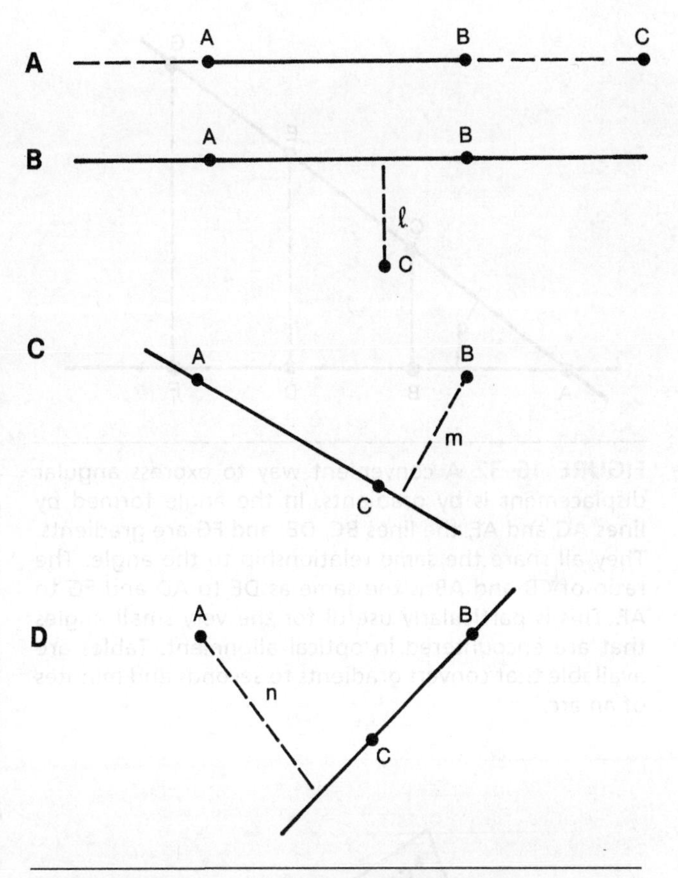

FIGURE 16–30 The shortest distance between two points is a straight line. All other points are either collinear with the line as point C in A above or may be expressed by one of three relationships as shown in B, C, and D. Any two of the points may be chosen to define the line, and the third point will be a displacement from it.

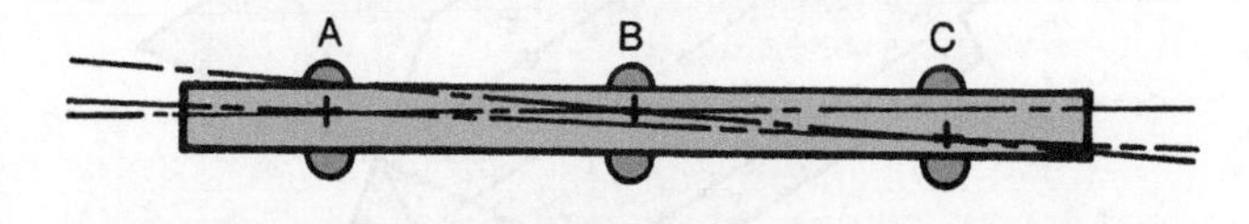

FIGURE 16–31 Consider a shaft supported by three bearings, where any two of the bearings could establish the desired axis. Then, the displacement of the third bearing is the error causing the malfunction. It is important to know which of the two are the true axes.

The basic unit of optical tooling is the telescope, which, in general, is an instrument that we can use to locate points in space and magnify them. The telescope is based on the conventions we accept about "rays" of light: that they are weightless, that they travel in straight lines, and that they can continue traveling infinitely.

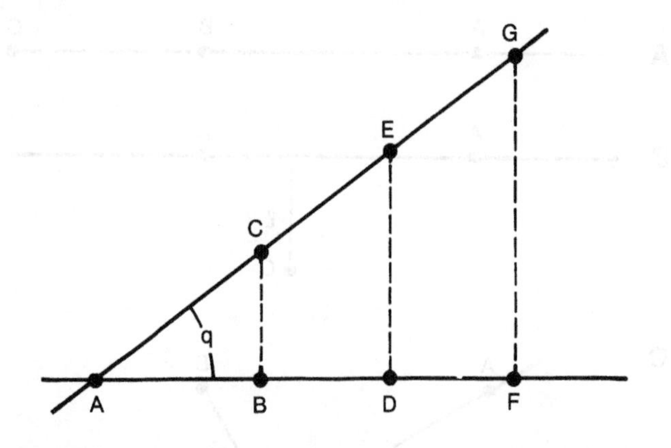

FIGURE 16–32 A convenient way to express angular displacement is by gradients. In the angle formed by lines AG and AF, the lines BC, DE, and FG are gradients. They all share the same relationship to the angle. The ratio of CB and AB is the same as DE to AD and FG to AF. This is particularly useful for the very small angles that are encountered in optical alignment. Tables are available that convert gradients to seconds and minutes of an arc.

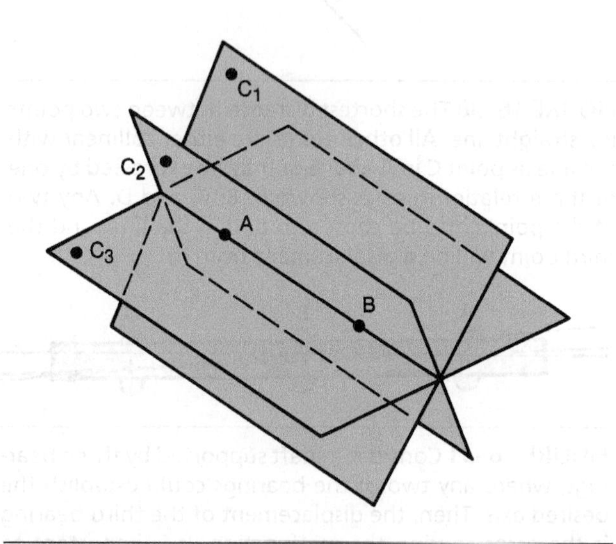

FIGURE 16–33 The straight line AB could lie in an infinite number of planes. Add any third point such as C_1, C_2, or C_3 and a specific plane is described. No other plane can contain the same three points.

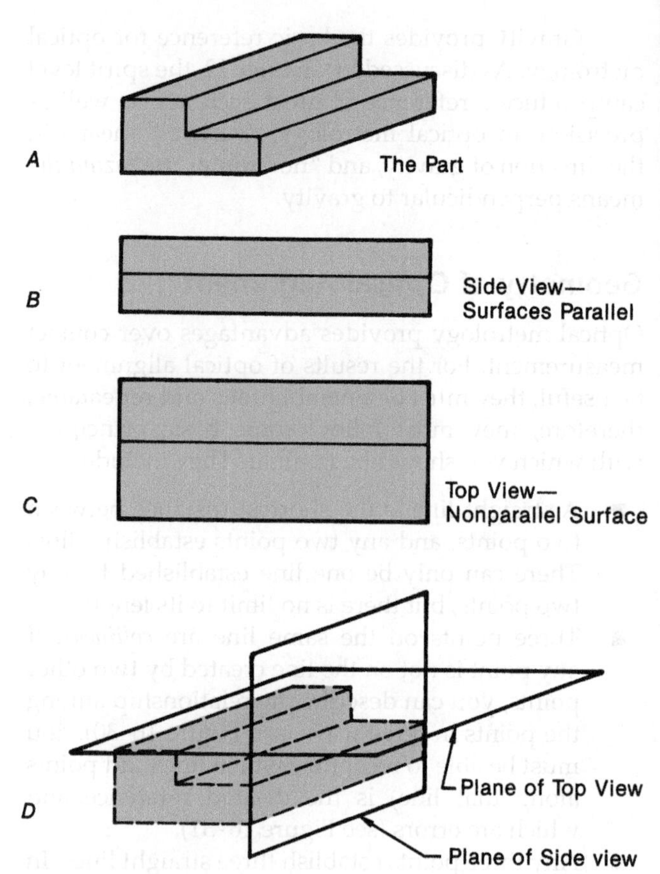

FIGURE 16–34 Whereas all of the horizontal edges of the part (A) may appear to be parallel, closer inspection may reveal that their parallelism depends on the choice of planes.

The telescopic instruments that are most widely used today are alignment telescopes, collimators, and survey-type instruments. Derived directly from surveying, the sight level and transit can be combined in a *jig transit* or the more specialized *theodolite*. The *line of sight telescope* is a simplified form of *alignment telescope*. Because alignment telescopes, collimators, and sight levels usually rest on the work you are examining, they establish their reference from the workpiece. The jig transit and theodolite are usually mounted apart from the work on separate stands. The various technologies of the optical systems and

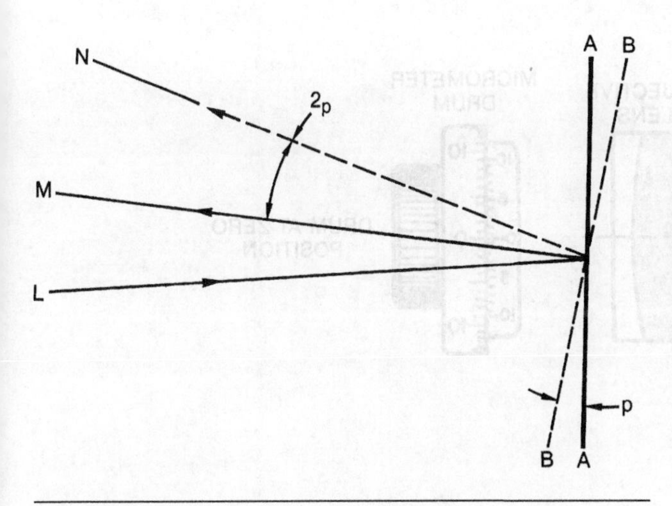

FIGURE 16–35 Incident light ray L on mirror AA is reflected as ray M. If the mirror is rotated by angle p to position BB, the reflected ray N rotates by 2p.

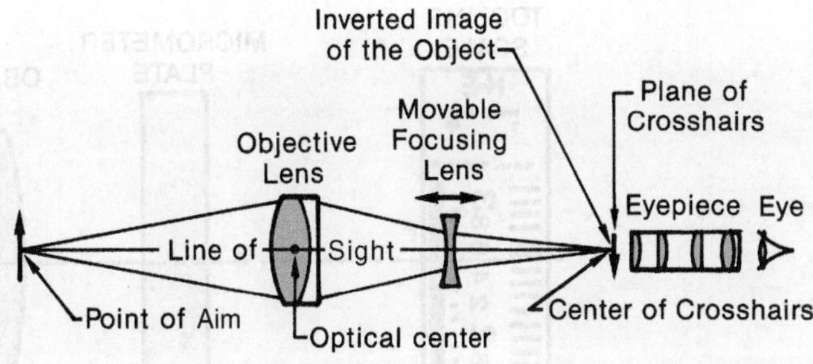

FIGURE 16–36 The optics of the alignment telescope place a crosshair reticle in the plane of the image of the object. In optical metrology, the object is usually a scale or a target.

applications give us a wider variety of instruments to choose from when we are making the important decision of which instrument is best for the job.

16-6 ALIGNMENT TELESCOPE—STRAIGHTNESS MEASUREMENT

Because it is similar to a riflescope, the alignment telescope is often called a "sight." The optics are similar for both instruments (see Figure 16–36). Riflescopes always have erecting systems so you can view the target in an upright position, and most optical metrology telescopes also have adopted this system.

Line of Sight Telescopes

The simplest form of alignment telescopes—line of sight telescopes—are most useful for establishing lines of reference. When you want to measure with them, you need to add an optical micrometer. The *optical micrometer* displaces the line of sight parallel to itself by means of an optical flat in the optical path (see Figure 16–37). The micrometer moves the flat through small angles.

When the flat is normal to the optical path, it does not affect the angle or the displacement of the rays passing through the micrometer plate. As you move the flat into a slight angle, however, the rays are displaced but the angle of the rays does not change. The range of displacement is usually 2.54 mm (0.100 in.), the discrimination is 2.54 mm (0.001 in), and the claimed accuracy is ±5 μm (±0.0002 in). In some instruments, you can use either a red or black scale, depending on the direction of displacement; one manufacturer has added an eyepiece for you to read the micrometer.

The Alignment Telescope

When we add an optical micrometer to a line of sight telescope, we get an alignment telescope. Most alignment telescopes have two micrometers: one for the x axis, and one for the y axis (see Figure 16–38). Typically, magnification varies automatically from 4X at zero focus to 46X at infinity focus.

We use the line of sight as the basic reference; therefore, we need to establish the most accurate line of sight possible in respect to the workpiece. To ingeniously solve this problem, manufacturers use two spheres arranged so that the line of sight passes through the center of each because, geometrically, you can turn a sphere without changing its center position. One sphere is aligned with the telescope, and the other

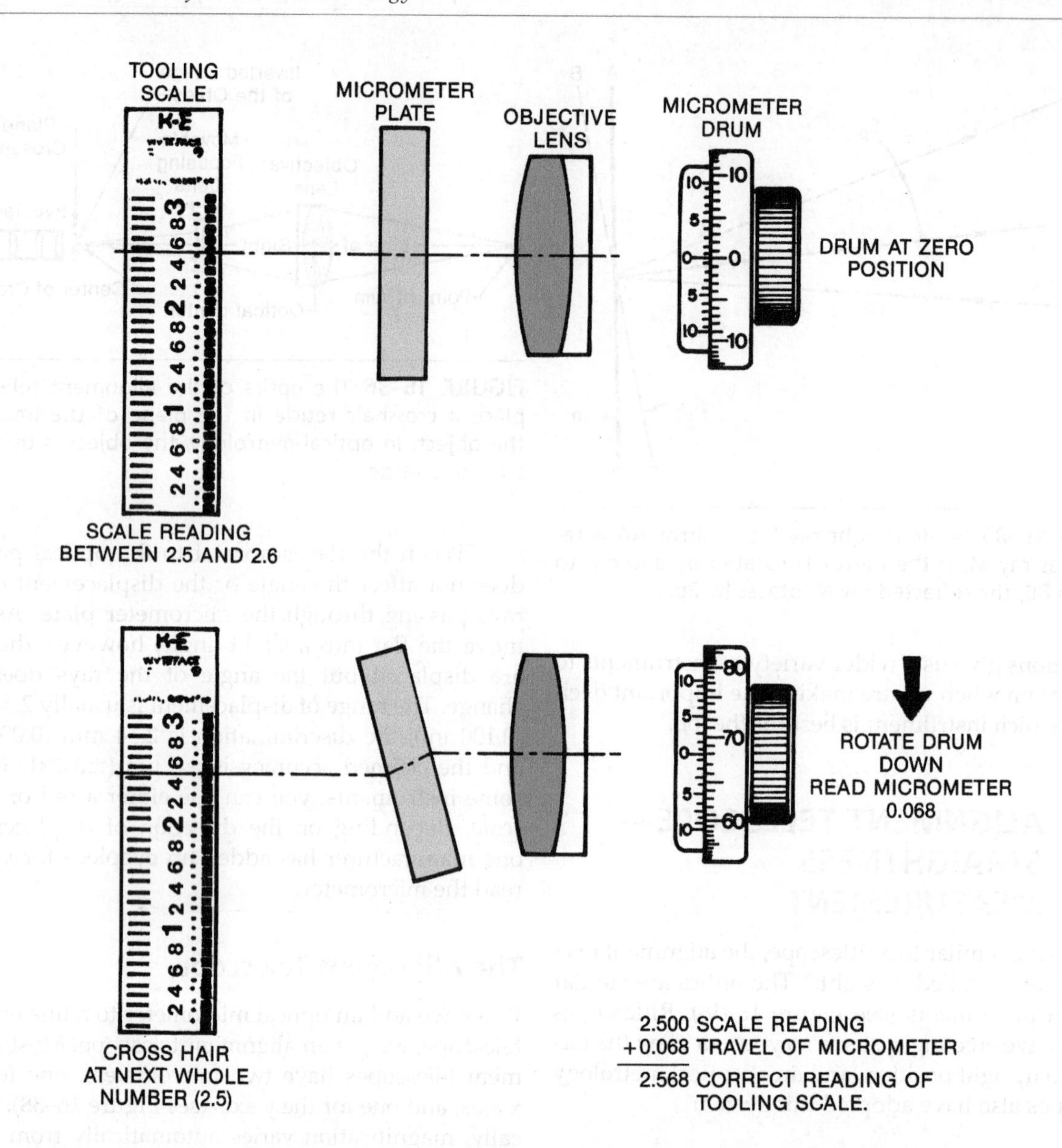

FIGURE 16–37 The optical micrometer displaces the crossline but does not change the angle of the line of sight. The displacement is read on a micrometer drum.

sphere is aligned with the target (see Figure 16–39). This mounting provides a constant datum (see Figure 16–40) from which we can take measurements. The socket for the sphere is known as a *mounting cup*, *flange cup*, or *cup mount*, and the telescope is mounted on a base (see Figure 16–41).

When a telescope and target are mounted in this manner and aligned with a line of sight, we can

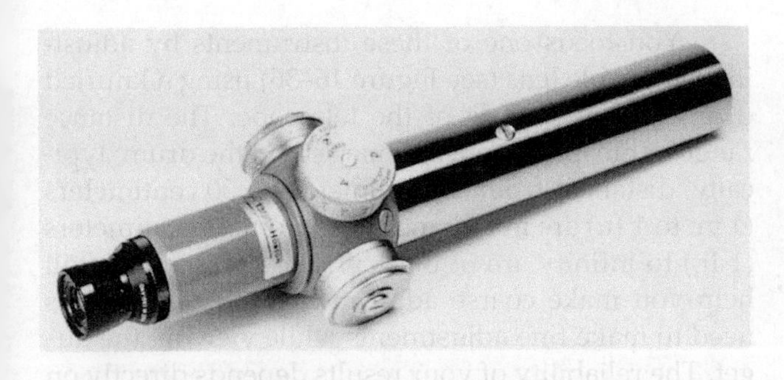

FIGURE 16–38 The alignment telescope is similar to the line of sight telescope but includes at least one optical micrometer. Some have optical micrometers in two axes. *(Courtesy of AMETEK Taylor Hobson.)*

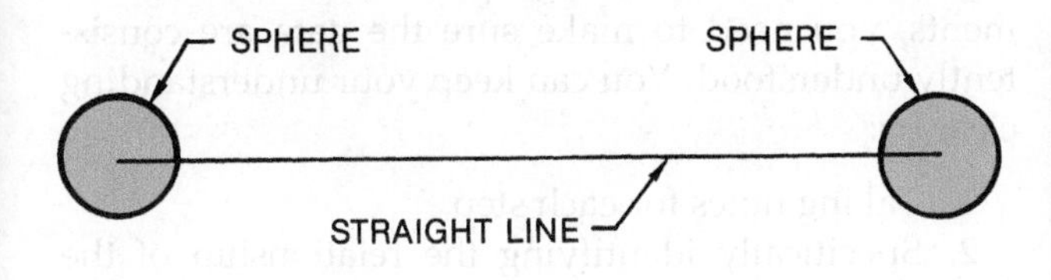

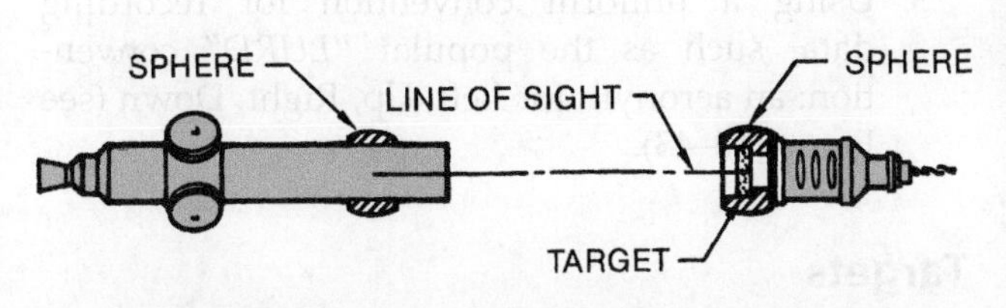

FIGURE 16–39 A straight line connecting the centers of two spheres will be undisturbed even if the spheres are rotated. By aligning the telescope's optical axis with one sphere and a target with the other sphere, a convenient line of sight is established.

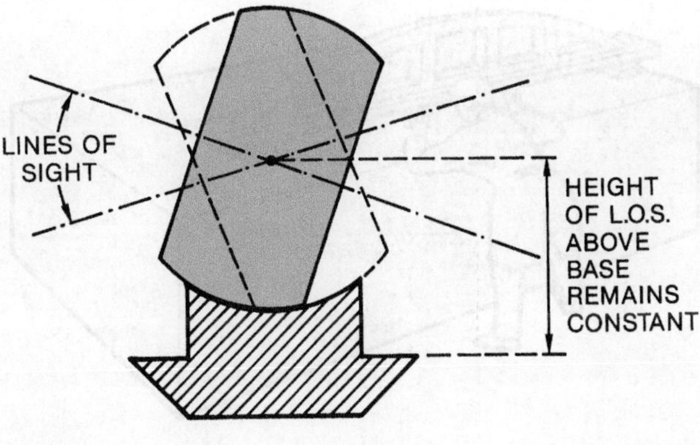

FIGURE 16–40 With the sphere mounted in a bearing, the line of sight always passes through its center. That provides a constant height datum for measurement.

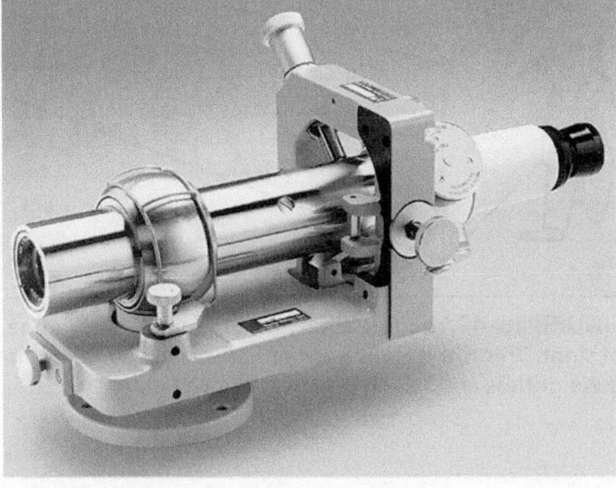

FIGURE 16–41 Alignment telescopes *(Courtesy of AMETEK Taylor Hobson)*

call it the optical equivalent of the straightedge (see Figure 16–42). We usually use it for a horizontal reference, but bases are available that make it possible for us to establish vertical line of sight alignment. Alignment telescopes are still widely used in aircraft, missile, and nuclear metrology.

Optical Limitations

The *field of view* is the largest diameter you can view through the telescope. As the distance from the telescope increases, the field of view increases and also varies by instrument.

Similarly, there is a maximum distance that we can have between two objects and still have both

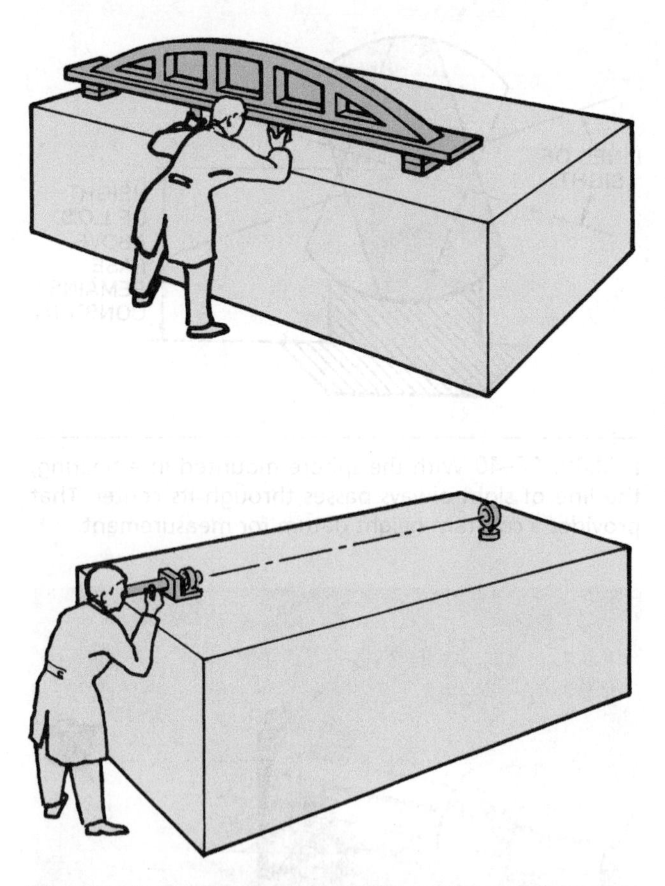

FIGURE 16–42 The telescope and target establish a line of sight. This is the equivalent of the straightedge, but it is weightless and has no practical length limits.

objects in focus. This is called *depth of focus.* Keeping two targets that are spaced widely apart in focus can be a problem, but depth of focus is mostly a subjective matter that each observer must control.

Distance from Telescope		Field of View	
1 ft.	0.3 m	0.7 in.	18 mm
10 ft.	3 m	2.9 in.	74 mm
50 ft.	15 m	12.5 in.	32 cm
100 ft.	30 m	24 in.	60 cm

You focus one of these instruments by adjusting a movable lens (see Figure 16–36) using a knurled drum on the outside of the telescope. The distance range of most telescopes is marked on the drum; typically, distances from 25 millimeters to 30 centimeters (1 in. to 1 ft.) are in red and those from 30 centimeters (1 ft.) to infinity are in black. These graduations will help you make coarse adjustments, but you always need to make fine adjustments while viewing the target. The reliability of your results depends directly on the precision of your focusing. You can identify incorrect focus by using parallax: If you move your head while looking through a properly focused telescope, the relationship of the crosslines of the telescope and the image of the target should not change. If it does, you need to focus more closely.

When you are using optical alignment instruments, you need to make sure the data are consistently understood. You can keep your understanding clear by:

1. Taking notes for each step
2. Specifically identifying the relationship of the telescope to the workpiece at the beginning
3. Using a uniform convention for recording data, such as the popular *"LURD"* convention: an acronym for Left, Up, Right, Down (see Figure 16–43).

Targets

Targets are available in a variety of patterns (see Figure 16–44), each with a specific benefit for different applications. Sometimes, you only need one target, but in some cases, you will need two or more to ensure the precision of your measurement. The intermediate targets must be transparent, and both targets must have precisely parallel surfaces. Typically, the tolerances of the faces of an intermediate target are within 2 seconds of arc.

In the standard circular target (see Figure 16–45), the vertical crossline of the telescope bisects the 0.6 annular ring, which means that the reading is 15.24 mm (0.6 in.) from the center of the target. Even though establishing this measurement requires visual approximation, you, as an experienced observer, should

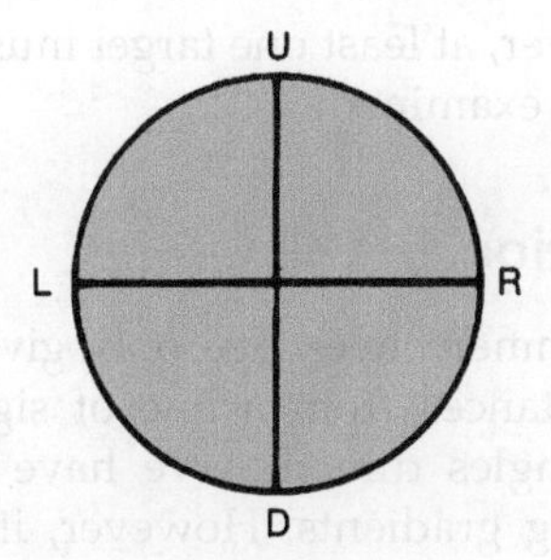

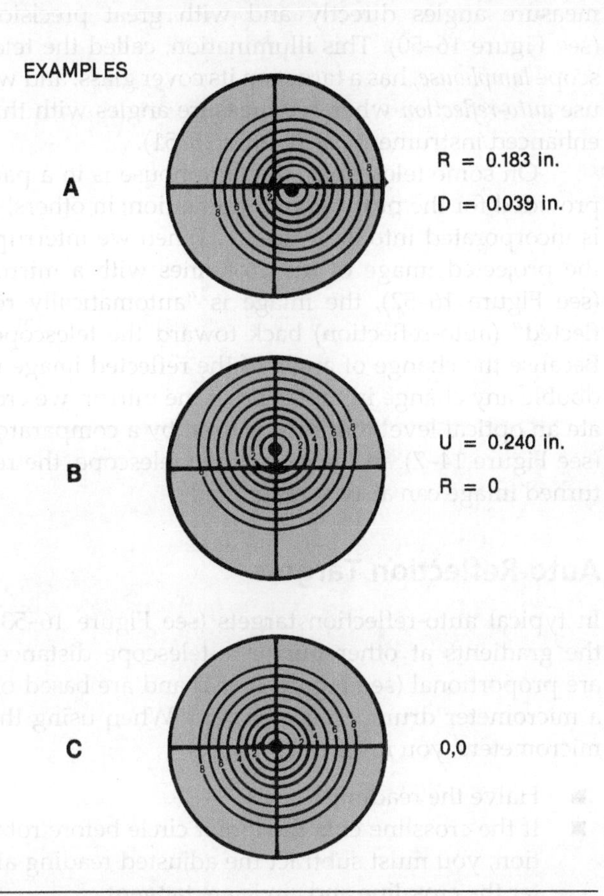

FIGURE 16–43 LURD is a convention for annotating measurements to prevent error. It stands for Left, Up, Right, Down. The three views above show a target in various positions with respect to the crosslines of the telescope. The displacements to the right are the micrometer readings. The letters "L.U.R.D." to their left show the way the directions of their displacements are positively identified.

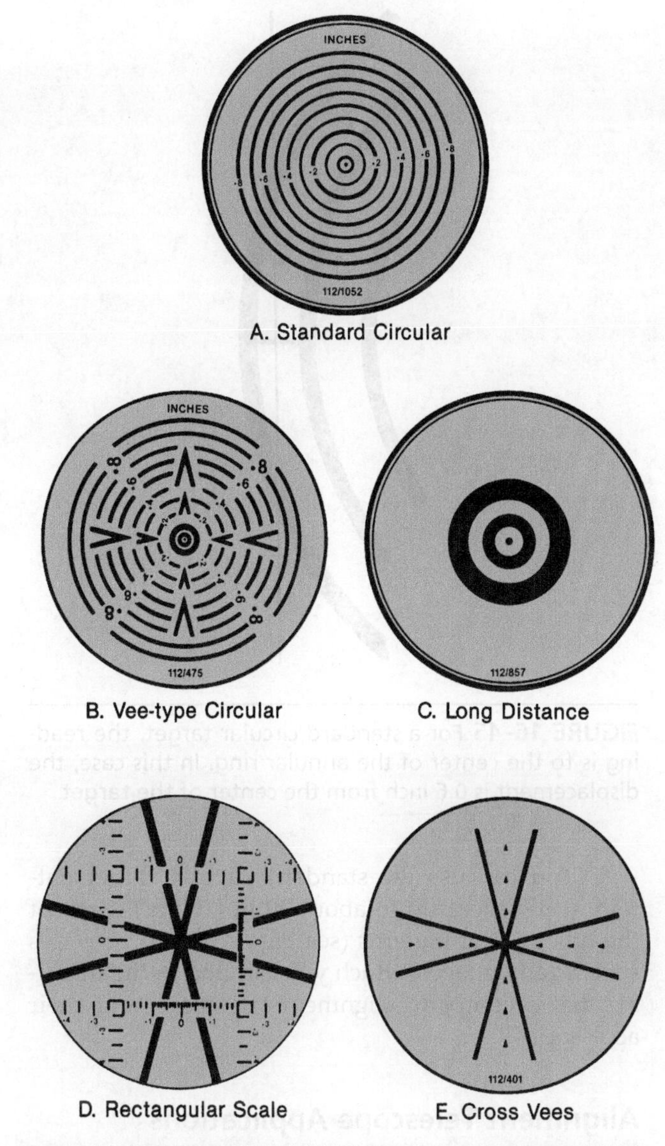

FIGURE 16–44 Typical target patterns. The graduations are helpful when the displacement is greater than the range of the micrometers.

be able to produce repeatable micrometer readings to within 5 μm (0.0002 in.) for targets up to 9 m (30 ft.) away. The general reading technique is shown in Figure 16–46.

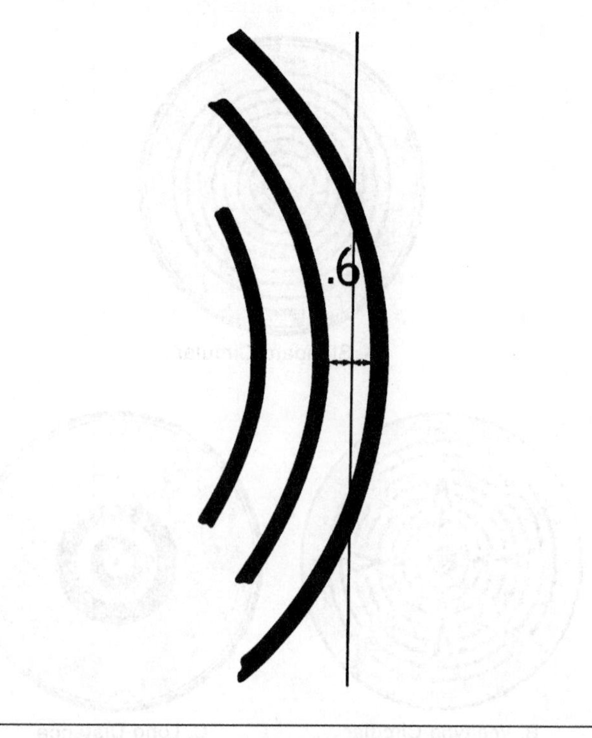

FIGURE 16–45 For a standard circular target, the reading is to the center of the annular ring. In this case, the displacement is 0.6 inch from the center of the target.

You can use the standard circular target pattern at distances up to about 30 m (100 ft.). Each of the other target patterns (see Figure 16–44) provides specific advantages, which you can find in the manuals that accompany alignment telescopes and their accessories.

Alignment Telescope Applications

In general, we use alignment telescopes because they can create a line of sight (see Figure 16–47). Then, we can use this established line to reveal any displacement from the line of sight (see Figure 16–48). We can also read vertical displacements directly by moving one target along the line of sight (L.O.S.) (see Figure 16–49).

For both of these general applications, the telescope and the datum target are supported independently of the workpiece, so you can detach them. In all cases, however, at least one target must rest on the feature(s) being examined.

Auto-Reflection

So far, the alignment telescope only gives us a way to measure distances from a line of sight; we cannot measure angles directly. We have to calculate the angles using gradients. However, if we add an internal illumination source to the telescope, we can measure angles directly and with great precision (see Figure 16–50). This illumination, called the telescope *lamphouse,* has a target on its cover glass, and we use *auto-reflection* when we measure angles with this enhanced instrument (see Figure 16–51).

On some telescopes, the lamphouse is in a part provided for the purpose of illumination; in others, it is incorporated into an eyepiece. When we interrupt the projected image of the crosslines with a mirror (see Figure 16–52), the image is "automatically reflected" (auto-reflection) back toward the telescope. Because the change of angle of the reflected image is double any change in the angle of the mirror, we create an optical level that can be used by a comparator (see Figure 14–7). In the case of the telescope, the returned image can also be magnified.

Auto-Reflection Targets

In typical auto-reflection targets (see Figure 16–53), the gradients at other mirror-to-telescope distances are proportional (see Figure 16–54) and are based on a micrometer drum setting of zero. When using the micrometers, you must be careful to:

- Halve the readings.
- If the crossline cuts the target circle before rotation, you must subtract the adjusted reading after the crossline and circle are tangent.
- If the crossline is outside of the circle before rotation, you must add the adjusted reading (see Figure 16–55).

The optical leverage provides the advantage of this method, but it also magnifies any errors in the mirror mounting. A variety of holders is available,

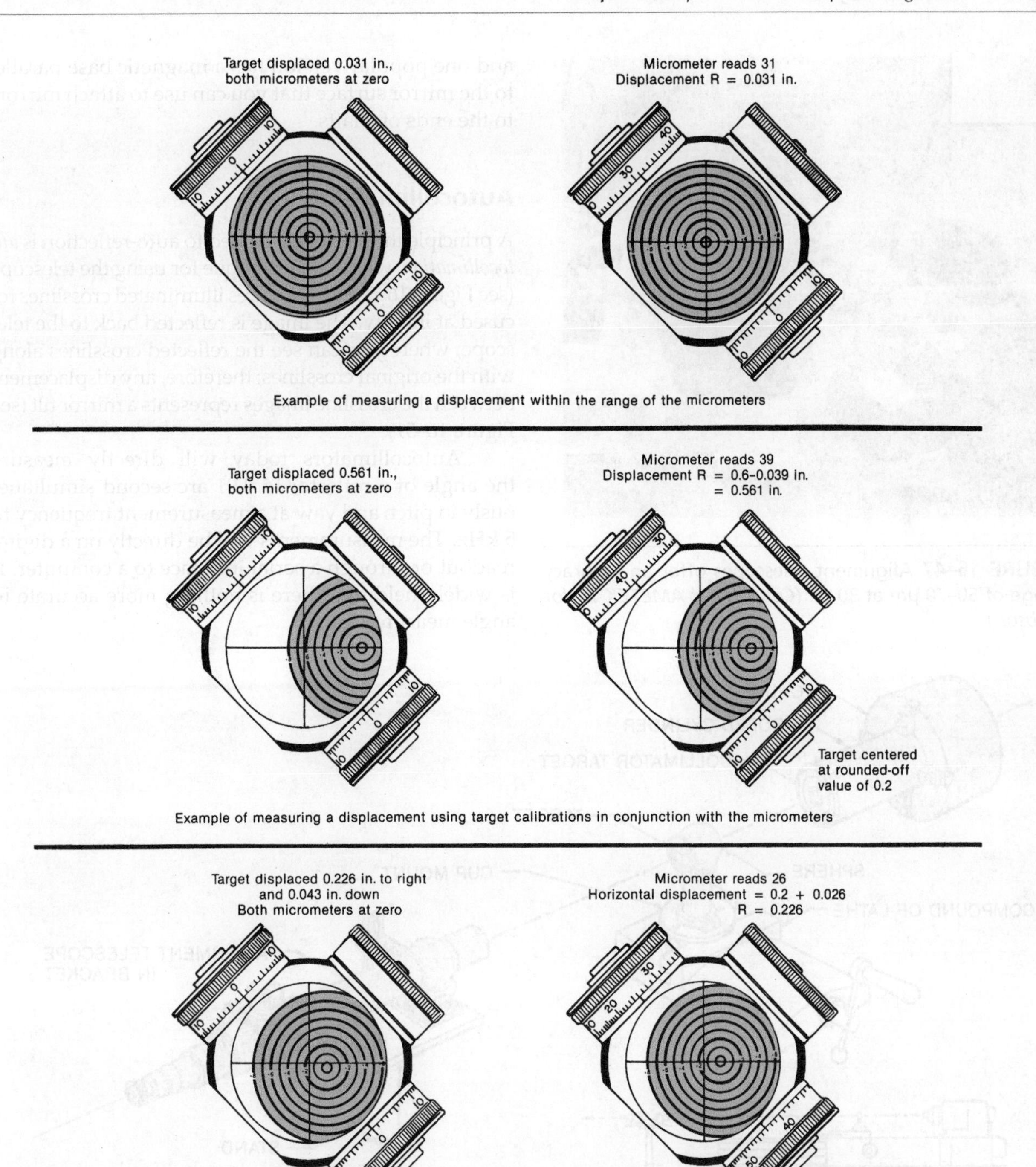

FIGURE 16–46 These examples show how the measurements are taken using the target pattern and micrometer.

FIGURE 16–47 Alignment telescopes offer an accuracy range of 50–70 μm at 30 m. *(Courtesy of AMETEK Taylor Hobson)*

and one popular version has a magnetic base parallel to the mirror surface that you can use to attach mirrors to the ends of shafts.

Autocollimation

A principle that is closely related to auto-reflection is *autocollimation,* which is a technique for using the telescope (see Figure 16–56) that features illuminated crosslines focused at infinity. The image is reflected back to the telescope, where you can see the reflected crosslines along with the original crosslines; therefore, any displacement between the crossline images represents a mirror tilt (see Figure 16–51).

Autocollimators today will directly measure the angle of a reflector to 0.01 arc second simultaneously in pitch and yaw at a measurement frequency to 5 kHz. The measurements will be directly on a digital readout or through a serial interface to a computer. It is widely held that there is nothing more accurate in angle measurement.

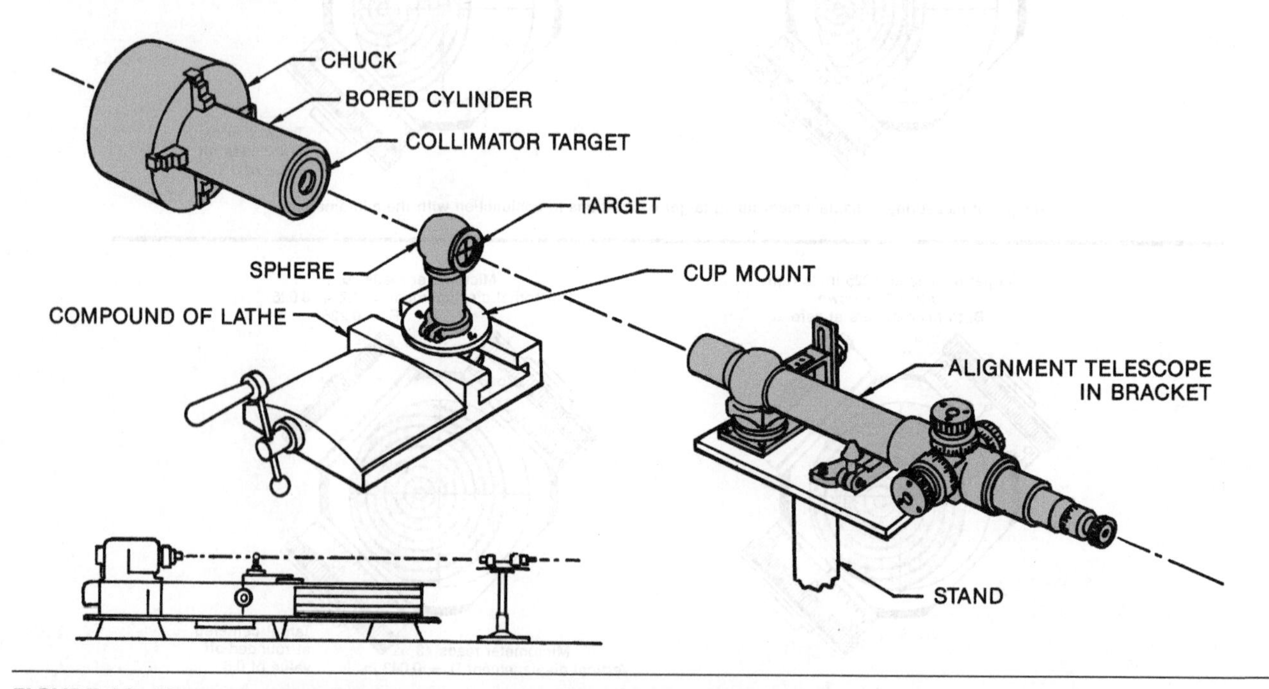

FIGURE 16–48 This is a typical application of the alignment telescope. A line of sight has been established along the center of rotation of a lathe spindle. An intermediate target attached to the compound allows the path of compound travel to be compared with the axis of the target spindle.

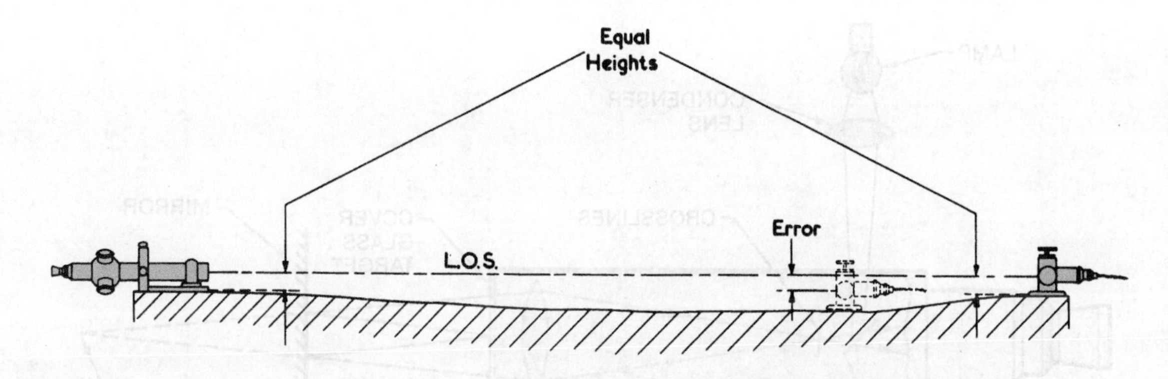

FIGURE 16–49 In addition to the measurement of deviations from straightness by means of one or more intermediate targets, the datum target may be used directly to measure vertical displacements.

FIGURE 16–50 A digital autocollimator used to precisely calibrate master angle gage blocks *(Courtesy of Micro-Radian Instruments)*

Industrial use of autocollimators ranges from angular feedback in military and aerospace environments to angle calibrations in controlled laboratories, including the calibration laboratory at NIST. Typical applications include flatness measurements of granite surface plates, straightness measurements of machine ways, measurement of positioning system repeatability, angular feedback for servo controllers, the alignment of optical elements, and the calibration of rotary and indexing tables (see Figure 16–57).

Modern autocollimators are compact in size. Most instruments fit in the palm of your hand and weigh only a few ounces. This makes them ideal for remote applications or in applications where the size and weight of measuring equipment is restricted.

Projection

You can use a telescope with crossline illumination as a projector (see Figure 16–58). In the example, the instrument is designed to drill a hole in plate C; however, the new hole must be in line with the holes in plates A and B. First, we place targets in the holes in plates A and B, then we shift the telescope or the workpiece (whichever is easier) until the line of sight passes through the centers of both targets. Because the crosslines are thin, the projected image will be faint and may require some reduction of the ambient light. Finally, we remove the targets so that the projected crosslines appear on plate C and show the centers for the new hole.

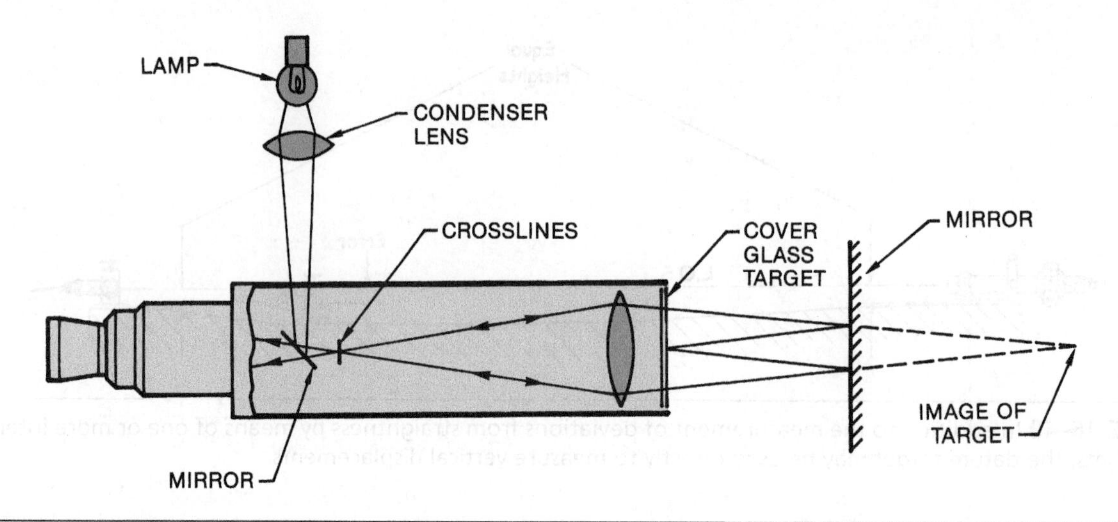

FIGURE 16–51 An internal illuminator may be used to illuminate the crosslines of the reticle. These are then projected with the line of sight.

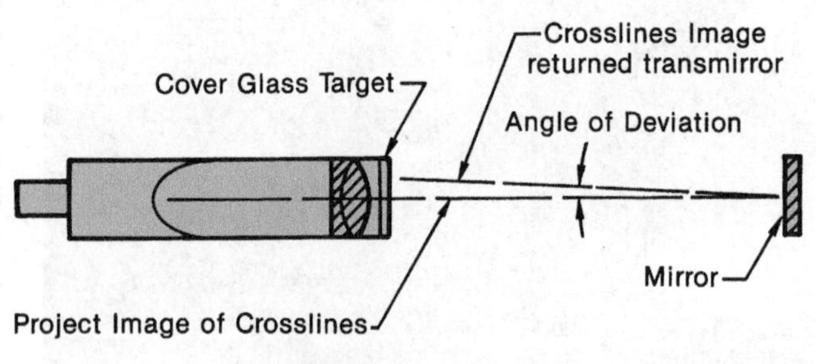

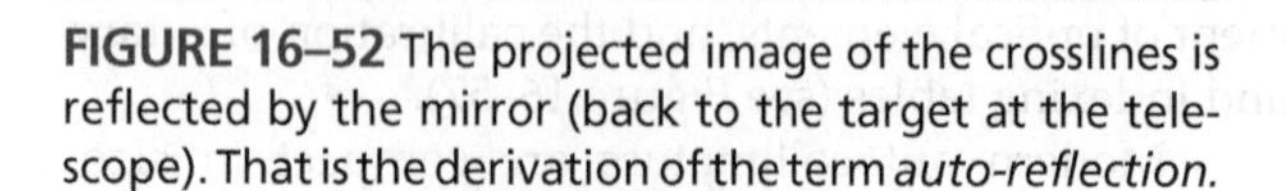
FIGURE 16–52 The projected image of the crosslines is reflected by the mirror (back to the target at the telescope). That is the derivation of the term *auto-reflection.*

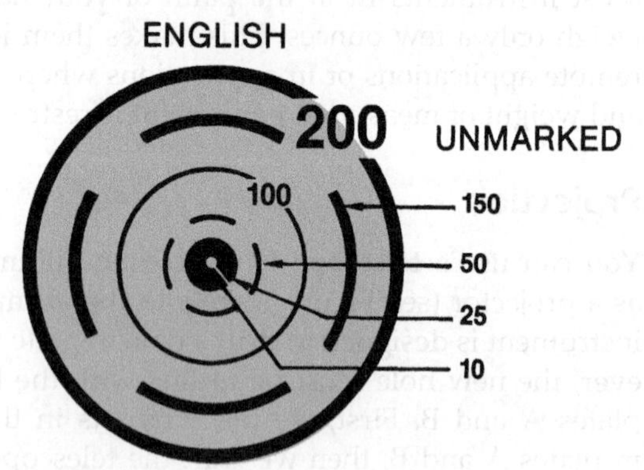

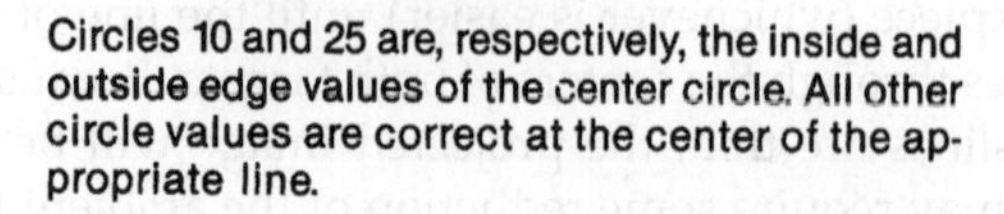
Circles 10 and 25 are, respectively, the inside and outside edge values of the center circle. All other circle values are correct at the center of the appropriate line.

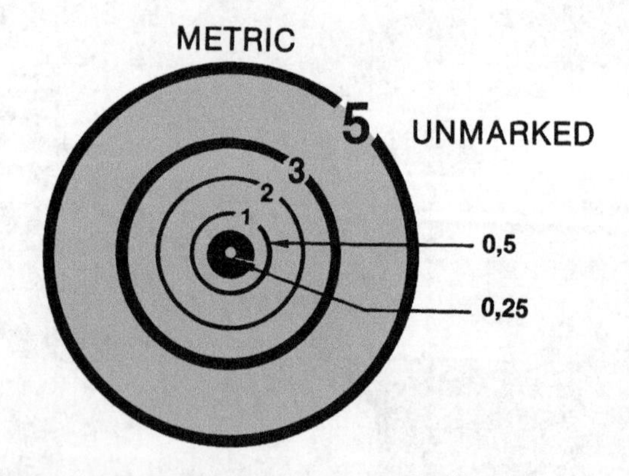

Circles 0,25 and 0,50 are, respectively, the inside and outside edge values of the center circle. All other circle values are correct at the center of the appropriate line.

FIGURE 16–53 Auto-reflection targets are read directly in gradients of the tilted mirror. In conventional units, the readings represent thousandths of an inch per foot when the mirror is 1 foot in front of the telescope. For SI measurements, the readings represent millimeters per meter when the mirror is 1 meter from the telescope.

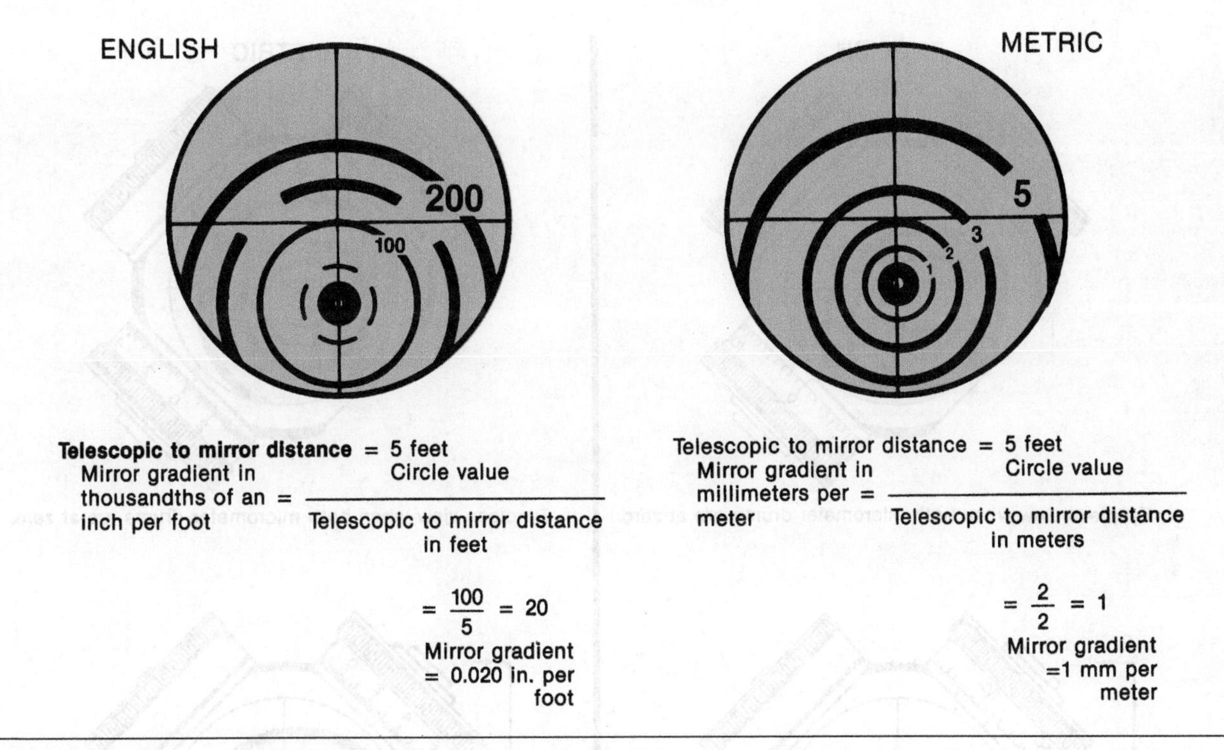

FIGURE 16–54 These examples are based on both micrometer drums being set at zero.

The projection feature is a great aid during setup, particularly when you are working over long distances and need to get things aligned within the field of view. With projected light, you can place your hand or a piece of paper in the beam to locate the crosslines, which can save setup time.

Measuring of Infinity

In all forms of optics, from photography to optical metrology, infinity means something different than it does in mathematics. In metrology, *infinity* designates the distance from which rays reaching the observer are parallel, or *collimated.*

In Figure 16–59, the *alignment collimator* has two targets: the reticle, also called the *tilt target,* permanently focused at infinity, and the target on the objective lens or its cover glass, the *alignment target* (see Figure 16–60).

When we tilt the alignment collimator with respect to a telescope focused at infinity, the center of the tilt target misses the crosslines of the telescope's reticle (see Figure 16–61A). If the crosslines are within the field of view, we can read the angle of tilt directly on the tilt target.

If you are trying to align items, you can adjust the collimator or workpiece so its rays are parallel to the line of sight of the telescope (see Figure 16–61B) and the crosslines and center of the tilt target coincide. The telescope and collimator must have parallel axes, but they need not be on the same line of sight axis (see Figures 16–61B and 16–61C).

Next, focus the telescope on the alignment target of the collimator, and the tilt target will disappear. You can then use the collimator in the conventional manner—that is, as a target to bring the telescope line of sight into coincidence with the optical axis of the collimator (see Figure 16–62).

ENGLISH

Eyepiece view when both micrometer drums are at zero.

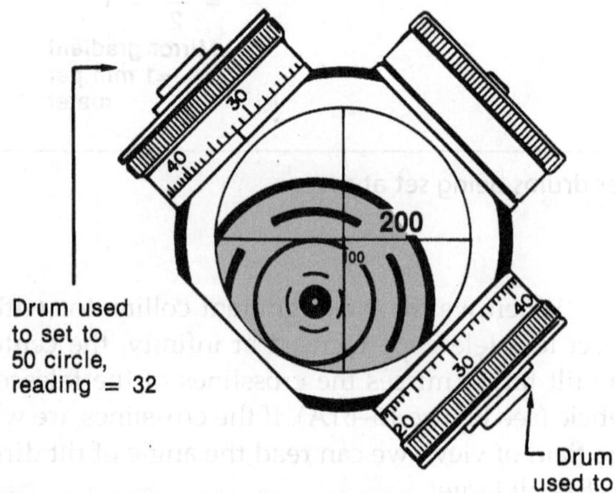

Drum used to set to 50 circle, reading = 32

Drum used to set to 100 circle, reading = 30

Eyepiece view when micrometer drums have been adjusted to bring 100 circle coincident with horizontal crossline and 50 circle coincident with verticle crossline.

Telescope to mirror distance = 10 feet.

Gradient of mirror in thous/foot about horizontal axis $= \frac{100 - 15}{10} = \frac{85}{10} = 8.5$
= 0.0085 in per foot.

Gradient of mirror in thous/foot about vertical axis $= \frac{50 + 16}{10} = \frac{66}{10} = 6.6$
= 0.0066 in per foot.

METRIC

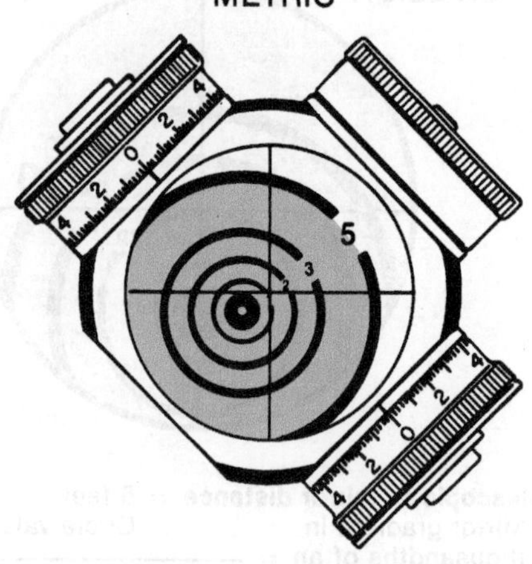

Eyepiece view when both micrometer drums are at zero.

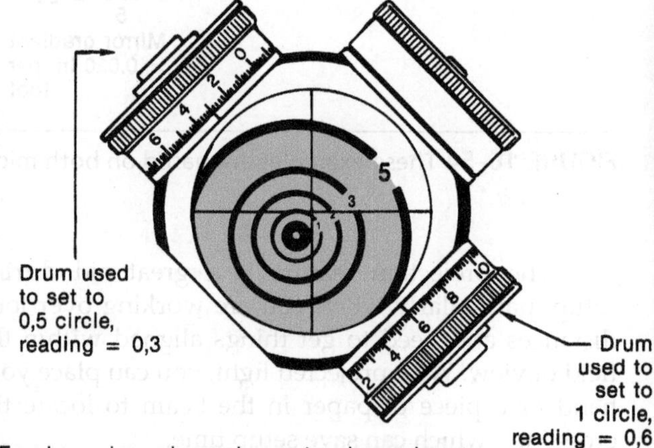

Drum used to set to 0,5 circle, reading = 0,3

Drum used to set to 1 circle, reading = 0,6

Eyepiece view when micrometer drums have been adjusted to bring 1 circle coincident with horizontal crossline and 0,5 circle coincident with vertical crossline.

Telescope to mirror distance = 3 meters.

Gradient of mirror in mm/meter about horizontal axis $= \frac{1 - 0{,}3}{3} = \frac{0{,}7}{3} = 0{,}23$
= 0,23 mm per meter.

Gradient of mirror in mm/meter about vertical axis $= \frac{0{,}5 + 0{,}15}{3} = \frac{0{,}65}{3} = 0{,}217$
= 0,217 mm per meter.

FIGURE 16–55 Examples of auto-reflection measurement using the micrometer drums.

FIGURE 16–56 Laser autocollimator for angle measurement of parts to 1 mm *(Courtesy of Micro-Radian Instruments)*

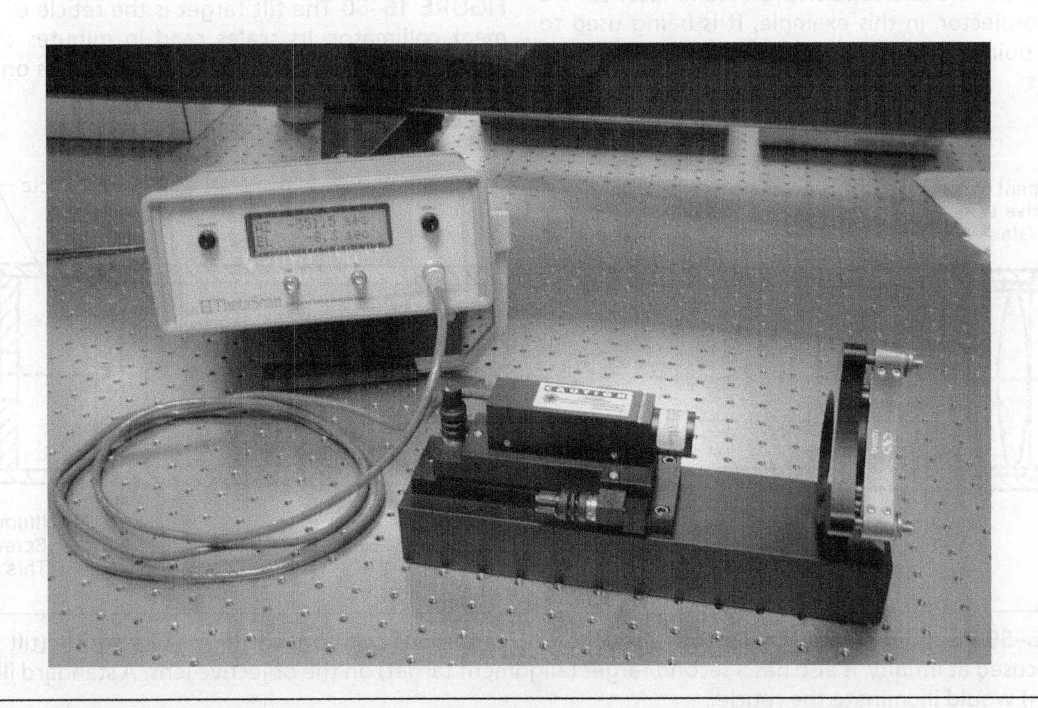

FIGURE 16–57 Laser autocollimator used to monitor angular movement of Newport gimbal mount positioner *(Courtesy of Micro-Radian Instruments)*

Instrument Calibration

You can use the alignment collimator in a variety of applications, most of which also involve the use of optical squares and other optical instruments (discussed later). One of the collimator's most important roles—one that is much older than modern optical metrology—is the calibration and adjustment of telescopes, binoculars, range finders, and similar devices. The straightness of a line of sight collimator has several reticle patterns so we can simulate targets at various distances. When you test a telescope, you "buck it in" at the shortest and longest distances and calibrate any deviations at intermediate distance reticles.

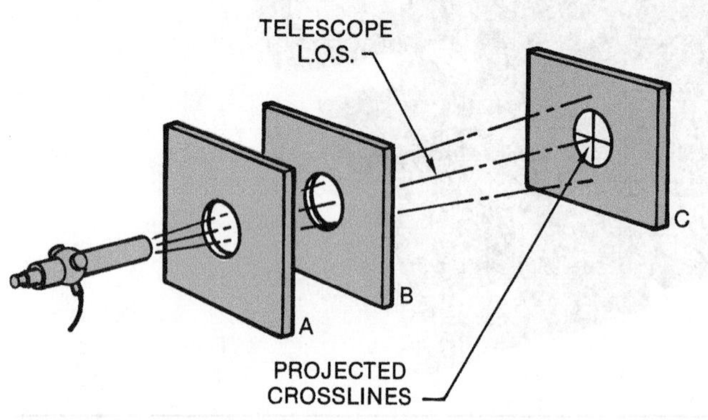

FIGURE 16–58 The telescope with an illuminator can be used as a projector. In this example, it is being used to locate the point in which to drill a hole in line with the other holes.

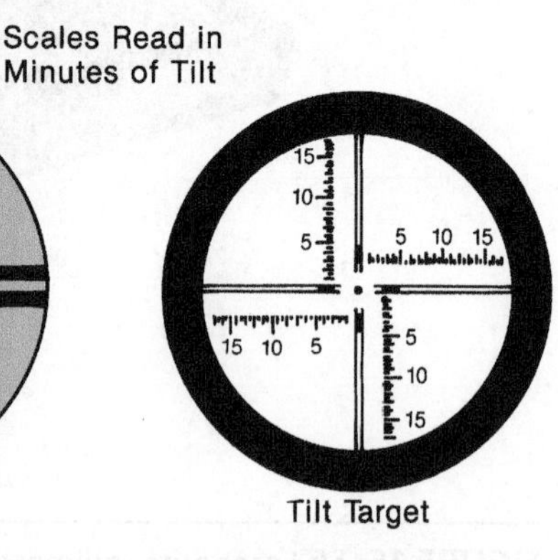

FIGURE 16–60 The tilt target is the reticle of the alignment collimator. Its scales read in minutes of tilt. The alignment target is a dispersion target. It is on the cover glass of the collimator.

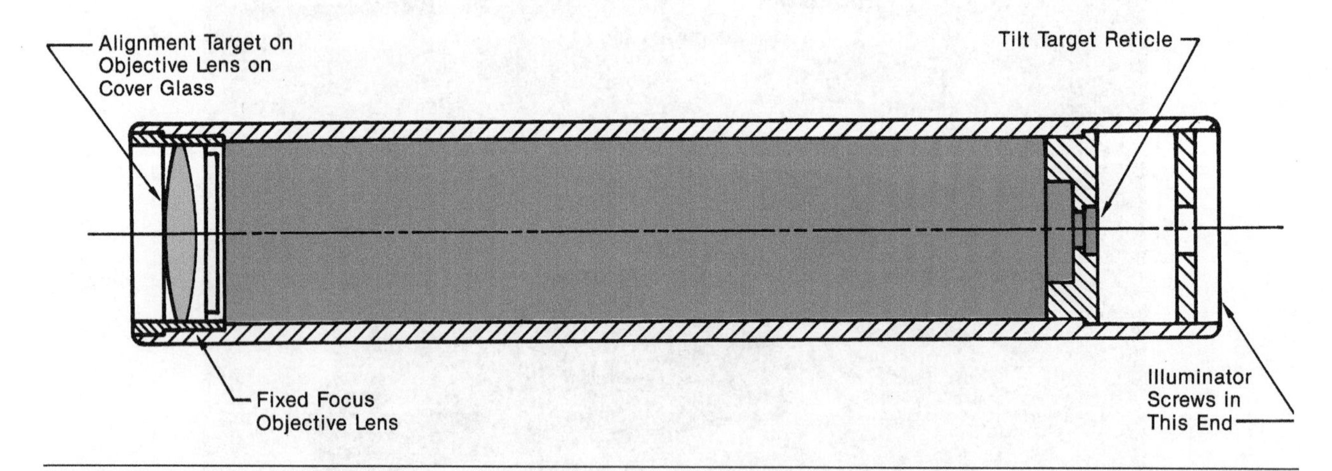

FIGURE 16–59 The alignment collimator consists of an objective lens positioned in front of a reticle (tilt target) so that it is focused at infinity. It also has a second target (alignment target) on the objective lens. A standard illuminator (not shown) would illuminate the reticle.

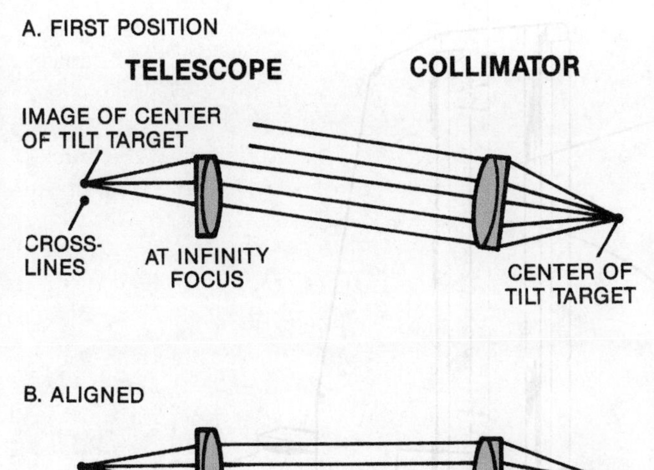

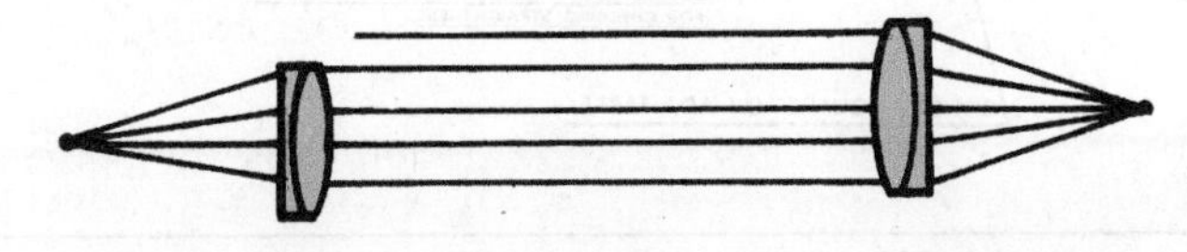

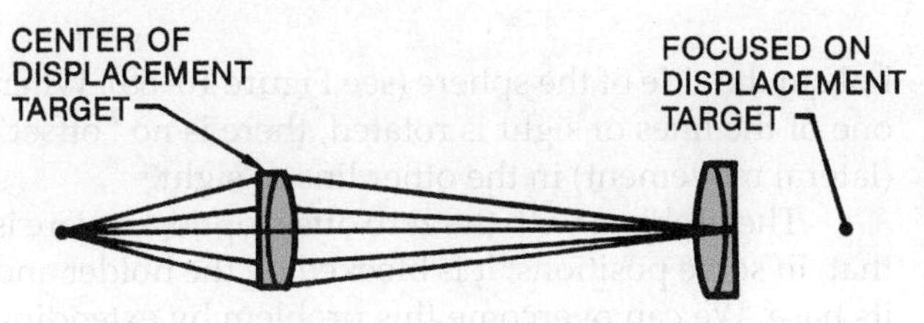

FIGURE 16–61 If the telescope and collimator axes are at an angle to each other, the crosslines of the telescope will be displaced from the center of the tilt target as in A. When the axes are parallel, as at B, the crosslines and center coincide. C shows that it is not necessary for the axes to be on the same line of sight-only parallel. When the axes are parallel, the telescope may then be focused on the alignment target.

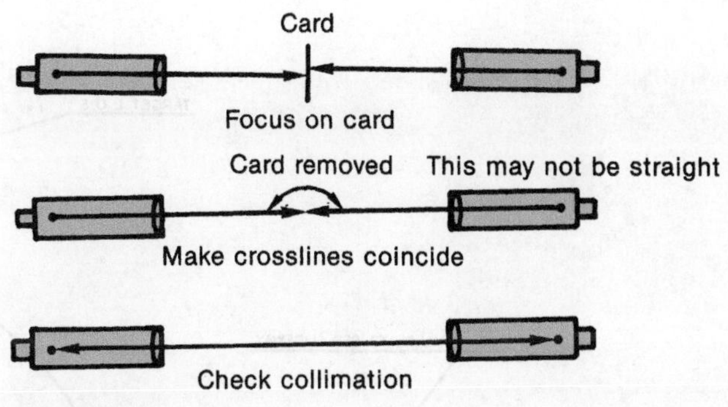

FIGURE 16–62 This technique saves time. The telescope and the collimator have their reticle images projected onto a card. They are then adjusted in the usual manner.

16–7 OPTICAL SQUARES—SQUARENESS

Even though some of the instruments we have discussed offer angle measurement capability (usually limited to reporting an amount of misalignment) most of these instruments only offer line of sight alignment. In order to align right angles (thereby extending line of sight to include planes), we need the *optical square* (see Figure 16–63). An optical square is any optical instrument that can turn the line of sight 90° from its original path.

If we set a mirror at 45°, we could easily divert a beam of light 90°; however, there are two problems (see Figure 16–64). Any error in mounting the mirror or maintaining its base parallel in a fixed beam reference will be magnified by the optical lever effect. In most situations, these errors would be greater than the workpiece squareness telescope.

Instead, we use a pentagonal prism, also called a "pentaprism," with such a unique optical path (see Figure 16–65) that we can move the prism through five of its six degrees of freedom (refer to Figure 10–46) without disturbing the right angle deviation of the beam of light. Movement in only one degree of freedom—rotating the prism so that the entering and emerging rays are not perpendicular to the imaginary

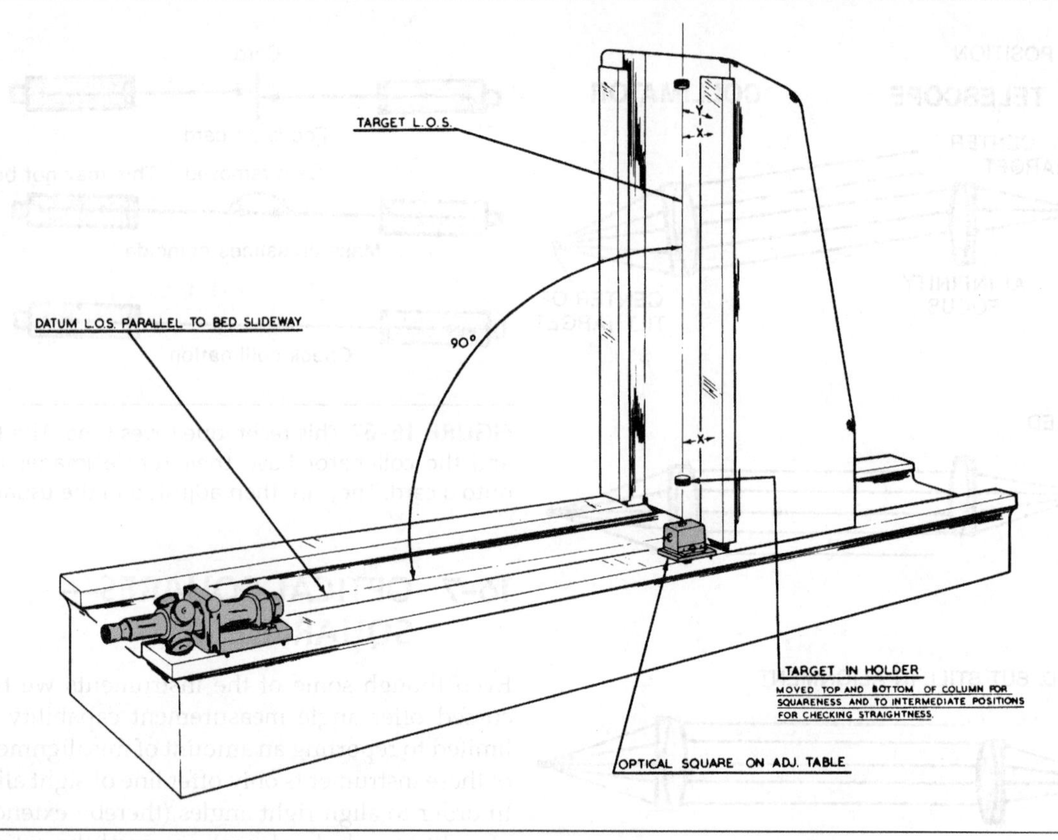

FIGURE 16–63 This is a typical situation in which an optical square is needed to determine the relationship of the vertical ways of the boring mill to the horizontal ways.

vertex of the 45° angle—affects the 90° angle (see Figure 16–66).

In optical metrology, we generally use two types of optical squares. In one, the square is fitted to another instrument like a telescope, and the prisms are factory aligned to ensure that the line of sight is perpendicular to the vertex. The second type of optical is used separately from the line of sight instrument and has provisions for adjustments of the line of sight plane.

Attached Optical Squares

Optical squares can be attached to the telescope (or attached to a telescope/base instrument (see Figure 16–67). The right angle beam in a telescope/base setup, a *zero offset optical square,* exits through the hole in the side of the sphere (see Figure 16–68). When one of the lines of sight is rotated, there is no "offset" (lateral movement) in the other line of sight.

The problem with the zero offset optical square is that, in some positions, it is blocked by the holder and its base. We can overcome this problem by extending the prism forward of the sphere, but then the prism is offset so that, when the telescope is turned in the vertical axis, the plane swept by the right angle line of sight moves laterally. The *double sphere optical square,* often called "offset optical square" (see Figure 16–69), can be used in either manner.

The Tooling Bar

When you use an optical square attached to an alignment telescope (see Figure 16–70), the telescope with

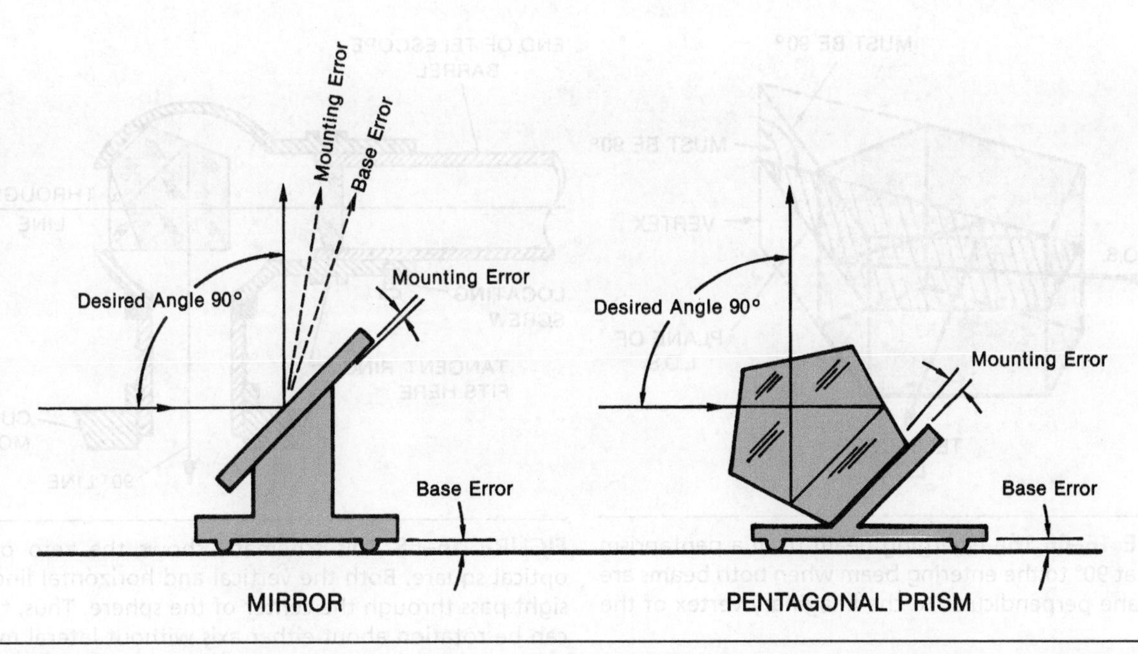

FIGURE 16–64 A mirror at a 45° angle to the incident ray of light should reflect the ray at 90°, but any errors in the mirror mounting or the parallelism of the base to the incident ray will be doubled in the reflected ray. If a pentagonal prism ("pentaprism") were used instead, the mounting error and base error would not affect the 90° reflection.

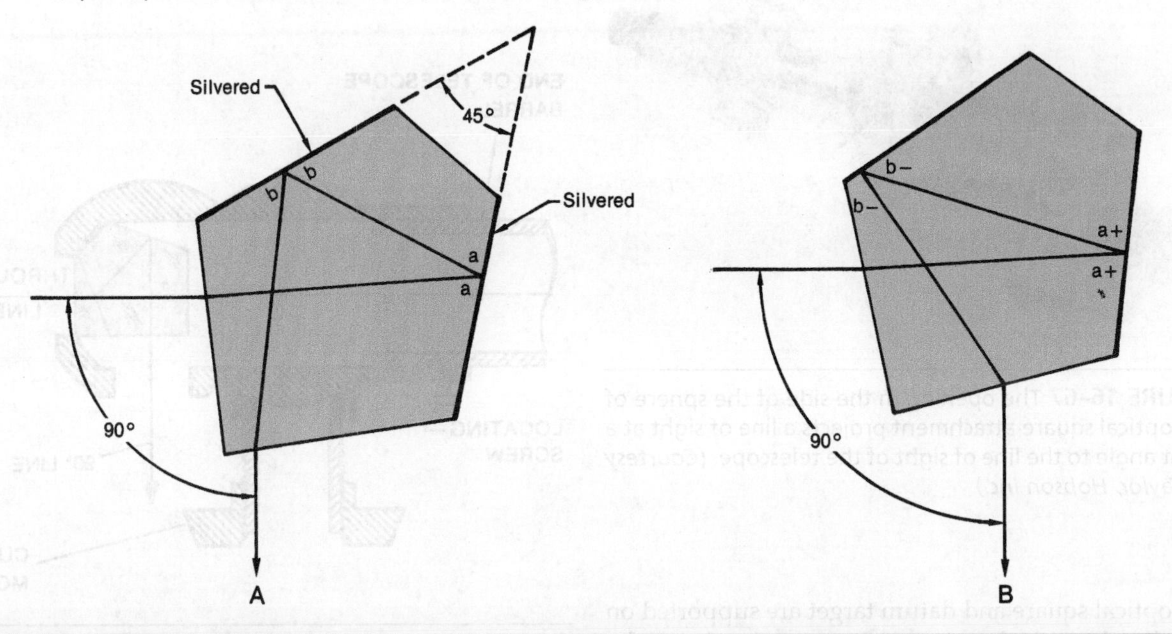

FIGURE 16–65 The pentaprism may be considered as two mirrors at precisely 45° to each other. Thus, each ray undergoes two reflections. In B, the prism has been rotated 5°. As a result, the first reflection increases from 75° in A to 80° in B. However, the second reflection is reduced by exactly the same amount. Therefore, the 90° displacement remains unchanged although the prism has turned.

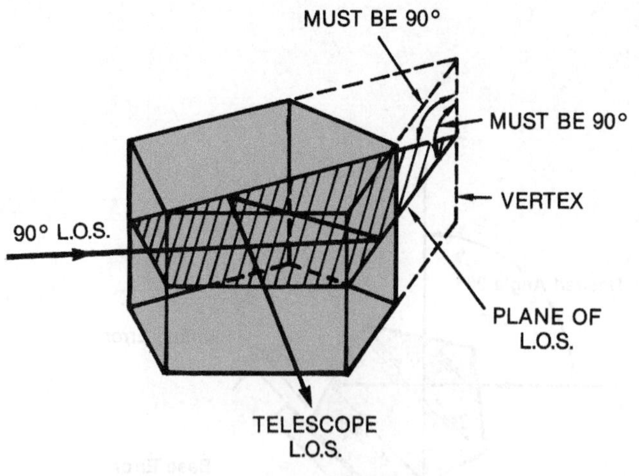

FIGURE 16–66 The emerging beam from a pentaprism is only at 90° to the entering beam when both beams are in a plane perpendicular to the imaginary vertex of the prism.

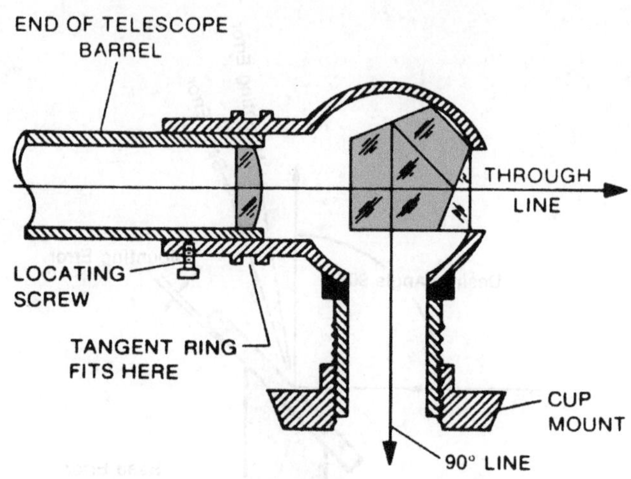

FIGURE 16–68 This schematic shows the zero offset optical square. Both the vertical and horizontal lines of sight pass through the center of the sphere. Thus, there can be rotation about either axis without lateral movement of ether axis.

FIGURE 16–67 The opening in the side of the sphere of the optical square attachment projects a line of sight at a right angle to the line of sight of the telescope. *(Courtesy of Taylor, Hobson Inc.)*

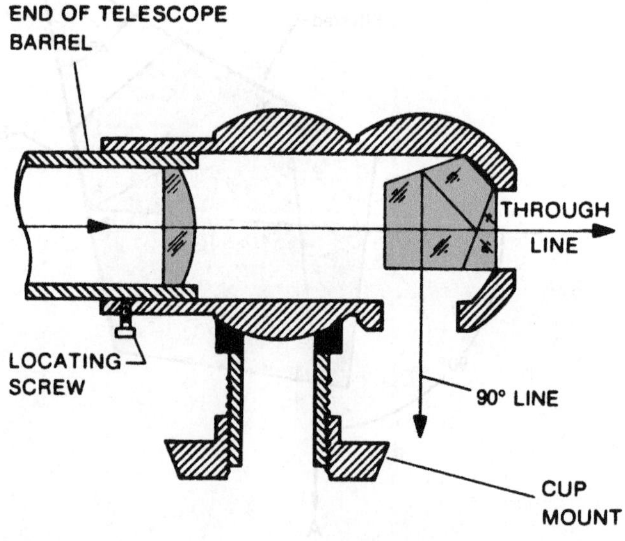

FIGURE 16–69 The offset optical square has two spheres, usually 4 inches apart. It may be used as a zero offset optical square or extended so that the right angle line of sight avoids obstructions.

the optical square and datum target are supported on a *tooling bar,* which is a rack that was developed for optical metrology so we can place optical instruments precisely along it. You use the tooling bar so you can erect two or more plates parallel to each other. The

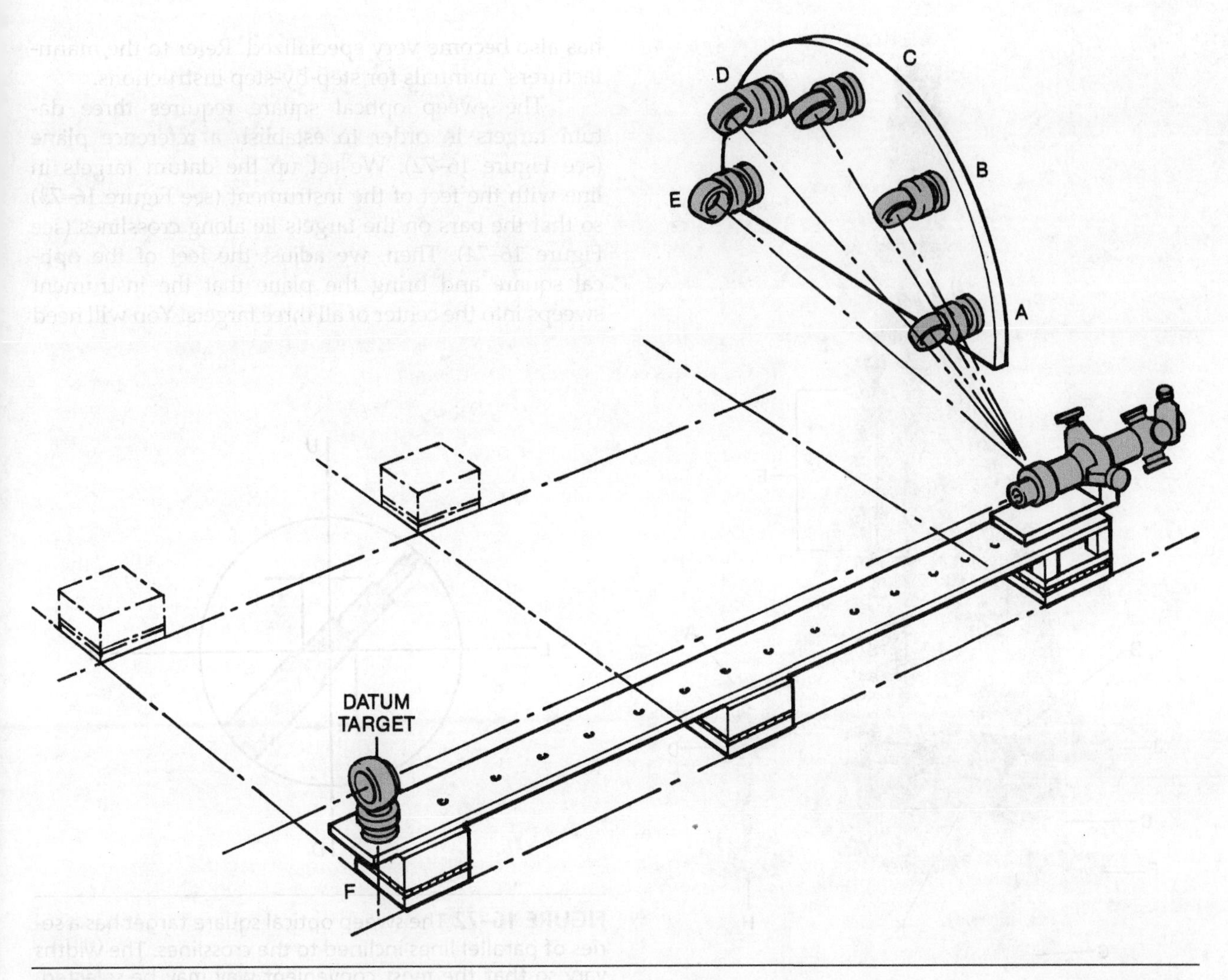

FIGURE 16–70 In this example, parallel plates are aligned by optical means. A tooling bar is used to simplify the support of the instruments. If that were not available, an intermediate target could be used to establish the datum line of sight.

telescope and datum target provide a line of sight, and by means of the optical square, you establish a right angle line of sight across the surface of the first plate you erect. Next, place a target at various locations on the plate, and move the plate until it is aligned with the right angle line of sight in each position. You really only need to check three points, but using more points can minimize errors. After the first plate is properly affixed, move the telescope along the tooling bar to the distance you want to erect the second plate, repeat the procedure, and affix the second plate. You can repeat this process for any number of plates.

The Sweep Optical Square

A particularly useful version of the attached optical square is the *sweep optical square* (see Figure 16–71), which can be used widely to check flatness. It consists of an alignment telescope with an optical square in a three-footed, vertical stand. The stand allows the line

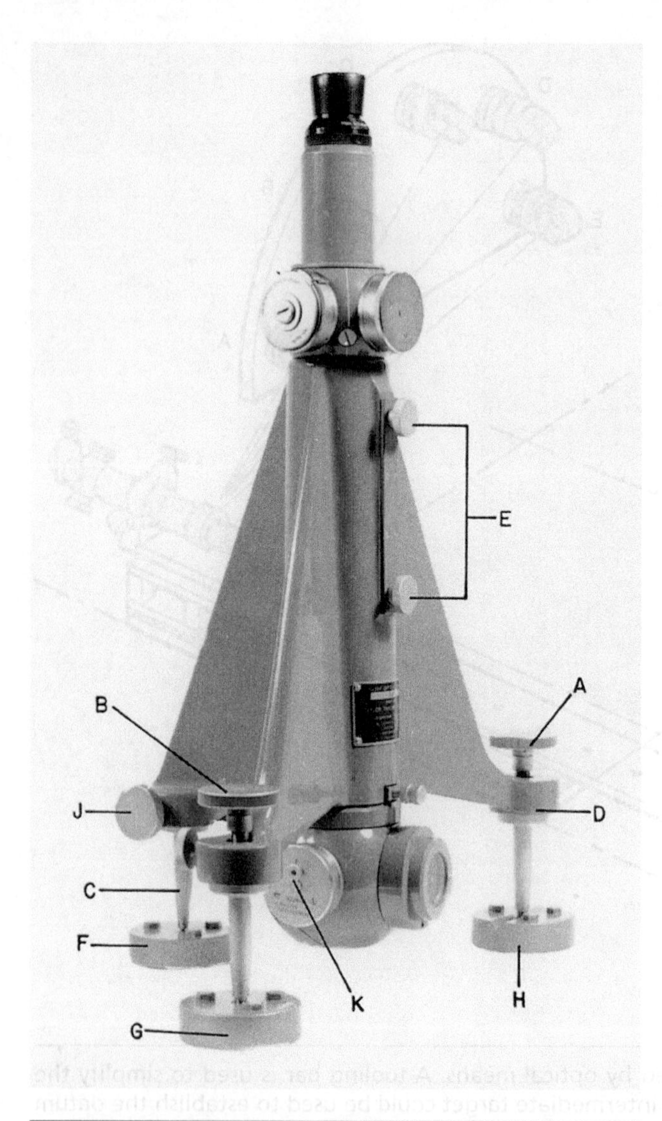

FIGURE 16–71 The sweep optical square is particularly useful in multiple measurements from one horizontal plane. Pad F is nonadjustable, whereas pads G and H are adjusted by knobs B and A. The knob J provides fine rotational movement. D is the lock against rotation. K is the locking screw for the optical micrometer on the horizontal line of sight. *(Courtesy of AMETEK Taylor Hobson.)*

of sight to sweep through a 360° horizontal plane, and the feet are adjustable so you can select which reference plane to use.

A special target has been developed for the sweep optical square, and the use of this instrument has also become very specialized. Refer to the manufacturers' manuals for step-by-step instructions.

The sweep optical square requires three datum targets in order to establish a reference plane (see Figure 16–72). We set up the datum targets in line with the feet of the instrument (see Figure 16–73) so that the bars on the targets lie along crosslines (see Figure 16–74). Then, we adjust the feet of the optical square and bring the plane that the instrument sweeps into the center of all three targets. You will need

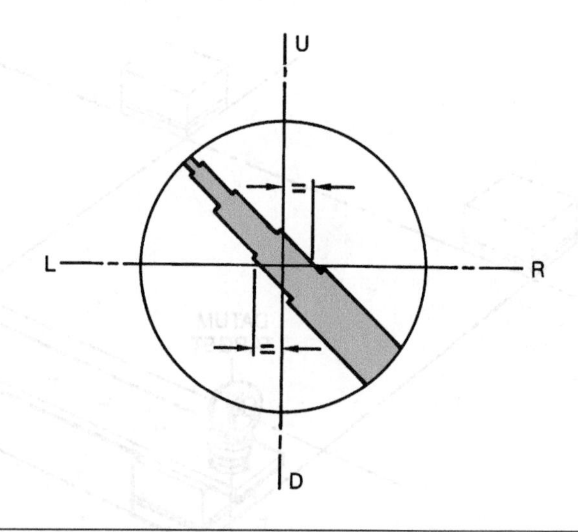

FIGURE 16–72 The sweep optical square target has a series of parallel lines inclined to the crosslines. The widths vary so that the most convenient way may be selected, the widest for the most distant. The setting is made to equalize the lengths of the crosslines or the areas.

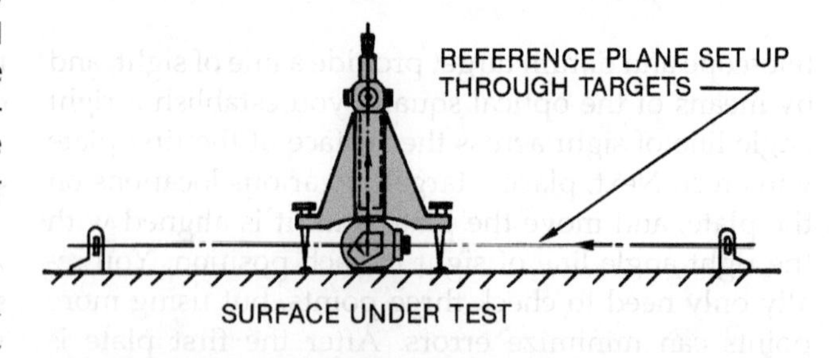

FIGURE 16–73 The sweep optical square rests on the surface under test. Targets resting on that surface are used to establish the reference.

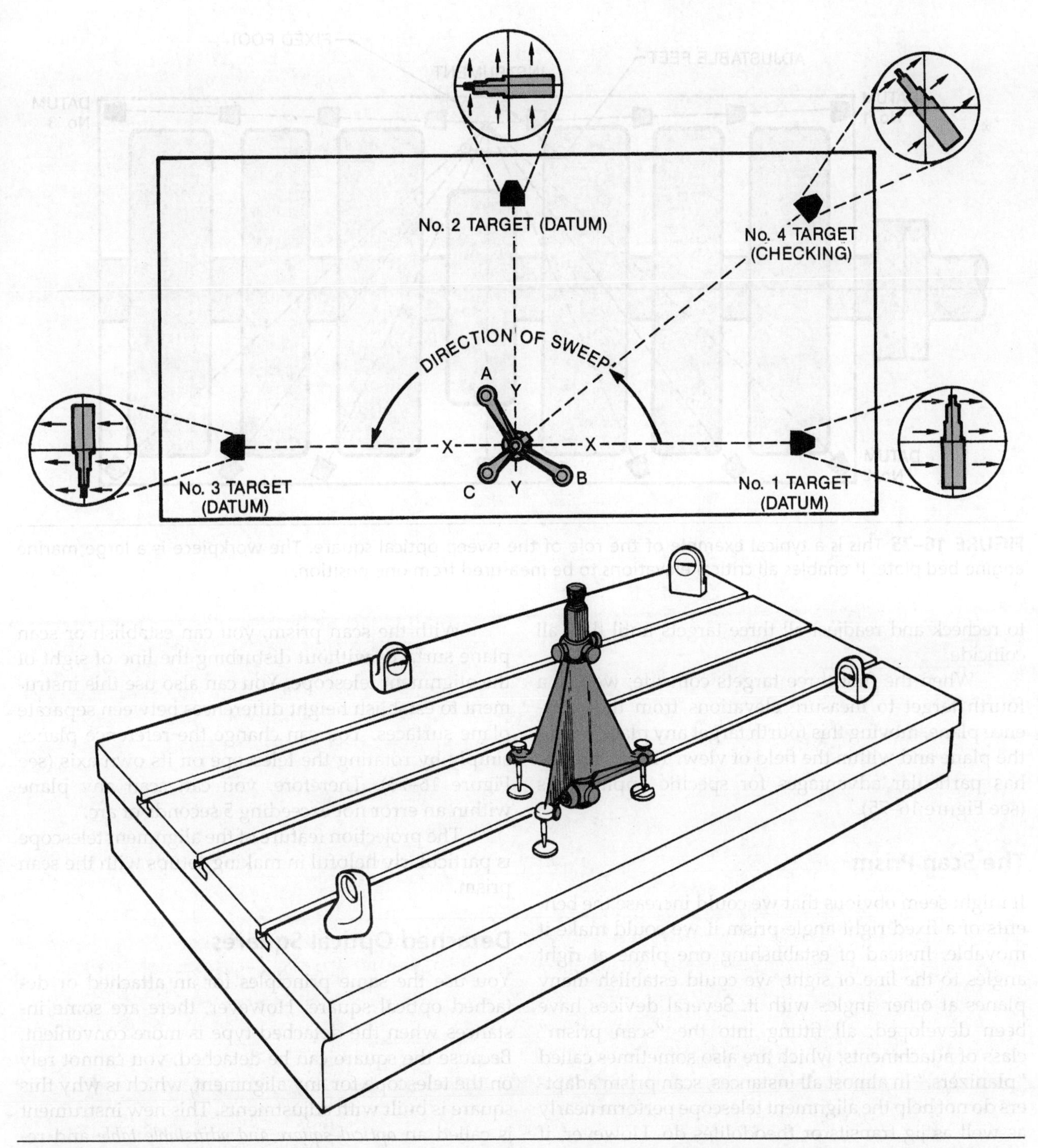

FIGURE 16–74 Three datum targets are used in conjunction with the sweep optical square. After they are set to establish the reference plane, a fourth target can be used for vertical measurements any place in that plane.

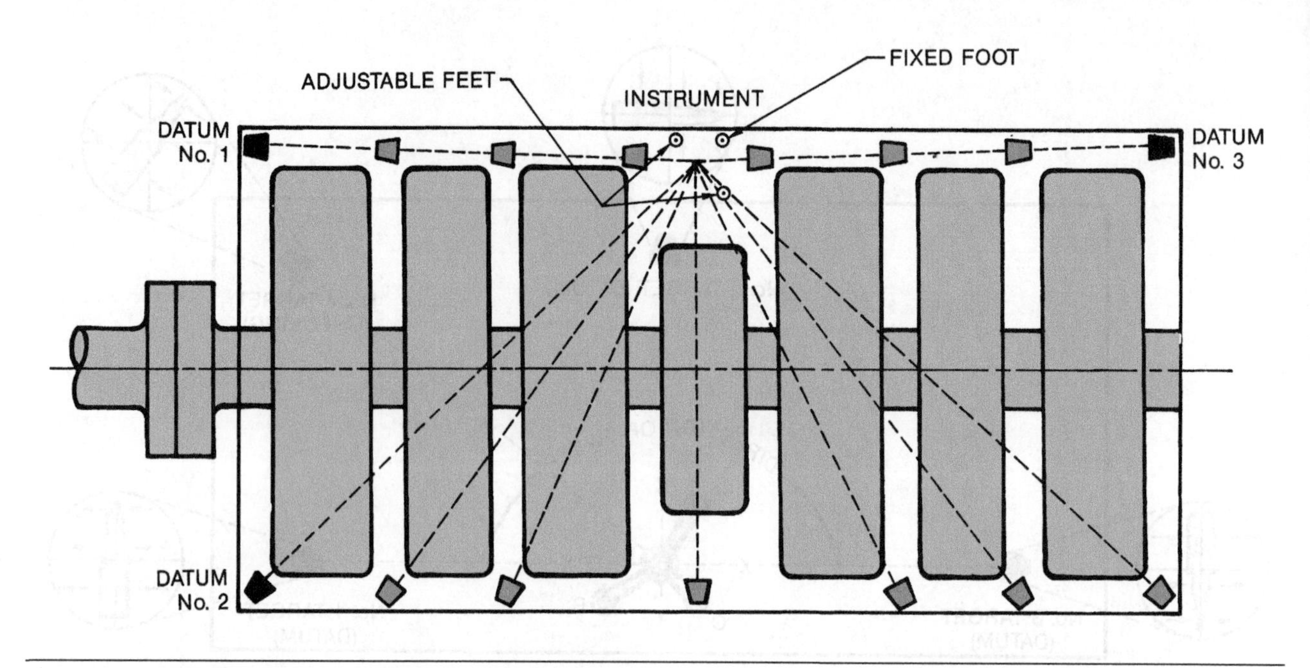

FIGURE 16–75 This is a typical example of the role of the sweep optical square. The workpiece is a large marine engine bed plate. It enables all critical elevations to be measured from one position.

to recheck and readjust all three targets until they all coincide.

When the first three targets coincide, we use a fourth target to measure elevations from the reference plane, moving this fourth target any place within the plane and within the field of view. This technique has particular advantages for specific applications (see Figure 16–75).

The Scan Prism

It might seem obvious that we could increase the benefits of a fixed right angle prism if we could make it movable. Instead of establishing one plane at right angles to the line of sight, we could establish many planes at other angles with it. Several devices have been developed, all fitting into the "scan prism" class of attachments, which are also sometimes called "planizers." In almost all instances, scan prism adapters do not help the alignment telescope perform nearly as well as jig transits or theodolites do. However, if you already have an alignment telescope, scan prisms are much less expensive.

With the scan prism, you can establish or scan plane surfaces without disturbing the line of sight of the alignment telescope. You can also use this instrument to establish height differences between separate plane surfaces. You can change the reference planes simply by rotating the telescope on its own axis (see Figure 16–76). Therefore, you can scan any plane within an error not exceeding 5 seconds of arc.

The projection feature of the alignment telescope is particularly helpful in making setups with the scan prism.

Detached Optical Squares

You use the same principles for an attached or detached optical square. However, there are some instances when the detached type is more convenient. Because the square can be detached, you cannot rely on the telescope for any alignment, which is why this square is built with adjustments. This new instrument is called an *optical square and adjustable table* and requires a series of adjustments to ensure the alignment (see Figure 16–77).

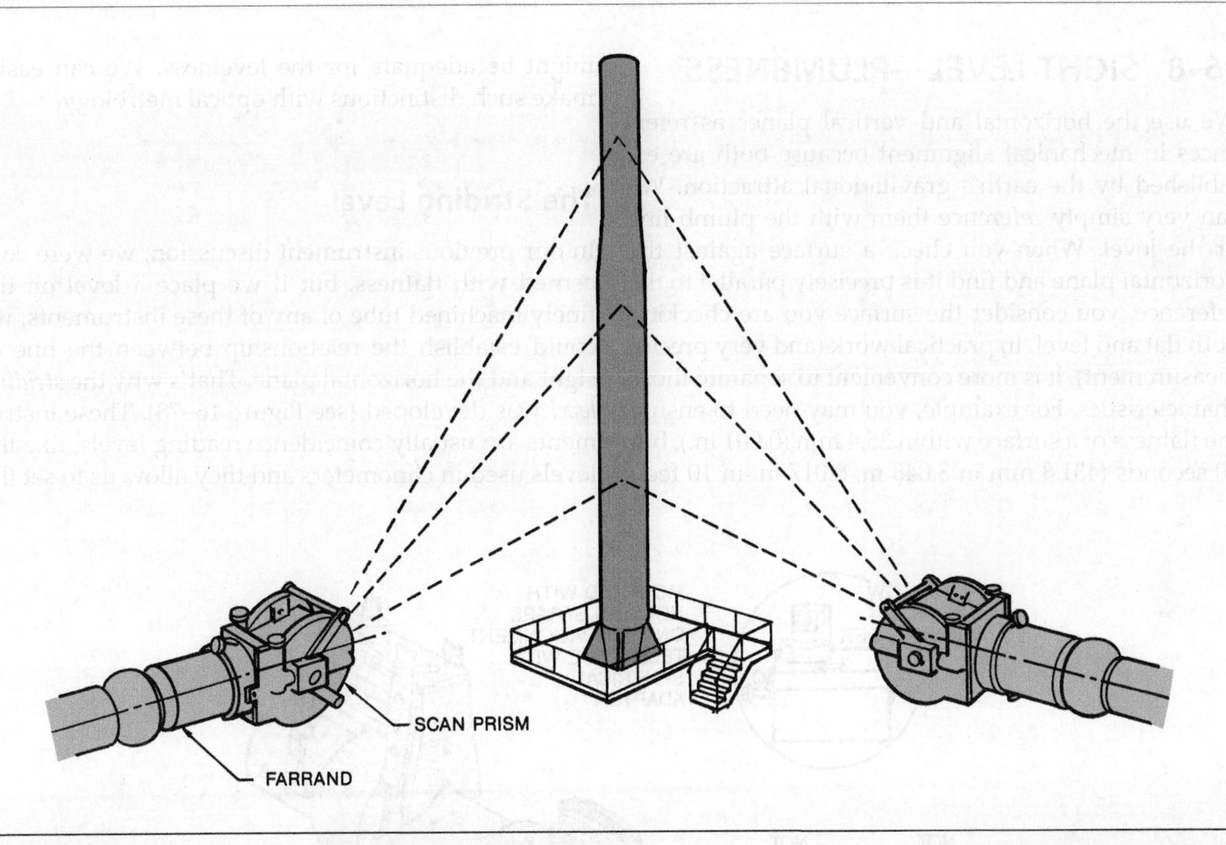

FIGURE 16–76 The reference plane can be selected by rotating the telescope. In this example, vertical planes are being established.

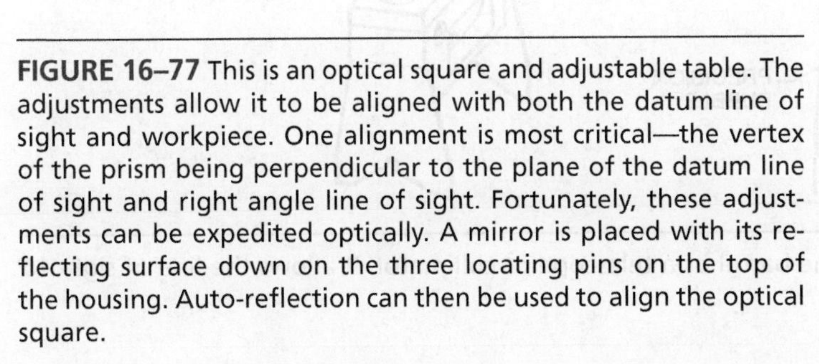

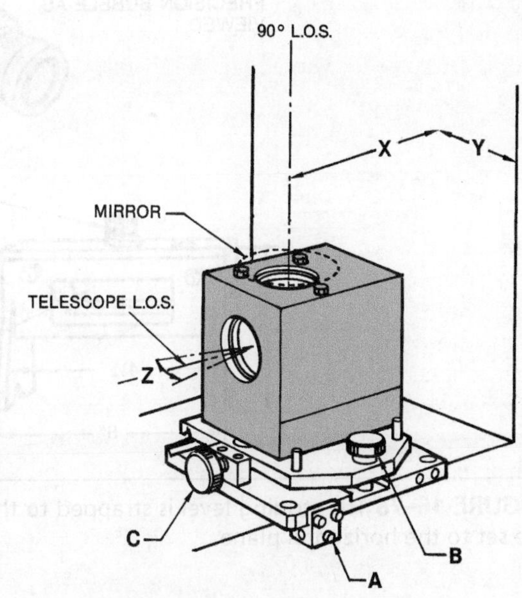

FIGURE 16–77 This is an optical square and adjustable table. The adjustments allow it to be aligned with both the datum line of sight and workpiece. One alignment is most critical—the vertex of the prism being perpendicular to the plane of the datum line of sight and right angle line of sight. Fortunately, these adjustments can be expedited optically. A mirror is placed with its reflecting surface down on the three locating pins on the top of the housing. Auto-reflection can then be used to align the optical square.

16-8 SIGHT LEVEL—PLUMBNESS

We use the horizontal and vertical planes as references in mechanical alignment because both are established by the earth's gravitational attraction. We can very simply reference them with the plumb line or the level. When you check a surface against the horizontal plane and find it is precisely parallel to the reference, you consider the surface you are checking both flat and level. In practical work (and very precise measurement), it is more convenient to separate those characteristics. For example, you may need to ensure the flatness of a surface within 25.4 mm (0.001 in.), but 30 seconds (431.8 mm in 3.048 m, 0.017 in. in 10 feet) might be adequate for the levelness. We can easily make such distinctions with optical metrology.

The Striding Level

In our previous instrument discussion, we were concerned with flatness, but if we place a level on the finely machined tube of any of these instruments, we could establish the relationship between the line of sight and the horizontal plane. That's why the *striding level* was developed (see Figure 16–78). These instruments are usually coincidence reading levels, like the levels used in clinometers and they allow us to set the

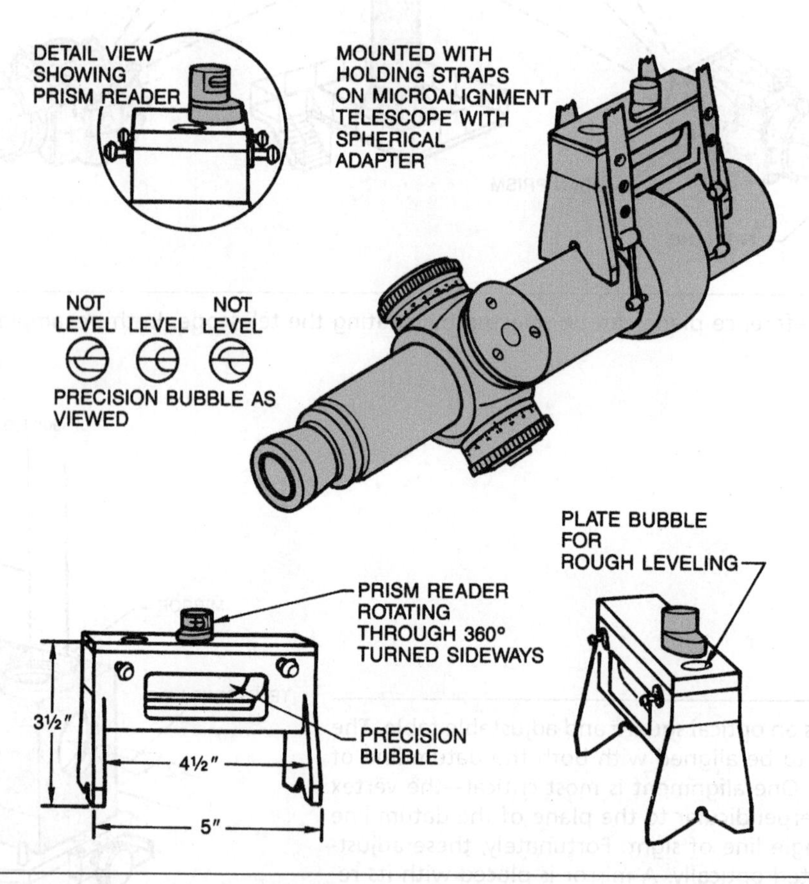

FIGURE 16–78 The striding level is strapped to the barrel of a telescope or collimator. It allows the line of sight to be set to the horizontal plane.

barrel within 2 seconds of arc. Electronic versions are available with even greater sensitivity.

The Cross Level

Another type of striding level, the *cross level,* mounts on the barrel at 90° from the line of sight and locates from a hole in the barrel. We use these levels to align the crosslines of the reticle with the horizontal and vertical planes.

Because striding levels reference to the barrel and not directly to the line of sight, the manufacturing tolerance may be as much as 3 seconds away from horizontal. You can measure this error by the reversal technique and correct the adjustment.

Figure 16–79 shows a typical application. In this example, we need to have eight mounting pads level and in the same plane; therefore, we use a datum target at each of seven pads to bring them into alignment.

The Sight Level

Because the need to level within a single plane is so common and the optical method is so much more expeditious than mechanical methods, metrologists have developed an instrument specifically for the task. The *sight level,* often called the *tilting level* or *optical tooling level,* can be supported on almost any surface on or off of a workpiece and swept through a 360° horizontal, level plane, unlike the telescope, which must be supported on the workpiece or a tooling bar.

Construction

Essentially, the sight level is a telescope with an integral level mounted on a tilting base, and when you align the instrument so that when the bubble of the level is centered, the line of sight is horizontal. Using the micrometer screw, you can tilt the telescope level assembly about the horizontal axis, and you can use

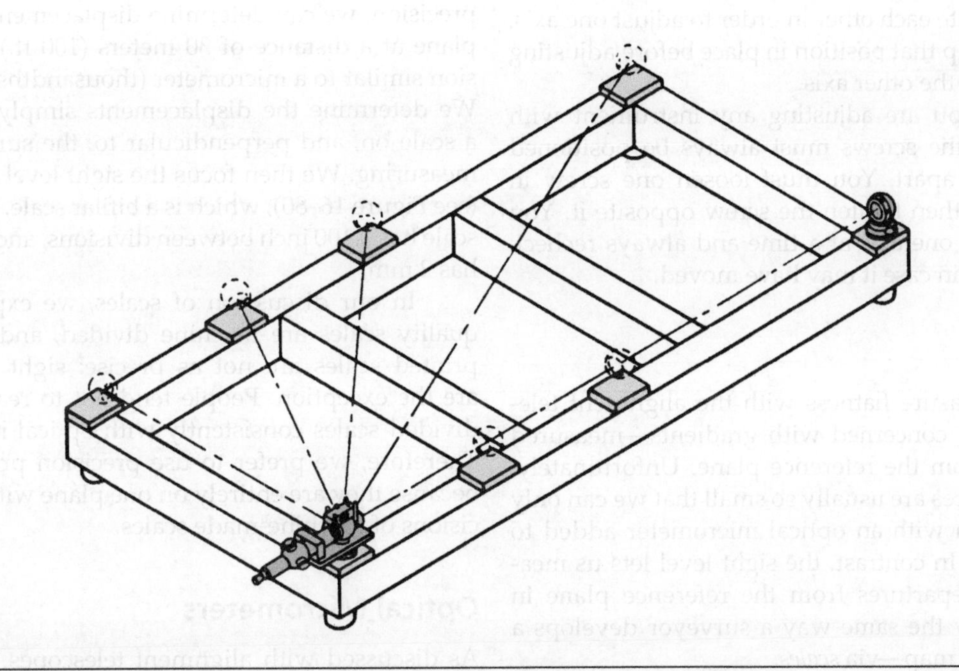

FIGURE 16–79 A typical application of the alignment telescope with striding level is this example. The requirement is to place eight mounting pads in the same plane and ensure that the plane is level.

the base to aim the telescope in any direction around the azimuth axis. The azimuth axis is set perpendicular to the horizontal axis by the circular level and the *leveling head.* You should use the reversal method to check the vertical axis: center the bubble and rotate the telescope 180°; if the bubble remains centered, the azimuth axis is vertical. When you rotate the telescope, it sweeps a horizontal plane from which you may use the scales to take measurements at any place.

The Four Leveling Screws

You may be wondering why we need four adjusting screws with the sight level. Using four screws seems to contradict the principles of kinematics that we have stressed throughout the text—particularly the principle of three-point suspension. We must have four screws because we must be able to adjust each axis independently of the other. If there were only three screws, one adjustment would affect the other two screw axes that are at an angle to the one we are adjusting. By using four screws, we can adjust the screws opposite each other in order to adjust one axis, and then clamp that position in place before adjusting and clamping the other axis.

When you are adjusting any instrument with four screws, the screws must always be positioned in pairs 180° apart. You must loosen one screw in the pair, and then tighten the screw opposite it. You should adjust one axis at a time and always recheck the other axis in case it may have moved.

Scales

When we measure flatness with the alignment telescope, we are concerned with gradients—measured departures from the reference plane. Unfortunately, these departures are usually so small that we can only measure them with an optical micrometer added to the telescope. In contrast, the sight level lets us measure major departures from the reference plane in almost exactly the same way a surveyor develops a topographical map—via *scales.*

The sight level sweeps a plane that is level and horizontal to within 1 second of arc. With such precision, we can determine displacements from that plane at a distance of 30 meters (100 ft.) with precision similar to a micrometer (thousandths of an inch). We determine the displacements simply by placing a scale on, and perpendicular to, the surface we are measuring. We then focus the sight level on the scale (see Figure 16–80), which is a bifilar scale. The English scale has 0.100 inch between divisions, and the SI scale has 2 mm.

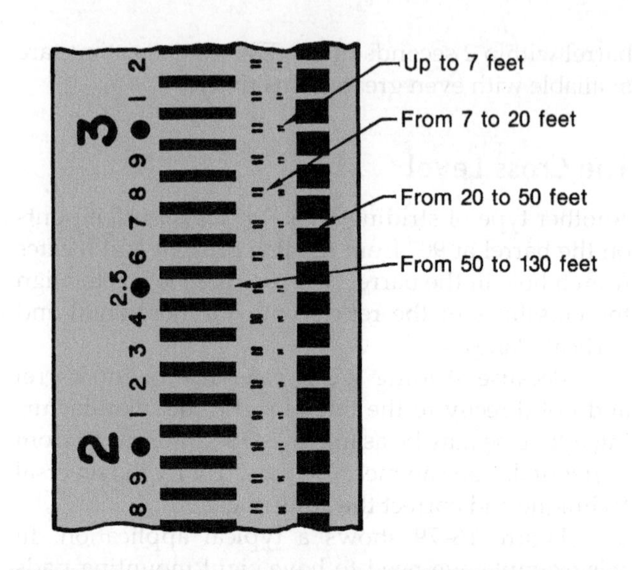

FIGURE 16–80 Typical sight level scale. Note that it has bifilar scales for use at four distances. The target is the center of the space between pairs of lines.

In our discussion of scales, we explained that quality scales are machine divided, and painted or printed scales are not as precise: sight level scales are the exception. People tend not to read machine-divided scales consistently with optical instruments. Therefore, we prefer to use precision printed scales because they are entirely on one plane without the incisions of machine-made scales.

Optical Micrometers

As discussed with alignment telescopes, optical micrometers are used with the sight level to divide the space between major scale divisions. For English scales, the micrometers have 100 divisions that may be read to 0.001 inch; the SI micrometer is divided into 0.02 mm divisions. Both scale systems have verniers that can read to one-tenth of the drum calibrations; however, there is no conclusive proof that anyone can make the bifilar alignment that closely. The vernier is useful if you make the reading to the finest decimal, then round up to the next larger one.

You should use one point as a base from which you measure the differences in height for a given workpiece. Measure top surfaces upward, and bottom surfaces downward. This technique does not improve your metrological accuracy or precision, but it provides consistency and minimizes computational errors. If the points are widely spaced, you can minimize instrument errors by centering the sight level approximately equidistant from the two extremes.

16-9 OPTICAL POLYGONS—ANGLES

We covered angle measurement in detail in Chapter 12 and came to the conclusion that optical instruments, like the ones we will now discuss, provide the fastest and most accurate precision measurement of angles. The alignment telescope, with its various attachments, can be one of the most valuable angle-measuring instruments when used in conjunction with *optical polygons.*

Construction

Optical polygons are physical representations of polygons with various numbers of faces, and we use them as standards for the exterior angles of polygons. Optical polygons are available in steel (see Figure 16–81) and glass with metal enclosures to protect them and have openings at the faces (see Figure 16–82). In both materials, the sides are flat, parallel, and precisely perpendicular to the reflective faces. Unlike divided circular scales, polygons are free from eccentricity error. Glass polygons are available with 5 to 36 faces, whereas steel polygons come with up to 72 faces. Polygons of both materials are considered accurate to within a few seconds of arc and are available with calibration certification to one-tenth second of arc.

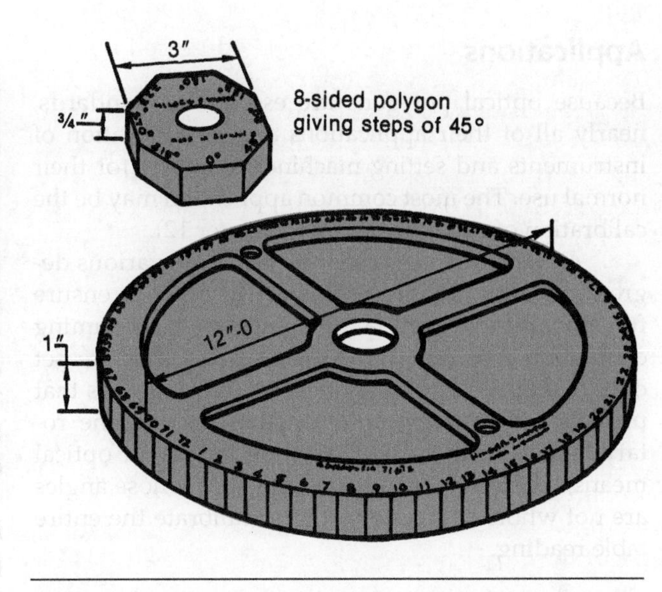

FIGURE 16–81 There are 68 steel optical polygons from 5 to 72 faces.

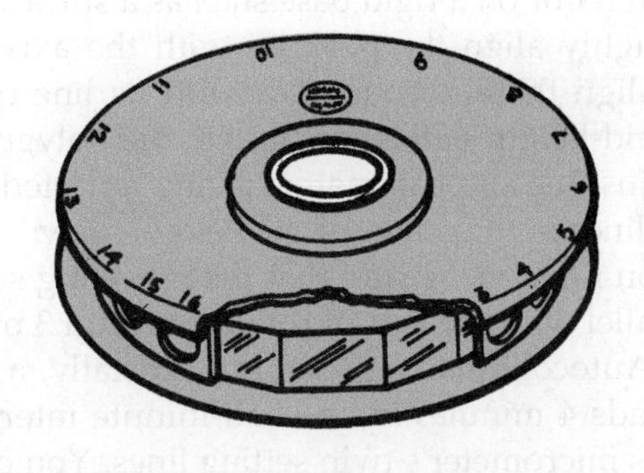

FIGURE 16–82 Glass optical polygons have steel enclosures to prevent injury to their reflective surfaces when handling. They are available in 32 sizes from 5 to 36 faces.

Manufacturers use two methods to identify the faces. Smaller polygons may have the angles engraved adjacent to the faces, and polygons with a large number of faces, or in which the nominal angles are not whole numbers of degrees, have their faces identified by consecutive numbers.

Applications

Because optical polygons are essentially standards, nearly all of their applications are for calibration of instruments and setting machines to angles for their normal use. The most common application may be the calibration of rotary tables (see Chapter 12).

As with other instruments, there are various degrees of calibration: a quick and easy test to ensure functionality through a time-and-labor consuming calibration. We can check rotary tables that are set only to degrees directly against optical polygons that provide whole degrees. When the scale on the rotary table is further divided by mechanical or optical means, we need to use optical polygons whose angles are not whole degrees in order to calibrate the entire table reading.

Rotary Table Calibration

If we need to calibrate a table that indexes to degrees, you could use an eight-face polygon. Set the table and autocollimator on a rigid base such as a surface plate, and roughly align the polygon with the axis of the table. Align the autocollimator with its line of sight at the mid-height of the zero face of the polygon, and then adjust the autocollimator for any reflected image of crosslines in the center of the field of view.

You must make sure that the reflecting surfaces are parallel with the axis of rotation within 3 minutes of arc. Autocollimators differ, but typically, a coarse scale reads 4 minutes in one-half-minute intervals at 90° to its micrometer's twin setting lines. You can use this scale to measure the setup error.

With the table at zero, set the reflected image of the crossline to the 4-minute scale (see Figure 16–83). Rotate the table 135° (see Figure 16–83) and note any change on the coarse scale in the horizontal image of the target wires. Continue rotating the table to the 225° face (see Figure 16–84C) and again note any change. In most cases, all three of these readings will be within the required 3 minutes, but if the deviation is greater, use shims to correct it (see Figure 16–84).

At this point, you can take your first angle reading. Take the autocollimator reading for the zero face (see Figure 16–85A) and rotate the table to the 45° position and take a reading for that face of the polygon (see Figure 16–85B). In each position, you should take two or more readings and average the results. Repeat this process at 45° intervals up to 360°, remembering that 360° is also 0°. Enter all your readings into a chart for the table (see Figure 16–86).

Rotate the table past its zero, then reverse your direction and return to zero in the opposite direction. Take the zero reading and repeat readings at 45° intervals until you reach 360° again, entering these data on the chart, too.

From these data, calculate the error, remembering to enter the calibration error furnished with the polygon. Be careful that you apply the error adjustments in the proper direction (refer to manuals for instructions).

Two-Polygon Calibration

If optical polygons are available with up to 72 sides, we must need a large polygon inventory. Fortunately, polygons add and subtract, just like angle gage blocks. For example, if we need to calibrate a rotary table at 5 degrees, the best choice for the calibration would be a 72-face polygon. However, we could also do the calibration with two polygons, one with eight faces and the other with nine faces:

	72 faces	$360 \div 72 = 5°$
	8 faces	$360 \div 8 = 45°$
less	9 faces	$360 \div 9 = 40°$
		$= 5°$

The two small polygons cost considerably less than the one large polygon and are easier to handle.

The Calibration of a Polygon

When you are using any standard as precise as an optical polygon, you must be just as suspicious of its precision as you are of any other instrument. Thanks to

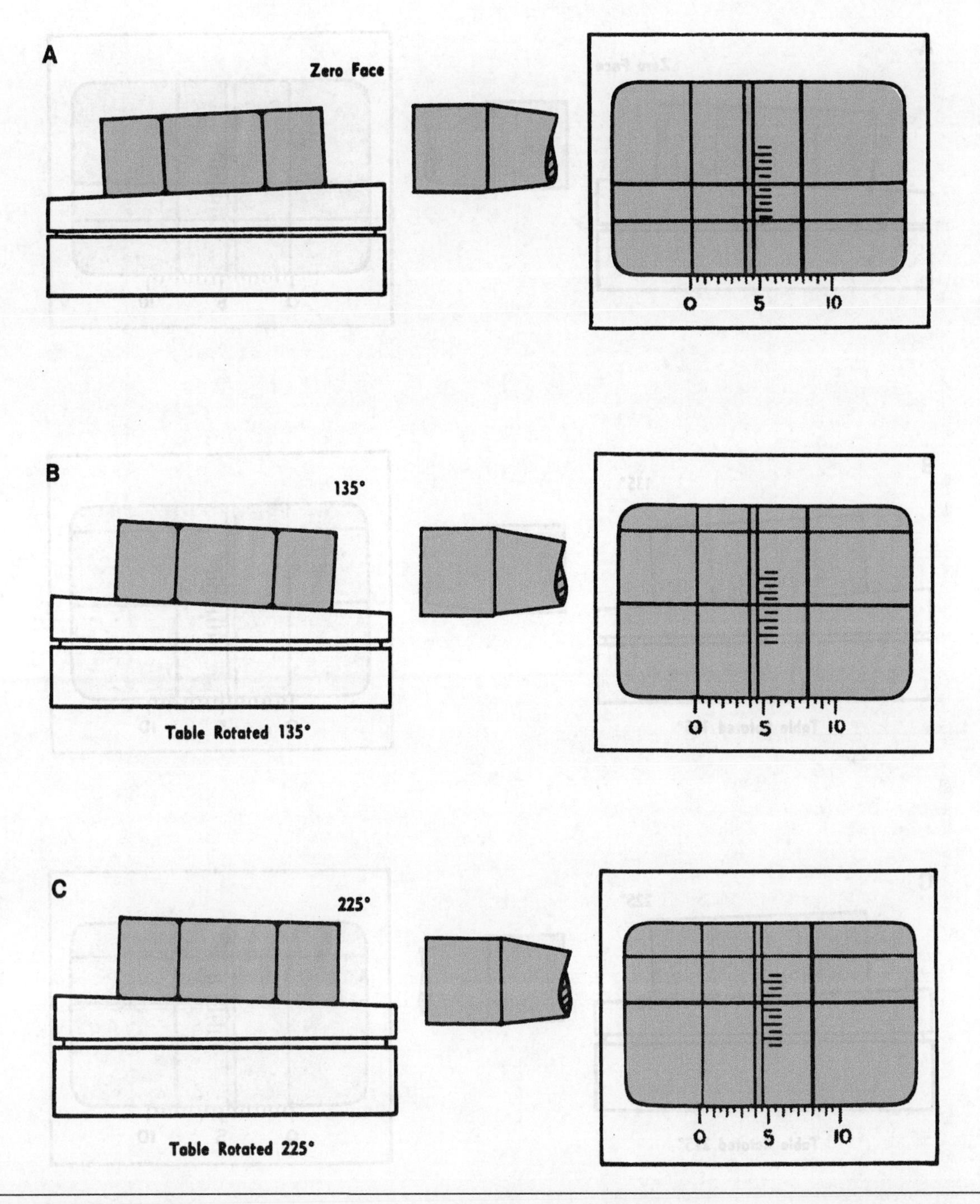

FIGURE 16–83 A course scale in the autocollimator simplifies alignment of the polygon to the axis of rotation. Without this scale, a single axis autocollimator would have been turned so that the drum would be vertical. For calibration, it is returned to this position.

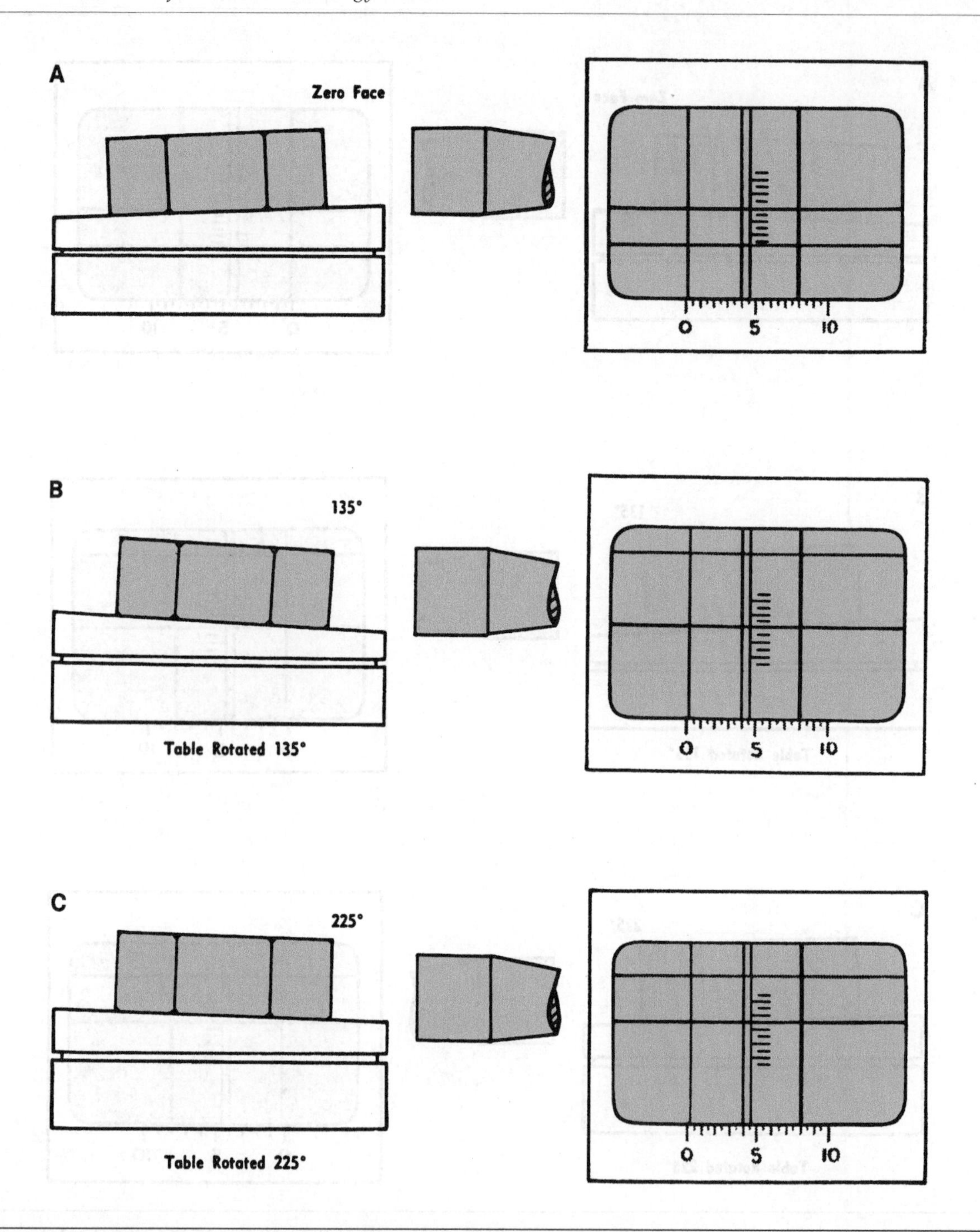

FIGURE 16–84 In Figure 16–83, greater deviation than 3° was discovered. Therefore, it is necessary to use spacers until the deviation is within 3° of all three checkpoints.

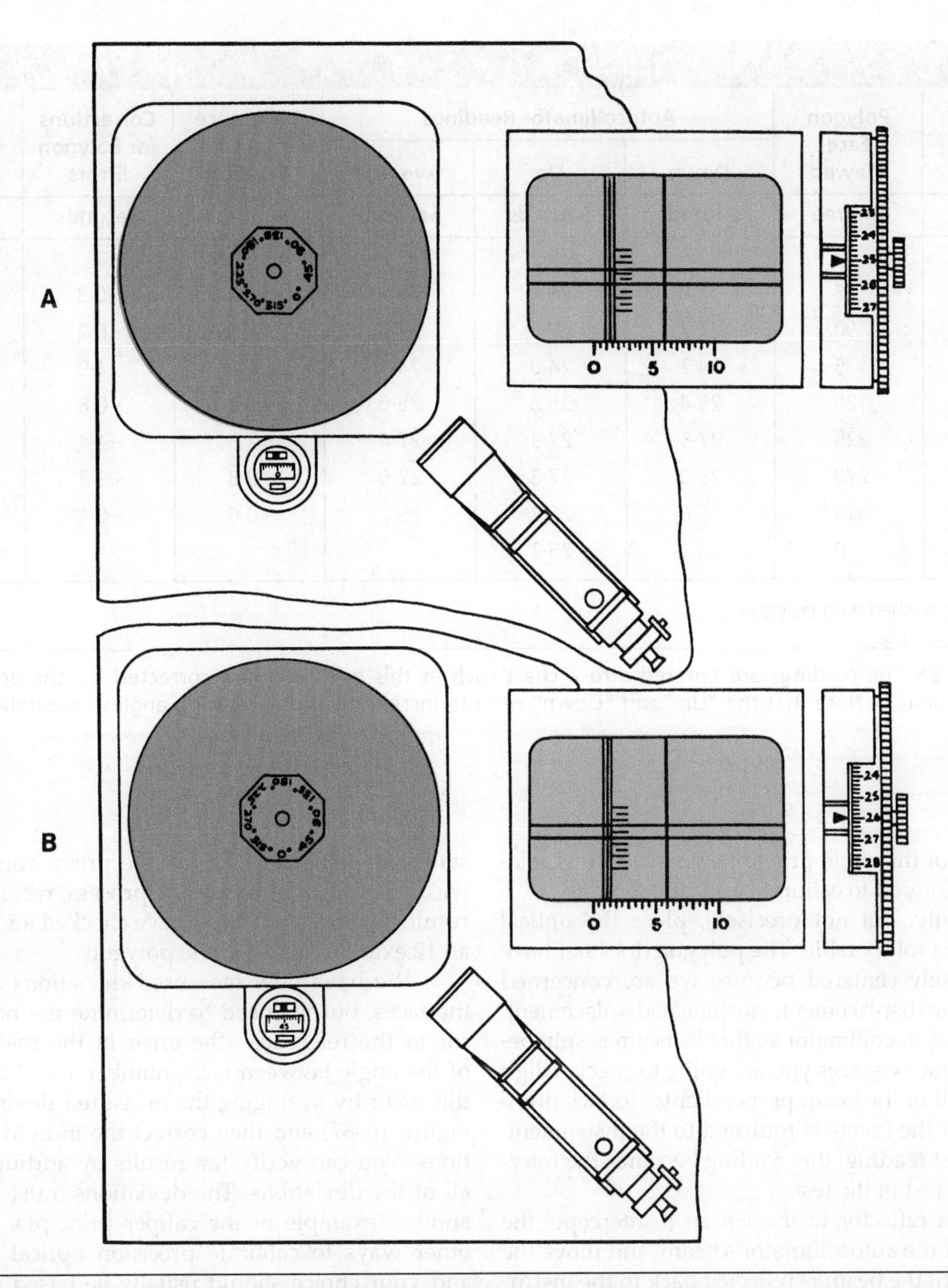

FIGURE 16–85 In A, a reading is taken for the zero face. Then, the table is rotated, and in B a reading is taken from the 45° face.

1	2	3	4	5	6	7*	8
Table Setting	Polygon Face Viewed	Autocollimator Readings			Difference from First Reading	Corrections for Polygon Errors	Errors in Table
		Down	Up	Average			
Degree	Degree	Seconds	Seconds	Seconds	Seconds	Seconds	Seconds
0	0	25·0	25·1	25·1	0	0	0
45	45	26·0	25·8	25·9	+0·8	−0·3	+0·5
90	90	27·7	27·5	27·6	+2·5	−0·2	+2·3
135	135	24·2	24·0	24·1	−1·0	−1·9	−2·9
180	180	25·4	25·8	25·6	+0·5	−0·8	−0·3
225	225	27·5	27·3	27·4	+2·3	−1·4	+0·9
270	270	28·0	27·8	27·9	+2·8	−3·3	−0·5
315	315	25·6	25·8	25·7	+0·6	+0·1	+0·7
0	0	25·1	25·2				

*From table furnished with polygon.

FIGURE 16–86 The readings are entered into a chart such as this one. They are corrected for the errors in the polygon (column 7). Note that the "Up" and "Down" refer to increasing and decreasing angles, respectively.

the closing of the circle principle, however, it is relatively easy for you to calibrate optical polygons.

Carefully, but not precisely, place the optical polygon on a rotary table. The polygon does not have to be precisely centered because we are concerned with angular displacement, not linear displacement. Position the autocollimator so that its beam is split between the first two faces you are going to check. Align the right half of the beam perpendicular to face number 1 so that the image is returned to the instrument. Take the first reading; this reading becomes the reference for the rest of the test.

Place a reflector to the left so it intercepts the other half of the autocollimator's beam, and move the reflector until the beam is reflected back to the instrument. Take the second reading, subtract it from the first reading, and divide the result by 2. (We need to divide because the angles have been doubled by the stationary reflector.) Rotate the prism counterclockwise one face, and repeat the process, recording your result. Continue until you have checked and recorded all 12 exterior angles of the polygon.

We have now measured deviations among all the faces, but we need to determine the nominal error in the reference—the error in the measurement of the angle between faces number 1 and 2. We find this error by averaging the measured deviations (see Figure 16–87) and then correct the individual deviations. You can verify the results by adding together all of the deviations. The deviations must total 360°, another example of the caliper principle. There are other ways to calibrate precision optical polygons, and your choice should usually be based on the instrumentation that is available to you. Other common calibrations use two autocollimators or an alignment telescope with a collimator.

Calibration of Twelve-Sided Optical Polygon with One Autocollimator					
Column 1	Column 2	Column 3	Column 4	Column 5	Column 6
Face	Angle (degree)	Recorded Differences	Deviation from Reference (second of arc)	Correction Factor (second of arc)	True Deviation from Nominal (second of arc)
1	0–30	8.0	0	−0.7	+0.7
2	30–60	7.1	−0.9	−0.7	−0.2
3	60–90	6.5	−1.5	−0.7	−0.8
4	90–120	6.3	−1.7	−0.7	−1.0
5	120–150	6.9	−1.1	−0.7	−0.4
6	150–180	8.4	+0.4	−0.7	+1.1
7	180–210	8.7	+0.7	−0.7	+1.4
8	210–240	8.4	+0.4	−0.7	+1.1
9	240–270	6.8	−1.2	−0.7	−0.5
10	270–300	6.4	−1.6	−0.7	−0.9
11	300–330	6.9	−1.1	−0.7	−0.4
12	330–360	7.2	0.8	−0.7	−0.1
total			−8.4		0

Note: Correction factor is −0.7 (−8.4/12)

FIGURE 16–87 The readings between the pairs of faces are in the recorded differences column (column 3). The "deviation from the reference" column (column 4) is derived from it. For example, 6.5 − 6.3 + (−1.5) = −1.7. We can measure all the differences between the faces. However, there is still the error in the initial measurement (face no. 1 to face no. 2), which is carried into all the subsequent measurements. This is compensated for by averaging all the deviations (see note at the bottom of the chart), and subtracting (adding algebraically) from the measured deviations (column 4 minus column 5). The results are the true deviations (column 6). Because a circle must have 360°, the test is that the summation of the deviations equals zero.

16–10 JIG TRANSIT—PLANES

Manufacturers call this instrument an *optical transit square, transit square,* or *jig level.* We usually use the jig transit to establish precise vertical planes and plumb lines. You can also use a right angle eyepiece to establish *zenith sights*—the point directly above the point of reference at infinity.

The jig transit looks much like a surveyor's transit, but it does not have circles for reading angles. It does have a telescope level, an optical micrometer, and accurate plate levels. We can extend its applications when we add an autocollimator and projector, as discussed with other instruments. A typical reticle pattern is shown in Figure 16–88.

The geometry of the jig transit (see Figure 16–89) lets us rotate the telescope 360° in both the horizontal and vertical planes. You can read straight down through an opening in the base and leveling head, which can be particularly useful for reading the scales on tooling bars.

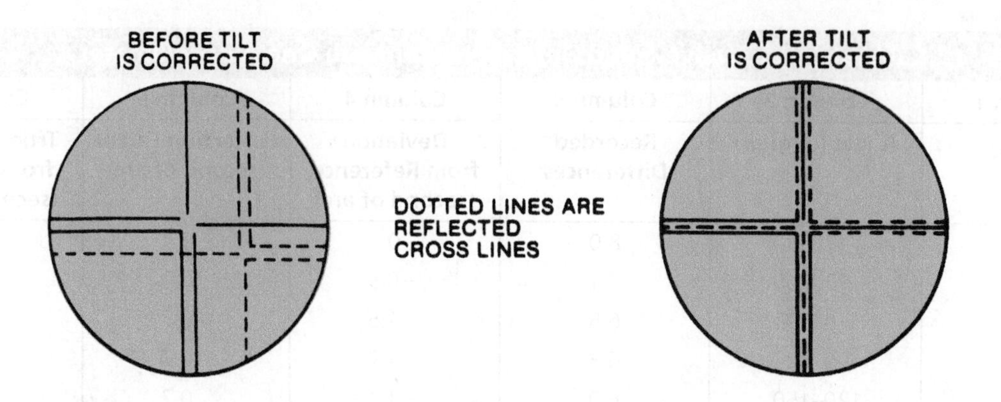

FIGURE 16–88 The reticle pattern for the jig transit is typically the crosspattern type but with paired lines in the two quadrants and single lines in the other two quadrants. The broken lines show how this appears when used with autocollimation.

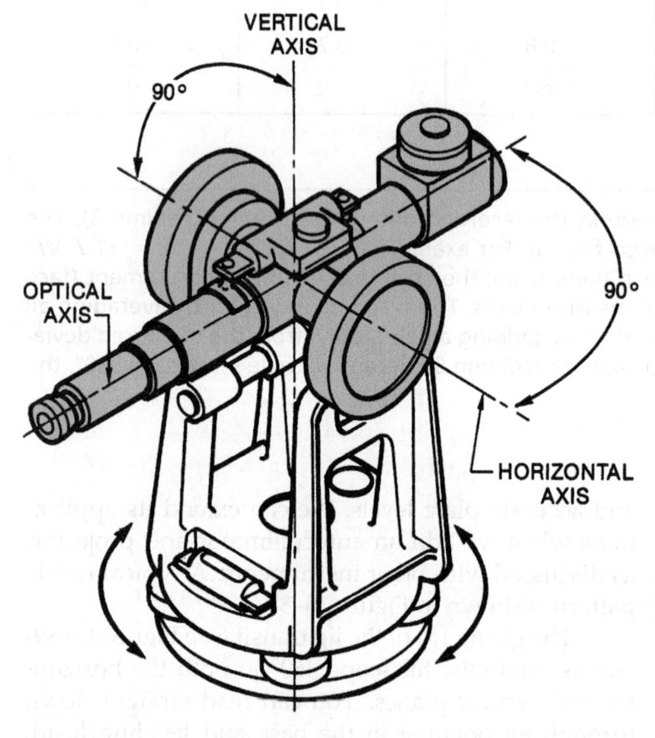

FIGURE 16–89 With the jig transit, all three axes intersect and rotation is possible about two of them.

The Cross-Axis Feature

The most important feature of the jig transit is the instrument's hollow elevation axis, which lets two perpendicular lines of sight intersect on the azimuth of the instrument. We can add a *telescope axle mirror* or a telescope along this axis. One mirror version has a partially silvered mirror on one end of the axis and a clear glass window at the other end. Another version uses retroreflective (nontransparent) mirrors, which can be attached at either or both ends of the elevation axis.

When you are using the mirror version, you do need a second person who will move the alignment telescope, which establishes the line of sight axis for the elevation axis, or you can use an instrument where the elevation axis has a built-in telescope. When the telescope is built-in, we call the instrument a *jig transit telescope square* (see Figure 16–90). The *cross-axis* (elevation axis) telescope has infinity focus and a crosslines pattern reticle.

Leveling

Always use the circular level (sometimes called a "bull's-eye" level) first. In some situations, that is all the leveling you will need to do. Adjust the four

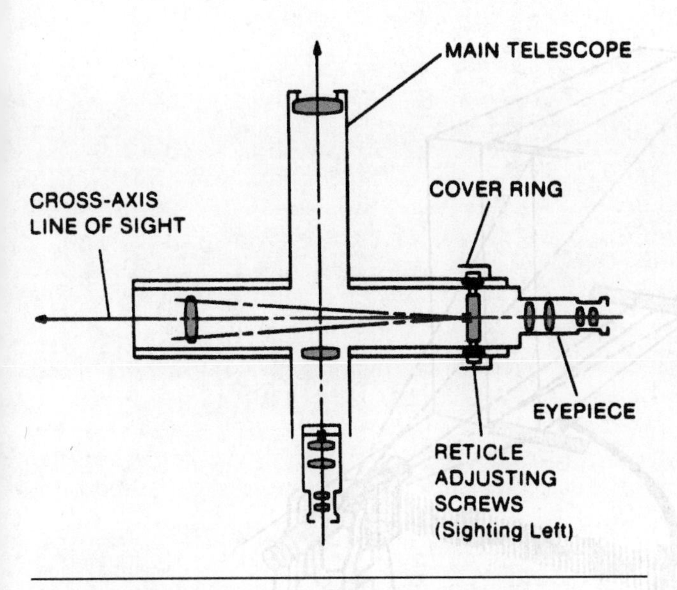

FIGURE 16–90 The two lines of sight intersect but do not optically interact.

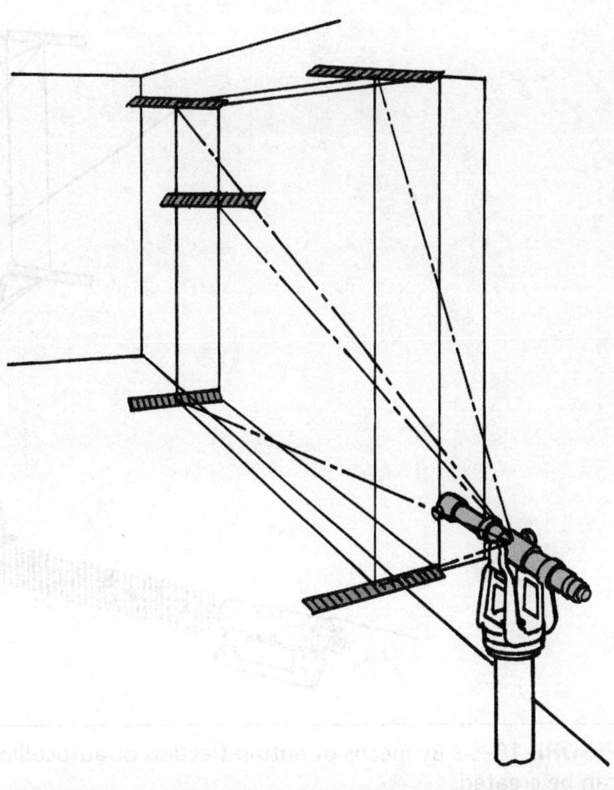

FIGURE 16–91 The telescope swings 360°, thereby establishing a vertical plane from which measurements may be taken by means of scales.

leveling screws according to the principles discussed for the sight level.

Next, rotate the alidade so that the long plate level comes into line with one pair of opposite adjusting screws. Use the screws to level it, and then rotate to the other pair of leveling screws. Repeat the process, and then use the reversal process of turning the alidade 180° to verify the results.

Applications

The jig transit is invaluable for establishing planes so that you can take measurements from them and use them for alignment references (see Figure 16–91), especially when:

- You need to take multiple measurements.
- It is not easy for you to access the reference from the features you are measuring or aligning.
- Long distances separate features and references.

In our example, the jig transit is supported by a stand so that swinging the head in its elevation axis establishes a vertical plane called the *station plane.* Using the same scales as the sight level (see Figure 16–67), you can take measurements from any feature to that plane, except a small area masked by parts of the instrument and its stand.

For precise results, you must be sure that the line of measurement is perpendicular to the plane of the features and references; if not, you are generating a cosine error. If you are only taking a few measurements, it may be easier to correct the data mathematically. However, if you are working with a large number of measurements, you should establish the station plane parallel to the plane of the features in one of a variety of ways, including some that are very simple (see Figure 16–92). You can use either auto-reflection or autocollimation to establish a line of sight at 90° to the station plane. The

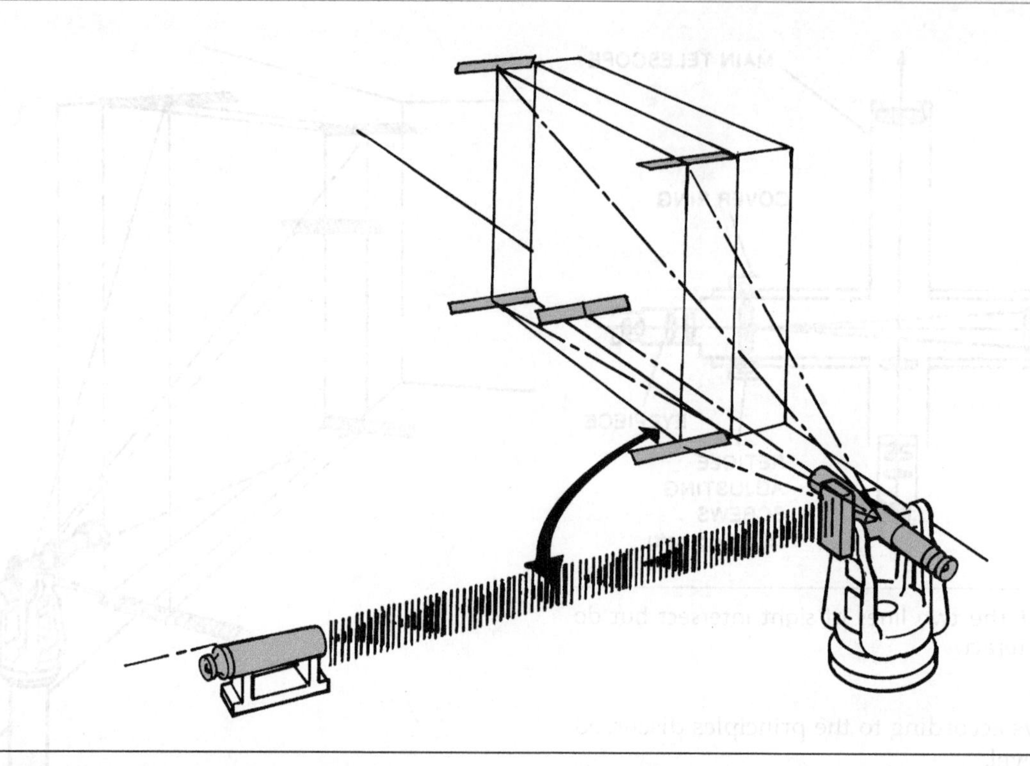

FIGURE 16–92 By means of auto-reflection or autocollimation, a plane at 90° to the measurement reference plane can be created.

station plane is not necessarily the plane of the features being investigated; therefore, you must align the workpiece so you create a relationship among three of the known features or reference areas and the station plane. You increase the accuracy of your results if features and reference areas are in the same plane—a plane parallel to the station plane. If they are not, you can measure the distances to the plane (see Figure 16–93).

In many situations, we are more concerned with alignment than measurement, such as with the alignment of rollers on large process industry machinery in paper mills (see Figure 16–94). When aligning the three rolls in our example, we establish two lines of sight, BD and CE, using alignment telescopes and targets. These lines must be level, parallel, in a vertical plane, and perpendicular to the axes of the rollers, all of which we can do optically.

We attach mirrors to the ends of the rollers so that the mirrors are perpendicular to the axes and their center marks are in line with the centers of rotation of the rollers. When we observe the end of a shaft with a telescope, we see that the center remains stationary even when the roller is rotated. We use jig transits to ascertain the axes of the rollers and lines of sight through them, and then establish the reference to determine whether or not those axes are parallel to each other.

The jig transit lets us use the reversal technique with even greater ease than we use the technique with other optical alignment instruments. For example, if we need to verify the right angle between the elevation axis and the line of sight, we first turn the telescope upside down and then aim it at a target. Reverse the telescope and read the scale in the opposite direction. Finally, with the telescope fixed in position, turn the azimuth to the target. You should obtain the same scale reading in both directions.

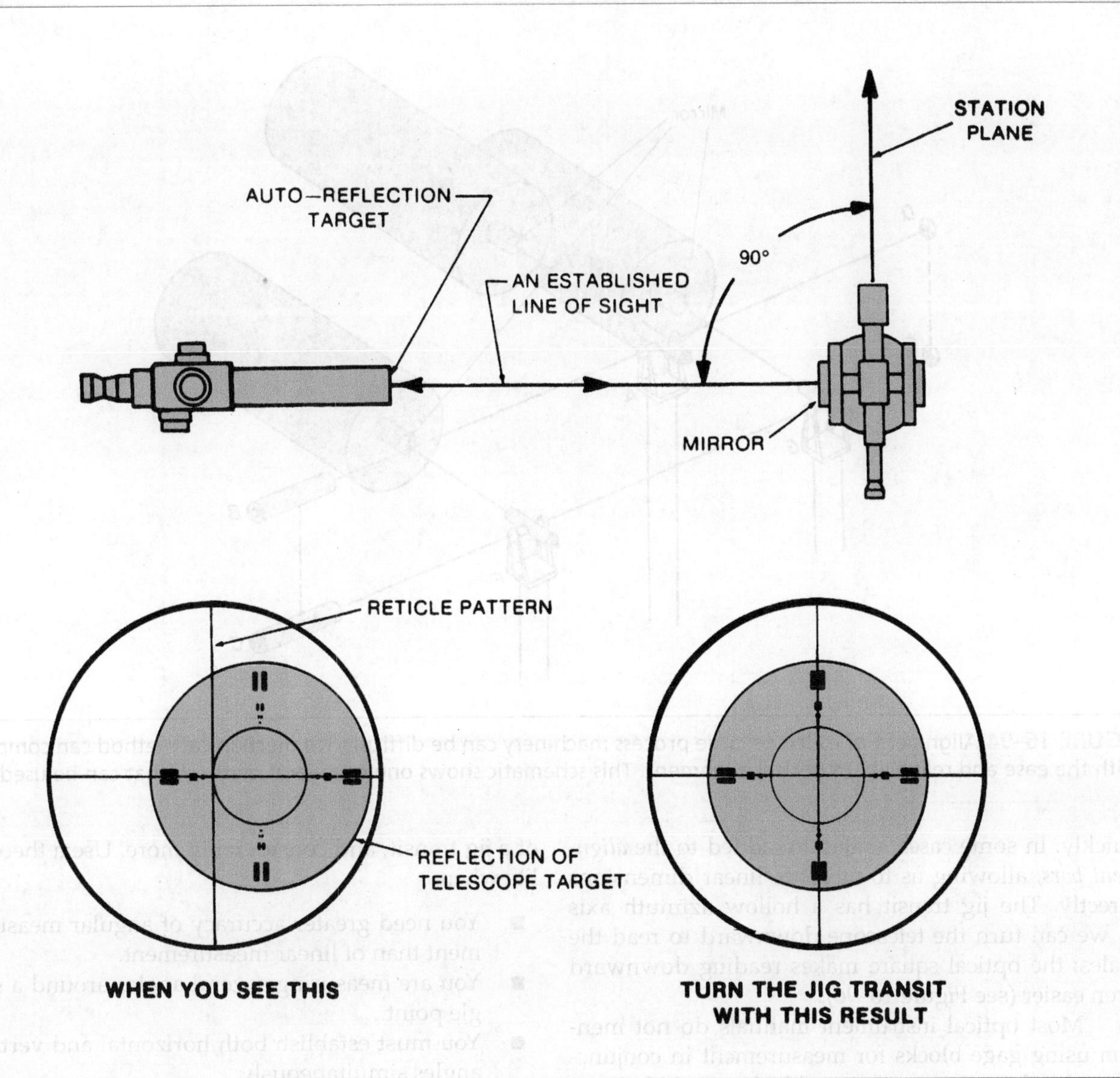

FIGURE 16–93 This schematic shows the way a line of sight can be created at 90° to the measurement plane (station plane) of the jig transit. The bottom drawings show the way the reticle pattern and the target are brought into coincidence.

Supports

We have, for the most part, ignored the supports for optical instruments because, from a metrological standpoint, the supports are less important than the supports we use for mechanical and electronic methods. For optical measurement, the most critical element is the beam of light, which is weightless, does not sag, and is not affected by distance.

Practically, the supports are very important (see Figure 16–95), because they allow us to set up

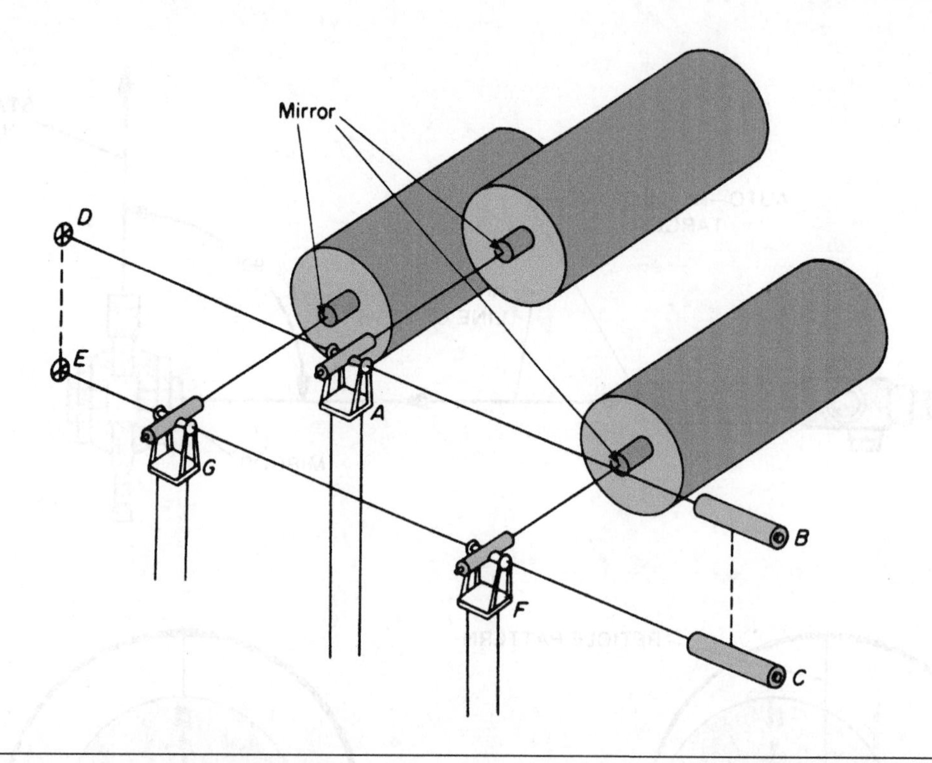

FIGURE 16–94 Alignment of rollers in large process machinery can be difficult. No mechanical method can compare with the ease and reliability of optical alignment. This schematic shows one of several methods that can be used.

quickly. In some cases, scales are added to the *alignment bars,* allowing us to measure linear dimensions directly. The jig transit has a hollow azimuth axis so we can turn the telescope downward to read the scales; the optical square makes reading downward even easier (see Figure 16–96).

Most optical instrument manuals do not mention using gage blocks for measurement in conjunction with them; however, gage blocks would be very useful with alignment bars.

16–11 THEODOLITE—ANGLES AND PLANES

A theodolite is basically a jig transit with graduated scales that you use to measure angles around its vertical and horizontal axes. It provides all the applications of a jig transit, and considerably more. Use a theodolite when:

- You need greater accuracy of angular measurement than of linear measurement.
- You are measuring several angles around a single point.
- You must establish both horizontal and vertical angles simultaneously.
- You must measure large angles.
- You must take an angular measurement at a distance from the reference line.
- You have difficulty reaching features by mechanical means because of obstructions.

Construction

The theodolite structurally resembles the jig transit (see Figure 16–89). You read the enclosed circular

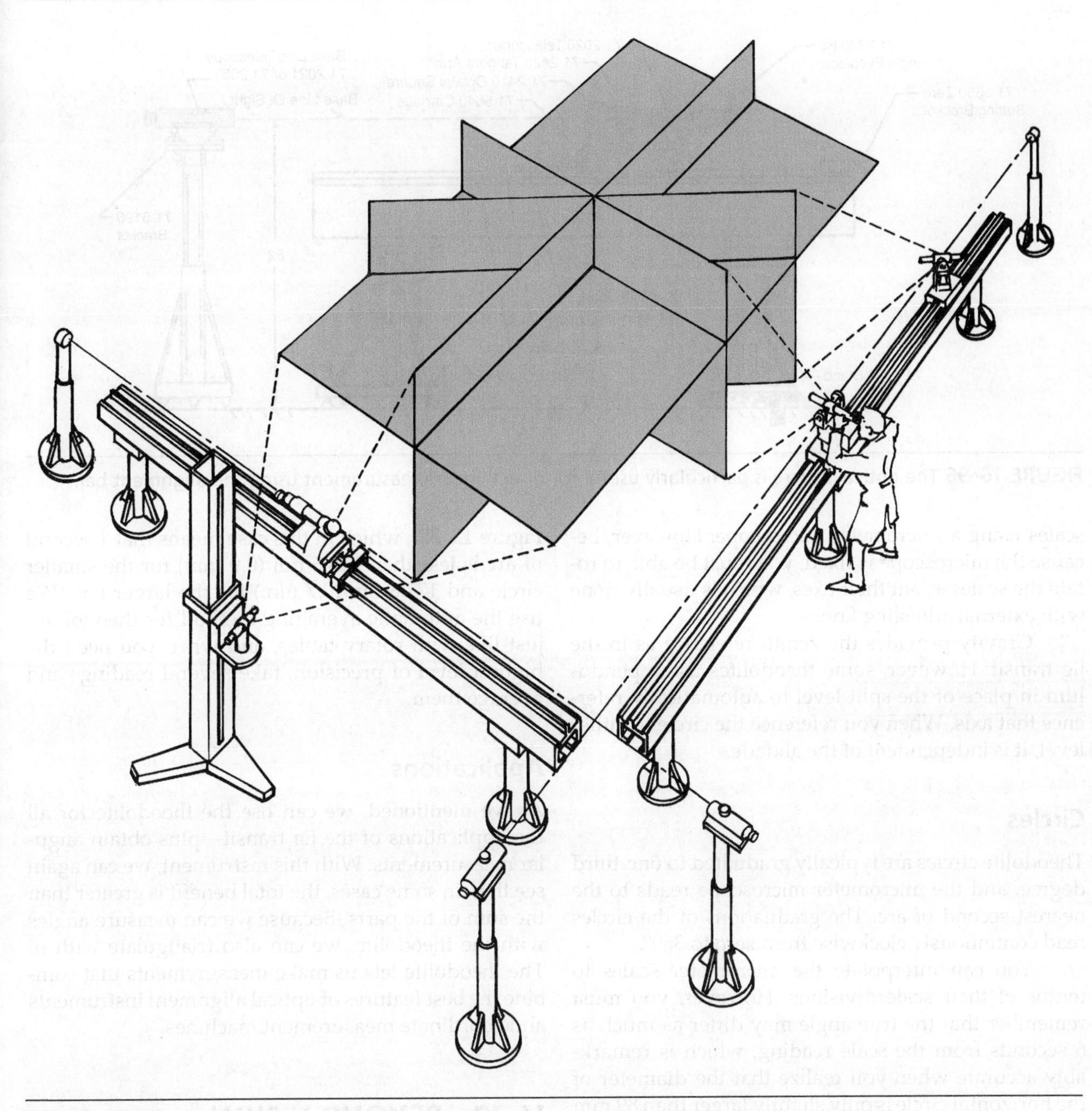

FIGURE 16–95 The use of standard supports facilitates optical alignment and measurement.

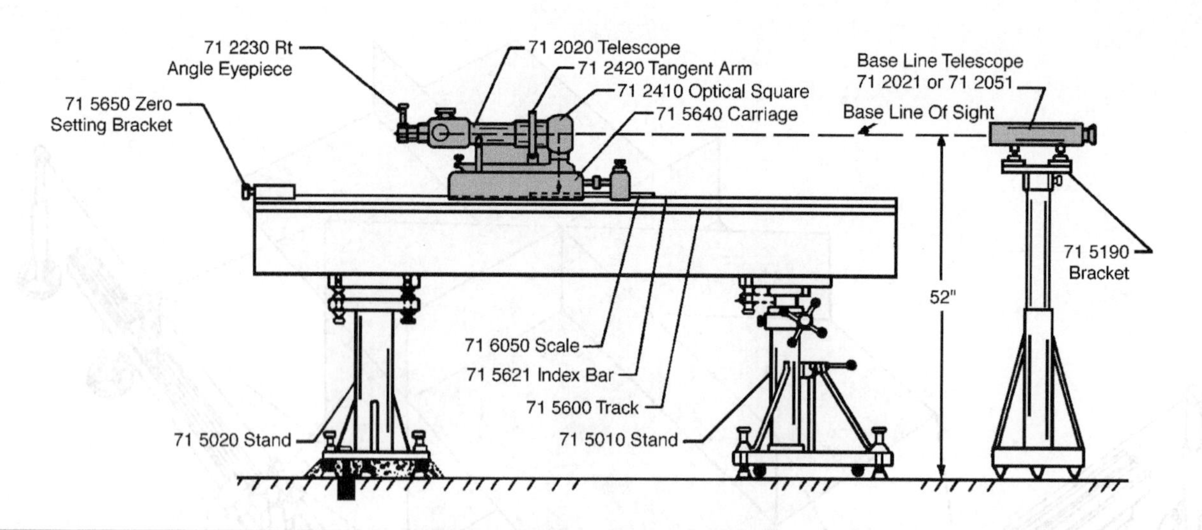

FIGURE 16–96 The optical square is particularly useful for direct linear measurement using the alignment bar.

scales using a micrometer microscope. However, because the microscope is fixed, you must be able to rotate the scales about their axes, which is usually done with external adjusting knobs.

Gravity provides the zenith reference as in the jig transit. However, some theodolites use a pendulum in place of the split level to automatically reference that axis. When you reference the circle to either level, it is independent of the alidade.

Circles

Theodolite circles are typically graduated to one-third degree, and the micrometer microscope reads to the nearest second of arc. The graduations of the circles read continuously clockwise from zero to 360°.

You can interpolate the micrometer scales to tenths of their scale divisions. However, you must remember that the true angle may differ as much as 6 seconds from the scale reading, which is remarkably accurate when you realize that the diameter of the horizontal circle is only slightly larger than 89 mm (3.5 in.) and the diameter of the vertical circle is slightly smaller than 76 mm (3 in.). There will be errors from the micrometer microscope, but the largest source of error is the eccentricity error (see Figure 12–87), which in this case means that 1 second of arc is less than 228.6 nm (0.9 μin.) for the smaller circle and 177.8 nm (0.7 μin.) for the larger one. We use the automatic averaging method for theodolites, just like with rotary tables, and when you need the highest level of precision, take several readings and average them.

Applications

As we mentioned, we can use the theodolite for all the applications of the jig transit—plus obtain angular measurements. With this instrument, we can again see that, in some cases, the total benefit is greater than the sum of the parts. Because we can measure angles with the theodolite, we can also triangulate with it. The theodolite lets us make measurements that combine the best features of optical alignment instruments and coordinate measurement machines.

16–12 BEYOND VISION

Optical measurement depends on another remarkable instrument—the human eye—but the eye is also optical measurement's greatest limitation. Because it is directly connected to the human brain, the eye's effectiveness is affected by the judgments and measurement decisions of the observer. Even though we can see the readings, we still have to interpret them. We cannot eliminate this variable, but we can lessen its effects by using interferometry, electronic sensors, and lasers.

Interferometry

We discussed the formation of visible bands of light by interferometry earlier. The *pointing interferometer*, an attachment mounted on an alignment telescope uses this principle to measure very small angular displacements. Inside the attachment, a prism divides a light beam into two paths, and these beams are directed at an external mirror. The mirror returns the beams, and the prism recombines them. When the two paths are nearly equal in length, you will see interference fringes in the eyepiece.

If you rotate the mirror, you change the return paths, creating fringes. One fringe (scale unit) represents a mirror rotation of 1 second of arc, and the optical micrometer of the telescope subdivides this scale unit into increments as small as 0.02 second of arc. The range is approximately 2,000 seconds of arc, and manufacturers claim that error does not exceed 0.1 percent of the measured angle or 0.02 second of arc, whichever is larger.

The interferometer has some other unique applications. For example, you can place any transparent material in one of the beams, and the material will affect that path length. From the distortion of the light path, you can measure optical properties such as density, refractory index, and so forth.

Photoelectric Instruments

Because photoelectric sensors have been developed to great sensitivity (defined in the Glossary), they have been added to autocollimators in several different versions. As varied as the instruments are, however, they all use somewhat the same principle. As with all autocollimators, the instrument directs a beam of light to a mirror and redirects the light to the instrument. One or more photoelectric sensor(s) measures the displacement of the returned beam in one of a variety of ways. The simplest method for reading the displacement uses two sensors to read the amount of light in each beam. Electronically, the output of one sensor is reversed with respect to the other one. When the sensors are aligned, the null-setting meter shows zero output—that is, the indicator on the meter points to zero in the center of the dial.

We roughly align the instrument using the eyepiece in the usual way, and then we use the optical micrometers to zero the photoelectric system. For each successive reading, you should zero the instrument with the meter and the deviations you recorded from the micrometers. The next most sophisticated versions provide limited angle readings directly from the meter. In addition, more sophisticated versions automatically align themselves, and you read them directly.

The two primary roles for *photoelectric instruments* are:

- The elimination of most eye fatigue from conventional instruments, which allows you to make repetitive measurements without risking additional reading errors
- Providing a practical method to continuously monitor an instrument—for example, the calibration of gyroscopes for navigational instruments

Some professionals also claim that it takes less skill to properly operate photoelectric autocollimators than conventional ones.

Laser Instruments

We did not go into great detail during our discussion of photoelectric instruments because of the laser. Laser is an acronym for "light amplification by simulated emission of radiation" and is produced by applying light to a solid or gas that is "excited" (in a highly energized state). Under proper conditions, an intense, narrow beam of single-color light results. Albert Einstein anticipated this phenomenon, but the laser was not developed until the proper material was developed (the first successful lasers used ruby).

The laser and the photoelectric phenomenon are natural partners; however, you should not look directly into even the so-called eye-safe lasers. If we use lasers for metrology, we need a better method of reading than the conventional eyepiece. On the other hand, although conventional optical metrology instruments have limited range, the laser beam is unlimited in length, for most practical purposes. In addition, you can conveniently see the line of sight beam with the laser instruments.

Laser Light

A laser beam is produced by a continuous wave plasma tube. Because it is coherent—unlike white light, the beam consists of light with only one wavelength—all the waves are in phase with each other. In most instruments, the beam emerges at 9.525 mm (0.375 in.) diameter; and even at a distance of 90 m (300 ft.), it will only have spread to 17.4625 mm (0.6875 in.).

A *laser alignment instrument* is constructed much like a conventional laser: All the components are inside a precisely machined barrel (see Figure 16–97). The line of sight—in this case, the laser beam—is precisely centered on and parallel to the axis of the barrel. You can use all of the holding and support fixtures for the alignment telescope with the laser instrument.

Equipment

The basic elements are the laser instrument and its support. Because lasers demand a greater amount of

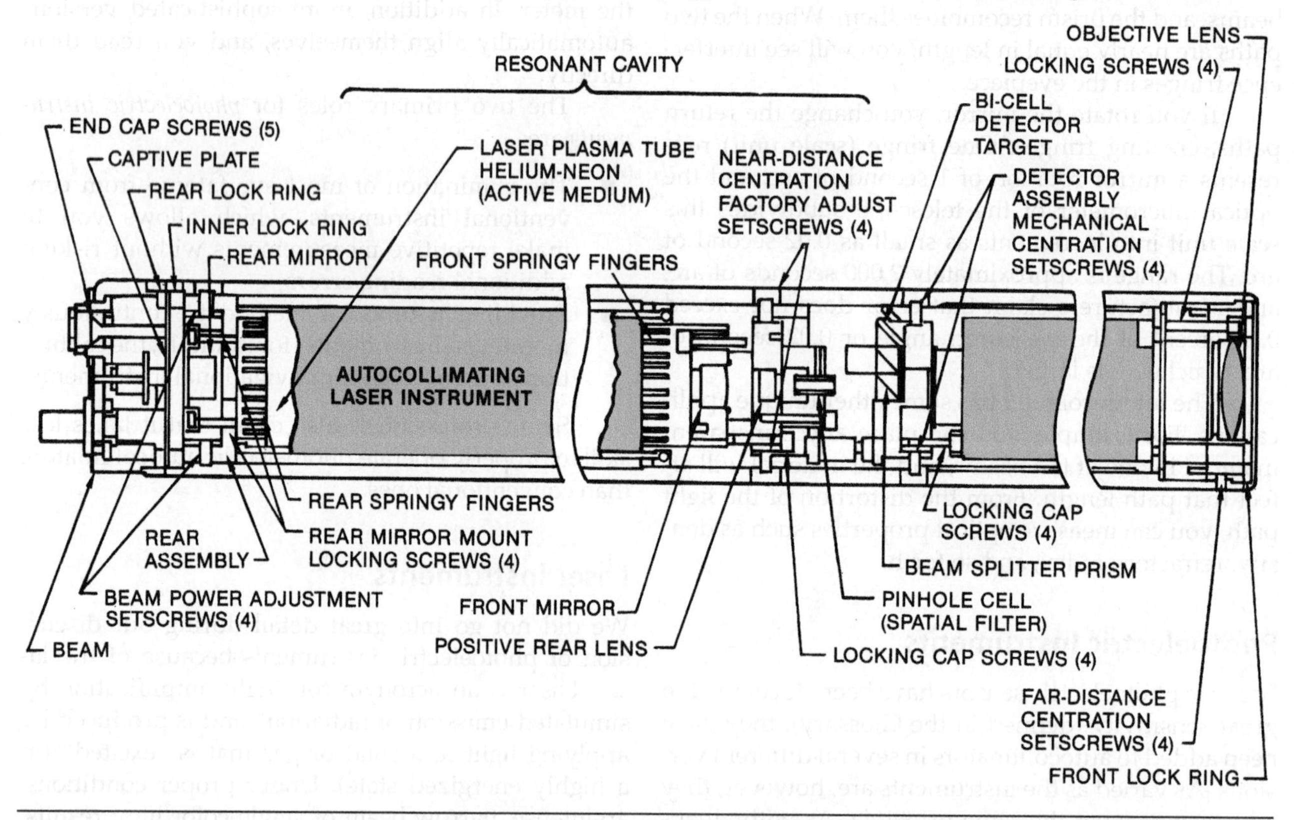

FIGURE 16–97 Externally, the laser instrument resembles the line of sight telescope. Internally, the only resemblance is the important one. Both have an optical axis exactly in line with the mechanical axis of their barrels.

power to operate, they have not revolutionized machining metals and warfare as many people predicted in the past. In addition to the power supply, you must remember to pay attention to the amount of heat dissipated by the laser. Finally, as with all measurement instruments, accessories are available.

When we discussed photoelectric autocollimators, we showed you that they retain their eyepieces and can be operated as conventional instruments. In contrast, the laser instrument has no eyepiece because we can see the beam in ambient light.

WARNING

Never look into the laser beam. It could injure your eyes. Only authorized personnel should use the equipment.

Alignment Role

With the addition of a target, you can use a laser alignment instrument like you use an alignment telescope. The specialized target uses photoelectric sensors to define the alignment. One target type is the *quad cell target,* because one cell in each quadrant detects changes in the light intensity on it. You can support the unit the same way you support conventional targets as long as the unit is held with its quadrants at 45° to the vertical and horizontal axes. You then plug it into the readout unit, which electronically changes the four measurements of light intensity into vertical and horizontal displacements and displays them digitally. If the displacement is to the right or down, the reading is preceded by a minus sign.

To roughly align the instrument, visually aim the laser beam at the target. You make fine adjustments with the instrument's optical micrometer, much like a conventional telescope. With the system illustrated, you may connect up to four targets at one time and then you can conveniently select their readings using the switches.

You can perform all of the techniques for the alignment telescope with a laser alignment instrument using readily available accessories like the see-through target (see Figure 16–98). With a see-through target, we divert a small amount of the laser beam 90° with a beam-splitting assembly. The diverted portion acts on a quad-cell array of sensors exactly like it acts on the sensors at the target. Manufacturers claim a resolution of 0.001 displacements from the center of the beam for both the quad cell target and the see-through target.

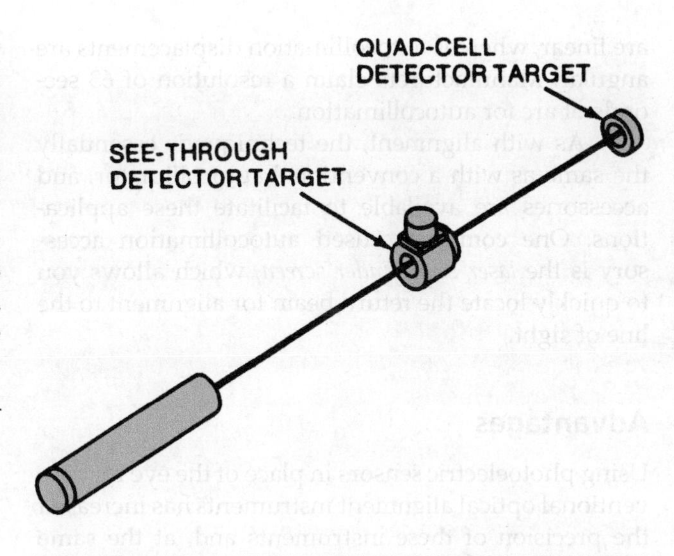

FIGURE 16–98 See-through targets may be placed along an established laser line of sight and displacements from that reference can be determined.

Autocollimation Role

The instrument barrel, in addition to the laser emitter, contains a beam splitter and an array of photoelectric sensors, which are arranged so that the sensors do not receive laser emissions. They receive the return beam from the mirror used in autocollimation.

Unlike the quad cell external targets, the internal target is a bi-cell array of sensors, and each of the two sensors responds to displacements in one axis. Because of this difference, you must switch the logic circuit of the readout unit from autoalignment to autocollimator operation, because autoalignment displacements are linear, whereas autocollimation displacements are angular. Manufacturers claim a resolution of 63 seconds of arc for autocollimation.

As with alignment, the technique is essentially the same as with a conventional autocollimator, and accessories are available to facilitate these applications. One commonly used autocollimation accessory is the *laser beam finder screen,* which allows you to quickly locate the return beam for alignment to the line of sight.

Advantages

Using photoelectric sensors in place of the eye for conventional optical alignment instruments has increased the precision of these instruments and, at the same time, has increased the ease of setup. When we use lasers to establish lines of sight, their benefits again increase, which is just one more of the many benefits of lasers:

- Hard-copy records—Because the instrument converts displacements to electrical signals, these instruments can input to printers as well as meters.
- Remote readout—In some setups, you may find it difficult to physically get into position to use the eyepiece. In other situations, such as in the nuclear industry, the intolerable environment keeps you from making direct observations of the readings.
- Continuous monitoring—With remote readouts, you can also continuously monitor the changing values of any operation and use a computer to control the processes of production directly.
- Multiple viewers—Two or more people can view the meter at the same time, aiding discussion and analysis.
- Vision errors—Variables in human eyesight, such as astigmatism, have less effect on meter reading than eyepiece measurement, which becomes particularly important when you and your co-workers need to make a number of observations or one of you is feeling a little tired.

SUMMARY

- There are four primary applications of optics to metrology: magnification, alignment, measurement, and standardization. Of these, magnification and alignment have the most frequent applications.
- Light is propagated in waves. These are not convenient for the study of optical systems. Therefore, an imaginary concept, the ray, has been devised.
- Rays allow us to follow the passage of light and account for the phenomena of reflection and refraction.
- Direct measurement with optical flats is restricted to an extremely short range. Beyond this range, the fringe pattern corresponds to the transducer stage only of the generalized measurement system (see Figure 14–10).
- Optical flat methods are unexcelled for the examination of surfaces. They are not as important for comparison measurement as they once were because of the development of the high-amplification, electronic comparator instruments. These instruments are more rapid and require relatively less skill than optical flats.
- The use of optical flats requires considerable manipulation. This results in a serious heat transfer problem. The flats themselves do not heat rapidly. Unfortunately, this also means they normalize very slowly. When measuring to a discrimination of one band or 11.6 μin., the heat transfer to steel parts can quickly render the measurement nearly valueless. On the other hand, there are fewer variables to cause errors with the optical flat method than with any other method of comparable precision and accuracy. If care is taken to ensure uniform wringing intervals, and if ample time is available for normalizing, this method provides greater reliability than comparable methods.
- The unique properties of light rays have been used since early man, but it took aircraft production to awaken industry to the potential for modern manufacturing. Measurement with

light rays is unique in that it does not depend on a physical object for the standard, utilizing ever-present gravity instead. This makes it valuable for evaluation of straightness, flatness, plumbness, and squareness, particularly when the workpieces are large. Optical alignment using conventional instruments nearly removed the limitation of size. Now with laser instruments that limitation is completely removed.

- The telescope is the basic element of optical alignment. It allows targets placed on the workpiece to be located in respect to its line of sight. A special type of telescope, the collimator, projects a line of sight, which is returned by a mirror. This permits very small angular changes to be measured.
- Specialized telescope mountings adapt optical alignment to a wide variety of applications. The alignment telescope is the simplest and the mainstay of straightness measurement. The optical square aligns two light paths at precisely 90°. It is the basis for squareness evaluation. Plumbness is evaluated with the sight level. This instrument incorporates a precision level to reference the gravitational attraction of the earth. The jig transit resembles a surveyor's transit both in appearance and use. It is used to establish precise vertical planes as well as plumb lines. Its most sophisticated version is the theodolite. Optical polygons are precise standard angles that may be used with these instruments to establish angle references.
- Optical alignment basically depends on the human eye. However, some of the limitations of the eye can be minimized. One of the methods is by interferometry in which fringe bands are the means for reading. Other instruments use photoelectric sensors to substitute for the eye.
- The use of laser light sources in optical metrology adds nothing to the basic principles. Its benefits are the greater distances it makes possible, and the greater discrimination when used with photoelectric sensors.
- The instruments with photoelectric sensors do not alter the metrology of their applications, but they do provide desirable convenience features. Among these are remote reading, hard-copy records, continuous monitoring, multiple viewers, and less observer fatigue.

END-OF-CHAPTER QUESTIONS

1. How many primary applications are there of optics to metrology?
   a. Infinite number
   b. 3
   c. 2
   d. 4
2. What characteristic of light makes it a standard?
   a. Its relative brightness can easily be expressed in analog values.
   b. Its color (wavelength) can be selected through a wide spectrum.
   c. The length of the waves is known and unvarying.
   d. It does not change throughout the usable range of temperature.
3. Which of the following typifies measurement with light waves?
   a. Gage blocks
   b. Precisely lapped surfaces
   c. Optical flats
   d. Steel reference flats
4. Which of the following best explains the need for increased discrimination?
   a. Government regulations
   b. Improved alloys
   c. The need to utilize the new digital readout measurement instruments
   d. Increase of speed, larger number of components, and need for reliability
5. How did Albert A. Michelson contribute to metrology?
   a. He did not. He was an astronomer.
   b. He used light waves to calibrate gage blocks.

c. He measured the wavelength of one particular spectral line in relation to the meter.
d. He divided by 2 the difference between the wave shift of incoming interference bands and outgoing bands, thereby arriving at their absolute value.

6. Which of the following characteristics of light is the most important one for its role as the master standard of length?
   a. Its unchanging feature
   b. Ease of reproducibility
   c. Convenience in using
   d. Not temperature sensitive

7. What is interferometry?
   a. The use of an interferometer to measure lengths
   b. The highest precision method generally available for length measurement
   c. The interpretation of fringe patterns formed when rays of light interfere with each other
   d. The systematic calibration of length differences by means of fringe patterns

8. For practical work with optical flats, which of the following is the most important requirement of the light?
   a. That it be directed downward
   b. That it be generated from isotope number 86 of krypton
   c. That it be monochromatic
   d. That it be at least 50% brighter than the ambient light

9. Which of the following commercially available light sources is the most practical for shop work?
   a. Helium
   b. Krypton
   c. Mercury
   d. Sodium

10. Why is cleanliness so important for measurement with optical flats?
   a. The fringe bands cannot be seen if the flat is dirty.
   b. Dirt may cause wringing when not intended.
   c. The higher amplification increases the magnitude of the error.
   d. The optical flat attracts dust.

11. The origins of optical alignment are from which of the following?
   a. Building construction
   b. Archeology
   c. Surveying
   d. Military science

12. What is macrometrology?
   a. Measurement with micrometer-type instruments
   b. Measurement of two or more features of the same workpiece
   c. Total measurement of an object
   d. Very large measurement

13. Which of these statements about microscopes and telescopes is correct?
   a. A microscope uses lenses, whereas a telescope uses diopters.
   b. They are basically the same but have different distances to their objectives, virtual images, real images, reticles, and so forth.
   c. A telescope may be handheld but a microscope must have a stand.
   d. A telescope may be collimated but a microscope cannot.

14. What is the name of the act required for checking instrument alignment?
   a. Reversal technique
   b. Interferometry
   c. Angular alignment
   d. Abbe's law

15. In optical metrology, what does "azimuth" mean?
   a. Perpendicular to gravity
   b. Aiming angle for naval guns
   c. Perpendicular to the reference surface
   d. Zero angle

16. What is established, if anything, by three points on one line?
    a. A plane
    b. A triangle
    c. Collinear
    d. Infinity

17. What is the role for line of sight telescopes?
    a. Astronomical observation
    b. Angle measurement
    c. Measurement of opaque workpieces
    d. Establishment of lines of reference

18. In which axes do the optical micrometers function?
    a. None
    b. X only
    c. X and Y
    d. X, Y, and Z

19. What is field of view?
    a. Largest diameter that can be viewed without repositioning the telescope
    b. The angle between the two extreme positions on the reticle
    c. The view of the workpiece as seen without the telescope
    d. The clearest portion of the center of the image

20. Which of the following best describes depth of focus?
    a. Maximum distance between targets in which both are in focus
    b. Distance behind eyepiece at which target can still be seen
    c. Distance at which target is at maximum size
    d. Z axis measurement

21. What is the meaning of collimation?
    a. Higher-grade instruments
    b. Projection of parallel rays of light
    c. Disposal of any unwanted data
    d. Light passing through a telescope

22. What is an optical square?
    a. A drafting square with a magnifier
    b. Four sets of reticles set at 90° to each other
    c. A prism or mirror system that turns light 90°
    d. A beam of light reflected from an orbiting space station

23. What is the reason for the double sphere offset square?
    a. Prevents the line of sight from being blocked
    b. Distribution of wear over two spheres instead of one
    c. Easier accessibility of the right-angle prism
    d. Easier adjustment

24. What is the chief role of the sweep optical square?
    a. Checking the flatness of large flat surfaces
    b. Leveling of reference surfaces
    c. Separating yaw from pitch of machine components
    d. Permanent monitoring of optical measurement setups

25. What is the maximum error that can be expected from the use of the scan prism if the setup has been made properly?
    a. Ten seconds of arc
    b. Five seconds of arc
    c. Depends on the temperature
    d. Varies among instrument manufacturers

26. What are the references most frequently used for mechanical alignment?
    a. External and internal
    b. Flat surfaces and curved surfaces
    c. Flat surfaces, curved surfaces, and spherical surfaces
    d. Horizontal and vertical planes

27. What is the role of the striding level?
    a. Measurement of levelness with respect to strides in surveying
    b. Any level that can be saddled to the tube of an optical instrument
    c. An instrument that compares line of sight with the horizontal plane
    d. A level that can be easily moved from instrument to instrument

28. What is the role of the cross level?
    a. Provides both x and y axis level readings at the same time
    b. Provides either x and y axis readings on call
    c. To survey applications only
    d. Alignment of crosslines of the reticle to the horizontal and vertical planes
29. What can be used to align the crosslines of the reticles in both the horizontal and vertical planes?
    a. The dispositive level
    b. The cross level
    c. Four conical adjustment cones, two for x and two for y axes
    d. The autocollimator
30. Which of the following groups all refer to the same general instruments?
    a. Sight level, tilting level, optical tooling level
    b. Sight level, tilting level, clinometer
    c. Sight level, scan prism, clinometer
    d. Sight level, scan prism, optical tooling level
31. What is the standard for optical polygon measurement?
    a. Triangulation
    b. Gravity
    c. Angle gage blocks
    d. The circle
32. What is the largest use for optical polygons?
    a. In NIST and other national standards laboratories
    b. In the setup of rotating machinery
    c. For the calibration of instruments and tooling that involves angles
    d. For checking angle gage blocks
33. What instrument is used most in conjunction with optical polygons?
    a. Autocollimator
    b. Autocollimator with scan prism
    c. Sine bar
    d. Rotary table
34. Which of the following all refer to the same type of instrument?
    a. Optical transit square, transit square, pointing interferometer
    b. Optical transit square, transit square, jig level, jig transit
    c. Transit square, jig level, pointing interferometer
    d. Transit square, jig level, autocollimator
35. Which of the following is the most important role of the jig transit?
    a. Establishment of vertical angles and plumb lines
    b. Surveying
    c. Measurement of flatness
    d. Measurement of inaccessible features of large structures
36. What is the most important feature of the jig transit?
    a. Portability and ease of handling
    b. Use of built-in levels
    c. Both horizontal and vertical axes intersect
    d. Illumination of the reticle
37. Which of the following best describes a theodolite?
    a. An electronic jig transit
    b. A lightweight jig transit
    c. A jig transit with scales for vertical and horizontal measurement
    d. A jig transit whose three axes can be locked in position
38. What establishes the zenith reference in both the jig transit and the theodolite?
    a. Careful setup on a stable surface
    b. The reversal process
    c. The intersection of the horizontal and vertical axes
    d. Gravity

39. How are vertical and horizontal angles read on a theodolite?
    a. Episcopic projection of circle scales
    b. Rotation of the circles
    c. Micrometer microscope viewing of the circles
    d. Vernier scales on the circles

40. Which of the following best describes interferometry?
    a. Use of lasers in measurement
    b. Use of optical flats in measurement
    c. The interference of light waves to form visible patterns
    d. The spectral analysis of white light

41. Which of the following best describes laser light?
    a. Intense light focused on a small spot
    b. Light rays that travel in the same direction
    c. Light that cannot be reflected
    d. Coherent with only one wavelength and all waves in phase

42. What is the first thing to know about lasers?
    a. They are expensive.
    b. They require much electrical power.
    c. They cannot be used except in temperature-controlled conditions.
    d. Never look into the laser beam.

43. Which of the following comes closest to the dispersion of a 9.525 mm (0.375 in.) laser beam at 91.44 m (300 ft.)?
    a. 9.525 mm (3/8 in.)
    b. 19.05 mm (3/4 in.)
    c. 38.1 mm (1 1/2 in.)
    d. 127 mm (5 in.)

44. Which of the following instruments is most likely to be adapted for laser use?
    a. Alignment telescopes
    b. Striding levels
    c. Sweep optical squares
    d. Theodolites

45. Which of the following best describes the advantages of laser metrology?
    a. Hard-copy records, remote readout, continuous monitoring, full color screen
    b. Hard-copy records, remote readout, multiple viewers, temperature insensitive
    c. Remote readout, continuous monitoring, multiple viewers, low cost
    d. Remote readout, continuous monitoring, multiple viewers, compensation for eyesight of operator

17

CHAPTER SEVENTEEN

OPTICAL METROLOGY

LEARNING OBJECTIVES

- Understand how the reticle makes a microscope a measurement instrument.
- Detail the chief roles of the microscope in applied metrology.
- Describe the optical systems required for comparators.
- State the significance of interchange and displacement measurement.
- List the advantages and disadvantages of machine vision systems.
- Identify applications for microscopes, comparators, and vision systems.

OVERVIEW

We employ two different basic principles in optical metrology. Unlike instruments that require optical alignment, we observe the projection of the part onto a screen. The applications we will discuss are extensions of magnifiers (see Chapter 16 and Appendix C) because they increase the apparent size of features so that we can study them more easily.

In practical metrology, there are two major divisions of microscopy: direct measurement and positioning. When the instrument is its own standard, direct measurement uses the interchange method (see Figure 17–1). For example, the reticle of the microscope is both the standard and the means for reading the standard; therefore, direct measurement with the microscope concerns:

- Linear measurement
- Surface topology and metallurgical measurement
- Measurement of features inaccessible by other means

It might be expected that we use microscopes chiefly for these purposes; however, *microscopes* are used most extensively for positioning. After we position a part with a microscope so we can see the part feature and scale, we actually use another instrument to make the measurements. When we use the microscope for placement, we are using the displacement method.

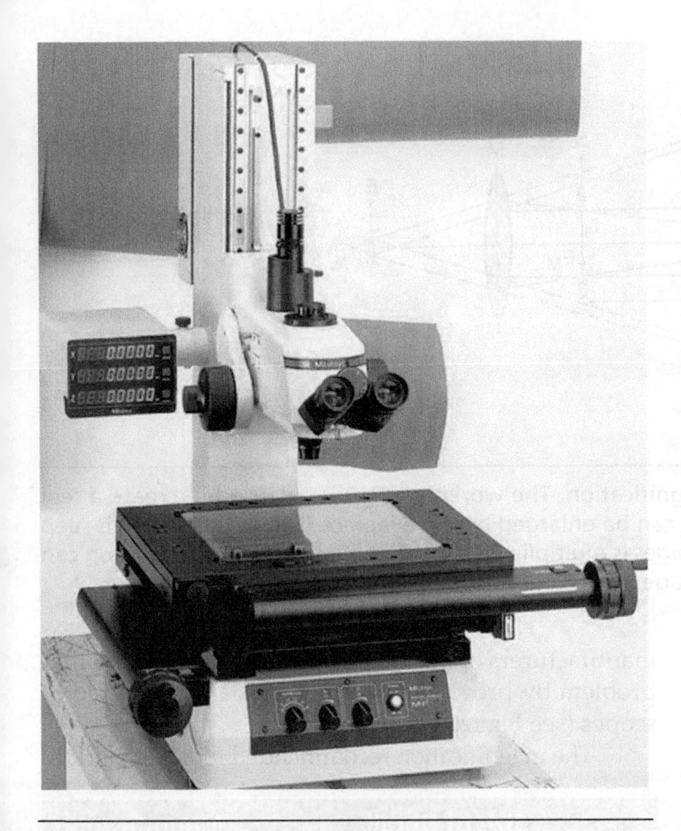

FIGURE 17–1 We associate the microscope, austere and pristine, with science and medicine. It is also a metrological tool of the most fundamental importance and greatest integrity. *(Courtesy of Mitutoyo American Corporation)*

FIGURE 17–2 Optical comparators are inexpensive and easy to use. *(Courtesy of Mitutoyo American Corporation)*

We use the optical instruments in this chapter to observe the magnified image and profile of the workpiece. Microscopes, optical comparators, and machine vision systems are by far the best-known optical instruments industry uses today, except for the simple magnifier (see Figure 17–2).

With optical comparators, we use the technique of *shadow projection* or *profile projection*, and we use many of the different types of instruments in much the same way. For example, you can equip a microscope for profile projection, and you can also adapt an optical comparator so you can observe surfaces. We will use the term *optical comparator* to designate an instrument and use *profile projection* to designate the technique you should employ. Comments about profile projection apply to microscopes as well as to optical comparators.

17–1 PRINCIPLES OF THE MICROSCOPE

The microscope is based on facts mentioned in Appendix C. After you magnify a part, you can then treat

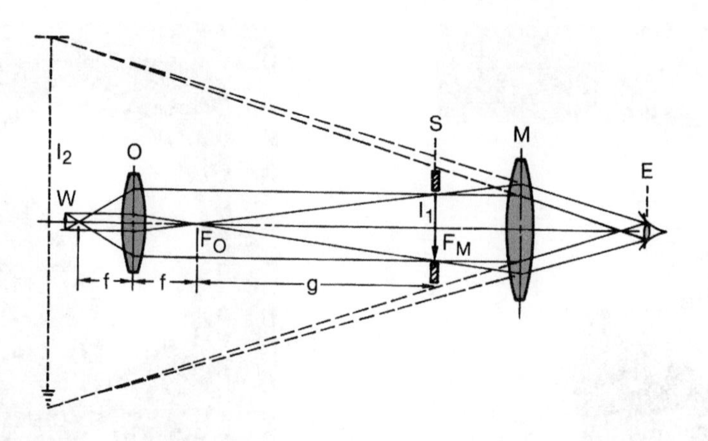

FIGURE 17–3 The microscope couples two stages of magnification. The workpiece (W) is enlarged to create a real image, I_1, at the stop (S). The stop frames the image so it can be enlarged by the eyepiece (M). As a result, it is seen as a virtual image, I_2. The magnification for each of the stages is multiplied. Thus, a great effective magnification can be achieved with only moderate magnification at each stage.

the optical image as a real object, which, in turn, can be magnified (see Figure 17–3).

We form an image using the objective lens O of the work W at I_1. We then view this image with the eyepiece magnifier E, producing the image I_2. Remember: I_1 is a *real* image; therefore, if we placed a screen at I_1, we would see an image on the screen. In contrast, I_2 is a *virtual* image—the image only appears to be located at I_2; therefore, if we placed a screen at I_2, we would not see the image on the screen. As a result, we must place any reticle in the real image plane.

The *effective magnification* is the product of the individual magnifications of the lenses. Eyepiece magnification is usually at 10X; therefore, when the objective lens has a magnification of 7X, the microscope magnifies 70 times. If you compare this magnification with the pocket comparator, you can see that the spacing would be read easily: 127 μm (0.005 in.) becomes 12.7 μm (0.0005 in.), which is a very fine measurement for such a rugged method. If you use an electronic instrument, you would need to be more careful in the setup. In addition, you must remember that none of these instruments measures directly. You would have to calibrate all of them against a standard.

In Appendix C, we mention that you must keep both eyes open when using a magnifier. The manufacturers of optical instruments have solved that problem by providing *binocular* or *stereoscopic* microscopes (see Figure 17–4).

The magnification recommended for most work is:

Size of Detail		Magnification
2.54 mm–254 μm	0.10–0.010 in.	6.6X–30X
254 μm–25.4 μm	0.010–0.001 in.	20X–60X
25.4 μm–12.7 μm	0.001–0.0005 in.	40X–90X
12.7 μm–254 nm	0.0005–0.00001 in.	80X–150X

The Measuring Microscope

As we stated, we can measure to such a high degree of reliability because measurement with a microscope is direct. As in all scaled instruments, the reticle itself is the standard—but it is also the microscope's limitation. As we increase the magnification, we decrease the microscope's *field of view* or the portion of the workpiece that we can view at one magnification. Manufacturers have developed sophisticated eyepieces and objective lenses to increase the field of view; however, even with the best of these enhancements, the field of view at magnifications over 50X is seldom as wide as the part feature you need to measure. As a result, we must use

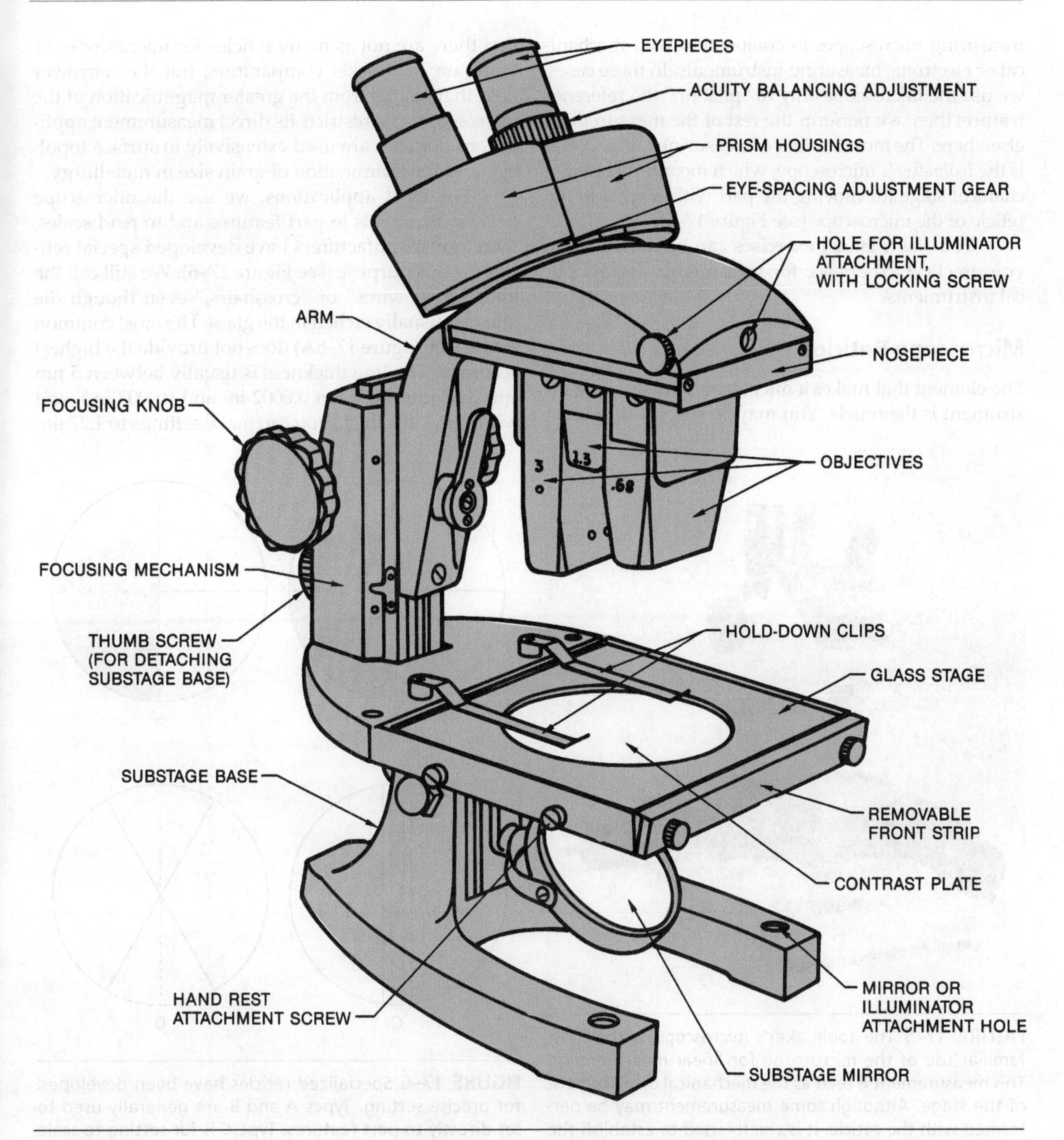

FIGURE 17–4 When both eyes are used, fatigue is reduced. Binocular or stereoscopic microscopes allow this to be done.

measuring microscopes in conjunction with mechanical or electronic measuring instruments. In these cases, we use the microscope only to "pick up" the reference feature; then, we perform the rest of the measurement elsewhere. The most familiar of measuring microscopes is the *toolmaker's* microscope, which incorporates a mechanical stage for moving the part with respect to the reticle of the microscope (see Figure 17–5).

Your lab manual exercises can be used to give you practical experience for measurement with optical instruments.

Microscope Reticles

The element that makes a microscope a measuring instrument is the reticle. You may be surprised to learn that there are not as many reticles for microscopes as there are for pocket comparators, but the narrower field that results from the greater magnification of the microscope also restricts its direct measurement applications. Reticles are used extensively in surface topology and the examination of grain size in metallurgy.

For most applications, we use the microscope to set instruments to part features and to read scales; therefore, manufacturers have developed special reticles for this purpose (see Figure 17–6). We still call the lines "cross wires" or "crosshairs," even though the lines are usually etched in the glass. The most common reticle (see Figure 17–6A) does not provide the highest accuracy. The line thickness is usually between 5 µm and 7.62 µm (between 0.0002 in. and 0.0003 in.), and with a line this thick, you can make settings to 1.27 µm

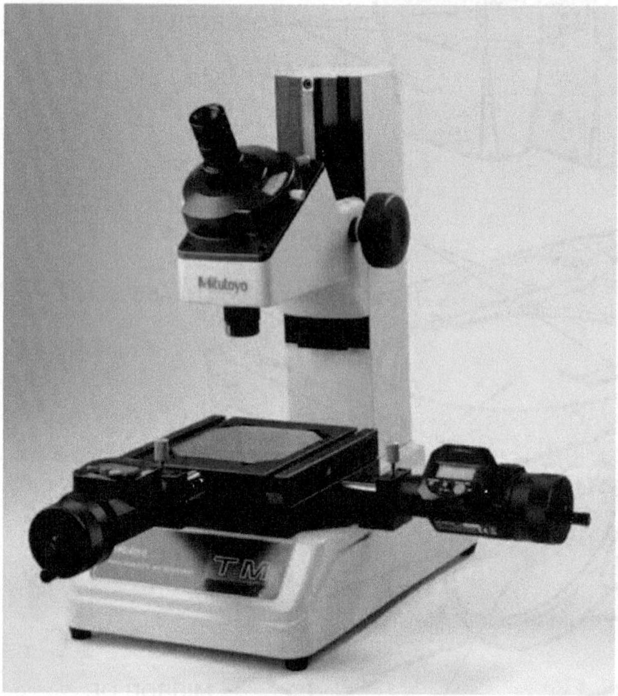

FIGURE 17–5 The toolmaker's microscope is the most familiar use of the microscope for linear measurement. This measurement is read as the mechanical displacement of the stage. Although some measurement may be performed with the reticle, it is chiefly used to establish the boundaries of the part features. *(Courtesy of Mitutoyo American Corporation)*

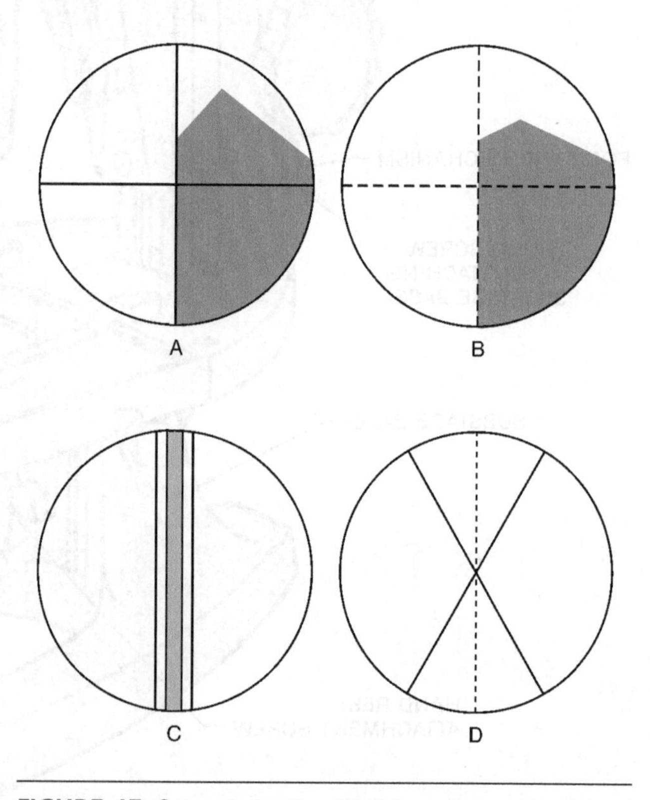

FIGURE 17–6 Specialized reticles have been developed for precise setting. Types A and B are generally used to set directly to part features. Type C is for setting to scale graduations and type D is used for ultraprecise work such as with semiconductor chips.

(50 μin.) using low magnifications, such as 2.5X or 5X for the objective and 10X for the eyepiece.

For greater reliability in your settings, use a reticle with broken lines. Broken lines are especially useful when the feature you need to measure is narrower than the reticle line (see Figure 17–6B).

For picking up scale lines, use a reticle, as shown in Figure 17–6C. Because your eye averages the slight irregularities of the edges of the scale lines when you see them in a clear space, the parallel lines are spaced slightly wider than the scale lines so you can make precise settings. This reticle is known as a *bifilar reticle.*

For the most precise work—work involving photographic plates, electronic workpieces, and other nonmetal workpieces—we use the reticle shown in Figure 17–6D. Two lines are at angles to each other, typically 30°. Your eye easily establishes symmetry among the four spaces that are created when the reticle is positioned over a fine line.

Micrometer Eyepiece

For reading vernier scales, the microscope offers unprecedented precision and can be equipped with a *micrometer eyepiece.* This eyepiece has a reticle in its principal image, which becomes superimposed over the workpiece. The reticle image is arranged so you can travel along one coordinate, usually the x axis (see Figure 17–7).

Stage Illumination

As we increase magnification, the distance between the objective lens and the workpiece decreases, which causes problems when we are trying to light the part adequately to see the feature we are measuring. Specifically, the distribution of light reflected from the workpiece is reduced by the square of the increase in magnification; therefore, a part feature clearly visible at 10X may be invisible at 50X. Because microscopy emerged from the natural sciences, not metrology, scientists could slice their specimens into thin sections and pass light through them. Surface illumination is not a problem for natural scientists, but it is a huge problem for a manufacturer trying to ensure the quality of its parts.

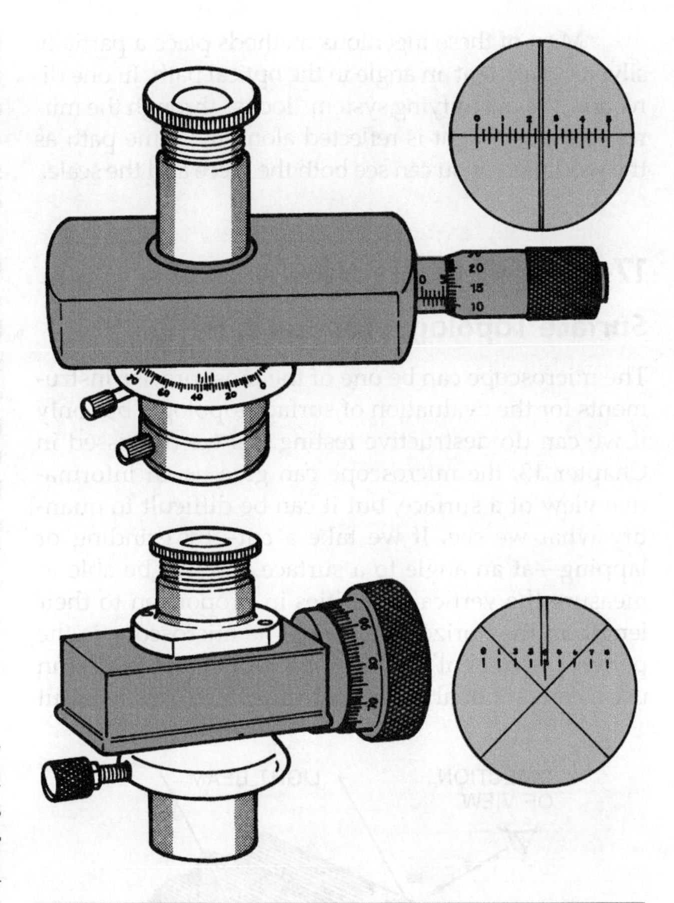

FIGURE 17–7 The micrometer eyepiece allows the scale of the reticle to be moved from one part feature to another. The displacement between the feature is read both by the micrometer and the reticle. The bottom version makes very precise linear measurements. With a magnification of 100X, the direct measurements are 5 μin., or 0.1 μm. The top version is called a *filar-goniometer* because it provides angular measurement to 20 minutes of an arc.

We use two general methods to alleviate the problem:

- Project focused beams of light onto the part of the workpiece being examined, which is inadequate even at the highest magnifications
- Break into the optical path and inject light toward the workpiece

Most of these ingenious methods place a partially silvered mirror at an angle in the optical path. In one direction, the magnifying system "looks" through the mirror, but when light is reflected along the same path as the workpiece, you can see both the piece and the scale.

17-2 APPLICATIONS

Surface Topology Applications

The microscope can be one of the most precise instruments for the evaluation of surface topology, but only if we can do destructive testing. As we discussed in Chapter 13, the microscope can give us an informative view of a surface, but it can be difficult to quantify what we see. If we take a cut—by grinding or lapping—at an angle to a surface, we will be able to measure the vertical asperities in proportion to their length in the horizontal plane. The microscope is the perfect instrument for this purpose; however, you can use a conventional stylus instrument to directly input the information into a computer for analysis. With a microscope, you have to manually record the data, and then input it into the computer for evaluation.

One method lets us evaluate a surface without destroying the piece we are working with. We use a plane of light instead of the angular cut (see Figure 17–8).

The microscopes we use for surface metrology (up to 400X) magnify more than magnifiers we use for scale reading and direct measurements (up to 50X), but the microscopes we use for metallurgical work magnify even more. Here, microscopes are used to examine and measure crystalline structures; therefore, they need to be able to magnify up to 2000X. Manufacturers have constructed specialized microscopes (see Figure 17–9), because metallurgical microscopes need such great magnification.

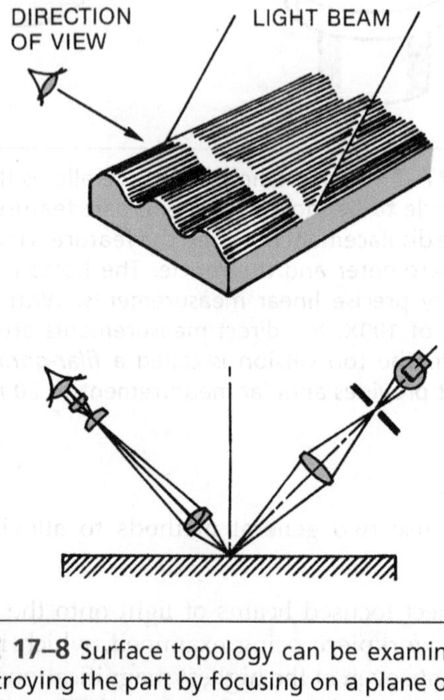

FIGURE 17–8 Surface topology can be examined without destroying the part by focusing on a plane of light at an angle to the surface and then taking measurements along the horizontal deviations.

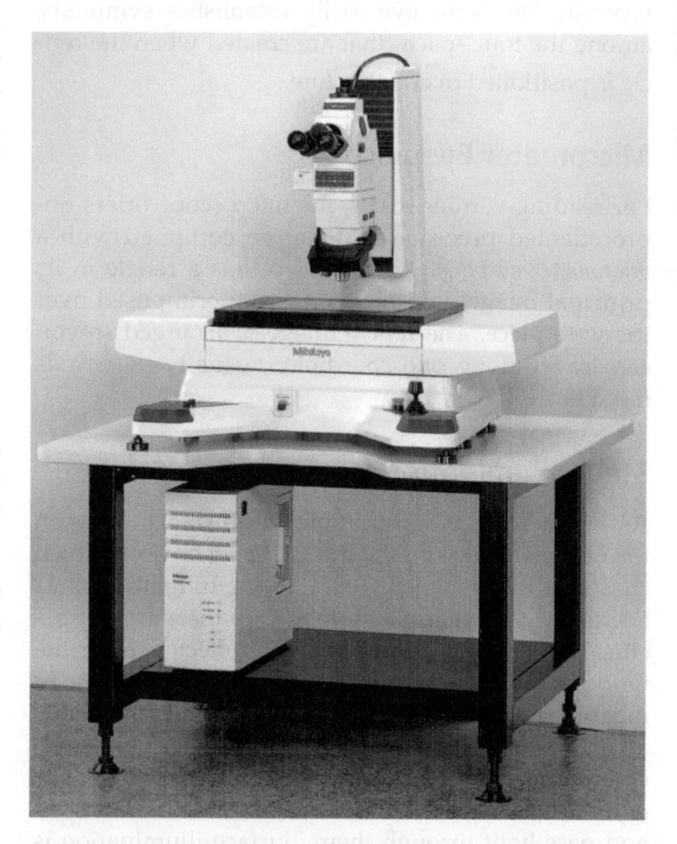

FIGURE 17–9 The highest power microscopes are used in metallurgical testing. *(Courtesy of Mitutoyo American Corporation)*

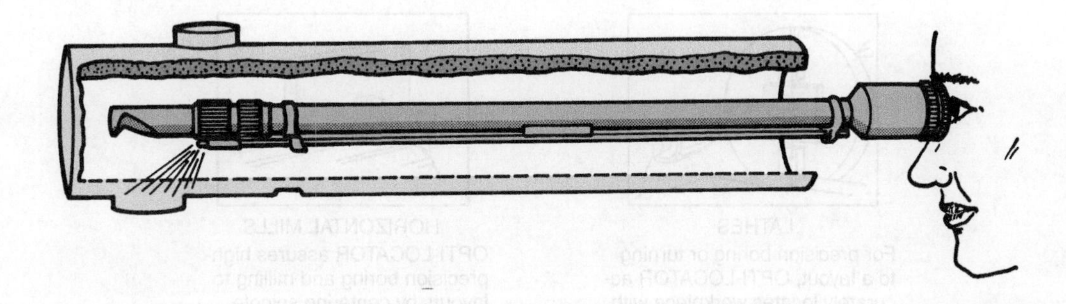

FIGURE 17–10 Optics offer the most practical way to inspect inaccessible part features. When coupled with the microscope, they provide reliable measurements as well.

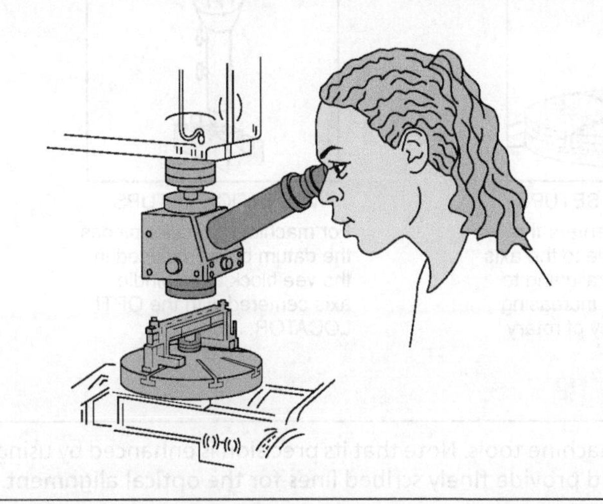

FIGURE 17–11 Instruments are available that attach to the spindles of conventional machine tools in order to facilitate alignment to part features.

Other Applications

You would probably not be surprised to learn that we also use optical instruments to inspect otherwise inaccessible features. We can argue whether we should actually call these instruments telescopes or microscopes (see Figure 17–10), but we can use most of them to both measure and inspect, as long as we use an eyepiece and the appropriate reticle.

On the other hand, you might be surprised to know that the microscope is most widely used in metrology almost as an accessory to other instruments.

When you are locating part features, you often cannot use a contact instrument. Sometimes, you cannot reach the boundary of a part feature; in other cases, the feature does not present a face at 90° to the line of measurement. Less often, you are working with extremely fine materials that cannot be distorted even by the slight contact of the stylus points. In all of these instances, you can use the microscope to pick up the feature.

Nearly all CMMs (see Chapter 18) have microscope adapters to locate part features. Microscopes are also widely used as the *center finder* and *edge finder* (see Figure 17–11), and they can be used with a wide variety of machine tools (see Figure 17–12).

17–3 COMPARISON OF OPTICAL COMPARATORS AND MICROSCOPES

Metrologists generally agree that optical comparators have five advantages over microscopes; however, they do not agree about the order of importance of the advantages:

- In general, the field of view may be much larger, so you can examine larger areas at one setting. The microscope also extends the range for direct comparison measurement, which in some cases is more reliable than displacement measurement.

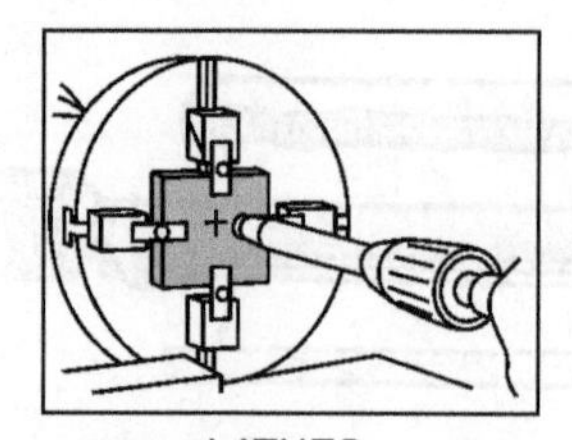

LATHES
For precision boring or turning to a layout, OPTI-LOCATOR accurately locates workpiece with axis of lathe spindle even though the tailstock is out of alignment.

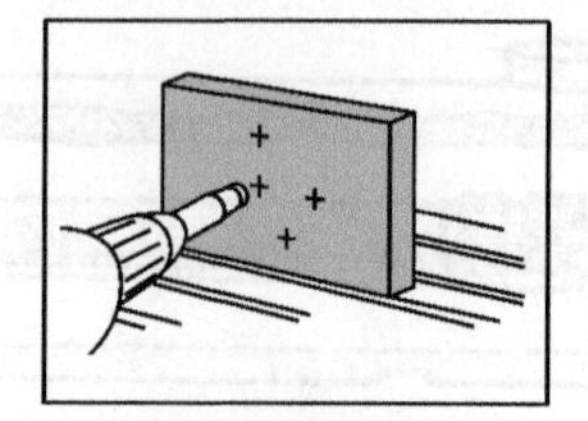

HORIZONTAL MILLS
OPTI-LOCATOR assures high-precision boring and milling to layouts by centering spindle axis to layout lines.

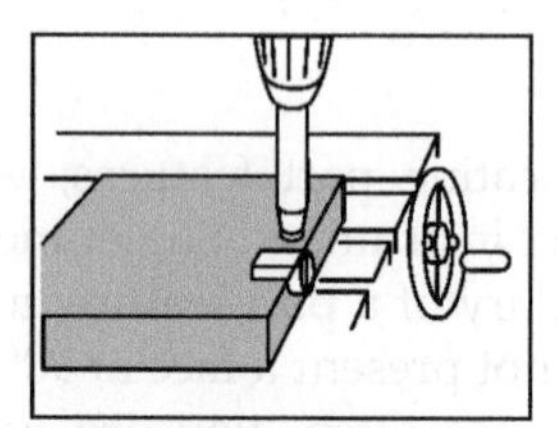

VERTICAL MILLS AND JIG BORERS
When used in combination with datum blocks, OPTI-LOCATOR aligns spindle the axis to any finished edge, furnishing a zero point for measurements.

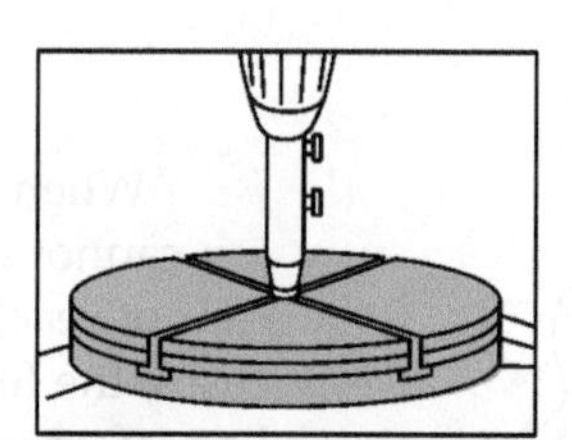

ROTARY TABLE SETUPS
OPTI-LOCATOR centers the machine tool spindle to the axis of a rotary table by aligning to the centering plug, increasing speed and accuracy of rotary table setups.

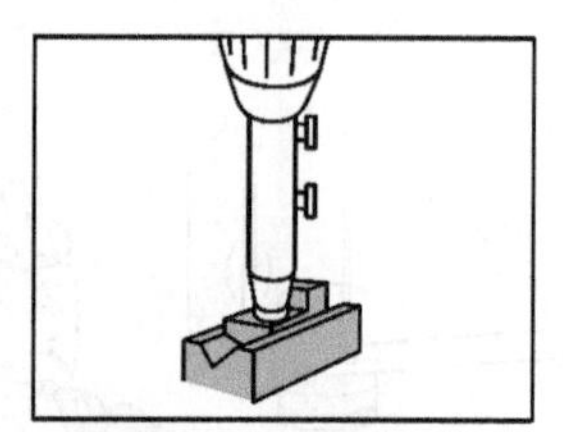

VEE BLOCK SETUPS
For machining circular pieces, the datum block is placed in the vee block and spindle axis centered with the OPTI-LOCATOR.

FIGURE 17–12 The optical attachment can be used in most machine tools. Note that its precision is enhanced by using various locating plugs. These locate from the part surface and provide finely scribed lines for the optical alignment.

- Because the size of the area being examined is larger, more than one person can be involved in measurement and analysis. You can point out particular features rather than try to explain where they are and risk having others misunderstand.
- You can make your measurements directly on the screen using ordinary drafting instruments; therefore, you rarely need special configurations. When you do need a special configuration, you can draw it on tracing paper and place it on the screen, with certain precautions, which we will discuss later.
- The photographic adapter is simpler than for a microscope. In fact, you can photograph the screen with an ordinary camera without special adapters.
- There is less eyestrain than with binocular microscopes, which leads to less operator fatigue.

One final advantage that is not usually mentioned but is particularly important in the United States is that optical comparators are widely used in industry, whereas the microscope is not a common instrument in hard goods manufacturing. Most people doing shop work are familiar with optical comparators, so there is little need for additional training.

The major disadvantage of the optical comparator is bulk. In order to project an image of an object

x wide at a magnification of m times, you need a screen at least mx in size: an 8.89 mm (0.35 in.) object at 25X requires a screen 22.23 cm (8.75 in.) across. If a microscope has an objective lens of 2.5X and a 10X eyepiece, the image would appear at the same 22.23 cm (8.75 in.) across, but it would actually exist in the reticle plane with a width of only 22.225 mm (0.875 in.), which is contained in the small barrel of a microscope. It is much easier to take a microscope into your work area than it is to take most optical comparators.

The other disadvantage of the optical comparator is its higher cost. Microscope prices begin very, very low for basic models and extend to very, very high for the most sophisticated versions. Prices for optical comparators do not start low, but they extend as high. If we compare only the microscopes and comparators we are likely to use for typical shop inspection, microscopes have the price advantage.

17–4 OPTICAL CONSIDERATIONS

In the basic optics system for a comparator (see Figure 17–14), the condensing lens (*C*) images the light source (*S*) in the plane of the object you are examining (*AB*). Of course, the image of the light must be larger than the object. The projection lens (*P*) receives the light that passes around or through the object and forms a real, inverted image (*A'B'*) on the projection screen (*G*). Figure 17–13 shows only two rays each from points *A* and *B;* however, all rays from these points will behave similarly, forming the image on the screen. The ratio of *AB* to *A'B'* is the magnification and is determined by the focal length of the projection lens.

If the magnification *A'B'*/*AB* equals m, we measure the object and image distances x and x' from the principal focus: $x' = mf$ and $x = f/m$. We add the focal length to each of these distances to obtain the object and image distances:

$$l = x + f = f(l \text{ to } l/m)$$
$$l' = x' + f = f(l + m)$$

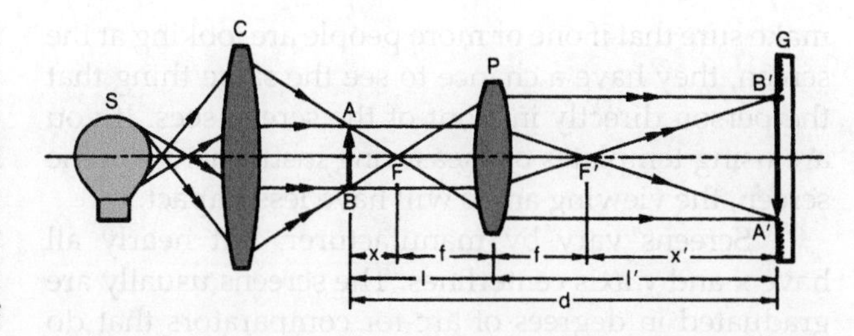

FIGURE 17–13 All projectors are based on this optical system. The condensing lens (C) images the light source (S) in the plane of the object (AB). The projection lens (P) then forms a real image (A' B') on the screen (G). The real image is inverted.

Therefore, the overall separation d of object and image is:

$$d = l + l' = f(2 + m + l/m)$$

We disregarded the nodal points in the lens, because we want to show that increases in magnification require increases in the length of the optical system.

The Projection Screen

There are two types of screens: translucent and opaque. A translucent screen displays magnified images projected on the back side of the screen. An opaque screen is similar to a moving picture screen in that the projection is from the front. Theoretically, opaque screens result in the highest accuracy. However, from a practical standpoint, opaque screens are used only for the largest optical comparators, while translucent screens are used for all others.

When you are using translucent screens, you need to remember that a phenomenon similar to parallax occurs. Because the amount of light transmitted is greatest in one line—the line to the back of the screen that those particular rays hit—the transmitted light falls off rapidly as your eye moves from side to side. In general, the maximum viewing angle should be 20° normal to the surface. Beyond this angle, you do not see all portions of the screen in the same relative relationships. For practical measurement, you should make sure that if one or more people are looking at the screen, they have a chance to see the same thing that the person directly in front of the screen sees. If you are using templates or measuring instruments on the screen, the viewing angle will have less impact.

Screens vary by manufacturer, but nearly all have x and y axis centerlines. The screens usually are graduated in degrees of arc for comparators that do not have bezels for screen rotation. Basic instrument screens do not look like the screens seen in many photographs—with multitudes of reference lines. These are *charts* placed over the screen, which we will discuss later.

Limitations of Profile Projection

The greatest limitation of profile projection is that we must ensure that the part feature we are examining lies in the plane of sharpness. This plane is the part of the screen that is normal to the optical axis and, generally, is between 127 μm and 254 μm (0.005 in. and 0.010 in.) thick, which can cause difficulties for many workpieces. For example, we most often use profile projectors—both optical comparators and microscopes—to inspect threads. However, threads are helixes, so their profiles lie at an angle to the workpiece (see Figure 17–14). Even when we use an angularly adjustable fixture to hold the workpiece, we do not solve the problem because portions of the workpiece can hide the feature we need to examine, and the errors are usually on the order of 2.54 μm (0.0001 in.).

The reflection from a workpiece can also cause problems in profile projection. We have the least difficulty with thin workpieces, such as plate gages and templates, and workpieces with beveled edges may generate no problems.

In most cases, workpieces are very long in the optical axis, and their smooth surfaces can result in specular reflection. If the reflected light falls on the screen outside of the dark profile, there is no problem, and sometimes we see this reflected light as a halo. If the reflected light falls in the dark area, however, we lose the contrast. When this reflection is severe, we may not be able to determine the edge of the profile with certainty.

FIGURE 17–14 Table movements of this optical comparator are interfaced to a dedicated computer and a digital display of the x and y axes. *(Courtesy of Optical Gaging Products Inc.)*

If a face of an object, like a rectangular block, is parallel to the optical axis, we will get a clear image of only the edge nearest the projection lens. In Figure 17–15, *p* is the focal plane of the projection lens, and in the top view, block *A* has a face in that plane. The block's top edge is defined by ray *ab* from the light source, so all lower rays are dark on the screen and all higher ones are bright. Any ray at a higher angle striking the top surface, like ray *cd,* will be reflected, and the reflected light will show in the bright area of the screen. If the opposite face of the block is in the focal plane as at *B,* ray *ab* still defines the top edge; however, the reflected light from *c* now forms an image on

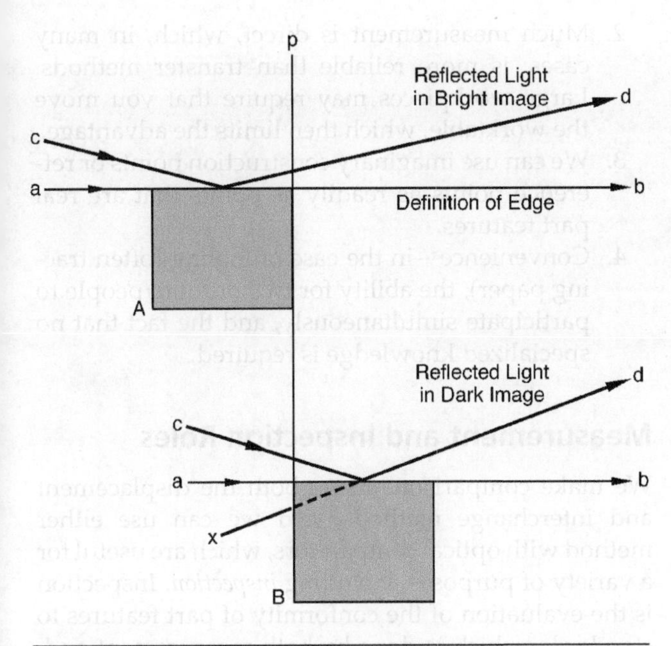

FIGURE 17–15 The face to be examined must be closest to the projection lens or the reflected light will fall in the dark portion of the image, thereby reducing its clarity.

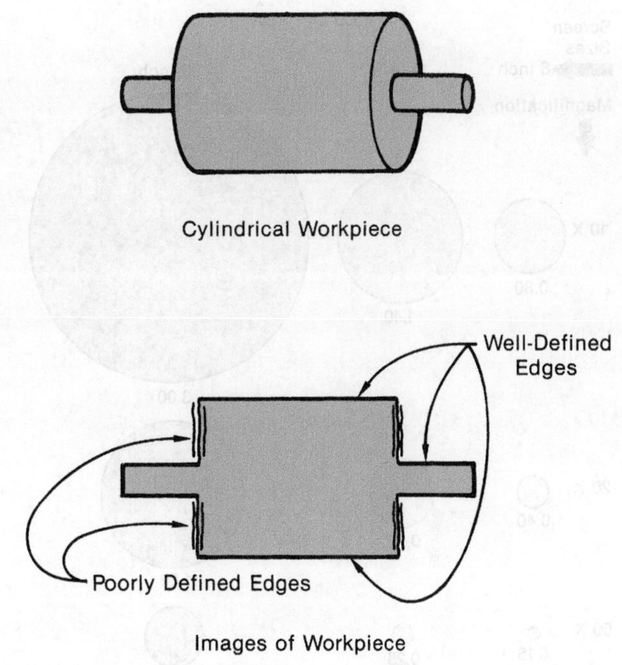

FIGURE 17–16 Reflection gives little problem for the rounded surfaces, but when they are in focus, the flat sides may be indistinct.

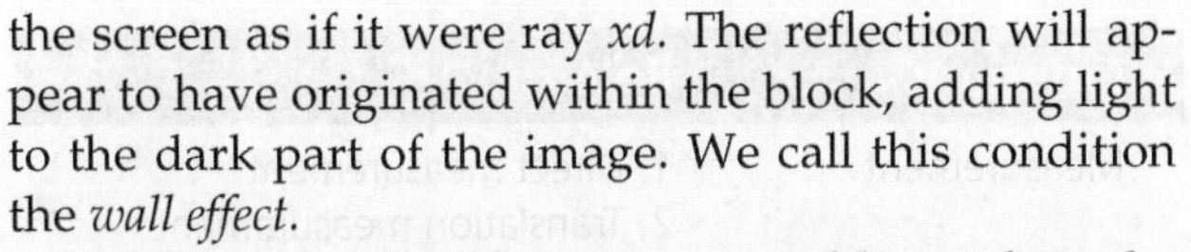
the screen as if it were ray *xd*. The reflection will appear to have originated within the block, adding light to the dark part of the image. We call this condition the *wall effect*.

The wall effect becomes a problem when the edges of the workpiece do not lie in the same plane—for example, in a cylindrical workpiece. If the top and bottom of the cylindrical portion are in the focal plane, the front and rear edges of the end faces must be equally spaced behind and in front of that plane, respectively (see Figure 17–16).

These limitations are important, but the limitation that users are most aware of is the *field of view*. The screen, of course, limits the amount of the workpiece that we can show at one time. The more we magnify the workpiece, the less of it we can display within the screen's borders. The amount of the piece we can show is determined by the screen diameter divided by the magnification (see Figure 17–17).

Episcopic and Diascopic Projection

Diascopic projection, also called *dispositive* projection, is projection through films, transparencies, and similar materials. *Episcopic projection* is projection of the surfaces of opaque objects, and we use it most often to examine engraved surfaces, embossed patterns, and typefaces.

Episcopic projection appears easy—simply illuminate the surface you are examining, and then turn off the illumination behind the surface—but in practice it is difficult. The image on the screen consists entirely of light reflected from the workpiece surface, and only a small fraction of the incident light can be collected for the projection lens; therefore, we must use an intense beam because most of the light is lost. In addition, the best lenses for profile projection have small apertures to minimize distortion, so some optical comparators have special lenses for surface inspection.

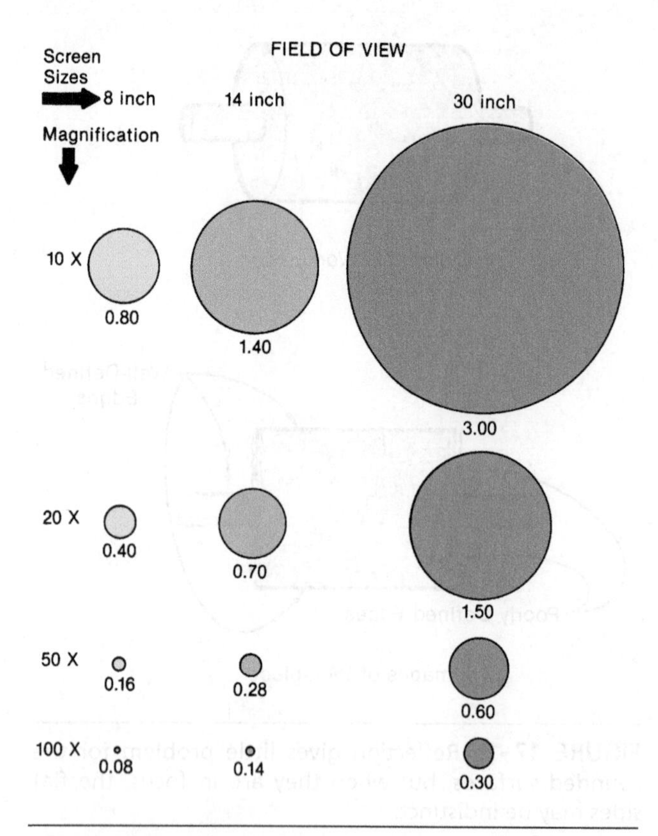

FIGURE 17–17 The field of view is the screen diameter divided by the magnification. The circles in this figure are the maximum areas in which workpiece features may be seen at one time.

The greater the angle of the illuminating beam, the less incident light enters the projection lens. If we "inject" the illuminating beam into the projection system, we can reduce incident light. We use a partially silvered mirror to inject the light in the opposite direction to the projected image.

17–5 APPLICATIONS OF THE OPTICAL COMPARATOR

The reasons that profile projection by means of the optical comparator is important (in order of importance) are:

1. We can examine a large number of part features in two dimensions simultaneously.
2. Much measurement is direct, which, in many cases, is more reliable than transfer methods. Large workpieces may require that you move the worktable, which then limits the advantage.
3. We can use imaginary construction points or reference points as readily as points that are real part features.
4. Convenience—in the ease of tooling (often tracing paper), the ability for two or more people to participate simultaneously, and the fact that no specialized knowledge is required.

Measurement and Inspection Roles

We make comparisons using both the displacement and interchange methods, and we can use either method with optical comparators, which are useful for a variety of purposes, including *inspection.* Inspection is the evaluation of the conformity of part features to standards, which is done by both measurement and, as in production inspection, *attribute inspection* (see Chapter 8). We usually perform attribute inspection with fixed, go/no-go gages. For attribute inspection, the optical comparator provides distinct advantages and an entirely new set of applications:

Category	Role
Measurement	1. Direct measurement
	2. Translation measurement
Inspection	3. Linear comparison
	4. Angle comparison
	5. Contour comparison

Direct Measurement

You should use direct measurement whenever possible, because it is inherently more reliable. This is especially when the following criteria apply:

- The workpieces are suitable for profile projection. (Remember the guidelines in "Limitations of Profile Projection.")
- You can display the longest feature of the workpiece entirely on the screen.
- Tolerances are within the range of the instrument.

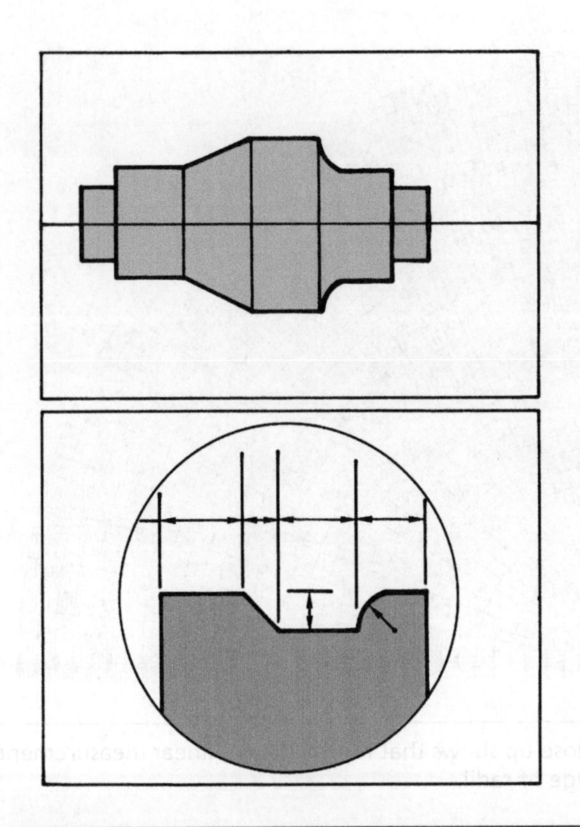

FIGURE 17–18 The top drawing is a screw machine part and the bottom drawing is the form tool for producing it. Both the part and the tool are typical examples of items that lend themselves to inspection by the optical comparator.

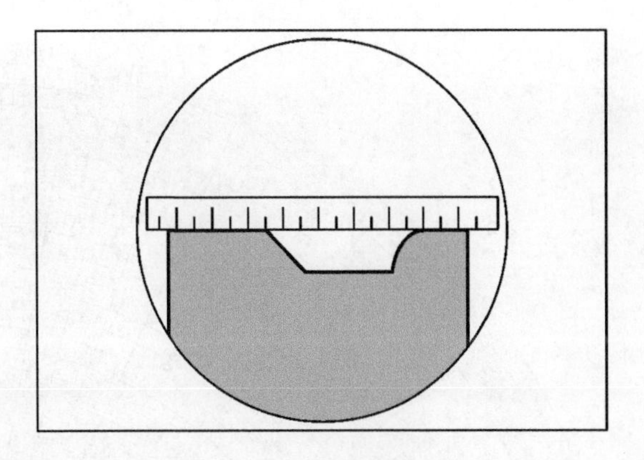

FIGURE 17–19 A drafting scale may be used directly to measure features from the screen. Clear plastic scales are available to facilitate this method.

For example, in Figure 17–18 (top), we have a screw machine workpiece and, at the bottom, we have a drawing of the form tool required to turn it. We could directly use a steel scale, but only if the tolerance is wider than any steel scale we would usually encounter. Remember, it is difficult for your eye to resolve small divisions on a steel rule (see Figure 17–21).

If we use the magnifying power of an optical comparator, we could use a 31.25X lens. If your eye can reliably read a scale to 793.75 μm (0.03125 in.), this lens would magnify the profile of the workpiece so that 793.75 μm (0.03125 in.) would represent 25.4 μm (0.001 in.) on the workpiece. You can readily measure this length on the screen using a drafting scale (see Figure 17–19), or any of the comparator "charts" available.

A chart, such as a *toolroom chart-gage* (see Figure 17–20), also provides markings so you can compare arcs and measure angles. You could also use it to measure the form cutting tool in Figure 17–18. Optical comparator manufacturers claim that their charts can accommodate 99.9% of all profiles you might encounter in the range of the screen size. However, charts only provide ways to measure three of the lines you may encounter: straight lines (both horizontal and vertical), angled lines, and arcs. The fourth type of line, the irregular contour line, is not predictable. With the chart, the cutting tool appears as in Figure 17–21.

The most important limitation for direct measurement with optical comparators is the size of the screen, and the best evidence that direct measurement is one of the fundamental advantages of optical comparators is the variety of large sizes that are available (see Figure 17–14). The related limitation is the magnification, because, in order to magnify some features adequately, we create an image larger than any screen—a larger field of view. In these cases, we measure by movement called *translation*.

Another case in which you would prefer to use an optical comparator is shown in Figure 17–22. You would choose the optical comparator for two

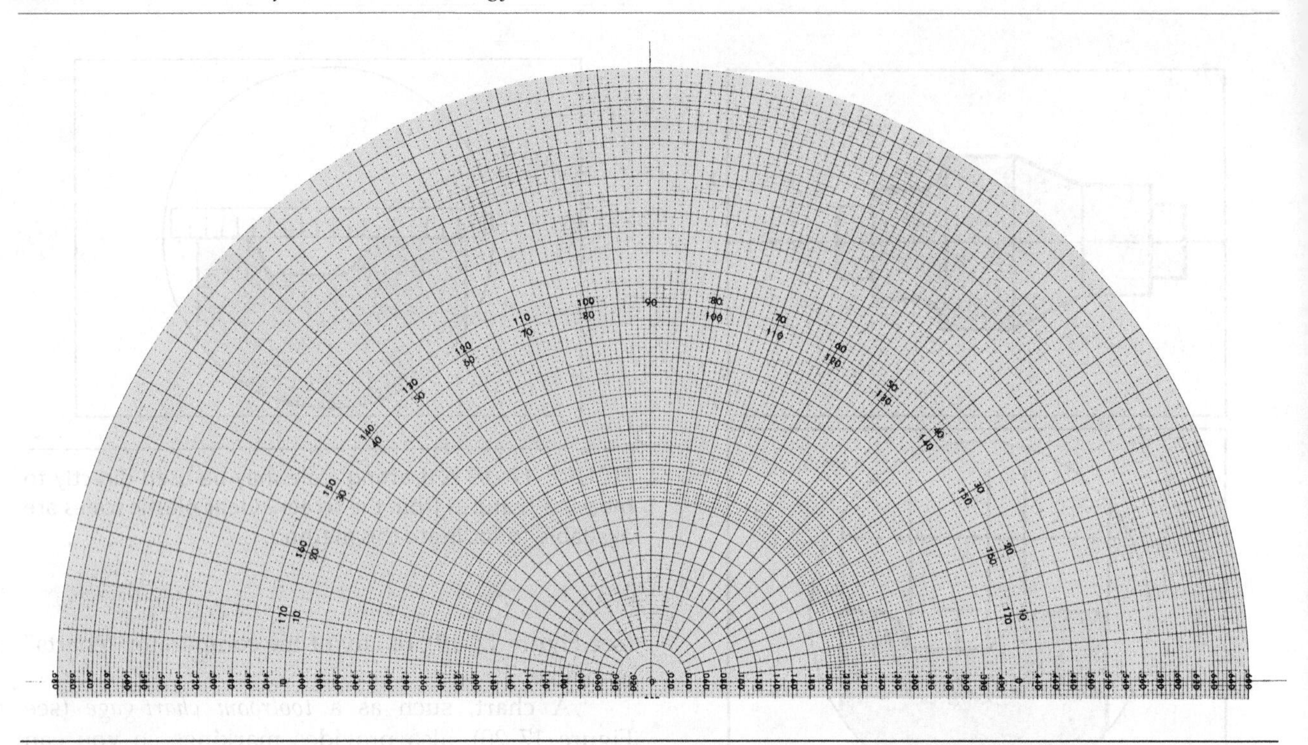

FIGURE 17–20 This is known as a toolroom chart-gage. The close-up shows that it provides for linear measurement, angle measurement, and the comparison of arcs of a wide range of radii.

reasons: You can check the geometry with the optical comparator—a micrometer cannot measure squareness—and you will be able to see a small burr or grain of dust with the optical comparator.

Measurement by Translation

Translation refers to motion along a line without rotation—any linear measurement. In common usage, translation refers to measurement in which the workpiece must be indexed. You must index a workpiece whenever the longest feature is longer than the width of the screen, and in order to index a workpiece, you must use the worktable or *stage* to move the workpiece through the length of the feature you are measuring. Indexing uses the displacement method. In these cases, you are using the optical comparator to "pick up" the reference ends of the part features.

The worktable moves the workpiece across the screen through the references, which are visually set. This method can be more practical than physical contact, but it also lacks the theoretical integrity of direct measurement. This process adds error inherent in the movement of the workpiece and the readings of the positions. Practically, when you are using a high-grade optical comparator, the error in optical referencing is greater than in table indexing.

Unlike all other measurement instruments, optical comparators do not distort the workpiece. There is no physical contact other than its support, and the size and angle of contact of a probe are not factors.

In the recommended technique for translation measurement, you closely watch the diminishing band of light between the shadow of the profile and the reference line on the screen as you move the worktable to bring the profile and the reference line together

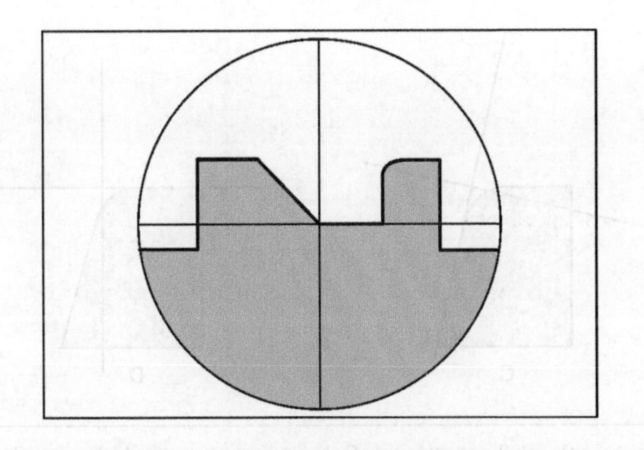

FIGURE 17–21 The cutting tool from Figure 17–18 is shown against the toolroom chart-gage.

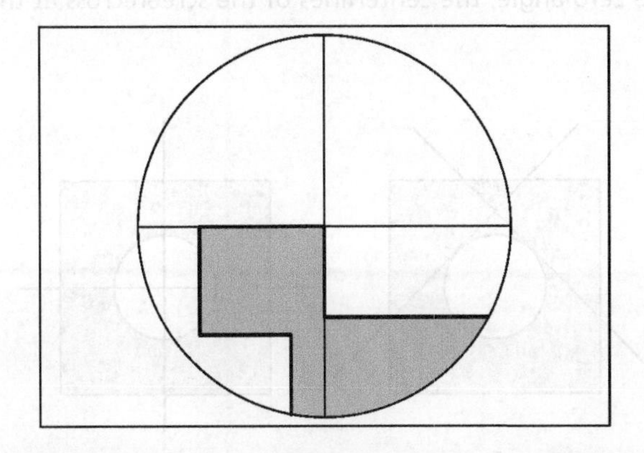

FIGURE 17–22 The optical comparator allows geometry to be checked at the same time measurements are taken.

(see Figure 17–23). You should stop moving the table when you cannot see any more light. With this technique, you can obtain a repeatability of 2.54 μm (0.0001 in.) at 50X magnification, and with practice, you can cut this value in half. You must remember, however, that the lines have thickness; therefore, you must always approach the lines from the same direction. In a more reliable method, you use a crenelated reference or a "hopscotch line" (see Figure 17–24), which consists of two interrupted lines whose inside edges lie along the same line.

Only optical instruments can measure directly from imaginary references—those "points in space."

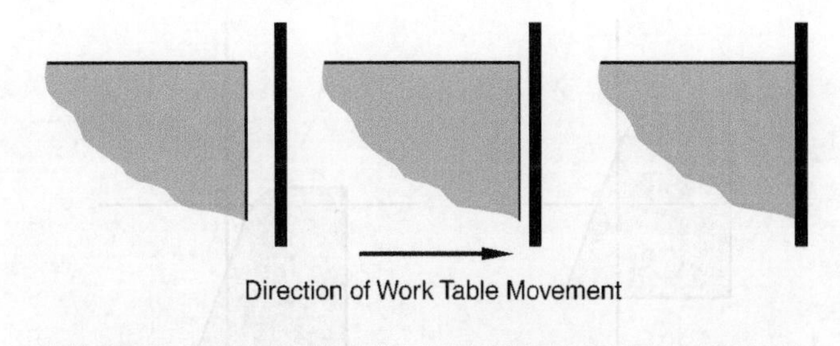

FIGURE 17–23 Table movement is slowed as the profile approaches the reference line on the screen. It is stopped the moment that the last vestige of light between the profile and the line disappears.

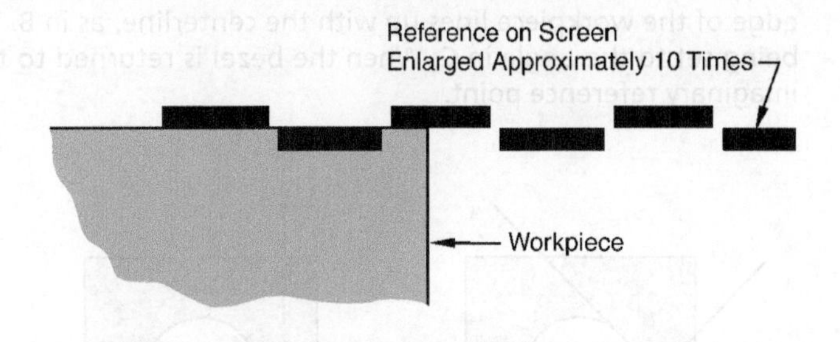

FIGURE 17–24 Very precise settings can be made with this type of reference line. The centerline created by the opposing rows of dashes is a reference of zero width.

For most shop work, the comparator makes it practical to use the translation technique with these imaginary references (see Figure 17–25).

Referencing to holes in workpieces is another instance when the optical comparator can be invaluable. Other instruments require plugs or other means to simulate the open space with real material, but the optical comparator uses the principle of the V-block (see Figure 17–26). Remember that, regardless of the size of the round part placed in a V-block, the centerline is always the same.

You can also combine direct measurement and measurement by indexing. In fact, you will commonly have a cluster of part features that all fit on the screen but are from some other feature greater than the screen width. You must measure the long distance by translation, but you can measure the cluster of small distances based on your own good judgment.

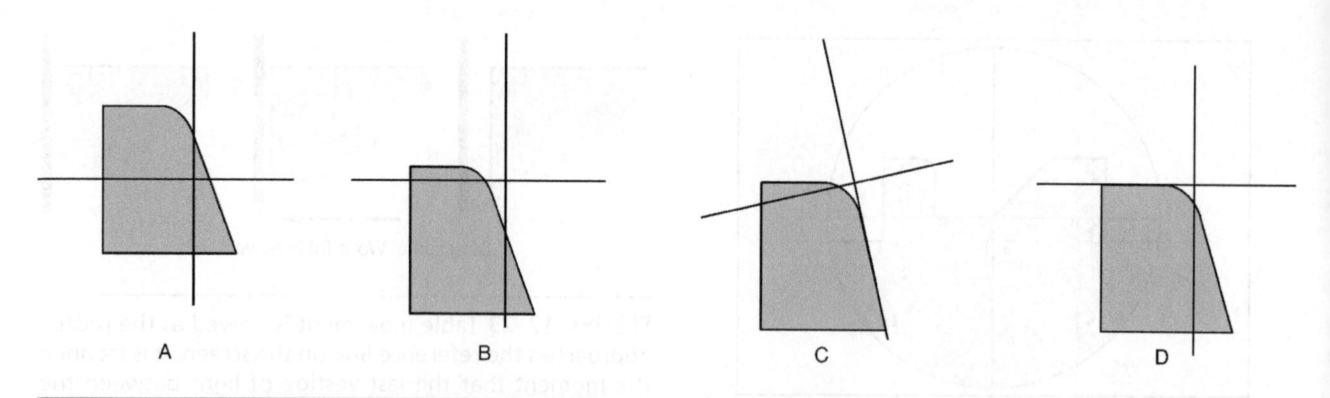

FIGURE 17–25 Only optical instruments can directly determine the intersection of the top edge and the angle, despite the radius on the corner of the workpiece. From the position at A, the workpiece is lowered so that the top edge of the workpiece lines up with the centerline, as in B. The workpiece is then moved to the right as the bezel is being set to the angle in C. When the bezel is returned to the zero angle, the centerlines of the screen cross at the imaginary reference point.

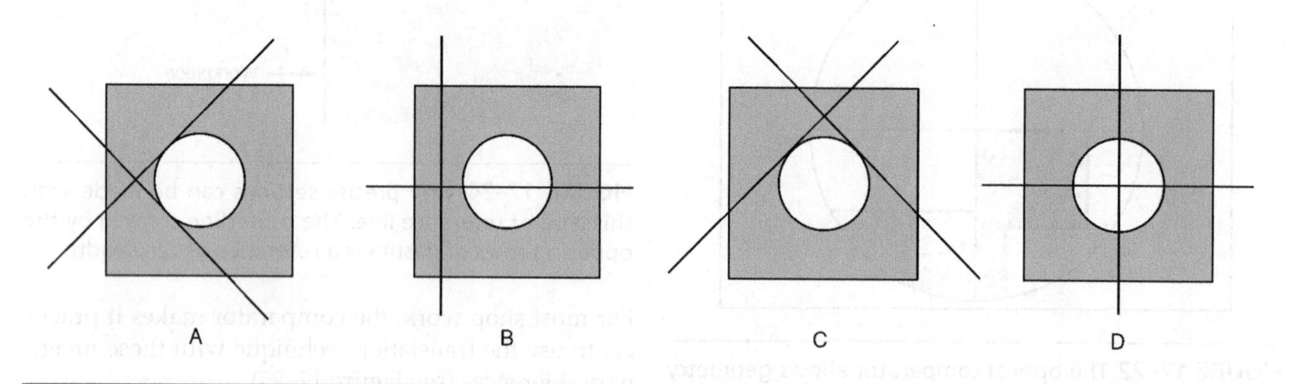

FIGURE 17–26 The center reference of a hole is easily set to the centerlines of the comparator screen. The centerlines are first set at 45°. The work table is moved horizontally until the centerlines are tangent to one side of the hole as at A. When the centerlines are returned to 0°–90°, the horizontal axis will be in line as at B. The bezel is again set to 45° and the work table moved until the hole is tangent in the vertical quadrant as at C. When the bezel is reset to 0°–90°, both centerlines will pass through the center of the hole.

Measurement by translation is limited by the distance the worktable will travel, and travel can be in the x-y axis for some instruments and in the x-z axis for others. The x axis has the longest travel, typically 20 cm (8 in.); when you use 100X magnification, the x axis provides nearly 20 m (67 ft.) of profile that you can examine.

Unfortunately, only a few comparators actually are calibrated and directly readable for the entire work table travel. In fact, most instruments are limited to 1 inch of travel along the two axes, which is achieved using large drum micrometers. Because of the size of the drum and the accuracy of the screw, you can read the micrometers directly to 2.54 μm (0.0001 in.), but you must resist the urge to interpolate between graduations. You may think you are gaining precision, but you are sacrificing accuracy. With an optical comparator—or an electronic comparator—you can switch to a higher magnification to increase the precision of your measurement.

The short length of the calibrated worktable travel is practical because a 203 mm (8 in.) calibrated screw would either be much less accurate or much more expensive. In addition, the micrometer-driven table travel encompasses the usual screen diameter; therefore, measurement by translation with no indexing has at least the same size range as by direct measurement.

We can also argue that there is a metrological justification for the short travel. On one hand, we can make direct measurements with only negligible errors that are limited to the comparison of the ends of part features to a scale on the screen. On the other hand, we can measure by translation, which means we have to align a succession of ends of part features with one fixed reference line on the screen. Although this method adds errors of the micrometer, worktable movement, and reading of the micrometer, it substitutes an easier and "more reliable" method of aligning the image and reference line.

When the part is larger than the 25 mm (1 in.) micrometer travel, we reposition the worktable by inserting measuring rods or gage blocks between the micrometer spindle and the worktable, which adds potential errors. This method makes the benefits of the optical comparator available for parts of considerable size, however, so it is well worth a little extra time and care.

When you are indexing, you must also keep accumulation of error in mind. You will find that it is set to zero and take each measurement separately; however, for best accuracy, you should take as many measurements as possible from the same reference, and then determine the individual feature sizes arithmetically. As we mentioned, some optical comparators have worktables calibrated for their full travel, and these tables are read digitally to four decimal places 2.54 μm (0.0001 in.). This is more reliable than the use of gage blocks or measuring rods.

One important form of measurement by translation is the tracer technique, which has its greatest application in contour inspection (discussed later).

Measurement by Comparison, Linear

When you use an optical comparator for linear measurement by comparison, it is very much like when you are doing attribute gaging using fixed gages. You can fairly easily prepare complete outlines of the workpiece, which you then place on the screen. The image of the workpiece is superimposed on the outlines, which can also show both high and low tolerance limits. Except for borderline situations, any inspector can readily accept or reject parts (see Figure 17–27).

Although this technique is used primarily for production inspection, notice that the technique would be meaningless without measurement: for part layout, for tooling, and for the development of the gaging, which in this case is the chart on the comparator screen.

Measurement by Comparison, Angles

Remember, an angle remains exactly the same regardless of its magnification. Increasing the magnification of an angle allows us to compare the angle on the workpiece more closely with the one on the screen. Optical comparators provide—by far—the easiest method of angle measurement, because most other methods of angle measurement require a computation of the trigonometric functions of the angle. Generally, you use the x and y coordinates of three

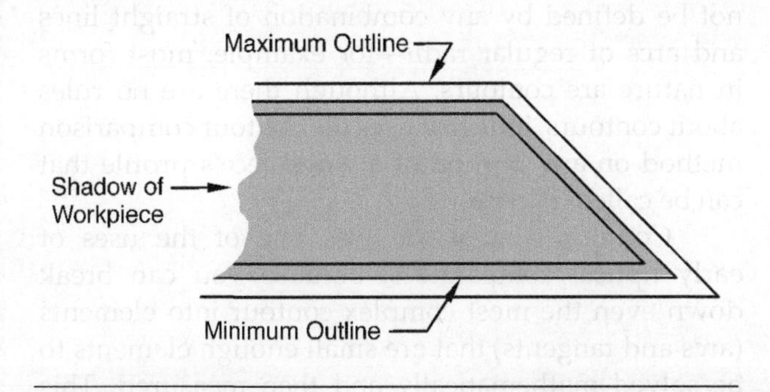

FIGURE 17–27 Maximum and minimum material conditions can be shown on the screen and the workpiece features quickly compared.

points along the angular feature to solve for the angle. In contrast, the optical comparator simply compares the unknown angle directly with a known one.

In addition, angle measurement using an optical comparator offers a variety of choices: You can directly compare angles by superimposing a drawing of the feature, including the angle, on the screen; prepared charts are available for those occasional, nonrepetitive angles you have to measure; and most comparators include a bezel, calibrated in degrees, that you can use to rotate the screen. Verniers on the bezels help you make readings to 5 minutes of arc for small comparators and 1 minute of arc for larger comparators. In the largest instruments, this information is provided on a digital readout. The bezel is particularly helpful when you need to examine compound angles—angles that do not lie in one plane. You must, of course, have all of the angles you are going to measure at one time in the same plane.

Angle measurement with an optical comparator is limited to the machine's screen size and accuracy. A great deal of angle measurement in industry involves large workpieces, such as parts for production machinery, where the angles are better determined by *optical alignment*, not comparators.

Measurement by Comparison, Contours

Literally, a *contour* is the outline of any shape; therefore, a square has a contour. In popular use, contour is used to refer to any irregular outside shape that cannot be defined by any combination of straight lines and arcs of regular radii—for example, most forms in nature are contours. Although there are no rules about contours, industry uses the contour comparison method on any portion of a workpiece's profile that can be called a "contour."

Contour comparison was one of the uses of early optical comparators, because you can break down even the most complex contour into elements (arcs and tangents) that are small enough elements to be solved mathematically and then measured. This method is very time consuming, so modern instruments, like certain CMMs, were designed to trace these essentially analog paths and express them digitally. With the optical comparator, you do not need to use these techniques for most workpieces.

Inspecting a contour is no more difficult than designing one. You select a comparator magnification that allows you to see the high and low tolerance limits clearly; then, you draw the profile to that scale and place the drawing on the comparator screen. You align the workpiece with the drawing and visually inspect the workpiece's profile (see Figure 17–27).

Unlike any precision measurement instrument (other than the microscope), you can see the entire profile at a glance. You will also be able to see a grain of dust or a minute burr and measure accurately in spite of it.

The Tracer Technique

This special role of measurement by translation is applied most often in contour inspection. We use the tracer technique for optical comparator inspection of internal features that cannot be projected directly.

In Figure 17–28, we assume that the contour of the workpiece is in a recess and that we cannot project its profile. We hold the workpiece on the worktable with a suitable fixture and mount the tracer directly to the comparator. We then see the image of the workpiece on the screen of the comparator (see Figure 17–29).

This method is limited by the contour length that you can inspect in the field area, which equals screen size divided by magnification. We can solve this problem by using multiple *follower probes* attached to the *tracer probe* (see Figure 17–30). As one probe runs out of screen, the next one comes into view (see Figure 17–31).

One variation of the *tracer technique* has no theoretical limitation—either in workpiece size or amplification. We use a *reticle-gage follower*—which is like a chart-gage, except that it is drawn to the same size as the workpiece features—instead of a probe follower (see Figure 17–32). We project its image on the screen. On the screen is a chart-gage with a reference circle proportional to the size of the trace probe

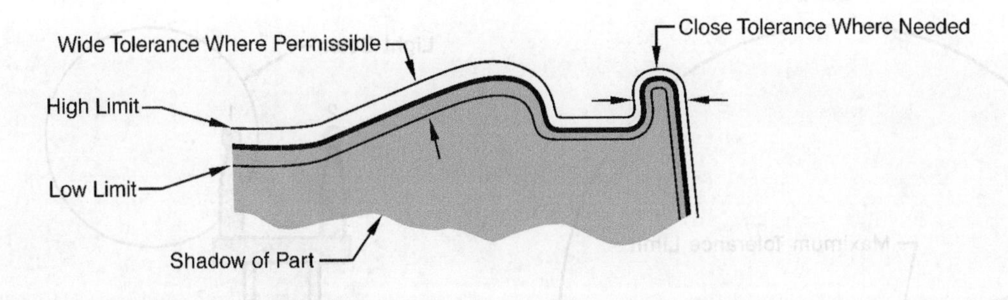

FIGURE 17–28 With an optical comparator, the entire profile can be examined at a glance. The tolerance can be opened or closed as required for each feature.

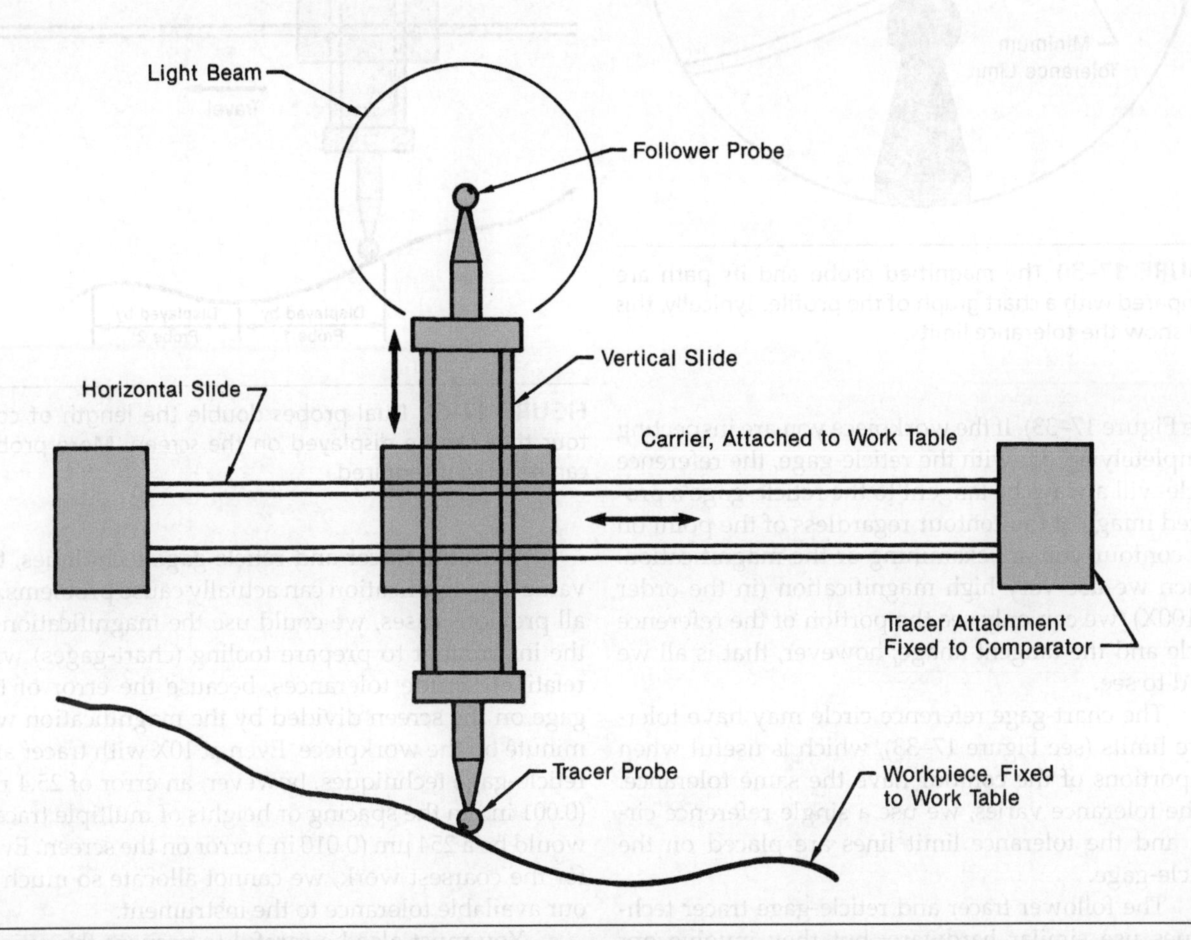

FIGURE 17–29 The schematic of the tracer attachment for inspecting inaccessible contours. The actual construction differs considerably. Unlike other translation measurement methods, the workpiece here is fixed. The work table moves the carrier which, thereby, traces the profile and exposes that trace in the optical path.

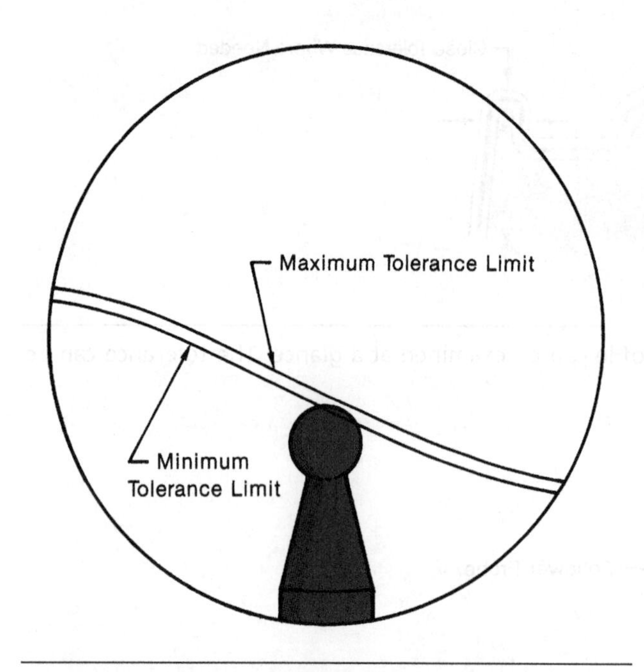

FIGURE 17–30 The magnified probe and its path are compared with a chart graph of the profile. Typically, this will show the tolerance limits.

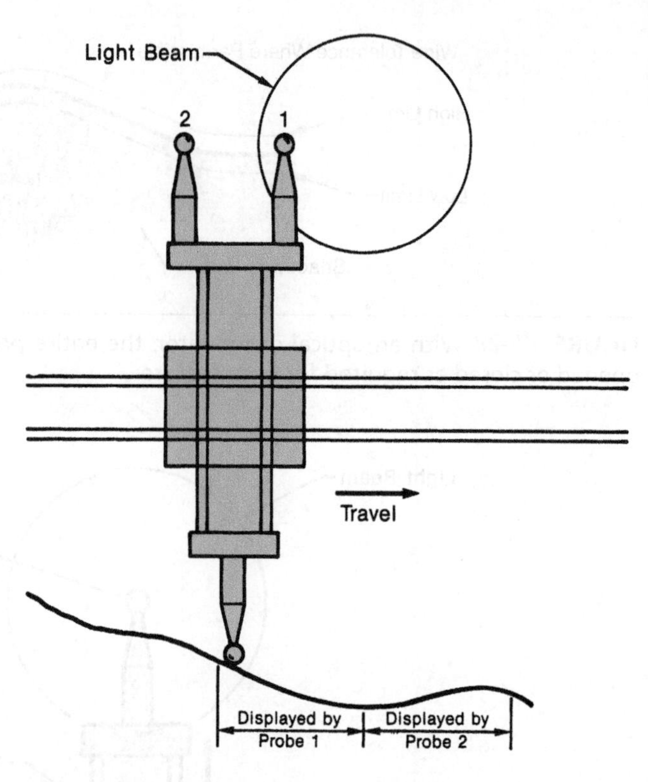

FIGURE 17–31 Dual probes double the length of contour that can be displayed on the screen. More probes can be used if required.

(see Figure 17–33). If the workpiece you are inspecting completely agrees with the reticle-gage, the reference circle will always be tangent to the reticle-gage's projected image of the contour regardless of the point on the contour you are examining or the magnification. When we use very high magnification (in the order of 100X), we can only see the portion of the reference circle and the tangent image; however, that is all we need to see.

The chart-gage reference circle may have tolerance limits (see Figure 17–33), which is useful when all portions of the contour have the same tolerance. If the tolerance varies, we use a single reference circle, and the tolerance limit lines are placed on the reticle-gage.

The follower tracer and reticle-gage tracer techniques use similar hardware, but they involve opposite principles. With the follower tracer, the tracer moves relative to a fixed "gage" drawn on the screen. With the reticle-gage, the "gage" drawn on the instrument moves relative to a fixed reference on the screen.

With the tracer and reticle-gage techniques, the value of magnification can actually cause problems. In all previous cases, we could use the magnification of the instrument to prepare tooling (chart-gages) with relatively crude tolerances, because the error of the gage on the screen divided by the magnification was minute on the workpiece. Even at 10X with tracer and reticle-gage techniques, however, an error of 25.4 μm (0.001 in.) in the spacing or heights of multiple tracers would be a 254 μm (0.010 in.) error on the screen. Even for the coarsest work, we cannot allocate so much of our available tolerance to the instrument.

You must also be careful to prepare the reticle-gages, and a variety of specialized techniques have been developed. These techniques are all based on producing the reticle-gage at a greatly enlarged scale, and then reducing it photographically to the final size.

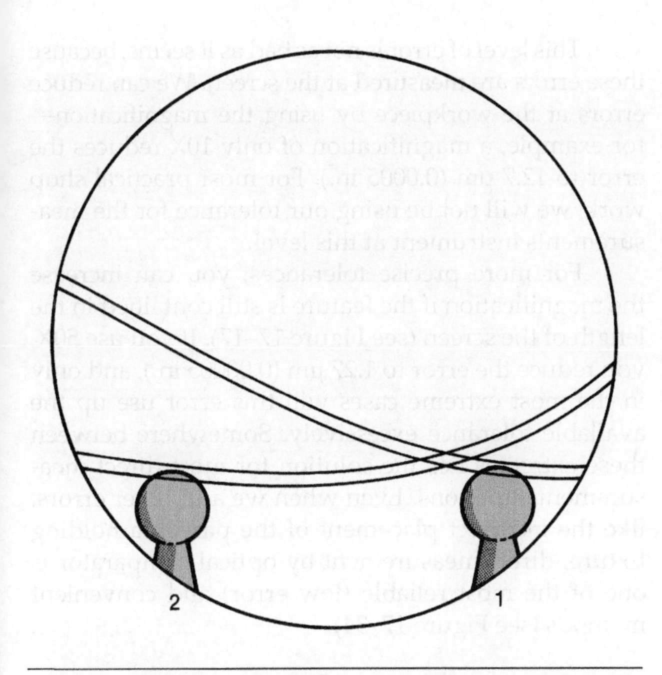

FIGURE 17–32 With multiple probes, multiple paths must be shown on the chart-gage. Probes 1 and 2 are at the same place on the contour because they are replications of the tracer probe.

17–6 THE ACCURACY OF OPTICAL COMPARATORS

There are two general classes of errors: instrument and operator. We must be most concerned about operator error, because these errors are most often simple mistakes. Although you will rarely make a mistake in a delicate process, you might use the wrong setup or make an arithmetic error in your calculations. Because optical comparators are so easy to use, they are also more vulnerable to errors; therefore, whenever something appears strange, check your arithmetic first. If arithmetic is not the problem, you can investigate some of the more subtle causes of errors, which we will discuss now.

Accuracy, Translation Method

Translation and direct measurement differ in relative accuracy. (Measurement by comparison is a special case of direct measurement and will be discussed separately.)

Accuracy for translation measurement comprises:

- The accuracy of the motion of the worktable—a function of the accuracy of the table's mechanical design and construction.
- The accuracy of the measuring provisions of the worktable—the built-in standard and the means for reading it.
- Operator accuracy—your ability to align the part feature on the workpiece to a reference line on the screen.

We do not include the accuracy of the optical system, because when you make repeated end-of-travel alignments from the same reference line, the optical system error does not affect the results.

For quality instruments, you can expect that the accuracy of the worktable—both the mechanical motion and the means for reading—will add no more

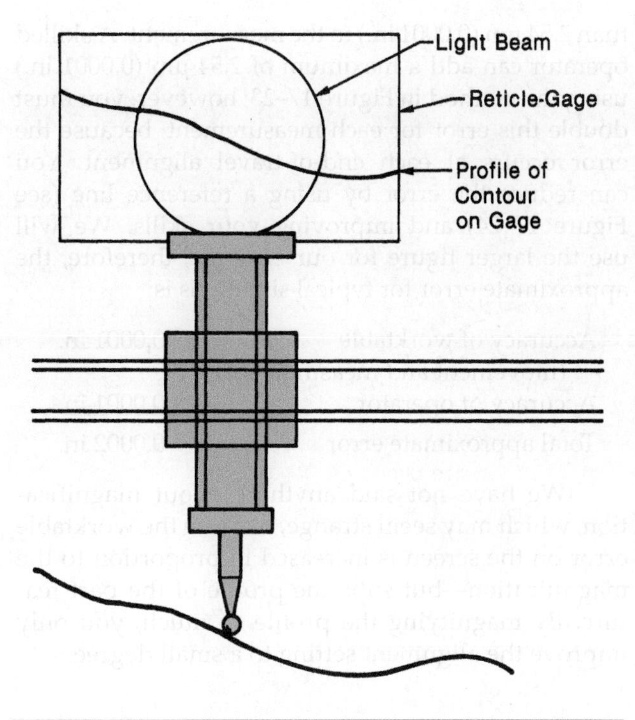

FIGURE 17–33 If a reticle-gage replaces the follower probe, a replica of the contour is projected on the screen.

than 2.54 μm (0.0001 in.) to the measurement. A skilled operator can add a maximum of 2.54 μm (0.0001 in.) using the method in Figure 17–23; however, you must double this error for each measurement, because the error occurs at each end-of-travel alignment. You can reduce the error by using a reference line (see Figure 17–24) and improving your skills. We will use the larger figure for our example; therefore, the approximate error for typical situations is:

Accuracy of worktable (movement and measurement)	0.0001 in.
Accuracy of operator	+0.0001 in.
Total approximate error	0.0002 in.

We have not said anything about magnification, which may seem strange, because the worktable error on the screen is increased in proportion to the magnification—but so is the profile of the part feature. By magnifying the profile so much, you only improve the alignment setting to a small degree.

Accuracy, Direct Measurement

Accuracy for direct measurement does not involve the worktable error, except when you combine direct measurement with translation measurement, which often happens. It does involve the optical system error, however, so you should remember that in direct measurement, you are examining both ends of the feature at the same time. Any optical distortion is reflected in the apparent separation of the ends of the part feature.

Even a 76.2 μm (0.003 in.) error across the screen is considered satisfactory for high-grade instruments; however, the error in the optical system is seldom linear. We measure this error from the screen center, but the error may return to zero before the outer edge is reached. You can quite easily calibrate the screen by projecting standards, usually spheres, onto it.

For a direct measurement, a typical error may be:

Accuracy of projection at the screen	0.003 in.
Accuracy of reading at the screen	+0.002 in.
Total approximate error at the screen	0.005 in.

This level of error is not as bad as it seems, because these errors are measured at the screen. We can reduce errors at the workpiece by using the magnification—for example, a magnification of only 10X reduces the error to 12.7 μm (0.0005 in.). For most practical shop work, we will not be using our tolerance for the measurements instrument at this level.

For more precise tolerances, you can increase the magnification if the feature is still contained in the length of the screen (see Figure 17–17). If you use 50X, you reduce the error to 1.27 μm (0.00005 in.), and only in the most extreme cases will this error use up the available tolerance excessively. Somewhere between these extremes lies the solution for most direct measurement situations. Even when we add other errors, like the incorrect placement of the part in a holding fixture, direct measurement by optical comparator is one of the most reliable (low error) and convenient methods (see Figure 17–34).

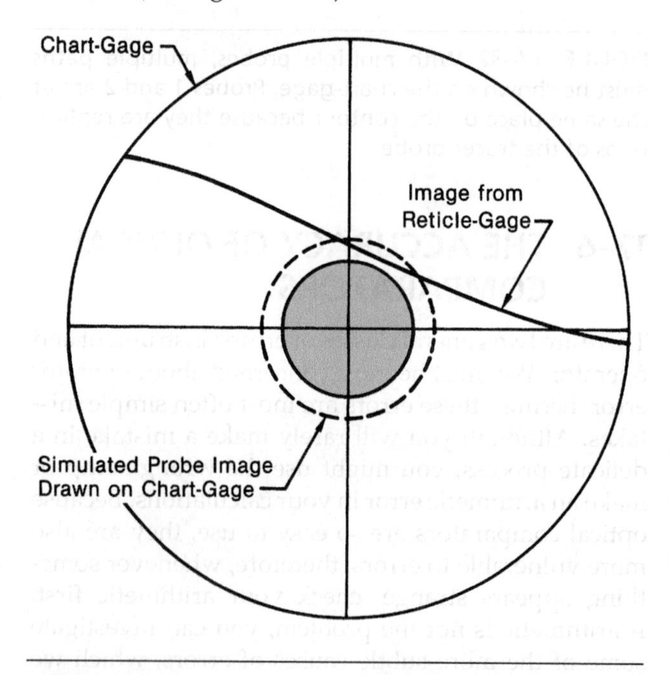

FIGURE 17–34 The contour image of the reticle-gage is projected on the chart-gage. The chart-gage has a circle which is the magnified size of the tracer probe. In this case, maximum and minimum tolerance limit circles are shown.

Accuracy, Chart-Gage Inspection

Chart-gage inspection, a special case of direct measurement, adds a chart-gage to the measurement process. The magnification simplifies measurement, because the chart-gages are superimposed on the screen; therefore, chart-gage errors on the workpiece are reduced by dividing by the magnification factor. At a sufficiently high magnification, error becomes insignificant.

The chart maker has an enormous advantage over the gage maker. Typically, a conventional gage manufacturer must subdivide the workpiece tolerance so that the instrument can obey the 10-to-1 rule (see Chapter 9). In contrast, the chart manufacturer preparing a chart-gage for 10X works with a gage tolerance as great as the workpiece tolerance.

In spite of this advantage, for the highest levels of precision inspection, we must still be aware of the chart-gage errors that must be taken into consideration, which depends on magnification, the method of gage layout, and the dimensional stability of the gage material. We can deal with the last factor by using glass, various plastics, and vellum drafting paper. Temperature changes have little effect, but you must be aware of changes in humidity. Glass chart-gages are unaffected by humidity but are the most expensive. Plastic ones include rigid plates and thin films of various materials and range from 18 μm to 17 μm (0.0007 in. to 0.0042 in.) distortion for a 35.6 cm (14 in.) screen at 50% humidity change.

The most convenient—and least expensive—material is vellum; however, it is greatly affected by humidity. A 50% humidity change may distort the chart 1 mm (0.040 in.) on a 36 cm (14 in.) screen. At 10X, this value is 10 mm (0.400 in.) on the part, which would be unacceptable in nearly every instance. This distortion does not eliminate vellum from consideration—for example, if you are comparing a feature that is 6 mm (0.250 in.) long, the distortion of 18 μm (0.0007 in.) will be acceptable in most cases.

You can use vellum even for large part features and higher magnification when its low cost or ease of preparation justifies its use. For large part features, you scale the chart-gage immediately after the

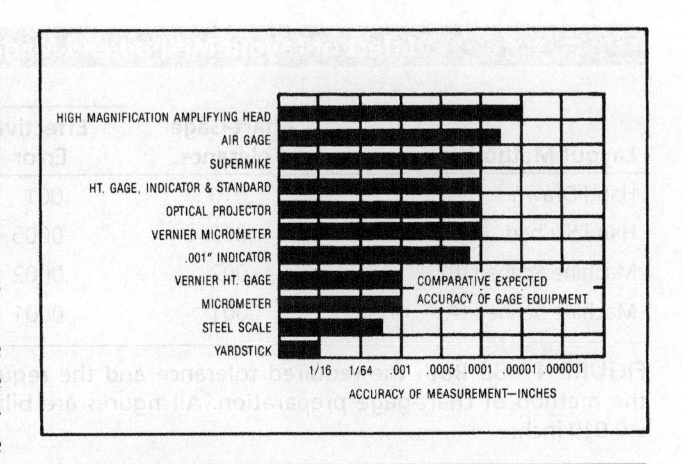

FIGURE 17–35 This graph compares measurement by optical comparator with other methods.

inspection and correct the results in proportion to the percentage of change between the desired dimension and the chart-gage representation of it.

You must also be aware of the effects of magnification and the method of layout on the chart-gage and how they interact (see Figure 17–35). We have shown two classes of machine scribed layouts: Class B is generally used because the higher cost of Class A layouts is not usually required by the workpiece tolerance (see Figure 17–36).

From all this information, you should conclude that when you are selecting an optical comparator, you should use the one with the highest magnification that will contain the feature on the screen. When you compare the capabilities of the optical comparator with the required tolerances of the workpiece feature, you will be able to determine the lowest cost method for the chart-gage.

17–7 MACHINE VISION SYSTEMS

The trend toward smaller, lighter, simpler products is forcing more manufacturers to deal with parts too fragile to measure with conventional contact systems. Although you can add the microscope or a video probe to a conventional coordinate measuring machine

Effect of Layout Method and Magnification on Gaging Tolerances					
		10x		50x	
Layout Method	Chart-Gage Tolerance	Effective Error	Part Tolerance	Effective Error	Part Tolerance
Hand Drawn	.010	.001	.010	.0002	.002
Hand Scribed	.005	.0005	.005	.0001	.001
Machine Scribed (B)	.002	.0002	.002	.00004	.0004
Machine Scribed (A)	.001	.0001	.001	.00002	.0002

FIGURE 17–36 Both the required tolerance and the required magnification must be considered when selecting the method of chart-gage preparation. All figures are bilateral; thus, a chart-gage tolerance of 0.010 is actually ±0.010 inch.

(CMM), vision systems are specifically designed for noncontact measurement (see Figure 17–37). Vision systems analyze an image of the part. The image is magnified so the system is similar to a microscope, with the eyepiece replaced by a camera. The vision system components include a high-resolution monochrome or color video camera, lighting, vision hardware (frame grabber and processor board, a motion controller for automated systems), and a computer system with vision-based-measurement software. The video camera captures the size and shape of the image on a pixel level. The frame grabber is a specialized converter that changes the analog signal into digital form.

FIGURE 17–37 This video measuring machine has manual X and Y stages with coarse and fine adjustments. *(Courtesy of Optical Gaging Products Inc.)*

Illumination

Vision systems work best under uniform, controlled lighting. Illumination of parts can affect measurement accuracy. The measurement data are taken, not from contact points, as with a CMM, but from reading the grayscale value of the pixelated image. The best measurements occur with a sharp focus and the highest contrast. Too much light can cause oversaturated pixels called "blooming." Too little light results in poor edge visibility. Vision systems are designed with several types of lighting control. Back or bottom lighting from below the part is best for profile types of measurements—edges and through holes. Top (or ring) lighting provides general surface lighting where features cannot be seen with light from underneath the part—top surfaces, slots, and blind holes. Auxiliary lighting is light sent through a beam splitter to provide a very intense collimated beam of light. This type of lighting works best for small deep grooves, slots, or blind holes where general surface lighting would not supply enough illumination to "see" the feature.

Most vision systems also offer some source of oblique LED lighting, where the light source skims across the part at an angle. Oblique lighting works well where surface features have a low profile, or where translucent plastic parts are difficult to illuminate.

Magnification

Systems can have zoom lenses that allow measurement of part features that vary greatly in size or fixed optics, which have the highest optical quality but trade off the range of magnification. Adding the zoom lens increases the flexibility of a vision system. The z axis moves along the optic sensor, perpendicular to the stage. A sharp image is critical to the measurement, so it is necessary to move the camera so that the feature is within the focus of the optics *at a particular magnification.* The *depth of focus* decreases as the magnification increases. At a maximum magnification, there is generally a very shallow depth of field. Most vision systems have an "autofocus" feature to assist in improving accuracy.

Vision System Measurement

Vision measurement parameters include movement or position in the x, y, and z axes, magnification, lighting source and intensity, and data acquisition. Manual systems generally use a crosshair tool, where the operator moves the stage to place the image of a crosshair over the feature to be measured and records that measurement point. Automated systems use sophisticated edge detection algorithms, where the software can create measurements from the grayscale values of a camera image.

Three-dimensional vision systems (see Figure 17–38) have the same advantages over microscopes and optical comparators that conventional CMMs have over single-purpose measurement tools. They offer all of the same automated measurement capabilities of traditional CMMs without touching the workpiece (see Figure 17–39). Vision systems can obtain measurement data quickly and easily, store automated part programs, and generate output reports. Measurement software can be linked with statistical analysis or 3-D modeling programs (see Figure 17–40).

Multisensor Systems

Many vision systems are adding touch and laser probes for increased functionality and accuracy (see

FIGURE 17–38 Computerized systems offer the advantage of powerful software algorithms. *(Courtesy of Optical Gaging Products Inc.)*

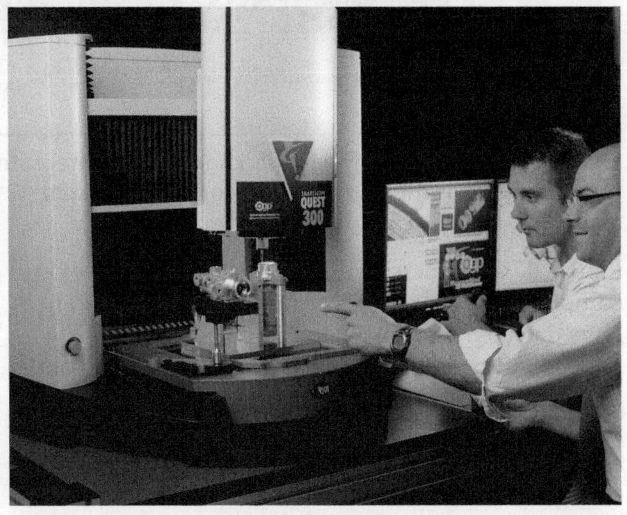

FIGURE 17–39 Fully automated vision systems offer all of the benefits of automated CMMs. This model also offers multisensor capabilities. *(Courtesy of Optical Gaging Products Inc.)*

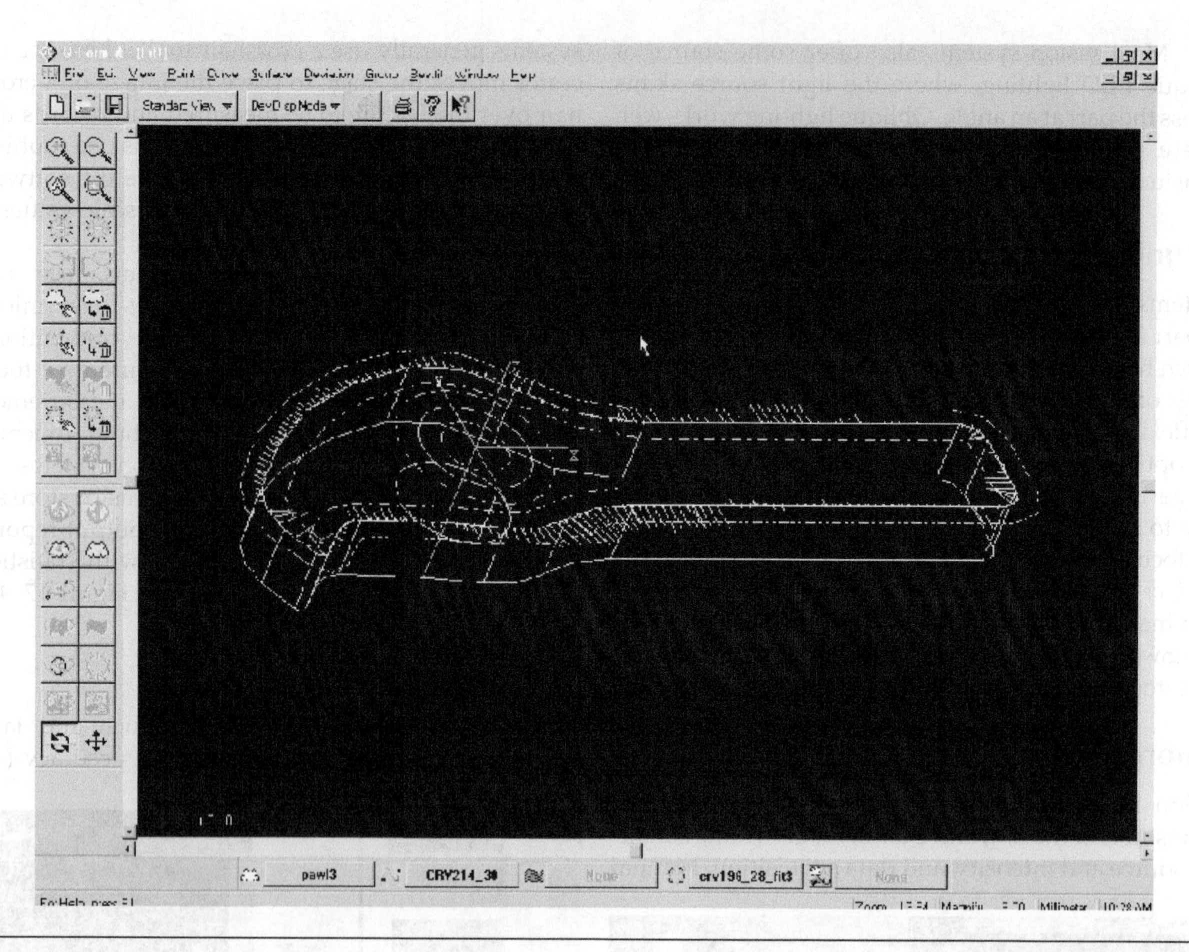

FIGURE 17–40 Many software packages can be integrated with statistical packages or three-dimensional modeling programs such as this one. *(Courtesy of CE Johansson)*

Figure 17–41). Increased manufacturing technology, complex part geometry, and tighter part tolerances have driven the need for multisensor systems.

A multisensor system has more overall capabilities than a single-purpose machine, but the dedicated system will provide more accurate results for the measurement for which it was built. Although software changes accommodate the integration of contact and noncontact measurements, one type of sensor will be the primary probe. It can be either a video probe or a touch probe, but one probe must be the primary sensor. Calibration routines must be performed for the offset position of the touch and laser probes. Adding a video sensor to a traditional CMM has limitations as well. Proper illumination, critical to measurement accuracy, is limited. CMM structure is not compatible with the same illumination as a video system; there are limited top lighting options, and there are no back lighting options.

Choosing the Best Inspection System

Choosing the best measurement system for your parts depends on a number of variables. Microscopes and optical comparators are generally manual systems and rely on human judgment. They can be simple and

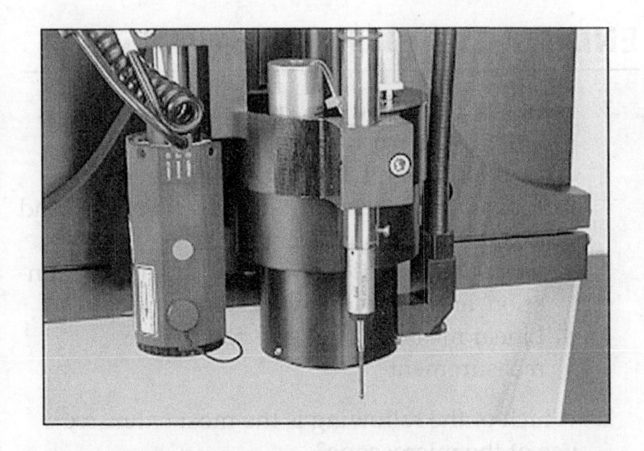

FIGURE 17–41 Multisensor systems combine contact and noncontact sensors with automated programming for fast, accurate inspection. *(Courtesy of CE Johansson)*

inexpensive but slow and limited to two axes. Optical comparators are easy to use and are cost-effective but better suited to flat, rather than contoured, surfaces. CMMs, vision machines, and multisensor systems can provide you with fast, accurate, automated three-dimensional inspection systems. The following are general factors that can help you evaluate system choices:

- **Size, weight, and geometry of the part:** For parts larger than 32 × 32 × 10 or heavier than 22 pounds, choose a CMM.
- **Contact or noncontact measurement:** For parts that can be damaged, distorted, or contaminated, a noncontact system is best.
- **Part tolerances required:** It used to be that the rule of thumb was if tolerances were less than ±0.0001 to 0.002 in. (25 to 50 mm), a noncontact system was a better choice. Today, CMM manufacturers claim volumetric accuracy in a range of 8 to 9 μm. Accuracy ranges for vision systems is ±3 μm.
- **Manufacturing volume or sampling rate:** For low-volume parts or short production runs, it is difficult to justify the higher costs of automated systems. Manual systems or hand tools are also less costly than dedicated hard gages.
- **Software complexity and flexibility:** The more flexible the measurement software, the more complex it is to operate. This will vary by manufacturer and machine type. Most measurement software today operate in a Microsoft Windows environment and are icon or menu driven, making them easy to learn.

Before purchasing any system, you should familiarize yourself with the appropriate ANSI standards, such as B89, that apply to CMMs.

SUMMARY

- Microscopy is important in two different aspects of metrology: direct measurement and positioning. Direct measurement has three principal roles: linear measurement, surface inspection, and measurement of inaccessible features.
- The microscope extends the magnification of the single lens magnifier by means of a phenomenon: an optical image may be treated as an object itself.
- A microscope by itself does not measure. Measurement is accomplished by a reticle. A reticle, like all scaled instruments, is both a standard and a means for reading measurement.
- At high magnification, the field of view, hence the measuring range, becomes very small. Therefore, a mechanical stage using micrometers may be required for most workpiece features.
- Reticles such as crosshairs are also used to line up an edge of a part feature so that a mechanical or electronic instrument may use it as a reference.
- The microscope makes easy the examination of surface finishes, but it does not measure surface asperities directly. In order to measure these, it is necessary to take an oblique cut across the surface or to use an oblique band of light to scan the surface.

- Optical comparators have five advantages over measuring microscopes: larger field of view, two or more simultaneous viewers, direct measurement from screen, simpler photographic records, and less eyestrain.
- The optics for the comparator differ from that of the slide projector in that a collimated beam of light is used for the comparator, whereas a converging beam is used for slide projection. The optical path may be very long, requiring the optics to be folded in order to get it into a reasonable size.
- Not all workpieces lend themselves to inspection by the optical comparator. It is necessary to have the edges of features in nearly the same plane for maximum sharpness. Reflection from the surface of the workpiece can also be a problem.
- For many workpieces, the large screen allows direct measurement. For larger workpieces, the worktable is traversed. That is displacement measurement. For comparators, it is often called measurement by translation or indexing.
- The optical comparator is a versatile measuring instrument. However, it finds even greater use for attribute inspection. Unlike many other methods, it allows two or more part features to be examined at the same time. Moreover, it enables geometry to be checked. This cannot be done directly by most other methods.
- One of the most useful capabilities of the optical comparator is to find and work from imaginary references. These exist only in space, such as the center of an internal diameter.
- Because the bezels of optical comparators may be rotated, they are one of the most effective methods to measure angles. Vernier reading bezel scales allow great precision.
- The inspection of irregular contours is extremely difficult by other methods, requiring multiple steps and calculations. With the optical comparator, it is easy and straightforward. The chart-gage is easily made and provides tolerance lines for comparison with the workpiece.
- Automated vision systems offer the same benefits of a CMM without touching the part.

END-OF-CHAPTER QUESTIONS

1. Which of the following are the major divisions of practical microscopy?
   a. Positioning and direct measurement
   b. Linear measurement, surface topology, and metallurgical measurement
   c. Direct examination, projected image examination, and virtual image examination
   d. Linear measurement and coordinate measurement

2. Which of the following is the most extensive use of the microscope?
   a. Metallurgical research
   b. Surface finish examination
   c. Examination of fringe bands
   d. Scale reading

3. What optical phenomenon makes possible the microscope?
   a. Aberration correction by achromatic lenses.
   b. The fact that a real image is inverted, whereas a virtual image is erect.
   c. Virtual images may be projected to infinity.
   d. An optical image may be treated as a real object.

4. What determines the effective magnification?
   a. Calibration of the instrument traceable to NIST
   b. The power of the objective lens
   c. The product of individual magnifications
   d. The sum of the individual magnifications

5. What is the most common eyepiece power?
   a. 5
   b. 7 1/2
   c. 10
   d. 50

6. What is the chief metrological difference between measuring with a microscope and with an electronic comparator?
   a. The microscope is limited to small workpieces.
   b. The comparator does not provide a view of the workpiece.

c. The comparator can only examine one point on the workpiece.
d. The microscope carries its own standard.

7. What is the relationship of the field of view?
a. It regulates the position of the observer's eye with respect to the eyepiece.
b. It determines the plane of the reticle.
c. It decreases with increased magnification.
d. It is the way the power of a microscope is designated.

8. How is decreased field of view compensated for in high-powered measuring microscopes?
a. By a mechanical stage
b. By multiple power eyepieces
c. By repetitive measurements
d. By monochromatic illumination of the object

9. Which element makes a microscope a measuring microscope?
a. The mechanical stage
b. The stand is larger and more rugged than on other microscopes
c. The interchangeable eyepieces
d. The reticle

10. What is the chief function of cross wires or crosshairs in a microscope?
a. The measurement of deviations from straightness
b. Finding the centers of circles
c. Alignment to part feature edges
d. As an aid in focusing

11. With increases in magnification, which of the following occur?
a. The field of view decreases.
b. The ambient illumination decreases.
c. The larger parts can be measured.
d. The eyepiece must be raised.

12. The short distance from the objective lens to the object causes problems at high magnification. Which of the following is the most serious?
a. Cleaning the object surface
b. Positioning the object
c. Reflection from the object surface
d. Illumination of the object

13. Which of the following is the most important advantage of a large field of view?
a. Allows larger areas to be examined
b. Relieves eyestrain
c. Minimizes parallax errors
d. Increases precision in proportion to the increase in field of view

14. The comparator allows more than one person at a time to examine the part features. Which of the following is the most important advantage of that?
a. Parallax error cancels out.
b. Particular features can be pointed to, thereby eliminating misunderstandings.
c. If one person leaves the company, a second one can still authenticate what that person saw.
d. It compensates for the differences in the resolving power of human vision when using microscopes.

15. Of the several disadvantages of the optical comparator in comparison with the microscope, which of the following is the most serious?
a. Ambient lighting control
b. Dependence on a mechanical stage
c. Sensitivity to temperature
d. Large size

16. What is the best point from which to view a feature on the comparator screen?
a. Within ±5° of normal in the horizontal plane
b. Above, looking down on the screen
c. Within ±20° in the horizontal plane
d. Within ±20° in any direction from normal

17. Which of the following observations about typical workpieces most affects the use of the optical comparator?
a. Most features have depth as well as length.
b. All workpieces have surface finish.
c. At suitable magnifications, the feature length may exceed the field of view.
d. Many features are internal rather than external.

18. What is the most common consideration caused by light reflected from the workpiece?
    a. Temporary blinding of observer
    b. Excessive heating
    c. Loss of contrast
    d. Ghost images
19. Which of the following provides the clearest image?
    a. Dark objects
    b. Objects with contrast between front surfaces and other surfaces
    c. The edge nearest the projection lens
    d. Rounded edges
20. What does diascopic projection refer to?
    a. Simultaneous projection of two features
    b. Simultaneous projection of two planes
    c. Projection through transparent material
    d. Projection through translucent material
21. What does episcopic projection mean?
    a. Same as slide projection
    b. Projection of rotating features
    c. Projection of internal features
    d. Projection of opaque surfaces
22. Which of these roles cannot be performed with the optical comparator?
    a. Clinometer measurement
    b. Attribute inspection
    c. Direct measurement
    d. Displacement measurement
23. To which of the following does a toolroom chart-gage correspond?
    a. Graduated reticle
    b. Vernier protractor
    c. Go/no-go attribute gage
    d. Mechanical worktable
24. Which of the following can be done with an optical comparator but not with micrometers, vernier calipers, height gages, or plain scales?
    a. Parallax can be eliminated.
    b. Heating the workpiece by handling is eliminated.
    c. Squareness can be checked.
    d. Multiple linear dimensions can be measured.
25. What is the most important limitation of direct measurement with the optical comparator?
    a. Sufficient illumination
    b. Field of view
    c. Length of the mechanical worktable
    d. Limited to episcopic projection
26. Which of the following is unique to measurement by optical comparators and, in some cases, microscopes?
    a. The workpiece is illuminated by the instrument.
    b. Only one plane may be observed.
    c. There is no distortion of the workpiece.
    d. The recognition of colors.
27. For which of the following does increased magnification make no change?
    a. Lengths approaching zero
    b. Negative dimensions
    c. Angles
    d. Circles
28. Which of the following is often used to inspect contours with an optical comparator?
    a. Solve the arcs and measure the tangents, then combine.
    b. Measure deviations in the y axis along the x axis.
    c. Compare with the chart graph.
    d. Match with the inverted contour projected on the screen.
29. What is the chief role of the tracer technique?
    a. Inspection of contours
    b. Measurement beyond the range of the work tables
    c. Conversion of translation measurement to direct measurement
    d. Examination of inaccessible features
30. What are the most common errors when using an optical comparator?
    a. Bezel calibration
    b. Scale readings
    c. Worktable calibration errors
    d. Arithmetic errors

31. Which of the following are not good applications for measurement on a visions system?
    a. Drill bits
    b. Circuit board
    c. Cam shaft
    d. Rubber gaskets

32. Which of the following has the greatest effect on measurement accuracy with a vision system?
    a. Magnification, illumination, and focus
    b. Operator, shift, and supervisor
    c. Part shape, location of machine, and software
    d. Position of part on the stage, material, and weight of part

18

CHAPTER EIGHTEEN

COORDINATE MEASURING MACHINES

LEARNING OBJECTIVES

- State reasons why coordinate measuring machines (CMMs) have potentiality not shared by individual measuring instruments.
- Explain the significance of the computer when coupled to a CMM.
- List the limitations as well as the opportunities for the CMM.

OVERVIEW

Coordinate measuring machines (CMMs) are extremely powerful metrological instruments (see Figure 18–1). They enable us to locate point coordinates on three-dimensional structures at the same time that they integrate both dimensions and the orthogonal relationships.

When we add a computer to the CMM, we create an instrument that can automatically perform complex analyses and that can learn measurement routines to compare how a piece conforms to its specifications. Most software programs operate within a Microsoft Windows environment, so good computer skills and a thorough understanding of basic geometry are both required to become a great CMM operator. Your lab manual has exercises to help you practice your CMM skills.

Instead of performing time-consuming measurement with traditional, single-axis instruments (micrometers, height gages, etc.) and cumbersome mathematics, you can dimensionally evaluate complex workpieces with precision and speed—and you can store the data for later analysis or comparisons. The greater the complexity of the piece, the greater the benefits from a CMM (Figure 18–2).

Even though a CMM provides many measurement benefits, it does not introduce a single new fundamental principle. Still, these machines are widely used for industrial applications. Even more importantly, for metrologists, a CMM combines all of the basic principles required for reliable measurement.

18–1 BACKGROUND

Development of Coordinate Measuring Machines

It is difficult to say when coordinate measuring machines emerged as a distinct class of instruments, but such metrology pioneers as C. E. Johansson and

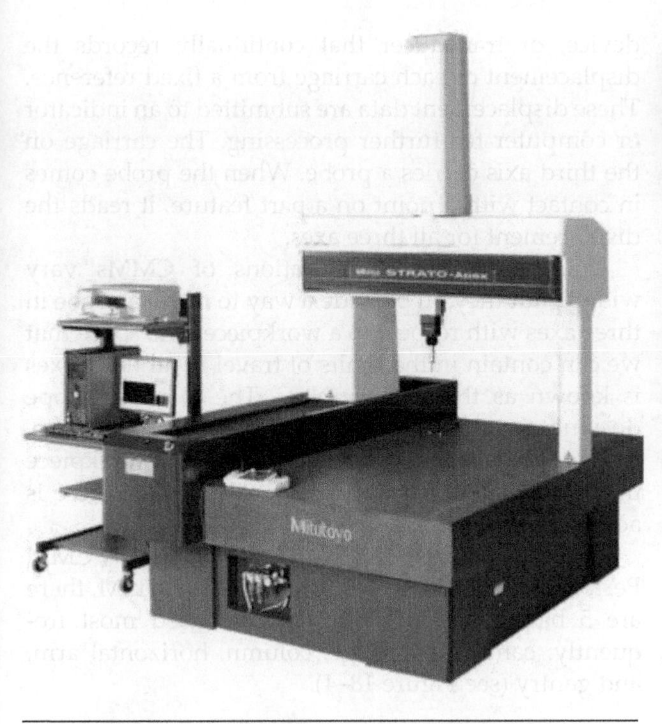

FIGURE 18–1 A bridge-type CMM. *(Courtesy of Mitutoyo American Corporation)*

F. H. Rolt wrote about machines that could measure in three axes.

The Atomic Energy Commission of the 1940s may have been among the first organizations to recognize the potential of CMMs, and their machines kept pace with the explosion in computer technology. The theorists and technicians in the field were quick to recognize the potential and to publicize their accomplishments. Unfortunately, the sophistication of CMMs—when compared with commonly used instruments—may have slowed their acceptance.

In the 1960s, the CMM began to emerge as a powerful tool; then, as microprocessors became the power in computing, manufacturers began to realize the real potential of CMM systems. The combination of the CMM and the computer's ability to process data allows us to apply these measurement systems in the most appropriate and effective ways.

FIGURE 18–2 Coordinate measuring machines can increase the speed and accuracy of the inspection process. *(Courtesy of Mitutoyo American Corporation)*

18–2 THE ROLE OF COORDINATE MEASURING MACHINES

Coordinate measuring machines play an important role in a large number of industries, including aerospace, automotive, electronics, food processing, health care, paper, pharmaceuticals, plastics, research and development, and semiconductors.

The ability to quickly and accurately capture and evaluate dimensional data distinguishes the CMM from other types of measurement processes. Sophisticated contact and noncontact sensors, combined with vast computer processing capabilities, make the CMM a practical, cost-effective solution. CMMs are especially suited for parts with complex geometry and assemblies (see Figure 18–3).

When you can reduce inspection time on the order of 80% to 90%, it is easy to see why manufacturers—as well as trade publications and professional journals—are enthusiastic about CMM systems.

CMMs are particularly well-suited for the following conditions:

- *Short runs*—We may be producing hundreds or even thousands of a part, but the production run

FIGURE 18–3 CMMs work well for complex part assemblies. *(Courtesy of CE Johansson)*

is not sufficient to justify the cost of production inspection tooling.

- *Multiple features*—When we have a number of features—both dimensional and geometric—to control, CMM is the instrument that makes control easy and economical.
- *Flexibility*—Because we can choose the application of the CMM system, we can also do short runs and measure multiple features.
- *High unit cost*—Because reworking or scrapping is costly, CMM systems significantly increase the production of acceptable parts.
- *Production interruption*—Whenever you have to inspect and pass one part before you can start machining on the next part, a machining center may actually be able to help a manufacturer save more money by reducing downtime than would be saved by inspection.

18–3 TYPES OF COORDINATE MEASURING MACHINES

The basic CMM has three perpendicular axes: x, y, and z. Each axis is fitted with a precision scale, measuring device, or transducer that continually records the displacement of each carriage from a fixed reference. These displacement data are submitted to an indicator or computer for further processing. The carriage on the third axis carries a probe. When the probe comes in contact with a point on a part feature, it reads the displacement for all three axes.

The physical configurations of CMMs vary widely, but they all provide a way to move a probe in three axes with respect to a workpiece. The space that we can contain in the limits of travel in all three axes is known as the *work envelope.* The work envelope does not necessarily limit the size of the workpiece. In many situations, some portions of the workpiece may be outside of the envelope—if the workpiece is adequately supported.

Although 10 configurations are covered by CMM Performance Standard ANSI/ASME B89.1.12M, there are 5 basic configurations that are used most frequently: cantilever, bridge, column, horizontal arm, and gantry (see Figure 18–4).

The Cantilever Type

In the cantilever-type CMM, a vertical probe moves in the z axis, carried by a cantilevered arm that moves in the y axis. This arm also moves laterally through the x axis. You have easy access to the work area, and the cantilever CMM provides a relatively large envelope without taking up too much floor space.

The Bridge Type

The most popular type of CMM, this machine has a moving bridge. It is similar to the cantilever type, because it has a support for the outer ends of the y-axis beam on the base. The bridge construction adds rigidity to the machine, but it also forces us to make sure that both ends of the y axis track at exactly the same rate. Loading the bridge CMM can be difficult because the bridge has two legs that both touch the base.

In a variation of the bridge CMM, the *fixed bridge* type, the bridge is fixed to the base. This setup eliminates potential tracking problems of the standard bridge CMM, providing superior accuracy and making it useful for gage room applications and production inspection.

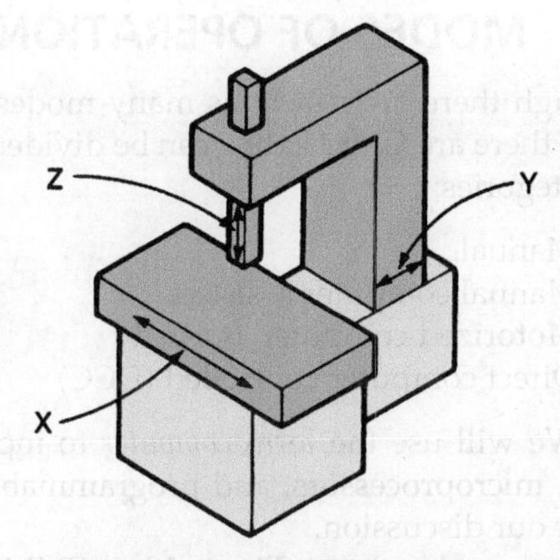

A. MOVING TABLE CANTILEVER ARM TYPE

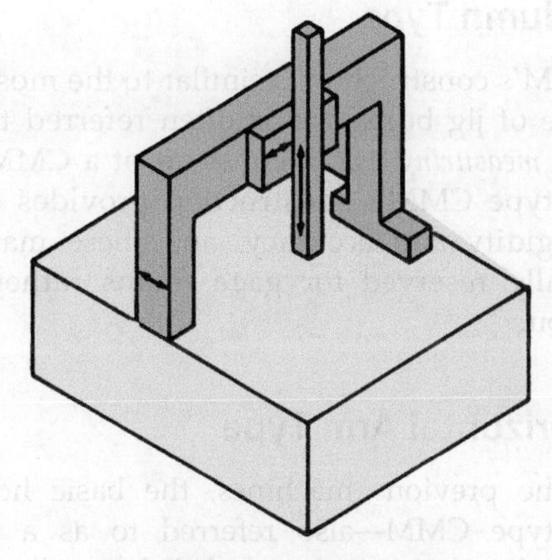

B. MOVING BRIDGE TYPE

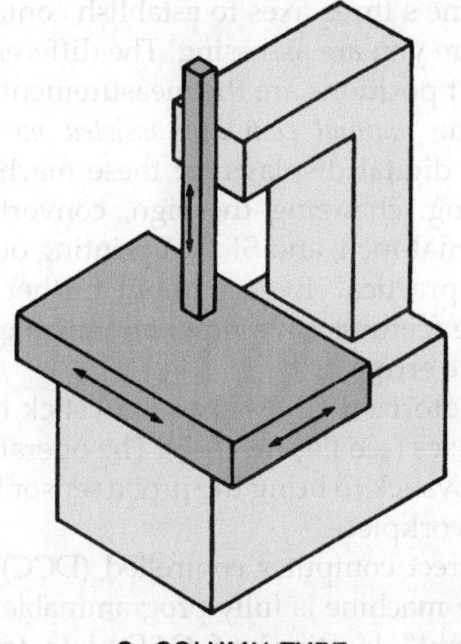

C. COLUMN TYPE

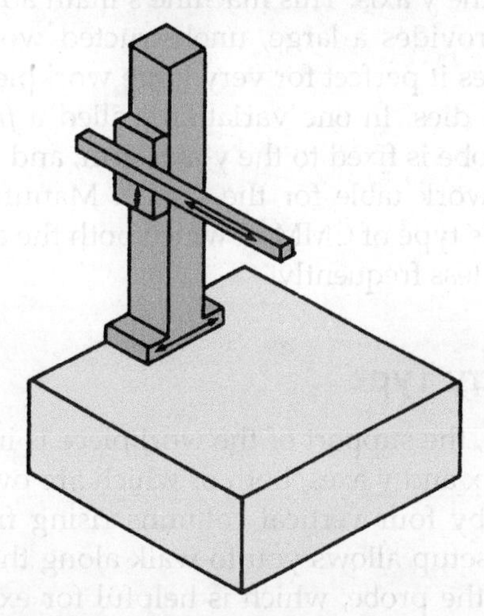

D. MOVING RAM HORIZONTAL ARM TYPE

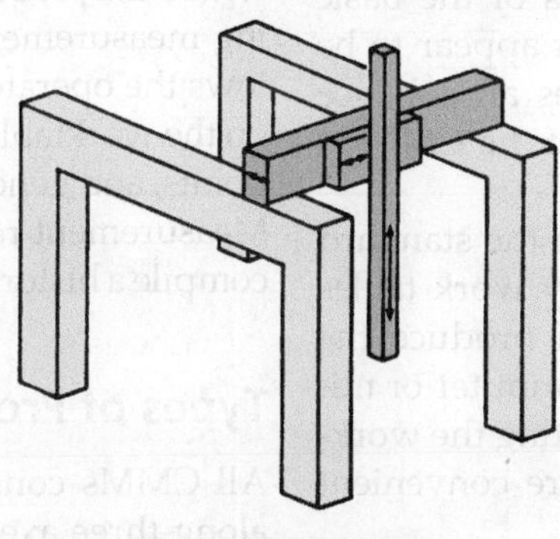

E. GANTRY TYPE

FIGURE 18–4 The five basic types of coordinate measuring machines.

The Column Type

This CMM's construction is similar to the most popular type of jig borer and is often referred to as a *universal measuring machine* instead of a CMM. The column type CMM's construction provides exceptional rigidity and accuracy, and these machines are usually reserved for gage rooms rather than inspection.

The Horizontal Arm Type

Unlike the previous machines, the basic horizontal arm-type CMM—also referred to as a layout machine—has a moving arm, and the probe is carried along the y axis. This machine's main advantage is that it provides a large, unobstructed work area, which makes it perfect for very large workpieces like automotive dies. In one variation, called a *fixed arm type,* the probe is fixed to the y-axis arm, and you use a moving work table for the y axis. Manufacturers produce this type of CMM in which both the arm and table move less frequently.

The Gantry Type

In this type, the support of the workpiece is independent of the x and y axes, both of which are overhead, supported by four vertical columns rising from the floor. This setup allows you to walk along the workpiece with the probe, which is helpful for extremely large pieces.

The five most common variations of the basic types (see Figure 18–5) may have what appear to be minor differences, but these differences all enhance the convenience of use and efficiency of the CMM in many situations.

Although they are not included in the standard as yet, CMMs are available with rotary work tables or probe spindles. In theory, you can produce the same data with any CMM that has a computer or microprocessor without mechanically rotating the workpiece; however, these rotary devices are convenient and can speed up inspection.

18–4 MODES OF OPERATION

Although there are nearly as many modes of operation as there are CMMs, they can be divided into general categories:

- Manual
- Manual computer assisted
- Motorized computer assisted
- Direct computer controlled (DCC)

We will use the term *computer* to include computers, microprocessors, and programmable controllers for our discussion.

A *manual mode* (see Figure 18–6) CMM has a free-floating probe that you, as the operator, move along the machine's three axes to establish contact with the part feature you are assessing. The differences among the contact positions are the measurements.

In the *manual computer-assisted mode,* we add electronic digital displays for these machines, doing zero setting, changing the sign, converting among inch, decimal-inch, and SI, and printing out data both easy and practical. Even without further sophistication, these features save time, minimize calculations, and reduce errors.

A motorized CMM uses a joystick to drive the machine axes (see Figure 18–6). The operator manipulates the joystick to bring the probe sensor into contact with the workpiece.

A direct computer controlled (DCC) coordinate measuring machine is fully programmable. The CMM uses "taught" locations of CAD data to determine where the probe sensor contacts the workpiece, collecting measurement data. The fully automated CMM allows the operator to place the workpiece in a fixture or on the worktable, run a stored program, collect the data points, and generate an output report (see Figure 18–7). Measurement reports can be saved in the computer to compile a historical record for statistical process control.

Types of Probes

All CMMs consist of a probe and a way to move it along three axes relative to a workpiece. Throughout

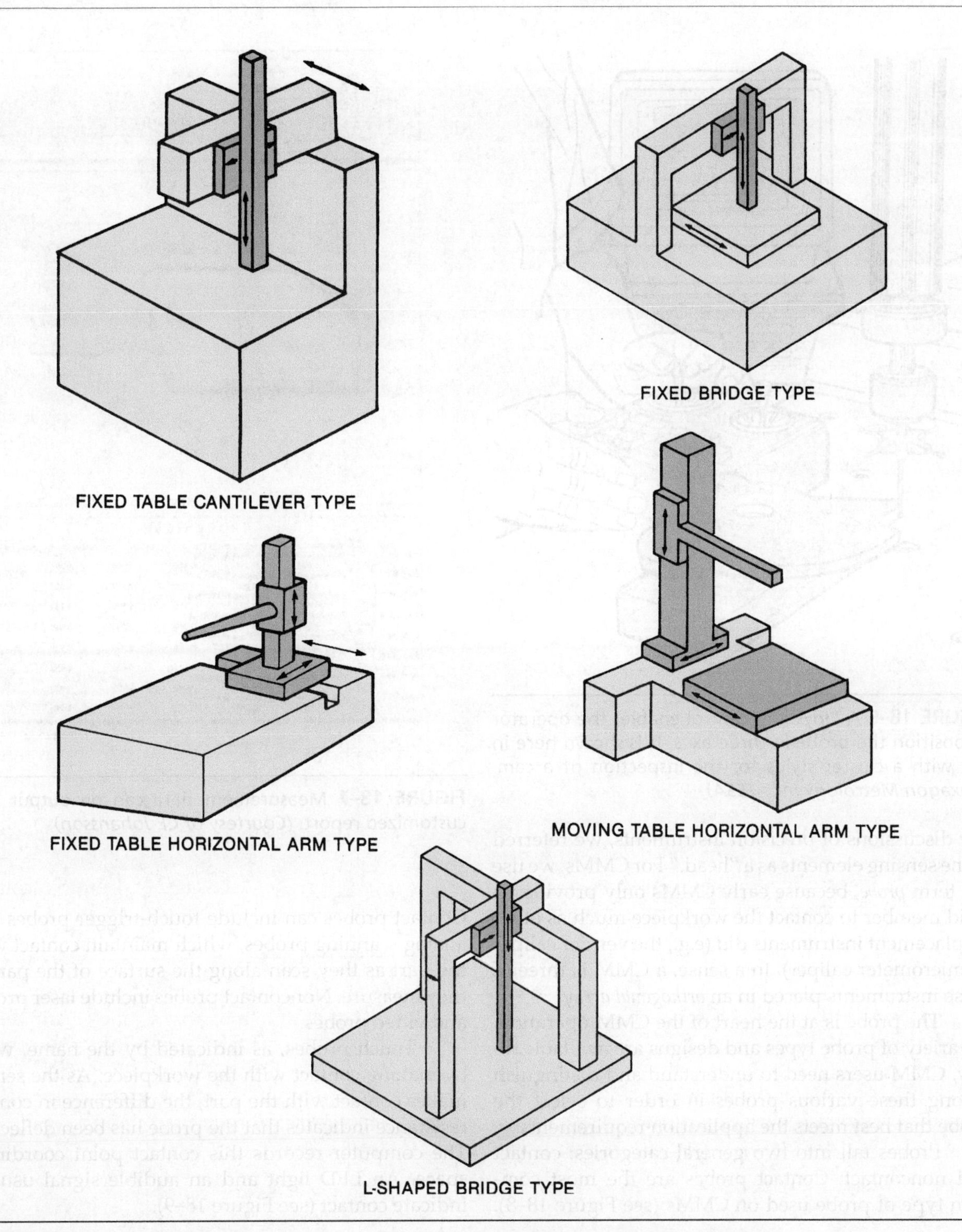

FIGURE 18–5 Popular variations of the five basic types of coordinate measuring machines.

FIGURE 18–6 A "joystick" control enables the operator to position the probe in three axes. It is shown here in use with a cluster stylus for the inspection of a cam. *(Hexagon Metrology Inc., TESA)*

our discussions of precision instruments, we referred to the sensing elements as a "head." For CMMs, we use the term *probe,* because early CMMs only provided a solid member to contact the workpiece much as other displacement instruments did (e.g., the vernier caliper or micrometer caliper). In a sense, a CMM is three of these instruments placed in an *orthogonal array.*

The probe is at the heart of the CMM operation. A variety of probe types and designs are available today. CMM users need to understand and distinguish among these various probes in order to select the probe that best meets the application requirements.

Probes fall into two general categories: contact and noncontact. Contact probes are the most common type of probe used on CMMs (see Figure 18–8).

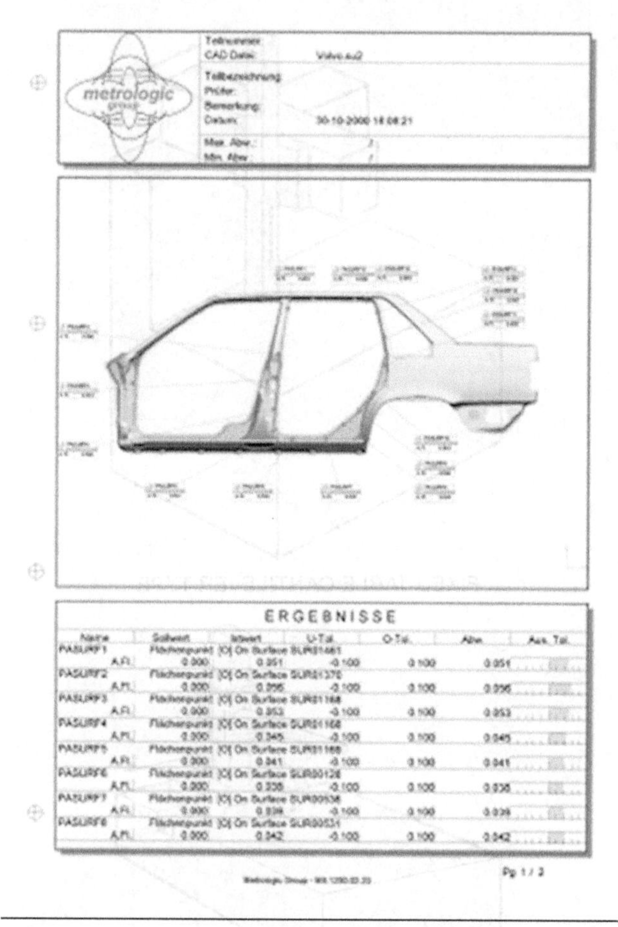

FIGURE 18–7 Measurement data can be output in a customized report. *(Courtesy of CE Johansson)*

Contact probes can include touch-trigger probes and analog scanning probes, which maintain contact with the part as they scan along the surface of the part as they measure. Noncontact probes include laser probes and video probes.

Touch probes, as indicated by the name, work by making contact with the workpiece. As the sensor makes contact with the part, the difference in contact resistance indicates that the probe has been deflected. The computer records this contact point coordinate space. An LED light and an audible signal usually indicate contact (see Figure 18–9).

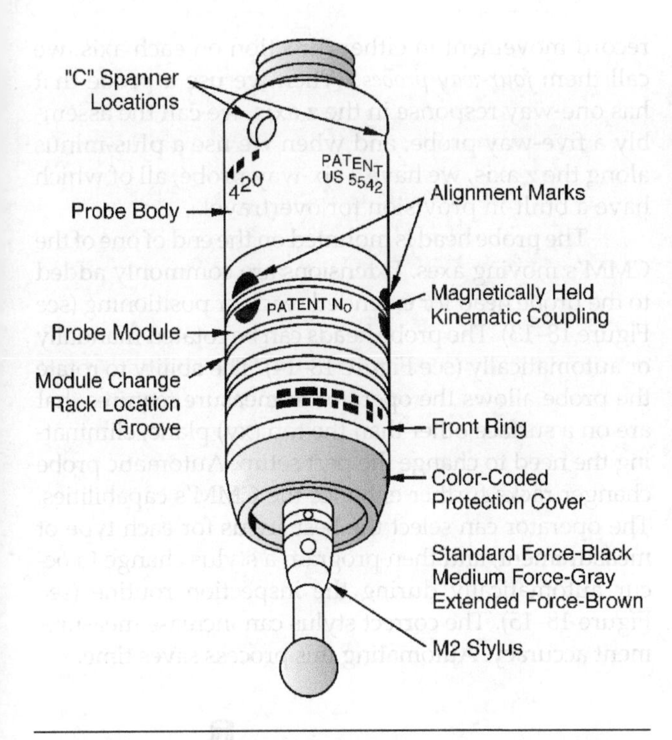

FIGURE 18–8 Contact probes, like this one made by Renishaw, are used to gather measurement data.

Analog scanning probes are a type of contact probe used to measure contoured surfaces, such as a turbine blade. The analog scanning probe remains in contact with the surface of the part as it moves, and produces analog readings rather than digital signals. Form measurement makes it necessary to collect and analyze large amounts of data quickly. Continuous analog scanning systems can acquire 10 to 50 times more data than a traditional touch-trigger probe in a given amount of time.

Analog probes improve the speed and accuracy at which measurement data are collected, particularly for surface mapping complex part shapes. They are often used for measurements of parts where the surface geometry is complex or irregular, such as a crankshaft or prosthetic device. Scanning probes are commonly used to gather data for reverse engineering applications.

One of the limitations of touch-trigger probes is the force required (a gram or more) to deflect the probe when it makes contact with the part. Measuring flexible or fragile parts can be difficult at best. Soft plastic, rubber, thin wire, or other flexible components can be distorted or damaged by the contact of a touch-trigger probe. Although noncontact probes are the most common solution, low-trigger-force, high-sensitivity-touch probes also are available (see Figure 18–10).

Advances in technology have made it possible to develop a new type of touch probe that uses a high-frequency stylus with a trigger force of less than

FIGURE 18–9 This probe head has an LED signal that is triggered when the probe sensor makes contact with the gear. *(Courtesy of CE Johansson)*

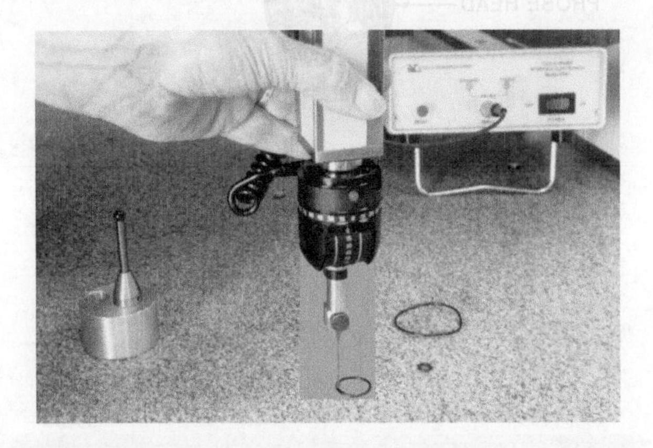

FIGURE 18–10 Low-trigger-force probes make it possible to measure small flexible parts like this rubber O-ring. *(Apollo Research Corp).*

10 milligrams to detect touch. The probe body contains a crystal that creates a vibration in the stylus. The probe resonates at 20 to 25 kilohertz. When the stylus makes contact with the part, the motion (vibration) changes. This change is detected by a microprocessor in the electronics box, which records this change as a touch. This touch registers as a data point before part deflection or damage occurs. The vibration in this probe also eliminates another common problem with low-force touch-trigger probes—"false triggers." False triggers occur when the machine movement, vibration, or acceleration causes a low trigger force probe to deflect, recording a false measurement point. The vibration in the new touch probe is not affected by changes in machine speed or movement.

Touch probe assemblies consist of three components: probe head, probe, and stylus (see Figures 18–11 and 18–12). The simplest touch probe assemblies are sensitive in the x and y axes, and because these probes record movement in either direction on each axis, we call them *four-way probes.* When we use a probe that has one-way response in the z axis, we call the assembly a five-way probe, and when we use a plus-minus along the z axis, we have a six-way probe, all of which have a built-in provision for overtravel.

The probe head is mounted on the end of one of the CMM's moving axes. Extensions are commonly added to the probe head for extended reach or positioning (see Figure 18–13). The probe heads can be rotated manually or automatically (see Figure 18–14). The ability to rotate the probe allows the operator to measure features that are on a surface other than the top (xy) plane, eliminating the need to change the part setup. Automatic probe changer racks further enhance the CMM's capabilities. The operator can select the best stylus for each type of measurement, and then program a stylus change to occur automatically during the inspection routine (see Figure 18–15). The correct stylus can increase measurement accuracy. Automating this process saves time.

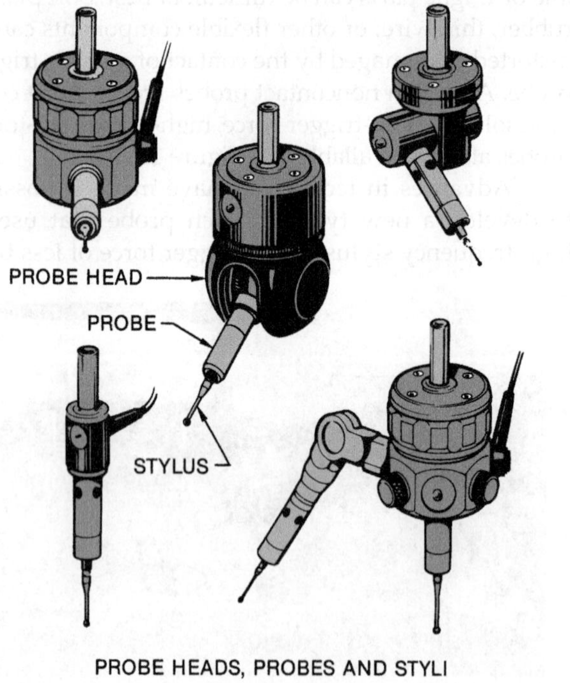

PROBE HEAD
PROBE
STYLUS
PROBE HEADS, PROBES AND STYLI

FIGURE 18–11 The probe is attached to the machine "quill" by means of a head. A variety of heads are available. Some are motorized and some carry two or more styli. *(Courtesy of Renishaw Plc)*

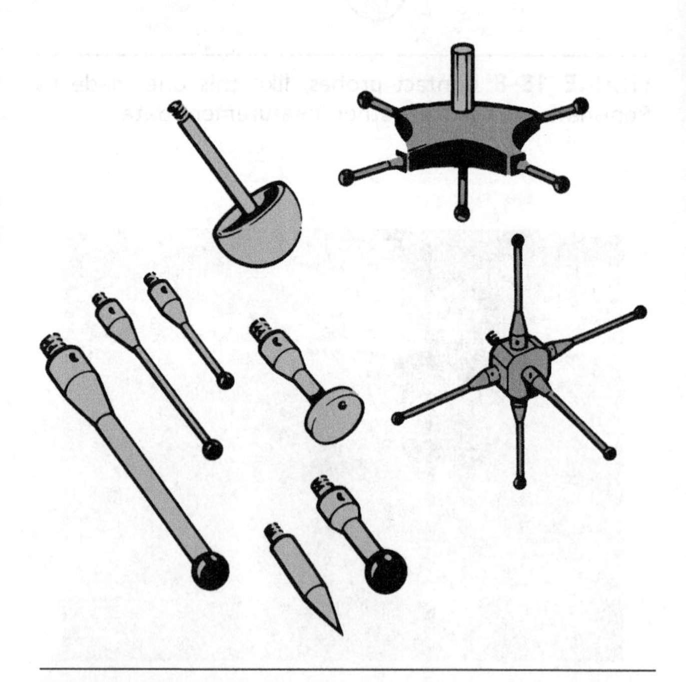

FIGURE 18–12 The ball styli are the most common, but other shapes are available, as well as multiple styli units. Ruby is often used for styli to extend their wear life. A multiple styli probe is often called a "cluster" probe. *(Courtesy of Renishaw Plc)*

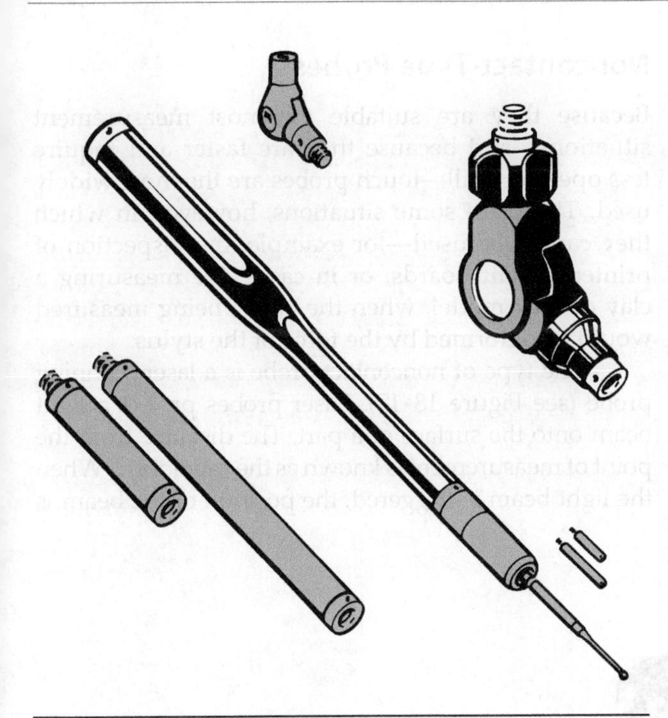

FIGURE 18–13 Combinations of extensions and joints enable the styli to reach otherwise inaccessible places. They may be used between heads and probes or between probes and styli. *(Courtesy of Renishaw Plc)*

FIGURE 18–14 Probe heads index in 7½ degree increments for increased measurement flexibility. *(Courtesy of CE Johansson)*

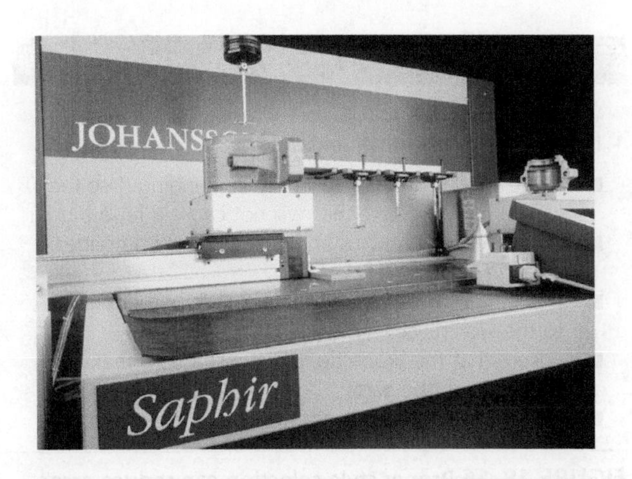

FIGURE 18–15 This CMM uses an automatic probe changer rack mounted on a system customized to operate on the shop floor. A conveyor system brings the part from the production line onto the CMM for in-process inspection. *(Courtesy of CE Johansson)*

Probe Calibration

By now, you should realize that when we measure in a sequence that progresses in the same direction, our result is more inherently reliable than if we make a sequence of measurements that change direction. One of the benefits of a CMM is that we can reverse directions without backlash, hysteresis, and other errors that destroy accuracy. Deflection of the probe is still a problem, but consistent calibration of the probe against a master standard by one of several methods helps keep this problem in check.

Probe calibration determines the effective working diameter of your stylus, which compensates for probe design. Measurement accuracy depends on careful calibration of the probe stylus. The probe stylus is used to measure a very accurate sphere to determine the probe tip's center and radius. When the probe contacts the workpiece, the coordinates of the tip are mathematically offset by the tip's radius. This calculation is performed automatically through the CMM's software.

The accuracy of your measurements with a touch probe also depends on using good probing techniques (see Figure 18–16). Effective probing techniques eliminate many common causes of measurement error.

Maximizing Probe Accuracy

Where possible, the following selections can increase the accuracy of measurements.

- Use a short stylus stiff to minimize bending. Use the lowest probing force that will not cause a false trigger.
- Calibrate to a known standard—sphere, ring gage, or gage blocks.
- Probe measurements should be made perpendicular to the workpiece.
- Ensure that the probe tip, not the shank, is making contact with the part.
- Make sure the stylus is not loose.

FIGURE 18–16 Proper style selection can reduce error.

Noncontact-Type Probes

Because they are suitable for most measurement situations—and because they are faster and require less operator skill—touch probes are the most widely used. There are some situations, however, in which they cannot be used—for example, the inspection of printed circuit boards, or in cases like measuring a clay or wax model, when the object being measured would be deformed by the force of the stylus.

One type of noncontact probe is a laser scanning probe (see Figure 18–17). Laser probes project a light beam onto the surface of a part. The distance from the point of measurement is known as the "standoff." When the light beam is triggered, the position of the beam is

FIGURE 18–17 The original objective of the noncontact probe was the measurement of soft or fragile surfaces, such as this wax prototype of an automobile. However, the great speeds at which it can operate make it valuable for inspection in which a large number of measurements must be taken, such as digitizing complex profiles. *(Courtesy of Renishaw Plc)*

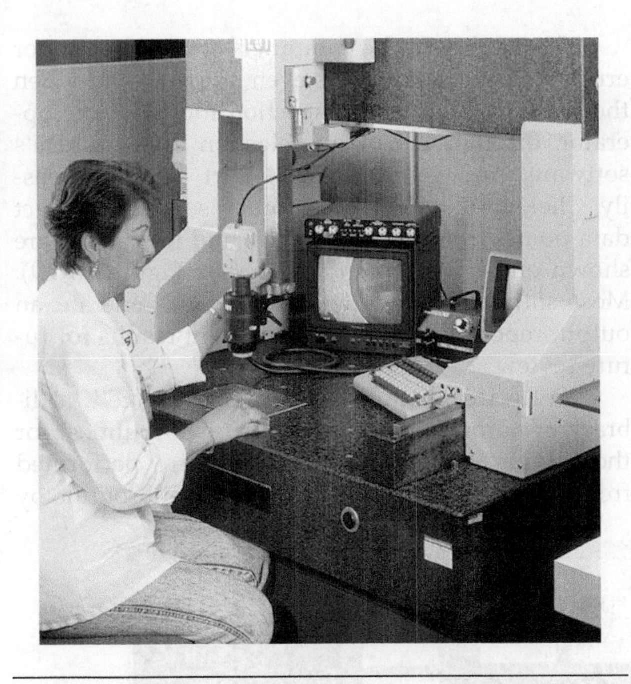

FIGURE 18–18 A video probe is added to the Z-axis arm for the inspection of printed circuit boards. *(Courtesy of CE Johansson)*

read by triangulation through a lens in the probe receptor. Laser tools have a high degree of speed and accuracy. A laser scanning probe is capable of 200 readings per second, when the surface has suitable contrast, and has a resolution of 1.27 μm (0.00005 in.), which is suitable for useful measurements to 2.54 μm (±0.0001 in.).

Another type of noncontact probe is the video probe (see Figure 18–18). Video probes rely on a high-resolution image of the part. This image is magnified and electronically digitized. The features are measured by computer "count" of the pixels of the electronic image. The camera is capable of generating a multitude of measurement points within a single video frame.

Direct Computer Control

You can increase versatility, convenience, and reliability of production when you let the computer control the processes. DCC, of course, is only applicable to motorized CMMs, and these machines are very similar to computer numerically controlled (CNC) machine tools. Because both the control and measurement cycles are handled by the computer, we refer to the direct computer control of manufacturing as CAM, or computer-assisted manufacturing.

The ability to provide this control was quickly recognized throughout industry. Today, CMM producers provide their own control packages, and computer manufacturers have also developed controls for CMMs. Software companies have also gotten into the CMM game: they provide for CMM suppliers, computer suppliers, and also specially designed programs for the manufacturers.

A program for DCC machines has three components:

- Move commands, which direct the probe to the data-collection points
- Measurement commands, which compare the distance traveled with the standard built into the machine for that axis
- Formatting commands, which translate the data into a form for display or printout

CMM Software

The programming of the machine or the software of the system enables a CMM to reach its full potential for accuracy, precision, and speed. Today's software is so highly developed that no one on the manufacturer's side has to do any programming, and because all CMM software is menu driven, the program basically will ask you, as the operator, what the end results you want are. In addition, software is available exclusively for, or that incorporates, statistical process analysis and control. Contour programs allow the CMM to quickly define detailed, complex nongeometric shapes without straight edges such as gears, cams, and injection molds. These programs also can be used to compare the measurement data with a computer-aided design (CAD) model.

A software package should allow any operator at any skill level to interface with the computer quickly and efficiently. Generally, software packages contain some or all of the following capabilities:

- Resolution selection
- Conversion between SI and English (millimeters and inches)

- Conversion of rectangular coordinates to polar coordinates
- Axis scaling
- Datum selection and reset
- Circle center and diameter solution
- Bolt-circle center and diameter
- Save and recall previous datum
- Nominal and tolerance entry
- Out-of-tolerance computation

Part Programs

Many of the benefits that manufacturers achieve when they use CMMs can actually be traced back to the computer that is part of the system.

The ability of the computer to reduce operator errors caused by fatigue is even more evident when there are programmed inspection routines. The operator, through the menu selection of the CMM's software, can generate CMM part programs easily. Click an icon, drive the probe sensor to collect data points from the workpiece, and the results are shown on the computer screen (see Figure 18–19). Most software allows an operator to generate an output report and store measurement results for future review.

Software manufacturers have created a library of automatic program steps (subroutines) for those types of measurements that are performed frequently. An operator can call up a subroutine by

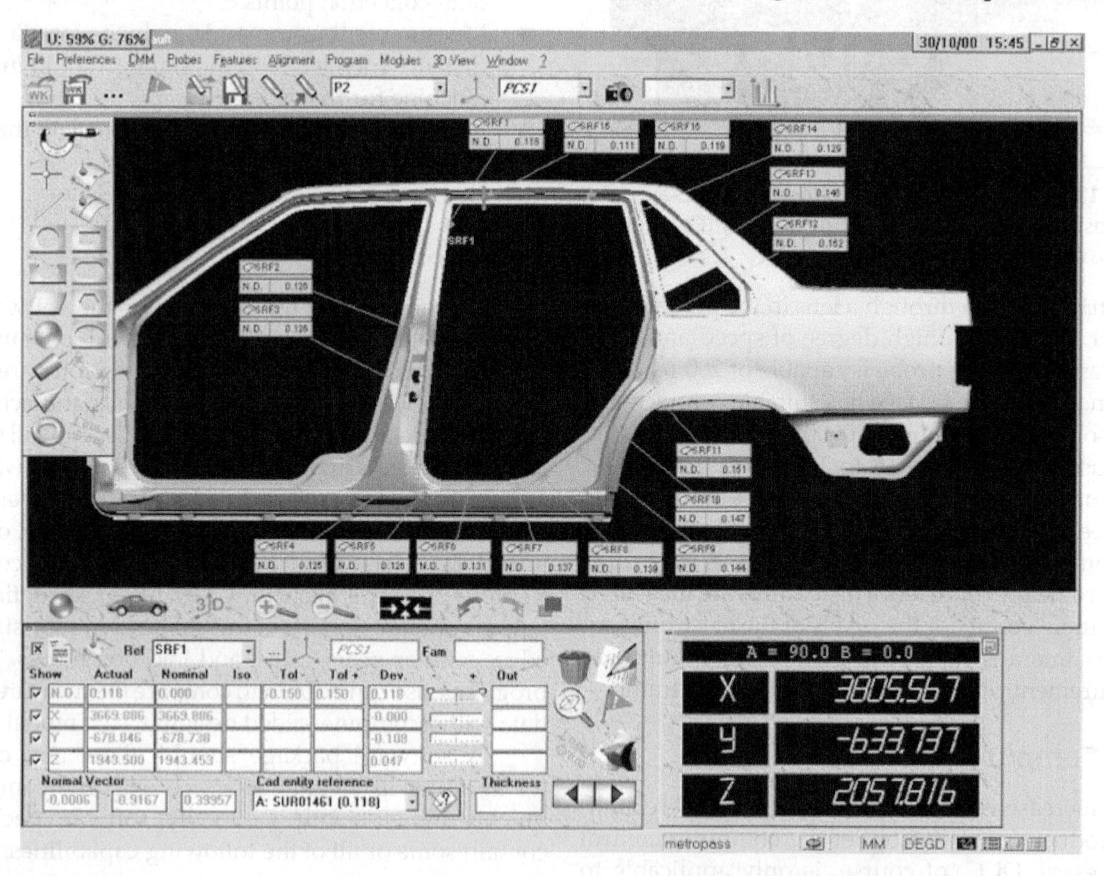

FIGURE 18–19 Software programs show a graphical image of the part and the measurement results on screen. *(Courtesy of CE Johansson)*

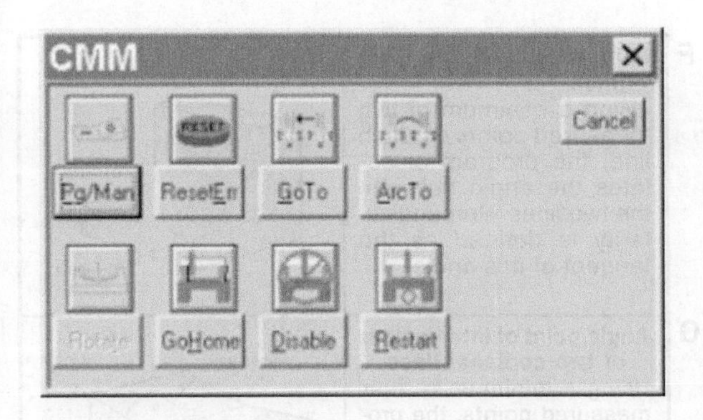

FIGURE 18–20 Measurements can be programmed with the click of an icon. *(Courtesy of Mitutoyo American Corporation)*

simply clicking the appropriate icon or menu choice (see Figures 18–20 and 18–21).

In the following examples of subroutines, you should first note that we only state the minimum number of points according to Euclidean geometry. By using so few points, we are assuming "perfection"—which never really exists; therefore, when we inspect, we are looking for the "best-fit" condition. For example, you only need to input three points into the computer to establish the center and diameter of a circle (see Figure 18–22A), but more are better because, to some degree, these points will depart from a true circle, which results in errors in both the center location and diameter size. Fortunately, it is easy to input data from other collection

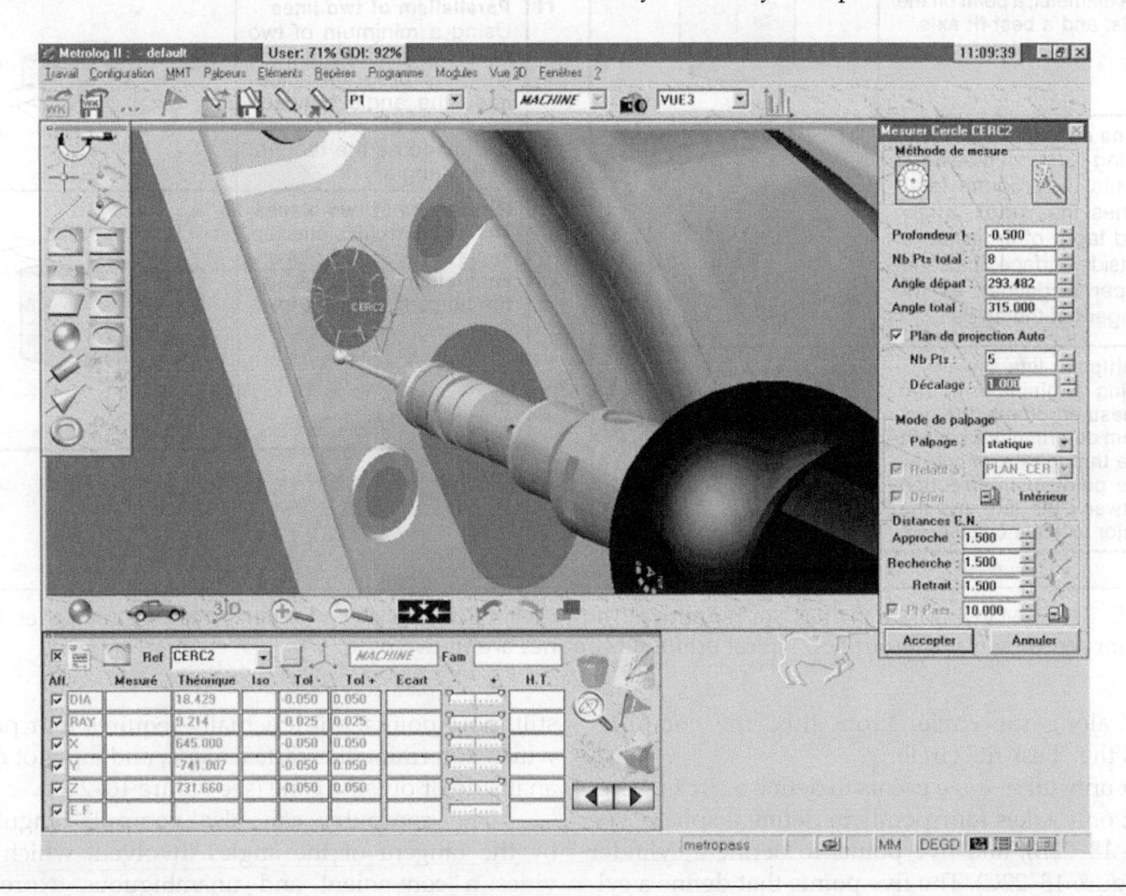

FIGURE 18–21 Inspection programs can be developed from CAD models, and run in "virtual" mode. An image of the probe movements is used to optimize the probe path and eliminate potential errors. *(Courtesy of CE Johansson)*

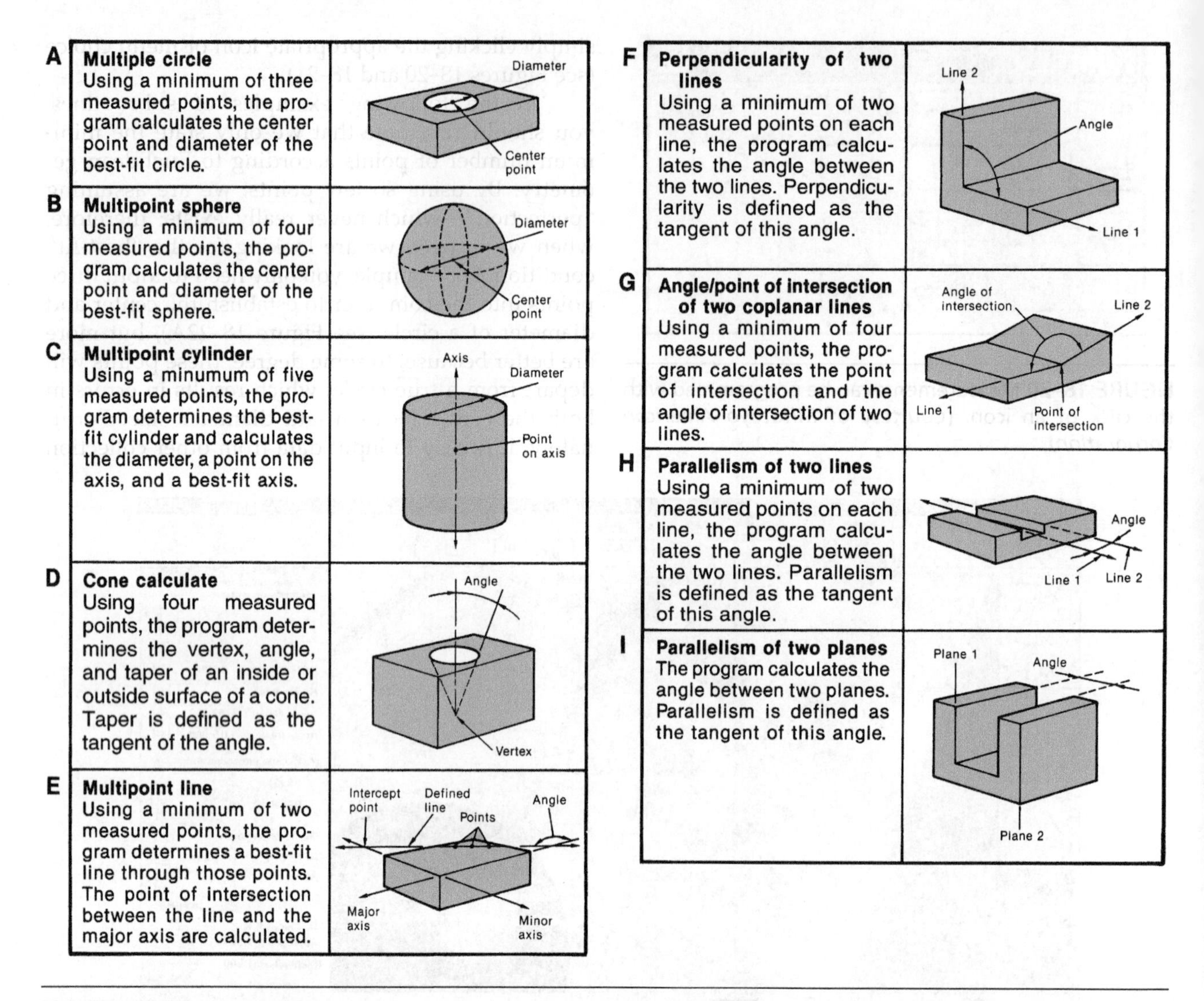

FIGURE 18–22 Many CMM activities are repetitive. Thus, their calculations can be performed by computer. These programs are known as subroutines. Typical useful subroutines are shown here.

points along the circle. From this, the computer selects the "best-fit" circle.

It only takes three points to define a circle; therefore, it only takes four points to define a sphere (see Figure 18–22B), and five points to define a cylinder (see Figure 18–22C). The five points that define a cylinder determine its size and its axis. You might expect that a complex figure such as a cone would require still more points, but it actually requires four points, which determine the vertex, angle, and taper of either an inside or outside cone (see Figure 18–22D).

The computer can also compute angularity by the tangent of the angles involved, which provides a convenient and unambiguous expression for such geometric relationships as parallelism and perpendicularity.

As you might expect, we only need two points to identify a line (see Figure 18–22E); however, any additional points allow the computer to identify the "best-fit" line, which can be very different than the two-point line we generate. Unlike plane geometry, the CMM also determines the direction of the line, which, in turn, provides the point of intersection between the line and the minor axis and the angle between the line and the major axis. If you were performing these calculations for a conventional plate, you might use several instruments and waste many hours and pages of paper.

When we are working with practical planes, a CMM with a computer can readily determine the "best-fit" plane from three points, and the computer can also determine the relationships among planes, which are common (see Figure 18–22F). The most common plane relationship—perpendicularity or squareness—requires only two points on each plane, and then the computer quickly compares the tangent of the angle with infinity. Another plane relationship—two planes that intersect (see Figure 18–22G)—is determined by four points, which establish the point of intersection and angle of intersection.

To determine the parallelism of two lines, we also need two points on each line, and we can use those points to define the deviation from parallelism in the plane of the lines using the tangent of the angle between them (see Figure 18–22H). You can use this principle to determine parallelism between planes (see Figure 18–22I). Other useful subroutines are shown in Figure 18–23.

Operational Modes

Manual Mode. In general, the manual mode lets us perform calculations required for single measurement applications—or "one-off" parts. You use the keyboard or mouse to select the program you need from a menu. The CMM obtains the measurement data, and a menu-driven program in the computer provides the measurements in a form you can use.

Automatic Mode. In the automatic mode, the CMM and its computer use a previously written program

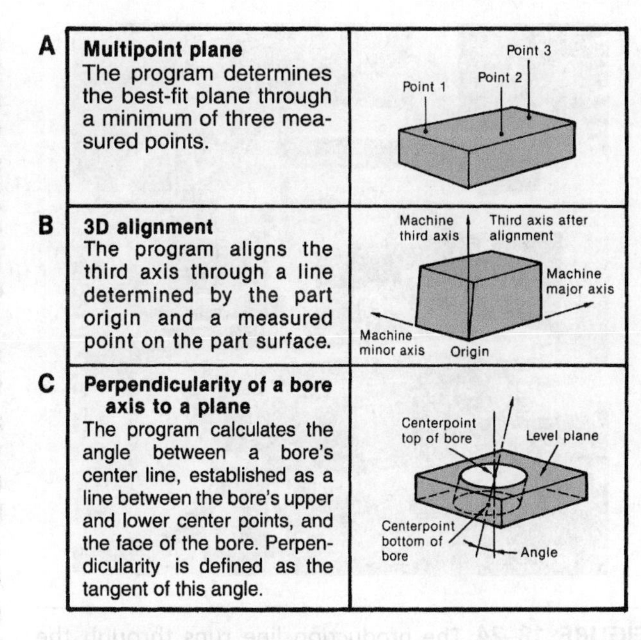

FIGURE 18–23 Subroutines can be prepared for nearly any measurement requirement such as these.

for the particular measurement task. The computer provides the measurements in the proper sequence, and then uses the data to compute the results. For a DCC machine, the program also drives the machine to the data collection points.

Programming Mode. In the programming mode, you can perform "real-time" programming, which is often referred to in this field as "self-teach." You "walk" the machine through the first inspection sequence, and then place that sequence in memory. You can repeat or correct this sequence as needed. You can also give the computer offline instructions in this mode, in much the same way you would program a numerically controlled (NC) machine tool.

Statistical Analysis Mode. You should only use the "batch" mode when you are producing large numbers of similar parts, because this mode only calls attention to trends that may lead to out-of-control conditions. We do not generally use CMMs for large production runs; however, a CMM's great speed does make them

FIGURE 18–24 The production line runs through the CMM, providing "real-time" inspection data that can be used to adjust the manufacturing processes. *(Courtesy of CE Johansson)*

useful for complex parts with many features that you must control. You can quickly adjust the machining center without continuing to produce faulty parts or leaving the machine idle while you are inspecting one part fully (see Figure 18–24).

18–5 METROLOGICAL FEATURES

CMM manufacturers claim repeatability to, or in excess of, +2.54 μm (+0.0001 in.) and accuracy of +10.16 μm (+0.0004 in.); however, these values decrease as the size of the CMM increases. These exceptional repeatability and accuracy capabilities might be surprising, considering that the concept of the coordinate measuring machine is in no way based in the principles of metrology. Designers and manufacturers take care that their machines, through their function, fulfill the fundamental principles, however, and features like system rigidity and the capability of the probe system make their success possible.

CMMs reduce computational errors, but this reduction must be balanced with any errors that we might find in the programming. Still, it is hard to beat the advantages of a CMM. When you manually measure, you add systematic error because of the method you use, even though you may not even recognize the errors are there. With the CMM, you can check the procedure without the workpiece, which will reveal errors before the actual measurements are made.

CMMs probably do not diminish thermal errors by being good heat sinks. In fact, they may add thermal errors because their mechanical operations generate heat. CMMs also use a great deal of granite, which is a poor conductor of heat. Though it takes a time to warm up, granite also does not release the heat buildup quickly.

CMM structure and part measurement can be affected by thermal expansion (the amount of change in a part due to variations in temperature). Thermal expansion can be caused by seasonal temperature changes or by day-to-day factors, such as sunlight, heating and cooling ducts, and the operation of nearby machining operations that produce heat. Because of this, conventional practice has been to limit the use of a CMM to a temperature- and humidity-controlled environment.

Manufacturing demands for higher quality and faster throughput have resulted in a desire for inspection equipment that works on the shop floor (see Figure 18–25). CMM manufacturers are addressing this need by using construction materials with a low coefficient of expansion and integrating temperature sensors in critical locations.

One material with a low coefficient of expansion that is being incorporated by manufacturers is a nickel alloy material called Invar, a composite of 64% iron and 36% nickel. Invar was developed by Swiss physicist Charles Edouard Guillaume in 1896 to improve the accuracy of precision clocks.

This is similar to the process used by civil engineers to compensate for the expansion and contraction of bridges. Transducers measure the growth of the axes due to temperature changes. The sensors measure the growth in microns, and feed back this information to the controller several times a minute. The mathematical algorithms can then compensate

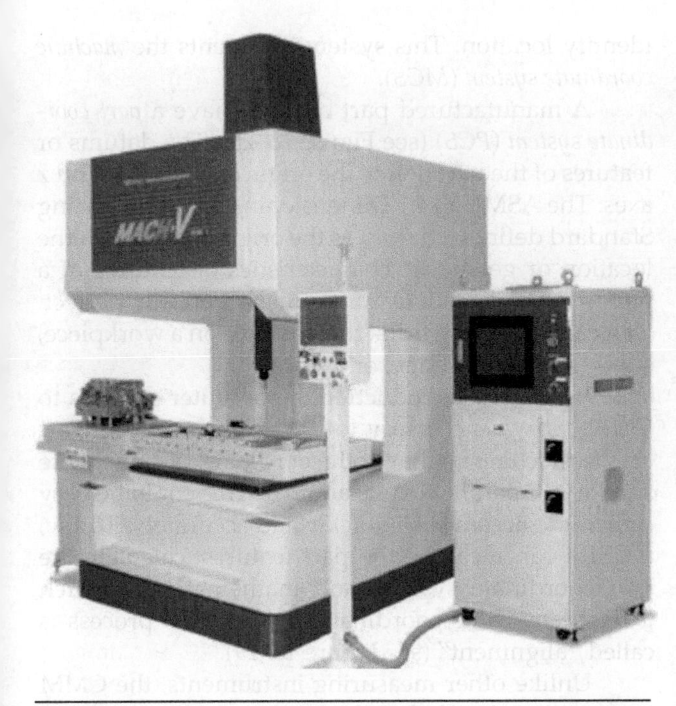

FIGURE 18–25 A high-speed CMM can be placed on the shop floor. *(Courtesy of Mitutoyo American Corporation)*

for these temperature changes as they occur (see Figure 18–26).

Similarly, you need to be concerned about vibration. Instead of damping vibrations, some CMMs may be receptive to them.

Coordinate measuring machines do not adhere to Abbe's law because the standard is never in line with the line of measurement. In fact, the standard is often far more removed from the line of measurement than it would be with conventional measurement instruments. The essential difference between using conventional instruments and a CMM is that you are responsible for making sure that the line of measurement is precisely parallel to the axis of the standard when you are using a conventional instrument. In contrast, this alignment is built into the CMM.

Coordinate Systems

A *coordinate system* allows the CMM to locate features on a workpiece relative to other features. The coordinate system is similar to a three-dimensional map, providing direction and location information.

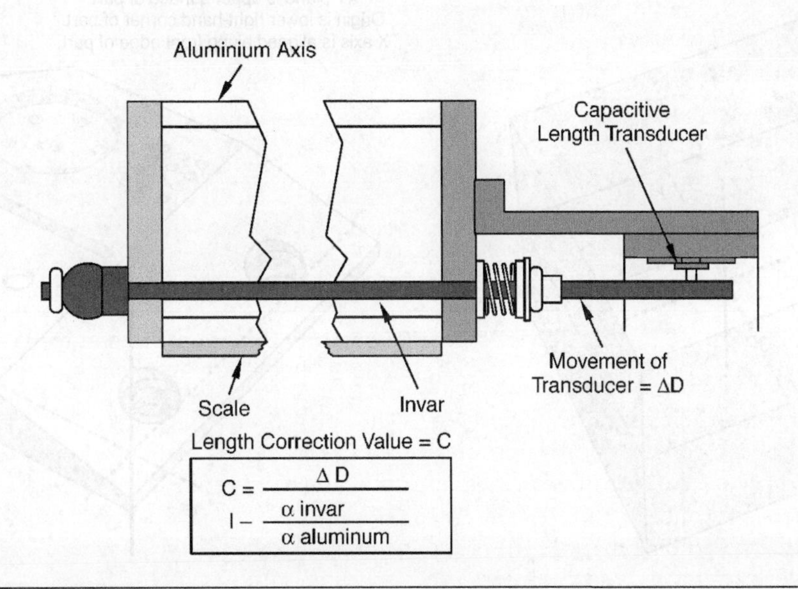

FIGURE 18–26 Temperature sensors and the incorporation of stable, low-expansion materials such as Invar reduce the effects of ambient temperature changes on the inspection process.

To eat dinner at Mark's World-Famous BarBQue, you would need to know which direction to travel and how far. If your starting point was home (the origin), you would need to travel to block 4, then turn and travel to block B, and then travel up to the 7th floor (see Figure 18–27).

A coordinate measuring machine uses these coordinates in much the same way. Each machine has a "home" position (an origin), and x, y, and z axes identify location. This system represents the *machine coordinate system* (*MCS*).

A manufactured part can also have a *part coordinate system* (*PCS*) (see Figure 18–28). The datums or features of the part define the origin and the x, y, and z axes. The ASME Y14.5 Dimensioning and Tolerancing Standard defines a datum as the origin from which the location or geometric characteristics of features of a part are established. Datums are theoretically perfect. Datum features are the actual features on a workpiece, such as a hole, an edge, or a surface.

Before the introduction of computer systems to CMMs, it was necessary to physically place the part on the machine so that both coordinate systems were square and parallel to one another. This could be very difficult to accomplish quickly and accurately. Today, a CMM can measure the part features, calculate the part coordinate system, and mathematically match it to the machine coordinate system. This process is called "alignment" (see Figure 18–29).

Unlike other measuring instruments, the CMM can measure itself; therefore, it has the inherent potential to check itself, too, which helps its reliability.

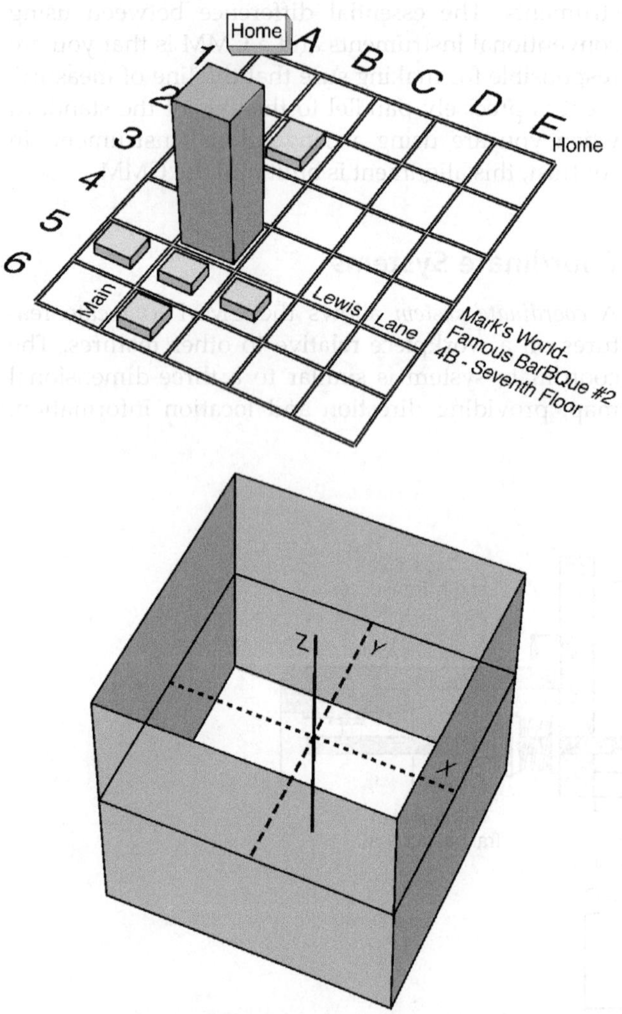

FIGURE 18–27 Maps give us direction and location, similar to a part coordinate system.

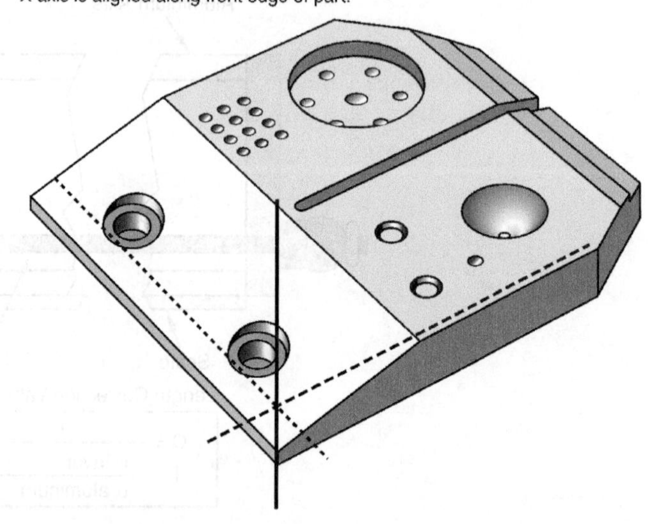

FIGURE 18–28 A part coordinate system identifies both the origin and the X, Y, and Z axes for a workpiece.

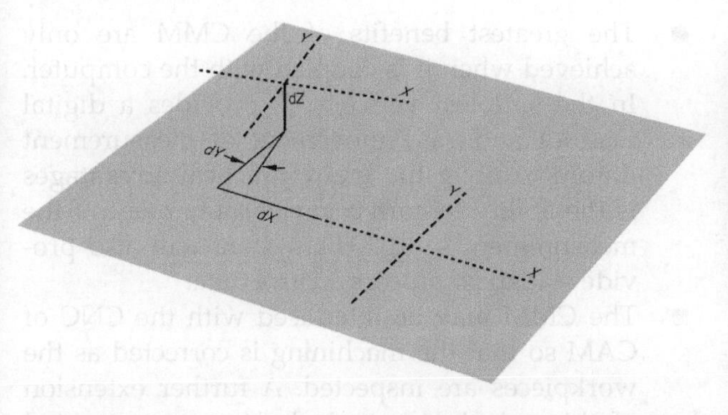

FIGURE 18–29 The alignment procedure matches our part coordinate system with the machine coordinates. Without this, we would need to perform complex mathematical calculations to determine the offsets from where our workpiece is in relation to the machine's origin.

You should not confuse self-checking with self-calibration, however. You can perform a systematic series of measurements on parts of the machine or use a test specimen to reveal errors in the standards and the alignment of the axes; however, even after meticulous analysis, there will always be some uncertainty in the amounts of the errors (calibration). You must compare these amounts with higher standards. From a practical standpoint, CMMs are usually calibrated against standards prepared specifically for their calibration.

Portable Arm Coordinate Measuring Machines

Sometimes a manufacturer will find it impractical or impossible to move a part or assembly to a stationary CMM. In those instances, a portable arm CMM can move to the part. Portable CMMs are an articulated measuring arm, with precision bearings and rotary transducers at each of its joints. At the arm's base is a mounting plate for direct attachment to a fixed surface or a portable tripod (see Figure 18–30). The arm's movement provides a basic spherical measurement envelope. The range of the arm can be extended through a process of "leapfrogging"—moving the arm to a new location and realigning to target datums.

FIGURE 18–30 The portable arm CMM can be taken to the workpiece or assembly. This CMM is combined with a laser scanning tool in order to inspect the geometry of the car door surface. *(Courtesy of CE Johansson)*

Manufacturers also offer a set of rails where the base can be attached. This extends the reach of the arm with little decrease in accuracy.

Portable arm CMMs are often used for reverse engineering applications, where prototype parts and assemblies are measured and the data collected are used to develop the CAD model or engineering documents. Although traditional CMMs are also used for reverse engineering, the portable arm has the additional advantage of portability. It can be taken to assemblies that are too large for the CMM's work envelope. Portable CMMs are not as accurate as the stationary CMM, measuring to an accuracy of 60.0008 inches. They can replace traditional, expensive hard gages on the shop floor, making them a very cost-effective solution.

18–6 FUTURE EXPECTATIONS

If we had to name one technological principle that is leading us into the next stage of the industrial revolution, we would probably choose "interactive response." Traditionally, we do something, then we analyze it, then we modify the original process before we move on to the next step. This system has been used in all endeavors from social science, to medicine, to educational systems, and, of course, throughout all manufacturing fields. Today, we use these same steps in total quality management (TQM) manufacturing systems. Interactive response lets us quickly discover problems and take action to deal with them before they grow larger.

Interaction, in short, means that we use the results to continuously monitor the process. Much of the development in this area is taking place in high production manufacturing. We can use interactive manufacturing and TQM—using continuous improvement techniques that modify actions based on their results—to produce high-quality products. The interactive combination of the CMM and CNC machining center will revolutionize any manufacturing operation where this combination is applied. Our ability to apply this combination determines the pace of progress.

The ultimate goal, however, is to interface CMM, CNC, and CAD. We could optimize the entire production process if we could create parts that could be adapted to the CAD-CNC-CMM team within functional parameters—which also could be computer controlled.

SUMMARY

- CMMs introduce no new principles of metrology. However, they utilize the fundamental principles to an extent not found in any other measurement instruments.
- The CMM is not a universal measurement device. It is applicable only to a relatively limited class of workpieces. These are chiefly those characterized by relatively small production, multiple features to be controlled, and high value.
- For workpieces that are applicable, the CMM reduces inspection time dramatically. However, in many cases, the greatest economic benefit is the reduction of downtime for machining while waiting for inspection to be completed.
- The greatest benefits of the CMM are only achieved when it is coupled with the computer. In the simplest versions, it provides a digital readout and easy movement of measurement datum. Among the many practical advantages is the ability to correct for misalignment of the measurement setup. The system will also provide statistical process control data.
- The CMM may be interfaced with the CNC of CAM so that the machining is corrected as the workpieces are inspected. A further extension of this principle may include computer-aided design and drafting (CADD).

END-OF-CHAPTER QUESTIONS

1. Which of the following is the most distinguishing feature of CMMs?
   a. Ability to measure large workpieces
   b. Ability to measure nearly all features for most workpieces at one setup
   c. Ability to both display and print the measurement digitally
   d. Ability to measure in three axes
2. Which of the following is most nearly correct?
   a. CMMs are based on an old concept that has now been rediscovered.
   b. CMMs are the latest innovation in metrology.
   c. CMMs extend back to the 1940s, but it has only been since the development of simplified computers that they have become prominent.
   d. CMMs were introduced in 1980 and reached prominence in 1985.
3. Which of the following is the most important advantage of the CMM?
   a. Greater accuracy than conventional instruments
   b. Greater versatility than conventional instruments
   c. Greater precision than conventional instruments
   d. Greater reliability than conventional instruments

4. For which of the following inspection situations does the CMM have the greatest economic benefits?
   a. Short production of relatively large workpieces with multiple features
   b. High production of parts that have both geometrical relationships and dimensional tolerances to be maintained
   c. All workpieces with multiple features in tolerance ranges between +0.001 and +0.0001 in.
   d. Workpieces of aluminum, magnesium, beryllium, and other lightweight metals

5. Which of the following characteristics are not used to define CMMs?
   a. Physical characteristics
   b. Type of built-in standard employed
   c. Size of measuring "envelope"
   d. Mode of operation

6. Which of the following best defines the envelope of a CMM?
   a. The sum of the x, y, and z axes
   b. The cubic space found by multiplying the maximum travel in the x, y, and z axes expressed in cubic centimeters
   c. The z axis times the area of the support surface
   d. The space bounded by the limits of travel in the x, y, and z axes

7. Which of the following CMMs is often called a universal measuring machine?
   a. Bridge type
   b. Bridge type with soft probe and computer assist
   c. Bridge type with soft probe, computer assist, and that is computer controlled
   d. Column type

8. Which of the following best describes the probe?
   a. The act of seeking the reference surfaces of features being measured
   b. The zero reference from which all measurements are taken
   c. The element that contacts the surfaces being examined
   d. The electronic gaging head

9. Which of the following best describes the difference between hard and soft probes?
   a. Soft probes have elastomer exteriors to prevent damage to work surfaces.
   b. Soft probes are spring mounted to the z axis.
   c. Hard probes provide direct readings; soft probes provide only indirect readings.
   d. Soft probes are electronic sensors.

10. Which of the following are the general modes for CMM operation?
    a. Manual, mechanically assisted manual, electrically powered, and computer controlled
    b. Manual, manual computer-assisted, motorized computer-assisted, and direct computer controlled
    c. Manual, manual digital, motorized, motorized digital, and direct computer controlled
    d. Manual, motorized, remote-controlled motorized digital readout, and computer-assisted digital readout

11. Which of the following can measure size and location simultaneously?
    a. Micrometers
    b. Coordinate measuring machines
    c. Vernier calipers
    d. Steel rule

12. CMM software packages include some or all of the following capabilities, except:
    a. Conversion between millimeters and inches
    b. Out-of-tolerance computation
    c. Datum selection
    d. Automatic detection of the part coordinate system

13. Which of the following is the accepted usage of "datum" in regard to CMMs?
    a. Data are the surfaces from which measurements are taken.
    b. Data are any numbers involved with the measurement act.

c. Datum is both ends of the part feature being measured.
d. Data refers to the angles between features.
e. Datum is the reference from which measurements are taken.

14. Which of the following is the most valid statement about the measurement-function library?
    a. It enables trigonometric problems to be solved.
    b. It is the storage facility for all of the software for one or more CMMs.
    c. It is the glossary section of the manual furnished with the particular CMM.
    d. It contains the programs that are in the machine's memory.

15. What type of geometry does a computer-assisted type of CMM utilize?
    a. Boolean
    b. Cartesian
    c. Euclidean
    d. Stochastic

16. What is the chief advantage of a computer built into a CMM in comparison to having separate units?
    a. Less floor space required
    b. Less error from data transference
    c. Separate computer operator not required
    d. No competition for computer time

APPENDIX A

DECIMAL EQUIVALENTS

Fractional: •Wire ••Metric xxLetter Sizes

Drill Size	Decimal	Drill Size	Decimal	Drill Size	Decimal	Drill Size	Decimal	Drill Size	Decimal
••0.10	0.0039	•70	0.0280	•52	0.0635	7/64	0.1094	•17	0.1730
•97	0.0059	•69	0.0292	••1.65	0.0650	•35	0.1100	••4.40	0.1732
•96	0.0063	••0.75	0.0295	••1.70	0.0669	••2.80	0.1102	•16	0.1770
•95	0.0067	•68	0.0310	•51	0.0670	•34	0.1110	••4.50	0.1772
•94	0.0071	1/32	0.0313	••1.75	0.0689	•33	0.1130	•15	0.1800
•93	0.0075	••0.80	0.0315	•50	0.0700	••2.90	0.1142	••4.60	0.1811
•92	0.0079	•67	0.0320	••1.80	0.0709	•32	0.1160	•14	0.1820
••0.20	0.0079	•66	0.0330	••1.85	0.0728	••3.00	0.1181	•13	0.1850
•91	0.0083	••0.85	0.0335	•49	0.0730	•31	0.1200	••4.70	0.1850
•90	0.0087	•65	0.0350	••1.90	0.0748	••3.10	0.1220	3/16	0.1875
•89	0.0091	••0.90	0.0354	•48	0.0760	1/8	0.1250	•12	0.1890
•88	0.0095	•64	0.0360	••1.95	0.0768	••3.20	0.1260	••4.80	0.1890
•87	0.0100	•63	0.0370	5/64	0.0781	•30	0.1285	•11	0.1910
•86	0.0105	••0.95	0.0374	•47	0.0785	••3.30	0.1299	••4.90	0.1929
•85	0.0110	•62	0.0380	••2.00	0.0787	••3.40	0.1339	•10	0.1935
•84	0.0115	•61	0.0390	••2.05	0.0807	•29	0.1360	•9	0.1960
••0.30	0.0118	••1.00	0.0394	•46	0.0810	••3.50	0.1378	••5.00	0.1968
•83	0.0120	•60	0.0400	•45	0.0820	•28	0.1405	•8	0.1990
•82	0.0125	•59	0.0410	••2.10	0.0827	9/64	0.1406	••5.10	0.2008
•81	0.0130	••1.05	0.0413	••2.15	0.0846	••3.60	0.1417	•7	0.2010
•80	0.0135	•58	0.0420	•44	0.0860	•27	0.1440	13/64	0.2031
••0.35	0.0138	•57	0.0430	••2.20	0.0866	••3.70	0.1457	•6	0.2040
•79	0.0145	••1.10	0.0433	••2.25	0.0886	•26	0.1470	••5.20	0.2047
1/64	0.0156	••1.15	0.0453	•43	0.0890	•25	0.1495	•5	0.2055
••0.40	0.0158	•56	0.0465	••2.30	0.0906	••3.80	0.1496	••5.30	0.2087
•78	0.0160	3/64	0.0469	••2.35	0.0925	•24	0.1520	•4	0.2090
••0.45	0.0177	••1.20	0.0472	•42	0.0935	••3.90	0.1535	••5.40	0.2126
•77	0.0180	••1.25	0.0492	3/32	0.0938	•23	0.1540	•3	0.2130
••0.50	0.0197	••1.30	0.0512	••2.40	0.0945	5/32	0.1562	••5.50	0.2165
•76	0.0200	•55	0.0520	•41	0.0960	•22	0.1570	7/32	0.2188
•75	0.0210	••1.35	0.0531	••2.45	0.0965	••4.00	0.1575	••5.60	0.2205
••0.55	0.0217	•54	0.0550	•40	0.0980	•21	0.1590	•2	0.2210
•74	0.0225	••1.40	0.0551	••2.50	0.0984	•20	0.1610	••5.70	0.2244
••0.60	0.0236	••1.45	0.0571	•39	0.0995	••4.10	0.1614	•1	0.2280
•73	0.0240	••1.50	0.0591	•38	0.1015	••4.20	0.1654	••5.80	0.2283
•72	0.0250	•53	0.0595	••2.60	0.1024	•19	0.1660	••5.90	0.2323

DECIMAL EQUIVALENTS (*CONTINUED*)

Fractional: •Wire ••Metric xxLetter Sizes

Drill Size	Decimal	Drill Size	Decimal	Drill Size	Decimal	Drill Size	Decimal	Drill Size	Decimal
••0.65	0.0256	••1.55	0.0610	•37	0.1040	••4.30	0.1693	xx A	0.2340
•71	0.0260	1/16	0.0625	••2.70	0.1063	•18	0.1695	15/64	0.2344
••0.70	0.0276	••1.60	0.0630	•36	0.1065	11/64	0.1719	••6.00	0.2362
xx B	0.2380	19/64	0.2969	••9.40	0.3701	••12.50	0.4921	3/4	0.7500
••6.10	0.2402	••7.60	0.2992	••9.50	0.3740	1/2	0.5000	49/64	0.7656
xx C	0.2420	xx N	0.3020	3/8	0.3750	••13.00	0.5118	••19.50	0.7677
••6.20	0.2441	••7.70	0.3031	xx V	0.3770	33/64	0.5156	25/32	0.7812
xx D	0.2460	••7.80	0.3071	••9.60	0.3780	17/32	0.5313	••20.00	0.7874
••6.30	0.2480	5/16	0.3125	••9.70	0.3819	••13.50	0.5315	51/64	0.7969
1/4	0.2500	••8.00	0.3150	••9.80	0.3858	35/64	0.5469	••20.50	0.8071
xx E	0.2500	xx O	0.3160	xx W	0.3860	••14.00	0.5512	13/16	0.8125
••6.40	0.2520	••8.10	0.3189	••9.90	0.3898	9/16	0.5625	••21.00	0.8268
••6.50	0.2559	••8.20	0.3228	25/64	0.3906	••14.50	0.5707	53/64	0.8281
xx F	0.2570	xx P	0.3230	••10.00	0.3937	37/64	0.5781	27/32	0.8438
••6.60	0.2598	••8.30	0.3268	xx X	0.3970	••15.00	0.5906	••21.50	0.8465
xx G	0.2610	21/64	0.3281	••10.20	0.4016	19/32	0.5938	55/64	0.8594
••6.70	0.2638	••8.40	0.3307	xx Y	0.4040	39/64	0.6094	••22.00	0.8661
17/64	0.2656	xx Q	0.3320	••10.30	0.4058	••15.50	0.6102	7/8	0.8750
xx H	0.2660	••8.50	0.3346	13/32	0.4062	5/8	0.6250	••22.50	0.8858
••6.80	0.2677	••8.60	0.3386	xx Z	0.4130	••16.00	0.6299	57/64	0.8906
••6.90	0.2717	xx R	0.3390	••10.50	0.4134	41/64	0.6406	••23.00	0.9055
xx I	0.2720	••8.70	0.3425	27/64	0.4219	••16.50	0.6496	29/32	0.9062
••7.00	0.2756	11/32	0.3438	••10.80	0.4252	21/32	0.6562	59/64	0.9219
xx J	0.2770	••8.80	0.3465	••11.00	0.4331	••17.00	0.6693	••23.50	0.9252
••7.10	0.2795	xx S	0.3480	7/16	0.4375	43/64	0.6719	15/16	0.9375
xx K	0.2810	••8.90	0.3504	••11.20	0.4409	11/16	0.6875	••24.00	0.9449
9/32	0.2812	••9.00	0.3543	••11.50	0.4528	••17.50	0.6890	61/64	0.9531
••7.20	0.2835	xx T	0.3580	29/64	0.4531	45/64	0.7031	••24.50	0.9646
••7.30	0.2874	••9.10	0.3583	••11.80	0.4646	••18.00	0.7087	31/32	0.9688
xx L	0.2900	23/64	0.3594	15/32	0.4688	23/32	0.7188	••25.00	0.9842
••7.40	0.2913	••9.20	0.3622	••12.00	0.4724	••18.50	0.7284	63/64	0.9844
xx M	0.2950	••9.30	0.3661	••12.20	0.4803	47/64	0.7344	1	1.0000
••7.50	0.2953	xx U	0.3680	31/64	0.4844	••19.00	0.7480		

APPENDIX B

CONVERSION TABLE

INCHES TO MILLIMETERS (mm)

Inches	mm	Inches	mm	Inches	mm	Inches	mm	Inches	mm
0.001	0.0254	0.041	1.0414	0.081	2.0574	0.310	7.8740	0.710	18.0340
0.002	0.0508	0.042	1.0668	0.082	2.0820	0.320	8.1280	0.720	18.2880
0.003	0.0762	0.043	1.0922	0.083	2.1082	0.330	8.3820	0.730	18.5420
0.004	0.1016	0.044	1.1176	0.084	2.1336	0.340	8.6360	0.740	18.7960
0.005	0.1270	0.045	1.1430	0.085	2.1590	0.350	8.8900	0.750	19.0500
0.006	0.1542	0.046	1.1684	0.086	2.1844	0.360	9.1440	0.760	19.3040
0.007	0.1778	0.047	1.1938	0.087	2.2098	0.370	9.3980	0.770	19.5580
0.008	0.2032	0.048	1.2192	0.088	2.2352	0.380	9.6520	0.780	19.8120
0.009	0.2286	0.049	1.2446	0.089	2.2606	0.390	9.9060	0.790	20.0660
0.010	0.2540	0.050	1.2700	0.090	2.2860	0.400	10.1600	0.800	20.3200
0.011	0.2794	0.051	1.2954	0.091	2.3114	0.410	10.4140	0.810	20.5740
0.012	0.3048	0.052	1.3208	0.092	2.3368	0.420	10.6680	0.820	20.8280
0.013	0.3302	0.053	1.3462	0.093	2.3622	0.430	10.9220	0.830	21.0820
0.014	0.3556	0.054	1.3716	0.094	2.3876	0.440	11.1760	0.840	21.3360
0.015	0.3810	0.055	1.3970	0.095	2.4130	0.450	11.4300	0.850	21.5900
0.016	0.4064	0.056	1.4224	0.096	2.4384	0.460	11.6840	0.860	21.8440
0.017	0.4318	0.057	1.4478	0.097	2.4638	0.470	11.9380	0.870	22.0980
0.018	0.4572	0.058	1.4732	0.098	2.4892	0.480	12.1920	0.880	22.3520
0.019	0.4826	0.059	1.4986	0.099	2.5146	0.490	12.4460	0.890	22.6060
0.020	0.5080	0.060	1.5240	0.100	2.5400	0.500	12.7000	0.900	22.8600
0.021	0.5334	0.061	1.5494	0.110	2.7940	0.510	12.9540	0.910	23.1140
0.022	0.5588	0.062	1.5748	0.120	3.0480	0.520	13.2080	0.920	23.3680
0.023	0.5842	0.063	1.6002	0.130	3.3020	0.530	13.4620	0.930	23.6220
0.024	0.6096	0.064	1.6256	0.140	3.5560	0.540	13.7160	0.940	23.8760
0.025	0.6350	0.065	1.6510	0.150	3.8100	0.550	13.9700	0.950	24.1300
0.026	0.6604	0.066	1.6764	0.160	4.0640	0.560	14.2240	0.960	24.3840
0.027	0.6858	0.067	1.7018	0.170	4.3180	0.570	14.4780	0.970	24.6380
0.028	0.7112	0.068	1.7272	0.180	4.5720	0.580	14.7320	0.980	24.8920
0.029	0.7366	0.069	1.7526	0.190	4.8260	0.590	14.9860	0.990	25.1460
0.030	0.7620	0.070	1.7780	0.200	5.0800	0.600	15.2400	1.000	25.4000
0.031	0.7874	0.071	1.8034	0.210	5.3340	0.610	15.4940	2.000	50.8000
0.032	0.8128	0.072	1.8288	0.220	5.5880	0.620	15.7734	3.000	76.2000
0.033	0.8382	0.073	1.8542	0.230	5.8420	0.630	16.0020	4.000	101.6000
0.034	0.8636	0.074	1.8796	0.240	6.0960	0.640	16.2560	5.000	127.0000
0.035	0.8890	0.075	1.9050	0.250	6.3500	0.650	16.5100	6.000	152.4000
0.036	0.9144	0.076	1.9304	0.260	6.6040	0.660	16.7640	7.000	177.8000
0.037	0.9398	0.077	1.9558	0.270	6.8580	0.670	17.0180	8.000	203.2000
0.038	0.9652	0.078	1.9812	0.280	7.1120	0.680	17.2720	9.000	228.6000
0.039	0.9906	0.079	2.0066	0.290	7.3660	0.690	17.5260	10.000	254.0000
0.040	1.0160	0.080	2.0320	0.300	7.6200	0.700	17.7800		

MILLIMETERS (mm) TO INCHES

mm	Inches	mm	Inches	mm	Inches	mm	Inches	mm	Inches
0.010	0.000394	0.530	0.020866	1.250	0.0492125	5.700	0.224409	10.900	0.429134
0.020	0.000787	0.540	0.021260	1.300	0.051181	5.800	0.228346	11.000	0.433071
0.030	0.001181	0.550	0.021654	1.350	0.053146	5.900	0.232283	11.100	0.437008
0.040	0.001575	0.560	0.022047	1.400	0.055118	6.000	0.236220	11.200	0.440945
0.050	0.001968	0.570	0.022441	1.450	0.0570865	6.100	0.240157	11.300	0.444882
0.060	0.002362	0.580	0.022835	1.500	0.059055	6.200	0.244094	11.400	0.448819
0.070	0.002756	0.590	0.023228	1.550	0.0610235	6.300	0.248031	11.500	0.452756
0.080	0.003150	0.600	0.023622	1.600	0.062992	6.400	0.251968	11.600	0.456693
0.090	0.003543	0.610	0.024016	1.650	0.0649605	6.500	0.255906	11.700	0.460630
0.100	0.003937	0.620	0.024409	1.700	0.066929	6.600	0.259842	11.800	0.464567
0.110	0.004331	0.630	0.024803	1.750	0.0688975	6.700	0.263780	11.900	0.468504
0.120	0.004724	0.640	0.025197	1.800	0.070866	6.800	0.267716	12.000	0.472441
0.130	0.005118	0.650	0.025591	1.850	0.0728345	6.900	0.271654	12.100	0.476378
0.140	0.005512	0.660	0.025984	1.900	0.074803	7.000	0.275591	12.200	0.480315
0.150	0.005906	0.670	0.026378	1.950	0.0767715	7.100	0.279528	12.300	0.484252
0.160	0.006299	0.680	0.026772	2.000	0.078740	7.200	0.283465	12.400	0.488189
0.170	0.006693	0.690	0.027166	2.100	0.082677	7.300	0.287402	12.500	0.492126
0.180	0.007087	0.700	0.027599	2.200	0.086614	7.400	0.281339	12.600	0.496063
0.190	0.007480	0.710	0.027953	2.300	0.090551	7.500	0.295276	12.700	0.500000
0.200	0.007874	0.720	0.028346	2.400	0.094488	7.600	0.299213	12.800	0.503937
0.210	0.008268	0.730	0.028740	2.500	0.098425	7.700	0.303150	12.900	0.507874
0.220	0.008661	0.740	0.029134	2.600	0.102362	7.800	0.307087	13.000	0.511811
0.230	0.009055	0.750	0.029528	2.700	0.106299	7.900	0.311024	13.200	0.519685
0.240	0.009449	0.760	0.029921	2.800	0.110236	8.000	0.314961	13.400	0.527559
0.250	0.009843	0.770	0.030315	2.900	0.114173	8.100	0.318898	13.600	0.535433
0.260	0.010236	0.780	0.030709	3.000	0.118110	8.200	0.322835	13.800	0.543307
0.270	0.010630	0.790	0.031102	3.100	0.122047	8.300	0.326772	14.000	0.551181
0.280	0.011024	0.800	0.031496	3.200	0.125984	8.400	0.330709	14.200	0.559055
0.290	0.011417	0.810	0.031890	3.300	0.129921	8.500	0.334646	14.400	0.566929
0.300	0.011811	0.820	0.032283	3.400	0.133858	8.600	0.338583	14.600	0.574803
0.310	0.012205	0.830	0.032677	3.500	0.137795	8.700	0.342520	14.800	0.582677
0.320	0.012598	0.840	0.033071	3.600	0.141732	8.800	0.346457	15.000	0.590551
0.330	0.012992	0.850	0.033465	3.700	0.145669	8.900	0.350394	15.200	0.598425
0.340	0.013386	0.860	0.033858	3.800	0.149606	9.000	0.354331	15.400	0.606299
0.350	0.013780	0.870	0.034252	3.900	0.153543	9.100	0.358268	15.600	0.614173
0.360	0.014173	0.880	0.034646	4.000	0.157480	9.200	0.362205	15.800	0.622047
0.370	0.014567	0.890	0.035039	4.100	0.161417	9.300	0.366142	16.000	0.629921
0.380	0.014961	0.900	0.035433	4.200	0.165354	9.400	0.370079	16.200	0.637795
0.390	0.015354	0.910	0.035827	4.300	0.169291	9.500	0.374016	16.400	0.645669
0.400	0.015748	0.920	0.036220	4.400	0.173228	9.600	0.377953	16.600	0.653543
0.410	0.016417	0.930	0.036614	4.500	0.177165	9.700	0.381890	16.800	0.661417
0.420	0.016535	0.940	0.037008	4.600	0.181102	9.800	0.385827	17.000	0.669291
0.430	0.016929	0.950	0.037402	4.700	0.185039	9.900	0.389764	17.200	0.677165
0.440	0.017323	0.960	0.037795	4.800	0.188976	10.000	0.393701	17.400	0.685039
0.450	0.017717	0.970	0.038189	4.900	0.192913	10.100	0.397638	17.600	0.692913
0.460	0.018110	0.980	0.038583	5.000	0.196850	10.200	0.401575	17.800	0.700787
0.470	0.018504	0.990	0.038976	5.100	0.200787	10.300	0.405512	18.000	0.708661
0.480	0.018898	1.000	0.039370	5.200	0.204724	10.400	0.409449	18.200	0.716535
0.490	0.019291	1.050	0.0413385	5.300	0.208661	10.500	0.413386	18.400	0.724409
0.500	0.019685	1.100	0.043307	5.400	0.212598	10.600	0.417323	18.600	0.732283
0.510	0.020079	1.150	0.0452755	5.500	0.216535	10.700	0.421260	18.800	0.740157
0.520	0.020472	1.200	0.047244	5.600	0.220472	10.800	0.425197	19.000	0.748031
19.200	0.755906	21.000	0.826772	22.800	0.897638	24.600	0.968504	26.400	1.039370
19.400	0.763780	21.200	0.834646	23.000	0.905512	24.800	0.976378	26.600	1.047244
19.600	0.711654	21.400	0.842520	23.200	0.913386	25.000	0.984252	26.800	1.055118
19.800	0.779528	21.600	0.850394	23.400	0.921260	25.200	0.992126	30.000	1.062992
20.000	0.787402	21.800	0.858268	23.600	0.929134	25.400	1.000000	30.200	1.070866
20.200	0.795276	22.000	0.866142	23.800	0.937008	25.600	1.007874	30.400	1.078740
20.400	0.803150	22.200	0.874016	24.000	0.944882	25.800	1.015748	30.600	1.086614
20.600	0.811024	22.400	0.881890	24.200	0.952756	26.000	1.023622	30.800	1.094488
20.800	0.818898	22.600	0.889764	24.400	0.960630	26.200	1.031496	31.000	1.102362

APPENDIX C

GEOMETRIC OPTICS

Note: This appendix is included as introductory and supplementary material for Chapters 13 and 18.

GEOMETRIC OPTICS

Waves and Rays

We have already discussed some aspects of optical metrology. This material is provided as additional information for the study of the optical lenses and optical phenomena.

We are concerned here with a certain portion of the wide spectrum of electromagnetic radiation. This radiation results from the movement of electrons from orbit to orbit. For each such migration, the frequency is constant but may range from very long wavelengths to very short. A portion of this spectrum can be perceived by living organisms. For humans and the higher animals this is achieved by the excitement of the retinas of their eyes.

Great contradictions separate our practical application of this phenomenon and our intellectual understanding of it. Physicists first recognized the wave characteristic of light but later became aware of its corpuscular character. The latter is embodied in *quantum mechanics.* It took Albert Einstein (1879–1955) to reconcile the two. These phenomena find practical application in interferometry and such beautifully simple techniques as measurement with optical flats (see Chapter 13).

In contrast to this ephemeral concept of light, optical magnification and optical alignment are based on the opposite characteristic. In these applications, we consider light to behave as *rays,* which are stable and true paths of *light* transmission. They do possess these characteristics, but they do not exist in nature. These characteristics are a human invention to aid in the practical applications of optics. Figure A shows that a ray is the line of travel of an advancing wave front. This concept provides a manageable tool for practical optics.

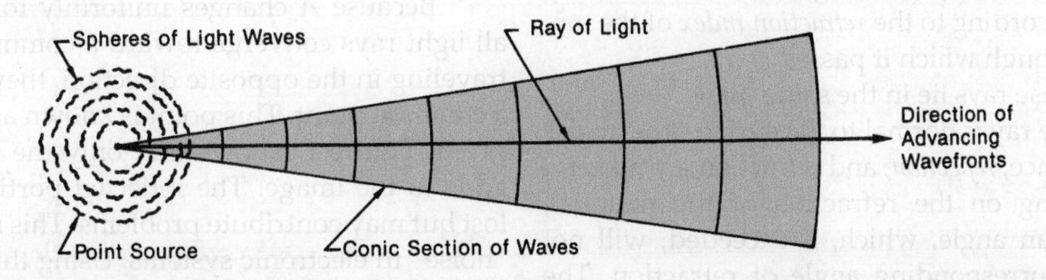

FIGURE A Light travels by wave fronts. Conic sections of the expanding spheres have such large radii they may be regarded as planes. A ray is the line of travel of the wave fronts.

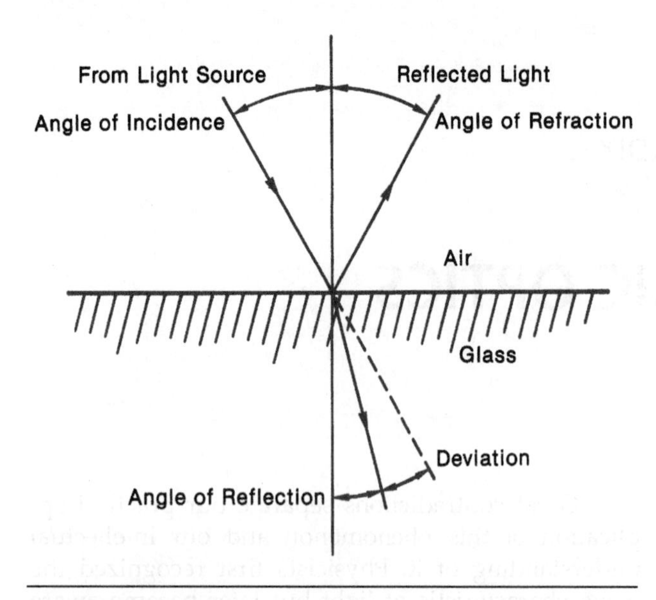

FIGURE B The angle of reflection is equal to the angle of incidence. The angle of refraction is governed by the difference in speed of light as it travels in each of the two media.

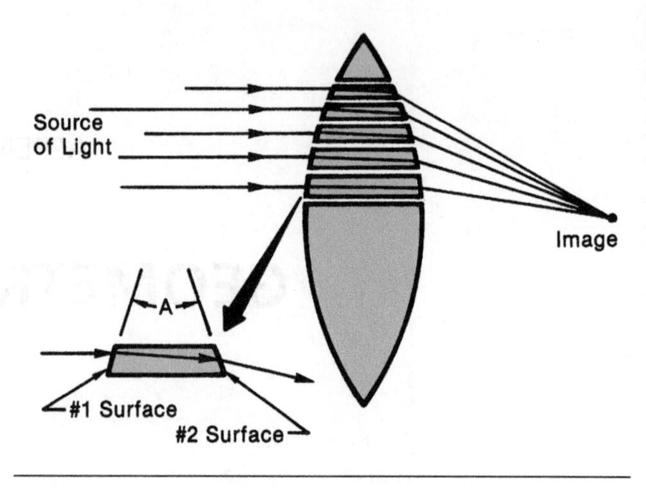

FIGURE C If a lens were sliced into very thin sections, the amount of curvature on each section would become negligible. Each section could then be considered as if it were a prism. As a result, the double refraction causes the rays from a distant source to converge to a point.

Geometrical Consideration

Light rays in a homogeneous medium such as air travel in straight lines. Their direction changes when they are reflected from a surface or refracted on entering another medium. They may also be scattered or absorbed. In many situations all of these actions occur concurrently. Figure B shows the relationship between reflection and refraction. It is important to note four facts:

1. Whereas the angle of incidence and the angle of reflection are equal, the angle of the refracted ray varies according to the *refraction index* of the material through which it passes.
2. All of these rays lie in the same plane.
3. When the ray is normal to the surface, the angles of incidence, *reflection*, and refraction are all zero.
4. Depending on the refraction of the material, there is an angle, which, if exceeded, will not have a corresponding angle of refraction. The ray will then be reflected internally. This is known as the *critical angle*.

For the sake of convenience let us consider *lens* to mean a piece of transparent material whose opposing surfaces are surfaces of revolution and highly polished. Let us now imagine that the lens is cut into very small slices as in Figure C. If these slices are thin enough, they can be considered to be without curvature, or to be prisms. Then if a light ray enters each of these sections, the following relationship expresses its path.

$$\delta = i_1 + i_2'$$

where:

δ (lowercase delta) is the deviation
i_1 is the angle of incidence of light on lens surface #1
i_2' is the angle of refraction of light on lens surface #2
A is the prism angle between surfaces #1 and #2

Because A changes uniformly for each section, all light rays converge toward a common point; or if traveling in the opposite direction, they diverge from a common point. This point is known as an *image*. The ray in Figure D shows that only the refracted light adds to the image. The reflected portion is not only lost but may contribute problems. This is analogous to "noise" in electronic systems. Using this basic behavior, the rays in even the most complex optical system may be plotted.

It should be noted that this discussion is based on a converging lens. This is a lens where one or both of the surfaces is convex. Lenses with concave surfaces

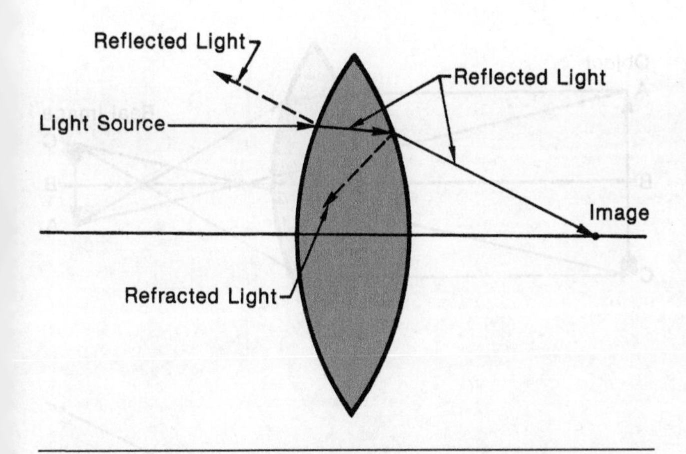

FIGURE D Only the refracted light contributes to the image. The reflected light is lost.

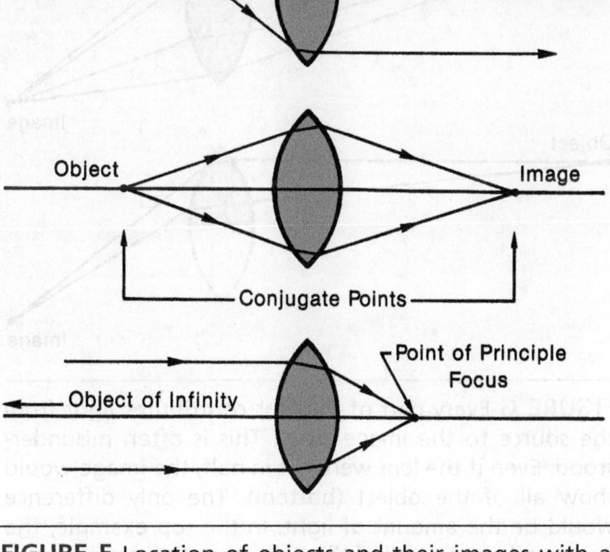

FIGURE F Location of objects and their images with a convergent lens. As the object becomes more distant, the image moves closer to the lens. The closest point is reached when the object is at infinity. That point is the primary focus.

are diverging lenses. Their ray paths are the reverse of those for converging systems, Figure E.

Figure F shows light rays starting at a common point and reuniting at another point on the opposite side of a convergent lens. The starting point is the *object* and the termination point is the image. Together these points are known as the *conjugate points.* For a convergent lens a *real image* is formed, whereas a divergent lens results in a *virtual image* (see Figure F). In both

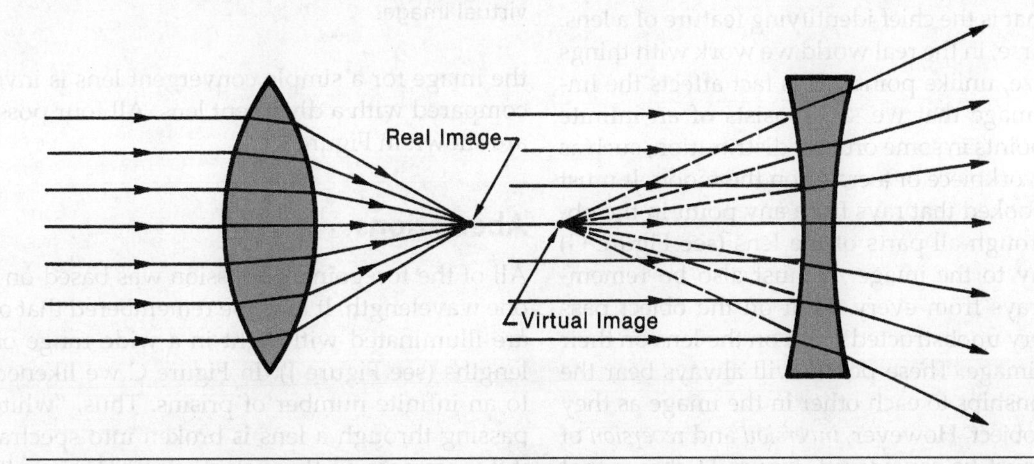

FIGURE E Converging lenses (left) have convex surfaces and cause the rays to converge into a real image. Divergent lenses (right) have concave surfaces and cause the rays to fan out. The point that they appear to come from is the virtual image.

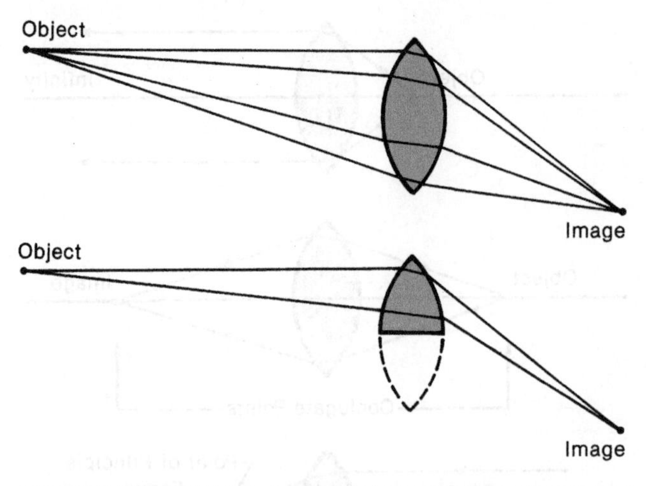

FIGURE G Every part of the lens contributes light from the source to the image (top). This is often misunderstood. Even if the lens were cut in half, the image would show all of the object (bottom). The only difference would be the amount of light. In the top example, the image will contain all of the light from the source that is received by its first surface less only the reflections, scattering, and absorption. In the bottom example, the image will receive only half of the light. That explains why you can cover half of a lens, but the image will still be a full circle.

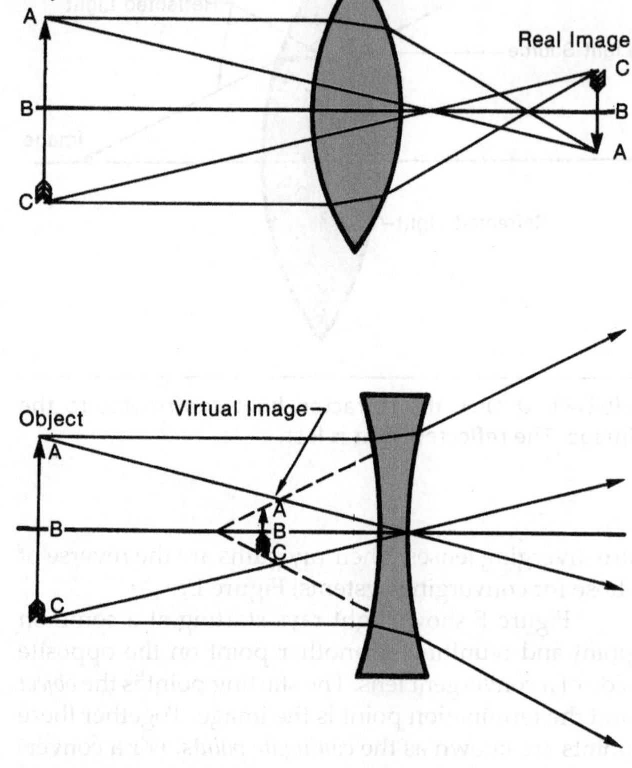

FIGURE H All real objects have size. Therefore, the images consist of an infinite series of points reduced here to A, B, and C for simplicity. When this is considered, the real image is found to be inverted with respect to the virtual image.

cases the image is closest to the lens for the most distant light sources. That point is known as the *primary focus* of the lens and its distance from the lens is the *focal length.* That is the chief identifying feature of a lens.

Of course, in the real world we work with things that have size, unlike points. This fact affects the images. Any image that we see consists of an infinite number of points in some orderly distribution, such as a line on a workpiece or a crater on the moon. It must not be overlooked that rays from any point in the object pass through all parts of the lens (see Figure G) on their way to the image. It must also be remembered that rays from every point on the object pass through every unobstructed point on the lens on their way to the image. These points will always bear the same relationships to each other in the image as they have in the object. However, *inversion* and *reversion* of the image must be considered. Figure H shows that the image for a simple convergent lens is inverted as compared with a divergent lens. All four possibilities are shown in Figure I.

Aberrations

All of the foregoing discussion was based on light of one wavelength. It must be remembered that our lives are illuminated with light in a wide range of wavelengths (see Figure J). In Figure C we likened a lens to an infinite number of prisms. Thus, "white" light passing through a lens is broken into spectra just as if it were passing through a prism. This results in the

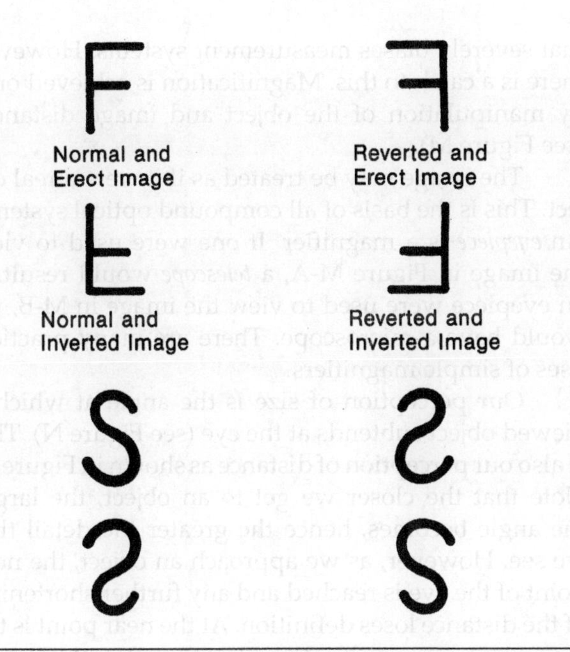

FIGURE I The four possible image orientations appear straightforward when the letter F is used to illustrate them. However, in practice they may be confusing as the example used illustrates.

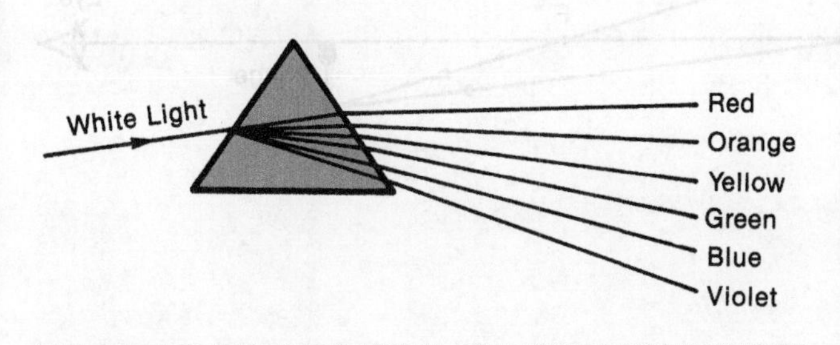

FIGURE J White light is made up of a combination of wavelengths. Of these, red is the longest wavelength and violet is the shortest. The amount of refraction varies with the wavelength. Therefore, a prism breaks white light into a spectrum.

rainbow fuzziness one sees around the image when using a simple lens. The effect is called *chromatic aberration.* At low magnifications it presents no problem. At high magnifications, however, instead of one image, images at all wavelengths are created.

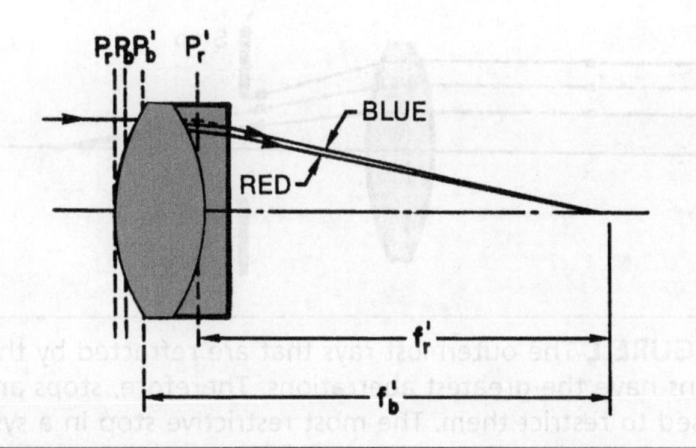

FIGURE K Chromatic aberrations are corrected by achromatic lenses. These lenses consist of two elements that have different indexes of refraction. In this example, the second element refracts the red wavelength less than the first element. Thus, the blue wavelength catches up with it and they recombine into white light.

Chromatic aberrations are corrected by multiple element lenses known as *achromatic* lenses, Figure K. They utilize lenses of different refraction indexes to bring back together the diverging wavelengths.

In addition to chromatic aberrations there are also *spherical aberrations.* This is an inherent defect of a simple lens. The rays passing through the center are focused in a different plane than those near the outer edge. This results in an unclear image.

This problem is also corrected by multiple element lenses. In most optical systems both aberrations must be corrected simultaneously. This requires tedious calculations for even a simple system. It explains why sophisticated optical systems such as those in cameras may have lenses with 10 or more elements. Prior to the computer, a lens designer could have devoted a lifetime to the perfection of only a small number of systems. Even with the computer, modern optical systems are among the wonders of the post-World War II period.

When examining lens systems, diaphragms known as "stops" will be found, Figure L. The outermost portions of lenses have the greatest aberrations. Thus, the stops aid in their control. The stops govern the "light gathering" capacity of the system. They are measured as *apertures* (a familiar term to photographers).

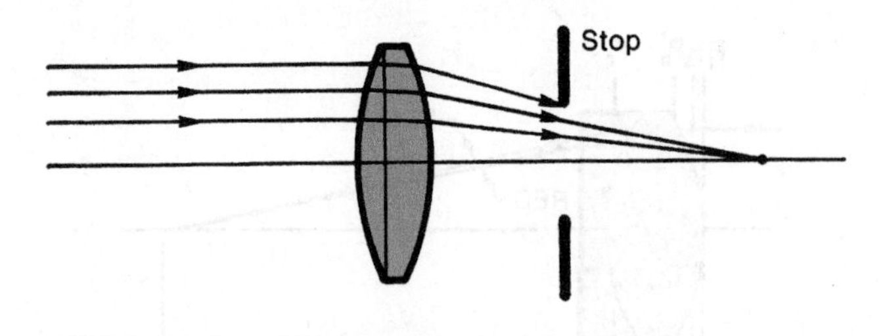

FIGURE L The outermost rays that are refracted by the lens have the greatest aberrations. Therefore, stops are used to restrict them. The most restrictive stop in a system determines the aperture of the system. The aperture is the amount of light from the source that forms the image.

Optical Magnification

The image produced by a lens may be made any size desired without the addition of external power as is required for most mechanical instruments. By way of review, it is the addition of such external power that severely biases measurement systems. However, there is a catch to this. Magnification is achieved only by manipulation of the object and image distances (see Figure M).

The image may be treated as if it were a real object. This is the basis of all compound optical systems. An *eyepiece* is a magnifier. If one were used to view the image in Figure M-A, a *telescope* would result. If an eyepiece were used to view the image in M-B, we would have a microscope. There are many practical uses of simple magnifiers.

Our perception of size is the angle at which a viewed object subtends at the eye (see Figure N). This is also our perception of distance as shown in Figure O. Note that the closer we get to an object, the larger the angle becomes, hence the greater the detail that we see. However, as we approach an object, the near point of the eye is reached and any further shortening of the distance loses definition. At the near point is the maximum size an object can be viewed with the unaided eye. For practical purposes, for adults 25.4 cm

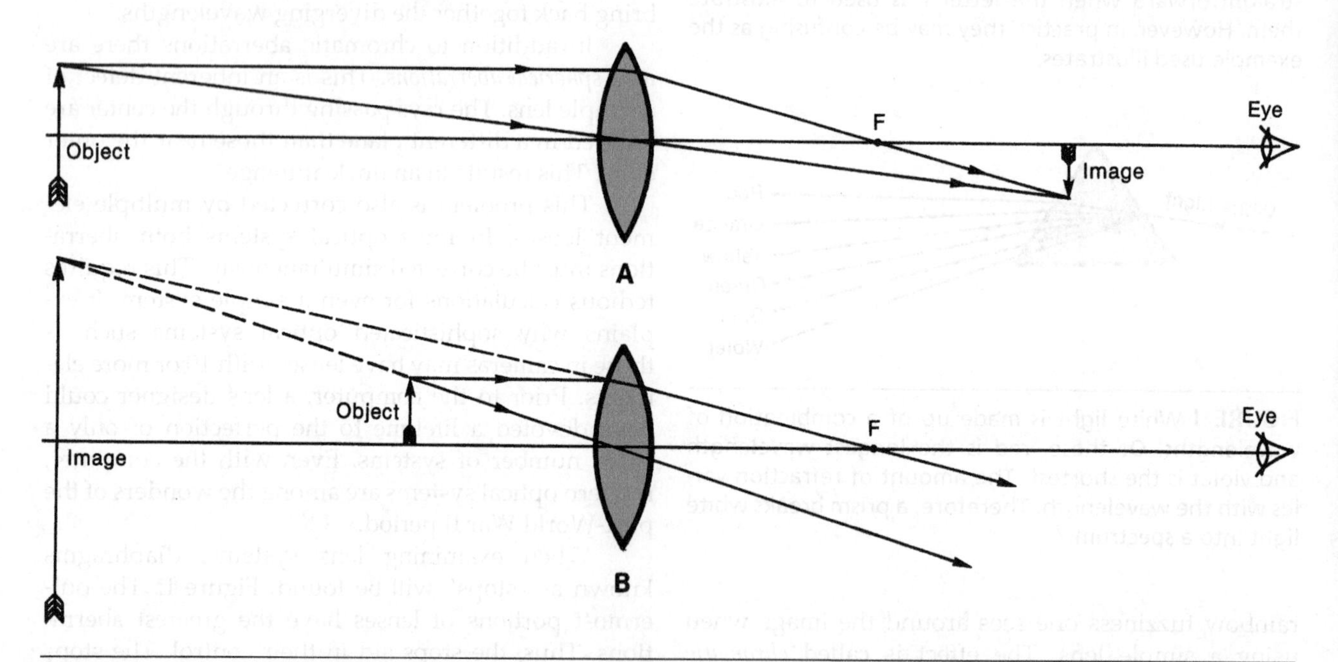

FIGURE M If one looks at a distant scene with a convergent lens, as in A, the image is inverted and smaller than the scene. In B, the object is close to the lens and the image is normal and erect but is larger than the object. This is the basis of simple optical magnification as used in metrology.

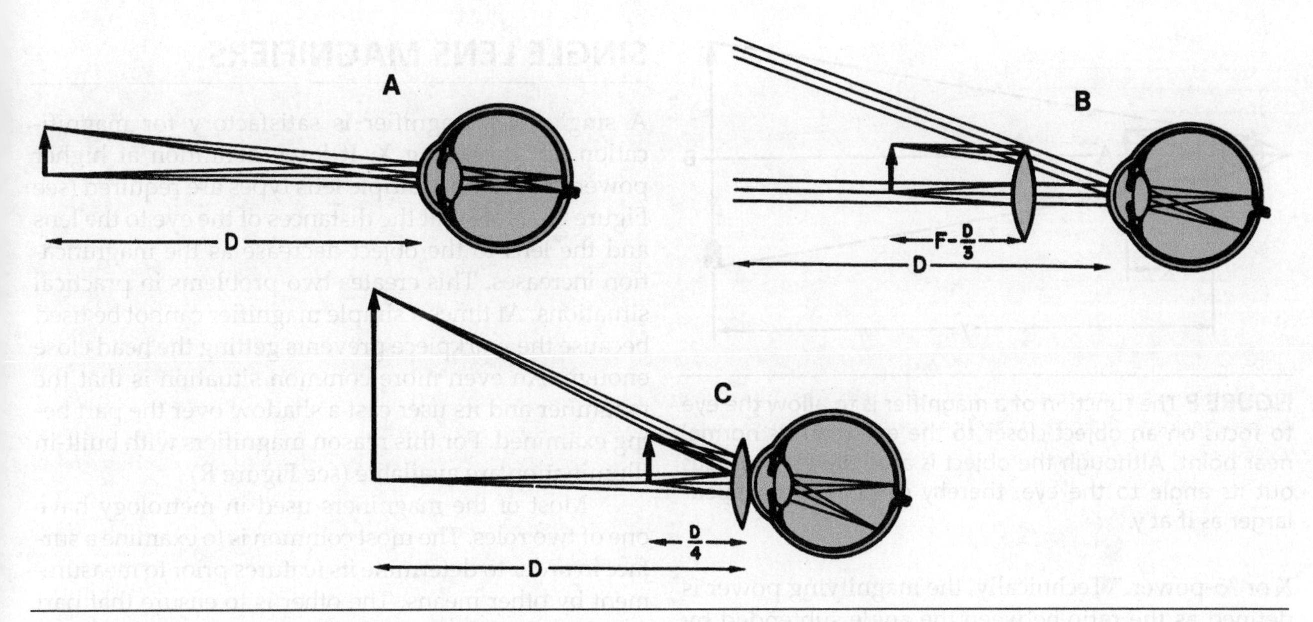

FIGURE N The ability to perceive size and distance is an organic function of the eye. In A, the "unaided" eye views a given object. The object subtends a small angle; thus, it is either far away or small in size. In B, a lens widens the perceived subtended angle. Although the object has not changed in size, it appears larger. If the same lens is brought closer to the eye as in C, the object appears still larger.

(10 in.) is the nearest distance of distinct vision for extended periods. The function of a magnifier is to enable the eye to remain relaxed while the object is viewed from shorter distances for greater clarity (see Figure P).

In Figure Q, the object is actually at distance x. If it could be seen, it would have a size A. The lens, however, spreads out the rays so that they appear to emanate from B at distance y. If, for example, x is 50 mm and y is 25 cm, the magnification is said to be

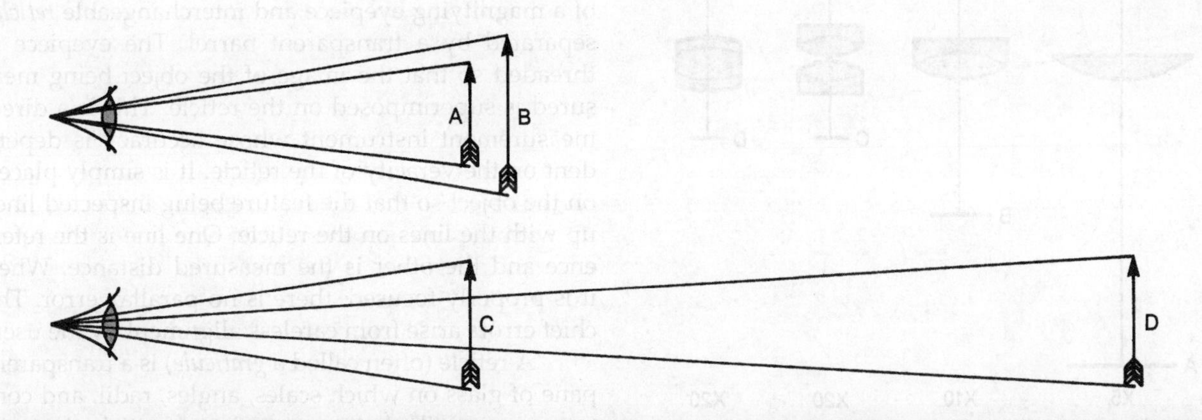

FIGURE O The perception of size and distance both depend on the angles that objects subtend on the eye. Although A and B lie nearly in the same plane (top), the wider angle from B makes us recognize that it is larger than A. Similarly, the smaller angle from D shows us that it is more distant than C (providing we know they are of the same size).

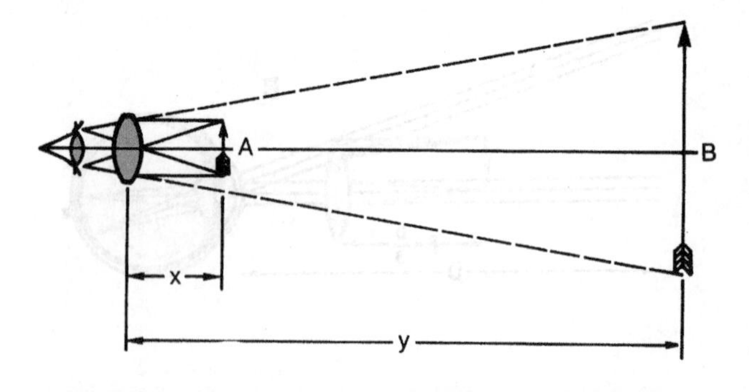

FIGURE P The function of a magnifier is to allow the eye to focus on an object closer to the eye than its normal near point. Although the object is at x, the lens spreads out its angle to the eye, thereby causing it to appear larger as if at y.

X or "5-power." Technically, the magnifying power is defined as the ratio between the angle subtended by the image at the eye and the angle subtended by the object if viewed directly at the nearest distance of distinct vision.

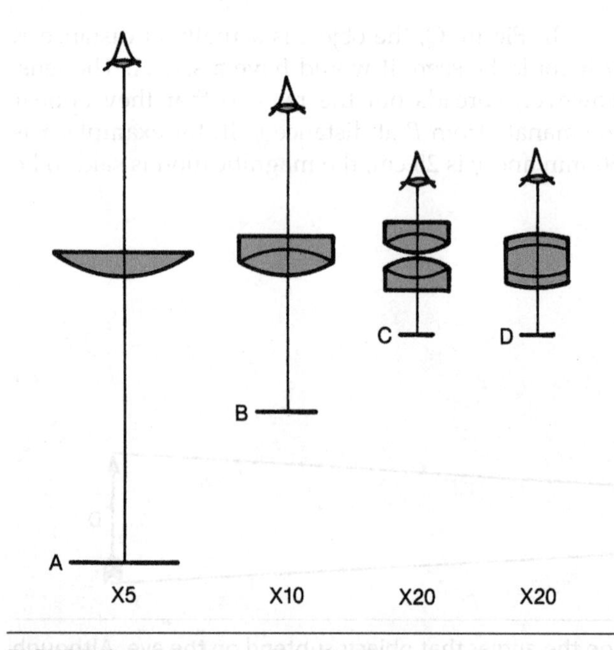

FIGURE Q As the size of the object decreases, it is necessary to use greater magnification for the same degree of clarity. At greater than 5X magnification, multiple-element lens systems are required as shown.

SINGLE LENS MAGNIFIERS

A single lens magnifier is satisfactory for magnification not exceeding X. It loses definition at higher power; therefore, multiple lens types are required (see Figure Q). Note that the distances of the eye to the lens and the lens to the object decrease as the magnification increases. This creates two problems in practical situations. At times a simple magnifier cannot be used because the workpiece prevents getting the head close enough. An even more common situation is that the magnifier and its user cast a shadow over the part being examined. For this reason magnifiers with built-in illumination are available (see Figure R).

Most of the magnifiers used in metrology have one of two roles. The most common is to examine a surface in order to determine its features prior to measurement by other means. The other is to ensure that part features line up properly with scale graduations and to assist in reading those graduations. This is less important today with digital reading instruments, but it is still important enough that everyone practicing precision measurement needs a good quality magnifier. In the foregoing roles the magnifier is a great assist, but it is not a metrological instrument. Not until a reticle is added does it become a measurement instrument.

The magnifier known as the *pocket comparator* (see Figure S) deserves to be better known. It consists of a magnifying eyepiece and interchangeable *reticles* separated by a transparent barrel. The eyepiece is threaded so that the image of the object being measured is superimposed on the reticle. This is a direct measurement instrument whose accuracy is dependent on the veracity of the reticle. It is simply placed on the object so that the feature being inspected lines up with the lines on the reticle. One line is the reference and the other is the measured distance. When it is properly focused, there is no parallax error. The chief errors arise from careless alignment by the user.

A reticle (often called a *graticule*) is a transparent pane of glass on which scales, angles, radii, and contours are inscribed. A great variety of reticles is available. A few are shown in Figure T. An illuminated version of the pocket comparator is shown in Figure U.

Seven power is the limit at which most pocket comparators are made. At this power reticles reading

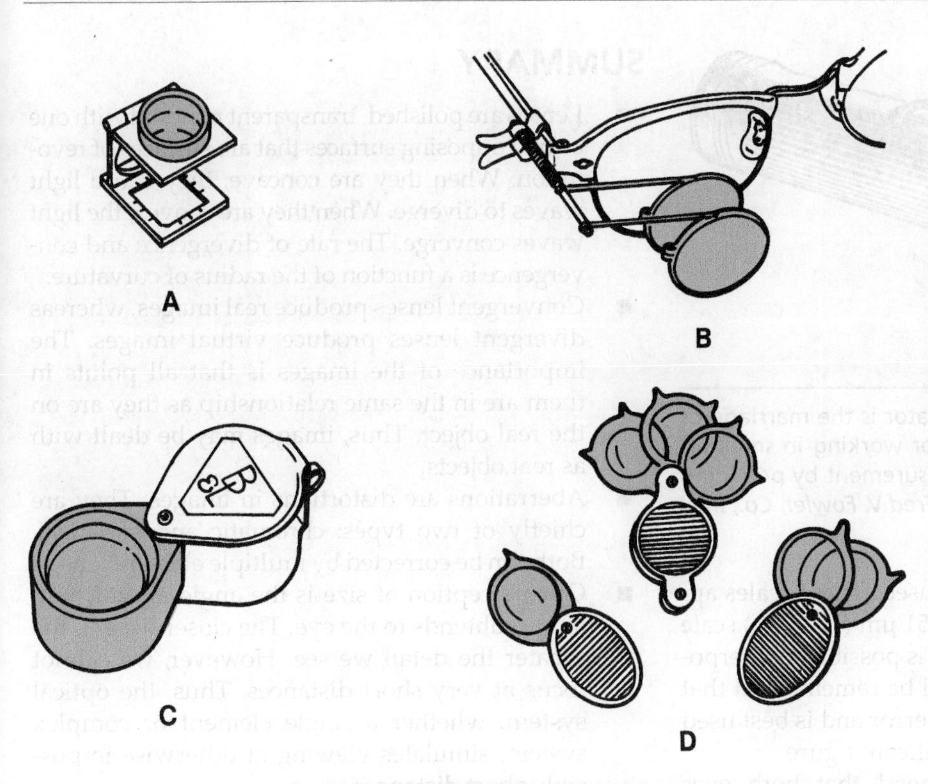

FIGURE R Many types of simple magnifiers are available. Those at the lowest magnification are called readers. There is no consensus about the power at which readers become magnifiers. A is a linen tester which is popular because it folds for easy carrying. B is an eyeglass loupe whose wide use by pawn shops shows its versatility. C is a pocket magnifier using a highly corrected Hasting Triplett lens. The pocket magnifiers at D are general-use items. Most of the magnifiers are in the 3X to 10X power range, but the Hastings Triplett goes to 20X.

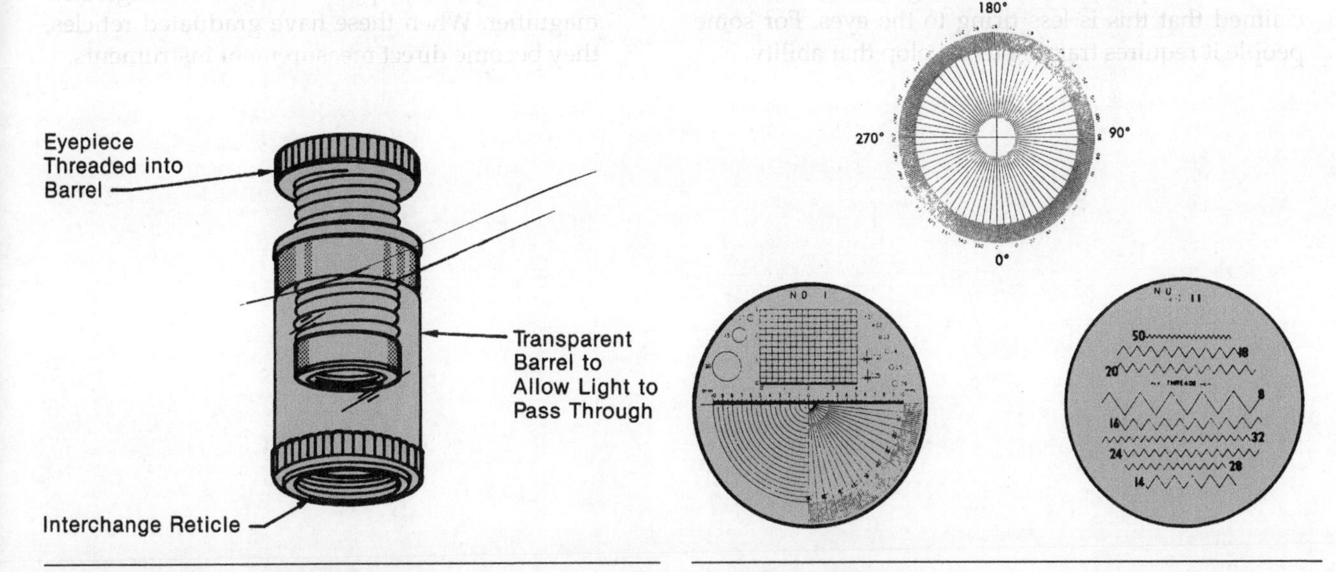

FIGURE S The pocket comparator is a direct measurement instrument. Because transfers and comparisons are not required, it is inherently an accurate instrument, as well as quick and easy to use.

FIGURE T Various reticles.

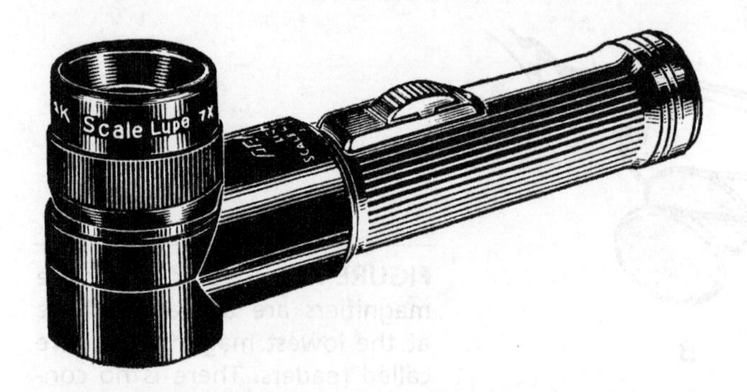

FIGURE U The flashlight comparator is the marriage of two useful instruments. Except for working in small recesses, it improves magnifier measurement by providing better illumination. *(Courtesy of Fred V. Fowler, Co., Inc.)*

in 0.1 mm (0.005 in.) are easily used. These scales appear roughly the same as a 793.51 μm (1/32 in.) scale to the unaided eye. Therefore, it is possible to interpolate fewer dimensions. It should be remembered that such interpolation increases the error and is best used for rounding off to a larger significant figure.

Some authorities recommend that both eyes should be kept open when using a magnifier. It is claimed that this is less tiring to the eyes. For some people it requires training to develop that ability.

SUMMARY

- Lenses are polished, transparent material with one or two opposing surfaces that are surfaces of revolution. When they are concave, they cause light waves to diverge. When they are convex, the light waves converge. The rate of divergence and convergence is a function of the radius of curvature.
- Convergent lenses produce real images, whereas divergent lenses produce virtual images. The importance of the images is that all points in them are in the same relationship as they are on the real object. Thus, images may be dealt with as real objects.
- Aberrations are distortions in images. They are chiefly of two types: chromatic and spherical. Both can be corrected by multiple element lenses.
- Our perception of size is the angle at which an angle subtends to the eye. The closer we get, the greater the detail we see. However, we cannot focus at very short distances. Thus, the optical system, whether a single element or complex system, simulates viewing at otherwise impossibly short distances.
- The most familiar optical device is the single lens magnifier. When these have graduated reticles, they become direct measurement instruments.

APPENDIX D

Prefixes Used As Multipliers

Prefix	Symbol	Multiplier	Decimal Value
yotta-	Y	10^{24}	1 000 000 000 000 000 000 000 000
zetta-	Z	10^{21}	1 000 000 000 000 000 000 000
exa-	E	10^{18}	1 000 000 000 000 000 000
peta-	P	10^{15}	1 000 000 000 000 000
tera-	T	10^{12}	1 000 000 000 000
giga-	G	10^{9}	1 000 000 000
mega-	M	10^{6}	1 000 000
kilo-	k	10^{3}	1 000
hecto-	h	10^{2}	100
deca-	da	10^{1}	10
deci-	d	10^{-1}	0.1
centi	c	10^{-2}	0.01
milli-	m	10^{-3}	0.001
micro-	µ	10^{-6}	0.000 001
nano-	n	10^{-9}	0.000 000 001
pico-	p	10^{-12}	0.000 000 000 001
femto-	f	10^{-15}	0.000 000 000 000 001
atto-	a	10^{-18}	0.000 000 000 000 000 001
zepto-	z	10^{-21}	0.000 000 000 000 000 000 001
yocto-	y	10^{-24}	0.000 000 000 000 000 000 000 001

Recommended Pronunciation of Prefixes

Prefix	Pronunciation
yotta	YOTT-a (a as in about)
zetta	ZETT-a (a as in about)
exa	EX-a (a as in about)
peta	PET-a (a as in petal)
tera	TERR-a (a as in terrace)
giga	GIG-a (gig as in giggle, a as in about)
mega	MEG-a (a as in megaphone)
kilo	KILL-oh (see NOTE)
hecto	HECH-toe
deca	DEK-a (as in decahedron)
deci	DESS-ih (as in decimal)
centi	SENT-ih (as in centipede)
milli	MILL-ih (as in military, but MILL-ee before a vowel)
micro	MIKE-roe (as in microbe)
nano	NAN-oh (a as in ant)
pico	PEEK-oh
femto	FEM-toe
atto	AT-toe (a as in hat)
yocto	YOCK-toe
zepto	ZEP-toe (E as in step)

NOTE: The first syllable of every prefix is accented to ensure that the prefix will retain its identity. Therefore, the preferred pronunciation of kilometer is KILL-oh-meter, not kil-LOM-muh-ter.

ANSI/IEEE Std 268-1992

APPENDIX E

TRIGONOMETRIC FUNCTIONS

Degrees	Radians	Sin	Cos	Tan	Cot	Sec	Csc		
6°00′	.1047	.1045	.9945	.1051	9.514	1.006	9.567	1.4661	84°00′
10	.1076	.1074	.9942	.1080	9.255	1.006	9.309	1.4632	50
20	.1105	.1103	.9939	.1110	9.010	1.006	9.065	1.4603	40
30	.1134	.1132	.9936	.1139	8.777	1.006	8.834	1.4574	30
40	.1164	.1161	.9932	.1169	8.556	1.007	8.614	1.4544	20
50	.1193	.1190	.9929	.1198	8.345	1.007	8.405	1.4515	10
7°00′	.1222	.1219	.9925	.1228	8.144	1.008	8.206	1.4486	83°00′
10	.1251	.1248	.9922	.1257	7.953	1.008	8.016	1.4457	50
20	.1280	.1276	.9918	.1287	7.771	1.008	7.835	1.4428	40
30	.1309	.1305	.9914	.1316	7.596	1.009	7.662	1.4399	30
40	.1338	.1334	.9911	.1346	7.429	1.009	7.496	1.4370	20
50	.1367	.1363	.9907	.1376	7.269	1.009	7.338	1.4341	10
8°00′	.1396	.1392	.9903	.1405	7.115	1.010	7.185	1.4312	82°00′
10	.1425	.1421	.9899	.1435	6.968	1.010	7.040	1.4283	50
20	.1454	.1449	.9894	.1465	6.827	1.011	6.900	1.4254	40
30	.1483	.1478	.9890	.1494	6.691	1.011	6.766	1.4225	30
40	.1513	.1507	.9886	.1524	6.561	1.012	6.637	1.4195	20
50	.1542	.1536	.9881	.1554	6.435	1.012	6.512	1.4166	10
9°00′	.1571	.1564	.9877	.1584	6.314	1.012	6.392	1.4137	81°00′
10	.1600	.1593	.9872	.1614	6.197	1.013	6.277	1.4108	50
20	.1629	.1622	.9868	.1644	6.085	1.013	6.166	1.4079	40
30	.1658	.1650	.9863	.1673	5.976	1.014	6.059	1.4050	30
40	.1687	.1679	.9858	.1703	5.871	1.014	5.956	1.4021	20
50	.1716	.1708	.9853	.1733	5.770	1.015	5.856	1.3992	10
10°00′	.1745	.1736	.9848	.1763	5.671	1.015	5.759	1.3963	80°00′
10	.1774	.1765	.9843	.1793	5.576	1.016	5.665	1.3934	50
20	.1803	.1794	.9838	.1823	5.485	1.016	5.575	1.3905	40
30	.1833	.1822	.9833	.1853	5.396	1.017	5.488	1.3875	30
40	.1862	.1851	.9827	.1883	5.309	1.018	5.403	1.3846	20
50	.1891	.1879	.9822	.1914	5.226	1.018	5.321	1.3817	10
		Cos	**Sin**	**Cot**	**Tan**	**Csc**	**Sec**	**Radians**	**Degrees**

Degrees	Radians	Sin	Cos	Tan	Cot	Sec	Csc		
11°00′	.1920	.1908	.9816	.1944	5.145	1.019	5.241	1.3788	79°00′
10	.1949	.1937	.9811	.1974	5.066	1.019	5.164	1.3759	50
20	.1978	.1965	.9805	.2004	4.990	1.020	5.089	1.3730	40
30	.2007	.1994	.9799	.2034	4.915	1.020	5.016	1.3701	30
40	.2036	.2022	.9793	.2065	4.843	1.021	4.945	1.3672	20
50	.2065	.2051	.9787	.2095	4.773	1.022	4.877	1.3643	10
12°00′	.2094	.2079	.9781	.2126	4.705	1.022	4.810	1.3614	78°00′
10	.2123	.2108	.9775	.2156	4.638	1.023	4.745	1.3585	50
20	.2153	.2136	.9769	.2186	4.574	1.024	4.682	1.3555	40
30	.2182	.2164	.9763	.2217	4.511	1.024	4.620	1.3526	30
40	.2211	.2193	.9757	.2247	4.450	1.025	4.561	1.3497	20
50	.2240	.2221	.9750	.2278	4.390	1.026	4.502	1.3468	10
13°00′	.2269	.2250	.9744	.2309	4.332	1.026	4.445	1.3439	77°00′
10	.2298	.2278	.9737	.2339	4.275	1.027	4.390	1.3410	50
20	.2327	.2306	.9730	.2370	4.219	1.028	4.336	1.3381	40
30	.2356	.2334	.9724	.2401	4.165	1.028	4.284	1.3352	30
40	.2385	.2363	.9717	.2432	4.113	1.029	4.232	1.3323	20
50	.2414	.2391	.9710	.2462	4.061	1.030	4.182	1.3294	10
14°00′	.2443	.2419	.9703	.2493	4.011	1.031	4.134	1.3265	76°00′
10	.2473	.2447	.9696	.2524	3.962	1.031	4.086	1.3235	50
20	.2502	.2476	.9689	.2555	3.914	1.032	4.039	1.3206	40
30	.2531	.2504	.9681	.2586	3.867	1.033	3.994	1.3177	30
40	.2560	.2532	.9674	.2617	3.821	1.034	3.950	1.3148	20
50	.2589	.2560	.9667	.2648	3.776	1.034	3.906	1.3119	10
15°00′	.2618	.2588	.9659	.2679	3.732	1.035	3.864	1.3090	75°00′
10	.2647	.2616	.9652	.2711	3.689	1.036	3.822	1.3061	50
20	.2676	.2644	.9644	.2742	3.647	1.037	3.782	1.3032	40
30	.2705	.2672	.9636	.2773	3.606	1.038	3.742	1.3003	30
40	.2734	.2700	.9629	.2805	3.566	1.039	3.703	1.2974	20
50	.2763	.2728	.9621	.2836	3.526	1.039	3.665	1.2945	10
16°00′	.2793	.2756	.9613	.2867	3.487	1.040	3.628	1.2915	74°00′
10	.2822	.2784	.9605	.2899	3.450	1.041	3.592	1.2886	50
20	.2851	.2812	.9596	.2930	3.412	1.042	3.556	1.2857	40
30	.2880	.2840	.9588	.2962	3.376	1.043	3.521	1.2828	30
40	.2909	.2868	.9580	.2994	3.340	1.044	3.487	1.2799	20
50	.2938	.2896	.9572	.3025	3.305	1.045	3.453	1.2770	10
17°00′	.2967	.2924	.9563	.3057	3.271	1.046	3.420	1.2741	73°00′
10	.2996	.2952	.9555	.3089	3.237	1.047	3.388	1.2712	50
20	.3025	.2979	.9546	.3121	3.204	1.048	3.357	1.2683	40
30	.3054	.3007	.9537	.3153	3.172	1.049	3.326	1.2654	30
40	.3083	.3035	.9528	.3185	3.140	1.049	3.295	1.2625	20
50	.3112	.3062	.9520	.3217	3.109	1.050	3.265	1.2596	10
18°00′	.3142	.3090	.9511	.3249	3.078	1.051	3.236	1.2566	72°00′
10	.3171	.3118	.9502	.3281	3.048	1.052	3.207	1.2537	50
20	.3200	.3145	.9492	.3314	3.018	1.053	3.179	1.2508	40
30	.3229	.3173	.9483	.3346	2.989	1.054	3.152	1.2479	30
40	.3258	.3201	.9474	.3378	2.960	1.056	3.124	1.2450	20
50	.3287	.3228	.9465	.3411	2.932	1.057	3.098	1.2421	10
		Cos	Sin	Cot	Tan	Csc	Sec	Radians	Degrees

Degrees	Radians	Sin	Cos	Tan	Cot	Sec	Csc	Radians	Degrees
19°00′	.3316	.3256	.9455	.3443	2.904	1.058	3.072	1.2392	71°00′
10	.3345	.3283	.9446	.3476	2.877	1.059	3.046	1.2363	50
20	.3374	.3311	.9436	.3508	2.850	1.060	3.021	1.2334	40
30	.3403	.3338	.9426	.3541	2.824	1.061	2.996	1.2305	30
40	.3432	.3365	.9417	.3574	2.798	1.062	2.971	1.2276	20
50	.3462	.3393	.9407	.3607	2.773	1.063	2.947	1.2246	10
20°00′	.3491	.3420	.9397	.3640	2.748	1.064	2.924	1.2217	70°00′
10	.3520	.3448	.9387	.3673	2.723	1.065	2.901	1.2188	50
20	.3549	.3475	.9377	.3706	2.699	1.066	2.878	1.2159	40
30	.3578	.3502	.9367	.3739	2.675	1.068	2.855	1.2130	30
40	.3607	.3529	.9357	.3772	2.651	1.069	2.833	1.2101	20
50	.3636	.3556	.9346	.3805	2.628	1.070	2.812	1.2072	10
21°00′	.3665	.3584	.9336	.3839	2.605	1.071	2.790	1.2043	69°00′
10	.3694	.3611	.9325	.3872	2.583	1.072	2.769	1.2014	50
20	.3723	.3638	.9315	.3906	2.561	1.074	2.749	1.1985	40
30	.3752	.3665	.9304	.3939	2.539	1.075	2.729	1.1956	30
40	.3782	.3692	.9293	.3973	2.517	1.076	2.709	1.1926	20
50	.3811	.3719	.9283	.4006	2.496	1.077	2.689	1.1897	10
22°00′	.3840	.3746	.9272	.4040	2.475	1.079	2.669	1.1868	68°00′
10	.3869	.3773	.9261	.4074	2.455	1.080	2.650	1.1839	50
20	.3898	.3800	.9250	.4108	2.434	1.081	2.632	1.1810	40
30	.3927	.3827	.9239	.4142	2.414	1.082	2.613	1.1781	30
40	.3956	.3854	.9228	.4176	2.395	1.084	2.595	1.1752	20
50	.3985	.3880	.9216	.4210	2.375	1.085	2.577	1.1723	10
23°00′	.4014	.3907	.9205	.4245	2.356	1.086	2.559	1.1694	67°00′
10	.4043	.3934	.9194	.4279	2.337	1.088	2.542	1.1665	50
20	.4072	.3961	.9182	.4314	2.318	1.089	2.525	1.1636	40
30	.4101	.3987	.9171	.4348	2.300	1.090	2.508	1.1607	30
40	.4131	.4014	.9159	.4383	2.282	1.092	2.491	1.1577	20
50	.4160	.4041	.9147	.4417	2.264	1.093	2.475	1.1548	10
24°00′	.4189	.4067	.9135	.4452	2.246	1.095	2.459	1.1519	66°00′
10	.4218	.4094	.9124	.4487	2.229	1.096	2.443	1.1490	50
20	.4247	.4120	.9112	.4522	2.211	1.097	2.427	1.1461	40
30	.4276	.4147	.9100	.4557	2.194	1.099	2.411	1.1432	30
40	.4305	.4173	.9088	.4592	2.178	1.100	2.396	1.1403	20
50	.4334	.4200	.9075	.4628	2.161	1.102	2.381	1.1374	10
25°00′	.4363	.4226	.9063	.4663	2.145	1.103	2.366	1.1345	65°00′
10	.4392	.4253	.9051	.4699	2.128	1.105	2.352	1.1316	50
20	.4421	.4279	.9038	.4734	2.112	1.106	2.337	1.1287	40
30	.4451	.4305	.9026	.4770	2.097	1.108	2.323	1.1257	30
40	.4480	.4331	.9013	.4805	2.081	1.109	2.309	1.1228	20
50	.4509	.4357	.9001	.4841	2.066	1.111	2.295	1.1199	10
26°00′	.4538	.4384	.8988	.4877	2.050	1.113	2.281	1.1170	64°00′
10	.4567	.4410	.8975	.4913	2.035	1.114	2.268	1.1141	50
20	.4596	.4436	.8962	.4950	2.020	1.116	2.254	1.1112	40
30	.4625	.4462	.8949	.4986	2.006	1.117	2.241	1.1083	30
40	.4654	.4488	.8936	.5022	1.991	1.119	2.228	1.1054	20
50	.4683	.4514	.8923	.5059	1.977	1.121	2.215	1.1025	10
		Cos	**Sin**	**Cot**	**Tan**	**Csc**	**Sec**	**Radians**	**Degrees**

Degrees	Radians	Sin	Cos	Tan	Cot	Sec	Csc		
27°00′	.4712	.4540	.8910	.5095	1.963	1.122	2.203	1.0996	63°00′
10	.4741	.4566	.8897	.5132	1.949	1.124	2.190	1.0967	50
20	.4771	.4592	.8884	.5169	1.935	1.126	2.178	1.0937	40
30	.4800	.4617	.8870	.5206	1.921	1.127	2.166	1.0908	30
40	.4829	.4643	.8857	.5243	1.907	1.129	2.154	1.0879	20
50	.4858	.4669	.8843	.5280	1.894	1.131	2.142	1.0850	10
28°00′	.4887	.4695	.8829	.5317	1.881	1.133	2.130	1.0821	62°00′
10	.4916	.4720	.8816	.5354	1.868	1.134	2.118	1.0792	50
20	.4945	.4746	.8802	.5392	1.855	1.136	2.107	1.0763	40
30	.4974	.4772	.8788	.5430	1.842	1.138	2.096	1.0734	30
40	.5003	.4797	.8774	.5467	1.829	1.140	2.085	1.0705	20
50	.5032	.4823	.8760	.5505	1.817	1.142	2.074	1.0676	10
29°00′	.5061	.4848	.8746	.5543	1.804	1.143	2.063	1.0647	61°00′
10	.5091	.4874	.8732	.5581	1.792	1.145	2.052	1.0617	50
20	.5120	.4899	.8718	.5619	1.780	1.147	2.041	1.0588	40
30	.5149	.4924	.8704	.5658	1.768	1.149	2.031	1.0559	30
40	.5178	.4949	.8689	.5696	1.756	1.151	2.020	1.0530	20
50	.5207	.4975	.8675	.5735	1.744	1.153	2.010	1.0501	10
30°00′	.5236	.5000	.8660	.5774	1.732	1.155	2.000	1.0472	60°00′
10	.5265	.5025	.8646	.5812	1.721	1.157	1.990	1.0443	50
20	.5294	.5050	.8631	.5851	1.709	1.159	1.980	1.0414	40
30	.5323	.5075	.8616	.5890	1.698	1.161	1.970	1.0385	30
40	.5352	.5100	.8602	.5930	1.686	1.163	1.961	1.0356	20
50	.5381	.5125	.8587	.5969	1.675	1.165	1.951	1.0327	10
31°00′	.5411	.5150	.8572	.6009	1.664	1.167	1.942	1.0297	59°00′
10	.5440	.5175	.8557	.6048	1.653	1.169	1.932	1.0268	50
20	.5469	.5200	.8542	.6088	1.643	1.171	1.923	1.0239	40
30	.5498	.5225	.8526	.6128	1.632	1.173	1.914	1.0210	30
40	.5527	.5250	.8511	.6168	1.621	1.175	1.905	1.0181	20
50	.5556	.5274	.8496	.6208	1.611	1.177	1.896	1.0152	10
32°00′	.5585	.5299	.8480	.6249	1.600	1.179	1.887	1.0123	58°00′
10	.5614	.5324	.8465	.6289	1.590	1.181	1.878	1.0094	50
20	.5643	.5348	.8450	.6330	1.580	1.183	1.870	1.0065	40
30	.5672	.5373	.8434	.6371	1.570	1.186	1.861	1.0036	30
40	.5701	.5397	.8418	.6412	1.560	1.188	1.853	1.0007	20
50	.5730	.5422	.8403	.6453	1.550	1.190	1.844	.9978	10
33°00′	.5760	.5446	.8387	.6494	1.540	1.192	1.836	.9948	57°00′
10	.5789	.5471	.8371	.6535	1.530	1.195	1.828	.9919	50
20	.5818	.5495	.8355	.6577	1.520	1.197	1.820	.9890	40
30	.5847	.5519	.8339	.6619	1.511	1.199	1.812	.9861	30
40	.5876	.5544	.8323	.6661	1.501	1.202	1.804	.9832	20
50	.5905	.5568	.8307	.6703	1.492	1.204	1.796	.9803	10
34°00′	.5934	.5592	.8290	.6745	1.483	1.206	1.788	.9774	56°00′
10	.5963	.5616	.8274	.6787	1.473	1.209	1.781	.9745	50
20	.5992	.5640	.8258	.6830	1.464	1.211	1.773	.9716	40
30	.6021	.5664	.8241	.6873	1.455	1.213	1.766	.9687	30
40	.6050	.5688	.8225	.6916	1.446	1.216	1.758	.9658	20
50	.6080	.5712	.8208	.6959	1.437	1.218	1.751	.9628	10
		Cos	**Sin**	**Cot**	**Tan**	**Csc**	**Sec**	**Radians**	**Degrees**

Degrees	Radians	Sin	Cos	Tan	Cot	Sec	Csc	Radians	Degrees
35°00′	.6109	.5736	.8192	.7002	1.428	1.221	1.743	.9599	55°00′
10	.6138	.5760	.8175	.7045	1.419	1.223	1.736	.9570	50
20	.6167	.5783	.8158	.7089	1.411	1.226	1.729	.9541	40
30	.6196	.5807	.8141	.7133	1.402	1.228	1.722	.9512	30
40	.6225	.5831	.8124	.7177	1.393	1.231	1.715	.9483	20
50	.6254	.5854	.8107	.7221	1.385	1.233	1.708	.9454	10
36°00′	.6283	.5878	.8090	.7265	1.376	1.236	1.701	.9425	54°00′
10	.6312	.5901	.8073	.7310	1.368	1.239	1.695	.9396	50
20	.6341	.5925	.8056	.7355	1.360	1.241	1.688	.9367	40
30	.6370	.5948	.8039	.7400	1.351	1.244	1.681	.9338	30
40	.6399	.5972	.8021	.7445	1.343	1.247	1.675	.9308	20
50	.6429	.5995	.8004	.7490	1.335	1.249	1.668	.9279	10
37°00′	.6458	.6018	.7986	.7536	1.327	1.252	1.662	.9250	53°00′
10	.6487	.6041	.7969	.7581	1.319	1.255	1.655	.9221	50
20	.6516	.6064	.7951	.7627	1.311	1.258	1.649	.9192	40
30	.6545	.6088	.7934	.7673	1.303	1.260	1.643	.9163	30
40	.6574	.6111	.7916	.7720	1.295	1.263	1.636	.9134	20
50	.6603	.6134	.7898	.7766	1.288	1.266	1.630	.9105	10
38°00′	.6632	.6157	.7880	.7813	1.280	1.269	1.624	.9076	52°00′
10	.6661	.6180	.7862	.7860	1.272	1.272	1.618	.9047	50
20	.6690	.6202	.7844	.7907	1.265	1.275	1.612	.9018	40
30	.6719	.6225	.7826	.7954	1.257	1.278	1.606	.8989	30
40	.6749	.6248	.7808	.8002	1.250	1.281	1.601	.8959	20
50	.6778	.6271	.7790	.8050	1.242	1.284	1.595	.8930	10
39°00′	.6807	.6293	.7771	.8098	1.235	1.287	1.589	.8901	51°00′
10	.6836	.6316	.7753	.8146	1.228	1.290	1.583	.8872	50
20	.6865	.6338	.7735	.8195	1.220	1.293	1.578	.8843	40
30	.6894	.6361	.7716	.8243	1.213	1.296	1.572	.8814	30
40	.6923	.6383	.7698	.8292	1.206	1.299	1.567	.8785	20
50	.6952	.6406	.7679	.8341	1.199	1.302	1.561	.8756	10
40°00′	.6981	.6428	.7660	.8391	1.192	1.305	1.556	.8727	50°00′
10	.7010	.6450	.7642	.8441	1.185	1.309	1.550	.8698	50
20	.7039	.6472	.7623	.8491	1.178	1.312	1.545	.8669	40
30	.7069	.6494	.7604	.8541	1.171	1.315	1.540	.8639	30
40	.7098	.6517	.7585	.8591	1.164	1.318	1.535	.8610	20
50	.7127	.6539	.7566	.8642	1.157	1.322	1.529	.8581	10
41°00′	.7156	.6561	.7547	.8693	1.150	1.325	1.524	.8552	49°00′
10	.7185	.6583	.7528	.8744	1.144	1.328	1.519	.8523	50
20	.7214	.6604	.7509	.8795	1.137	1.332	1.514	.8494	40
30	.7243	.6626	.7490	.8847	1.130	1.335	1.509	.8465	30
40	.7272	.6648	.7470	.8899	1.124	1.339	1.504	.8436	20
50	.7301	.6670	.7451	.8951	1.117	1.342	1.499	.8407	10
42°00′	.7330	.6691	.7431	.9004	1.111	1.346	1.494	.8378	48°00′
10	.7359	.6713	.7412	.9057	1.104	1.349	1.490	.8349	50
20	.7389	.6734	.7392	.9110	1.098	1.353	1.485	.8319	40
30	.7418	.6756	.7373	.9163	1.091	1.356	1.480	.8290	30
40	.7447	.6777	.7353	.9217	1.085	1.360	1.476	.8261	20
50	.7476	.6799	.7333	.9271	1.079	1.364	1.471	.8232	10
		Cos	**Sin**	**Cot**	**Tan**	**Csc**	**Sec**	**Radians**	**Degrees**

Degrees	**Radians**	**Sin**	**Cos**	**Tan**	**Cot**	**Sec**	**Csc**		
43°00′	.7505	.6820	.7314	.9325	1.072	1.367	1.466	.8203	47°00′
10	.7534	.6841	.7294	.9380	1.066	1.371	1.462	.8174	50
20	.7563	.6862	.7274	.9434	1.060	1.375	1.457	.8145	40
30	.7592	.6884	.7254	.9490	1.054	1.379	1.453	.8116	30
40	.7621	.6905	.7234	.9545	1.048	1.382	1.448	.8087	20
50	.7650	.6926	.7214	.9601	1.042	1.386	1.444	.8058	10
44°00′	.7679	.6947	.7193	.9657	1.036	1.390	1.440	.8029	46°00′
10	.7709	.6967	.7173	.9713	1.030	1.394	1.435	.7999	50
20	.7738	.6988	.7153	.9770	1.024	1.398	1.431	.7970	40
30	.7767	.7009	.7133	.9827	1.018	1.402	1.427	.7941	30
40	.7796	.7030	.7112	.9884	1.012	1.406	1.423	.7912	20
50	.7825	.7050	.7092	.9942	1.006	1.410	1.418	.7883	10
45°00′	.7854	.7071	.7071	1.0000	1.0000	1.414	1.414	.7854	45°00′
		Cos	**Sin**	**Cot**	**Tan**	**Csc**	**Sec**	**Radians**	**Degrees**

π radians = 180° $\pi = 3.14159$

1 radian = 57° 17′44″

1° = 0.01745 32925 radian = 60′ = 3600″

$\frac{\pi}{2} = 1.5708$ $4\pi = 12.5664$

$\pi = 3.1416$ $5\pi = 15.7080$

$2\pi = 6.2832$ $6\pi = 18.8496$

$3\pi = 9.4248$ $7\pi = 21.9911$

APPENDIX F

METROLOGY WEB SITES

Web Site	Description
a2la.org	The American Association for Laboratory Accreditation (A2LA) is a nonprofit, nongovernmental membership organization. Find information on proficiency testing, inspection bodies, and reference material producers.
ASME.org	The American Society of Mechanical Engineers administers voluntary U.S. standardization. Search here for quality publications, codes, and standards.
ASQ.org/knowledge-center	The American Society for Quality. Use the Tools tab to see examples and links to fishbone diagrams, and Gantt and Pareto charts. Use the Topics tab to review six sigma, cost of quality, and ISO and ANSI standards.
BIPM.org	The International Bureau of Weights and Measures. This organization in Severes, France, was created to ensure worldwide uniformity of measurements and their traceability. Find common measurement abbreviations and acronyms, along with links to metrology institutes, organizations, and accreditation bodies.
Metric-conversions.org	Use online conversion tables, calculators, and metric formulas on this site.
Nist.gov/traceability	National Institute of Standards and Technology offers information on laboratory accreditation and standards. The traceability link provides a thorough examination of traceability.
Techstreet.com	Tech Street lets you track an unlimited number of standards. Keep up-to-date on the latest version of standards by industry, or organization.
Wisc-online.com	The Wisconsin Online Resource Center contains a digital library of web-based learning objects. Measurement lessons, drills, and simulations can be found here.

APPENDIX G

BIBLIOGRAPHY

American Society of Mechanical Engineers (ASME) *ASME Y14.5 – 2009 Dimensioning and Tolerancing*, American Society of Mechanical Engineers, New York, 2009

BIPM *International Vocabulary of Metrology – Basic and General Concepts, (VIM)* JCGM 200: 2012, bipm.org/en/publications/guides/vim

Box, G. E. P., W. G. Hunter, and J. S. Hunter *Statistics for Experimenters*, 2nd ed., John Wiley, New York, 2005

Grant, E. L., and R. S. Leavenworth *Statistical Quality Control*, 7th ed., McGraw-Hill, New York, 1996

Gryna, F. R., R. Chua, J. Defeo *Juran's Quality Planning and Analysis for Enterprise Quality*, 5th ed., McGraw-Hill, New York, 2005

Hines, W. W., D. C. Montgomery, D. M. Goldsman, C. M. Borror *Probability and Statistics in Engineering*, 4th ed., John Wiley, New York, 2003

Hoffman, P., E. Hopewell, B. James *Precision Machining Technology*, 2nd ed., Cengage Learning, New York, 2012

Johnson, R. R., P. J. Kuby *Elementary Statistics*, 11th ed., Cengage Learning, New York, 2011

Montgomery, D. C. *Design and Analysis of Experiments*, 8th ed., John Wiley, New York, 2012

Montgomery, D. C. *Introduction to Statistical Control*, 7th ed., John Wiley, New York, 2012

Taguchi, G. *Introduction to Quality Engineering*, Asian Productivity Organization, UNIPUB, White Plains, New York, 1986

Western Electric *Statistical Quality Control Handbook*, 2nd ed., Western Electric Corporation, Indianapolis, Indiana, 1982

GLOSSARY

4:1 rule (Chapter 9).The *4:1 rule* is an ANSI/NCSL-Z-540 *calibration* requirement. If you are calibrating a gage, you need to use a check standard that has an uncertainty of measurement that is less than or equal to 25% of the total gage tolerance.

10-to-1 Rule (Chapter 9). The *10-to-1 rule* is a rule for enhancing measurement reliability. It states that the instrument should be able to divide the part tolerance into ten parts. It is often written as "one-to-ten" rule and spoken of as the "rule of ten." Modern statistical quality control procedures provide better methods for instrument selection and production control. However, the 10-to-1 rule is recommended whenever statistical methods are not available. See Figures 9–2, 9–3, and 9–4.

AA (Chapter 8). AA is the abbreviation for *arithmetical average.* The AA is the mean between a series of values. Compare with *RMS.*

Abbe's Law (Chapter 6). *Abbe's law* is the principle that for maximum reliability the axis of the standard must lie along the line of measurement. See Figure 6–11.

Aberrations (Appendix C). *Aberrations* is the term used in optics to designate what we know as errors in linear measurement. See Appendix C, Figure L.

Abstraction (or abstract) (Chapter 12). Is the quality of dealing with ideas rather than events.

Acceptable Quality Level (AQL) (Chapter 4). The maximum percentage of defectives that can be considered acceptable.

Acceptance Number (Chapter 4). The largest number of defects in the sample being considered that will permit acceptance of the inspection lot.

Accidental Errors (Chapter 9). See *random errors.*

Accumulation of Errors (Chapter 5). The compounding of errors in serial measurements is known as *accumulation of errors.* Serial measurements are interconnected measurements whose errors are compounded. See Figure 5–34.

Accuracy (Chapter 2). *Accuracy* is adherence to a prescribed standard. Accuracy is comparative, whereas precision is a positive matter, and reliability is relative. The opposite of accuracy is "uncertainty." See Figures 2–11 and 7–12.

Actual Mating Envelope (Chapter 3). An imaginary envelope that forms the boundary around a part. The actual mating envelope is defined according to feature type. For an external feature (a block or a pin), it is the smallest envelope of perfect form that can be circumscribed about the feature so that it just contacts the surface at the highest points. For an internal feature (a hole or a slot), it is the largest envelope of perfect form that can be inscribed within the feature. See Figures 3–10 and 3–11.

Acute Angle (Chapter 12). An angle that measures less than 90°. See Figure 12–3.

Additive Tolerances (Chapter 3). Tolerances that are combined by their sums. Note that they are combined algebraically with both minus and plus values.

Air Film (Chapter 8). A very thin layer of air that prevents gage blocks from complete contact. See Figure 8–21.

Alignment Collimator (Chapter 16). An *alignment collimator* is a special target in which the image of the target emerges in parallel rays. See also *collimation* and *infinity*. See Figure 16–61.

Alignment Telescope (Chapter 16). See *line of sight telescope.*

Ambient Temperature (Chapter 14). See *temperature.*

Amplification (Chapters 10, 14). *Amplification* is the amount by which the senses are multiplied in the perception of precision. It is an increase in the output of a system as compared with the input. Power of a system is the ratio of the output to the input.

Angle (Chapter 12). An *angle* is the rotation necessary to bring one line into coincidence with (or parallel to) another line in the same plane. *Complementary angles* total 90°. *Supplementary angles* total 180°. See Figures 12–5 and 12–6.

Angular displacement (Chapter 12). *Angular displacement* is the angle through which a point, line, or body is rotated about a specific axis.

Angularity (Chapter 3). A tolerance zone defined by two parallel planes at a specified angle, other than 90°, from one or more datums. See Figure 3–27. Geometric symbol: ∠

AOQ (Chapter 4). See *average outgoing quality.*

AOQL (Chapter 4). See *average outgoing quality limit.*

AQL (Chapter 4). See *acceptable quality level.*

Asperities (Chapter 13). Deviations from perfection in surfaces are termed *asperities.* The four most common types are *roughness, waviness, error of form,* and *flaws.* The first three of these recur in a given part. Their spacings are expressed as wavelengths. Flaws, however, are random asperities. The roughness is considered to be the *primary* texture. See Figures 13–2 and 13–3.

Attribute (Chapter 4). A characteristic that either does or does not exist. It implies gaging.

Autocollimator (Chapter 12). The autocollimator combines a collimator and a telescope for measuring angles.

Auto-Reflection (Chapter 16). The optical micrometer provides displacement of the optical path but does not affect the angle. For angle deflection, mirrors and prisms are used. With *auto-reflection* the image of a reticle is projected onto a target, which is a mirror. That image is directed back to the telescope and the displacement between the reticle and its reflected image is examined. This will be a measure of angular displacement. It requires that the telescope be provided with an illuminator. The illuminator is often called the "lamphouse." See also *collimation.* See Figures 16–53 and 16–55.

Average Outgoing Quality (AOQ) (Chapter 4). This includes all accepted lots together with rejected lots after they have been 100% inspected and defectives replaced.

Average Outgoing Quality Limit (AOQL) (Chapter 4). Maximum AOQ (q.v.) for all possible incoming qualities for a given sampling plan.

Axes (Chapter 10). See *vertical, horizontal.*

Azimuth (Chapter 16). For optical metrology, *azimuth* is a horizontal angle measured in a clockwise direction from the true north position.

Bias Error (Chapter 5). *Bias error* is the conscious or unconscious influencing of measurement. See also *error.*

Bilateral Tolerance (Chapter 3). See *tolerance.*

Binary Numeral (Base Two) **System** (Chapter 12). A numeral system using two symbols, typically 0 and 1. Binary systems are commonly used in electronic circuitry and in computer language codes.

Blade (Chapter 5). A part of the combination square, it is the steel rule or straightedge. It can be used by itself or in with a protractor head, a center head, or a square head. See Figure 5-19.

Bridging (Chapter 8). *Bridging* is the finding of a dimension by the testing of measurements spanning an unknown dimension and successively closing the distance between them until the unknown dimension is approximated.

Calibration (Chapters 7, 9). *Calibration* determines the accuracy relationship of a gage or an instrument to a higher standard. In the United States, the calibration traces the accuracy to the National Institute of Standards and Technology (NIST), successor to the National Bureau of Standards (NBS). The NIST, in turn, coordinates the U.S. standards to the international standards.

Caliper Measurement (Chapters 5, Chapters 6). The use of any instrument that can be set to the part feature being measured and can then transfer that setting to a standard for comparison. See Figures 5–40 and 5–41.

Center head (Chapter 5). A part of the combination square, it can be used with a steel rule (blade) to locate or lay out the exact center of round stock. See Figure 5-24.

Centerline Average (CLA) (Chapter 13). See *roughness average.*

Centralize (Chapter 6, 7). Movement of an instrument by feel of contact so that its axis is oriented parallel to the line of measurement. This is sometimes called "rocking." See Figure 7–34.

Circle (Chapter 12). The *circle* is the standard for angular measurement. It consists of a closed, plane curve, all of whose points are equidistant from one enclosed point, the center. See also *lobing* and *roundness.*

Circular Runout (Chapter 3). The composite deviation from the desired form through a full rotation (360°) of the part on a datum axis. See Figure 3–38. Geometric symbol: ↗

Circularity (Chapter 3). A condition in which all points of the surface intersected by a plane perpendicular to the axis are equidistant from that axis. Circularity is also called roundness. See Figure 3–23. Geometric symbol: ○

Collimation (Chapter 16). *Collimation* means to render parallel. Light rays emanating from a source diverge. A *collimator* directs them into parallel rays. See also *infinity.* See Figure 16–60. An alignment collimator is a target using collimation.

Collimator (Chapter 12). A collimator is a device for producing a parallel beam of rays. A collimator narrows a beam of particles, causing the beam to become more aligned is a specific direction (collimated, or parallel) See also autocollimator.

Collinear (Chapter 16). Three or more points passing through or lying on the same straight line. See Figure 16–30.

Comparator (Chapter 10). Both indicators and comparators are sensing elements in systems that measure by comparison. The unknown is compared to a standard. *Indicator* is the term applied to the lower precision instruments such as dial indicators. *Comparator* is the term applied to the higher precision instruments. These may be either mechanical or electronic but are chiefly electronic. For greatest reliability, these instruments should be used only to compare, not to measure directly. Their greatest accuracy is at the zero reading. The further from zero, the greater the error they add. A *test indicator* is the least precise version and is used only for rough setup, etc. See Figures 10–10, 10–13, and 10–14.

Comparison Measurement (Chapter 10). *Comparison measurement* is measurement with an instrument that detects the difference between the part and a separate standard. See also *direct measurement.*

Complementary Angle (Chapter 12). Two angles whose combined measurement equals 90°. See Figure 12–6.

Compound Angle (Chapter 12). *Compound angles* are formed by the edges of triangles that lie in different planes. *Face angles* are formed by the intersections of edges of solids. The *Dihedral angle* is the opening between planes.

Concentricity (Chapter 3). A condition in which two or more features (such as cylinders, cones, and spheres) have a common axis. Concentricity is measured by determining the median of all opposed points. See Figure 3–36. Geometric symbol: ◎

Contour (Chapter 17). The literal meaning of *contour* is the outside outline of any shape. In practice, however, it has come to mean the irregular outlines that cannot be conveniently defined by combinations of straight lines and arcs of regular radii.

Control Charts (Chapter 4). Charts that facilitate visual analysis of production trends. Each contains separate average chart and range chart.

Coordinate measurement (Chapter 4). Coordinate measurement is a 3-dimensional measurement system where dimensions from three rectangular coordinates: x, y, and z.

Coordinate Measuring Machine (CMM) (Chapters 5 and 18). *Coordinate measuring machines* are instruments that locate point coordinates on three-dimensional structures while simultaneously integrating both the dimensional and orthogonal relationships. CMMs are chiefly used for *plate work* (q.v.) on relatively large parts. They are available in a large variety of sizes and configurations. See Figures 18–4 and 18–5.

Cosine Error, Sine Error (Chapter 10). The *cosine error* is induced by lack of squareness between the feature of the part and the measurement instrument. The *sine error* results from misalignment of a flat contact with the part feature. See Figures 10–52, 10–52, and 14–19.

Cylindricity (Chapter 3). A condition where all circular elements of a surface are within the specified size tolerance. See Figure 3–24. Geometric symbol: ⌭

Datum (Chapter 3). A theoretically exact point, axis, or plane. A datum is the origin from which the location or geometric characteristic features are established.

Datum Feature (Chapter 3). An actual physical feature of a part that is used to establish a datum. A feature can be a surface, pin, tab, hole, or slot.

Datum Reference Frame (Chapter 3). A theoretical reference frame made up of three mutually perpendicular planes. See Figure 3–9.

Decimal Numeral (Base Ten) **System** (Chapter 12). The decimal system has a base of ten. It is the most commonly used numeral system, using digits of 0, 1, 2, 3, 4, 5, 6, 7, 8, and 9.

Defect (Chapter 4). A deviation from some standard. The term covers a wide range of possible severity; from merely a minor flaw to a fault sufficient to cause product failure.

Degradation of Workmanship (Chapter 11). *Degradation of workmanship* is the generalization that no part feature is as accurate as the process or machine that produced it. Thus, the less dependent a part feature is on previous operations, the greater the probability of accuracy.

Depth of Focus (Chapter 16 and 17). *Depth of focus* depends upon the numerical aperture (NA) as well as the magnification and is inversely proportional to both. The higher the magnification, the shorter the depth of focus for any given numerical aperture. Also known as *depth of field.*

Descriptive Statistics (Chapter 4). *Descriptive statistics* summarize patterns in the data. Examples of descriptive statistics are measures of central tendency, and measures of spread.

Determinate errors (Chapter 9). *Determinate errors* are errors know and controllable. Determinate errors are systemic.

Dial Calipers (Chapter 6). A caliper measurement instrument with a dial readout. Dial calipers are used for inside and outside and depth measurements. They are quick and easy to read, but are subject to jaw wear. See Figure 6–23.

Dial Indicators (Chapter 10). A precision measuring tool utilizing a dial face. A plunger moves in and out from the body of the indicator and rotates a needle on the dial face. The dial face is generally marked in increments of 0.001″. A smaller dial reads each revolution of the larger dial in increments of 0.100″, providing an overall range of 1–2″. See Figure 10–9.

Diascopic Projection (Chapter 17). *Diascopic projection* is the term for projection of transparent objects, such as film. See also *episcopic projection.*

Digital/Analog (Chapter 14). A digital instrument displays the measurement in numerals. In an analog instrument, a moving member (a "hand") moves in proportion to the change in measurement; therefore, it replicates the measurement, but greatly amplified. In general, digital readouts are desirable to show limits, whereas analog readouts are best when trends are to be followed. Most analog instruments use dial scales for readout.

Dihedral Angle (Chapter 12). The *dihedral angle* is the opening between planes. See also *compound angle*.

Dimension (Chapter 2). A *dimension* is an expression of length in a standards system. See Figure 2–5.

Dimensional Metrology (Chapter 1). See *linear metrology*.

Direct Measurement (Chapters 5 and 10). *Direct measurement* is measurement with an instrument that incorporates its standard. See also *comparison measurement*.

Discrepancy (Chapter 9). See *uncertainty*.

Discrimination (Chapter 5). *Discrimination* is the fineness of the scale divisions of an instrument. It is the smallest division of the scale that can be read reliably. As discrimination is increased, the range is decreased. The ratio for this varies among the types of instruments. See Figures 5–5, 7–7, 7–12, and 10–15.

Displacement Method (Chapter 6). The *displacement method* is a method of measurement in which each end of the part feature is separately compared with a standard. There is a displacement in getting from one end to the other. It is the relationship between the distance displaced and the standard that constitutes the measurement. See Figure 2–6 and Figure 6–10.

Distortion Ratio (Chapter 13). The distortion ratio is the disparity in the aspect ratio of axes. Surface measurement software utilizes this discrepancy to exaggerate surface roughness to better see the variation. Each surface measurement machine has its own particular distortion ratio. Graphs of the same surface measured on different machines will not look the same because they may use different distortion ratios.

Dividing Heads (Chapter 12). *Dividing heads* are mechanical devices for dividing a circle into angular portions.

Dynamic Measurement (Chapter 14). Most linear measurement is *static*. It is performed on something that is stable and unchanging, at least over a short period of time. *Dynamic measurement*, in contrast, is the continuous monitoring of a changing quantity. Although there are many exceptions, digital readouts are associated with static measurement and analog readouts are associated with dynamic measurement.

Eccentricity Error (Chapter 12). *Eccentricity error* is discrepancy in angular measurement caused by the measurement arc not being concentric with the center of rotation. See Figure 12–87.

Elasticity (Chapter 13). The ability of a solid to return to its original shape and size after being deformed by force.

Electronic Digital Calipers (Chapter 6). A caliper measurement instrument with an electronic readout. They add the advantage of easy conversion from inch to metric and some offer the ability to send readings to data collectors or a computer. See Figure 6–24.

End Standards (Chapter 8). *End standards* are special gage blocks that are placed on ends of stacks of gage blocks to convert to other measurements than the normal external end measurements of the stacks.

Engine engraved (Chapter 5). A process of engraving graduated marks on a rule, rather than stamping. Today, quality steel rules are engraved with a photo-engraving process.

Episcopic Projection (Chapter 17). *Episcopic projection* is the term for optical projection of opaque objects. See also *diascopic projection*.

Error (Chapters 5, 9). *Error* is the difference between the measured value and the true value. Error always exists although it may not be measurable. *Observational error* is formed during the reading of an instrument. *Parallax error* is caused by the apparent shifting of objects when the viewing position is changed. *Manipulative error* is caused by the handling of the instrument and the part. *Bias error* is the conscious or unconscious influencing of measurement. See also *systematic error, random error,* and *illegitimate error*. See Figure 7–12.

Face Angle (Chapter 12). *Face angles* are formed by the intersections of edges of solids. See also *compound angle.*

Feature (Chapter 3). A *feature* is a definable portion of a workpiece.

Feature Control Frame (Chapter 3). A rectangular box that contains geometric symbols, modifiers, and datum references. The feature control frame communicates design intent, indicating part feature size, geometry, and location. See Figure 3–18.

Feature of Size (Chapter 3). A feature associated with a size dimension, such as one cylindrical or spherical surface, or a set of two opposed elements or opposed parallel surfaces. See Figure 3–1.

Feel (Chapters 5, 6, 7). *Feel* is the perception of the distortion that results from physical contact between the instrument and the part or standard.

Field of View (Chapter 17). In an optical system, the *field of view* is the screen diameter divided by the magnification. See Figure 17–14.

Fixed Bridge Coordinate Measuring Machine (Chapter 18). A fixed bridge CMM has each end of the bridge fixed to the machine base. The machine table moves horizontally. See Figure 18-5.

Flatness (Chapter 3). *Flatness* is the degree to which the reference plane corresponds to the theoretically perfect plane. See Figure 3–22. Geometric symbol: ⏥

Freedom (Chapter 10). There are six possible degrees of *freedom*. Three are translational and involve movement in the x, y, and z axes. The other three are rotational about these axes. If movement in any of these degrees of freedom is prevented, it is said to be constrained. See also *orthogonal.* See Figure 10–47.

Fringe Bands (Chapter 16). *Fringe bands* are the alternate light and dark stripes that result when two rays combine and interference takes place. See Figure 16–5.

Fringe Patterns (Chapter 16). *Fringe patterns* are groups of fringe bands.

Gage Blocks (Chapter 8). *Gage blocks* are the mass produced end standards that combine arithmetically to form usable length combinations. They are standards that are traceable to the national bureau of standards in the countries of their use.

Gage, Gauge (Chapter 3). *Gauge* is the spelling used in Great Britain but finds some use in the United States. It is chiefly used for nonlinear measurement instruments (fluid gages, etc.). However, it is finding increased use for coordinate measuring machines, where it designates a device with a proportional range and some form of indicator.

Gages, Snap (Chapter 5). A *snap gage* is a caliper type device that has been preset to a given dimension. It is often made with two settings, one to admit acceptable parts, the other to reject out-of-tolerance parts. This general class is known as *attribute* gages and are frequently called *go and no-go* gages or *go/no-go* gages. They are also called *limit* gages. See also *go/no-go gaging.* See Figures 5–51 and 8–46.

Gaging (Chapter 5). *Gaging* is single-purpose measurement to determine if objects are between size limits or to sort objects into size categories.

Gaging Force (Chapter 5). *Gaging force* is the force holding the instrument to the part. It is often incorrectly called gaging "pressure." See Figure 7–26.

Geometric Optics (Chapter 16, Appendix C). *Geometric optics* is the technique for ray tracing. See also *ray tracing.*

Gon (Chapter 12). In angular measurement, the term gon is used in Europe and SI units, while *grad* is used in the United states. A gon is an angular measurement where the arc is divided by the circumference and multiplied by 400.

Go/No-Go Gaging (Chapters 5, 8, 9, 14). A *go* gage accepts all properly sized parts. If a female part is not accepted, it is too small. If a female part is accepted by the *no-go gage,* it is too large. The reverse applies to male parts. See Figures 5–51 and 8–46.

Grad (Chapter 12). In angular measurement, the term gon is used in Europe and SI units, while *grad* is used in the United States. A grad is an angular measurement where the arc is divided by the circumference and multiplied by 400.

Graduated Scales (Chapter 5). *Graduated scales*: a series of marks laid down at determinate distances, along a straight edge, for purposes of measurement (rulers).

Granite (Chapter 11). *Granite* is a lightweight, very hard granular, igneous rock consisting mainly of quartz, mica and feldspar. For granite surface plates, the quartz content varies from stone to stone, and is a deciding factor for surface plate material.

Heat Sink (Chapter 8). A *heat sink* is a body with rapid heat transfer on which workpieces and gages can be placed to arrive at the same temperature before measurement. The process is known as *normalizing*. See also *temperature.*

Heisenberg's Uncertainty Principle (Chapter 9). Refers to a limit to the precision any measurement can be made. The act of measurement alone alters the measurement.

Hermaphrodite Caliper (Chapter 5). A measurement instrument similar to dividers, with one curved leg and one straight leg. Hermaphrodite calipers are widely used to scribe lines parallel to an edge or to find the center of a circle. See Figures 5–45 and 5–46.

Histogram (Chapter 4). Graphical representation of a frequency distribution in which one dimension is proportional to the range of frequencies and the other dimension is proportional to the number of frequencies appearing within the range. See also *normal distribution curve.*

Horizontal (Chapter 12). See *vertical.*

Hysteresis (Chapter 15). *Hysteresis* is when the value of a physical property lags behind changes in the effect causing the change, as when magnetic induction lags behind the magnetizing force.

Illegitimate Error (Chapter 9). Whereas the other common errors must be assumed to exist to some extent, *illegitimate errors* should never occur. They are accidents, the avoidable mishandling of some portion of the measurement act. See also *errors, systematic errors,* and *random errors.*

Image (Chapter 16, Appendix C). Light rays passing through a lens converge on a point known as an *image.* Such rays originate at an object. Together they are known as *conjugate points.* For a convergent lens (convex), a *real image* is formed. For a divergent lens (concave), a *virtual image* results. The real image lies on the side of the lens away from the object and is inverted. The virtual image lies between the lens and the object and is erect. See Appendix C, Figures D, E, F, and H.

Inch-Pound System (Chapter 2). The *inch-pound system* is the conventional measurement system for the United States and Great Britain. It is based on the foot, pound, second, degree Fahrenheit, ampere, and candela.

Indeterminate errors (Chapter 9). *Indeterminate errors* are random or uncontrollable errors.

Index Heads (Chapter 12). *Index heads* are mechanical devices for dividing a circle into angular portions. They are also known as dividing heads.

Indexing (Chapter 12). *Indexing* is the act of measurement of angles or setting out angles for machining by use of a rotary table.

Indicators (Chapter 10). See *comparators.*

Inferential Statistics (Chapter 4). *Inferential statistics* is the process of drawing information from sampled observations of a population and making conclusions about the population. Inferential statistics use a portion of the population to deduce or infer a hypothesis.

Infinity (Chapter 16). In metrology, *infinity* has a special meaning: it is the distance from which rays reaching the observer are parallel. See also *collimation* and *zenith sight*.

Inspection (Chapter 1). *Inspection* is the verification of conformity to a standard, generally applied to the production of goods, but also used for remedial examination.

Interchange Method (Chapter 6). The *interchange method* is a method of measurement in which both ends of the unknown feature are compared with both ends of a standard of the desired dimension, or with a ruled scale. It is known as *measurement by comparison.*

Interference Fit (Chapter 3). *Interference* fit refers to parts that must be compressed to mate.

Interferometer (Chapter 8). An *interferometer* is any instrument making an application of interferometry. See also *interferometry.*

Interferometry (Chapter 8). *Interferometry* is the formation of visible bands of light by interacting wave fronts. Interferometry is used for precise comparisons using gage blocks (Chapter 16) and for surface evaluation (Chapter 16). Interferometry is also the basis for precise optical metrology instruments such as the *pointing interferometer* (Chapter 16).

International Inch (Chapter 2). The *international inch* is the reconciliation of the inch to the meter. One inch equals 25.4 mm.

International Vocabulary of Basic and General Terms in Metrology (VIM) (Chapter 9). Provides a set of definitions for a system of basic and general concepts used in metrology. It is intended to be a tool for better communication, providing a common reference for scientists, engineers, teachers, and practitioners, irrespective of the field of application.

Interpolation (Chapters 5, 7). *Interpolation* is the selection of the nearest graduation when a measurement lies between two identifiable values. The observational equivalent to the rounding off (q.v.) process in computation.

Iron Ulna (Chapter 2). One of the oldest units of length measurement used was the 'cubit' which was the length of the arm from the tip of the finger to the elbow. Around 1196, as a move toward "standardization", it was decreed that "Throughout the realm there shall be the same yard of the same size and it should be of **iron**". During King Edward I's reign, (1272-1307) the yard (or *Ulna*) and its divisions were further defined as "three feet and no more."

Jig Transit (Chapter 16). A *jig transit* is similar to a surveyor's transit, but it does not have scales for measuring angles. It has, however, provisions for both autocollimation (q.v.) and optical projection. Its most important feature is that the elevation axis is hollow. This allows two lines of sight perpendicular to each other to intersect on the azimuth (q.v.) of the instrument. Also known as *optical transit square* and *jig level*. See also *tachymetry.*

Johansson, C. E. (Chapter 8). Johansson was a pioneer in metrology and clearly anticipated modern, interchangeable gage blocks. More importantly, he understood the significance of this development.

Joystick Control (Chapter 18). A *joystick control* allows the operator to control the direction of travel of the moving member in two orthogonal axes, as well as the rate of travel. It is regularly used for manual operation of CMMs. See also *coordinate measuring machine.* See Figure 18–6.

Kinematics (Chapter 10). The branch of mechanics dealing with the study of motion without consideration of mass or force. Kinematic science describes the motion of objects using words, diagrams, numbers, graphs, and equations.

Knee (Chapter 5). A straight edge or square block, used as a reference point for measurement.

Krypton 86 (Chapter 16). *Krypton* 86 is a gas which, when electrically excited, emits the wavelength of light that is the basis for the international standard. It was selected because of its stability. However, it is more practical to use helium gas to emit light for most practical work.

Laser (Chapter 17). A *laser* emission is an intense, narrow beam of light that is a single wavelength. The beam will travel great distances with little loss of energy and only small dispersion. **Warning:** A laser beam can damage eyesight or cause blindness. Never look directly into the beam. Laser instruments are similar to conventional ones except for the laser light source and the use of photoelectric sensors. See also *photoelectric instruments.*

Layout (Chapter 5). *Layout* is the use of measurement instruments to mark features on workpieces for subsequent machining. Limited to rough machining, the final machining is controlled by machine control settings and verified by measurement.

Lead (Chapter 9). The distance a thread will advance in one revolution relative to its mating thread.

Least Count (Chapter 6). The *least count* is the discrimination of a vernier instrument.

Least Material Condition (LMC, Ⓛ) (Chapter 3). The condition in which a feature of size contains the least amount of material within its permissible limits. For a pin or a block, Least Material Condition exists when the feature is at the smallest size allowed by the tolerance limits. For a hole, the Least Material Condition results in the largest hole size allowed within the tolerance limits.

Level (Chapter 12). The *level* is an instrument for establishing a horizontal line. By means of two or more such lines intersecting, a horizontal plane may be established. The standard for a level is gravity. Very precise levels are known as *clinometers.*

Light (Chapter 16). *Light* is the electromagnetic energy radiation in frequencies to which the human eye is sensitive, as well as those of slightly longer and slightly shorter wavelengths. See also *light rays, light wave,* and *interferometry.* See Figure 16–4, and Appendix C, Figures A and J.

Light Rays (Chapter 16). *Light rays* are imaginary lines perpendicular to wave fronts. Although they do not physically exist, they are a convenience in tracing the passage of light through an optical system. See also *ray* and *ray tracing.*

Light Wave (Chapter 16). A *light wave* is the pulsation in space that transmits light energy. A wave front is an advancing pulse of energy. See also *light.*

Line of Collimation (Chapter 16). See *line of sight.*

Line of Sight Telescope (Chapter 16). The simplest optical alignment instrument is the *line of sight telescope.* This type of telescope establishes optical lines of reference. With optical micrometers they provide a limited measuring range. With the addition of the micrometers these instruments become known as *alignment telescopes.* See also *optical micrometer.* See Figures 16-37, 16-38.

Line references (Chapter 5). Reference line, an arbitrary fixed line (such as an axis, edge, or shoulder) from which coordinated of a point are computed, or from which a measurement is made.

Linear Metrology (Chapter 2). *Linear metrology* is related to measurements taken along a line. In general use, it includes measurement along curved lines as well as straight lines and angles. It applies to areas and volumes because these can be reduced to their component lines. It is often referred to as dimensional metrology.

Linear Variable Differential Transformer (LVDT). (Chapter 14). A type of electrical transformer used to measure linear displacement. The transformer has three solenoidal coils placed end to end in a tube. An alternating current is driven through the primary (center coil), causing the voltage induced in the secondary coils on either end to change. See Figure 14–12.

Lobing (Chapter 13). *Lobing* is deviation from roundness. It is to a round surface what *waviness* is to a flat surface. See also *roundness.* See Figure 13–18.

Logarithms (Chapter 12). The term *logarithms* is generally used to mean "common logarithms." These are the exponents to which ten must be raised to equal the number. For example, the logarithm of 100 is 2 because $10^2 = 100$. The logarithm of 23 is 1.36173 because $10^{1.36173} = 23$. Logarithms facilitate the expression and calculation of complex values.

Lot (Chapter 4). A definitive quantity (Number) of some manufactured product that is considered to be uniform.

LURD (Chapter 16). *LURD* is a convention for expressing measurement changes to minimize misunderstandings. It stands for: left, up, right, down. See Figure 16–43.

Machine Coordinate System (Chapter 18). A three-dimensional mapping system used by numerically controlled machines to identify coordinate points. See Figure 18–29.

Machining centers (Chapter 5). Manufacturing equipment used to remove metal and shape workpieces. In a machining center, the tool moves while the workpiece remains stationary.

Macrogeometry (Chapter 8). *Macrogeometry* means large-scale consideration of form or shape. In popular usage, macrogeometry is concerned with size and shape that can be expressed with standard inspection instruments. Evaluation of squareness in shop machines usually is macrogeometry.

Magnification (Chapters 16, 17). The total *magnification* is the product of the eyepiece magnification and the objective magnification. The eyepiece magnification is usually X10. Thus, coupled with an objective of X43, the total magnification would be 430 power.

Manipulative Error (Chapter 5). *Manipulative error* is error caused by the handling of the instrument and the part. See also *error*.

Margin of Safety (Chapter 5). In a measurement, it means the amount that the measurement process exceeds the required minimum accuracy and/or precision.

Master Flat (Chapter 16). A *master flat* is an optical flat having surfaces of maximum precision. It is used to calibrate *working flats*. See also *optical flat*.

Mastering (Chapter 15). *Mastering* means setting to standards. Although it can apply to all types of measuring systems, it is chiefly applied to pneumatic metrology.

Material Condition (Chapter 3). An outside edge is less than 180° of material. An inside edge is more than 180°. These are the *material conditions*. When material is dimensioned, the part feature is *male*. When the space between material is dimensioned the part feature is *female*.

Maximum Material Condition (MMC) (Chapter 3). *Maximum material condition* occurs when the male feature is at its largest size and the female feature at its minimum size.

Mean ($\bar{X}$) (Chapter 4). The average of a sample of *n* units. It is the sum (Σ) of the values divided by the number of units in the sample.

Measured Point (Chapter 2). The *measured point* is the point on a scale that coincides with the end of the feature being measured opposite from the reference point. See also *reference point*.

Measurement (Chapter 2). *Measurement* means to proportion by measured lots. The measurement may be any quality (length, weight, hardness, etc.) and lots may be in any units (inches, volts, picas, etc.). See also *metrology*. See Figure 2–5.

Measurement Function Library (Chapter 18). The *measurement function library* is the program incorporated in a direct computer controlled CMM. These standard programs are extended by additional software called *subroutines*.

Median (Med) (Chapter 4). The middle measurement when an odd number of units is arranged in order of size. When an even number is arranged, the median is the average of the two middle units.

Mensuration (Chapter 1). The process of measuring geometric quantities.

Metric System (Chapter 2). The familiar name for Le Système International d'Unités, the most recent of many so-called metric systems. Often abbreviated "SI." Based on the meter, kilogram, second, degree Celsius, ampere, and candela.

Metrology (Chapter 1). *Metrology* is the science of measurement. See also *measurement*.

Metrology Laboratory (Chapter 8). A *metrology laboratory* is a laboratory for the calibration of standards. It may be a department within a company or an outside service.

Microgeometry (Chapter 8). *Microgeometry* means small-scale considerations of form or shape. The precision in microgeometry is expressed in mikes. Evaluation of surface finish and flatness of gages is microgeometry.

Micrometer Instruments (Chapter 7). *Micrometer instruments* are measurement instruments that achieve their amplification as a result of the resolution of screw threads. See Figure 7–3.

Microscope (Chapter 17). A *microscope* is an instrument for greatly enlarging the view of a part feature. That capability is known as amplification (q.v.). Whereas most measurement systems alter their measurements to a considerable degree, optical systems do not impose those limits, but they have other problems. See Figure 17–3.

Midrange (Chapter 4). *Midrange* is the value midway between the highest and lowest values in a dataset.

Mike (micro) (Chapter 2). In the decimal-inch system of terminology, *micro* means millionths. However, in popular use "microinch" (with or without the hyphen) is the most common designation for millionths of an inch. See Figure 14–3 and Appendix D.

Mil (Chapter 2). In the decimal-inch system of terminology, *mil* means thousandths.

MIL-STD-105 (Chapter 4). The standard for sampling required for most Department of Defense applications.

Mode (Chapter 4). The most frequent value of the set of units.

Monochromatic Light (Chapter 16). *Monochromatic light* consists of one wavelength (color) only.

National Bureau of Standards (NBS). Former name for the National Institute of Standards and Technology, the official standardization agency for the government of the United States.

National Institute of Standards and Technology (NIST) (Chapter 2). New name for the National Bureau of Standards.

Natural Functions (Chapter 12). See *trigonometry*.

Normal Distribution Curve (Chapter 4). The *normal distribution curve* is the bell-shaped curve that results when a variable is changing by chance, such as the throws of dice. It is called a *histogram* because any production process may be graphed. See also *standard deviation*.

Normalize (Chapter 8). See *heat sink* and *temperature*.

Numerical Aperture (NA) (Chapter 17). The higher the *NA*, the greater the resolving power (q.v.). The NA is = n sin u, where n is the lowest refractive index that appears between the objective and the front lens of the objective and is half of the angular aperture of the objective. Fortunately, it is marked on lens mounts.

Observational Error (Chapter 9). Error formed during the reading of an instrument. See also *error*.

Obtuse Angle (Chapter 12). An angle that measures greater than 90° but less than 180°. See Figure 12–3.

Operating Characteristic Curves (Chapter 4). *Operating characteristic curves* are a graphical means for comparing sampling plans. They enable SQC plans to be compared and show the relationship between lot size and number of defectives.

Optical Alignment (Chapter 17). *Optical alignment* is that branch of metrology that uses light waves to examine sizes and geometrical relationships over larger distances than ordinarily required in shop and laboratory work.

Optical Comparator (Chapter 17). An *optical comparator* is an instrument that projects a greatly magnified image of a part feature onto a screen for examination.

Optical Flat (Chapter 16). An *optical flat* is a glass or quartz reference surface with one or both sides finished to be precisely flat and to have minimum surface imperfections. *Flats* is the general term for such devices made from nontransparent materials such as steel or ceramics. A *master flat* has surfaces of maximum precision. It is used to calibrate *working flats*. See Figure 16–15.

Optical Lever (Chapter 14). A device used to magnify a small displacement, making it possible to obtain an accurate measurement of the displacement. See Figure 14–7.

Optical Micrometer (Chapter 16). An *optical micrometer* is a mechanically adjustable means to offset an optical path in one axis. It is calibrated so that the offset may be known. Optical micrometers are frequently used in pairs at right angles to each other. They are not used alone but are incorporated into instruments such as the line-of-sight telescope. Note that the optical micrometer does not affect the angle of the optical path, only its displacement. See Figure 16–37.

Optical Polygons (Chapter 16). *Optical polygons* are polygons with precisely spaced faces that represent divisions of a circle. In other words, they represent angles. They are used with optical alignment instruments to set out and to measure angles. See Figures 16-81 through 16-87.

Optical Square (Chapter 16). An *optical square* is a means for turning all or some of the rays in the optical path to 90° from that path. See Figure 16-67.

Optical Tooling (Chapter 12). *Optical Tooling* is a technique of establishing precise reference lines and planes using telescopic lenses.

Orthogonal (Chapter 18). *Orthogonal* literally means rectangular or right angled. In popular use, it has come to mean movement in two or three axes at right angles to each other. See also *freedom, vertical,* and *horizontal.*

Orthogonal Array (Chapter 18). *Orthogonal array* refers to a device in which the functional parts act at right angles to each other.

Parallax Error (Chapter 5). *Parallax* error is the apparent shifting of objects when the viewing position is changed. See also *error.*

Parallel Separation Planes (Chapter 16). The concept of *parallel separation planes* simplifies measurement with optical flats. The concept consists of a set of planes parallel to the working surface and one-half wavelength apart. Their intersection with the part is seen as fringe bands. See Figure 16–7.

Parallelism (Chapter 3). The condition of a surface or centerplane that is equidistant at all points from a datum or axis. See Figures 3–31 and 3–32. Geometric symbol: //

Part Coordinate System (PCS) (Chapter 18). A three-dimensional coordinate system for identifying the location or coordinates of a manufactured part. See Figure 18–28.

Perpendicularity (Chapter 3). A tolerance zone defined by two parallel planes at a 90° angle from one or more datums. See Figure 3–28. Geometric symbol: ⊥

Photoelectric Instruments (Chapter 16). In order to avoid the subjective bias of the observer, optical instruments have been developed that use photoelectric sensors to ascertain the reference and measured points. The most advanced of these use lasers as their light source. See also *laser.*

Pitch (Chapter 19). *Pitch* is the distance from one point to an adjacent point. For example, the distance between teeth on a saw blade or threads on a screw.

Plane Angle (Chapter 12). An angle formed by two straight lines in the same plane.

Plasticity (Chapter 13). The ability to be formed or shaped through pressure without failure.

Plate Work (Chapter 5). See *surface plate.* See Figure 5–2.

Pneumatic Metrology (Chapter 15). *Pneumatic metrology* refers to any measurement system that uses pressure changes in a fluid (air or other gases) system to amplify changes in part sizes or shapes. Pneumatic gages are basically analog devices. See Figures 15–24 and 15–25.

Pointing interferometer (Chapter 12). A pointing interferometer is a highly precise instrument for measuring angles.

Polar Chart Method (Chapter 13). See *V-block method.*

Population, Population Size (N) (Chapter 4). The totality of the set of units or measurements under consideration.

Positional Tolerancing (Chapter 3). *Positional tolerancing* recognizes that the limits of clearances and fits between parts occur in circles, not the rectangles that result from tolerancing with orthogonal relationships (q.v.). This allows tolerances to be widened without loss of reliability. See also *tolerance.* See Figure 3–16.

Precision (Chapter 1). *Precision* refers to the fineness of readings or dispersion of measurements. It is the degree of mutual agreement among individual measurements. Precision is a positive matter, whereas accuracy is comparative and reliability is relative. See Figure 2–11.

Precision Instruments (Chapter 5). *Precision instruments* are those providing amplification of the natural senses of sight and touch.

Premise (Chapter 12). *Premise* is a proposition, argument, or theory previously proposed; a basic assertion.

Probability (Chapter 4). The chance of an occurrence over a number of trials. Probability can range from no chance (0) to full certainty (1).

Probe (Chapter 18). The probe is part of the coordinate measuring machine that gathers data by "touching" or sensing the workpiece. Probes can be contact or non-contact. They include hard probes, touch trigger probes, optical and laser probes. See Figures 18-8 through 18-11.

Profile of a Line (Chapter 3). A two-dimensional tolerance zone extending along the length of a feature. See Figure 3–25. Geometric symbol: ⌒

Profile of a Surface (Chapter 3). A three-dimensional tolerance zone extending along the length and width (or circumference) of a feature. See Figure 3–26. Geometric symbol: ⌓

Proof (Chapter 12). Is a deductive mathematical argument. The structure of such argument goes that if one thing is true, something that necessarily follows is also true.

Protractor (Chapter 12). A *protractor* is an instrument for measuring angles by the displacement method. They usually have calibrated arcs as standards. See Figure 12–38.

Protractor head (Chapter 5). A part of the combination square, it can be used with a steel rule (blade) to measure the angle between the base and the ruler. See Figure 5-22.

Quadrant (Chapter 12). In geometrical measurements, a sector equal to a quarter of a circle, or any of the 4 equal areas made by dividing a plane into 4 parts by an x- and y-axis.

Quality (Chapter 1). The totality of features and characteristics of a product or service that make it able to satisfy the customer's needs.

Quality Assurance (Chapter 1). A system of procedures whose purpose is to ensure that the overall quality control function is in fact accomplished. This is accomplished by verifications, audits and evaluations of the specifications, production, and end use of the product or service.

Quality Control (Chapter 1). The overall system of activities employed to provide a quality product or service that meets the needs of the customer. The aim of quality control is to provide quality products that are satisfactory, dependable, and economic. The overall system integrates several steps, including proper specifications, production processes to meet the specifications, and review of the end product to assure compliance with the customer's needs.

Quartiles and percentiles (Chapter 4). *Quartiles* split data into quarters: the first quartile at the 25th percentile, the second quartile at the 50th, and the third at the 75th. A percentile is the value of a variable that has a certain percentage of other values that fall beneath it. For example, a score in the 45th percentile means 45% of scores fall below that score.

R Chart (Chapter 3). Pronounced R-bar chart. See also *range chart.*

Radian (Chapter 15). The angle subtended by an arc of circle that is equal to the radius of the circle.

Random Errors (Chapter 9). *Random errors* result from erratic malfunctions of any part of the measurement system, including the observer. For example, a loose element that may unpredictably change positions could add errors. Random errors are often found by the discrepancies when a measurement act is repeated. Random errors are also known as accidental errors. See also *errors, systematic errors,* and *illegitimate errors.*

Random Number Table (Chapter 3). A *random number table* is a table of numbers in which there is no discernable pattern of occurrences.

Random Sample (Chapter 4). A *random sample* is the selection of a sample without bias. See also *sampling.*

Range (R) (Chapter 4). The range of a set of n numbers is the difference between the largest and the smallest number.

Ray Tracing (Chapter 16). *Ray tracing* is the study of optical systems by means of rays.

Readability (Chapter 7). *Readability* is the relative ease with which the measurement can be distinguished. For example, both a plain micrometer and a vernier caliper have the same discrimination, but the plain micrometer is more readable. See Figure 7–7.

Real Image (Appendix C). *A real image* is the image formed by light rays from an object passing through a convergent (convex) lens. The real image lies on the side of the lens away from the object and is inverted. See also *virtual image.*

Reference Plane (Chapter 11). The *reference plane* is the plane in which the reference points lie. It is always perpendicular to the line of reference. The reference plane by definition must be flat. The reference plane is also known as the *datum plane.* See also *flatness* and *surface plate.* See Figure 11–17.

Reference Point (Chapter 2). The *reference point* is the position on a part feature from which a dimension is expressed. Usually, but not necessarily, it is expressed as zero. See also *measured point.* See Figure 5–7.

Reflection (Appendix C). *Reflection* is the change of direction of light when it is directed upon a suitable surface. See Appendix C, Figure D.

Refraction (Appendix C). *Refraction* is the change in path when light passes from one transparent medium into a different medium. See Appendix C, Figure D.

Reliability (Chapter 1, 2). The *reliability* is the probability of achieving desired results. Reliability is a relative matter, whereas accuracy is comparative and precision is positive. See Figure 2–13.

Repeatability (Chapter 6). *Repeatability* is the variation among several measurements taken with one instrument on one part feature. It is a test of precision, not of accuracy. See Figure 6–20.

Resolution (Chapter 10). *Resolution* is the ability to visually distinguish with the unaided eye between separate items, usually lines. It is the ratio of the width of one scale division (one output unit) to the width of the band (the readout element). Resolution is not the same as readability. See Figures 10–19 and 10–20.

Resolving Power (Chapter 17). The *resolving power* is the property by which small elements are distinctly separated. For practical work, this is more important than the magnification. The measure of resolving power is numerical aperture (q.v.).

Reversal Process (Chapter 12). A method of reversing measurement doubling apparent error and subsequently halving the uncertainty. See Figure 12–32.

Reversal Technique (Chapters 11, 12). The *reversal technique* or *reversal process* is a method for detecting or canceling small changes by comparing a variable with itself but with a reversed algebraic sign. See Figures 12–21, 12–32, and 12–88.

Right Angle (Chapter 12). An angle equal to 90°. See Figure 12–3.

RMS (Chapter 8). *RMS* is the abbreviation for root mean square. The RMS is the average of the squares of the deviations of the high and low values. Compare with *AA*.

Roughness Average (Chapter 13). The *roughness average* is the arithmetic average (AA) of the absolute values of the profile height deviations. It is sometimes referred to by its earlier term, *centerline average* (CLA). This is the method standardized for use in the United States. Many other methods also are used and about 10 have been standardized for use in other countries. See also *lobing* and *V-block method.* See Figure 13–15.

Round Down (Chapter 2). *Round down* is to leave intact the next remaining decimal place when rounding off a numeral five in last place. See Figure 2–25.

Round Off (Chapter 2). *Round off* is to arbitrarily eliminate the last decimal place. See Figure 2–25.

Round Up (Chapter 2). *Round up* is to raise the next remaining decimal place when rounding off a numeral five in last place. See Figure 2–25.

Roundness (Chapters 12, 13). *Roundness* is the characteristic that occurs when all parts of a circle are identical. See also *lobing*.

Rule (Chapter 5). See *scale*.

Ruling engine (Chapter 5). A highly precise device employed to mark graduations on measuring instruments.

Sample (Chapter 4). A *sample* is a representative portion of a group. The larger the sample, the greater the chance that it will be representative of the lot.

Sampling (Chapter 4). *Sampling* is the procedure for removing a representative cross section of parts from a larger quantity. *Random sampling* refers to the fact that to be useful for statistical quality control (q.v.), the sample must not be biased. The method must ensure that any part has as much or as little chance as any other one of being selected. *Sampling plan* is a formal adoption of a sampling schedule. See also *statistics*.

Sampling length (Chapter 13). *Sampling length* is determined by the cut-off criteria. The sample lengths have the same numeric value as the cut-off, typically 0.8 mm. Choose the appropriate number of sample lengths to provide good data analysis. This can range from 5 to sixty samples lengths.

Sampling Plan (Chapter 4). See *sampling*.

Sampling Table (Chapter 4). A *sampling table* is a table that enables the sampling to be planned based on lot size and AQL.

Scale (Chapter 5). A *scale* is a measuring device whose graduations are proportional to a unit of length for either reduction or enlargement. A *rule,* in contrast, has graduations that are as near as practical to the designated divisions of its unit of length. In addition, the word *scale* designates that portion of an instrument that has rulings. For example, the readings from a rule are taken from its scale. See Figure 5–4.

Scriber (Chapter 5). A sharp pointed hand tool used to mark lines on a workpiece.

Sensitivity (Chapter 10). *Sensitivity* is the minimum input that results in a discernible output.

Setting standard (Chapter 7). A setting standard is a measurement standard, such as a gage block, used as a reference standard.

Sexagesimal (Chapter 12). The base-sixty (sexagesimal) system is a numeral system with sixty at its base. It is used, in modified form, for measuring time, angles, and geographic coordinates.

Sexagesimal Numeral (Base 60) System (Chapter 12). The *sexagesimal system* is a measurement system based on the number 60. Although there are other measurement systems for angles, the sexagesimal system predominates. Example: Circle = 360 degrees (360°); Degree = 60 minutes (60′); Minute = 60 seconds (60″).

Shadow Projection (Chapter 17). The mathematical projection of an object onto a plane through a light source. See Figure 17–2.

Shewhart Control Charts (Chapter 4). *Shewhart control charts* are a formalized method for collection and presentation of production data.

Side Adjacent (Chapter 12). A trigonometry term, side adjacent is a means to identify one of three legs of a right triangle. It is the adjacent, touching or included side of a referenced angle, rather than the side opposite of the referenced angle. The third leg of the triangle is the hypotenuse (opposite of the right angle).

Side Opposite (Chapter 12). A trigonometry term, side opposite is a means to identify one leg of three legs of a right triangle. It is the leg opposite of the referenced angle, rather than the side touching or included (side adjacent) to the referenced angle. The third leg of the triangle is the hypotenuse (opposite of the right angle).

Significant Figures (Chapter 2). *Significant figures* or digits are digits that begin with the first digit to the left that is not zero and end with the last digit to the right that is correct.

0.000543

significant figures 54

approximation 3

Sine Bar (Chapter 12). A *sine bar* is a bar to which two identical cylinders are attached with a known separation and a known relation to the reference surface of the bar. The bar becomes the hypotenuse of a triangle for angle measurement. See also *sine plate* and *sine table.* See Figures 12–58, 12–64, and 12–65.

Sine Plate (Chapter 12). A *sine plate* is a wide sine bar. *Sine block* is another name for sine plate.

Sine Table (Chapter 12). A *sine table* is a sine device incorporated into the worktable portion of an inspection device or machine tool. See Figures 12–58, 12–64, and 15–65.

Slide Caliper (Chapter 5). A tool used for small measurements. Slide calipers come in a wide range of sizes for both inside and outside measurements. See Figure 5–50.

Solid Angle (Chapter 12). An angle formed by three or more planes intersecting at a common point.

SQC (Chapter 4). See *statistical quality control.*

Square head (Chapter 5). A part of the combination square, it is used with a steel rule (blade) to check squareness. It measures 90º, 45 º, and depth measurements. See Figure 5-21.

Squareness (Chapter 11). *Squareness* refers to a condition of being at a right angle to a line or plane. Also, a right angle in material form. See also *perpendicularity.*

Stage (Chapter 17). For optical instruments such as the microscope and optical projector, *stage* is the term that corresponds to *worktable* for most other instruments.

Standard (Chapter 8). A *standard* is an agreed-on unit for use in measurement such as meter, pound, lumen, etc. The standard is the ultimate embodiment of the unit of length. A standard is a copy that is traceable back to the national standard.

Standard Deviation (Chapter 4). The *standard deviation* is obtained by mathematically adjusting the data. It is expressed by the Greek letter sigma (σ). It is the square root of the mean average of the squares of the deviations of all of the individual measurements from the mean average value x. See also *normal distribution curve.*

Statistical Process Control (SPC) (Chapter 4). The application of statistical methods to identify and control variation in the manufacturing process.

Statistical Quality Control (SQC) (Chapter 4). *Statistical quality control,* abbreviated SQC, refers to control of production by statistical methods (q.v.). SQC by attributes is based on go/no-go principles (q.v.). SQC by variables is the reverse. It is based on the natural capability (q.v.) of the process.

Statistics (Chapter 4). *Statistics* is a branch of mathematics. It is concerned with the collection of data involving a number of occurrences, the analysis of the data, and the presentation of conclusions based on the analysis.

Steradian (Chapter 12). The angle subtended at the center of a sphere by an area on the surface of the sphere that is equal to the radius squared.

Stereoscopic microscope (Chapter 17). *Stereoscopic microscope* is a microscope with double eyepieces and independent light paths, producing three-dimensional image. See also, microscope.

Straightness (Chapter 3). *Straightness* is the condition where one line element of an axis, or a surface, is in a straight line.

Striding level (Chapter 16). *Striding level* is a tube shaped level fastened to a frame. It allows the line of sight to be set on the horizontal plane. See Figure 16-78.

Stylus Method (Chapter 13). The *stylus method* for surface examination resembles playing a phonograph record. A very small, smooth member (the stylus) is drawn along the surface. Its movement is amplified and recorded. The recording is usually a paper tape which can subsequently be analyzed. See Figures 13–7 and 13–8.

Supplementary Angle (Chapter 12). Two angles whose combined measurement equals 180°. See Figure 12–5.

Surface Datum Instruments (Chapter 13). See *true-datum instruments.*

Surface Gage (Chapter 5). A gage consisting of a scriber mounted on an adjustable stand used in the measurement of plane surfaces.

Surface Metrology (Chapter 13). *Surface metrology,* also known as topography, refers to surface finish and geometry (flatness, roundness, etc.). See Figure 13–1.

Surface Plate (Chapter 11). The primary purpose of a *surface plate* is to provide a reference plane. It must also provide an adequate and convenient surface for the support of the part and the instrumentation. Plate work is layout and inspection performed from a surface plate.

Surface Texture Specimens (Chapter 13). Two types of *specimens* are in use. The most familiar is used to make visual comparisons with surface being examined. These are called *roughness comparison specimens* or *pilot specimens.* There are also precision reference specimens. These are used to calibrate the inspection instruments by comparison. See Figure 13–5.

Surface Topography (Chapter 8) The surface finish, including roughness, waviness, and finish characteristics. See Figures 8–8 through 8–11.

Systematic Error (Chapter 9). A *systematic* or *constant error* is an error that occurs in all readings uniformly. It may thereby go unnoticed. It may be caused by any element in the measurement system, including the observer. For example, measurements to mikes in any temperature other than 68° will contain systematic errors. Systematic errors are not revealed by repetition as are most other errors. See also *errors, random errors,* and *illegitimate errors.*

Tachymetry (Chapter 16). *Tachymetry* refers to the determination of distances and angles at distances by remote means, as in surveying. The theodolite is used for tachymetry. It is essentially a jig transit with scales for angle measurement. See also *jig transit.*

Target (Chapter 16). In optical metrology, a *target* is literally the feature that is viewed. This is often a graduated scale so that differences may be detected and measured. See Figures 16-45 and 16-48.

Telescope (Appendix C). The *telescope* and *microscope* are similar in that a real image (q.v.) formed by a lens can be treated as if it were the actual object being observed; that is, it can be magnified or reduced and its shape and dimensions examined.

Temperature (Chapter 14). Nearly all engineering materials change size when their *temperature* changes. Most increase their size. When measuring to tenth-mil or finer increments, this becomes an important consideration. The measurement should be performed at 68°F or the results corrected for the ambient temperature. *Ambient temperature* means surrounding temperature, that is, air temperature. It must be remembered that it takes some time for items of different temperatures to normalize when brought together in ambient air. *Normalize* refers to arriving at the same temperature. See also *heat sink.* See Figure 14-30.

Theodolite (Chapter 12). A theodolite is a precision instrument for measuring angles in the horizontal and vertical planes.

Tolerance (Chapter 3). A *tolerance* is an unambiguous statement about the size and shape limits that are acceptable. A *bilateral tolerance* expresses these limits on both sides of a nominal dimension (the ideal dimension). A *unilateral tolerance* has one limit at the nominal dimension and expresses the other limit entirely in a deviation in one direction from the nominal. See also *positional tolerancing* and *material condition.* See Figure 3–1.

Toolroom Chart-Gage (Chapter 17). *Toolroom chart-gages* are transparent overlays that are placed on the screens of optical projectors. They provide various combinations of angles, arcs, and radii for direct measurement of part features projected onto the screens.

Total Runout (Chapter 3). The composite deviation from the desired form through a full rotation (360°) of the part on a datum axis, and along the length of the part. See Figure 3–39. Geometric symbol: ⌰

Traceability (Chapter 14). *Traceability* refers to the ability to verify each successively higher step of calibration until the NIST (q.v.) international standards are reached. See Figure 9–1.

Tracer Technique (Chapter 17). The *tracer technique* is a technique for optical comparator measurement that utilizes a tracer that is moved along a feature to be inspected. The tracer simultaneously moves a similar member in the optical path so that its travel can be compared with the ideal path on a chart. This is useful when the feature to be examined cannot be placed directly in the optical path. That often occurs for inside features.

Translation (Chapter 14). *Translation* refers to motion without rotation, hence to linear motion. In measurement, it refers to the comparison of an unknown dimension to a known standard.

Traverse comparators (Chapters 6 and 14). Refer to a measurement where two microscopes point simultaneously at two extremes of a standard. The extremities of a second standard are then viewed under the microscope using transverse displacement.

Trigonometry (Chapter 12). *Trigonometry* is the branch of mathematics dealing with the relationships of sides of triangles to the angles formed by them. *Trigonometric functions* are the ratios of pairs of triangle sides. They are useful to designate angles. *Natural functions* are tables of trigonometric functions divided out into decimals. See Figures 12–56 and 12–57.

True Position (Chapter 3). The theoretically exact location of a feature. See Figure 3–33. Geometric symbol:

True-Datum Instruments (Chapter 13). With a *true-datum* surface inspection instrument, a stylus is drawn across the surface, guided by a true line. This line is incorporated in the instrument and is the standard. The graph shows deviations from that line. With a *surface datum* inspection instrument, the stylus is guided by a member that rests on the surface undergoing inspection. The member resting on the surface may be either a skid or a shoe. See Figures 13–8 and 13–9.

Uncertainty (Chapter 9). *Uncertainty* is the reverse of accuracy. *Discrepancy* is the difference between repeated measurements. See Figure 9–8.

Unilateral Tolerance (Chapter 10). See *tolerance.*

Unit (Chapter 1). An object on which a measurement or observation may be made.

Unit of Length (Chapter 2). The *unit of length* is the arbitrary smallest whole division of a standard of length. See also *standard.*

Universal bevel protractor (Chapter 12). *Universal bevel protractor* is a graduated semicircular protractor with a pivoted arm for measuring or marking angles.

Variable (Chapter 9). A *variable* is a characteristic that can be expressed in numerical values from a continuous scale. It implies measurement.

V-Block Method (Chapter 13). The most common method of measuring round parts is to support them in a *V-block,* which also provides the reference points. Unfortunately, this is affected by any lobing that may exist. A more precise method is to rotate the part around a center and take readings of the absolute distances from the center. When plotted, these readings are a polar chart. See Figure 13–19.

Vernier (Chapter 6). A *vernier* is an auxiliary scale or device used to make fine adjustments to the main scale on precision instruments.

Vernier Instruments (Chapter 6). *Vernier instruments* are only those instruments in which the vernier is the outstanding characteristic. Many instruments not so classified use vernier scales to increase their amplification.

Vernier protractor (Chapter 12). *Vernier protractor* is a precision instrument for measuring angles.

Vernier Scale (Chapter 6). A *vernier scale* is the amplifying scale that has one division more in a given length than the main scale. See Figure 6–4.

Vertex (Chapter 12). The point at which the sides of an angle intersect. See Figure 12–2.

Vertical (Chapter 12). *Vertical* literally means at right angles to the plane of the horizon. In metrology, this is altered to state that it is in the direction of the pull of gravity. *Horizontal* is any line or plane at right angles to the vertical axis. For convenience, the x and y axes are usually considered to be in the horizontal plane and the z axis in the vertical. See also *orthogonal array.*

Virtual Image (Appendix C). A *virtual image* is the image formed by light rays from an object passing through a divergent (concave) lens. The virtual image lies between the lens and the object and is erect. See also *image.*

Wave Front (Chapter 16). A *wave front* is an advancing pulse of energy.

Wear allowance (Chapter 9). Wear is the damage that inevitably occurs during the use of measurement instruments. Wear allowance takes wear into account during the design of gages. It is generally 10% of the gage tolerance and is applied in the direction opposite to wear. In a plug gage, this allowance would be added. In a snap gage, wear allowance would be subtracted.

Wear Blocks (Chapter 8). A thin block used on both ends of gage blocks to prevent wear. See Figure 8–17.

White Light (Chapter 16). *White light* is made up of a combination of colored light wavelengths. It varies according to its source. See Figure 16–17.

Wring (Chapter 8). To *wring* is to bring two surfaces of microinch flatness together so that they adhere with only a microinch of separation. *Wringing interval* is the space by which two wrung surfaces are separated. It varies but is in microinches. The air or liquid separating wrung surfaces is known as *film*. See Figures 8–21 and 8–24.

Wringing Interval (Chapter 8). See *wring*.

Zenith Sight (Chapter 16). A *zenith sight* is that point directly above the point of reference at infinity. See also *infinity*.

Zero of Ignorance (Chapter 2). The *zero of ignorance* is an unnecessary division of a dimension. The additional zeroes usually result from ignorance, having been added to make all values in a table or discussion come out neatly to the same length. However, they are sometimes added fraudulently to make work appear more precise than it is. For example, in 0.57000 the last two zeroes are the zeroes of ignorance when the actual value is 0.570.

INDEX

A

B

C

E

F

G

H

I